T0135322

Lecture Notes in Computer Science 13024

More information about this subseries at https://link.springer.com/bookseries/7412

Christian Bauckhage · Juergen Gall ·
Alexander Schwing (Eds.)

Pattern Recognition

43rd DAGM German Conference, DAGM GCPR 2021
Bonn, Germany, September 28 – October 1, 2021
Proceedings

Editors
Christian Bauckhage
Fraunhofer IAIS
Sankt Augustin, Germany

Juergen Gall
University of Bonn
Bonn, Germany

Alexander Schwing
University of Illinois at Urbana-Champaign
Urbana, IL, USA

ISSN 0302-9743 ISSN 1611-3349 (electronic)
Lecture Notes in Computer Science
ISBN 978-3-030-92658-8 ISBN 978-3-030-92659-5 (eBook)
https://doi.org/10.1007/978-3-030-92659-5

LNCS Sublibrary: SL6 – Image Processing, Computer Vision, Pattern Recognition, and Graphics

This Springer imprint is published by the registered company Springer Nature Switzerland AG
The registered company address is: Gewerbestrasse 11, 6330 Cham, Switzerland

Preface

It was a pleasure to organize the 43rd DAGM German Conference on Pattern Recognition (DAGM GCPR 2021), which was held as a virtual conference from September 28 to October 1, 2021. The conference included four Special Tracks on Computer Vision Systems and Applications, Pattern Recognition in the Life and Natural Sciences, Photogrammetry and Remote Sensing, and Robot Vision.

We received 116 submissions. Out of the 116 submissions, 9 were desk rejected due to violation of the submission policy. The remaining papers underwent a rigorous double-blind reviewing procedure with three Program Committee (PC) members assigned to each paper. Afterward, one of the involved PC members served as moderator for a discussion among the reviewers and prepared a consolidation report that was also forwarded to the authors in addition to the reviews. As a result of this rigorous reviewing procedure, 46 out of 116 submissions were accepted, which corresponds to an acceptance rate of 39.7%. All accepted papers are published in these proceedings and they cover the entire spectrum of pattern recognition, machine learning, optimization, action recognition, video segmentation, generative models, self-supervised learning, 3D modeling, and applications.

The conference was attended by over 720 registered participants. It included a Workshop on Scene Understanding in Unstructured Environments organized by Abhinav Valada (University of Freiburg), Peter Mortimer (Bundeswehr University Munich), Nina Heide (Fraunhofer IOSB), and Jens Behley (University of Bonn), a Tutorial on Geometric Deep Learning held by Emanuele Rodolà (Sapienza University of Rome), a Nectar Track on Machine Learning, and a Nectar Track on Pattern Recognition and Computer Vision. In the Nectar Tracks, 28 publications, which have been recently published in other conference proceedings or journals, were presented by the authors. In addition, we were very happy to have four internationally renowned researchers as our invited speakers to present their work in four fascinating areas: David Forsyth (University of Illinois at Urbana-Champaign, USA), Kristen Grauman (University of Texas at Austin and Facebook AI Research, USA), Thorsten Joachims (Cornell University, USA), and Jiri Matas (Czech Technical University in Prague, Czech Republic). The program was complemented by a Special Session on Unsolved Problems in Pattern Recognition where major challenges of pattern recognition were discussed. The presentations are available at the DAGM GCPR 2021 YouTube Channel[1].

The success of DAGM GCPR 2021 would not have been possible without the support of many people. First of all, we would like to thank all authors of the submitted papers and the speakers for their contributions. All PC members and Special Track Chairs deserve great thanks for their timely and competent reviews. We are grateful to our sponsors for their support as well. Finally, special thanks go to the support staff. We would like to

[1] https://www.youtube.com/channel/UCHm7nVTgWdCuUguSaJQV-_A/playlists.

thank Springer for giving us the opportunity of continuing to publish the DAGM GCPR proceedings in the LNCS series.

October 2021

Christian Bauckhage
Juergen Gall
Alexander Schwing

Organization

General Chair

Juergen Gall	University of Bonn, Germany

Program Chairs

Christian Bauckhage	University of Bonn, Germany
Alexander Schwing	University of Illinois at Urbana-Champaign, USA

Workshop/Tutorial Chair

Michael Möller	University of Siegen, Germany

Honorary Chair

Wolfgang Förstner	University of Bonn, Germany

Special Track Chairs

Joachim Denzler	Friedrich Schiller University Jena, Germany
Friedrich Fraundorfer	Graz University of Technology, Austria
Xiaoyi Jiang	University of Münster, Germany
Helmut Mayer	Bundeswehr University Munich, Germany
Bodo Rosenhahn	Leibniz University Hannover, Germany
Uwe Sörgel	University of Stuttgart, Germany
Carsten Steger	MVTec Software GmbH and TU Munich, Germany
Jörg Stückler	Max Planck Institute for Intelligent Systems, Germany

Award Committee

Thomas Brox	University of Freiburg, Germany
Wolfgang Förstner	University of Bonn, Germany
Simone Frintrop	University of Hamburg, Germany
Christoph Lampert	Institute of Science and Technology Austria, Austria

Program Committee

Abhinav Valada	University of Freiburg, Germany
Andrés Bruhn	University of Stuttgart, Germany
Angela Yao	National University of Singapore, Singapore
Arjan Kuijper	Fraunhofer Institute for Computer Graphics Research IGD and TU Darmstadt, Germany
Bastian Leibe	RWTH Aachen University, Germany
Bastian Goldlücke	University of Konstanz, Germany
Bernhard Rinner	University of Klagenfurt, Austria
Bernt Schiele	Max Planck Institute for Informatics, Germany
Björn Menze	TU Munich, Germany
Björn Andres	TU Dresden, Germany
Bodo Rosenhahn	Leibniz University Hannover, Germany
Boris Flach	Czech Technical University in Prague, Czech Republic
Carsten Steger	MVTec Software GmbH and TU Munich, Germany
Christian Riess	Friedrich-Alexander University Erlangen-Nuremberg, Germany
Christoph Lampert	Institute of Science and Technology Austria, Austria
Cyrill Stachniss	University of Bonn, Germany
Daniel Cremers	TU Munich, Germany
Danijel Skocaj	University of Ljubljana, Slovenia
Diego Marcos	Wageningen University, The Netherlands
Dietrich Paulus	Universtiy of Koblenz-Landau, Germany
Dorota Iwaszczuk	TU Darmstadt, Germany
Florian Kluger	Leibniz University Hannover, Germany
Friedrich Fraundorfer	Graz University of Technology, Austria
Gerard Pons-Moll	University of Tübingen, Germany
Gernot Stübl	PROFACTOR GmbH, Germany
Gernot Fink	TU Dortmund, Germany
Guillermo Gallego	TU Berlin, Germany
Hanno Scharr	Forschungszentrum Jülich, Germany
Helmut Mayer	Bundeswehr University Munich, Germany
Hendrik Lensch	University of Tübingen, Germany
Horst Bischof	Graz University of Technology, Austria
Jan Dirk Wegner	ETH Zurich, Switzerland
Jens Behley	University of Bonn, Germany
Joachim Denzler	Friedrich Schiller University Jena, Germany
Joachim Buhmann	ETH Zurich, Switzerland
Jörg Stückler	Max Planck Institute for Intelligent Systems, Germany
Josef Pauli	University of Duisburg-Essen, Germany
Julia Vogt	ETH Zurich, Switzerland
Juergen Gall	University of Bonn, Germany
Kevin Koeser	GEOMAR Helmholtz Centre for Ocean Research Kiel, Germany
Lionel Ott	ETH Zurich, Switzerland

Lucas Zimmer	University of Zurich, Switzerland
Marcello Pelillo	University of Venice, Italy
Marco Rudolph	Leibniz University Hannover, Germany
Marcus Rohrbach	Facebook AI Research, USA
Margret Keuper	University of Mannheim, Germany
Margrit Gelautz	Vienna University of Technology, Austria
Mario Fritz	CISPA Helmholtz Center for Information Security, Germany
Markus Ulrich	Karlsruhe Institute of Technology, Germany
Martin Humenberger	Naver Labs Europe, France
Martin Weinmann	Karlsruhe Institute of Technology, Germany
Matthias Hein	University of Tübingen, Germany
Michael Schmitt	German Aerospace Center, Germany
Misha Andriluka	Google, Switzerland
Olaf Hellwich	TU Berlin, Germany
Peter Ochs	University of Tübingen, Germany
Rainer Stiefelhagen	Karlsruhe Institute of Technology, Germany
Reinhard Koch	Kiel University, Germany
Ribana Roscher	University of Bonn, Germany
Ronny Hänsch	German Aerospace Center, Germany
Rudolph Triebel	TU Munich, Germany
Simone Frintrop	University of Hamburg, Germany
Stefan Leutenegger	Imperial College London, UK
Timo Bolkart	Max Planck Institute for Intelligent Systems, Germany
Torsten Sattler	Czech Technical University in Prague, Czech Republic
Ullrich Köthe	University of Heidelberg, Germany
Uwe Sörgel	University of Stuttgart, Germany
Vaclav Hlavac	Czech Technical University in Prague, Czech Republic
Vladyslav Usenko	TU Munich, Germany
Volker Roth	University of Basel, Switzerland
Wolfgang Förstner	University of Bonn, Germany
Xiaoyi Jiang	University of Münster, Germany
Zeynep Akata	University of Tübingen, Germany
Ziad Al-Halah	University of Texas at Austin, USA

Support Staffs

Alberto Martinez
Andreas Torsten Kofer
Beatriz Perez Cancer
Heidi Georges-Hecking
Jifeng Wang
Jinhui Yi
Julian Tanke
Mian Ahsan Iqbal
Olga Zatsarynna
Peter Lachart
Yasaman Abbasi
Yaser Souri
Yazan Abu Farha

Awards

DAGM GCPR 2021 Best Paper Award

InfoSeg: Unsupervised Semantic Image Segmentation with Mutual Information Maximization

Robert Harb	TU Graz, Austria
Patrick Knöbelreiter	TU Graz, Austria

DAGM GCPR 2021 Honorable Mentions

TxT: Crossmodal End-to-End Learning with Transformers

Jan-Martin O. Steitz	TU Darmstadt, Germany
Jonas Pfeiffer	TU Darmstadt, Germany
Iryna Gurevych	TU Darmstadt, Germany
Stefan Roth	TU Darmstadt, Germany

Video Instance Segmentation with Recurrent Graph Neural Networks

Joakim Johnander	Linköping University, Sweden
Emil Brissman	Linköping University, Sweden
Martin Danelljan	ETH Zurich, Switzerland
Michael Felsberg	Linköping University, Sweden

German Pattern Recognition Award

Zeynep Akata	University of Tübingen, Germany

DAGM MVTec Dissertation Award

Stability and Expressiveness of Deep Generative Models

Lars Morten Mescheder	University of Tübingen, Germany

DAGM Best Master's Thesis Award

$(SP)^2$ Net for Generalized Zero-Label Semantic Segmentation

Anurag Das	Saarland University, Germany

Contents

Machine Learning and Optimization

Sublabel-Accurate Multilabeling Meets Product Label Spaces 3
Zhenzhang Ye, Bjoern Haefner, Yvain Quéau, Thomas Möllenhoff, and Daniel Cremers

InfoSeg: Unsupervised Semantic Image Segmentation with Mutual Information Maximization 18
Robert Harb and Patrick Knöbelreiter

Sampling-Free Variational Inference for Neural Networks with Multiplicative Activation Noise 33
Jannik Schmitt and Stefan Roth

Conditional Adversarial Debiasing: Towards Learning Unbiased Classifiers from Biased Data 48
Christian Reimers, Paul Bodesheim, Jakob Runge, and Joachim Denzler

Revisiting Consistency Regularization for Semi-supervised Learning 63
Yue Fan, Anna Kukleva, and Bernt Schiele

Learning Robust Models Using the Principle of Independent Causal Mechanisms 79
Jens Müller, Robert Schmier, Lynton Ardizzone, Carsten Rother, and Ullrich Köthe

Reintroducing Straight-Through Estimators as Principled Methods for Stochastic Binary Networks 111
Alexander Shekhovtsov and Viktor Yanush

Bias-Variance Tradeoffs in Single-Sample Binary Gradient Estimators 127
Alexander Shekhovtsov

End-to-End Learning of Fisher Vector Encodings for Part Features in Fine-Grained Recognition 142
Dimitri Korsch, Paul Bodesheim, and Joachim Denzler

Investigating the Consistency of Uncertainty Sampling in Deep Active Learning 159
Niklas Penzel, Christian Reimers, Clemens-Alexander Brust, and Joachim Denzler

ScaleNet: An Unsupervised Representation Learning Method for Limited Information 174
Huili Huang and M. Mahdi Roozbahani

Actions, Events, and Segmentation

A New Split for Evaluating True Zero-Shot Action Recognition 191
Shreyank N. Gowda, Laura Sevilla-Lara, Kiyoon Kim, Frank Keller, and Marcus Rohrbach

Video Instance Segmentation with Recurrent Graph Neural Networks 206
Joakim Johnander, Emil Brissman, Martin Danelljan, and Michael Felsberg

Distractor-Aware Video Object Segmentation 222
Andreas Robinson, Abdelrahman Eldesokey, and Michael Felsberg

$(SP)^2$Net for Generalized Zero-Label Semantic Segmentation 235
Anurag Das, Yongqin Xian, Yang He, Bernt Schiele, and Zeynep Akata

Contrastive Representation Learning for Hand Shape Estimation 250
Christian Zimmermann, Max Argus, and Thomas Brox

Fusion-GCN: Multimodal Action Recognition Using Graph Convolutional Networks 265
Michael Duhme, Raphael Memmesheimer, and Dietrich Paulus

FIFA: Fast Inference Approximation for Action Segmentation 282
Yaser Souri, Yazan Abu Farha, Fabien Despinoy, Gianpiero Francesca, and Juergen Gall

Hybrid SNN-ANN: Energy-Efficient Classification and Object Detection for Event-Based Vision 297
Alexander Kugele, Thomas Pfeil, Michael Pfeiffer, and Elisabetta Chicca

A Comparative Study of PnP and Learning Approaches to Super-Resolution in a Real-World Setting 313
Samim Zahoor Taray, Sunil Prasad Jaiswal, Shivam Sharma, Noshaba Cheema, Klaus Illgner-Fehns, Philipp Slusallek, and Ivo Ihrke

Merging-ISP: Multi-exposure High Dynamic Range Image Signal Processing 328
Prashant Chaudhari, Franziska Schirrmacher, Andreas Maier, Christian Riess, and Thomas Köhler

Spatiotemporal Outdoor Lighting Aggregation on Image Sequences 343
Haebom Lee, Robert Herzog, Jan Rexilius, and Carsten Rother

Generative Models and Multimodal Data

AttrLostGAN: Attribute Controlled Image Synthesis from Reconfigurable Layout and Style 361
Stanislav Frolov, Avneesh Sharma, Jörn Hees, Tushar Karayil, Federico Raue, and Andreas Dengel

Learning Conditional Invariance Through Cycle Consistency 376
Maxim Samarin, Vitali Nesterov, Mario Wieser, Aleksander Wieczorek, Sonali Parbhoo, and Volker Roth

CAGAN: Text-To-Image Generation with Combined Attention Generative Adversarial Networks 392
Henning Schulze, Dogucan Yaman, and Alexander Waibel

TxT: Crossmodal End-to-End Learning with Transformers 405
Jan-Martin O. Steitz, Jonas Pfeiffer, Iryna Gurevych, and Stefan Roth

Diverse Image Captioning with Grounded Style 421
Franz Klein, Shweta Mahajan, and Stefan Roth

Labeling and Self-Supervised Learning

Leveraging Group Annotations in Object Detection Using Graph-Based Pseudo-labeling 439
Daniel Pototzky, Matthias Kirschner, and Lars Schmidt-Thieme

Quantifying Uncertainty of Image Labelings Using Assignment Flows 453
Daniel Gonzalez-Alvarado, Alexander Zeilmann, and Christoph Schnörr

Implicit and Explicit Attention for Zero-Shot Learning 467
Faisal Alamri and Anjan Dutta

Self-supervised Learning for Object Detection in Autonomous Driving 484
Daniel Pototzky, Azhar Sultan, Matthias Kirschner, and Lars Schmidt-Thieme

Assignment Flows and Nonlocal PDEs on Graphs 498
Dmitrij Sitenko, Bastian Boll, and Christoph Schnörr

Applications

Viewpoint-Tolerant Semantic Segmentation for Aerial Logistics 515
Shiming Wang, Fabiola Maffra, Ruben Mascaro, Lucas Teixeira, and Margarita Chli

T6D-Direct: Transformers for Multi-object 6D Pose Direct Regression 530
Arash Amini, Arul Selvam Periyasamy, and Sven Behnke

TetraPackNet: Four-Corner-Based Object Detection in Logistics Use-Cases 545
Laura Dörr, Felix Brandt, Alexander Naumann, and Martin Pouls

Detecting Slag Formations with Deep Convolutional Neural Networks 559
Christian von Koch, William Anzén, Max Fischer, and Raazesh Sainudiin

Virtual Temporal Samples for Recurrent Neural Networks: Applied to Semantic Segmentation in Agriculture 574
Alireza Ahmadi, Michael Halstead, and Chris McCool

Weakly Supervised Segmentation Pretraining for Plant Cover Prediction 589
Matthias Körschens, Paul Bodesheim, Christine Römermann, Solveig Franziska Bucher, Mirco Migliavacca, Josephine Ulrich, and Joachim Denzler

How Reliable Are Out-of-Distribution Generalization Methods for Medical Image Segmentation? 604
Antoine Sanner, Camila González, and Anirban Mukhopadhyay

3D Modeling and Reconstruction

Clustering Persistent Scatterer Points Based on a Hybrid Distance Metric 621
Philipp J. Schneider and Uwe Soergel

CATEGORISE: An Automated Framework for Utilizing the Workforce of the Crowd for Semantic Segmentation of 3D Point Clouds 633
Michael Kölle, Volker Walter, Ivan Shiller, and Uwe Soergel

Zero-Shot Remote Sensing Image Super-Resolution Based on Image Continuity and Self Tessellations 649
Rupak Bose, Vikrant Rangnekar, Biplab Banerjee, and Subhasis Chaudhuri

A Comparative Survey of Geometric Light Source Calibration Methods 663
Mariya Kaisheva and Volker Rodehorst

Quantifying Point Cloud Realism Through Adversarially Learned Latent Representations 681
Larissa T. Triess, David Peter, Stefan A. Baur, and J. Marius Zöllner

Full-Glow: Fully Conditional Glow for More Realistic Image Generation 697
Moein Sorkhei, Gustav Eje Henter, and Hedvig Kjellström

Multidirectional Conjugate Gradients for Scalable Bundle Adjustment 712
Simon Weber, Nikolaus Demmel, and Daniel Cremers

Author Index 725

Machine Learning and Optimization

Sublabel-Accurate Multilabeling Meets Product Label Spaces

Zhenzhang Ye[1(✉)], Bjoern Haefner[1], Yvain Quéau[2], Thomas Möllenhoff[3], and Daniel Cremers[1]

[1] TU Munich, Garching, Germany
yez@in.tum.de
[2] GREYC, UMR CNRS 6072, Caen, France
[3] RIKEN Center for AI Project, Tokyo, Japan

Abstract. Functional lifting methods are a promising approach to determine optimal or near-optimal solutions to difficult nonconvex variational problems. Yet, they come with increased memory demands, limiting their practicability. To overcome this drawback, this paper presents a combination of two approaches designed to make liftings more scalable, namely product-space relaxations and sublabel-accurate discretizations. Our main contribution is a simple way to solve the resulting semi-infinite optimization problem with a sampling strategy. We show that despite its simplicity, our approach significantly outperforms baseline methods, in the sense that it finds solutions with lower energies given the same amount of memory. We demonstrate our empirical findings on the nonconvex optical flow and manifold-valued denoising problems.

Keywords: Variational methods · Manifold-valued problems · Convex relaxation · Global optimization

1 Introduction

Many tasks in imaging and low-level computer vision can be transparently modeled as a variational problem. In practice, the resulting energy functionals are often nonconvex, for example due to data terms based on image-matching costs or manifold-valued constraints. The goal of this work is to develop a convex optimization approach to total variation-regularized problems of the form

$$\inf_{u:\Omega\to\Gamma} \int_{\Omega} c(x, u_1(x), \ldots, u_k(x))\,\mathrm{d}x + \sum_{i=1}^{k} \lambda_i \mathrm{TV}(u_i). \tag{1}$$

Here, $\Gamma = \{(\gamma_1, \ldots, \gamma_k) \in \mathbf{R}^N : \gamma_i \in \Gamma_i, i = 1 \ldots k\}$ is based on compact, embedded manifolds $\Gamma_i \subset \mathbf{R}^{N_i}$ with $N = N_1 + \ldots + N_k$. Throughout this paper we only consider imaging applications and pick $\Omega \subset \mathbf{R}^2$ to be a rectangular image domain. The cost function $c : \Omega \times \Gamma \to \mathbf{R}_{\geq 0}$ in (1) can be a general *nonconvex* function. Notably, we only assume that we can *evaluate* the cost function

C. Bauckhage et al. (Eds.): DAGM GCPR 2021, LNCS 13024, pp. 3–17, 2021.
https://doi.org/10.1007/978-3-030-92659-5_1

$c(x, u(x))$ but no gradient information or projection operators are available. This allows us to consider degenerate costs that are out of reach for gradient-based approaches.

As a regularization term in (1) we consider a simple separable *total variation regularization* $\text{TV}(u_i)$ on the individual components $u_i : \Omega \to \mathbf{R}^{N_i}$ weighted by a tunable hyper-parameter $\lambda_i > 0$. The total variation (TV) encourages a spatially smooth but edge-preserving solution. It is defined as

$$\text{TV}(u_i) := \sup_{\substack{p:\Omega \to \mathbf{R}^{N_i \times 2} \\ \|p(x)\|_* \leq 1}} \int_\Omega \langle \text{Div}_x\, p(x), u_i(x) \rangle \, \mathrm{d}x = \int_\Omega \|\nabla u_i(x)\| \, \mathrm{d}x, \tag{2}$$

where the last equality holds for sufficiently smooth u_i. We denote by $\nabla u_i(x) \in \mathbf{R}^{N_i \times 2}$ the Jacobian matrix in the Euclidean sense and by $\|\cdot\|_*$ the dual norm. Since our focus is on the data cost c, we consider only this separable TV case.

Problems of the form (1) find applications in low-level vision and signal processing. An example is the *optical flow* estimation between two RGB images $I_1, I_2 : \Omega \to \mathbf{R}^3$, where $\Gamma_1 = \Gamma_2 = [a, b] \subset \mathbf{R}$ are intervals and the cost function is given by $c(x, u_1(x), u_2(x)) = |I_1(x + (u_1(x), u_2(x))) - I_2(x)|$. In many applications, Γ_i is a curved manifold, see [24,37]. Examples include $\Gamma_i = \mathbb{S}^2$ in the case of normal field processing [24], SO(3) in the case of motion estimation [16] or the circle $\mathbb{S}^1$ for processing of cyclic data [11,33].

As one often wishes to estimate multiple quantities in a joint fashion, one naturally arrives at the product space formulation as considered in (1). A popular approach to address such joint optimization problems are expectation maximization procedures [12] or block-coordinate descent and alternating direction-type methods [6], where one estimates a single quantity while holding the other ones fixed. Sometimes, such approaches depend on a good initialization and can be prone to getting stuck in bad local minima. Our goal is to devise a convex relaxation of Problem (1) that can be directly solved to global optimality with standard proximal methods (possibly implemented on GPUs) such as the primal dual algorithm [29]. To achieve this, we offer the following contributions:

- To tackle relaxations of (1) in a memory-efficient manner, we propose a sublabel-accurate implementation of the product-space lifting [15]. This implementation is enabled by building on ideas from [26], which views sublabel-accurate multilabeling as a finite-element discretization.
- Our main contribution presented in Sect. 4 is a simple way to implement the resulting optimization problem with a sampling strategy. Unlike previous liftings [25,26,36], our approach does not require epigraphical projections and can therefore be applied in a black-box fashion to any cost $c(x, u(x))$.
- We show that our sublabel-accurate implementation attains a lower energy than the product-space lifting [15] on optical flow estimation and manifold-valued denoising problems.

The following Sect. 2 is aimed to provide an introduction to relaxation methods for (1) while also reviewing existing works and our contributions relative to them. We present the relaxation for (1) and its discretization in Sect. 3. In Sect. 4, we show how to implement the discretized relaxation with the proposed sampling strategy. Section 5 presents numerical results on optical flow and manifold-valued denoising and our conclusions are drawn in Sect. 6.

2 Related Work: Convex Relaxation Methods

Let us first consider a simplified version of problem (1) where Ω consists only of a single point, i.e., the nonconvex minimization of one data term:

$$\min_{\gamma \in \Gamma} c(\gamma_1, \ldots, \gamma_k). \tag{3}$$

A well-known approach to the global optimization of (3) is a *lifting* or *stochastic relaxation* procedure, which has been considered in diverse fields such as polynomial optimization [19], continuous Markov random fields [3,13,28], variational methods [30], and black-box optimization [5,27,32]. The idea is to relax the search space in (3) from $\gamma \in \Gamma$ to *probability distributions*[1] $\mathbf{u} \in \mathcal{P}(\Gamma)$ and solve

$$\min_{\mathbf{u} \in \mathcal{P}(\Gamma)} \int_{\Gamma} c(\gamma_1, \ldots, \gamma_k) \, \mathrm{d}\mathbf{u}(\gamma_1, \ldots, \gamma_k). \tag{4}$$

Due to linearity of the integral wrt. $\mathbf{u}$ and convexity of the relaxed search space, this is a convex problem for any c. Moreover, the minimizers of (4) concentrate at the optima of c and can hence be identified with solutions to (3). If Γ is a continuum, this problem is infinite-dimensional and therefore challenging.

Discrete/Traditional Multilabeling. In the context of Markov random fields [17,18] and multilabel optimization [9,21,22,39] one typically discretizes Γ into a finite set of points (called the *labels*) $\Gamma = \{\mathbf{v}_1, \ldots, \mathbf{v}_\ell\}$. This turns (4) into a finite-dimensional linear program $\min_{\mathbf{u} \in \Delta^\ell} \langle c', \mathbf{u} \rangle$ where $c' \in \mathbf{R}^\ell$ denotes the label cost and $\Delta^\ell \subset \mathbf{R}^\ell$ is the $(\ell - 1)$-dimensional unit simplex.. If we evaluate the cost at the labels, this program *upper bounds* the continuous problem (3), since instead of all possible solutions, one considers a restricted subset determined by the labels. Since the solution will be attained at one of the labels, typically a fine meshing is needed. Similar to black-box and zero-order optimization methods, this strategy suffers from the *curse of dimensionality.* When each Γ_i is discretized into ℓ labels, the overall number is ℓ^k which quickly becomes intractable since many labels are required for a smooth solution. Additionally, for pairwise or regularizing terms, often a large number of dual constraints has to be implemented. In that context, the work [23] considers a constraint pruning strategy as an offline-preprocessing.

Sublabel-Accurate Multilabeling. The discrete-continuous MRF [13,38,40] and lifting methods [20,25,26] attempt to find a more label-efficient convex formulation. These approaches can be understood through duality [13,26]. Applied to (3), the idea is to replace the cost $c : \Gamma \to \mathbf{R}$ with a dual variable $\mathbf{q} : \Gamma \to \mathbf{R}$:

[1] $\mathcal{P}(\Gamma)$ is the set of nonnegative Radon measures on Γ with total mass $\mathbf{u}(\Gamma) = 1$.

$$\min_{\mathbf{u} \in \mathcal{P}(\Gamma)} \sup_{\mathbf{q}:\Gamma \to \mathbf{R}} \int_{\Gamma} \mathbf{q}(\gamma_1, \ldots, \gamma_k) \, \mathrm{d}\mathbf{u}(\gamma_1, \ldots, \gamma_k), \text{ s.t. } \mathbf{q}(\gamma) \leq c(\gamma) \; \forall \gamma \in \Gamma. \quad (5)$$

The inner supremum in the formulation (5) *maximizes* the lower-bound $\mathbf{q}$ and if the dual variable is sufficiently expressive, this problem is equivalent to (4).

Approximating $\mathbf{q}$, for example with piecewise linear functions on Γ, one arrives at a *lower-bound* to the nonconvex problem (3). It has been observed in a recent series of works [20,25,26,36,40] that piecewise linear dual variables can lead to smooth solutions even when $\mathbf{q}$ (and therefore also $\mathbf{u}$) is defined on a rather coarse mesh. As remarked in [13,20,25], for an *affine* dual variable this strategy corresponds to minimizing the convex envelope of the cost, $\min_{\gamma \in \Gamma} c^{**}(\gamma)$, where c^{**} denotes the Fenchel biconjugate of c.

The implementation of the constraints in (5) can be challenging even in the case of piecewise-linear $\mathbf{q}$. This is partly due to the fact that the problem (5) is a semi-infinite optimization problem [4], i.e., an optimization problem with *infinitely many constraints*. The works [25,40] implement the constraints via projections onto the epigraph of the (restricted) conjugate function of the cost within a proximal optimization framework. Such projections are only available in closed form for some choices of c and expensive to compute if the dimension is larger than one [20]. This limits the applicability in a "plug-and-play" fashion.

Product-Space Liftings. The product-space lifting approach [15] attempts to overcome the aforementioned exponential memory requirements of labeling methods in an orthogonal way to the sublabel-based methods. The main idea is to exploit the product-space structure in (1) and optimize over k marginal distributions of the probability measure $\mathbf{u} \in \mathcal{P}(\Gamma)$, which we denote by $\mathbf{u}_i \in \mathcal{P}(\Gamma_i)$. Applying [15] to the single data term (3) one arrives at the following relaxation:

$$\min_{\{\mathbf{u}_i \in \mathcal{P}(\Gamma_i)\}} \sup_{\{\mathbf{q}_i:\Gamma_i \to \mathbf{R}\}} \sum_{i=1}^{k} \int_{\Gamma_i} \mathbf{q}_i(\gamma) \, \mathrm{d}\mathbf{u}_i(\gamma) \;\; \text{s.t.} \; \sum_{i=1}^{k} \mathbf{q}_i(\gamma) \leq c(\gamma) \; \forall \gamma \in \Gamma. \quad (6)$$

Since one only has to discretize the individual Γ_i this substantially reduces the memory requirements from $\mathcal{O}(\ell^N)$ to $\mathcal{O}(\sum_{i=1}^{k} \ell^{N_i})$. While at first glance it seems that the curse of dimensionality is lifted, the difficulties are moved to the dual, where we still have a large (or even infinite) number of constraints. A global implementation of the constraints with Lagrange multipliers as proposed in [15] again leads to the same exponential dependancy on the dimension.

As a side note, readers familiar with optimal transport may notice that the supremum in (6) is a multi-marginal transportation problem [8,35] with transportation cost c. This view is mentioned in [1] where relaxations of form (6) are analyzed under submodularity assumptions.

In summary, the sublabel-accurate lifting methods, discrete-continuous MRFs [25,40] and product-space liftings [15] all share a common difficulty: *implementation of an exponential or even infinite number of constraints on the dual variables.*

Summary of Contribution. Our main contribution is a simple way to implement the dual constraints in an online fashion with a random sampling strategy which we present in Sect. 4. This allows a black-box implementation, which only requires an *evaluation* of the cost c and no epigraphical projection operations as in [25, 40]. Moreover, the sampling approach allows us to propose and implement a sublabel-accurate variant of the product-space relaxation [15] which we describe in the following section.

3 Product-Space Relaxation

Our starting point is the convex relaxation of (1) presented in [15, 34]. In these works, $\Gamma_i \subset \mathbf{R}$ is chosen to be an interval. Following [36] we consider a generalization to manifolds $\Gamma_i \subset \mathbf{R}^{N_i}$ which leads us to the following relaxation:

$$\min_{\{\mathbf{u}_i:\Omega\to\mathcal{P}(\Gamma_i)\}} \sup_{\substack{\{\mathbf{q}_i:\Omega\times\Gamma_i\to\mathbf{R}\}\\ \{\mathbf{p}_i:\Omega\times\Gamma_i\to\mathbf{R}^2\}}} \sum_{i=1}^{k} \int_{\Omega}\int_{\Gamma_i} \mathbf{q}_i(x,\gamma_i) - \operatorname{Div}_x \mathbf{p}_i(x,\gamma_i)\, \mathrm{d}\mathbf{u}_i^x(\gamma)\, \mathrm{d}x, \tag{7}$$

$$\text{s.t.} \quad \|P_{T_{\gamma_i}} \nabla_{\gamma_i} \mathbf{p}_i(x,\gamma_i)\|_* \leq \lambda_i, \text{ for all } 1 \leq i \leq k,\ (x,\gamma_i) \in \Omega \times \Gamma_i, \tag{8}$$

$$\sum_{i=1}^{k} \mathbf{q}_i(x,\gamma_i) \leq c(x,\gamma), \text{ for all } (x,\gamma) \in \Omega \times \Gamma. \tag{9}$$

This cost function appears similar to (6) explained in the previous section, but with two differences. First, we now have marginal distributions $\mathbf{u}_i(x)$ for every $x \in \Omega$ since we do not consider only a single data term anymore. The notation $\mathrm{d}\mathbf{u}_i^x$ in (7) denotes the integration against the probability measure $\mathbf{u}_i(x) \in \mathcal{P}(\Gamma_i)$. The variables $\mathbf{q}_i$ play the same role as in (6) and lower-bound the cost under constraint (9). The second difference is the introduction of additional dual variables $\mathbf{p}_i$ and the term $-\operatorname{Div}_x \mathbf{p}_i$ in (7). Together with the constraint (8), this can be shown to implement the total variation regularization [24, 36]. Following [36], the derivative $\nabla_{\gamma_i} \mathbf{p}_i(x,\gamma_i)$ in (8) denotes the $(N_i \times 2)$-dimensional Jacobian considered in the Euclidean sense and $P_{T_{\gamma_i}}$ the projection onto the tangent space of Γ_i at the point γ_i. Next, we describe a finite-element discretization of (7).

3.1 Finite-Element Discretization

We approximate the infinite-dimensional problem (7) by restricting $\mathbf{u}_i$, $\mathbf{p}_i$ and $\mathbf{q}_i$ to be piecewise functions on a discrete meshing of $\Omega \times \Gamma_i$. The considered discretization is a standard finite-element approach and largely follows [36]. Unlike the forward-differences considered in [36] we use lowest-order Raviart-Thomas elements (see, e.g., [7, Section 5]) in Ω, which are specifically tailored towards the considered total variation regularization.

Discrete Mesh. We approximate each d_i-dimensional manifold $\Gamma_i \subset \mathbf{R}^{N_i}$ with a simplicial manifold Γ_i^h, given by the union of a collection of d_i-dimensional simplices $\mathcal{T}_i$. We denote the number of vertices ("labels") in the triangulation of Γ_i as ℓ_i. The set of labels is denoted by $\mathcal{L}_i = \{\mathbf{v}_{i,1}, \ldots, \mathbf{v}_{i,\ell_i}\}$. As assumed,

$\Omega \subset \mathbf{R}^2$ is a rectangle which we split into a set of faces $\mathcal{F}$ of edge-length h_x with edge set $\mathcal{E}$. The number of faces and edges are denoted by $F = |\mathcal{F}|$, $E = |\mathcal{E}|$.

Data Term and the $\mathbf{u}_i$, $\mathbf{q}_i$ *Variables.* We assume the cost $c : \Omega \times \Gamma \to \mathbf{R}_{\geq 0}$ is constant in $x \in \Omega$ on each face and denote its value as $c(x(f), \gamma)$ for $f \in \mathcal{F}$, where $x(f) \in \Omega$ denotes the midpoint of the face f. Similarly, we also assume the variables $\mathbf{u}_i$ and $\mathbf{q}_i$ to be constant in $x \in \Omega$ on each face but continuous piecewise linear functions in γ_i. They are represented by coefficient functions $\mathbf{u}_i^h, \mathbf{q}_i^h \in \mathbf{R}^{F \cdot \ell_i}$, i.e., we specify the values on the labels and linearly interpolate inbetween. This is done by the interpolation operator $\mathbf{W}_{i,f,\gamma_i} : \mathbf{R}^{F \cdot \ell_i} \to \mathbf{R}$ which given an index $1 \leq i \leq k$, face f, and (continuous) label position $\gamma_i \in \Gamma_i$ computes the function value: $\mathbf{W}_{i,f,\gamma_i} \mathbf{u}_i^h = \mathbf{u}_i(x(f), \gamma_i)$. Note that after discretization, $\mathbf{u}_i$ is only defined on Γ_i^h but we can uniquely associate to each $\gamma_i \in \Gamma_i^h$ a point on Γ_i.

Divergence and $\mathbf{p}_i$ *variables.* Our variable $\mathbf{p}_i$ is represented by coefficients $\mathbf{p}_i^h \in \mathbf{R}^{E \cdot \ell_i}$ which live on the edges in Ω and the labels in Γ_i. The vector $\mathbf{p}_i(x, \gamma_i) \in \mathbf{R}^2$ is obtained by linearly interpolating the coefficients on the vertical and horizontal edges of the face and using the interpolated coefficients to evaluate the piecewise-linear function on Γ_i^h. Under this approximation, the discrete divergence $\mathrm{Div}_x^h : \mathbf{R}^{E \cdot \ell_i} \to \mathbf{R}^{F \cdot \ell_i}$ is given by $(\mathrm{Div}_x^h \mathbf{p}_i^h)(f) = \left(\mathbf{p}_i^h(e_r) + \mathbf{p}_i^h(e_t) - \mathbf{p}_i^h(e_l) - \mathbf{p}_i^h(e_b)\right)/h_x$ where e_r, e_t, e_l, e_b are the right, top, left and bottom edges of f, respectively.

Total Variation Constraint. Computing the operator $P_{T_{\gamma_i}} \nabla_{\gamma_i}$ is largely inspired by [36, Section 2.2]. It is implemented by a linear map $\mathbf{D}_{i,f,\alpha,t} : \mathbf{R}^{E \cdot \ell_i} \to \mathbf{R}^{d_i \times 2}$. Here, $f \in \mathcal{F}$ and $\alpha \in [0,1]^2$ correspond to a point $x \in \Omega$ while $t \in \mathcal{T}_i$ is the simplex containing the point corresponding to $\gamma_i \in \Gamma_i$. First, the operator computes coefficients in $\mathbf{R}^{\ell_i}$ of two piecewise-linear functions on the manifold by linearly interpolating the values on the edges based on the face index $f \in \mathcal{F}$ and $\alpha \in [0,1]^2$. For each function, the derivative in simplex $t \in \mathcal{T}_i$ on the triangulated manifold is given by the gradient of an affine extension. Projecting the resulting vector into the d_i-dimensional tangent space for both functions leads to a $d_i \times 2$-matrix which approximates $P_{T_{\gamma_i}} \nabla_{\gamma_i} \mathbf{p}_i(x, \gamma_i)$.

Final Discretized Problem. Plugging our discretized $\mathbf{u}_i$, $\mathbf{q}_i$, $\mathbf{p}_i$ into (7), we arrive at the following finite-dimensional optimization problem:

$$\min_{\{\mathbf{u}_i^h \in \mathbf{R}^{F \cdot \ell_i}\}} \max_{\substack{\{\mathbf{p}_i^h \in \mathbf{R}^{E \cdot \ell_i}\},\\ \{\mathbf{q}_i^h \in \mathbf{R}^{F \cdot \ell_i}\}}} h_x^2 \cdot \sum_{i=1}^{k} \langle \mathbf{u}_i^h, \mathbf{q}_i^h - \mathrm{Div}_x^h \mathbf{p}_i^h \rangle + \sum_{f \in \mathcal{F}} \mathbf{i}\{\mathbf{u}_i^h(f) \in \Delta^{\ell_i}\}, \tag{10}$$

$$\text{s.t.} \quad \|\mathbf{D}_{i,f,\alpha,t} \mathbf{p}_i^h\|_* \leq \lambda_i, \ \forall 1 \leq i \leq k, f \in \mathcal{F}, \alpha \in \{0,1\}^2, t \in \mathcal{T}_i, \tag{11}$$

$$\sum_{i=1}^{k} \mathbf{W}_{i,f,\gamma_i} \mathbf{q}_i^h \leq c\left(x(f), \gamma\right), \ \forall f \in \mathcal{F}, \gamma \in \Gamma, \tag{12}$$

where $\mathbf{i}\{\cdot\}$ is the indicator function. In our applications, we found it sufficient to enforce the constraint (11) at the corners of each face which corresponds to choosing $\alpha \in \{0,1\}^2$. Apart from the infinitely many constraints in (12), this is a finite-dimensional convex-concave saddle-point problem.

3.2 Solution Recovery

Before presenting in the next section our proposed way to implement the constraints (12), we briefly discuss how a primal solution $\{\mathbf{u}_i^h\}$ of the above problem is turned into an approximate solution to (1). To that end, we follow [24,36] and compute the Riemannian center of mass via an iteration $\tau = 1, \ldots, T$:

$$V_j^\tau = \log_{u_i^\tau}(\mathbf{v}_{i,j}), \quad v^\tau = \sum_{j=1}^{\ell_i} \mathbf{u}_i^h(f,j) V_j^\tau, \quad u_i^{\tau+1} = \exp_{u_i^\tau}(v^\tau). \tag{13}$$

Here, $u_i^0 \in \Gamma_i$ is initialized by the label with the highest probability according to $\mathbf{u}_i^h(f,\cdot)$. $\log_{u_i^\tau}$ and $\exp_{u_i^\tau}$ denote the logarithmic and exponential mapping between Γ_i^h and it's tangent space at $u_i^\tau \in \Gamma_i$, which are both available in closed-form for the manifolds we consider here. In our case $T = 20$ was enough to reach convergence. For flat manifolds, $T = 1$ is enough, as both mappings boil down to the identity and (13) computes a weighted Euclidean mean.

In general, there is no theory which shows that $u^T(x) = (u_1^T(x), \ldots, u_k^T(x))$ from (13) is a global minimizer of (1). Tightness of the relaxation in the special case $k = 1$ and $\Gamma \subset \mathbf{R}$ is shown in [31]. For higher dimensional Γ, the tightness of related relaxations is ongoing research; see [14] for results on the Dirichlet energy. By computing a-posteriori optimality gaps, solutions of (7) were shown to be typically near the global optimum of the problem (1); see, e.g., [15].

4 Implementation of the Constraints

Though the optimization variables in (10) are finite-dimensional, the energy is still difficult to optimize because of the infinite constraints in (12).

Before we present our approach, let us first describe what we refer to as the *baseline method* in the rest of this paper. For the baseline approach, we consider the direct solution of (10) where we implemented the constraints only at the label/discretization points $\mathcal{L}_1 \times \ldots \times \mathcal{L}_k$ via Lagrange multipliers. This strategy is also employed by the (global variant) of the product-space approach [15].

We aim for a framework that allows for solving a better approximation of (12) than the above baseline while being of similar memory complexity. To achieve this, our algorithm alternates the following two steps in an iterative way.

1) Sampling. Based on the current solution we prune previously considered but feasible constraints and sample a new subset of the infinite constraints in (12). From all current sampled constraints, we consider the most violated constraints for each face, add one sample at the current solution and discard the rest.

2) Solving the subsampled problem. Considering the current finite subset of constraints, we solve problem (10) using a primal-dual algorithm.

These two phases are performed alternatingly, with the aim to eventually approach the solution of the continuous problem (10). The details of our constraint sampling strategy are shown in Algorithm 1. For each face in $\mathcal{F}$, the

Algorithm 1: Sampling strategy at face $f \in \mathcal{F}$.

Inputs: Solution $u = (u_1, \dots, u_k)$ at face f, sublabel-set $\mathcal{S}_f$, n, δ, r

1 $\mathcal{S}'_f \leftarrow$ uniformSample(Γ, n) `/* global exploration */`

2 $\mathcal{S}'_f \leftarrow \mathcal{S}'_f \cup$ localPerturb(u, δ, n) `/* local exploration around sol. */`

3 $\mathcal{S}_f \leftarrow \{\gamma \in \mathcal{S}_f : \sum_{i=1}^{k} \mathbf{q}_i(f,\gamma) > c(f,\gamma)\}$ `/* remove feas. cons. */`

4 $\mathcal{S}_f \leftarrow$ top-k($\mathcal{S}'_f$, r) $\cup \mathcal{S}_f$ `/* add the most violated r samples */`

5 $\mathcal{S}_f \leftarrow \mathcal{S}_f \cup \{u\}$. `/* have one sample at cur. sol. */`

6 **return** $\mathcal{S}_f$

Algorithm 2: Proposed algorithm for problem (1).

Inputs: $c : \Omega \times \Gamma \to \mathbf{R}$, $\lambda_i > 0$, $N_{it} > 0$, $M_{it} > 0$, $n > 0$, $\delta > 0$, $r > 0$

1 $\mathbf{u}_i^{h,0} = \mathbf{1}/\ell_i$, $\mathbf{q}_i^{h,0} = \mathbf{0}$, $\mathbf{p}_i^{h,0} = \mathbf{0}$, $\mathcal{S}_f^0 = \mathcal{L}_1 \times \ldots \times \mathcal{L}_k$.

2 **for** $it = 0$ **to** N_{it} **do**

3 Obtain $\mathbf{u}_i^{h,it+1}, \mathbf{q}_i^{h,it+1}, \mathbf{p}_i^{h,it+1}$ by running M_{it} iterations of the primal-dual method [29] on (10), with constraints (12) implemented at $\mathcal{S}_f^{it}$ for each $f \in \mathcal{F}$ via Lagrange multipliers; warmstart at $\mathbf{u}_i^{h,it}, \mathbf{q}_i^{h,it}, \mathbf{p}_i^{h,it}$. Compute current approximate solution u^{it+1} to (1) via (13).

4 Get $\mathcal{S}_f^{it+1}$ by calling Alg. 1 with ($u^{it+1}(f)$, $\mathcal{S}_f^{it}$, n, δ, r) for each $f \in \mathcal{F}$.

5 **end**

algorithm generates a finite set of "sublabels" $\mathcal{S}_f \subset \Gamma$ at which we implement the constraints (12). In the following, we provide the motivation behind each line in the algorithm.

Random Uniform Sampling (Line 1). To have a global view of the cost function, we consider a uniform sampling on the label space Γ. The parameter $n > 0$ determines the number of the samples for each face.

Local Perturbation Around the Mean (Line 2). Besides the global information, we apply local perturbation around the current solution u. In case the current solution is close to the optimal one, this strategy allows us to refine it with these samples. The parameter $\delta > 0$ determines the size of the local neighbourhood. In experiments, we always used a Gaussian perturbation with $\delta = 0.1$.

Pruning Strategy (Lines 3–4). Most samples from previous iterations are discarded because the corresponding constraints are already satisfied. We prune all current feasible constraints as in [4]. Similarly, the two random sampling strategies (Lines 1 and 2) might return some samples for which the constraints are already fulfilled. Therefore, we only consider the samples with violated constraints and pick the r most violated from them. This pruning strategy is essential for a memory efficient implementation as shown later.

Sampling at u (Line 5). Finally, we add one sample which is exactly at the current solution $u \in \Gamma$ to have at least one guaranteed sample per face.

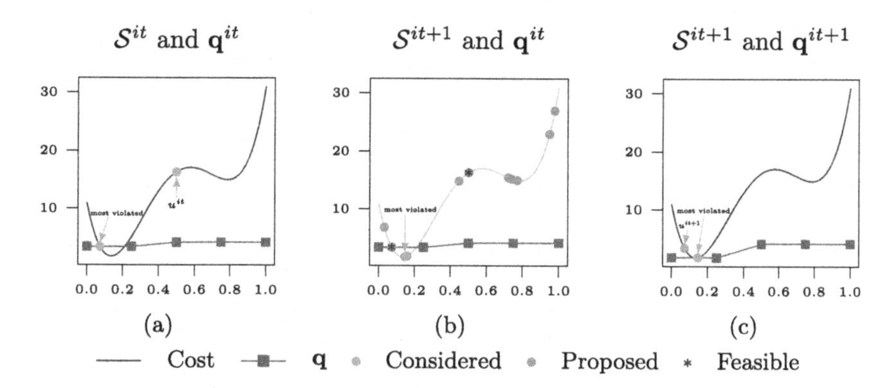

Fig. 1. Illustration of sampling strategies. (a) Two samples (red dots) are considered leading to the shown optimal dual variable $\mathbf{q}$ after running primal-dual iterations. (b) The two samples are pruned because the constraints are feasible. Several random samples are proposed (gray dots) and only one of them is picked (red dot). (c) One more sample on u^{it} is added and the $\mathbf{q}$ is refined. (Color figure online)

Overall Algorithm. After implementing the constraints at the finite set determined by Algorithm 1, we apply a primal-dual method [10] with diagonal preconditioning [29] to solve (10). Both constraints (11) and (12) are implemented using Lagrange multipliers. Based on the obtained solution, a new set of samples is determined.

This scheme is alternated for a fixed number of outer iterations N_{it} and we have summarized the overall algorithm in Algorithm 2. While we do not prove convergence of the overall algorithm, convergence results for related procedures exist; see, e.g., [4, Theorem 2.4].

Finally, let us note that a single outer iteration of Algorithm 2 with large number of M_{it} corresponds to the baseline method.

5 Numerical Validation

Our approach and the baseline are implemented in PyTorch. Code for reproducing the following experiments can be found here: https://github.com/zhenzhangye/sublabel_meets_product_space. Note that a specialized implementation as in [15] will allow the method to scale by factor $10 - 100\times$.

5.1 Illustration of Our Sampling Idea

To illustrate the effect of the sampling strategies, we consider the minimization of a single nonconvex data term. The cost c and the corresponding dual variable $\mathbf{q}$ are plotted in Fig. 1. As shown in (a), the primal-dual method can obtain the optimal $\mathbf{q}^h$ for the sampled subproblem only. Our sampling strategy can provide necessary samples and prune the feasible ones as, cp. (b). These few but necessary samples lead the $\mathbf{q}^h$ to achieve global optimality, cp. (c). If one more iteration is performed, the sampling at u^{it} can stabilize the optimal $\mathbf{q}$.

Table 1. Ablation study indicating the effect of individual lines in Algorithm 1. Numbers in parentheses indicate the standard deviation across 20 runs.

	Labels	Baseline	+Line 1&4	+Line 2	+Line 3	+Line 5
Energy	3	4589 (±0.00)	2305 (±3.73)	2291 (±3.6)	8585 (±130.4)	2051 (±10.7)
Time [s]		8.98	22.77	23.22	23.22	23.33
Mem. [Mb]		11.21	13.94	15.53	11.65	12.05
Energy	7	2582 (±0.00)	2020 (±2.68)	2012 (±1.3)	7209 (±116.7)	1969 (±3.6)
Time [s]		74.13	16.02	16.61	15.56	18.38
Mem. [Mb]		28.35	32.96	33.49	28.356	28.68
Energy	13	2029 (±0.00)	1935 (±1.14)	1926 (±0.7)	5976 (±75.7)	1901 (±3.7)
Time [s]		183.80	37.65	38.84	38.29	38.22
Mem. [Mb]		52.85	60.55	60.94	54.35	54.73

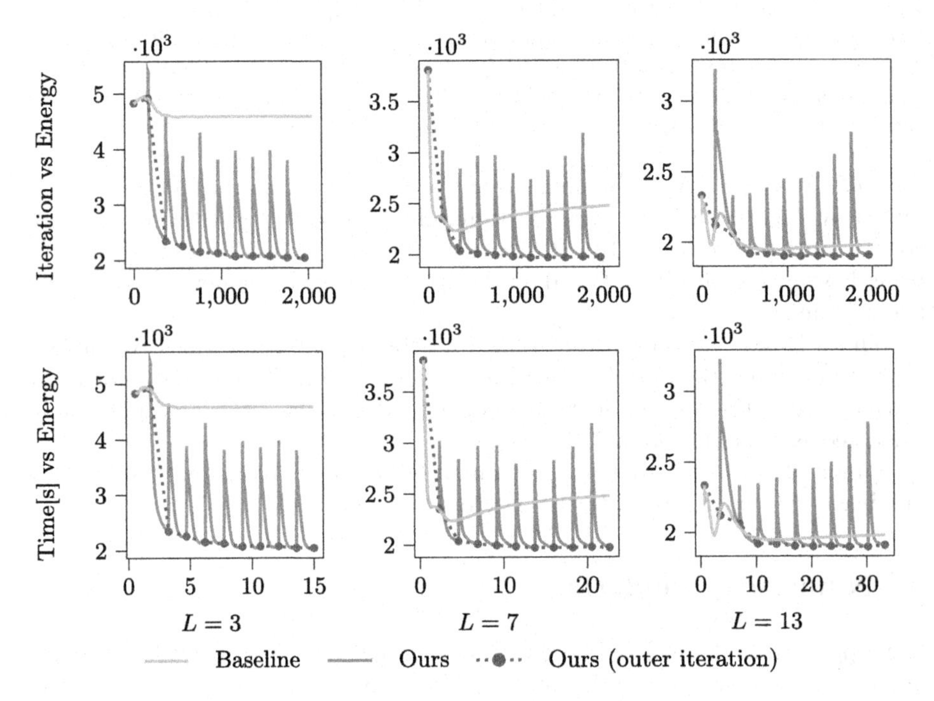

Fig. 2. Comparison between the baseline and our approach on a 64×64 image, degraded with Gaussian and salt-and-papper noise. Our approach finds lower energies in fewer iterations and time, which implements the constraints only at the label points.

5.2 Ablation Study

Next, we study the effect of each line in Algorithm 1. We evaluate our method on the truncated quadratic energy $c(x, u(x)) = \min\{(u(x) - f(x))^2, \nu\}$. where $f : \Omega \to \mathbf{R}$ is the input data. For this specific experiment, the parameters are chosen as $\nu = 0.025$, $\lambda = 0.25$, $N_{it} = 10$, $M_{it} = 200$, $n = 10$ and $r = 1$. To reduce

Baseline						
#Labels	3	7	11	15	19	23
Energy	56730.6	35902.5	25943.1	18923.1	16402.7	15297.1
Memory [Mb]	334.9	1166.6	2427.3	4111.3	6220.6	8748.51
aep [px]	3.08	2.43	2.19	2.10	2.09	2.07
aae [°]	0.79	0.51	0.37	0.32	0.31	0.30
Ours						
#Labels	3	7	11	15	19	23
Energy	**31917.5**	**18843.4**	**12891.6**	**11238.0**	**10451.8**	**9867.4**
Memory [Mb]	334.9	1167.3	2427.3	4111.2	6213.1	8748.51
aep [px]	2.58	2.32	2.09	2.08	2.07	2.07
aae [°]	0.65	0.44	0.31	0.30	0.30	0.29

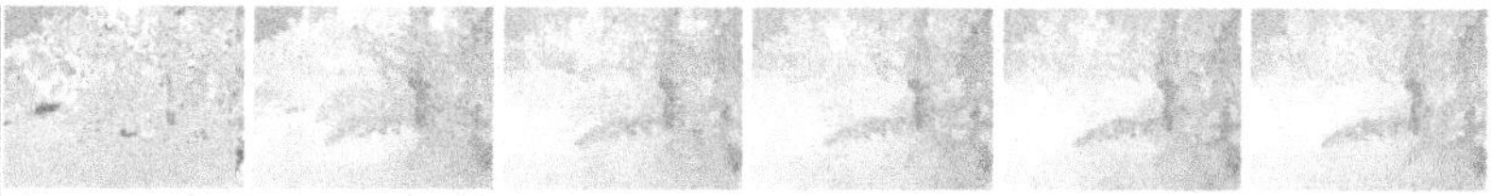

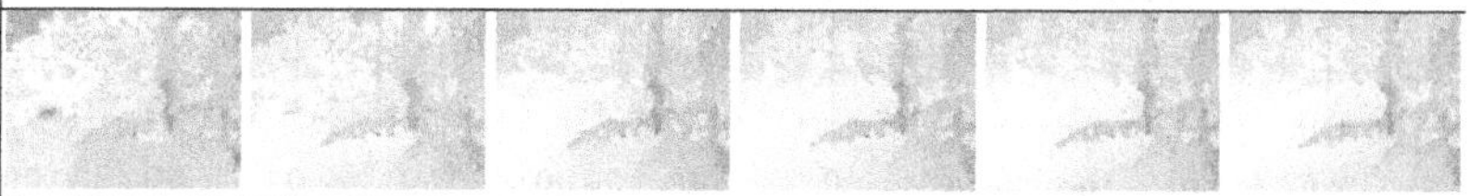

Fig. 3. We compute the optical flow on Grove3 [2] using our method and our baseline for a varying amount of labels. Given an equal number of labels/memory, our sampling strategy performs favorably to an implementation of the constraints at the labels.

the effect of randomness, we run each algorithm 20 times and report mean and standard deviation of the final energy for different number of labels in Table 1.

Adding uniform sampling and picking the most violated constraint per face already decreases the final energy significantly, i.e. Line 1 and Line 4 of Algorithm 1. We also consider local exploration around the current solution, cf. Line 2, which helps to find better energies at the expense of higher memory requirements. To circumvent that, we introduce our pruning strategy in Line 3 of Algorithm 1. However, the energy deteriorates dramatically because some faces could end up having no samples after pruning. Therefore, keeping the current solution as a sample (Line 5) per face prevents the energy from degrading.

Including all the sampling strategies, the proposed method can achieve the best energy and run-time, at comparable memory usage to the baseline method. We further illustrate the comparison on the number of iterations and time between the baseline and our proposed method in Fig. 2. Due to the replacement on the samples, we have a peak right after each sampling phase. The energy however converges immediately, leading to an overall decreasing trend.

5.3 Optical Flow

Given two input images I_1, I_2, we compute the optical flow $u : \Omega \to \mathbf{R}^2$. The label space $\Gamma = [a, b]^2$ in our case is chosen as $a = -2.5$ and $b = 7.5$. We use a simple ℓ_2-norm for the data term, i.e. $c(x, u(x)) = ||I_2(x) - I_1(x + u(x))||$

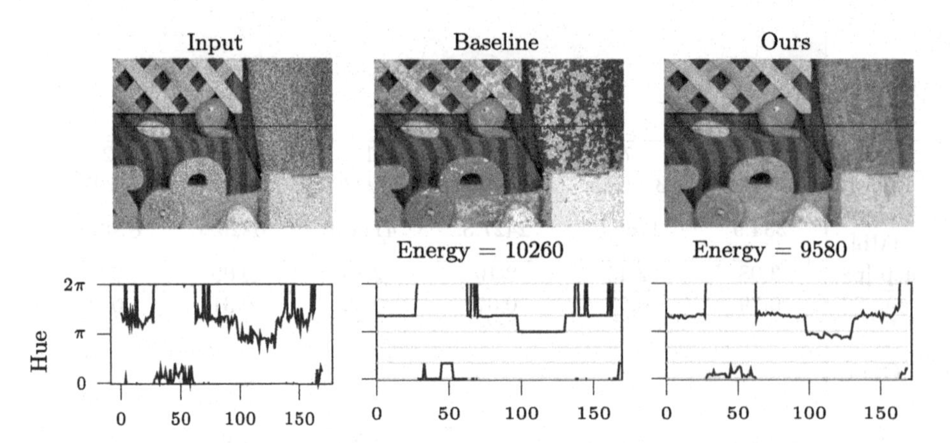

Fig. 4. Denoising of an image in HSV color space ($\Gamma_1 = \mathbb{S}^1$) using our method and the baseline. Since our approach implements the constraints adaptively inbetween the labels it reaches a lower energy with less label bias.

and set the regularization weight as $\lambda = 0.04$. The baseline approach runs for 50K iterations, while we set $N_{it} = 50$ and $M_{it} = 1000$ for a fair comparison. Additionally, we choose $n = 20$ and $r = 1$ in Algorithm 1.

The results are shown in Fig. 3. Our method outperforms the baseline approach regarding energy under the same number of labels and requires the same amount of memory. We can achieve lower energy with about half number of labels.

5.4 Denoising in HSV Color Space

In our final application, we evaluate on a manifold-valued denoising problem in HSV color space. The hue component of this space is a circle, i.e., $\Gamma_1 = \mathbb{S}^1$, $\Gamma_2, \Gamma_3 = [0, 1]$. The data term of this experiment is a truncated quadratic distance, where for the hue component the distance is taken on the circle $\mathbb{S}^1$.

Both the baseline and our method are implemented with 7 labels. 30K iterations are performed on the baseline and $N_{it} = 100$ outer iterations for our method with 300 inner primal-dual steps are used to get an equal number of total iterations. Other parameters are chosen as $\lambda = 0.015$, $n = 30$ and $r = 5$. As shown in Fig. 4, our method can achieve a lower energy than the baseline. Qualitatively, since our method implements the constraints not only at the labels but also inbetween, there is less bias compared to the baseline.

6 Conclusion

In this paper we made functional lifting methods more scalable by combining two advances, namely product-space relaxations [15] and sublabel-accurate discretizations [26,36]. This combination is enabled by adapting a cutting-plane method from semi-infinite programming [4]. This allows an implementation of sublabel-accurate methods without difficult epigraphical projections.

Moreover, our approach makes sublabel-accurate functional-lifting methods applicable to any cost function in a simple black-box fashion. In experiments, we demonstrate the effectiveness of the approach over a baseline based on the product-space relaxation [15] and provided a proof-of-concept experiment showcasing the method in the manifold-valued setting.

References

1. Bach, F.: Submodular functions: from discrete to continuous domains. Math. Program. **175**, 419–459 (2018). https://doi.org/10.1007/s10107-018-1248-6
2. Baker, S., Scharstein, D., Lewis, J.P., Roth, S., Black, M.J., Szeliski, R.: A database and evaluation methodology for optical flow. Int. J. Comput. Vis. (IJCV) **92**(1), 1–31 (2011)
3. Bauermeister, H., Laude, E., Möllenhoff, T., Moeller, M., Cremers, D.: Lifting the convex conjugate in Lagrangian relaxations: a tractable approach for continuous Markov random fields. arXiv:2107.06028 (2021)
4. Blankenship, J.W., Falk, J.E.: Infinitely constrained optimization problems. J. Optim. Theory Appl. **19**(2), 261–281 (1976)
5. de Boer, P., Kroese, D.P., Mannor, S., Rubinstein, R.Y.: A tutorial on the cross-entropy method. Ann. Oper. Res. **134**(1), 19–67 (2005)
6. Boyd, S.P., Parikh, N., Chu, E., Peleato, B., Eckstein, J.: Distributed optimization and statistical learning via the alternating direction method of multipliers. Found. Trends Mach. Learn. **3**(1), 1–122 (2011)
7. Caillaud, C., Chambolle, A.: Error estimates for finite differences approximations of the total variation. Preprint hal-02539136 (2020)
8. Carlier, G.: On a class of multidimensional optimal transportation problems. J. Convex Anal. **10**(2), 517–530 (2003)
9. Chambolle, A., Cremers, D., Pock, T.: A convex approach to minimal partitions. SIAM J. Imaging Sci. **5**(4), 1113–1158 (2012)
10. Chambolle, A., Pock, T.: A first-order primal-dual algorithm for convex problems with applications to imaging. J. Math. Imaging Vis. **40**, 120–145 (2011)
11. Cremers, D., Strekalovskiy, E.: Total cyclic variation and generalizations. J. Math. Imaging Vis. **47**(3), 258–277 (2013)
12. Dempster, A.P., Laird, N.M., Rubin, D.B.: Maximum likelihood from incomplete data via the EM algorithm. J. Roy. Stat. Soc.: Ser. B (Methodol.) **39**(1), 1–22 (1977)
13. Fix, A., Agarwal, S.: Duality and the continuous graphical model. In: Fleet, D., Pajdla, T., Schiele, B., Tuytelaars, T. (eds.) ECCV 2014. LNCS, vol. 8691, pp. 266–281. Springer, Cham (2014). https://doi.org/10.1007/978-3-319-10578-9_18
14. Ghoussoub, N., Kim, Y.H., Lavenant, H., Palmer, A.Z.: Hidden convexity in a problem of nonlinear elasticity. SIAM J. Math. Anal. **53**(1), 1070–1087 (2021)
15. Goldluecke, B., Strekalovskiy, E., Cremers, D.: Tight convex relaxations for vector-valued labeling. SIAM J. Imaging Sci. **6**(3), 1626–1664 (2013)
16. Görlitz, A., Geiping, J., Kolb, A.: Piecewise rigid scene flow with implicit motion segmentation. In: International Conference on Intelligent Robots and Systems (IROS) (2019)

17. Ishikawa, H.: Exact optimization for Markov random fields with convex priors. IEEE Trans. Pattern Anal. Mach. Intell. (PAMI) **25**(10), 1333–1336 (2003)
18. Kappes, J., et al.: A comparative study of modern inference techniques for discrete energy minimization problems. In: IEEE Conference on Computer Vision and Pattern Recognition (CVPR) (2013)
19. Lasserre, J.B.: Global optimization with polynomials and the problem of moments. SIAM J. Optim. **11**(3), 796–817 (2000)
20. Laude, E., Möllenhoff, T., Moeller, M., Lellmann, J., Cremers, D.: Sublabel-accurate convex relaxation of vectorial multilabel energies. In: Leibe, B., Matas, J., Sebe, N., Welling, M. (eds.) ECCV 2016. LNCS, vol. 9905, pp. 614–627. Springer, Cham (2016). https://doi.org/10.1007/978-3-319-46448-0_37
21. Lellmann, J., Schnörr, C.: Continuous multiclass labeling approaches and algorithms. SIAM J. Imaging Sci. **4**(4), 1049–1096 (2011)
22. Lellmann, J., Kappes, J., Yuan, J., Becker, F., Schnörr, C.: Convex multi-class image labeling by simplex-constrained total variation. In: Tai, X.-C., Mørken, K., Lysaker, M., Lie, K.-A. (eds.) SSVM 2009. LNCS, vol. 5567, pp. 150–162. Springer, Heidelberg (2009). https://doi.org/10.1007/978-3-642-02256-2_13
23. Lellmann, J., Lellmann, B., Widmann, F., Schnörr, C.: Discrete and continuous models for partitioning problems. Int. J. Comput. Vis. (IJCV) **104**(3), 241–269 (2013)
24. Lellmann, J., Strekalovskiy, E., Koetter, S., Cremers, D.: Total variation regularization for functions with values in a manifold. In: International Conference on Computer Vision (ICCV) (2013)
25. Möllenhoff, T., Laude, E., Moeller, M., Lellmann, J., Cremers, D.: Sublabel-accurate relaxation of nonconvex energies. In: IEEE Conference on Computer Vision and Pattern Recognition (CVPR) (2016)
26. Möllenhoff, T., Cremers, D.: Sublabel-accurate discretization of nonconvex free-discontinuity problems. In: International Conference on Computer Vision (ICCV) (2017)
27. Ollivier, Y., Arnold, L., Auger, A., Hansen, N.: Information-geometric optimization algorithms: a unifying picture via invariance principles. J. Mach. Learn. Res. **18**, 18:1–18:65 (2017)
28. Peng, J., Hazan, T., McAllester, D., Urtasun, R.: Convex max-product algorithms for continuous MRFs with applications to protein folding. In: International Conference on Machine Learning (ICML) (2011)
29. Pock, T., Chambolle, A.: Diagonal preconditioning for first order primal-dual algorithms in convex optimization. In: International Conference on Computer Vision (ICCV) (2011)
30. Pock, T., Schoenemann, T., Graber, G., Bischof, H., Cremers, D.: A convex formulation of continuous multi-label problems. In: Forsyth, D., Torr, P., Zisserman, A. (eds.) ECCV 2008. LNCS, vol. 5304, pp. 792–805. Springer, Heidelberg (2008). https://doi.org/10.1007/978-3-540-88690-7_59
31. Pock, T., Cremers, D., Bischof, H., Chambolle, A.: Global solutions of variational models with convex regularization. SIAM J. Imaging Sci. **3**(4), 1122–1145 (2010)
32. Schaul, T.: Studies in continuous black-box optimization. Ph.D. thesis, Technische Universität München (2011)
33. Steinke, F., Hein, M., Schölkopf, B.: Nonparametric regression between general Riemannian manifolds. SIAM J. Imaging Sci. **3**(3), 527–563 (2010)
34. Strekalovskiy, E., Chambolle, A., Cremers, D.: Convex relaxation of vectorial problems with coupled regularization. SIAM J. Imaging Sci. **7**(1), 294–336 (2014)

35. Villani, C.: Optimal Transport: Old and New. Springer, Heidelberg (2008)
36. Vogt, T., Strekalovskiy, E., Cremers, D., Lellmann, J.: Lifting methods for manifold-valued variational problems. In: Grohs, P., Holler, M., Weinmann, A. (eds.) Handbook of Variational Methods for Nonlinear Geometric Data, pp. 95–119. Springer, Cham (2020). https://doi.org/10.1007/978-3-030-31351-7_3
37. Weinmann, A., Demaret, L., Storath, M.: Total variation regularization for manifold-valued data. SIAM J. Imaging Sci. **7**(4), 2226–2257 (2014)
38. Zach, C.: Dual decomposition for joint discrete-continuous optimization. In: International Conference on Artificial Intelligence and Statistics (AISTATS) (2013)
39. Zach, C., Gallup, D., Frahm, J.M., Niethammer, M.: Fast global labeling for real-time stereo using multiple plane sweeps. In: Proceedings of the Vision, Modeling and Visualization Workshop (VMV) (2008)
40. Zach, C., Kohli, P.: A convex discrete-continuous approach for Markov random fields. In: Fitzgibbon, A., Lazebnik, S., Perona, P., Sato, Y., Schmid, C. (eds.) ECCV 2012. LNCS, vol. 7577, pp. 386–399. Springer, Heidelberg (2012). https://doi.org/10.1007/978-3-642-33783-3_28

InfoSeg: Unsupervised Semantic Image Segmentation with Mutual Information Maximization

Robert Harb(✉) and Patrick Knöbelreiter

Institute of Computer Graphics and Vision, Graz University of Technology, Graz, Austria
robert.harb@icg.tugraz.at

Abstract. We propose a novel method for unsupervised semantic image segmentation based on mutual information maximization between local and global high-level image features. The core idea of our work is to leverage recent progress in self-supervised image representation learning. Representation learning methods compute a single high-level feature capturing an entire image. In contrast, we compute multiple high-level features, each capturing image segments of one particular semantic class. To this end, we propose a novel two-step learning procedure comprising a segmentation and a mutual information maximization step. In the first step, we segment images based on local and global features. In the second step, we maximize the mutual information between local features and high-level features of their respective class. For training, we provide solely unlabeled images and start from random network initialization. For quantitative and qualitative evaluation, we use established benchmarks, and COCO-Persons, whereby we introduce the latter in this paper as a challenging novel benchmark. InfoSeg significantly outperforms the current state-of-the-art, e.g., we achieve a relative increase of 26% in the Pixel Accuracy metric on the COCO-Stuff dataset.

Keywords: Unsupervised semantic segmentation · Representation learning

1 Introduction

Semantic image segmentation is the task of assigning a class label to each pixel of an image. Various applications make use of it, including autonomous driving, augmented reality, or medical imaging. As a result, a lot of research was dedicated to semantic segmentation in the past. However, the vast majority of research focused on supervised methods. A major drawback of supervised methods is that they require large labeled training datasets containing images

Supplementary Information The online version contains supplementary material available at https://doi.org/10.1007/978-3-030-92659-5_2.

C. Bauckhage et al. (Eds.): DAGM GCPR 2021, LNCS 13024, pp. 18–32, 2021.
https://doi.org/10.1007/978-3-030-92659-5_2

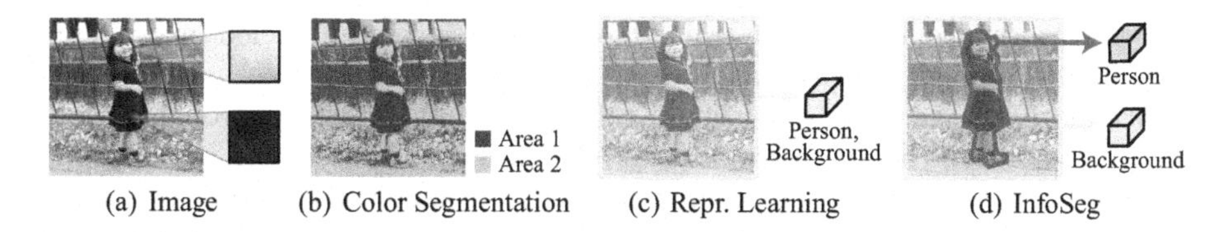

Fig. 1. **(a)** Input image. The two magnified image patches have vastly different low-level appearance despite covering the same semantic object: a person. **(b)** Color based segmentation fails to capture any high-level structure of the image. **(c)** Representation learning captures high-level information of the entire image in a single feature. **(d)** InfoSeg captures semantically similar image areas in separate features.

together with pixel-wise class labels. These datasets have to be created manually by humans with great effort. For example, annotating a single image of the Cityscapes [7] dataset required 90 min of human labor on average. This dependence of supervised methods on large human-annotated training datasets limits practical applications. We tackle this problem by introducing a novel approach on semantic image segmentation that does not require any labeled training data.

The major challenge of semantic image segmentation is to identify high-level structures in images. State-of-the-art methods approach this by learning from labeled data. While extensive research exists in segmentation without labeled data, it mainly focuses on non-learning based methods using low-level features such as color or edges [1,6,11,18]. In general, low-level features are insufficient for semantic segmentation. They are not homogenous across high-level structures. Figure 1(a–b) illustrate this problem. An image depicting a person is segmented based on color. Color changes vastly across image areas, even if they are semantically correlated. Consequently, the resulting segmentation does not capture any high-level structures. Contrarily, Fig. 1(d) illustrates how InfoSeg maps unlabeled images to segmentations that capture high-level structures. These segmentations often directly capture the semantic classes of labeled datasets.

The core idea of our method is to leverage image-level representation learning for pixel-level segmentation. Only recently, self-supervised representation learning methods [5,14,26] showed how to extract high-level features from images without any annotated training data. However, they compute features that capture the *entire* content of images. Therefore, they are not suitable for segmentation. To enable segmentation, we instead use multiple high-level features, each capturing semantically similar image areas. This allows us to assign pixels to classes based on their attribution to each of these features. Figure 1(c–d) illustrate how our approach differs from image-level representation learning. We learn high-level features with a mutual information (MI) maximization approach, inspired by Local Deep InfoMax [14]. However, unlike Local Deep InfoMax, we follow a novel two-step learning procedure enabling segmentation. At each iteration, we perform a Segmentation and Mutual Information Maximization step. In the first step, we segment images using the current features. In the second step, we update the features based on the segmentation from the first step. This two-step procedure allows us to train InfoSeg using solely unlabeled images and without pre-trained network backbones.

We motivate the exact structure of InfoSeg by first giving a thorough review of current-state-of-the-art methods [17,27], followed by a discussion of their limitations and how we approach them in InfoSeg. Our qualitative and quantitative evaluation show that InfoSeg significantly outperforms all compared methods. For example, we achieve a relative increase of 26% in Pixel-Accuracy (PA) on the COCO-Stuff dataset [4]. Even though we follow the standard evaluation protocol for quantitative evaluation, we provide a critical discussion of it and uncover problems left undiscussed by recent work [17,27]. Furthermore, in addition to established datasets, we introduce COCO-Persons as a novel benchmark. COCO-Persons contains complex scenes requiring high-level interpretation for segmentation. Our experiments show that InfoSeg handles the challenging scenes of COCO-Persons significantly better than compared methods. Finally, we perform an ablation study.

2 Related Work

Self-supervised Image Representation Learning. aims to capture high-level content of images without using any labeled training data. State-of-the-art methods follow a contrastive learning framework [2,5,10,12–14,26,30]. In contrastive learning, one computes multiple representations of differently augmented versions of the same input image. Augmentations can include photometric or geometric image transformations. During training, one enforces similarity on representations computed from the same image and dissimilarity on representations of different images. To this end, various objectives exist, such as the normalized cross entropy [26] or MI [14].

Unsupervised Semantic Image Segmentation. Invariant Information Clustering (IIC) [17] is a clustering approach also applicable for semantic segmentation. Briefly, IIC uses a MI objective that enforces the same prediction for differently augmented image patches. The authors of IIC proposed to use photometric or geometric image transformations to compute augmentations. For example, one can create augmentations by random color jittering, rotation, or scaling. Ouali *et al.* [27] did a follow-up work on IIC. In addition to standard image transformations, they proposed to process image patches through various masked convolutions. We further discuss these two methods and its differences to InfoSeg in Sect. 3.2. Concurrent to our work, Mirsadeghi *et al.* proposed InMARS [23]. InMARS is also related to IIC. However, instead of operating on each pixel individually, InMARS utilizes a superpixel representation. Furthermore, a novel adversarial training scheme is introduced.

Another recently introduced method that states to perform unsupervised semantic segmentation is SegSort [15]. However, we note that SegSort still uses supervised learning at multiple stages. First, they initialize parts of their network architecture with pre-trained weights obtained by supervised training of a classifier on the ImageNet [8] dataset. Second, they use pseudo ground truth masks generated by a HED contour detector [31], which is trained supervised using the BSDS500 [1] dataset. Therefore, we do not consider SegSort as an unsupervised method.

3 Motivation

In this section, we first review how recent work [17,27] uses MI for unsupervised semantic image segmentation. Then, we discuss limitations of these methods, and how we tackle them in InfoSeg.

3.1 Unsupervised Semantic Image Segmentation

State-of-the-art methods [17,27] adapt the MI based image clustering approach of IIC [17] for segmentation. In the following, we introduce IICs' approach on image clustering and then the proposed modifications for segmentation.

For clustering, one creates two versions x and x' of the same image. These versions show the same semantic content, but alter low-level appearance by using random photometric or geometric transformations. Consequently, semantic class predictions y and y' of the two images x and x' should be the same. To achieve this, one maximizes the MI between y and y'

$$\max_{\psi} \quad I(\Phi_{\psi}(x); \Phi_{\psi}(x')) = I(y; y'), \tag{1}$$

where Φ is a CNN parametrized by ψ. Considering we can express the MI between y and y' as

$$I(y; y') = H(y) - H(y|y'), \tag{2}$$

Equation (1) maximizes the entropy $H(y)$ while minimizing the conditional entropy $H(y|y')$. Minimizing $H(y|y')$ pushes predictions of the two images x and x' together. Therefore, the network has to compute predictions invariant to the different low-level transformations. This should encourage class predictions to depend on high-level image content instead. While sole minimization of $H(y|y')$ can trivially be done by assigning the same class to all images. Additional maximization of $H(y)$ has a regularization effect against such degenerate solutions. Since maximizing $H(y)$ encourages predictions that put equal probability mass on all classes. Consequently, predictions for all images can not collapse to a single class.

For segmentation, Ji *et al.* [17] proposed to use the previously introduced clustering approach on image patches rather than entire images. Two image versions are pushed through a network that computes dense pixel-wise class predictions. The objective given in Eq. (1) is now applied on the pixel-wise class predictions. Therefore, each prediction depends on an image patch rather than an entire image. Patches are defined by the receptive field for each output pixel of the network. Additionally, one enforces local spatial invariance by maximizing MI of predictions from adjacent image patches. This approach on unsupervised semantic segmentation was initially proposed by IIC [17]. Furthermore, Ouali *et al.* [27] proposed an extension by generating views using different masked convolutions [25]. In the following, we discuss three major limitations of these two works, and how we tackle them in InfoSeg.

3.2 Limitations of Current Methods

The first limitation of discussed methods is that they do not incorporate global image context. Global context is essential to capture high-level structures, since they often cover large image areas having diverse local appearance. Therefore observing only small image patches is often not sufficient to identify them. Ideally, each pixel-wise prediction should depend on the entire image. Nevertheless, the discussed approaches make pixel-wise predictions based on image patches. The receptive field of the network Φ determines the size of these patches. In general, one could enlarge the receptive field by changing the network architecture. However, adapting IIC from clustering to segmentation is based on restricting each prediction's receptive field from entire images to patches. By making each pixel-wise prediction dependent on the entire image again, one would fall back to clustering. In InfoSeg we capture global context in global high-level features that cover the entire image. We make pixel-wise predictions based on the MI between these global features and local patch-wise features. This allows each pixel-wise prediction to depend on the entire image.

A second limitation of discussed methods is that they fail to leverage recent advances in image representation learning [2,5,14,26]. These methods are effective at capturing high-level image content, but only at the image-level. Adapting them for pixel-level segmentation is not trivial. Ouali *et al.* [27] attempted this with their Autoregressive Representation Learning (ARL) loss, but failed to increase segmentation performance. Despite high-level information is constant across large image areas, ARL computes for *each pixel* a separate high-level feature. Contrarily, in InfoSeg, we share high-level features over the *entire image*. To still allow pixel-wise segmentation, we compute multiple high-level features. Each high-level feature encodes only image areas depicting one class. We then assign pixels to classes based on their attribution to each of these features.

Finally, discussed methods jointly learn features and segmentations. They use intermediate feature representations to assign pixels to class labels. At the beginning of training, features depend on random initialization and contain no high-level information. This can lead to classes that latch onto low-level features instead of capturing high-level information. This issue was first discussed for image classification by SCAN [29]. Instead, we decouple feature learning and segmentation. Therefore, we perform two steps at each iteration. First, we compute features that are explicitly trained to encode high-level information. Then, we use them for segmentation.

4 InfoSeg

In InfoSeg, we tackle unsupervised semantic image segmentation. We take a set $\{\mathbf{X}^{(n)} \in \mathcal{X}\}_{n=1}^{N}$ of N unlabeled images and assign a label $\mathcal{Z} = \{z_1, \ldots, z_K\}$ to every pixel of each image. Importantly, for one particular image, we do not specify which nor how many labels should be assigned. We only provide the total number of labels K in all images. After training, we follow the standard

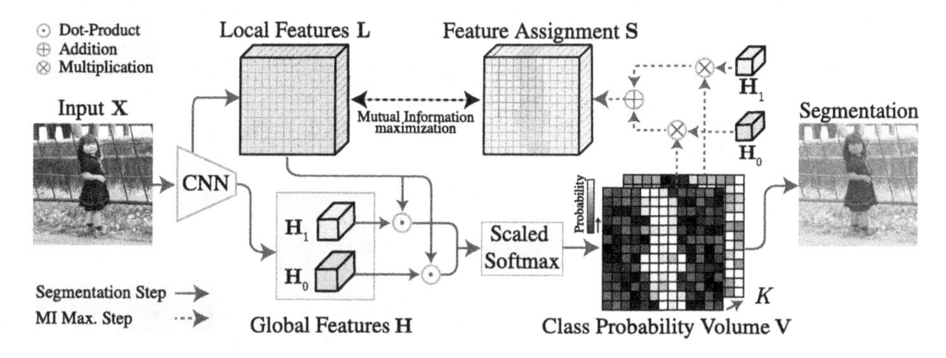

Fig. 2. Overview of InfoSeg for $K = 2$ classes. At each training iteration, we alternate the following two steps. **Segmentation Step** (solid lines): An input image $\mathbf{X}$ is passed through a CNN to compute local patch-wise features $\mathbf{L}$ and for each class k a global image-level feature $\mathbf{H}_k$. We then score local $\mathbf{L}$ with global $\mathbf{H}$ features using a dot-product. The result is passed through a scaled softmax function to compute the class probability volume $\mathbf{V}$. Finally, we obtain a segmentation by assigning each pixel to the class with the largest probability. **Mutual Information Maximization Step** (dashed lines): The global feature assignment $\mathbf{S}$ is computed as a sum of global features, weighted by their respective class probabilities at each spatial position. Finally, we maximize Mutual Information between local features $\mathbf{L}$ and their respective feature assignment $\mathbf{S}$.

evaluation protocol and map the learned labels of InfoSeg directly to the semantic classes of an annotated dataset.

InfoSeg is designed to tackle the three limitations of state-of-the-art methods discussed in Sect. 3.2. Figure 2 shows an overview of InfoSeg. In the following, we first discuss how we leverage recent progress in representation learning for semantic segmentation in Sect. 4.1. Then we provide further details of our method in Sect. 4.2 and Sect. 4.3.

4.1 Representation Learning for Segmentation

We first review how Local Deep InfoMax [14] captures high-level information of entire images, and then how InfoSeg adapts this approach to target image segmentation.

Local Deep InfoMax [14] learns global high-level features of images by maximizing their average MI with local features. Local features cover image patches, and the global feature covers the entire image. If the global feature has limited capacity, the network cannot simply copy all local features' content into the global feature to maximize MI. Instead, the network has to encode a compact representation that shares information with as many image patches as possible. Hjelm *et al.* [14] showed that the resulting global features encode high-level image information. They motivated this by the idea that high-level information is often constant over an entire image, while low-level information such as pixel-

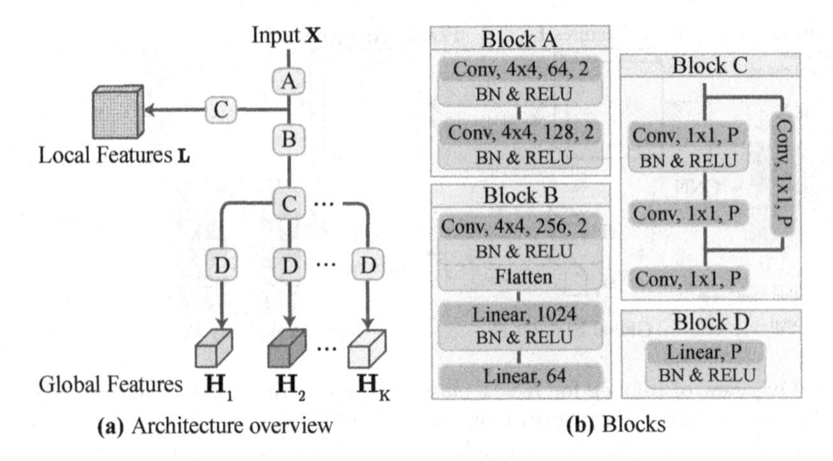

Fig. 3. Feature computation for K classes. **(a)** Overview of network architecture. **(b)** Used blocks. **Legend**: *Conv,* $W \times W, C, d$: Convolution with filter size $W \times W$, C channels and stride d. Blocks that are used multiple times, each have their own set of parameters.

level noise varies. Consequently, the global feature is encouraged to encode the former while disregarding the latter.

To enable pixel-wise segmentation, we compute for each image multiple global features instead of a single one. Each global feature only encodes image areas that depict a particular class. This allows us to segment images by assigning pixels to classes based on their attribution to each global feature. During training, we maximize for each global feature MI only with local features covering its respective class. Therefore, we learn high-level features in a similar way as Local Deep InfoMax [14], but target segments instead of entire images. This requires us to learn high-level features together with segmentations. To this end, we alternate two steps at each iteration. In the *Segmentation Step*, we assign local to global features based on their content, *i.e.*, we segment images. In the *Mutual Information Maximization step*, we maximize the MI between all global features and assigned local features, *i.e.*, we learn the features. We describe both steps in the following.

4.2 Segmentation Step

Given an input image $\mathbf{X} \in \mathcal{X} = \mathbb{R}^{M \times N \times C}$, we compute P-dimensional global $\mathbf{H} \in \mathbb{R}^{K \times P}$ and patch-wise local $\mathbf{L} \in \mathbb{R}^{U \times V \times P}$ features. The k-th global feature $\mathbf{H}_k \in \mathbb{R}^P$ encodes a high-level representation for the k-th class and covers the entire image. The local feature $\mathbf{L}_{i,j} \in \mathbb{R}^P$ at the spatial position (i, j) encodes an image patch. Furthermore, the spatial resolution of $\mathbf{L}$ is downsampled by a rate of d from the input resolution, *i.e.* $U = M/d$ and $V = N/d$.

Figure 3 shows the architecture of our feature computation network. First, the input image is processed by Block A, resulting in a grid of patch-wise image features. To compute the local features $\mathbf{L}$, we further process these patch-wise features by Block C. Adding this additional residual block of pointwise convolutions led to better performance, than using Block A's output directly for the local features. To compute the global features $\mathbf{H}_k$, we first process the output of Block A to image-level features using Block B. Then, similarly, as for the local features, we add a residual block of pointwise convolutions using Block C. Finally, each global feature is computed using a separate linear layer using Block D.

To compute an image segmentation, we use the dot-product of a local and global feature pair $\langle \mathbf{L}_{i,j}, \mathbf{H}_k \rangle$ as a class score. A high score indicates that the k-th class is shown at the position (i, j). We elaborate in Sect. 4.3 how MI maximization increases the dot-product of a local feature and the global feature of its corresponding class. After computing the class scores, we apply a pixel-wise scaled softmax to compute a class-probability volume $\mathbf{V} \in \mathbb{R}^{U \times V \times K}$ with elements

$$V_{i,j,k} = \frac{\exp\left(\tau \cdot \langle \mathbf{L}_{i,j}, \mathbf{H}_k \rangle\right)}{\sum_{\hat{k}} \exp\left(\tau \cdot \langle \mathbf{L}_{i,j}, \mathbf{H}_{\hat{k}} \rangle\right)}, \tag{3}$$

where τ is a hyper-parameter that controls the smoothness of the resulting distribution. Using the probability volume, we compute the low-res segmentation $\mathbf{K}$ for every pixel (i, j) with

$$k_{i,j} = \arg\max_{k \in \mathcal{Z}} V_{i,j,k}, \tag{4}$$

by taking the class with the largest probability. We can then compute a full-res segmentation $\mathbf{Z}$ by upsampling the low-res segmentation $\mathbf{K}$ to the input image resolution.

4.3 Mutual Information Maximization Step

We first need to assign each local feature to its corresponding class's global feature. We could do this using the segmentation $\mathbf{K}$. However, this disregards class probabilities, instead of utilizing their exact values, *e.g.* to account for uncertainty. Especially at the beginning of training, segmentations are uncertain and depend on random network initialization. Reinforcing possibly incorrect predictions can lead to degenerate solutions. To alleviate this problem, we do not make hard class assignments using $\mathbf{K}$, but soft assignments using class-probabilities $\mathbf{V}$. Instead of assigning a single global feature, we weight each global feature by its respective class probability. To this end, we define the function $S_{\boldsymbol{\theta}}^{(i,j)}$ that computes a soft global feature assignment for the local feature $\mathbf{L}_{(i,j)}$ as follows

$$S_{\boldsymbol{\theta}}^{(i,j)}(\mathbf{X}) = \sum_k V_{i,j,k} \cdot H_{\boldsymbol{\theta}}^{(k)}(\mathbf{X}), \tag{5}$$

where the function $H_{\boldsymbol{\theta}}^{(k)}(\mathbf{X})$ computes the k-th global feature $\mathbf{H}_k$ for an image $\mathbf{X}$, and $\boldsymbol{\theta}$ denotes the learnable parameters of our network.

During training, we maximize the MI between the output of $S_\theta^{(i,j)}(\mathbf{X})$ and the corresponding local feature $\mathbf{L}_{(i,j)}$ for all spatial positions (i,j). Hence our objective is given as

$$\max_\theta \mathbb{E}_\mathbf{X}\left[\frac{1}{UV}\sum_{i,j} I\left(L_\theta^{(i,j)}(\mathbf{X}); S_\theta^{(i,j)}(\mathbf{X})\right)\right], \tag{6}$$

where $\mathbb{E}_\mathbf{X}$ denotes the expectation over all training images $\mathbf{X}$ and the function $L_\theta^{(i,j)}(\mathbf{X})$ computes the local features $\mathbf{L}_{i,j}$ given an input image $\mathbf{X}$.

To evaluate our objective Eq. 6, consider that local and global features are high-dimensional continuous random variables. MI computation of such variables is challenging. Contrarily to discrete variables as in the objective of IIC Eq. 1, where exact computation is possible. For continuous variables, Belghazi *et al.* [3] proposed MI estimation by maximizing lower bounds parametrized by neural networks. They used a bound based on the Donsker & Varadhan (DV) representation of the Kullback-Leibler (KL) divergence. While several other bounds exist [28], we use a bound based on the Jensen-Shannon Divergence (JSD). Mainly because Hjelm *et al.* [14] showed favorable properties of the JSD bound compared to others in their representation learning setting. This includes increased training stability and better performance with smaller batch sizes. Nevertheless, we also perform experiments using the DV bound in our ablation studies. A JSD based MI estimator $\widehat{I}_{JSD}(X;Y)$ for two random variables X and Y can be defined as follows [24]

$$I(X;Y) \geq \widehat{I}_{JSD}(X;Y) := \mathbb{E}_{p(x,y)}[-\operatorname{sp}(-T(x,y))] - \mathbb{E}_{p(x)p(y)}[\operatorname{sp}(T(x,y))], \tag{7}$$

where $\operatorname{sp}(x) = \log(1+e^x)$ and T is a discriminator mapping sample pairs from X and Y to a real valued score. The first and second expectations are taken over samples from the joint $p(x,y)$ and marginal $p(x)p(y)$ distributions. Consequently, to tighten the bound, the discriminator T needs to discriminate samples from the joint and marginal distributions by assigning high or low scores, respectively.

To use the JSD estimator Eq. 7 in our objective Eq. 6, we have to define the discriminator T and a sampling strategy. Following recent work [2,14], we create joint and marginal samples by combining feature pairs computed from the same image $\mathbf{X}$ and two randomly paired images $\mathbf{X}$ and $\mathbf{X}'$, respectively. The discriminator T can be implemented using any arbitrary function that maps feature pairs to a discrimination score, *e.g.*, a neural network. For efficiency, we use the dot-product to compute discrimination scores *i.e.*, $T(x,y) := \langle x, y\rangle$. This requires only a single expensive forward pass through our network to compute the features, while we can then score any arbitrary combination with a cheap dot-product. Omitting the spatial indices (i,j) to avoid notational clutter, this leads to the MI estimator

$$\widehat{I}_{JSD}\left(L_\theta(\mathbf{X}); S_\theta(\mathbf{X})\right) := \mathbb{E}_\mathbb{P}[-\operatorname{sp}(-\langle L_\theta(x), S_\theta(x)\rangle)] - \mathbb{E}_{\mathbb{P}\times\tilde{\mathbb{P}}}[\operatorname{sp}(\langle L_\theta(x); S_\theta(x')\rangle)], \tag{8}$$

where $\mathbb{P}$ is the empirical distribution of our dataset, x is an image sampled from $\mathbb{P}$ and x' is an image sampled from $\tilde{\mathbb{P}} = \mathbb{P}$. We can now simply insert the estimator of Eq. 8 into our objective Eq. 6. Maximizing the resulting objective increases the dot-product of local features with the global feature of their assigned class. Consequently, we use the dot-product as a class score, as described in Sect. 4.2.

5 Experiments

We first introduce our experimental setup and discuss challenges at the quantitative evaluation of unsupervised segmentation. Then we perform an evaluation using established benchmarks [4,22], and COCO-Persons, a novel dataset introduced in this work. On all datasets, InfoSeg significantly outperforms compared methods. Finally, we perform ablation studies.

5.1 Setup

We start training from random network initialization and provide solely unlabeled images. We set $P = 1024$, $\tau = 0.8$ and use the ADAM optimizer [19] with a learning rate of 10^{-4} and a batch size of 64. Furthermore, the network architecture we use results in a downsampling rate of $d = 4$, and we set the number of classes K to be equal to the number of classes in each dataset.

Note that InfoSeg requires a network with a different structure as Invariant Information Clustering (IIC) and Autoregressive Clustering (AC). For InfoSeg, the final outputs are 1×1 sized global image features. Contrarily, in IIC and AC, the final outputs are pixel-wise class predictions downscaled from the input image resolution. This impedes a comparison with these methods using the exact same architecture. Nevertheless, we provide an experiment in our ablation study where we apply the objective of IIC on the output of our Segmentation Step.

5.2 Quantitative Evaluation

Meaningful quantitative evaluation of unsupervised semantic segmentation is challenging. Recent work used the PA for quantitative evaluation. The PA is defined as the percentage of pixels assigned to the same class as in a given annotation. However, in unsupervised semantic segmentation, one does not specify which classes should be used for segmentation. Instead, many different segmentations can be considered as equally valid. Nevertheless, quantitative evaluation metrics, such as the PA or mean Intersection-Over-Union (mIoU), evaluate all pixel-wise predictions as incorrect that do not exactly match the given annotations. While this has been left undiscussed by previous work [17,27], we emphasize this has to be considered when interpreting quantitative metrics of unsupervised methods.

We can further illustrate problems at quantitative evaluation using the COCO-Stuff [4] dataset as an example. The dataset contains the class *rawmaterial* that labels image areas depicting metal, plastic, paper, or cardboard. We argue that this is a very specific class and aggregating these four materials in one class is an arbitrary design choice of the dataset. It is unfeasible to expect an unsupervised method to come up with this specific solution. Nevertheless, recent methods [17,27] reported significant increases over baseline models on the PA. We attribute this to the dataset's vast class imbalance. Besides very specific classes such as *rawmaterial*, the dataset also contains more generic classes such as *water*, or *plant*. These classes are overrepresented and make up more than 50% of all pixels. Therefore, an algorithm can achieve high PA by focusing mainly on these few overrepresented classes. To illustrate this effect, we provide a confusion matrix of our predictions in the supplementary material.

Despite the discussed problems, we follow prior work and use the PA to evaluate all of our results quantitatively. Following the standard evaluation protocol [17,27], we map each of the predicted classes in $\mathcal{Z}$ to one of the annotated classes in $\mathcal{Z}'$ before computing the PA. This is necessary because class ordering is unknown without providing labeled data during training. We find the one-to-one mapping between $\mathcal{Z}$ and $\mathcal{Z}'$ by solving the linear assignment problem using the Hungarian method [20]. We compute this mapping once after training and use the same mapping for all images in the dataset.

5.3 Data

Recent work [17,27] established the COCO-Stuff [4] and Potsdam [22] datasets as benchmarks. COCO-Stuff contains 15 classes and Potsdam 6 classes. Additionally, for both datasets, a reduced 3-class variation exists. We use the same pre-processing as in the compared methods, resulting in 128×128 sized RGB images for COCO-Stuff, and 200×200 sized RGBIR images for Potsdam.

While unsupervised segmentation of COCO-Stuff and Potsdam is challenging, most classes in these datasets still have a homogeneous low-level appearance. For example, low-level features such as color and texture are often sufficient to segment areas labeled as *water* in COCO-Stuff or *road* in Potsdam. To show that InfoSeg can go one step further, we evaluate on an additional dataset where segmentation is more reliant on high-level image features. To this end, we introduce the COCO-Persons dataset, which we will provide publicly. Each image depicts one or multiple persons and is annotated with a person and a non-person class. Face, hair, and clothing of persons vary vastly in color, texture, and shape, and the non-person areas cover a variety of complex indoor and outdoor scenes. The dataset is a subset of the COCO [21] dataset and contains 15 399 images having 128×128 pixels.

5.4 Results

We provide quantitative and qualitative results in Table 1 and Fig. 4, respectively. To compute results for COCO-Persons, we used publicly available implementations, if available. In our experiments, InfoSeg significantly outperformed all compared methods [9,16,17,23,27]. We discuss qualitative results in the following.

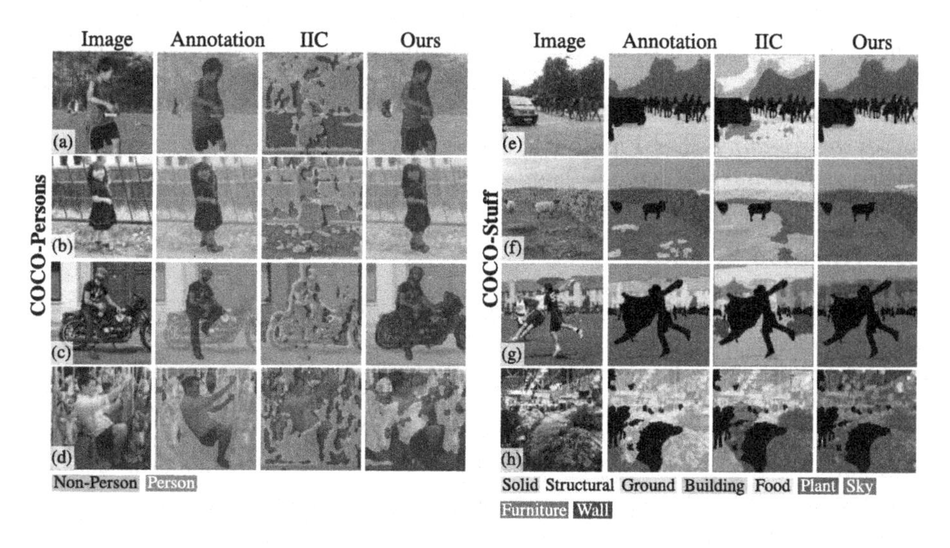

Fig. 4. Qualitative comparison. Non-stuff areas in COCO-Stuff are masked in black.

Table 1. Pixel-Accuracy of InfoSeg and compared methods. *Clustering of features from methods that are not specifically designed for image segmentation.

Method	COCO-Persons	COCO-Stuff	COCO-Stuff-3	Potsdam	Potsdam-3
Random CNN	52.3	19.4	37.3	28.3	38.2
K-Means	54.3	14.1	52.2	35.3	45.7
Doersch* [9]	55.6	23.1	47.5	37.2	49.6
Isola* [16]	57.5	24.3	54.0	44.9	63.9
IIC [17]	57.1	27.7	72.3	45.4	65.1
AC [27]	–	30.8	72.9	49.3	66.5
InMARS [23]	–	31.0	73.1	47.3	70.1
InfoSeg (ours)	**69.6**	**38.8**	**73.8**	**57.3**	**71.6**

COCO-Persons. In Fig. 4(a–c), we show successful segmentation of images with vastly inhomogenous low-level appearance. InfoSeg even captures the two small persons in the background of Fig. 4(a). In Fig. 4(c), the motorbike is assigned to the same class as the person. The dataset contains several images where persons are shown together with motorbikes. Therefore, without supervision, it is challenging to disentangle these two semantic concepts. In Fig. 4(d), we show a challenging example yielding a failure case.

COCO-Stuff. Figure 4(e–f) show examples where our predictions are close to the annotations. Figure 4(g–h) provide reasonable segmentations, even though large portions differ from the annotations. These examples demonstrate challenges at the evaluation of COCO-Stuff due to overly specific classes. Matching the annotations requires precise distinction of similar high-level concepts, which is difficult without supervision. The example in Fig. 4(g) shows multiple houses that are assigned to the same class as the stone wall in Fig. 4(f). However, the ground truth of COCO-Stuff assigns the stone wall to a *wall* class and the houses to a *building* class. Figure 4(h) shows a market scene containing vegetables labeled as *food* but predicted as *plants*. Arguably, vegetables are food and plants.

5.5 Ablation Studies

To examine the influence of individual components, we perform the following ablation studies. First, we evaluate the effectiveness of soft assignments by replacing them with hard assignments. Therefore, we change our objective Eq. 6 to maximize the MI at each spatial position between the local feature and the global feature of the assigned class according to the segmentation $\mathbf{K}$. Second, we replace the JSD MI estimator with a DV one. Finally, in the last ablation study, we omit our Mutual Information Maximization step and solely perform our Segmentation Step. As a replacement for our Mutual Information Maximization step we apply the MI maximization objective of IIC, referred to as IIC-MI. To create the two image versions required by IIC-MI, we use the same transformations as in IIC.

Table 2. Ablation studies on COCO-Stuff-3.

Measure	MI Max. Step	PA
IIC-MI	–	60.2
DV	Hard Assignment	53.9
DV	Soft Assignment	55.1
JSD	Hard Assignment	67.3
JSD	Soft Assignment	**73.8**

Table 2 shows the results of our ablation studies, whereby we performed all experiments using the COCO-Stuff-3 dataset. We can observe the following: Using soft assignments increases performance over hard assignments. A JSD-based MI estimator performs better than a DV-based, which aligns with the results of Hjelm *et al.* [14]. And replacing our Mutual Information Maximization step with the objective of IIC leads to a decline in performance.

6 Conclusion

We proposed a novel approach for unsupervised semantic image segmentation. Our experiments showed that our method yields semantically meaningful predictions and significantly outperforms related methods. We used the established datasets for evaluation and introduced a novel challenging benchmark COCO-Person. Furthermore, we discussed several problems making the quantitative evaluation of unsupervised semantic segmentation challenging. Finally, we performed ablation studies on our model.

Acknowledgements. This work was partly funded by the Austrian Research Promotion Agency (FFG) under project 874065.

References

1. Arbelaez, P., Maire, M., Fowlkes, C., Malik, J.: Contour detection and hierarchical image segmentation. IEEE Trans. Pattern Anal. Mach. Intell. **33**, 898–916 (2010)
2. Bachman, P., Hjelm, R.D., Buchwalter, W.: Learning representations by maximizing mutual information across views. In: Conference on Neural Information Processing Systems, pp. 15535–15545 (2019)
3. Belghazi, M.I., et al.: Mutual information neural estimation. In: International Conference on Learning Representations, pp. 530–539 (2018)
4. Caesar, H., Uijlings, J., Ferrari, V.: COCO-stuff: thing and stuff classes in context. In: Computer Vision and Pattern Recognition, pp. 1209–1218 (2018)
5. Chen, T., Kornblith, S., Norouzi, M., Hinton, G.: A simple framework for contrastive learning of visual representations. arXiv preprint arXiv:2002.05709 (2020)
6. Comaniciu, D., Meer, P.: Mean shift: a robust approach toward feature space analysis. IEEE Trans. Pattern Anal. Mach. Intell. **24**(5), 603–619 (2002)
7. Cordts, M., et al.: The cityscapes dataset for semantic urban scene understanding. Technical report. www.cityscapes-dataset.net
8. Deng, J., Dong, W., Socher, R., Li, L.J., Li, K., Fei-Fei, L.: ImageNet: a large-scale hierarchical image database. In: Computer Vision and Pattern Recognition (2009)
9. Doersch, C., Gupta, A., Efros, A.A.: Unsupervised visual representation learning by context prediction. In: International Conference on Computer Vision, pp. 1422–1430 (2015)
10. Dosovitskiy, A., Springenberg, J.T., Riedmiller, M., Brox, T.: Discriminative unsupervised feature learning with convolutional neural networks. In: Conference on Neural Information Processing Systems, pp. 766–774 (2014)
11. Felzenszwalb, P.F., Huttenlocher, D.P.: Efficient graph-based image segmentation. Int. J. Comput. Vis. **59**, 167–181 (2004)
12. He, K., Fan, H., Wu, Y., Xie, S., Girshick, R.: Momentum contrast for unsupervised visual representation learning. In: Computer Vision and Pattern Recognition (2020)
13. Henaff, O.: Data-efficient image recognition with contrastive predictive coding. In: International Conference on Machine Learning, pp. 4182–4192 (2020)
14. Hjelm, R.D., et al.: Learning deep representations by mutual information estimation and maximization. In: International Conference on Learning Representations (2019)
15. Hwang, J.J., et al.: SegSort: segmentation by discriminative sorting of segments. In: International Conference on Computer Vision (2019)
16. Isola, P., Zoran, D., Krishnan, D., Adelson, E.H.: Learning visual groups from co-occurrences in space and time. arXiv preprint arXiv:1511.06811 (2015)
17. Ji, X., Henriques, J.F., Vedaldi, A.: Invariant information clustering for unsupervised image classification and segmentation. In: International Conference on Computer Vision, October 2019
18. Shi, J., Malik, J.: Normalized cuts and image segmentation. IEEE Trans. Pattern Anal. Mach. Intell. **22**, 888–905 (2000)
19. Kingma, D.P., Ba, J.: Adam: a method for stochastic optimization. In: International Conference on Learning Representations (2015)
20. Kuhn, H.W.: The Hungarian method for the assignment problem. In: Naval Research Logistics Quarterly, vol. 2, pp. 83–97. Wiley Online Library (1955)
21. Lin, T.-Y., et al.: Microsoft COCO: common objects in context. In: Fleet, D., Pajdla, T., Schiele, B., Tuytelaars, T. (eds.) ECCV 2014. LNCS, vol. 8693, pp. 740–755. Springer, Cham (2014). https://doi.org/10.1007/978-3-319-10602-1_48

22. Markus Gerke, I.: Use of the stair vision library within the ISPRS 2D semantic labeling benchmark (Vaihingen) (2015)
23. Mirsadeghi, S.E., Royat, A., Rezatofighi, H.: Unsupervised image segmentation by mutual information maximization and adversarial regularization. IEEE Robot. Autom. Lett. **6**(4), 6931–6938 (2021)
24. Nowozin, S., Cseke, B., Tomioka, R.: F-GAN: training generative neural samplers using variational divergence minimization. In: Conference on Neural Information Processing Systems, pp. 271–279 (2016)
25. Van den Oord, A., Kalchbrenner, N., Espeholt, L., Vinyals, O., Graves, A., et al.: Conditional image generation with PixelCNN decoders. In: Conference on Neural Information Processing Systems, pp. 4790–4798 (2016)
26. den Oord, A.V., Li, Y., Vinyals, O.: Representation learning with contrastive predictive coding. arXiv preprint arXiv:1807.03748. vol. abs/1807.03748 (2018)
27. Ouali, Y., Hudelot, C., Tami, M.: Autoregressive unsupervised image segmentation. In: Vedaldi, A., Bischof, H., Brox, T., Frahm, J.-M. (eds.) ECCV 2020. LNCS, vol. 12352, pp. 142–158. Springer, Cham (2020). https://doi.org/10.1007/978-3-030-58571-6_9
28. Poole, B., Ozair, S., van den Oord, A., Alemi, A., Tucker, G.: On variational bounds of mutual information. In: International Conference on Machine Learning, pp. 5171–5180 (2019)
29. Van Gansbeke, W., Vandenhende, S., Georgoulis, S., Proesmans, M., Van Gool, L.: SCAN: learning to classify images without labels. In: Vedaldi, A., Bischof, H., Brox, T., Frahm, J.-M. (eds.) ECCV 2020. LNCS, vol. 12355, pp. 268–285. Springer, Cham (2020). https://doi.org/10.1007/978-3-030-58607-2_16
30. Wu, Z., Xiong, Y., Yu, S.X., Lin, D.: Unsupervised feature learning via non-parametric instance discrimination. IEEE Trans. Pattern Anal. Mach. Intell. **37**, 3733–3742 (2018)
31. Xie, S., Tu, Z.: Holistically-nested edge detection. In: International Conference on Computer Vision, vol. 125, pp. 3–18. Kluwer Academic Publishers, Hingham, December 2017

Sampling-Free Variational Inference for Neural Networks with Multiplicative Activation Noise

Jannik Schmitt[1(✉)] and Stefan Roth[1,2]

[1] Department of Computer Science, TU Darmstadt, Darmstadt, Germany
jannik.schmitt@visinf.tu-darmstadt.de
[2] hessian.AI, Darmstadt, Germany

Abstract. To adopt neural networks in safety critical domains, knowing whether we can trust their predictions is crucial. Bayesian neural networks (BNNs) provide uncertainty estimates by averaging predictions with respect to the posterior weight distribution. Variational inference methods for BNNs approximate the intractable weight posterior with a tractable distribution, yet mostly rely on sampling from the variational distribution during training and inference. Recent sampling-free approaches offer an alternative, but incur a significant parameter overhead. We here propose a more efficient parameterization of the posterior approximation for sampling-free variational inference that relies on the distribution induced by multiplicative Gaussian activation noise. This allows us to combine parameter efficiency with the benefits of sampling-free variational inference. Our approach yields competitive results for standard regression problems and scales well to large-scale image classification tasks, e.g. ImageNet.

Keywords: Bayesian deep learning · Variational inference · Uncertainty estimation

1 Introduction

When applying deep networks to safety critical problems, uncertainty estimates for their predictions are paramount. Bayesian inference is a theoretically well-founded framework for estimating the model-inherent uncertainty by computing the posterior distribution of the parameters. While sampling from the posterior of a Bayesian neural network is possible with different Markov chain Monte Carlo (MCMC) methods, often an explicit approximation of the posterior can be beneficial, for example for continual learning [34]. Variational inference (VI) can be used to approximate the posterior with a simpler distribution. Since the

---

Supplementary Information The online version contains supplementary material available at https://doi.org/10.1007/978-3-030-92659-5_3.

C. Bauckhage et al. (Eds.): DAGM GCPR 2021, LNCS 13024, pp. 33–47, 2021.
https://doi.org/10.1007/978-3-030-92659-5_3

variational objective as well as the predictive distribution cannot be calculated analytically, they are often approximated through Monte Carlo integration using samples from the approximate posterior [1,9]. During training, this introduces additional gradient variance, which can be a problem when training large BNNs. Further, multiple forward-passes are required to compute the predictive distribution, which makes deployment in time-critical systems difficult.

Recently, sampling-free variational inference methods [11,38,50] have been proposed, with similar predictive performance to sampling-based VI methods on small-scale tasks. They may also be able to remedy the gradient variance problem of sampling-based VI for larger-scale tasks. Still, sampling-free methods incur a significant parameter overhead. To address this, we here propose a sampling-free variational inference scheme for BNNs – termed MNVI – where the approximate posterior can be induced by *multiplicative Gaussian activation noise*. This helps us *decrease the number of parameters of the Bayesian network to almost half*, while still being able to analytically compute the Kullback-Leibler (KL) divergence with regard to an isotropic Gaussian prior. Further, assuming multiplicative activation noise allows to reduce the computational cost of variance propagation in Bayesian networks compared to a Gaussian mean-field approximate posterior. We then discuss how our MNVI method can be applied to modern network architectures with batch normalization and max-pooling layers. Finally, we describe how regularization by the KL-divergence term differs for networks with the induced variational posterior from networks with a mean-field variational posterior.

In experiments on standard regression tasks [3], our proposed sampling-free variational inference approach achieves competitive results while being more lightweight than other sampling-free methods. We further apply our method to large-scale image classification problems using modern convolutional networks including ResNet [12], obtaining well-calibrated uncertainty estimates while also improving the prediction accuracy compared to standard deterministic networks.

We make the following contributions: *(i)* We propose to reduce the number of parameters and computations for sampling-free variational inference by using the distribution induced by multiplicative Gaussian activation noise in neural networks as a variational posterior (cf. Fig. 1); *(ii)* we show how our MNVI method can be applied to common network architectures that are used in practice; *(iii)* we demonstrate experimentally that our method retains the accuracy of sampling-free VI with a mean-field Gaussian variational posterior while performing better with regard to uncertainty specific evaluation metrics despite fewer parameters; and *(iv)* we successfully apply it to various large-scale image classification problems such as classification on ImageNet.

2 Related Work

Bayesian neural networks (BNNs) allow for a principled quantification of model uncertainty in neural networks. Early approaches for learning the posterior distribution of BNNs rely on MCMC methods [33] or a Laplace approximation [30] and are difficult to scale to modern neural networks. More modern

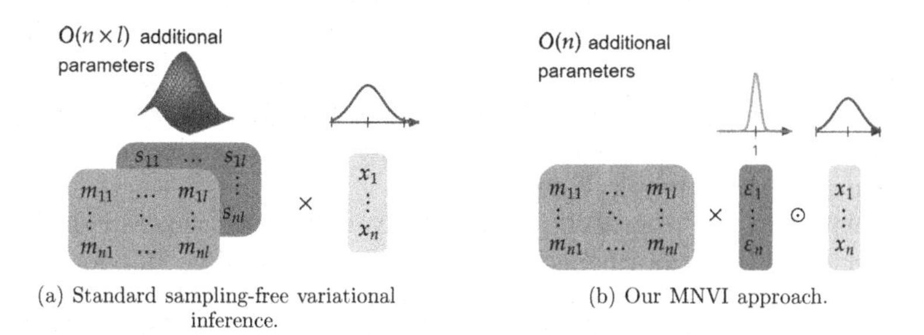

(a) Standard sampling-free variational inference.

(b) Our MNVI approach.

Fig. 1. Both standard sampling-free mean-field VI *(a)* and our MNVI approach *(b)* assume that the input x to a linear layer with weight matrix M as well as the weight uncertainty of the posterior can be approximated by Gaussian distributions. We additionally assume that the variational posterior is induced by multiplicative Gaussian activation noise ε with mean 1. This lets us represent the weight uncertainty *compactly* as an n-dimensional vector, storing the variance of the activation noise. In contrast, a mean-field variational posterior needs an $n \times l$ matrix S to store the weight variances.

approaches use a variety of Bayesian inference methods, such as assumed density filtering for probabilistic backpropagation [8,14], factorized Laplace approximation [37], stochastic gradient MCMC methods like SGLD [46], IASG [31], or more recent methods [13,51], as well as variational inference.

Variational inference for neural networks was first proposed by Hinton and Van Camp [16], motivated from an information-theoretic perspective. Further stochastic descent-based VI algorithms were developed by Graves [9] and Blundell et al. [1]. Latter utilizes the local re-parametrization trick [23] and can be extended to variational versions [21] of adaptive gradient-based optimization with preconditioning depending on the gradient noise, such as Adam [22] and RMSProp [15]. Swiatkowski et al. [41] recently proposed an efficient low-rank parameterization of the mean-field posterior.

Multiplicative activation noise, such as Bernoulli or Gaussian noise, is well-known from the regularization methods Dropout and Gaussian Dropout [40]. Gal and Ghahramani [6] and Kingma et al. [24] showed that Dropout and Gaussian Dropout, respectively, can be interpreted as approximate VI. However, regarding the respective priors that are chosen to result in simple L_2-regularization of the network weights for Dropout and a regularization term that depends only on the variance of the noise variables for Gaussian Dropout, the evidence lower bound (ELBO) is not well-defined [17]. Rank-1 BNNs [4] use multiplicative noise with hierarchical priors to perform sampling-based VI. More complex distributions for the multiplicative noise modeled by normalizing flows were studied by Louizos and Welling [29]. Multiplicative noise has further been considered for reducing sampling variance by generating pseudo-independent weight samples [48] and to address the increase of parameters for ensembles [47].

Variance propagation for sampling-free VI has been proposed by Roth and Pernkopf [38], Haussmannn et al. [11], and Wu et al. [50]. Jankowiak [19] computes closed form objectives in the case of a single hidden layer. Variance propagation has also been used in various uncertainty estimation algorithms for estimating epistemic uncertainty by Bayesian principles, such as Probabilistic Back Propagation [14], Stochastic Expectation Propagation [28], and Neural Parameter Networks [44], or aleatoric uncertainty by propagating an input noise estimate through the network [7,20].

Most related to our approach, multiplicative activation noise and variance propagation have been jointly used by Wang and Manning [45] for training networks with dropout noise without sampling, as well as by Postels et al. [36] for sampling-free uncertainty estimation for networks that are trained by optimizing the approximate ELBO of [6]. However, [45] estimates a different objective than the ELBO in variational inference and does not estimate predictive uncertainty, while [36] uses Monte Carlo Dropout during training. Further, both methods assume Bernoulli noise on the network's activation that is not adapted during training. Our approach not only retains the low parameter overhead of dropout-based methods, but offers a well-defined ELBO. To the best of our knowledge, we are the first to perform sampling-free VI for networks with adaptive Gaussian activation noise.

3 MNVI – Variational Inference in Neural Networks with Multiplicative Gaussian Activation Noise

3.1 Variational Inference and the Evidence Lower Bound

Given a likelihood function $p(y|x, w)$ parameterized by a neural network with weights w, data $\mathcal{D} = (X, Y)$, and a prior on the network's weights $p(w)$, we are interested in the posterior distribution of the network weights

$$p(w|X, Y) = \frac{1}{p(Y|X)} p(Y|X, w)p(w). \tag{1}$$

For neural networks, however, computing $p(Y|X)$ and the exact posterior $p(w|X, Y)$ is intractable. A popular method to approximate the posterior is variational inference. Let $q(w|\theta)$ be a family of tractable distributions parameterized by θ. The objective of variational inference [42] is to choose a distribution $q(w|\theta^*)$ such that θ^* minimizes $\mathrm{KL}\big[q(w|\theta)\big|\big|p(w|X, Y)\big]$, where $\mathrm{KL}[\cdot||\cdot]$ denotes the Kullback-Leibler divergence. Minimizing the divergence is equivalent to maximizing the evidence lower bound

$$\mathrm{ELBO}(\theta) = \mathbb{E}_{q(w|\theta)}\big[\log p(Y|X, w)\big] - \mathrm{KL}\big[q(w|\theta)\big|\big|p(w)\big], \tag{2}$$

where $\mathbb{E}_{q(w|\theta)}[\log p(Y|X, w)]$ is the expected log-likelihood with respect to the variational distribution and $\mathrm{KL}[q(w|\theta)||p(w)]$ is the divergence of the variational distribution from the prior. Finding a distribution that maximizes the ELBO can, therefore, be understood as a trade-off between fitting to a data-dependent term while not diverging too far from the prior belief about the weight distribution.

3.2 The Implicit Posterior of BNNs with Adaptive Gaussian Activation Noise

For variational inference in Bayesian neural networks, the expected log-likelihood cannot be calculated exactly and has to be approximated, commonly through Monte Carlo integration using samples from the variational distribution [1,9]:

$$\mathbb{E}_{q(w|\theta)}\left[\log p(Y|X,w)\right] \approx \frac{1}{S}\sum_{s=1}^{S}\log p(Y|X,w^s), \quad w^s \sim q(w|\theta). \tag{3}$$

Since w is a random variable, for any given X and neural network h_w depending on the weights w, the output $h_w(X)$ is a random variable with respect to the same probability space that defines a distribution for the network output $h_w(X)$. Wu et al. [50] showed that in case the output distribution $q(h_w(X))$ is known, the expected value can instead be computed by integration with respect to the output distribution:

$$\mathbb{E}_{q(w|\theta)}\left[\log p(Y|X,w)\right] = \mathbb{E}_{q(h_w(X))}\left[\log p(Y|h_w(X))\right]. \tag{4}$$

An alternative way to approximate the expected log-likelihood is, therefore, to approximate the output distribution using variance propagation.

The variational family used to approximate the weight posterior of a BNN is often chosen to be Gaussian. However, even a naive Gaussian mean-field approximation with a diagonal covariance matrix doubles the number of parameters compared to a deterministic network, which is problematic for large-scale networks that are relevant in various application domains.

Assumption 1. *To reduce the parameter overhead, we here propose to use the induced Gaussian variational distribution of a network with multiplicative Gaussian activation noise* [24].

Given the output x_j of an activation function with independent multiplicative $\mathcal{N}(1,\alpha_j)$-distributed noise ε_{ij} and a deterministic weight m_{ij}, the product $m_{ij}\varepsilon_{ij}x_j$ is distributed equally to the product of x_j and a stochastic $\mathcal{N}(m_{ij},\alpha_j m_{ij}^2)$-distributed weight w_{ij}. Therefore, the implicit variational distribution of networks with independent multiplicative Gaussian activation noise is given by

$$q(w|m,\alpha) = \prod_{i,j}\mathcal{N}(w_{ij}|m_{ij},\alpha_j m_{ij}^2). \tag{5}$$

This distribution is parameterized by the weight means m_{ij} and activation noise α_j, hence for a linear layer with M_1 input units and M_2 output units, it introduces only M_1 new parameters in addition to the $M_1 \times M_2$ parameters of a deterministic model. This way, the quadratic scaling of the number of additional parameters with respect to the network width for a traditional mean-field variational posterior with a diagonal covariance matrix approximation [9,50] can be *reduced to linear scaling* in our case, as visualized in Fig. 1. Thus our assumption significantly reduces the parameter overhead. As we will see, this not only benefits practicality, but in fact even improves uncertainty estimates.

Further for convolutional layers, we assume that the variance parameter of the multiplicative noise is shared within a channel. For a convolutional layer with C_1 input channels, C_2 output channels, and filter width W, only C_1 new parameters are introduced in addition to the $C_1 \times C_2 \times W^2$ parameters of the deterministic layer. While the number of additional parameters for a mean-field variational posterior scales quadratically with respect to the number of channels $C_1 \times C_2$ and filter width W, the number of additional parameters for the distribution induced by our multiplicative activation noise assumption scales only *linearly* in C_1, again significantly reducing the parameter overhead.

3.3 Variance Propagation

Assuming we can compute the distribution of a network's output $q(h_w(x))$, which is induced by parameter uncertainties, as well as $\mathbb{E}_{q(h_w(X))}[\log p(Y|h_w(X))]$ and $\mathrm{KL}[q(w|\theta)||p(w)]$ for the chosen variational family and prior distribution, we can perform sampling-free variational inference [50]. To approximate the output distribution $q(h_w(x))$, variance propagation can be utilized. Variance propagation can be made tractable and efficient by two assumptions:

Assumption 2. *The output of a linear or convolutional layer is approximately Gaussian.*

Since a multivariate Gaussian is characterized by its mean vector and covariance matrix, the approximation of the output distribution of a linear layer can be calculated by just computing the first two moments. For sufficiently wide networks this assumption can be justified by the central limit theorem.

Assumption 3. *The correlation of the outputs of linear and convolutional layers can be neglected.*

This assumption is crucial for reducing the computational cost, since it allows our MNVI approach to propagate only the variance vectors instead of full covariance matrices, thus avoiding quadratic scaling with regards to layer width. Without it, applying variance propagation to modern network architecture would not be practical. While multiplication with independent noise reduces the correlation of the outputs of linear layers, the correlation will generally be non-zero. Empirically, Wu et al. [50] verified that for shallow networks the predictive performance of a BNN is mostly retained under this assumption. In the supplemental material, we show that, while the output variance is somewhat underestimated, the predictive performance is retained also for deeper networks and the mean of the output distribution is estimated well by variance propagation.

Reducing the Computation. For two independent random variables A and X with finite second moment, mean and variance of their product can be calculated by

$$\mathbb{E}[AX] = \mathbb{E}[A]\mathbb{E}[X], \quad \mathbb{V}[AX] = \mathbb{V}[A]\mathbb{V}[X] + \mathbb{V}[A]\mathbb{E}[X]^2 + \mathbb{E}[A]^2\mathbb{V}[X]. \tag{6}$$

Given a linear layer with independent $\mathcal{N}(m_{ij}, s_{ij})$-distributed weights and independent inputs with means x_j and variances v_j, we can calculate the mean and variance of the outputs Z as

$$\mathbb{E}[Z_i] = \sum_j m_{ij}x_j, \qquad \mathbb{V}[Z_i] = \sum_j s_{ij}v_j + s_{ij}x_j^2 + m_{ij}^2 v_j, \tag{7}$$

which can be vectorized as

$$\mathbb{E}[Z] = Mx, \qquad \mathbb{V}[Z] = S(v + x \circ x) + (M \circ M)v, \tag{8}$$

where $\circ$ is the Hadamard-product. Hence, in addition to one matrix-vector multiplication for the mean propagation, two more matrix-vector multiplications are needed for variance propagation through a linear layer, tripling the computational cost compared to its deterministic counterpart.

Importantly, in the case of our proposed posterior weight distribution for networks with multiplicative activation noise, however, the weight variance $s_{ij} = \alpha_j m_{ij}^2$ can be linked to the weight mean, so that the output variance for our MNVI approach can be computed as

$$\mathbb{V}[Z] = (M \circ M)(\alpha \circ (v + x \circ x)) + (M \circ M)v = (M \circ M)((1 + \alpha) \circ v + \alpha \circ x \circ x). \tag{9}$$

Therefore, variance propagation through linear layers for multiplicative noise networks only requires computing one matrix-vector product and the cost for computing the variance can be reduced by half.

Similarly, for convolutional layers with $\mathcal{N}(m_{ij}, s_{ij})$-distributed weights in addition to one convolutional operation for mean propagation, generally two additional convolutional operations are needed for variance propagation. By linking the weight variance to the weight mean with multiplicative activation noise, i.e. $w_{ij} \sim \mathcal{N}(m_{ij}, \alpha_j m_{ij}^2)$, the cost of variance propagation through a convolutional layer can also be reduced to only one additional convolutional operation.

Variance Propagation Through Activation Layers. For non-linear activation functions f, we assume that the input Z can be approximated Gaussian with mean μ and variance σ^2. We thus calculate the mean and variance of the output distribution as

$$\begin{aligned} \mathbb{E}[f(Z)] &= \int_{\mathbb{R}} f(z)\phi\left(\frac{z-\mu}{\sigma}\right) \mathrm{d}z, \\ \mathbb{V}[f(Z)] &= \int_{\mathbb{R}} f(z)^2\phi\left(\frac{z-\mu}{\sigma}\right) \mathrm{d}z - \mathbb{E}[f(Z)]^2, \end{aligned} \tag{10}$$

where $\phi(z)$ is the standard Gaussian probability density function.

As an example, for the rectifier function, mean and variance can be computed analytically [5] as

$$\begin{aligned} \mathbb{E}[\mathrm{ReLU}(Z)] &= \mu\Phi\left(\frac{\mu}{\sigma}\right) + \sigma\phi\left(\frac{\mu}{\sigma}\right), \\ \mathbb{V}[\mathrm{ReLU}(Z)] &= (\mu^2 + \sigma^2)\Phi\left(\frac{\mu}{\sigma}\right) + \mu\sigma\phi\left(\frac{\mu}{\sigma}\right) - \mathbb{E}[\mathrm{ReLU}(Z)]^2, \end{aligned} \tag{11}$$

where Φ is the cumulative density function of a standard normal distribution. If these integrals cannot be computed analytically for some other activation function, they can instead be approximated by first-order Taylor expansion, yielding

$$\mathbb{E}[f(Z)] \approx f(\mathbb{E}[Z]), \quad \mathbb{V}[f(Z)] \approx \left.\frac{\mathrm{d}f(z)}{\mathrm{d}z}\right|_{\mathbb{E}[Z]}^{2} \cdot \mathbb{V}[Z]. \tag{12}$$

Batch Normalization and Max-Pooling Layers. To apply our approach to modern neural networks, variance estimates have to be propagated through batch-normalization [18] and pooling layers. Similar to Osawa et al. [35], we aim to retain the stabilizing effects of batch normalization during training and, therefore, do not model a distribution for the batch-normalization parameters and do not change the update rule for the input mean x. For propagating the variance v, we view the batch-normalization layer as a linear layer, obtaining

$$x' = \alpha \cdot \frac{x - \mathrm{mean(x)}}{\sqrt{\mathrm{var}(x) + \varepsilon}} + \beta, \quad v' = \frac{\alpha^2 v}{\mathrm{var}(x) + \varepsilon}, \tag{13}$$

where $\mathrm{mean}(x)$ and $\mathrm{var}(x)$ are the batch mean and variance, α, β are the affine batch-normalization parameters, and ε is a stabilization constant.

The output mean and variance of the max-pooling layer under the assumption of independent Gaussian inputs cannot be calculated analytically for more then two variables. Therefore, we use a simple approximation similar to [11] that preserves the sparse activations of the deterministic max-pooling layer. Given an input window of variables with means (x_{ij}) and variances (v_{ij}), the max-pooling layer outputs $x_{ij^*} = \max_{ij} x_{ij}$ and the variance of the respective entry v_{ij^*}.

3.4 The NLLH Term

Regression problems are often formulated as L_2-loss minimization, which is equivalent to maximizing the log-likelihood under the assumption of additive homoscedastic Gaussian noise. We obtain a more general loss function, if we allow for heteroscedastic noise, which can be modeled by a network predicting an input-dependent estimate of the log-variance $c(x)$ in addition to the predicted mean $\mu(x)$ of the Gaussian. The log-likelihood of a data point (x, y) under these assumptions is

$$\log p\big(y|\mu(x), c(x)\big) = -\frac{1}{2}\left(\log 2\pi + c(x) + e^{-c(x)}(\mu(x) - y)^2\right). \tag{14}$$

Given a Gaussian approximation for the output distribution of the network, the expected log-likelihood can now be computed analytically. To represent heteroscedastic aleatoric uncertainty, we assume that the network predicts the mean $\mu(x)$ and the log-variance $c(x)$ of the Gaussian conditional distribution and additionally outputs the variances $v_\mu(x)$ and $v_c(x)$ of those estimates due to the weight uncertainties represented by the multiplicative Gaussian activation noise.

Note that $\mu(x)$, $c(x)$, $v_\mu(x)$, and $v_c(x)$ thus depend on the learnable parameters m and α. The expected log-likelihood can then be computed by integrating the log-likelihood with respect to an approximation $\tilde{q}(h_w(x))$ of the output distribution of the network and is given by [50]

$$\begin{aligned}\mathbb{E}_{\tilde{q}(w|\theta)}\big[\log p(y|x,w)\big] = -\frac{1}{2}\big(&\log 2\pi + c(x) \\ &+ e^{-c(x)+v_c(x)}\big(v_\mu(x) + (\mu(x)-y)^2\big)\big).\end{aligned} \tag{15}$$

For classification, however, the expected log-likelihood cannot be calculated analytically. Instead, it can be approximated by Monte Carlo integration by sampling logits from the Gaussian approximation $\tilde{q}(h_w(x)) = \mathcal{N}(\mu(x), v_\mu(x))$ of the network's output distribution:

$$\begin{aligned}\mathbb{E}_{q(w|\theta)}\big[\log p(y|x,w)\big] &\approx \frac{1}{S}\sum_{s=1}^{S} \log p\big(y|\operatorname{softmax}(h_w^s(x))\big), \\ h_w^s(x) &\sim \tilde{q}(h_w(x)).\end{aligned} \tag{16}$$

Note that computing this Monte Carlo estimate only requires sampling from the output distribution, which has negligible computational costs compared to sampling predictions by drawing from the weight distribution and computing multiple forward-passes. Thus our MNVI variational inference scheme is efficient despite this sample approximation. This basic approach can also be used to compute the expected log-likelihood for a general probability density function.

3.5 The KL Term

Assumption 4. *We choose a Gaussian weight prior* $p(w) = \prod \mathcal{N}(w_{ij}|0, \sigma^2)$.

This allows us to analytically compute the KL-divergence of the variational distribution and the prior:

$$\operatorname{KL}\big[q(w|\theta)\|p(w)\big] = \sum_{i,j}\frac{1}{2}\left(\log\frac{\sigma^2}{\alpha_j w_{ij}^2} + \frac{(1+\alpha_j)w_{ij}^2}{\sigma^2} - 1\right). \tag{17}$$

This leads to a different implicit prior for the weight means, as can be seen in Fig. 2. While for a mean-field variational distribution the Kullback-Leibler divergence encourages weight means close to zero, the variational distribution induced by multiplicative activation noise favors small but non-zero weight means where the optimal size is dependent on the variance of the activation noise α and prior variance σ^2 and discourages sign changes, leading to different regularization of the network's weights.

Note that $\text{KL}[q(w|\theta)||p(w)]$ converges to zero for $\alpha_j \to \infty$, $w_{ij}^2 \to 0$, and $\alpha_j w_{ij}^2 \to \sigma^2$ for all i, j. Therefore, by minimizing the KL divergence term the activation noise is increased while the influence on the weights resembles that of weight decay. Further, for fixed $\alpha > 0$ the KL divergence is strongly convex with respect to $w^2 > 0$ and obtains its minimum at $\frac{\sigma^2}{1+\alpha}$, so the weight decay effect is stronger for weights where the activation noise α_j is high.

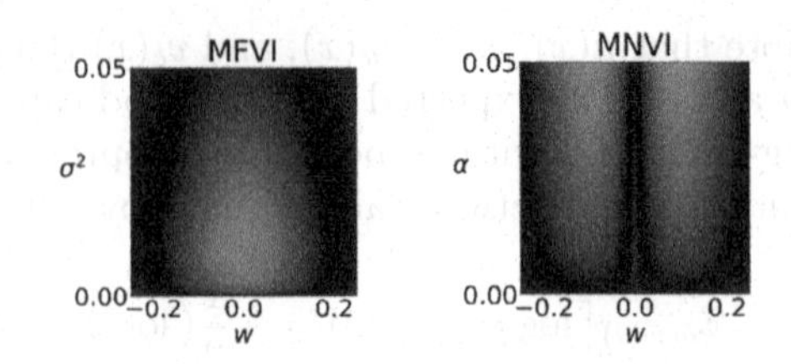

Fig. 2. KL divergence from a Gaussian prior for a one-dimensional mean-field Gaussian variational posterior *(left)* and a variational posterior induced by our multiplicative activation noise *(right)*. Brighter areas correspond to a lower divergence.

For learning a tempered posterior [43,52], we will allow rescaling of the Kullback-Leibler divergence term, so the generalized objective becomes

$$\min_{\theta} -\mathbb{E}_{q(w|\theta)}\left[\log p(Y|X,w)\right] + \kappa \text{KL}\left[q(w|\theta)\middle|\middle|p(w)\right] \tag{18}$$

for some $\kappa > 0$. Bayesian inference algorithms for deep networks often use $\kappa \in (0,1)$, which corresponds to a cold posterior [49].

3.6 Predictive Distribution

Similar to the log-likelihood, the predictive distribution can be computed analytically given the output distribution of the network for a Gaussian predictive distribution, while for a categorical distribution it has to be approximated. Following the calculations in [50] for the Gaussian predictive distribution, we obtain

$$p(y|x) \approx \mathcal{N}\left(\mu(x), v_{\mu(x)} + e^{c(x)+\frac{1}{2}v_c(x)}\right). \tag{19}$$

For classification, the class probability vector $\mathbf{p}$ of the categorical distribution can be estimated by sampling from the output distribution:

$$\mathbf{p}(y|x) \approx \frac{1}{S}\sum_{s=1}^{S} \text{softmax}(h_w^s(x)), \quad h_w^s(x) \sim \tilde{q}(h_w(x)). \tag{20}$$

4 Experiments

4.1 Regression on the UCI Datasets

We first test our approach on the UCI regression datasets [3] in the standard setting of a fully-connected network with one hidden layer of width 50 and 10-fold cross validation. This is the same training and evaluation setup used by [50]. Please refer to the supplemental material for details on our implementation

Table 1. *Comparison of the average log-likelihood and standard deviation for multiple Bayesian inference methods on the UCI regression datasets.* Higher is better. We use the MC-MFVI implementation of [50].

	boston	concrete	energy	kin8	power	wine	yacht
MC-MFVI [1,9]	-2.43±0.03	**-3.04±0.02**	-2.38±0.02	**2.40±0.05**	**-2.66±0.01**	**-0.78±0.02**	-1.68±0.04
MC Dropout [6]	-2.46±0.25	**-3.04±0.09**	-1.99±0.09	0.95±0.03	**-2.66±0.01**	**-0.78±0.02**	-1.68±0.04
Ensemble [26]	**-2.41±0.25**	-3.06±0.18	-1.38±0.22	1.20±0.02	-2.79±0.04	-0.94±0.12	-1.18±0.21
DVI [50]	**-2.41±0.02**	-3.06±0.02	**-1.01±0.06**	1.13±0.00	-2.80±0.00	-0.90±0.01	-0.47±0.03
dDVI [50]	-2.42±0.02	-3.07±0.02	-1.06±0.06	1.13±0.00	-2.80±0.00	-0.91±0.02	-0.47±0.03
SMFVI	-3.51±0.03	-3.42±0.04	-1.11±0.04	1.17±0.00	-2.88±0.00	-2.01±0.01	**-0.37±0.03**
MNVI (ours)	-2.43±0.02	-3.05±0.01	-1.33±0.05	1.15±0.01	-2.86±0.00	-0.96±0.01	**-0.37±0.02**

of MNVI and information on the training settings. As can be seen in Table 1, our MNVI method retains most of the predictive performance of the recent sampling-free DVI and dDVI [50], which propagate the full covariance matrix or only variances in a Bayesian network with a mean-field Gaussian variational posterior, while ours is more lightweight. When comparing against other VI approaches, such as Monte Carlo mean-field VI [1,9] and Monte Carlo dropout [6], as well as ensembling [26], the best performing method is dataset dependent, but our proposed MNVI generally obtains competitive results.

We also compare to a sampling-free mean-field variational inference scheme that shares the variance parameter for weights connected to the same input channel (SMFVI) and thus has the same parameter overhead as MNVI. While it is competitive on three of the six benchmark datasets, a significant drop in log-likelihood can be observed for the boston, concrete, and wine datasets.

4.2 Image Classification

We train a LeNet on MNIST [27], an All Convolutional Network [39] on CIFAR-10 [25], and a ResNet-18 on CIFAR-10, CIFAR-100 [25], and ImageNet [2]. All models based the same network architecture use an identical training protocol across the different BNN realizations. Details on the training setup can be found in the supplemental. We compare the predictive performance of our proposed sampling-free variational inference scheme for networks with multiplicative activation noise to the deterministic baseline network, Monte Carlo Dropout [6] with 8 samples computed at inference time, as well as sampling-free variational inference for a Gaussian mean field posterior (MFVI) similar to [11,38,50] and a Gaussian mean-field posterior with shared variance parameter for weights connected to the same input node (SMFVI).

To evaluate uncertainty estimates with respect to the expected calibration error (ECE) [10], we divide the interval $[0, 1]$ into $K = 20$ bins of equal length and assign predictions on the test set of size N to bins based on their confidence. For each bin of size B_i, we calculate the average confidence conf_i and accuracy acc_i of predictions within the bin. The ECE can then be computed as

Table 2. *Experimental results on the image classification datasets MNIST, CIFAR-10, CIFAR-100, and ImageNet.* Lower is better for all metrics. Our proposed sampling-free MNVI approach consistently performs as well or better than sampling-free MFVI despite requiring significantly fewer additional parameters.

		Misclass. [%]	NLLH	ECE [%]	AUMRC	Inference [ms]	Parameters [$\times 10^6$]
MNIST LeNet	Deterministic	$\mathbf{0.55}_{\pm 0.06}$	$0.027_{\pm 0.001}$	$0.40_{\pm 0.08}$	$9.91_{\pm 1.23} \times 10^{-5}$	$0.86_{\pm 0.16}$	1.111
	MC Dropout	$0.59_{\pm 0.01}$	$0.021_{\pm 0.002}$	$0.26_{\pm 0.02}$	$10.42_{\pm 3.03} \times 10^{-5}$	$7.35_{\pm 0.36}$	1.111
	MFVI	$0.57_{\pm 0.04}$	$\mathbf{0.017}_{\pm 0.001}$	$0.21_{\pm 0.07}$	$8.30_{\pm 0.62} \times 10^{-5}$	$6.28_{\pm 0.45}$	2.224
	SMFVI	$0.60_{\pm 0.04}$	$\mathbf{0.017}_{\pm 0.004}$	$0.20_{\pm 0.02}$	$\mathbf{8.27}_{\pm 0.30} \times \mathbf{10^{-5}}$	$6.27_{\pm 0.50}$	1.114
	MNVI (ours)	$\mathbf{0.55}_{\pm 0.03}$	$0.018_{\pm 0.001}$	$\mathbf{0.19}_{\pm 0.03}$	$8.33_{\pm 0.64} \times 10^{-5}$	$5.97_{\pm 0.63}$	1.114
CIFAR-10 AllCNN	Deterministic	$7.97_{\pm 0.20}$	$0.428_{\pm 0.009}$	$5.74_{\pm 0.25}$	$9.42_{\pm 0.26} \times 10^{-3}$	$1.84_{\pm 0.11}$	1.370
	MC Dropout	$\mathbf{7.16}_{\pm 0.25}$	$\mathbf{0.257}_{\pm 0.003}$	$\mathbf{2.76}_{\pm 0.14}$	$\mathbf{8.12}_{\pm 0.14} \times \mathbf{10^{-3}}$	$11.86_{\pm 0.55}$	1.370
	MFVI	$7.72_{\pm 0.14}$	$0.348_{\pm 0.006}$	$4.95_{\pm 0.07}$	$8.98_{\pm 0.44} \times 10^{-3}$	$12.75_{\pm 0.64}$	2.739
	SMFVI	$8.39_{\pm 0.17}$	$0.482_{\pm 0.009}$	$5.86_{\pm 0.17}$	$10.34_{\pm 0.18} \times 10^{-3}$	$12.69_{\pm 0.20}$	1.372
	MNVI (ours)	$7.62_{\pm 0.35}$	$0.352_{\pm 0.015}$	$4.92_{\pm 0.33}$	$8.95_{\pm 0.57} \times 10^{-3}$	$9.96_{\pm 0.52}$	1.372
CIFAR-10 ResNet18	Deterministic	$5.94_{\pm 0.26}$	$0.261_{\pm 0.006}$	$3.86_{\pm 0.21}$	$6.23_{\pm 0.66} \times 10^{-3}$	$5.64_{\pm 0.23}$	11.17
	MC Dropout	$5.70_{\pm 0.09}$	$\mathbf{0.219}_{\pm 0.006}$	$\mathbf{2.78}_{\pm 0.06}$	$5.76_{\pm 0.21} \times 10^{-3}$	$47.34_{\pm 2.12}$	11.17
	MFVI	$5.63_{\pm 0.21}$	$0.256_{\pm 0.006}$	$3.72_{\pm 0.11}$	$5.64_{\pm 0.25} \times 10^{-3}$	$34.92_{\pm 1.18}$	22.34
	SMFVI	$5.84_{\pm 0.28}$	$0.233_{\pm 0.023}$	$3.04_{\pm 0.06}$	$7.50_{\pm 0.69} \times 10^{-3}$	$35.89_{\pm 1.17}$	11.18
	MNVI (ours)	$\mathbf{5.60}_{\pm 0.14}$	$0.246_{\pm 0.007}$	$3.46_{\pm 0.10}$	$\mathbf{5.53}_{\pm 0.04} \times \mathbf{10^{-3}}$	$33.82_{\pm 0.66}$	11.18
CIFAR-100 ResNet18	Deterministic	$27.38_{\pm 0.57}$	$1.266_{\pm 0.019}$	$13.3_{\pm 0.4}$	$8.23_{\pm 0.14} \times 10^{-2}$	$5.75_{\pm 0.25}$	11.22
	MC Dropout	$27.87_{\pm 0.37}$	$1.240_{\pm 0.005}$	$11.6_{\pm 0.5}$	$8.30_{\pm 0.02} \times 10^{-2}$	$46.75_{\pm 2.21}$	11.22
	MFVI	$26.91_{\pm 0.10}$	$1.271_{\pm 0.016}$	$13.1_{\pm 0.3}$	$7.87_{\pm 0.03} \times 10^{-2}$	$34.71_{\pm 0.86}$	22.43
	SMFVI	$27.18_{\pm 0.18}$	$1.297_{\pm 0.023}$	$13.6_{\pm 0.2}$	$8.03_{\pm 0.07} \times 10^{-2}$	$35.81_{\pm 0.77}$	11.23
	MNVI (ours)	$\mathbf{25.30}_{\pm 0.50}$	$\mathbf{1.085}_{\pm 0.011}$	$\mathbf{10.5}_{\pm 0.5}$	$\mathbf{7.40}_{\pm 0.17} \times \mathbf{10^{-2}}$	$34.02_{\pm 0.83}$	11.23
ImageNet ResNet18	Deterministic	31.09	1.282	**3.13**	0.1106	$8.63_{\pm 0.59}$	11.69
	MNVI (ours)	**31.05**	**1.276**	3.88	**0.1092**	$43.75_{\pm 1.06}$	11.69

$$\mathrm{ECE} = \sum_{i=1}^{K} \frac{B_i}{N} |\mathrm{conf}_i - \mathrm{acc}_i|. \tag{21}$$

Further, we examine how well misclassifications can be identified by reporting the area under the misclassification-rejection curve (AUMRC) [32]. To compute the AUMRC, we sort predictions on the test set by their predictive entropy in descending order and iteratively remove the remaining data point with the highest predictive entropy from the evaluation set, adding the point (x_i, y_i) to the misclassification-rejection-curve, where x_i is the percentage of data removed from the evaluation set and y_i is the misclassification rate for the remaining data points. The AUMRC can then be calculated as the area under the resulting step function. We also report the average required time for computing a single forward-pass with a batch size of 100 for all methods on a NVIDIA GTX 1080 Ti GPU as well as the number of parameters.

As can be seen in Table 2, our proposed MNVI can be applied to various image classification problems, scaling up to ImageNet [2]. Our approach is able to substantially improve both classification accuracy and uncertainty metrics compared to the deterministic baseline in all settings but classification on MNIST, where it matches the classification accuracy of a deterministic network. Compared to the popular approximate Bayesian inference method MC Dropout [6],

MNVI matches the accuracy on MNIST while significantly increasing the accuracy for ResNet18 on CIFAR-10 and -100. While MC-Dropout produces better ECE for both architectures on CIFAR-10, MNVI achieves lower calibration error on MNIST and CIFAR-100 as well as lower AUMRC in three out of four settings. Sampling-free variational inference with a mean field posterior approximation (MFVI), while having almost double the number of parameters and slightly higher inference time, only achieves a slightly worse misclassification rate than MNVI. More importantly, for all settings MNVI has better calibrated predictions as indicated by the lower ECE and better separates false from correct predictions resulting in a lower AUMRC in three out of four setting despite being clearly more efficient, especially for deep networks. Compared to this, the additive weight-sharing scheme of SMFVI leads to a drop in classification performance for all datasets and to worse calibration for the AllCNN on CIFAR-10 and ResNet18 on CIFAR-100. Thus, our proposed method has significantly fewer parameters than MFVI and is faster than MC Dropout, scaling up all the way to ImageNet, while matching or even surpassing the accuracy and uncertainty estimation of these methods.

5 Conclusion

In this paper, we proposed to use the distribution induced by multiplicative Gaussian activation noise as a posterior approximation for sampling-free variational inference in Bayesian neural networks. The benefits of this variational posterior are a reduction of the number of parameters and required computation. Our experiments show that the suggested posterior approximation retains or even improves over the accuracy of the Gaussian mean-field posterior approximation, while requiring a negligible amount of additional parameters compared to a deterministic network or MC dropout. Our approach can be successfully applied to train Bayesian neural networks for various image classification tasks, matching or even surpassing the predictive accuracy of deterministic neural networks while producing better calibrated uncertainty estimates. Because of these promising results, we hope that this as well as further research on efficient sampling-free variational inference methods will lead to a more widespread adoption of Bayesian neural networks in practice.

Acknowledgements. This project has received funding from the European Research Council (ERC) under the European Union's Horizon 2020 research and innovation programme (grant agreement No. 866008).

References

1. Blundell, C., Cornebise, J., Kavukcuoglu, K., Wierstra, D.: Weight uncertainty in neural networks. In: ICML, pp. 1613–1622 (2015)
2. Deng, J., Dong, W., Socher, R., Li, L., Li, K., Li, F.: ImageNet: a large-scale hierarchical image database. In: CVPR, pp. 248–255 (2009)
3. Dua, D., Graff, C.: UCI machine learning repository (2017). http://archive.ics.uci.edu/ml

4. Dusenberry, M.W., et al.: Efficient and scalable Bayesian neural nets with rank-1 factors. In: ICML, pp. 2782–2792 (2020)
5. Frey, B.J., Hinton, G.E.: Variational learning in nonlinear Gaussian belief networks. Neural Comput. **11**(1), 193–213 (1999)
6. Gal, Y., Ghahramani, Z.: Dropout as a Bayesian approximation: representing model uncertainty in deep learning. In: ICML, pp. 1050–1059 (2016)
7. Gast, J., Roth, S.: Lightweight probabilistic deep networks. In: CVPR, pp. 3369–3378 (2018)
8. Ghosh, S., Delle Fave, F.M., Yedidia, J.: Assumed density filtering methods for learning Bayesian neural networks. In: AAAI, pp. 1589–1595 (2016)
9. Graves, A.: Practical variational inference for neural networks. In: NIPS*2011, pp. 2348–2356 (2011)
10. Guo, C., Pleiss, G., Sun, Y., Weinberger, K.Q.: On calibration of modern neural networks. In: ICML, pp. 1321–1330 (2017)
11. Haussmann, M., Kandemir, M., Hamprecht, F.A.: Sampling-free variational inference of Bayesian neural nets. In: UAI, pp. 563–573 (2019)
12. He, K., Zhang, X., Ren, S., Sun, J.: Deep residual learning for image recognition. In: CVPR, pp. 770–778 (2016)
13. Heek, J., Kalchbrenner, N.: Bayesian inference for large scale image classification. arXiv:1908.03491 [cs.LG] (2019)
14. Hernández-Lobato, J.M., Adams, R.: Probabilistic backpropagation for scalable learning of Bayesian neural networks. In: ICML, pp. 1861–1869 (2015)
15. Hinton, G.E., Srivastava, N., Swersky, K.: Neural networks for machine learning: lecture 6a - overview of mini-batch gradient descent (2012)
16. Hinton, G.E., Van Camp, D.: Keeping the neural networks simple by minimizing the description length of the weights. In: COLT, pp. 5–13 (1993)
17. Hron, J., Matthews, A., Ghahramani, Z.: Variational Bayesian dropout: pitfalls and fixes. In: ICML, pp. 2019–2028 (2018)
18. Ioffe, S., Szegedy, C.: Batch normalization: accelerating deep network training by reducing internal covariate shift. In: ICML, pp. 448–456 (2015)
19. Jankowiak, M.: Closed form variational objectives for Bayesian neural networks with a single hidden layer. arXiv:2002.02655 [stat.ML] (2018)
20. Jin, J., Dundar, A., Culurciello, E.: Robust convolutional neural networks under adversarial noise. arXiv:1511.06306 [cs.LG] (2015)
21. Khan, M., Nielsen, D., Tangkaratt, V., Lin, W., Gal, Y., Srivastava, A.: Fast and scalable Bayesian deep learning by weight-perturbation in Adam. In: ICML, pp. 2611–2620 (2018)
22. Kingma, D.P., Ba, J.: Adam: a method for stochastic optimization. In: ICLR (2015)
23. Kingma, D.P., Welling, M.: Auto-encoding variational Bayes. In: ICLR (2014)
24. Kingma, D.P., Salimans, T., Welling, M.: Variational dropout and the local reparameterization trick. In: NIPS*2015, pp. 2575–2583 (2015)
25. Krizhevsky, A., Hinton, G.E.: Learning multiple layers of features from tiny images. Technical report, University of Toronto (2009)
26. Lakshminarayanan, B., Pritzel, A., Blundell, C.: Simple and scalable predictive uncertainty estimation using deep ensembles. In: NIPS*2017, pp. 6402–6413 (2017)
27. LeCun, Y., Bottou, L., Bengio, Y., Haffner, P.: Gradient-based learning applied to document recognition. Proc. IEEE **86**(11), 2278–2324 (1998)
28. Li, Y., Hernández-Lobato, J.M., Turner, R.E.: Stochastic expectation propagation. In: NIPS*2015, pp. 2323–2331 (2015)
29. Louizos, C., Welling, M.: Multiplicative normalizing flows for variational Bayesian neural networks. In: ICML (2017)

30. MacKay, D.J.: A practical Bayesian framework for backpropagation networks. Neural Comput. **4**(3), 448–472 (1992)
31. Mandt, S., Hoffman, M.D., Blei, D.M.: Stochastic gradient descent as approximate Bayesian inference. J. Mach. Learn. Res. **18**, 134:1–134:35 (2017)
32. Nadeem, M.S.A., Zucker, J.D., Hanczar, B.: Accuracy-rejection curves (ARCs) for comparing classification methods with a reject option. In: Machine Learning in Systems Biology, pp. 65–81 (2009)
33. Neal, R.M.: Bayesian learning for neural networks. Ph.D. thesis, University of Toronto (1995)
34. Nguyen, C.V., Li, Y., Bui, T.D., Turner, R.E.: Variational continual learning. In: ICLR (2018)
35. Osawa, K., et al.: Practical deep learning with Bayesian principles. In: NeurIPS*2019, pp. 4289–4301 (2019)
36. Postels, J., Ferroni, F., Coskun, H., Navab, N., Tombari, F.: Sampling-free epistemic uncertainty estimation using approximated variance propagation. In: ICCV, pp. 2931–2940 (2019)
37. Ritter, H., Botev, A., Barber, D.: A scalable Laplace approximation for neural networks. In: ICLR (2018)
38. Roth, W., Pernkopf, F.: Variational inference in neural networks using an approximate closed-form objective. In: Proceedings of the NIPS 2016 Workshop on Bayesian Deep Learning (2016)
39. Springenberg, J., Dosovitskiy, A., Brox, T., Riedmiller, M.: Striving for simplicity: the all convolutional net. In: ICLR (Workshop Track) (2015)
40. Srivastava, N., Hinton, G.E., Krizhevsky, A., Sutskever, I., Salakhutdinov, R.: Dropout: a simple way to prevent neural networks from overfitting. J. Mach. Learn. Res. **15**(1), 1929–1958 (2014)
41. Swiatkowski, J., et al.: The k-tied normal distribution: a compact parameterization of Gaussian mean field posteriors in Bayesian neural networks. In: ICML (2020)
42. Wainwright, M.J., Jordan, M.I.: Graphical models, exponential families, and variational inference. Found. Trends Mach. Learn. **1**(1–2), 1–305 (2008)
43. Walker, S., Hjort, N.L.: On Bayesian consistency. J. R. Stat. Soc. Ser. B (Stat. Methodol.) **63**(4), 811–821 (2001)
44. Wang, H., Xingjian, S., Yeung, D.Y.: Natural-parameter networks: a class of probabilistic neural networks. In: NIPS*2016, pp. 118–126 (2016)
45. Wang, S., Manning, C.: Fast dropout training. In: ICML, pp. 118–126 (2013)
46. Welling, M., Teh, Y.W.: Bayesian learning via stochastic gradient Langevin dynamics. In: ICML, pp. 681–688 (2011)
47. Wen, Y., Tran, D., Ba, J.: BatchEnsemble: an alternative approach to efficient ensemble and lifelong learning. In: ICLR (2020)
48. Wen, Y., Vicol, P., Ba, J., Tran, D., Grosse, R.: Flipout: efficient pseudo-independent weight perturbations on mini-batches. In: ICLR (2018)
49. Wenzel, F., et al.: How good is the Bayes posterior in deep neural networks really? In: ICML, pp. 10248–10259 (2020)
50. Wu, A., Nowozin, S., Meeds, E., Turner, R., Hernández-Lobato, J., Gaunt, A.: Deterministic variational inference for robust Bayesian neural networks. In: ICLR (2019)
51. Zhang, R., Li, C., Zhang, J., Chen, C., Wilson, A.G.: Cyclical stochastic gradient MCMC for Bayesian deep learning. In: ICLR (2020)
52. Zhang, T.: From ε-entropy to KL-entropy: analysis of minimum information complexity density estimation. Ann. Stat. **34**(5), 2180–2210 (2006)

Conditional Adversarial Debiasing: Towards Learning Unbiased Classifiers from Biased Data

Christian Reimers[1,2](✉), Paul Bodesheim[1], Jakob Runge[2,3], and Joachim Denzler[1,2]

[1] Computer Vision Group, Friedrich Schiller University Jena, 07743 Jena, Germany
creimers@bgc-jena.mpg.de
[2] Institute of Data Science, German Aerospace Center (DLR), 07745 Jena, Germany
[3] Technische Universität Berlin, 10623 Berlin, Germany

Abstract. Bias in classifiers is a severe issue of modern deep learning methods, especially for their application in safety- and security-critical areas. Often, the bias of a classifier is a direct consequence of a bias in the training set, frequently caused by the co-occurrence of relevant features and irrelevant ones. To mitigate this issue, we require learning algorithms that prevent the propagation of known bias from the dataset into the classifier. We present a novel adversarial debiasing method, which addresses a feature of which we know that it is spuriously connected to the labels of training images but statistically independent of the labels for test images. The debiasing stops the classifier from falsly identifying this irrelevant feature as important. Irrelevant features co-occur with important features in a wide range of bias-related problems for many computer vision tasks, such as automatic skin cancer detection or driver assistance. We argue by a mathematical proof that our approach is superior to existing techniques for the abovementioned bias. Our experiments show that our approach performs better than the state-of-the-art on a well-known benchmark dataset with real-world images of cats and dogs.

Keywords: Adversarial debiasing · Causality · Conditional dependence

1 Introduction

Deep neural networks have demonstrated impressive performances in many computer vision and machine learning tasks, including safety- and security-critical applications such as skin cancer detection [15] or predicting recidivism [4]. However, many people and domain experts advise against employing deep learning in those applications, even if classifiers outperform human experts, as in skin lesion

Supplementary Information The online version contains supplementary material available at https://doi.org/10.1007/978-3-030-92659-5_4.

C. Bauckhage et al. (Eds.): DAGM GCPR 2021, LNCS 13024, pp. 48–62, 2021.
https://doi.org/10.1007/978-3-030-92659-5_4

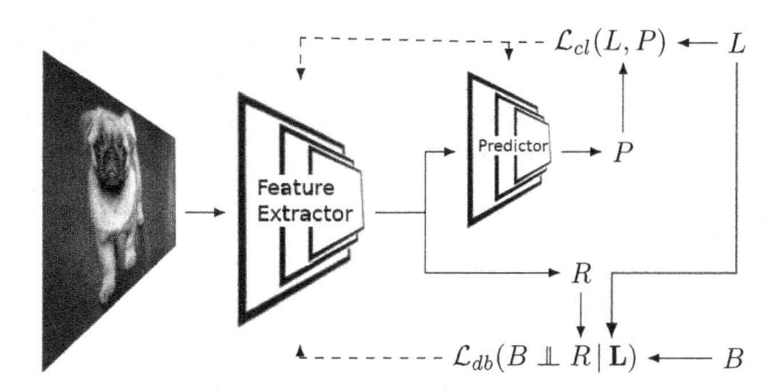

Fig. 1. In adversarial debiasing, a debiasing loss $\mathcal{L}_{db}$ is often used to enforce independence between the bias variable B and a representation R. In this work, we show that it is beneficial to condition this independence on the label L.

classification [20]. One reason for their concerns is bias in the classifiers. Indeed, almost all image datasets contain some kind of bias [22] and, consequently, the performance of classifiers varies significantly across subgroups [4,13].

One major reason for bias in classifiers is dataset bias. Every dataset is a unique slice through the visual world [19]. Therefore, an image dataset does not represent the real world perfectly but contains unwanted dependencies between meaningless features and the labels of its samples. This spurious connection can be caused by incautious data collection or by justified concerns, if, for example, the acquirement of particular examples is dangerous. A classifier trained on such a dataset might pick the spuriously dependent feature to predict the label and is, thus, biased. Mitigating such a bias is challenging even if the spurious connection is known. To this end, it is important to understand the nature of the spurious dependence. Therefore, we start our investigation at the data generation process. We provide a formal description of the data generation model for a common computer vision bias in Sect. 3.1. In contrast to other approaches that do not provide a model for the data generation process and, hence, rely solely on empirical evaluations, this allows us to investigate our proposed method theoretically.

The main contribution of our work is a novel adversarial debiasing strategy. The basic concept of adversarial debiasing and the idea of our improvement can be observed in Fig. 1. For adversarial debiasing, a second loss $\mathcal{L}_{db}$ is used in addition to the regular training loss $\mathcal{L}_{cl}$ of a neural network classifier. This second loss penalizes the dependence between the bias variable B and an intermediate representation R from the neural network. The main difference we propose in this paper is replacing this dependence $B \not\perp\!\!\!\perp R$ by the conditional dependence $B \not\perp\!\!\!\perp R \,|\, L$ with L being the label. In fact, it turns out that this conditional dependence is better suited than the unconditional dependence for the considered kind of bias. The motivation for this replacement, and a mathematical proof for its suitability can be found in Sect. 3.2.

To use our new conditional independence criterion for adversarial debiasing, we have to implement it as a differentiable loss. We provide three possible implementations in Sect. 3.3. We demonstrate that these new loss functions lead to an increase in accuracy on unbiased test sets. In Sect. 4, we provide results of experiments on a synthetic dataset, a dataset with real-world images of cats and dogs that is used by previous work to evaluate adversarial debiasing, and an ablation study to show that the proposed change of the criterion causes the increase in accuracy.

2 Related Work

Traditionally, adversarial debiasing aims to learn a feature representation that is informative for a task but independent of the bias. Hence, a second neural network that should predict the bias from the feature representation is introduced to enforce this independence. The original network for classification and this second network are then trained in an adversarial fashion. To this end, different loss functions for the original network are suggested to decrease the performance of the second network for predicting the bias. Previous work aims at minimizing the cross-entropy between bias prediction and a uniform distribution [3] or maximizing the mean squared error between the reconstruction and the bias [23]. Further approaches enclose the joint maximization of the cross-entropy between the predicted and true distribution of the bias variable and the entropy of the distribution of the predicted bias [11] or the minimization of the correlation between the ground-truth bias and the prediction of the bias [1]. However, as shown in another study [17], independence is too restrictive as a criterion for determining whether a deep neural network uses a certain feature. This fact is also reflected in the experimental results of the abovementioned papers. The resulting classifiers are less biased, but this often leads to decreasing performance on unbiased test sets. As one example, significantly less bias in an age classifier trained on a dataset biased by gender has been reported [3], but the classification performance on an unbiased test set decreased from 0.789 to 0.781. Our work is fundamentally different. Instead of a different loss, we suggest a different criterion to determine whether a neural network uses a feature. We use the conditional independence criterion proposed by [17] rather than independence between the representation and the bias.

While the vast majority of adversarial debiasing methods acknowledge that bias has many forms, they rarely link the suggested solutions to the processes that generate the biased data. Instead, they rely exclusively on empirical evaluations. In contrast, we provide a specific model for a specific kind of bias as well as a theoretical proof that our approach is better suited for this case.

3 Proposed Conditional Adversarial Debiasing Approach

In this section, we motivate our novel approach for adversarial debiasing and introduce our novel adversarial debiasing criterion. We prove mathematically that the new criterion fits our specific bias model. Finally, we provide three possible implementations for loss functions that realize this criterion.

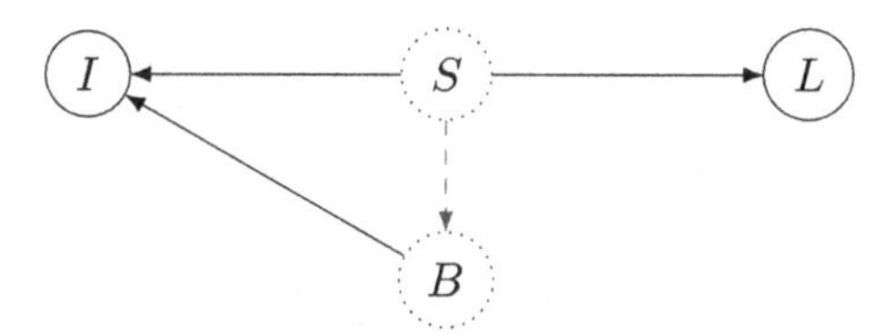

Fig. 2. A graphical representation of the specific bias. Circles represent variables, dotted circles represent unobserved variables. The label L is only dependent on a signal S, while the input I is also dependent on some variable B. In the training set, the signal S influences the variable B due to bias. This is indicated by the red dashed arrow. In contrast, in the test set, the two are independent.

3.1 Bias Model

Different kinds of bias influence visual datasets in various ways [22]. We consider a specific kind of bias and the corresponding model covers many relevant tasks in computer vision. To describe this bias model, we start with a graphical model of the underlying data generation process displayed in Fig. 2.

For classification tasks, like separating cats from dogs, we assume that a signal S is contained in and can be extracted from the input I that contains the relevant information. Following the graphical model, a labeling process generates the label L ("cat" or "dog") from this signal S. However, the input I is a mixture of multiple signals. Besides S, another signal B influences I. In the cat/dog example, B might relate to the fur's color. Since the fur's color is not meaningful for distinguishing cats from dogs, B is independent of S and L during the application of the classifier in practice, i.e., on an unbiased test set, $B \perp\!\!\!\perp L$. In contrast, if taking images of dark-furred dogs would be bad luck, we might find an unwanted dependence between the signal S and the signal B in a training set, leading to $B \not\perp\!\!\!\perp L$, where we call B the bias variable. This dependence can be utilized by a classifier to reach a higher accuracy on the training set, resulting in a biased classifier and a lower accuracy on the test set.

To better understand the direction of the arrow from S to B, we want to emphasize, that data for a task is selected with a purpose. Images are included in the dataset because they show cats or dogs and one will, if necessary, deliberately accept imbalances in variables like fur-color. In contrast, if one find that our dataset misrepresents fur-color one would never accept a major misrepresentation of the ratio of cats and dogs to compensate for this problem. This demonstrates, that S influences B through the dataset creation while B does not influence S.

This bias model covers many, but not all, relevant situations in computer vision. To this end, we present one example for a situations where the bias model fits the data, and one example where it does not.

The first example is a driver assistance system that uses a camera to estimate aquaplaning risk [9]. To train such a system, example images of safe conditions and aquaplaning conditions are required. While the former can easily be collected in the wild, it is dangerous to drive a car under aquaplaning conditions. Thus, images for aquaplaning are collected in a specific facility. In this example, the signal S is the standing water, and the bias variable B is the location that determines the background of the image. Because of the safety risk, they are dependent in the training set, but not at the time of application.

The second example is a system that predicts absenteeism in the workplace [2]. If an automated system predicts absenteeism, it should ignore pregnancy. Here, the bias variable B is the sex and the signal S is the chance of an employee to be absent from work. In this situation, similar to many others in algorithmic fairness, our bias model does not apply because the sex (B) does also influence the chance to be absent from work (L) during the application.

3.2 Conditional Independence for Debiasing

Deep neural networks unite a feature extractor and a predictor [16]. For adversarial debiasing, we separate the two at some intermediate layer. We denote the output of the feature extractor with R. Note that it is valid to use the whole network for feature extraction such that R contains the class predictions. Both networks are trained using a classification loss $\mathcal{L}_{cl}$, e.g., cross-entropy loss. Additionally, a debiasing loss $\mathcal{L}_{db}$ is used to prevent the extraction of the bias variable B. For a visualization, see Fig. 1. Most approaches for adversarial debiasing [1,3,11,23] aim to find a representation R of I that is independent of the bias variable B while still being informative for the label L, i.e.,

$$R \perp\!\!\!\perp B \quad \wedge \quad R \not\perp\!\!\!\perp L. \tag{1}$$

In this work, we propose a novel strategy: Instead of independence, we aim for conditional independence of R and B, given the label L, i.e.,

$$R \perp\!\!\!\perp B \,|\, L \quad \wedge \quad R \not\perp\!\!\!\perp L. \tag{2}$$

First, we show that our strategy agrees with state-of-the-art results in explaining deep neural networks [17] and second, that an optimal classifier fulfills the conditional independence (2) but not the independence (1). We prove this statement for the case that all data generation processes are linear and L is scalar. Thus, loss functions that enforce the independence (1) will decrease the classifier's performance, while loss functions that ensure the conditional independence (2) will not.

The goal of debiasing is to prevent a deep neural network from using a biased feature. To reach this goal, we first need to determine whether a classifier uses a feature. So far, most approaches for adversarial debiasing use the dependence between a feature and the classifier's prediction to measure whether a classifier is using a feature. In contrast, we build on previous work for understanding deep neural networks [17]. While the independence criterion (1) obviously ensures that a bias variable B is not used for classification, the authors of [17] reveal that independence is too restrictive to determine whether a deep neural network uses a certain feature. They employ the framework of causal inference [14] to show that the ground-truth labels are a confounding variable for features of the input and the predictions of a deep neural network. In theoretical considerations and empirical experiments, they further demonstrate that the prediction of a neural network and a feature of the input can be dependent even though the feature is not used by the network. The authors, therefore, suggest using the conditional independence (2) to determine whether a feature is used by a classifier.

Thus, using the independence criterion (1) is too strict. Even if the deep neural network ignores the bias, it might not satisfy (1) and, hence, not minimize a corresponding loss. Furthermore, minimizing such a loss based on the independence criterion will likely result in a less accurate classifier. Therefore, we use the conditional dependence criterion (2) for adversarial debiasing.

To corroborate this claim, we present a mathematical proof for the following statement. If the bias can be modeled as explained in Sect. 3.1, the optimal classifier, which recovers the signal and calculates the correct label for every input image, fulfills the conditional independence (2) but not the independence (1). In this work, we only include the proof for the linear, uni-variate case, i.e., all data generating processes (Φ, Ψ, Ξ) are linear and L is scalar. However, this proof can further be extended to the nonlinear case by using a kernel space in which the data generation processes are linear and replacing covariances with the inner product of that space.

Theorem 1. *If the bias can be modeled as described in Sect. 3.1, the optimal classifier fulfills the conditional independence (2) but not the independence (1).*

Proof. For this proof, we denote all variables with capital Latin letters. Capital Greek letters are used for processes, and lower-case Greek letters for their linear coefficients. The only exception is the optimal classifier denoted by F^*. First, we define all functions involved in the model. Afterward, since dependence results in correlation in the linear case, a simple calculation proves the claim.

Let S denote the signal according to the bias model, as explained in Sect. 3.1. In the linear case, the bias variable B can be split into a part that is fully determined by S and a part that is independent of S. Let B^* be the part of the bias variable that is independent of S, e.g., noise. According to the bias model, the bias variable B, the image I, and the label L are given by:

$$B = \alpha_1 S + \alpha_2 B^* =: \Phi(S, B^*), \tag{3}$$

$$I =: \Psi(S, B) = \Psi(S, \Phi(S, B^*)), \tag{4}$$

$$L = \zeta_1 S =: \Xi(S). \tag{5}$$

Thus, the label L can be calculated from the signal S only. The optimal solution F^* of the machine learning problem will recover the signal and calculate the label. By the assumptions of the bias model, the signal can be recovered from the input. Thus, there exists a function $\Psi^\dagger$ such that $\Psi^\dagger(\Psi(S, B)) = S$ holds. Therefore, F^* is given by

$$F^* := \Xi \Psi^\dagger. \tag{6}$$

Now, we have defined all functions appearing in the model. The rest of the proof are two straightforward calculations. In the linear case, the independence of variables is equivalent to variables being uncorrelated. We denote the covariance of two variables A, B with $\langle A, B \rangle$. To prove that (1) does not hold, we calculate

$$\langle F^*(I), B \rangle = \left\langle \Xi \Psi^\dagger \Psi \left(S, \Phi\left(S, B^*\right)\right), \Phi\left(S, B^*\right) \right\rangle = \zeta_1 \alpha_1 \langle S, S \rangle . \quad (7)$$

This is equal to zero if and only if either all inputs contain an identical signal ($\langle S, S \rangle = 0$), the dataset is unbiased ($\alpha_1 = 0$), or the label does not depend on the signal ($\zeta_1 = 0$). For conditional independence, we can use partial correlation. Using its definition we obtain

$$\langle F^*(I), B \rangle | L = \left\langle F^*(I) - \frac{\langle F^*(I), L \rangle}{\langle L, L \rangle} L, B - \frac{\langle B, L \rangle}{\langle L, L \rangle} L \right\rangle . \quad (8)$$

We substitute L by (5) and use the properties of the inner product to arrive at

$$\langle F^*(I), B \rangle - \frac{\langle \zeta_1 S, \zeta_1 S \rangle \langle \zeta_1 S, B \rangle}{\langle \zeta_1 S, \zeta_1 S \rangle} = \zeta_1 \alpha_1 \langle S, S \rangle - \frac{\alpha_1 \zeta_1^3 \langle S, S \rangle^2}{\zeta_1^2 \langle S, S \rangle} = 0. \quad (9)$$

This completes the proof for the linear case. For more detailed calculations we refer to the supplementary material. □

The optimal classifier does not minimize loss criteria based on the independence (1). Further, from (7), we see that the dependence contains ζ_1, which is the correlation between the signal S and the neural network's prediction. Loss functions based on that criterion aim to reduce this parameter and, hence, will negatively affect the classifier's performance. We demonstrate this effect using a synthetic dataset in Sect. 4. In contrast, loss terms based on our new criterion (2) are minimized by the optimal classifier. Thus, corresponding loss functions do not reduce the accuracy to minimize bias.

3.3 Implementation Details

In practice, we are faced with the problem of integrating our criterion into the end-to-end learning framework of deep neural networks. Hence, we provide three possibilities to realize (2) as a loss function. Turning an independence criterion into a loss function is not straightforward. First, the result of an independence test is binary and, hence, non-differentiable. Second, we need to consider distributions of variables to perform an independence test. However, we only see one mini-batch at a time during the training of a deep neural network. Nevertheless, multiple solutions exist for the unconditional case.

In this section, we describe three possible solutions, namely: mutual information (MI), the Hilbert-Schmidt independence criterion (HSIC) and the maximum correlation criterion (MCC). We adapt the corresponding solutions from the unconditional case and extend them to conditional independence criteria.

The first solution makes use of the MI of R and B as suggested in [11]. Here, the criterion for independence is $\mathrm{MI}(R; B) = 0$, and the MI is the differentiable loss function. In contrast, we use conditional independence. Our criterion is $\mathrm{MI}(R;B|L)=0$, and the loss is given by the conditional MI:

$$\mathrm{MI}(R;B|L)=\sum_{l\in\mathcal{L},b\in\mathcal{B}r\in\mathcal{R}} p_{R,B,L}(r,b,l)\log\frac{p_L(l)p_{R,B,L}(r,b,l)}{p_{R,L}(r,l)p_{B,L}(b,l)}. \tag{10}$$

To incorporate this loss, we need to estimate the densities $p_{R,B,L}, p_{R,L}$ and $p_{B,L}$ in every step. We use kernel density estimation on the mini-batches with a Gaussian kernel and a variance of one fourth of the mean pairwise distance within a batch. This setting proved best in preliminary experiments on reconstructing densities.

As a second solution, we extend the Hilbert Schmidt independence criterion [8]. The variables are independent if and only if $\mathrm{HSIC}(R,B)=0$ holds for a sufficiently large kernel space. The HSIC was extended to a conditional independence criterion by [7]. Multiple numerical approximations exist, we use

$$\mathrm{tr}\, G_R S_L G_B S_L = 0. \tag{11}$$

Here, S_L is given by $(\mathbb{I}+1/mG_L)^{-1}$, where $\mathbb{I}$ is the identity matrix and $G_X = HK_XH$ with K_X the kernel matrix for $X\in\{B,R,L\}$ and $H_{ij}=\delta_{ij}-m^{-2}$ for δ_{ij} the Kronecker-Delta and m the number of examples. For the relation to HSIC and further explanations, we refer to [7]. We use the same kernel as above and estimate the loss on every mini-batch independently.

The third idea we extend is the predictability criterion from [1]

$$\max_f \quad \mathrm{Corr}(f(R),B)=0. \tag{12}$$

To use this criterion within a loss function, they parametrize f by a neural network. However, this is not an independence criterion as it can be equal to zero, even if R and B are dependent. Therefore, it is unclear how to incorporate the conditioning on L. As a consequence, we decided to extend the proposed criterion in two ways. First, we use the maximum correlation coefficient (MCC)

$$\mathrm{MCC}(R,B)=\max_{f,g} \quad \mathrm{Corr}(f(R),g(B))=0, \tag{13}$$

which is equal to zero if and only if the two variables are independent [18]. Second, we use the partial correlation conditioned on the label L, which leads to

$$\max_{f,g} \quad \mathrm{PC}(f(R),g(B)\,|\,L)=0. \tag{14}$$

To parametrize both functions f and g, we use neural networks. The individual effects of the two extensions can be observed through our ablation study in Sect. 4.2. Note that all three implementations can be used for vector-valued variables and, therefore, also for multiple bias variables in parallel.

3.4 Limitations

Debiasing with our method is only possible if the bias is known and a numerical value can be assigned to each image. This is, however, true for all adversarial debiasing methods, e.g. the methods described in Sect. 2. Further, we assume the bias model described in Sect. 3.1. As described in the same section, the model covers many, but not all biases in computer vision.

One drawback of our method is that testing for conditional dependence is more complicated than testing for dependence. This is less a problem of calculation time and more of stability. Since the time complexity of both tests scales with the batch size, it can be ignored compared to the time complexity of backpropagation. However, the final data effects are stronger in the conditional dependence tests compared to their unconditional counterparts.

4 Experiments and Results

This section contains empirical results that confirm our theoretical claims. We first present experiments on a synthetic dataset that is designed to maximize the difference between the independence criterion (1) and the conditional independence criterion (2). Afterward, we report the results of an ablation study demonstrating that the gain in performance can be credited to the change of the independence criterion. Finally, we show that our findings also apply to a real-world dataset. For this purpose, we present experiments on different biased subsets of the cats and dogs dataset [12]. To evaluate our experiments, we measure the accuracy on an unbiased test set. Our debiasing approach is designed for applications in which the training set is biased, but where the classifier is used in an unbiased, real-world situation. Hence, the accuracy on an unbiased test set is our goal and therefore the most precise measure in this case. Further evaluations are included in the supplementary material.

4.1 Synthetic Data

If a feature is independent of the label for a given classification task, the independence criterion (1) and the conditional independence criterion (2) agree. Since we aim to maximize the difference between the two criteria, we use a dataset with a strong dependence between the label L and the variable B. We create a dataset of eight-by-eight pixels images that combine two signals. The first signal S, determines the shape of high-intensity pixels in the image, either a cross or a square, both consisting of the same number of pixels. The second signal B is the color of the image. To maximize the dependence between the label L and the bias variable B, every training image of a cross is green and every training image of a square is violet. In the test set, these two signals are independent. Example images from the training and test set can be seen in Fig. 3.

For our first experiment (Setup I), we use the shape as the signal S to determine the label L and the mean color of the image as the bias variable B. To avoid any influence of shape- or color-preference of neural networks, in a second experiment, we use the inverse setting (Setup II). For this second experiment, we use the color as the signal S to determine the label L. Here, the bias variable B is calculated as the difference between the values of pixels in the square

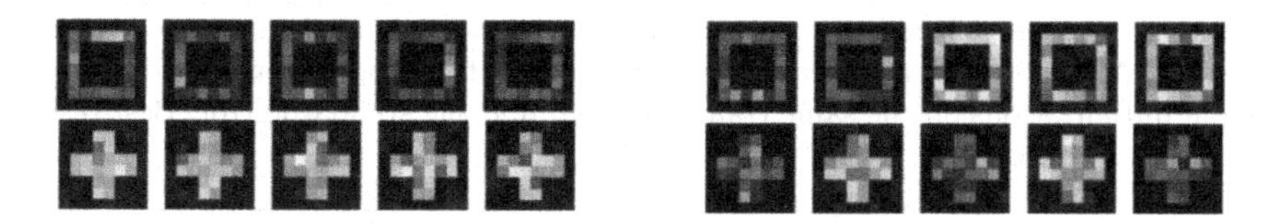

Fig. 3. Example images of the synthetic dataset, left: training set, right: test set.

Table 1. Results our method and all baselines on synthetic data. For both setups, we report mean accuracy ± standard error from 100 different random initializations on the same train/test-split. Best results in **bold**

	Baseline model	Adeli et al. [1]	Zhang et al. I [23]	Zhang et al. II [23]	Kim et al. [11]	Ours (MI)	Ours (HSIC)	Ours (MCC)
Setup I	0.819 ±0.016	0.747 ±0.015	0.736 ±0.018	0.747 ±0.016	0.771 ±0.012	0.840 ±0.014	0.846 ±0.021	**0.854** ±0.013
Setup II	0.791 ±0.016	0.776 ±0.014	0.837 ±0.017	0.750 ±0.013	0.767 ±0.016	**0.871** ±0.012	0.868 ±0.013	0.867 ±0.013

and those in the cross. We use a neural network with two convolutional layers (each with 16 filters of size 3×3) and two dense layers (128 neurons with ReLU activations and 2 neurons with softmax) as our backbone for classification. As a baseline, we use this network without any debiasing method. We have reimplemented four existing methods listed in Table 1. Two methods are proposed by Zhang et al. [23]. The first one called *Zhang et al. I* penalizes the predictability of B from R, the second one called *Zhang et al. II* penalizes predictability of B from R and L. Average accuracies of the competing approaches and from all three implementations of our proposed criterion are shown in Table 1.

Note that hyperparameter selection influences which feature (color or shape) the neural network uses for classifications, and we mitigate this by rigorous hyperparameter optimization. We use grid search for hyperparameters of our implementations as well as for general parameters like learning rates, and set hyperparameters for competing methods to values reported in the corresponding papers. We trained ten neural networks for each combination of different hyperparameters and evaluated them on an unbiased validation set. A list of the hyperparameters for every method is included in the supplementary material.

In Setup I, we see that differences between methods from the literature using the unconditional independence criterion (1) and the methods using our new conditional independence criterion (2) are much larger than the differences between methods within these groups. For the first experiment, the worst method using conditional independence performs 6.9 % points better than the best method using unconditional independence. The differences between the best and worst method within these groups is 1.4 % points and 3.5 % points, respectively. In the second setup, the results are similar. One competing method is surprisingly

Table 2. The results of the ablation study. Every method is trained on a biased training set and evaluated on an unbiased test set. We report the accuracy averaged over 100 random initializations on the same train/test-split and the standard error. Best results are marked in **bold**

	Uncond. MI	Cond. MI	Uncond. HSIC	Cond. HSIC	Adeli et al. [1]	Uncond. MCC	Only PC	Ours (MCC)
Setup I	0.583	**0.840**	0.744	**0.846**	0.747	0.757	0.836	**0.854**
	±0.010	±0.014	±0.011	±0.021	±0.015	±0.016	±0.014	±0.013
Setup II	0.833	**0.871**	0.590	**0.868**	0.776	0.807	0.830	**0.867**
	±0.011	±0.012	±0.011	±0.013	±0.014	±0.015	±0.015	±0.013

good in this experiment, but our models using conditional debiasing still perform much better with a gain of at least 3 % points.

We draw two conclusions from these experiments. First, none of the existing methods was able to improve the results of the baseline in Setup I, and only one approach did so for Setup II. This coincides with previous observations from the literature that adversarial debiasing methods are challenged in situations with strong bias, and reducing bias leads to decreased accuracy [3]. It also agrees with our findings discussed in Sect. 3.2. Second, we observe that all methods that use our new debiasing criterion with conditional independence reach higher accuracies than the baseline and, consequently, also a higher accuracies than existing methods. Note that we were able to reach the baseline performance for every method by allowing hyperparameters that deactivate the debiasing completely, e.g., by setting the weight of the debiasing loss to zero. To avoid this, we have limited the hyperparameter search to the range used in the respective publications.

4.2 Ablation Study

In the previous experiments, debiasing with our new criterion achieved higher accuracies than existing debiasing methods. We now conduct an ablation study for the three implementations from Sect. 3.3 to show that this increase can be attributed to our conditional independence criterion. More specifically, we report results for a method using unconditional mutual information and for a method using the unconditional HSIC as a loss. Furthermore, we present two methods that investigate the gap between the method of Adeli et al. [1] and our approach using the conditional maximal correlation coefficient. The first one uses the unconditional maximum correlation coefficient, and the second one incorporates the partial correlation (PC) instead of the correlation in (12). We use the settings and evaluation protocols of Setup I and Setup II from the previous section. The results are presented in Table 2.

The unconditional versions of MI and HSIC perform at least 3.8% points worse than our conditional counterparts in both setups. For the third method, we observe that the change from the predictability criterion to the maximum correlation coefficient ("Adeli et al. [1]" vs. "Uncond. MCC" and "Only PC" vs. "Ours(MCC)") increases the accuracy by at most 3.7% points. In contrast, the change from correlation to partial correlation ("Adeli et al. [1]" vs. "Only PC" and "Uncond. MCC" vs. "Ours(MCC)") increases the accuracy by at least 5.4% points. These observations indicate that the improvements found in Sect. 4.1 can be attributed to the difference between unconditional and conditional independence.

4.3 Real-World Data

Finally, we want to investigate whether increasing accuracies can also be observed for real-world image data. To evaluate the performance of our debiasing method, we require an unbiased test set.

Hence, we use a dataset with labels for multiple signals per image. This allows us to introduce a bias in the training set but not in the test set. We choose the cats and dogs dataset introduced by [12] for the same purpose. This dataset contains images of cats and dogs that are additionally labeled as dark-furred or light-furred. We first remove 20% of each class/fur combination as an unbiased test set. Then, we create eleven training sets with different levels of bias. We start with a training set that contains only light-furred dogs and only dark-furred cats. For each of the other sets, we increase the fraction of dark-furred dogs and light-furred cats by ten percent. Therefore, the last dataset contains only dark-furred dogs and light-furred cats. All training sets are created to have the same size for a fair evaluation. Hence, the number of training images is restricted by the rarest class/fur combination, leading to only less than 2500 training images.

We use a ResNet-18 [10] as a classifier. Details about this network and the selected hyperparameters for this experiment can be found in the supplementary material. To solely focus on the bias in our training sets, we refrain from pretraining on ImageNet [6] because ImageNet already contains thousands of dog images, and train from scratch instead. We only report accuracies for our approach with HSIC because it was most robust for different hyperparameters.

The results shown in Table 3 are averaged across three runs. Since the labels L and the bias variable B are binary, the two signals are indistinguishable for 0% and 100% dark-furred dogs, respectively. Furthermore, we obtain an unbiased training set for 50% dark-furred dogs. Our method reaches the highest accuracies in seven out of the remaining eight biased scenarios and the highest overall accuracy of 0.875 for 40% dark-furred dogs in the training set. For six out of these seven scenarios, the baseline was outside of our method's 95% confidence interval. We observe that competing methods only outperform the baseline in situations with little bias. This result supports our finding that existing methods are not suited for the bias model described in Sect. 3.1.

To further investigate the effectiveness of our approach, we compare the conditional and unconditional HSIC in Table 3 as well. We see that the conditional HSIC outperforms the unconditional HSIC in all biased scenarios. The stronger the bias, the bigger is the difference between the two methods. The correlation

Table 3. Experimental results on the cats and dogs dataset. All methods were trained on training sets in which p% of all dogs are dark-furred dogs and p% of all cats are light-furred. The first column indicates the fraction p, the others contain the accuracies on an unbiased test set. Best results in **bold**

Frac. p	Baseline model	Adeli et al. [1]	Zhang et al. I [23]	Zhang et al. II [23]	Uncond. HSIC	Ours (HSIC)
0%	**0.627** ±0.004	0.597 ±0.004	0.590 ±0.002	0.617 ±0.001	0.611 ±0.003	0.615 ±0.005
10%	0.800 ±0.001	0.774 ±0.002	0.779 ±0.005	0.785 ±0.007	0.759 ±0.012	**0.801** ±0.001
20%	0.845 ±0.003	0.829 ±0.000	0.812 ±0.002	0.809 ±0.005	0.816 ±0.002	**0.855** ±0.004
30%	0.852 ±0.007	0.842 ±0.003	0.837 ±0.004	0.834 ±0.003	0.834 ±0.002	**0.863** ±0.002
40%	0.859 ±0.007	0.855 ±0.004	0.870 ±0.002	0.850 ±0.001	0.861 ±0.003	**0.875** ±0.003
50%	0.859 ±0.006	**0.866** ±0.003	0.856 ±0.001	0.853 ±0.001	0.863 ±0.004	0.860 ±0.002
60%	**0.866** ±0.006	0.837 ±0.001	0.850 ±0.003	0.860 ±0.004	0.844 ±0.001	0.856 ±0.005
70%	0.844 ±0.003	0.854 ±0.003	0.835 ±0.005	0.841 ±0.005	0.835 ±0.003	**0.859** ±0.000
80%	0.829 ±0.002	0.822 ±0.005	0.820 ±0.005	0.826 ±0.003	0.820 ±0.007	**0.836** ±0.002
90%	0.773 ±0.010	0.743 ±0.001	0.758 ±0.001	0.731 ±0.002	0.757 ±0.003	**0.791** ±0.004
100%	0.612 ±0.001	0.612 ±0.004	0.604 ±0.001	0.609 ±0.001	0.606 ±0.002	**0.616** ±0.002

between the bias, measured as the absolute value between the difference of fractions of dark- and light-furred dogs, and the difference in accuracy between the conditional and unconditional HSIC method is 0.858.

5 Conclusion

In this work, we investigated a specific kind of dataset bias with a graphical model for data generation. Our exact model formulation allowed us to provide a mathematical proof to confirm our proposed conditional adversarial debiasing approach. Hence, our work differs from related work on adversarial debiasing, which solely relies on empirical evaluations. Our experimental results also support our theoretical claims. If a bias can be modeled with the investigated bias model, our conditional independence criterion is a better choice compared to an unconditional one. This is confirmed by our experiments. On synthetic data, the difference between conditional and unconditional debiasing criteria has been maximized. We further demonstrated in an ablation study that the conditional independence criterion is the reason for an increase in accuracy on unbiased test data, and improved accuracies have also been observed for real-world data.

In the future we aim to extend these empirical evaluations to get a better practical understanding of the method. This includes more detailed investigations in synthetic datasets, experiments on biased real-world datasets, e.g. HAM10000[21] or CAMELYON17[5], and experiments on datasets that do not fit the bias model in Sect. 3.1.

References

1. Adeli, E., et al.: Representation learning with statistical independence to mitigate bias. In: Proceedings of the IEEE/CVF Winter Conference on Applications of Computer Vision, pp. 2513–2523 (2021)
2. Ali Shah, S.A., Uddin, I., Aziz, F., Ahmad, S., Al-Khasawneh, M.A., Sharaf, M.: An enhanced deep neural network for predicting workplace absenteeism. Complexity **2020**, 1–12 (2020)
3. Alvi, M., Zisserman, A., Nellåker, C.: Turning a blind eye: explicit removal of biases and variation from deep neural network embeddings. In: Leal-Taixé, L., Roth, S. (eds.) ECCV 2018. LNCS, vol. 11129, pp. 556–572. Springer, Cham (2019). https://doi.org/10.1007/978-3-030-11009-3_34
4. Angwin, J., Larson, J., Mattu, S., Kirchner, L.: Machine Bias. PropPblica (2016). https://www.propublica.org/article/machine-bias-risk-assessments-in-criminal-sentencing
5. Bandi, P., et al.: From detection of individual metastases to classification of lymph node status at the patient level: the camelyon17 challenge. IEEE Trans. Med. Imaging **38**(2), 550–560 (2018)
6. Deng, J., Dong, W., Socher, R., Li, L.J., Li, K., Fei-Fei, L.: ImageNet: a large-scale hierarchical image database. In: 2009 IEEE Conference on Computer Vision and Pattern Recognition, pp. 248–255. IEEE (2009)
7. Fukumizu, K., Gretton, A., Sun, X., Schölkopf, B.: Kernel measures of conditional dependence. In: Advances in Neural Information Processing Systems, pp. 489–496 (2008)
8. Gretton, A., Fukumizu, K., Teo, C.H., Song, L., Schölkopf, B., Smola, A.J.: A kernel statistical test of independence. In: Advances in Neural Information Processing Systems, pp. 585–592 (2008)
9. Hartmann, B., Raste, T., Kretschmann, M., Amthor, M., Schneider, F., Denzler, J.: Aquaplaning - a potential hazard also for automated driving. In: ITS Automotive Nord e.V. (Hrsg.) Braunschweig (2018)
10. He, K., Zhang, X., Ren, S., Sun, J.: Deep residual learning for image recognition. In: Proceedings of the IEEE Conference on Computer Vision and Pattern Recognition, pp. 770–778 (2016)
11. Kim, B., Kim, H., Kim, K., Kim, S., Kim, J.: Learning not to learn: training deep neural networks with biased data. In: Proceedings of the IEEE Conference on Computer Vision and Pattern Recognition, pp. 9012–9020 (2019)
12. Lakkaraju, H., Kamar, E., Caruana, R., Horvitz, E.: Discovering blind spots of predictive models: representations and policies for guided exploration. arXiv preprint arXiv:1610.09064 (2016)
13. Muckatira, S.: Properties of winning tickets on skin lesion classification. arXiv preprint arXiv:2008.12141 (2020)
14. Pearl, J.: Causality. Cambridge University Press, Cambridge (2009)

15. Perez, F., Vasconcelos, C., Avila, S., Valle, E.: Data augmentation for skin lesion analysis. In: Stoyanov, D., et al. (eds.) CARE/CLIP/OR 2.0/ISIC -2018. LNCS, vol. 11041, pp. 303–311. Springer, Cham (2018). https://doi.org/10.1007/978-3-030-01201-4_33
16. Reimers, C., Requena-Mesa, C.: Deep learning-an opportunity and a challenge for geo-and astrophysics. In: Knowledge Discovery in Big Data from Astronomy and Earth Observation, pp. 251–265. Elsevier (2020)
17. Reimers, C., Runge, J., Denzler, J.: Determining the relevance of features for deep neural networks. In: Vedaldi, A., Bischof, H., Brox, T., Frahm, J.-M. (eds.) ECCV 2020. LNCS, vol. 12371, pp. 330–346. Springer, Cham (2020). https://doi.org/10.1007/978-3-030-58574-7_20
18. Sarmanov, O.V.: The maximum correlation coefficient (symmetrical case). In: Doklady Akademii Nauk, vol. 120, pp. 715–718. Russian Academy of Sciences (1958)
19. Torralba, A., Efros, A.A.: Unbiased look at dataset bias. In: CVPR 2011, pp. 1521–1528. IEEE (2011)
20. Tschandl, P., et al.: Comparison of the accuracy of human readers versus machine-learning algorithms for pigmented skin lesion classification: an open, web-based, international, diagnostic study. Lancet Oncol. **20**(7), 938–947 (2019)
21. Tschandl, P., Rosendahl, C., Kittler, H.: The HAM10000 dataset, a large collection of multi-source dermatoscopic images of common pigmented skin lesions. Sci. data **5**, 180161 (2018)
22. Wang, A., Narayanan, A., Russakovsky, O.: REVISE: a tool for measuring and mitigating bias in visual datasets. In: Vedaldi, A., Bischof, H., Brox, T., Frahm, J.-M. (eds.) ECCV 2020. LNCS, vol. 12348, pp. 733–751. Springer, Cham (2020). https://doi.org/10.1007/978-3-030-58580-8_43
23. Zhang, B.H., Lemoine, B., Mitchell, M.: Mitigating unwanted biases with adversarial learning. In: Proceedings of the 2018 AAAI/ACM Conference on AI, Ethics, and Society, pp. 335–340 (2018)

Revisiting Consistency Regularization for Semi-supervised Learning

Yue Fan(✉), Anna Kukleva, and Bernt Schiele

Max Planck Institute for Informatics, Saarland Informatics Campus, Saarbrücken, Germany
{yfan,akukleva,schiele}@mpi-inf.mpg.de

Abstract. Consistency regularization is one of the most widely-used techniques for semi-supervised learning (SSL). Generally, the aim is to train a model that is invariant to various data augmentations. In this paper, we revisit this idea and find that enforcing invariance by decreasing distances between features from differently augmented images leads to improved performance. However, encouraging equivariance instead, by increasing the feature distance, further improves performance. To this end, we propose an improved consistency regularization framework by a simple yet effective technique, FeatDistLoss, that imposes consistency and equivariance on the classifier and the feature level, respectively. Experimental results show that our model defines a new state of the art for various datasets and settings and outperforms previous work by a significant margin, particularly in low data regimes. Extensive experiments are conducted to analyze the method, and the code will be published.

1 Introduction

Deep learning requires large-scale and annotated datasets to reach state-of-the-art performance [30,42]. As labels are not always available or expensive to acquire a wide range of semi-supervised learning (SSL) methods have been proposed to leverage unlabeled data [1,3,6,7,10,18,28,29,33,38,46,48,49,51].

Consistency regularization [2,28,44] is one of the most widely-used SSL methods. Recent work [27,46,51] achieves strong performance by utilizing unlabeled data in a way that model predictions should be invariant to input perturbations. However, when using advanced and strong data augmentation schemes, we question if the model should be invariant to such strong perturbations. On the right of Fig. 1 we illustrate that strong data augmentation leads to perceptually highly diverse images. Thus, we argue that improving equivariance on such strongly augmented images can provide even better performance rather than making the model invariant to all kinds of augmentations. To this end, we propose a simple yet effective technique, Feature Distance Loss (FeatDistLoss), to improve data-augmentation-based consistency regularization.

Supplementary Information The online version contains supplementary material available at https://doi.org/10.1007/978-3-030-92659-5_5.

C. Bauckhage et al. (Eds.): DAGM GCPR 2021, LNCS 13024, pp. 63–78, 2021.
https://doi.org/10.1007/978-3-030-92659-5_5

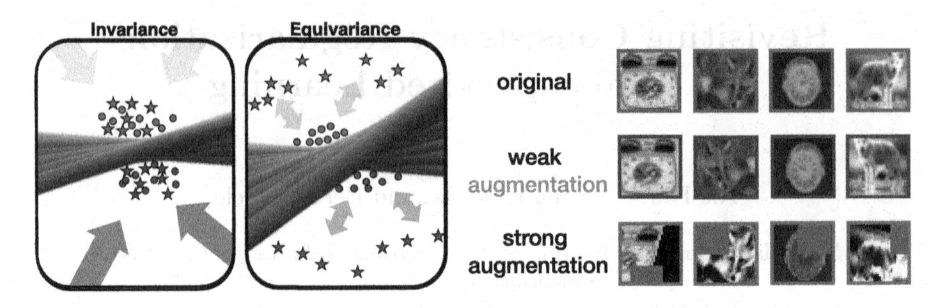

Fig. 1. Left: Binary classification task. Stars are features of strongly augmented images and circles are of weakly augmented images. While encouraging invariance by decreasing distance between features from differently augmented images gives good performance (left), encouraging equivariant representations by increasing the distance regularizes the feature space more, leading to even better generalization performance. **Right:** Examples of strongly and weakly augmented images from CIFAR-100. The visually large difference between them indicates that it can be more beneficial if they are treated differently.

We formulate our FeatDistLoss as to explicitly encourage invariance or equivariance between features from different augmentations while enforcing the same semantic class label. Figure 1 left shows the intuition behind the idea. Specifically, encouragement of equivariance for the same image but different augmentations (increase distance between stars and circles of the same color) pushes representations apart from each other, thus, covering more space for the class. Imposing invariance, on the contrary, makes the representations of the same semantic class more compact. In this work we empirically find that increasing equivariance to differently augmented versions of the same image can lead to better performance especially when rather few labels are available per class (see Sect. 4.3).

This paper introduces the method *CR-Match* which combines FeatDistLoss with other strong techniques defining a new state-of-the-art across a wide range of settings of standard SSL benchmarks, including CIFAR-10, CIFAR-100, SVHN, STL-10, and Mini-Imagenet. More specifically, our contribution is fourfold. (1) We improve data-augmentation-based consistency regularization by a simple yet effective technique for SSL called *FeatDistLoss* which regularizes the distance between feature representations from differently augmented images of the same class. (2) We show that while encouraging invariance results in good performance, encouraging equivariance to differently augmented versions of the same image consistently results in even better generalization performance. (3) We provide comprehensive ablation studies on different distance functions and different augmentations with respect to the proposed FeatDistLoss. (4) In combination with other strong techniques, we achieve *new state-of-the-art results* across a variety of standard semi-supervised learning benchmarks, specifically in low data regimes.

2 Related Work

SSL is a broad field aiming to exploit both labeled and unlabeled data. Consistency regularization is a powerful method for SSL [2,39,44]. The idea is that the model should output consistent predictions for perturbed versions of the same input. Many works explored different ways to generate such perturbations. For example, [48] uses an exponential moving average of the trained model to produce another input; [28,44] use random max-pooling and Dropout [6,27,46,47,51] use advanced data augmentation; [6,7,49] use MixUp regularization [54], which encourages convex behavior "between" examples. Another spectrum of popular approaches is pseudo-labeling [29,35,45], where the model is trained with artificial labels. [1] trained the model with "soft" pseudo-labels from network predictions; [38] proposed a meta learning method that deploys a teacher model to adjust the pseudo-label alongside the training of the student; [29,46] learn from "hard" pseudo-labels and only retain a pseudo-label if the largest class probability is above a predefined threshold. Furthermore, there are many excellent works around generative models [16,25,37] and graph-based methods [5,24,31,32]. We refer to [8,55,56] for a more comprehensive introduction of SSL methods.

Noise injection plays a crucial role in consistency regularization [51]. Thus advanced data augmentation, especially combined with weak data augmentation, introduces stronger noise to unlabeled data and brings substantial improvements [6,46]. [46] proposes to integrate pseudo-labeling into the pipeline by computing pseudo-labels from weakly augmented images, and then uses the cross-entropy loss between the pseudo-labels and strongly augmented images. Besides the classifier level consistency, our model also introduces consistency on the feature level, which explicitly regularizes representation learning and shows improved generalization performance. Moreover, self-supervised learning is known to be beneficial in the context of SSL. In [9,10,20,41], self-supervised pre-training is used to initialize SSL. However, these methods normally have several training phases, where many hyper-parameters are involved. We follow the trend of [6,53] to incorporate an auxiliary self-supervised loss alongside training. Specifically, we optimizes a rotation prediction loss [19].

Equivariant representations are recently explored by capsule networks [22, 43]. They replaced max-pooling layers with convolutional strides and dynamic routing to preserve more information about the input, allowing for preservation of part-whole relationships in the data. It has been shown, that the input can be reconstructed from the output capsule vectors. Another stream of work on group equivariant networks [12,13,50] explores various equivariant architectures that produce transform in a predictable linear manner under transformations of the input. Different from previous work, our work explores equivariant representations in the sense that differently augmented versions of the same image are represented by different points in the feature space despite the same semantic label. As we will show in Sect. 4.3, information like object location or orientation is more predictable from our model when features are pushed apart from each other.

3 CR-Match

Consistency regularization is highly-successful and widely-adopted technique in SSL [2,27,28,44,46,51]. In this work, we aim to leverage and improve it by even further regularizing the feature space. To this end, we present a simple yet effective technique FeatDistLoss to explicitly regularize representation learning and classifier learning at the same time. We describe our SSL method, called CR-Match, which shows improved performance across many different settings, especially in scenarios with few labels. In this section, we first describe our technique FeatDistLoss and then present CR-Match that combines FeatDistLoss with other regularization techniques inspired from the literature.

3.1 Feature Distance Loss

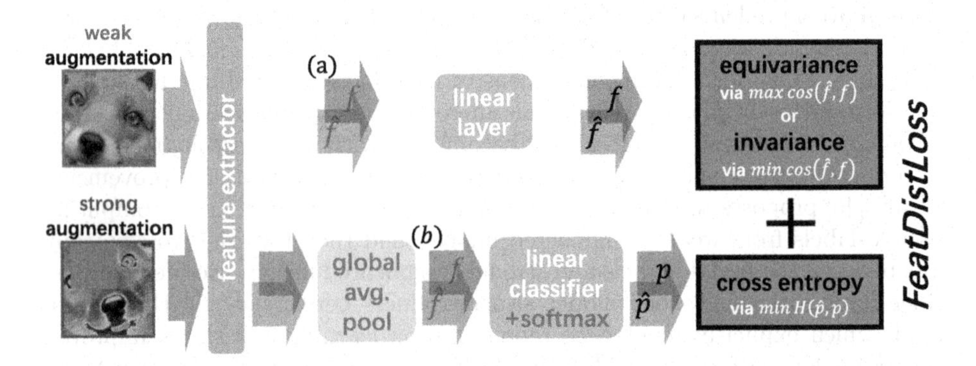

Fig. 2. The proposed FeatDistLoss utilizes unlabeled images in two ways: On the classifier level, different versions of the same image should generate the same class label, whereas on the feature level, representations are encouraged to become either more equivariant (pushing away) or invariant (pulling together). f and $\hat{f}$ denote strong and weak features; p and $\hat{p}$ are predicted class distributions from strong and weak features; a) and b) denote features before and after the global average pooling layer. Our final model takes features from a) and encourages equivariance to differently augmented versions of the same image. An ablation study of other choices is in Sect. 4.3.

Background: The idea of consistency regularization [2,28,44] is to encourage the model predictions to be invariant to input perturbations. Given a batch of n unlabeled images $\mathbf{u}_i, i \in (1, ..., n)$, consistency regularization can be formulated as the following loss function:

$$\frac{1}{n}\sum_{i=1}^{n} ||f(\mathcal{A}(\mathbf{u}_i)) - f(\alpha(\mathbf{u}_i))||_2^2 \tag{1}$$

where f is an encoder network that maps an input image to a d-dimensional feature space; $\mathcal{A}$ and α are two stochastic functions which are, in our case, strong and weak augmentations, respectively (details in Sect. 3.3). By minimizing the L_2

distance between perturbed images, the representation is therefore encouraged to become more invariant with respect to different augmentations, which helps generalization. The intuition behind this is that a good model should be robust to data augmentations of the images.

FeatDistLoss: As shown in Fig. 2, we extend the above consistency regularization idea by introducing consistency on the classifier level and invariance or equivariance on the feature level. FeatDistLoss thus allows to apply different types of control for these levels. In particular, when encouraging to reduce the feature distance, it becomes similar to classic consistency regularization, and encourages invariance between differently augmented images. As argued above, making the model predictions invariant to input perturbations gives good generalization performance. Instead, in this work we find it is more beneficial to treat images from different augmentations differently because some distorted images are largely different from their original images as demonstrated visually in Fig. 1 right. Therefore, the final model (CR-Match) uses FeatDistLoss to increase the distance between image features from augmentations of different intensities while at the same time enforcing the same semantic label for them. Note that in Sect. 4.3, we conduct an ablation study on the choice of distance function, where we denote CR-Match as CR-Equiv, and the model that encourages invariance as CR-Inv.

The final objective for the FeatDistLoss consists of two terms: $\mathcal{L}_{Dist}$ (on the feature level), that explicitly regularizes feature distances between embeddings, and a standard cross-entropy loss $\mathcal{L}_{PseudoLabel}$ (on the classifier level) based on pseudo-labeling.

With $\mathcal{L}_{Dist}$ we either decrease or increase the feature distance between weakly and strongly augmented versions of the same image in a low-dimensional space projected from the original feature space to overcome the curse of dimensionality [4]. Let $d(\cdot,\cdot)$ be a distance metric and z be a linear layer that maps the high-dimensional feature into a low-dimensional space. Given an unlabeled image $\mathbf{u}_i$, we first extract features with strong and weak augmentations by $f(\mathcal{A}(\mathbf{u}_i))$ and $f(\alpha(\mathbf{u}_i))$ as shown in Fig. 2 (a), and then FeatDistLoss is computed as:

$$\mathcal{L}_{Dist}(\mathbf{u}_i) = d(z(f(\mathcal{A}(\mathbf{u}_i))), z(f(\alpha(\mathbf{u}_i)))) \tag{2}$$

Different choices of performing $\mathcal{L}_{Dist}$ are studied in Sect. 4.3, where we find empirically that applying $\mathcal{L}_{Dist}$ at (a) in Fig. 2 gives the best performance.

At the same time, images from strong and weak augmentations should have the same class label because they are essentially generated from the same original image. Inspired by [46], given an unlabeled image $\mathbf{u}_i$, a pseudo-label distribution is first generated from the weakly augmented image by $\hat{\mathbf{p}}_i = g(f(\alpha(\mathbf{u}_i)))$, and then a cross-entropy loss is computed between the pseudo-label and the prediction for the corresponding strongly augmented version as:

$$\mathcal{L}_{PseudoLabel}(\mathbf{u}_i) = \ell_{CE}(\hat{\mathbf{p}}_i, g(f(\mathcal{A}(\mathbf{u}_i)))) \tag{3}$$

where ℓ_{CE} is the cross-entropy, g is a linear classifier that maps a feature representation to a class distribution, and $\mathcal{A}(\mathbf{u}_i)$ denotes the operator for strong augmentations.

Putting it all together, FeatDistLoss processes a batch of unlabeled data $\mathbf{u}_i, i \in (1, ..., B_u)$ with the following loss:

$$\mathcal{L}_U = \frac{1}{B_u} \sum_{i=1}^{B_u} \mathbb{1}\{c_i > \tau\}(\mathcal{L}_{Dist}(\mathbf{u}_i) + \mathcal{L}_{PseudoLabel}(\mathbf{u}_i)) \tag{4}$$

where $c_i = max\ \hat{\mathbf{p}}_i$ is the confidence score, and $\mathbb{1}\{\cdot\}$ is the indicator function which outputs 1 when the confidence score is above a threshold. This confidence thresholding mechanism ensures that the loss is only computed for unlabeled images for which the model generates a high-confidence prediction, which gives a natural curriculum to balance labeled and unlabeled losses [46].

As mentioned before, depending on the function d, FeatDistLoss can decrease the distance between features from different data augmentation schemes (when d is a distance function, thus pulling the representations together), or increase it (when d is a similarity function, thus pushing the representations apart). As shown in Table 4, we find that both cases results in an improved performance. However, increasing the distance between weakly and strongly augmented examples consistently results in better generalization performance. We conjecture that the reason lies in the fact that FeatDistLoss by increasing the feature distance explores equivariance properties (differently augmented versions of the same image having distinct features but the same label) of the representations. It encourages the model to have more distinct weakly and strongly augmented images while still imposing the same label, which leads to both more expressive representation and more powerful classifier. As we will show in Sect. 4.3, information like object location or orientation is more predictable from models trained with FeatDistLoss that pushes the representations apart. Additional ablation studies of other design choices such as the distance function and the linear projection z are also provided in Sect. 4.3.

3.2 Overall CR-Match

Now we describe our SSL method called CR-Match leveraging the above FeatDistLoss. Pseudo-code for processing a batch of labeled and unlabeled examples is shown in algorithm 1 in supplementary material.

Given a batch of labeled images with their labels as $\mathcal{X} = \{(\mathbf{x}_i, \mathbf{p}_i) : i \in (1, ..., B_s)\}$ and a batch of unlabeled images as $\mathcal{U} = \{\mathbf{u}_i : i \in (1, ..., B_u)\}$. CR-Match minimizes the following learning objective:

$$\mathcal{L}_S(\mathcal{X}) + \lambda_u \mathcal{L}_U(\mathcal{U}) + \lambda_r \mathcal{L}_{Rot}(\mathcal{X} \cup \mathcal{U}) \tag{5}$$

where $\mathcal{L}_S$ is the supervised cross-entropy loss for labeled images with weak data augmentation regularization; $\mathcal{L}_U$ is our novel feature distance loss for unlabeled images which explicitly regularizes the distance between weakly and strongly augmented images in the feature space; and $\mathcal{L}_{Rot}$ is a self-supervised loss for unlabeled images and stands for rotation prediction from [19] to provide an additional supervisory and regularizing signal.

Fully Supervised Loss for Labeled Data: We use cross-entropy loss with weak data augmentation regularization for labeled data:

$$\mathcal{L}_S = \frac{1}{B_s} \sum_{i=1}^{B_s} \ell_{CE}(\mathbf{p}_i, g(f(\alpha(\mathbf{x}_i)))) \tag{6}$$

where ℓ_{CE} is the cross-entropy loss, $\alpha(\mathbf{x}_i)$ is the extracted feature from a weakly augmented image $\mathbf{x}_i$, g is the same linear classifier as in Eq. 2, and $\mathbf{p}_i$ is the corresponding label for $\mathbf{x}_i$.

Self-supervised Loss for Unlabeled Data: Rotation prediction [19] (RotNet) is one of the most successful self-supervised learning methods, and has been shown to be complementary to SSL methods [7,41,53]. Here, we create four rotated images by 0°, 90°, 180°, and 270° for each unlabeled image $\mathbf{u}_i$ for $i \in (1, ..., \mu B)$. Then, classification loss is applied to train the model predicting the rotation as a four-class classification task:

$$\mathcal{L}_{Rot} = \frac{1}{4B_u} \sum_{i=1}^{B_u} \sum_{r \in \mathbb{R}} \ell_{CE}(r, h(\alpha(Rotate(\mathbf{u}_i, r)))) \tag{7}$$

where $\mathbb{R}$ is $\{0°, 90°, 180°, 270°\}$, and h is a predictor head.

3.3 Implementation Details

Data Augmentation: As mentioned above, CR-Match adopts two types of data augmentations: weak augmentation and strong augmentation from [46]. Specifically, the weak augmentation α corresponds to a standard random cropping and random mirroring with probability 0.5, and the strong augmentation $\mathcal{A}$ is a combination of RandAugment [14] and CutOut [17]. At each training step, we uniformly sample two operations for the strong augmentation from a collection of transformations and apply them with a randomly sampled magnitude from a predefined range. The complete table of transformation operations for the strong augmentation is provided in the supplementary material.

Other Implementation Details: For our results in Sect. 4, we minimize the cosine similarity in FeatDistLoss, and use a fully-connected layer for the projection layer z. The predictor head h in rotation prediction loss consists of two fully-connected layers and a ReLU [34] as non-linearity. We use the same $\lambda_u = \lambda_r = 1$ in all experiments. As a common practice, we repeat each experiment with five different data splits and report the mean and the standard deviation of the error rate. We refer to supplementary material for further details and ablation of the hyper-parameters.

4 Experimental Results

Following protocols from previous work [7,46], we conduct experiments on several commonly used SSL image classification benchmarks to test the efficacy of CR-Match. We show our main results in Sect. 4.1, where we achieve state-of-the-art error rates across all settings on SVHN [36], CIFAR-10 [26], CIFAR-100 [26],

STL-10 [11], and mini-ImageNet [40]. In our ablation study in Sect. 4.2 we analyze the effect of FeatDistLoss and RotNet across different settings. Finally, in Sect. 4.3 we extensively analyse various design choices for our FeatDistLoss.

Table 1. Error rates on CIFAR-10, and CIFAR-100. A Wide ResNet-28-2 [52] is used for CIFAR-10 and a Wide ResNet-28-8 with 135 filters per layer [7] is used for CIFAR-100. We use the same code base as [46] (i.e., same network architecture and training protocol) to make the results directly comparable. *Numbers are generated by [46]. †Numbers are produced without CutOut. The best number is in bold and the second best number is in italic.

	CIFAR-10			CIFAR-100		
Per class labels	4 labels	25 labels	400 labels	4 labels	25 labels	100 labels
Mean Teacher [48]	–	32.32 ± 2.30*	9.19 ± 0.19*	–	53.91 ± 0.57*	35.83 ± 0.24*
MixMatch [7]	47.54 ± 11.50*	11.08 ± 0.87	6.24 ± 0.06	67.61 ± 1.32*	39.94 ± 0.37*	25.88 ± 0.30
UDA [51]	29.05 ± 5.93*	5.43 ± 0.96	4.32 ± 0.08*	59.28 ± 0.88*	33.13 ± 0.22*	24.50 ± 0.25*
ReMixMatch [6]	19.10 ± 9.64*	6.27 ± 0.34	5.14 ± 0.04	*44.28* ± 2.06*	*27.43* ± 0.31*	23.03 ± 0.56*
FixMatch (RA) [46]	13.81 ± 3.37	*5.07* ± 0.65	4.26 ± 0.05	48.85 ± 1.75	28.29 ± 0.11	*22.60* ± 0.12
FixMatch (CTA) [46]	*11.39* ± 3.35	*5.07* ± 0.33	*4.31* ± 0.15	49.95 ± 3.01	28.64 ± 0.24	23.18 ± 0.11
FeatMatch [27]	–	6.00 ± 0.41	4.64 ± 0.11	–	–	–
CR-Match	**10.70** ± 2.91	**5.05** ± 0.12	**3.96** ± 0.16	**39.45** ± 1.69	**25.43** ± 0.14	**20.40** ± 0.08

Table 2. Left: Error rates on STL-10 and SVHN. A Wide ResNet-28-2 and a Wide ResNet-37-2 [52] is used for SVHN and STL-10, repectively. The same code base is adopted as [46] to make the results directly comparable. Notations follow Table 1 **Right:** Error rates on Mini-ImageNet with 40 labels and 100 labels per class. All methods are evaluated on the same ResNet-18 architecture.

	STL-10	SVHN		
Per class labels	100 labels	4 labels	25 labels	100 labels
Mean Teacher [48]	21.34±2.39*	-	3.57±0.11*	3.42±0.07*
MixMatch [7]	10.18±1.46	42.55±14.53*	3.78±0.26	3.27±0.31
UDA [51]	7.66±0.56*	52.63±20.51*	2.72±0.40	*2.23*±0.07
ReMixMatch [6]	6.18±1.24	*3.34*±0.20*	3.10±0.50	2.83±0.30
FixMatch (RA) [46]	7.98±1.50	3.96±2.17	*2.48*±0.38	2.28±0.11
FixMatch (CTA) [46]	*5.17*±0.63	7.65±7.65	2.64±0.64	2.36±0.19
FeatMatch [27]	-	-	3.34±0.19†	3.10±0.06†
CR-Match	**4.89**±0.17	**2.79**±0.93	**2.35**±0.29	**2.08**±0.07

	Mini-ImageNet	
Per class labels	40 labels	100 labels
Mean Teacher [48]	72.51±0.22	57.55±1.11
Label Propagation [23]	70.29±0.81	57.58±1.47
PLCB [1]	56.49±0.51	46.08±0.11
FeatMatch [27]	*39.05*±0.06	*34.79*±0.22
CR-Match	**34.87**±0.99	**32.58**±1.60

4.1 Main Results

In the following, each dataset subsection includes two paragraphs. The first provides technical details and the second discusses experimental results.

CIFAR-10, CIFAR-100, and SVHN. We follow prior work [46] and use 4, 25, and 100 labels per class on CIFAR-100 and SVHN without extra data. For CIFAR-10, we experiment with settings of 4, 25, and 400 labels per class. We create labeled data by random sampling, and the remaining images are regarded as unlabeled by discarding their labels. Following [6,7,46], we use a Wide ResNet-28-2 [52] with 1.5M parameters on CIFAR-10 and SVHN, and a Wide ResNet-28-8 with 135 filters per layer (26M parameters) on CIFAR-100.

As shown in Table 1 and Table 2, our method improves over previous methods across all settings, and defines a new state-of-the-art. Most importantly, we improve error rates in low data regimes by a large margin (e.g., with 4 labeled examples per class on CIFAR-100, we outperform FixMatch and the second best method by 9.40% and 4.83% in absolute value respectively). Prior works [6,7,46] have reported results using a larger network architecture on CIFAR-100 to obtain better performance. On the contrary, we additionally evaluate our method on the small network used in CIFAR-10 and find that our method is more than 17 times ($17 \approx 26/1.5$) parameter-efficient than FixMatch. We reach 46.05% error rate on CIFAR-100 with 4 labels per class using the small model, which is still slightly better than the result of FixMatch using a larger model.

STL-10. STL-10 contains 5,000 labeled images of size 96-by-96 from 10 classes and 100,000 unlabeled images. The dataset pre-defines ten folds of 1,000 labeled examples from the training data, and we evaluate our method on five of these ten folds as in [6,46]. Following [7], we use the same Wide ResNet-37-2 model (comprising 5.9M parameters), and report error rates in Table 2.

Our method achieves state-of-the-art performance with 4.89% error rate. Note that FixMatch with error rate 5.17% used the more advanced CTAugment [6], which learns augmentation policies alongside model training. When evaluated with the same data augmentation (RandAugment) as we use in CR-Match, our result surpasses FixMatch by 3.09% (3.09% = 7.98% − 4.89%), which indicates that CR-Match itself induces a strong regularization effect.

Mini-ImageNet. We follow [1,23,27] to construct the mini-ImageNet training set. Specifically, 50,000 training examples and 10,000 test examples are randomly selected for a predefined list of 100 classes [40] from ILSVRC [15]. Following [27], we use a ResNet-18 network [21] as our model and experiment with settings of 40 labels per class and 100 labels per class.

As shown in Table 2, our method consistently improves over previous methods and achieves a new state-of-the-art in both the 40-label and 100-label settings. Especially in the 40-label case, CR-Match achieves an error rate of 34.87% which is 4.18% higher than the second best result. Note that our method is 2 times more data efficient than the second best method FeatMatch [27] (FeatMatch, using 100 labels per class, reaches a similar error rate as our method with 40 labeled examples per class).

4.2 Ablation Study

Table 3. Ablation studies across different settings. Error rates are reported for a single split.

RotNet	FeatDistLoss	MiniImageNet@40	CIFAR10@4	CIFAR100@4	SVHN@4
		35.13	11.86	46.22	2.42
	✓	34.14	**10.33**	43.48	2.34
✓		34.64	11.27	41.48	2.21
✓	✓	**33.82**	10.92	**39.22**	**2.09**

In this section, we analyze how FeatDistLoss and RotNet influence the performance across different settings, particularly when there are few labeled samples. We conduct experiments on a single split on CIFAR-10, CIFAR-100, and SVHN with 4 labeled examples per class, and on MiniImageNet with 40 labels per class. Specifically, we remove the $\mathcal{L}_{Dist}$ from Eq. 4 and train the model again using the same training scheme for each setting. We do not ablate $\mathcal{L}_{PseudoLabel}$ and $\mathcal{L}_S$ due to the fact that removing one of them leads to a divergence of training. We present a more detailed analysis of the pseudo-label error rate, the error rate of contributing unlabeled images, and the percentage of contributing unlabeled images during training on CIFAR-100 in the supplementary material.

We report final test error rates in Table 3. We see that both RotNet and FeatDistLoss contribute to the final performance while their proportions can be different depending on the setting and dataset. For MiniImageNet, CIFAR-100 and SVHN, the combination of both outperforms the individual losses. For CIFAR-10, FeatDistLoss even outperforms the combination of both. This suggests that RotNet and FeatDistLoss are both important components for CR-Match to achieve the state-of-the-art performance.

4.3 Influence of Feature Distance Loss

In this section, we analyze different design choices for FeatDistLoss to provide additional insights of how it helps generalization. We focus on a single split with 4 labeled examples from CIFAR-100 and report results for a Wide ResNet-28-2 [52]. For fair comparison, the same 4 random labeled examples for each class are used across all experiments in this section.

Different Distance Metrics for FeatDistLoss. Here we discuss the effect of different metric functions d for FeatDistLoss. Specifically, we compare two groups of functions in Table 4 left: metrics that increase the distance between features, including cosine similarity, negative JS divergence, and L2 similarity (i.e. normalized negative L2 distance); metrics that decrease the distance between features, including cosine distance, JS divergence, and L2 distance. We find that both increasing and decreasing distance between features of different augmentations give reasonable performance. However, increasing the distance always performs better than the counterpart (e.g., cosine similarity is better than cosine distance). We conjecture that decreasing the feature distance corresponds to an increase of the invariance to data augmentation and leads to ignorance of information like rotation or translation of the object. In contrast, increasing the feature distance while still imposing the same label makes the representation equivariant to these augmentations, resulting in more descriptive and expressive representation with respect to augmentation. Moreover, a classifier has to cover a broader space in the feature space to recognize rather dissimilar images from the same class, which leads to improved generalization. In summary, we found that both increasing and decreasing feature distance improve over the model which only applies consistency on the classifier level, whereas increasing distances shows better performance by making representations more equivariant.

Table 4. Left: Effect of different distance functions for FeatDistLoss. The same split on CIFAR-100 with 4 labels per class and a Wide ResNet-28-2 is used for all experiments. Metrics that pull features together performs worse than those that push features apart. The error rate of CR-Match without FeatDistLoss is shown at the bottom. **Right:** Error rates of binary classification (whether a specific augmentation is applied) on the features from CR-Equiv (increasing the cosine distance) and CR-Inv (decreasing the cosine distance). We evaluate translation, scaling, rotation, and color jittering. Lower error rate indicates more equivariant features. Results are averaged over 10 runs.

Metric		Error rate
Impose equi-variance	cosine similarity	**45.52**
	L_2 similarity	46.22
	negative JS div.	46.46
Impose invariance	cosine distance	46.98
	L_2 distance	48.74
	JS divergence	47.48
CR-Match w/o FeatDistLoss		48.89

Transformations	Feature extractor	
	CR-Equiv	CR-Inv
Translation	33.22±0.28	36.80±0.30
Scaling	11.09±0.66	14.87±0.40
Rotation	15.05±0.33	21.92±0.32
ColorJittering	31.04±0.50	35.99±0.27

Invariance and Equivariance. Here we provide an additional analysis to demonstrate that increasing the feature distance provides equivariant features while the other provides invariant features. Based on the intuition that specific transformations of the input image should be more predictable from equivariant representations, we quantify the equivariance by how accurate a linear classifier can distinguish between features from augmented and original images. Specifically, we compare two models from Table 4 left: the model trained with cosine similarity denoted as *CR-Equiv* and the model trained with cosine distance denoted as *CR-Inv*. We train a linear SVM to predict whether a certain transformation is applied for the input image. 1000 test images from CIFAR-100 are used for training and the rest (9000) for validation. The binary classifier is trained by an SGD optimizer with an initial learning rate of 0.001 for 50 epochs, and the feature extractor is fixed during training. We evaluate translation, scaling, rotation, and color jittering in Table 4 right. All augmentations are from the standard PyTorch library. The SVM has a better error rate across all augmentations when trained on CR-Equiv features, which means information like object location or orientation is more predictable from CR-Equiv features, suggesting that CR-Equiv produces more equivariant features than CR-Inv. Furthermore, if the SVM is trained to classify strongly and weakly augmented image features, CR-Equiv achieves a 0.27% test error while CR-Inv is 46.18%.

Different Data Augmentations for FeatDistLoss. In our main results in Sect. 4.1, FeatDistLoss is computed between features generated by weak augmentation and strong augmentation. Here we investigate the impact of FeatDistLoss with respect to different types of data augmentations. Specifically, we evaluate the error rate of CR-Inv and CR-Equiv under three augmentation strategies: weak-weak pair indicates that FeatDistLoss uses two weakly augmented images, weak-strong pair indicates that FeatDistLoss uses a weak augmentation and a strong augmentation, and strong-strong pair indicates that FeatDistLoss uses two strongly augmented images.

As shown in Table 5, using either CR-Inv or CR-Equiv using weak-strong pairs conistently outperforms the other augmentation settings (weak-weak and strong-strong).

Table 5. Effect of combinations of weak and strong augmentation in FeatDistLoss on a Wide ResNet-28-2 for CR-Inv and CR-Equiv.

Error rate	CR-Inv	CR-Equiv
Weak-Weak	48.88	48.51
Weak-Strong	**46.98**	**45.52**
Strong-Strong	48.57	48.05

Additionally, CR-Equiv consistently achieves better generalization performance across all three settings. In particular, in the case advocated in this paper, namely using weak-strong pairs, CR-Equiv outperforms CR-Inv by 1.46%. Even in the other two settings, CR-Equiv leads to improved performance even though only by a small margin. This suggests that, on the one hand, that it is important to use different types of augmentations for our FeatDistLoss. And on the other hand, maximizing distances between images that are inherently different while still imposing the same class label makes the model more robust against changes in the feature space and thus gives better generalization performance.

Linear Projection and Confidence Threshold in FeatDistLoss. As mentioned in Sect. 3, we apply $\mathcal{L}_{Dist}$ at (a) in Fig. 2 with a linear layer mapping the feature from the encoder to a low-dimensional space before computing the loss, to alleviate the curse of dimensionality. Also, the loss only takes effect when the model's prediction has a confidence score above a predefined threshold τ. Here we study the effect of other design choices in Table 6. While features after the global average pooling (i.e. (b)) gives a better result than the ones directly from the feature extractor, (b) performs worse than (a) when additional projection heads are added. Thus, we use features from the feature extractor in CR-Match.

Table 6. Effect of the projection head z, and the place to apply $\mathcal{L}_{Dist}$. (a) denotes un-flattened features taken from the feature extractor directly. (b) denotes features after the global average pooling. MLP has 2 FC layers and a ReLU. Removing the linear projection head harms the test error, and a non-linear projection head does not improve the performance further.

Features taken from Fig. 2 at	Feature	Feature + linear	Feature + MLP
(a)	48.37	**45.52**	47.52
(b)	47.37	46.10	47.15

The error rate increases from 45.52% to 48.37% and 47.52% when removing the linear layer and replacing the linear layer by a MLP (two fully-connected layers and a ReLU activation function), respectively. This suggests that a lower dimensional space serves better for comparing distances, but a non-linear mapping does not give further improvement. Moreover, when we apply FeatDistLoss for all pairs of input images by removing the confidence threshold, the test error increases from 45.52% to 46.94%, which suggests that regularization should be only performed on features that are actually used to update the model parameters, and ignoring those that are also ignored by the model.

5 Conclusion

The idea of consistency regularization gives rise to many successful works for SSL [2,27,28,44,46,51]. While making the model invariant against input perturbations induced by data augmentation gives improved performance, the scheme tends to be suboptimal when augmentations of different intensities are used. In this work, we propose a simple yet effective improvement, called FeatDistLoss. It introduces consistency regularization on both the classifier level, where the same class label is imposed for versions of the same image, and the feature level, where distances between features from augmentations of different intensities is increased. By encouraging the representation to distinguish between weakly and strongly augmented images, FeatDistLoss encourages more equivariant representations, leading to improved classification boundaries, and a more robust model.

Through extensive experiments we show the superiority of our training framework, and define a new state-of-the-art on CIFAR-10, CIFAR-100, SVHN, STL-10 and Mini-ImageNet. Particularly, our method outperforms previous methods in low data regimes by significant margins, e.g., on CIFAR-100 with 4 annotated examples per class, our error rate (39.45%) is 4.83% better than the second best (44.28%). In future work, we are interested in integrating more prior knowledge and stronger regularization into SSL to further push the performance in low data regimes.

References

1. Arazo, E., Ortego, D., Albert, P., O'Connor, N.E., McGuinness, K.: Pseudo-labeling and confirmation bias in deep semi-supervised learning. In: International Joint Conference on Neural Networks (IJCNN). IEEE (2020)
2. Bachman, P., Alsharif, O., Precup, D.: Learning with pseudo-ensembles. In: Advances in Neural Information Processing Systems (2014)
3. Bachman, P., Hjelm, R.D., Buchwalter, W.: Learning representations by maximizing mutual information across views. In: Advances in Neural Information Processing Systems (2019)
4. Bellman, R.: Dynamic programming. Science **153**(3731), 34–37 (1966)
5. Bengio, Y., Delalleau, O., Le Roux, N.: 11 label propagation and quadratic criterion (2006)
6. Berthelot, D., et al.: ReMixMatch: semi-supervised learning with distribution matching and augmentation anchoring. In: 8th International Conference on Learning Representations, ICLR (2020)
7. Berthelot, D., Carlini, N., Goodfellow, I., Papernot, N., Oliver, A., Raffel, C.A.: MixMatch: a holistic approach to semi-supervised learning. In: Advances in Neural Information Processing Systems (2019)
8. Chapelle, O., Scholkopf, B., Zien, A.: Semi-supervised learning (Chapelle, O. et al. eds. 2006) [Book Reviews]. IEEE Trans. Neural Netw. **20**(3) (2009)
9. Chen, T., Kornblith, S., Norouzi, M., Hinton, G.: A simple framework for contrastive learning of visual representations. arXiv preprint arXiv:2002.05709 (2020)
10. Chen, T., Kornblith, S., Swersky, K., Norouzi, M., Hinton, G.: Big self-supervised models are strong semi-supervised learners. arXiv preprint arXiv:2006.10029 (2020)

11. Coates, A., Ng, A., Lee, H.: An analysis of single-layer networks in unsupervised feature learning. In: Proceedings of the Fourteenth International Conference on Artificial Intelligence and Statistics (2011)
12. Cohen, T., Welling, M.: Group equivariant convolutional networks. In: International Conference on Machine Learning (2016)
13. Cohen, T.S., Welling, M.: Steerable CNNs. arXiv preprint arXiv:1612.08498 (2016)
14. Cubuk, E.D., Zoph, B., Shlens, J., Le, Q.V.: RandAugment: practical automated data augmentation with a reduced search space. In: Proceedings of the IEEE Conference on Computer Vision and Pattern Recognition Workshops (2020)
15. Deng, J., Dong, W., Socher, R., Li, L.J., Li, K., Fei-Fei, L.: ImageNet: a large-scale hierarchical image database. In: IEEE Conference on Computer Vision and Pattern Recognition (2009)
16. Denton, E., Gross, S., Fergus, R.: Semi-supervised learning with context-conditional generative adversarial networks. arXiv preprint arXiv:1611.06430 (2016)
17. DeVries, T., Taylor, G.W.: Improved regularization of convolutional neural networks with cutout. arXiv preprint arXiv:1708.04552 (2017)
18. French, G., Oliver, A., Salimans, T.: Milking CowMask for semi-supervised image classification. CoRR abs/2003.12022 (2020)
19. Gidaris, S., Singh, P., Komodakis, N.: Unsupervised representation learning by predicting image rotations. In: International Conference on Learning Representations (2018)
20. He, K., Fan, H., Wu, Y., Xie, S., Girshick, R.: Momentum contrast for unsupervised visual representation learning. In: Proceedings of the IEEE/CVF Conference on Computer Vision and Pattern Recognition (2020)
21. He, K., Zhang, X., Ren, S., Sun, J.: Deep residual learning for image recognition. In: Proceedings of the IEEE Conference on Computer Vision and Pattern Recognition (2016)
22. Hinton, G.E., Sabour, S., Frosst, N.: Matrix capsules with EM routing. In: International Conference on Learning Representations (2018)
23. Iscen, A., Tolias, G., Avrithis, Y., Chum, O.: Label propagation for deep semi-supervised learning. In: Proceedings of the IEEE Conference on Computer Vision and Pattern Recognition (2019)
24. Joachims, T.: Transductive learning via spectral graph partitioning. In: Proceedings of the 20th International Conference on Machine Learning (ICML) (2003)
25. Kingma, D.P., Mohamed, S., Rezende, D.J., Welling, M.: Semi-supervised learning with deep generative models. In: Advances in Neural Information Processing Systems (2014)
26. Krizhevsky, A., Hinton, G., et al.: Learning multiple layers of features from tiny images. Technical report (2009)
27. Kuo, C.-W., Ma, C.-Y., Huang, J.-B., Kira, Z.: FeatMatch: feature-based augmentation for semi-supervised learning. In: Vedaldi, A., Bischof, H., Brox, T., Frahm, J.-M. (eds.) ECCV 2020. LNCS, vol. 12363, pp. 479–495. Springer, Cham (2020). https://doi.org/10.1007/978-3-030-58523-5_28
28. Laine, S., Aila, T.: Temporal ensembling for semi-supervised learning. In: 5th International Conference on Learning Representations, ICLR (2017)
29. Lee, D.H.: Pseudo-label: the simple and efficient semi-supervised learning method for deep neural networks. In: Workshop on Challenges in Representation Learning, ICML (2013)

30. Lin, T.-Y., et al.: Microsoft COCO: common objects in context. In: Fleet, D., Pajdla, T., Schiele, B., Tuytelaars, T. (eds.) ECCV 2014. LNCS, vol. 8693, pp. 740–755. Springer, Cham (2014). https://doi.org/10.1007/978-3-319-10602-1_48
31. Liu, B., Wu, Z., Hu, H., Lin, S.: Deep metric transfer for label propagation with limited annotated data. In: Proceedings of the IEEE International Conference on Computer Vision Workshops (2019)
32. Luo, Y., Zhu, J., Li, M., Ren, Y., Zhang, B.: Smooth neighbors on teacher graphs for semi-supervised learning. In: Proceedings of the IEEE Conference on Computer Vision and Pattern Recognition (2018)
33. Miyato, T., Maeda, S., Koyama, M., Ishii, S.: Virtual adversarial training: a regularization method for supervised and semi-supervised learning. IEEE Trans. Pattern Anal. Mach. Intell. **41**(8), 1979–1993 (2018)
34. Nair, V., Hinton, G.E.: Rectified linear units improve restricted Boltzmann machines. In: ICML (2010)
35. Nesterov, Y.: A method of solving a convex programming problem with convergence rate $o(k^2)$. Doklady Akademii Nauk (1983)
36. Netzer, Y., Wang, T., Coates, A., Bissacco, A., Wu, B., Ng, A.Y.: Reading digits in natural images with unsupervised feature learning. In: NIPS Workshop on Deep Learning and Unsupervised Feature Learning (2011)
37. Odena, A.: Semi-supervised learning with generative adversarial networks. arXiv preprint arXiv:1606.01583 (2016)
38. Pham, H., Xie, Q., Dai, Z., Le, Q.V.: Meta pseudo labels. arXiv preprint arXiv:2003.10580 (2020)
39. Rasmus, A., Berglund, M., Honkala, M., Valpola, H., Raiko, T.: Semi-supervised learning with ladder networks. In: Advances in Neural Information Processing Systems (2015)
40. Ravi, S., Larochelle, H.: Optimization as a model for few-shot learning. In: 5th International Conference on Learning Representations, ICLR (2017)
41. Rebuffi, S.A., Ehrhardt, S., Han, K., Vedaldi, A., Zisserman, A.: Semi-supervised learning with scarce annotations. In: Proceedings of the IEEE/CVF Conference on Computer Vision and Pattern Recognition Workshops (2020)
42. Russakovsky, O., et al.: ImageNet large scale visual recognition challenge. Int. J. Comput. Vis. **115**, 211–252 (2015)
43. Sabour, S., Frosst, N., Hinton, G.E.: Dynamic routing between capsules. In: Advances in Neural Information Processing Systems (2017)
44. Sajjadi, M., Javanmardi, M., Tasdizen, T.: Regularization with stochastic transformations and perturbations for deep semi-supervised learning. In: Advances in Neural Information Processing Systems (2016)
45. Scudder, H.: Probability of error of some adaptive pattern-recognition machines. IEEE Trans. Inf. Theory **11**, 363–371 (1965)
46. Sohn, K., et al.: FixMatch: simplifying semi-supervised learning with consistency and confidence. In: Advances in Neural Information Processing Systems (2020)
47. Srivastava, N., Hinton, G., Krizhevsky, A., Sutskever, I., Salakhutdinov, R.: Dropout: a simple way to prevent neural networks from overfitting. J. Mach. Learn. Res. **15**, 1929–1958 (2014)
48. Tarvainen, A., Valpola, H.: Mean teachers are better role models: weight-averaged consistency targets improve semi-supervised deep learning results. In: Advances in Neural Information Processing Systems (2017)

49. Verma, V., Lamb, A., Kannala, J., Bengio, Y., Lopez-Paz, D.: Interpolation consistency training for semi-supervised learning. In: Proceedings of the Twenty-Eighth International Joint Conference on Artificial Intelligence, IJCAI. International Joint Conferences on Artificial Intelligence Organization (2019)
50. Weiler, M., Cesa, G.: Advances in Neural Information Processing Systems (2019)
51. Xie, Q., Dai, Z., Hovy, E., Luong, T., Le, Q.: Unsupervised data augmentation for consistency training. In: Larochelle, H., Ranzato, M., Hadsell, R., Balcan, M.F., Lin, H. (eds.) Advances in Neural Information Processing Systems, vol. 33, pp. 6256–6268. Curran Associates, Inc. (2020). https://proceedings.neurips.cc/paper/2020/file/44feb0096faa8326192570788b38c1d1-Paper.pdf
52. Zagoruyko, S., Komodakis, N.: Wide residual networks. In: Proceedings of the British Machine Vision Conference (BMVC) (2016)
53. Zhai, X., Oliver, A., Kolesnikov, A., Beyer, L.: S4L: self-supervised semi-supervised learning. In: Proceedings of the IEEE International Conference on Computer Vision (2019)
54. Zhang, H., Cisse, M., Dauphin, Y.N., Lopez-Paz, D.: Mixup: beyond empirical risk minimization. In: 6th International Conference on Learning Representations, ICLR (2018)
55. Zhu, X.: Semi-supervised learning literature survey. Technical report. 1530, Computer Sciences, University of Wisconsin-Madison (2005)
56. Zhu, X., Goldberg, A.B.: Introduction to semi-supervised learning. Synth. Lect. Artif. Intell. Mach. Learn. **3**(1), 1–130 (2009)

Learning Robust Models Using the Principle of Independent Causal Mechanisms

Jens Müller[1,2(✉)], Robert Schmier[2,3], Lynton Ardizzone[1,2], Carsten Rother[1,2], and Ullrich Köthe[1,2]

[1] Heidelberg Collaboratory for Image Processing, Heidelberg University, Heidelberg, Germany
jens.mueller@iwr.uni-heidelberg.de
[2] Computer Vision and Learning Lab, Heidelberg University, Heidelberg, Germany
[3] Bosch Center for Artificial Intelligence, Renningen, Germany

Abstract. Standard supervised learning breaks down under data distribution shift. However, the principle of independent causal mechanisms (ICM, [31]) can turn this weakness into an opportunity: one can take advantage of distribution shift between different environments during training in order to obtain more robust models. We propose a new gradient-based learning framework whose objective function is derived from the ICM principle. We show theoretically and experimentally that neural networks trained in this framework focus on relations remaining invariant across environments and ignore unstable ones. Moreover, we prove that the recovered stable relations correspond to the true causal mechanisms under certain conditions, turning domain generalization into a causal discovery problem. In both regression and classification, the resulting models generalize well to unseen scenarios where traditionally trained models fail.

Keywords: Domain generalization · Principle of independent causal mechanisms

1 Introduction

Standard supervised learning has shown impressive results when training and test samples follow the same distribution. However, many real world applications do not conform to this setting, so that research successes do not readily translate into practice [20]. *Domain Generalization* (DG) addresses this problem: it aims at training models that generalize well under domain shift. In contrast to domain *adaption*, where a few labeled and/or many unlabeled examples are provided for each target test domain, in DG absolutely no data is available from the test domains' distributions making the problem unsolvable in general.

In this work, we view the problem of DG specifically using ideas from causal discovery. This viewpoint makes the problem of DG well-posed: we assume that

C. Bauckhage et al. (Eds.): DAGM GCPR 2021, LNCS 13024, pp. 79–110, 2021.
https://doi.org/10.1007/978-3-030-92659-5_6

there exists a feature vector $h^\star(\mathbf{X})$ whose relation to the target variable Y is invariant across all environments. Consequently, the conditional probability $p(Y \mid h^\star(\mathbf{X}))$ has predictive power in each environment. From a causal perspective, changes between domains or environments can be described as interventions; and causal relationships – unlike purely statistical ones – remain invariant across environments unless explicitly changed under intervention. This is due to the fundamental principle of "Independent Causal Mechanisms" which will be discussed in Sect. 3. From a causal standpoint, finding robust models is therefore a *causal discovery* task [4,24]. Taking a causal perspective on DG, we aim at identifying features which (i) have an invariant relationship to the target variable Y and (ii) are maximally informative about Y. This problem has already been addressed with some simplifying assumptions and a discrete combinatorial search by [22,35], but we make weaker assumptions and enable gradient based optimization. The later is attractive because it readily scales to high dimensions and offers the possibility to *learn* very informative features, instead of merely selecting among predefined ones. Approaches to invariant relations similar to ours were taken by [10], who restrict themselves to linear relations, and [2,19], who consider a weaker notion of invariance. Problems (i) and (ii) are quite intricate because the search space has combinatorial complexity and testing for conditional independence in high dimensions is notoriously difficult. Our main contributions to this problem are the following: *First*, by connecting invariant (causal) relations with normalizing flows, we propose a differentiable two-part objective of the form $I(Y; h(\mathbf{X})) + \lambda_I \mathcal{L}_I$, where I is the mutual information and $\mathcal{L}_I$ enforces the invariance of the relation between $h(\mathbf{X})$ and Y across all environments. This objective operationalizes the ICM principle with a trade-off between feature informativeness and invariance controlled by parameter λ_I. Our formulation generalizes existing work because our objective is not restricted to linear models. *Second*, we take advantage of the continuous objective in three important ways: (1) We can learn invariant new features, whereas graph-based methods as in e.g. [22] can only select features from a pre-defined set. (2) Our approach does not suffer from the scalability problems of combinatorial optimization methods as proposed in e.g. [30] and [35]. (3) Our optimization via normalizing flows, i.e. in the form of a density estimation task, facilitates accurate maximization of the mutual information. *Third*, we show how our objective simplifies in important special cases and under which conditions its optimal solution identifies the true causal parents of the target variable Y. We empirically demonstrate that the new method achieves good results on two datasets proposed in the literature.

2 Related Work

Different types of invariances have been considered in the field of DG. One type is defined on the feature level, i.e. features $h(\mathbf{X})$ are invariant across environments if they follow the same distribution in all environments (e.g. [5,8,27]). However, this form of invariance is problematic since the distribution of the target variable might change between environments, which induces a corresponding change in the distribution of $h(\mathbf{X})$. A more plausible and theoretically justified assumption is the invariance of *relations* [22,30,35]. The relation between a target Y and features $h(\mathbf{X})$ is invariant across environments, if the conditional distribution $p(Y \mid h(\mathbf{X}))$ remains unchanged in all environments. Existing approaches exhaustively model conditional distributions for all possible feature selections and check for the invariance property [22,30,35], which scales poorly for large feature spaces. We derive a theoretical result connecting *normalizing flows* and *invariant relations*, which enables gradient-based learning of an invariant solution. In order to exploit our formulation, we also use the Hilbert-Schmidt-Independence Criterion that has been used for robust learning by [11] in the one environment setting. [2,19,38] also propose gradient-based learning frameworks, which exploit a weaker notion of invariance: They aim to match the conditional expectations across environments, whereas we address the harder problem of matching the entire conditional distributions. The connection between DG, invariance and causality has been pointed out for instance by [24,35,39]. From a causal perspective, DG is a causal discovery task [24]. For studies on causal discovery in the purely observational setting see e.g. [6,29,36], but they do not take advantage of variations across environments. The case of different environments has been studied by [4,9,15,16,22,26,30,37]. Most of these approaches rely on combinatorial optimization or are restricted to linear mechanisms, whereas our continuous objective efficiently optimizes very general non-linear models. The distinctive property of causal relations to remain invariant across environments in the absence of direct interventions has been known since at least the 1930s [7,13]. However, its crucial role as a tool for causal discovery was – to the best of our knowledge– only recently recognized by [30]. Their estimator – *Invariant Causal Prediction* (ICP) – returns the intersection of all subsets of variables that have an invariant relation w.r.t. Y. The output is shown to be the set of the direct causes of Y under suitable conditions. Again, this approach requires linear models and exhaustive search over all possible variable sets $\mathbf{X}_S$. Extensions to time series and non-linear additive noise models were studied in [14,33]. Our treatment of invariance is inspired by these papers and also discusses identifiability results, i.e. conditions when the identified variables are indeed the direct causes, with two key differences: Firstly, we propose a formulation that allows for a gradient-based learning and does not need strong assumptions on the underlying causal model. Second, while ICP tends to exclude features from the parent set when in doubt, our algorithm prefers to err towards best predictive performance in this situation.

3 Preliminaries

In the following we introduce the basics of this work as well as the connection between DG and causality. Basics on causality are presented in Appendix A. We first define our notation as follows: We denote the set of all variables describing the system under study as $\widetilde{\mathbf{X}} = \{X_1, \ldots, X_D\}$. One of these variables will be singled out as our prediction target, whereas the remaining ones are observed and may serve as predictors. To clarify notation, we call the target variable $Y \equiv X_i$ for some $i \in \{1, \ldots, D\}$, and the remaining observations are $\mathbf{X} = \widetilde{\mathbf{X}} \setminus \{Y\}$. Realizations of a random variable (RV) are denoted with lower case letters, e.g. x_i. We assume that observations can be obtained in different environments $e \in \mathcal{E}$. Symbols with superscript, e.g. Y^e, refer to a specific environment, whereas symbols without refer to data pooled over all environments. We distinguish known environments $e \in \mathcal{E}_{\text{seen}}$, where training data are available, from unknown ones $e \in \mathcal{E}_{\text{unseen}}$, where we wish our models to generalize to. The set of all environments is $\mathcal{E} = \mathcal{E}_{\text{seen}} \cup \mathcal{E}_{\text{unseen}}$. We assume that all RVs have a density p_A with probability distribution P_A (for some variable or set A). We consider the environment to be a RV E and therefore a system variable similar to [26]. This gives an additional view on causal discovery and the DG problem. Independence and dependence of two variables A and B is written as $A \perp B$ and $A \not\perp B$ respectively. Two RVs A, B are conditionally independent given C if $P(A, B \mid C) = P(A \mid C)P(B \mid C)$. This is denoted with $A \perp B \mid C$. It means A does not contain any information about B if C is known (see e.g. [31]). Similarly, one can define independence and conditional independence for sets of RVs.

3.1 Invariance and the Principle of ICM

DG is in general unsolvable because distributions between seen and unseen environments could differ arbitrarily. In order to transfer knowledge from $\mathcal{E}_{\text{seen}}$ to $\mathcal{E}_{\text{unseen}}$, we have to make assumptions on how seen and unseen environments relate. These assumptions have a close link to causality. We assume certain relations between variables remain invariant across all environments. A subset $\mathbf{X}_S \subset \mathbf{X}$ of variables *elicits an invariant relation* or *satisfies the invariance property* w.r.t. Y over a subset $W \subset \mathcal{E}$ of environments if

$$\forall e, e' \in W: \quad P(Y^e \mid \mathbf{X}_S^e = u) = P(Y^{e'} \mid \mathbf{X}_S^{e'} = u) \tag{1}$$

for all u where both conditional distributions are well-defined. Equivalently, we can define the invariance property by $Y \perp E \mid \mathbf{X}_S$ and $I(Y; E \mid \mathbf{X}_S) = 0$ for E restricted to W. The *invariance property* for computed features $h(\mathbf{X})$ is defined analogously by the relation $Y \perp E \mid h(\mathbf{X})$. Although we can only test for Eq. 1 in $\mathcal{E}_{\text{seen}}$, taking a causal perspective allows us to derive plausible conditions for an invariance to remain valid in all environments $\mathcal{E}$. In brief, we assume that environments correspond to interventions in the system and invariance arises from the principle of *independent causal mechanisms* [31]. We specify these conditions later in Assumption 1 and 2. At first, consider the joint density $p_{\widetilde{\mathbf{X}}}(\widetilde{\mathbf{X}})$.The chain rule offers a combinatorial number of ways to decompose this distribution into a product of conditionals. Among those, the *causal factorization*

$$p_{\widetilde{\mathbf{X}}}(x_1, \ldots, x_D) = \prod_{i=1}^{D} p_i(x_i \mid \mathbf{x}_{pa(i)}) \tag{2}$$

is singled out by conditioning each X_i onto its *direct causes* or *causal parents* $\mathbf{X}_{pa(i)}$, where $pa(i)$ denotes the appropriate index set. The special properties of this factorization are discussed in [31]. The conditionals p_i of the causal factorization are called *causal mechanisms*. An *intervention* onto the system is defined by replacing one or several factors in the decomposition with different (conditional) densities $\overline{p}$. Here, we distinguish *soft-interventions* where $\overline{p}_j(x_j \mid \mathbf{x}_{pa(j)}) \neq p_j(x_j \mid \mathbf{x}_{pa(j)})$ and *hard-interventions* where $\overline{p}_j(x_j \mid \mathbf{x}_{pa(j)}) = \overline{p}_j(x_j)$ is a density which does not depend on $x_{pa(j)}$ (e.g. an atomic intervention where x_j is set to a specific value $\overline{x}$). The resulting joint distribution for a single intervention is

$$\overline{p}_{\tilde{\mathbf{X}}}(x_1, \ldots, x_D) = \overline{p}_j(x_j \mid \mathbf{x}_{pa(j)}) \prod_{i=1, i \neq j}^{D} p_i(x_i \mid \mathbf{x}_{pa(i)}) \tag{3}$$

and extends to multiple simultaneous interventions in the obvious way. The principle of *independent causal mechanisms* (ICM) states that every mechanism acts independently of the others [31]. Consequently, an intervention replacing p_j with $\overline{p}_j$ has no effect on the other factors $p_{i \neq j}$, as indicated by Eq. 3. This is a crucial property of the causal decomposition – alternative factorizations do not exhibit this behavior. Instead, a coordinated modification of several factors is generally required to model the effect of an intervention in a non-causal decomposition. We utilize this principle as a tool to train *robust* models. To do so, we make two additional assumptions, similar to [30] and [14]:

Assumption *(1) Any differences in the joint distributions $p^e_{\tilde{\mathbf{X}}}$ from one environment to the other are fully explainable as interventions: replacing factors $p^e_i(x_i \mid \mathbf{x}_{pa(i)})$ in environment e with factors $p^{e'}_i(x_i \mid \mathbf{x}_{pa(i)})$ in environment e' (for some subset of the variables) is the only admissible change.* ***(2)*** *The mechanism $p(y \mid \mathbf{x}_{pa(Y)})$ for the target variable Y is invariant under changes of environment, i.e. we require conditional independence $Y \perp E \mid \mathbf{X}_{pa(Y)}$.*

Assumption 2 implies that Y must not directly depend on E. Consequences in case of omitted variables are discussed in Appendix B. If we knew the causal decomposition, we could use these assumptions directly to train a robust model for Y – we would simply regress Y on its parents $\mathbf{X}_{pa(Y)}$. However, we only require that a causal decomposition with these properties exists, but do not assume that it is known. Instead, our method uses the assumptions indirectly – by simultaneously considering data from different environments – to identify a stable regressor for Y. We call a regressor stable if it solely relies on predictors whose relationship to Y remains invariant across environments, i.e. is not influenced by any intervention. By assumption 2, such a regressor always exists. However, predictor variables beyond $\mathbf{X}_{pa(Y)}$, e.g. children of Y or parents of children, may be included into our model as long as their relationship to Y remains invariant across all environments. We discuss this and further illustrate Assumption 2 in Appendix B. In general, prediction accuracy will be maximized when all suitable predictor variables are included into the model. Accordingly, our algorithm will asymptotically identify the full set of stable predictors for Y. In addition, we will prove under which conditions this set contains exactly the parents of Y.

3.2 Domain Generalization

To exploit the principle of ICM for DG, we formulate the DG problem as follows

$$h^\star := \underset{h \in \mathcal{H}}{\arg\max} \left\{ \min_{e \in \mathcal{E}} I(Y^e; h(\mathbf{X}^e)) \right\} \qquad \text{s.t.} \quad Y \perp E \mid h(\mathbf{X}) \tag{4}$$

The optimization problem in Eq. 4 asks to find features $h(\mathbf{X})$ which are maximally informative in the worst environment subject to the invariance constraint. where $h \in \mathcal{H}$ denotes a learnable feature extraction function $h\colon \mathbb{R}^D \to \mathbb{R}^M$ where M is a hyperparameter. This optimization problem defines a maximin objective: The features $h(\mathbf{X})$ should be as informative as possible about the response Y even in the most difficult environment, while conforming to the ICM constraint that the relationship between features and response must remain invariant across all environments. In principle, our approach can also optimize related objectives like the average mutual information over environments. However, very good performance in a majority of the environments could then mask failure in a single (outlier) environment. We opted for the maximin formulation to avoid this. On the other hand there might be scenarios where the maxmin formulation is limited. For instance when the training signal is very noisy in one environment, the classifier might discard valuable information from the other environments. As it stands, Eq. 4 is hard to optimize, because traditional independence tests for the constraint $Y \perp E \mid h(\mathbf{X})$ cannot cope with conditioning variables selected from a potentially infinitely large space $\mathcal{H}$. A re-formulation of the DG problem to circumvent these issues is our main theoretical contribution.

3.3 Normalizing Flows

Normalizing flows form a class of probabilistic models that has recently received considerable attention, see e.g. [28]. They model complex distributions by means of invertible functions T (chosen from some model space $\mathcal{T}$), which map the densities of interest to latent normal distributions. Normalizing flows are typically built with specialized neural networks that are invertible by construction and have tractable Jacobian determinants. We represent the conditional distribution $P(Y \mid h(\mathbf{X}))$ by a *conditional* normalizing flow (see e.g. [1]). The literature typically deals with Structural Causal Models restricted to additive noise. With normalizing flows, we are able to lift this restriction to the much broader setting of arbitrary distributions (for details see Appendix C). The corresponding loss is the negative log-likelihood (NLL) of Y under T, given by

$$\mathcal{L}_{\text{NLL}}(T, h) := \mathbb{E}_{h(\mathbf{X}),Y}\left[\|T(Y; h(\mathbf{X})\|^2/2 - \log|\det \nabla_y T(Y; h(\mathbf{X}))|\right] + C \quad (5)$$

where $\det \nabla_y T$ is the Jacobian determinant and $C = \dim(Y)\log(\sqrt{2\pi})$ is a constant that can be dropped [28]. Equation 5 can be derived from the change of variables formula and the assumption that T maps to a standard normal distribution [28]. If we consider the NLL on a particular environment $e \in \mathcal{E}$, we denote this with $\mathcal{L}^e_{\text{NLL}}$. Lemma 1 shows that normalizing flows optimized by NLL are indeed applicable to our problem:

Lemma 1. *(proof in Appendix C) Let $h^\star, T^\star := \arg\min_{h\in\mathcal{H}, T\in\mathcal{T}} \mathcal{L}_{\text{NLL}}(T, h)$ be the solution of the NLL minimization problem on a sufficiently rich function space $\mathcal{T}$. Then the following properties hold for any set $\mathcal{H}$ of feature extractors:*

(a) $h^\star$ also maximizes the mutual information, i.e. $h^\star = \arg\max_{g\in\mathcal{H}} I(g(\mathbf{X}); Y)$

(b) $h^\star$ and the latent variables $R = T^\star(Y; h^\star(\mathbf{X}))$ are independent: $h^\star(\mathbf{X}) \perp R$

Statement (a) guarantees that $h^\star$ extracts as much information about Y as possible. Hence, the objective (4) becomes equivalent to optimizing (5) when we restrict the space $\mathcal{H}$ of admissible feature extractors to the subspace $\mathcal{H}_\perp$ satisfying the invariance constraint $Y \perp E \mid h(\mathbf{X})$: $\arg\min_{h\in\mathcal{H}_\perp} \max_{e\in\mathcal{E}} \min_{T\in\mathcal{T}} \mathcal{L}^e_{\text{NLL}}(T; h) = \arg\max_{h\in\mathcal{H}_\perp} \min_{e\in\mathcal{E}} I(Y^e; h(\mathbf{X}^e))$ (Appendix C). Statement (b) ensures that the flow indeed implements a valid structural equation, which requires that R can be sampled independently of the features $h(\mathbf{X})$.

4 Method

In the following we propose a way of indirectly expressing the constraint in Eq. 4 via normalizing flows. Thereafter, we combine this result with Lemma 1 to obtain a differentiable objective for solving the DG problem. We also present important simplifications for least squares regression and softmax classification and discuss relations of our approach with causal discovery.

4.1 Learning the Invariance Property

The following theorem establishes a connection between invariant relations, prediction residuals and normalizing flows. The key consequence is that a suitably trained normalizing flow translates the statistical independence of the latent variable R from the features and environment $(h(\mathbf{X}), E)$ into the desired invariance of the mechanism $P(Y \mid h(\mathbf{X}))$ under changes of E. We will exploit this for an elegant reformulation of the DG problem (4) into the objective (7) below.

Theorem 1. *Let h be a differentiable function and $Y, \mathbf{X}, E$ be RVs. Furthermore, let $R = T(Y; h(\mathbf{X}))$ be a continuous, differentiable function that is a diffeomorphism in Y. Suppose that $R \perp (h(\mathbf{X}), E)$. Then, it holds that $Y \perp E \mid h(\mathbf{X})$.*

Proof. The decomposition rule for the assumption (i) $R \perp (h(\mathbf{X}), E)$ implies (ii) $R \perp h(\mathbf{X})$. To simplify notation, we define $Z := h(\mathbf{X})$. Because T is invertible in Y and due to the change of variables (c.o.v.) formula, we obtain

$$p_{Y|Z,E}(y \mid z, e) \overset{(c.o.v.)}{=} p_{R|Z,E}(T(y,z) \mid z, e) \left|\det \frac{\partial T}{\partial y}(y,z)\right|$$
$$\overset{(i)}{=} p_R(r) \left|\det \frac{\partial T}{\partial y}(y,z)\right| \overset{(ii)}{=} p_{R|Z}(r \mid z) \left|\det \frac{\partial T}{\partial y}(y,z)\right| \overset{(c.o.v.)}{=} p_{Y|Z}(y \mid z).$$

This implies $Y \perp E \mid Z$. The theorem states in particular that if there exists a suitable diffeomorphism T such that $R \perp (h(\mathbf{X}), E)$, then $h(\mathbf{X})$ satisfies the invariance property w.r.t. Y. Note that if Assumption 2 is violated, the condition $R \perp (h(\mathbf{X}), E)$ is unachievable in general and therefore the theorem is not applicable (see Appendix B). We use Theorem 1 in order to *learn* features

h that meet this requirement. In the following, we denote a conditional normalizing flow parameterized via θ with T_θ. Furthermore, h_ϕ denotes a feature extractor implemented as a neural network parameterized via ϕ. We can relax condition $R \perp (h_\phi(\mathbf{X}), E)$ by means of the Hilbert Schmidt Independence Criterion (HSIC), a kernel-based independence measure (see Appendix D for the definition and [12] for details). This loss, denoted as $\mathcal{L}_I$, penalizes dependence between the distributions of R and $(h_\phi(\mathbf{X}), E)$. The HSIC guarantees that

$$\mathcal{L}_I\big(P_R, P_{h_\phi(\mathbf{X}),E}\big) = 0 \quad \iff \quad R \perp (h_\phi(\mathbf{X}), E) \tag{6}$$

where $R = T_\theta(Y; h_\phi(\mathbf{X}))$ and $P_R, P_{h_\phi(\mathbf{X}),E}$ are the distributions implied by the parameter choices ϕ and θ. Due to Theorem 1, minimization of $\mathcal{L}_I(P_R, P_{h_\phi(\mathbf{X}),E})$ w.r.t. ϕ and θ will thus approximate the desired invariance property $Y \perp E \mid h_\phi(\mathbf{X})$, with exact validity upon perfect convergence. When $R \perp (h_\phi(\mathbf{X}), E)$ is fulfilled, the decomposition rule implies $R \perp E$ as well. However, if the differences between environments are small, empirical convergence is accelerated by adding a Wasserstein loss which enforces the latter (see Appendix D and Sect. 5.2).

4.2 Exploiting Invariances for Prediction

Equation 4 can be re-formulated as a differentiable loss using a Lagrange multiplier λ_I on the HSIC loss. λ_I acts as a hyperparameter to adjust the trade-off between the invariance property of $h_\phi(\mathbf{X})$ w.r.t. Y and the mutual information between $h_\phi(\mathbf{X})$ and Y. See Appendix F for algorithm details. In the following, we consider normalizing flows in order to optimize Eq. 4. Using Lemma 1(a), we maximize $\min_{e\in\mathcal{E}} I(Y^e; h_\phi(\mathbf{X}^e))$ by minimizing $\max_{e\in\mathcal{E}}\{\mathcal{L}_{\text{NLL}}(T_\theta; h_\phi)\}$ w.r.t. ϕ, θ. To achieve the described trade-off between goodness-of-fit and invariance, we therefore optimize

$$\arg\min_{\theta,\phi} \left(\max_{e\in\mathcal{E}} \left\{ \mathcal{L}^e_{\text{NLL}}(T_\theta, h_\phi) \right\} + \lambda_I \mathcal{L}_I(P_R, P_{h_\phi(\mathbf{X}),E}) \right) \tag{7}$$

where $R^e = T_\theta(Y^e, h_\phi(\mathbf{X}^e))$ and $\lambda_I > 0$. The first term maximizes the mutual information between $h_\phi(\mathbf{X})$ and Y in the environment where the features are least informative about Y and the second term aims to ensure an invariant relation. In the special case that the data is governed by additive noise, Eq. 7 simplifies: Let f_θ be a regression function, then solving for the noise term gives $Y - f_\theta(\mathbf{X})$ which corresponds to a diffeomorphism in Y, namely $T_\theta(Y; X) = Y - f_\theta(\mathbf{X})$. Under certain assumptions (see Appendix E) we obtain an approximation of Eq. 7 via

$$\arg\min_{\theta} \left(\max_{e\in\mathcal{E}_{\text{seen}}} \left\{ \mathbb{E}\big[(Y^e - f_\theta(\mathbf{X}^e))^2\big] \right\} + \lambda_I \mathcal{L}_I(P_R, P_{f_\theta(\mathbf{X}),E}) \right) \tag{8}$$

where $R^e = Y^e - f_\theta(\mathbf{X}^e)$ and $\lambda_I > 0$. Here, $\arg\max_\theta I(f_\theta(\mathbf{X}^e), Y^e)$ corresponds to the argmin of the L2-Loss in the corresponding environment. Alternatively we can view the problem as to find features $h_\phi \colon \mathbb{R}^D \to \mathbb{R}^m$ such that $I(h_\phi(\mathbf{X}), Y)$ gets maximized under the assumption that there exists a model $f_\theta(h_\phi(\mathbf{X})) + R = Y$ where R is independent of $h_\phi(\mathbf{X})$ and is Gaussian. In this case we obtain the learning objective

$$\arg\min_{\theta,\phi}\left(\max_{e\in\mathcal{E}_{\text{seen}}}\left\{\mathbb{E}\left[(Y^e - f_\theta(h_\phi(\mathbf{X}^e)))^2\right]\right\} + \lambda_I\mathcal{L}_I(P_R, P_{h_\phi(\mathbf{X}),E})\right) \tag{9}$$

For the classification case, we consider the expected cross-entropy loss

$$-\mathbb{E}_{\mathbf{X},Y}\left[f(\mathbf{X})_Y - \log\left(\sum_c \exp\left(f(\mathbf{X})_c\right)\right)\right] \tag{10}$$

where $f\colon \mathcal{X} \to \mathbb{R}^m$ returns the logits. Minimizing the expected cross-entropy loss amounts to maximizing the mutual information between $f(\mathbf{X})$ and Y [3,34, Eq. 3]. We set $T(Y; f(\mathbf{X})) = Y \cdot \text{softmax}(f(\mathbf{X}))$ with component-wise multiplication. Then T is invertible in Y conditioned on the softmax output and therefore Theorem 1 is applicable. Now we can apply the same invariance loss as above in order to obtain a solution to Eq. 4.

4.3 Relation to Causal Discovery

Under certain conditions, solving Eq. 4 leads to features which correspond to the direct causes of Y (identifiability). In this case we obtain the causal mechanism by computing the conditional distribution of Y given the direct causes. Hence Eq. 4 can be seen as an approximation of the causal mechanism when the identifiability conditions are met. The following Proposition states the conditions when the direct causes of Y can be found by exploiting Theorem 1.

Proposition 1. *We assume that the underlying causal graph G is faithful with respect to $P_{\tilde{\mathbf{X}},E}$. We further assume that every child of Y in G is also a child of E in G. A variable selection $h(\mathbf{X}) = \mathbf{X}_S$ corresponds to the direct causes if the following conditions are met: (i) $T(Y; h(\mathbf{X})) \perp E, h(\mathbf{X})$ is satisfied for a diffeomorphism $T(\cdot; h(\mathbf{X}))$, (ii) $h(\mathbf{X})$ is maximally informative about Y and (iii) $h(\mathbf{X})$ contains only variables from the Markov blanket of Y.*

The Markov blanket of Y is the only set of vertices which are necessary to predict Y (see Appendix A). We give a proof of Proposition 1 as well as a discussion in Appendix G. To facilitate explainability and explicit causal discovery, we employ the same gating function and complexity loss as in [17]. The gating function h_ϕ is a 0-1 mask that marks the selected variables, and the complexity loss $\mathcal{L}(h_\phi)$ is a soft counter of the selected variables. Intuitively speaking, if we search for a variable selection that conforms to the conditions in Proposition 1, the complexity loss will exclude all non-task relevant variables. Therefore, if $\mathcal{H}$ is the set of gating functions, then $h^\star$ in Eq. 4 corresponds to the direct causes of Y under the conditions listed in Proposition 1. The complexity loss as well as the gating function can be optimized by gradient descent.

5 Experiments

The main focus of this work is on the theoretical and methodological improvements of causality-based domain generalization using information theoretical concepts. A complete and rigorous quantitative evaluation is beyond the scope of this work. In the following we demonstrate proof-of-concept experiments.

5.1 Synthetic Causal Graphs

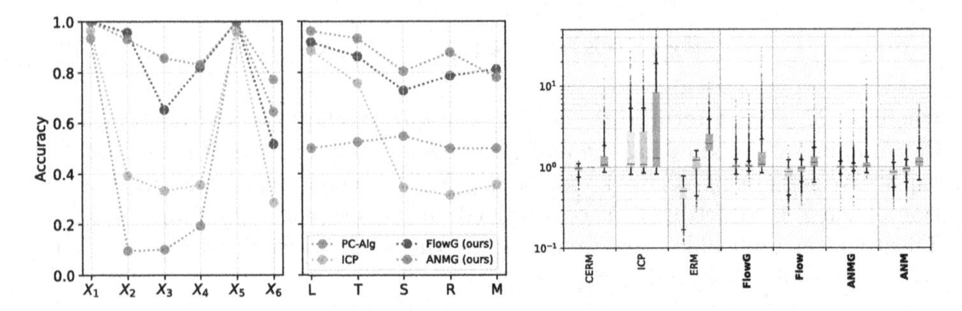

Fig. 1. (a) Detection accuracy of the direct causes for baselines and our gating architectures, broken down for different target variables (left) and mechanisms (right: **L**inear, **T**anhshrink, **S**oftplus, **R**eLU, **M**ultipl. Noise) (b) Logarithmic plot of L2 errors, normalized by CERM test error. For each method (ours in bold) from left to right: training error, test error on seen environments, domain generalization error on unseen environments.

To evaluate our methods for the regression case, we follow the experimental design of [14]. It rests on the causal graph in Fig. 2. Each variable $X_1, ..., X_6$ is chosen as the regression target Y in turn, so that a rich variety of local configurations around Y is tested. The corresponding structural equations are selected among four model types of the form $f(\mathbf{X}_{pa(i)}, N_i) = \sum_{j \in pa(i)} \mathtt{mech}(a_j X_j) + N_i$, where $\mathtt{mech}$ is either the identity (hence we get a linear Structural Causal Model (SCM)), Tanhshrink, Softplus or ReLU, and one multiplicative noise mechanism of the form $f_i(\mathbf{X}_{pa(i)}, N_i) = (\sum_{j \in pa(i)} a_j X_j) \cdot (1 + (1/4)N_i) + N_i$, resulting in 1365 different settings. For each setting, we define one observational environment (using exactly the selected mechanisms) and three interventional ones, where soft or do-interventions are applied to non-target variables according to Assumptions 1 and 2 (full details in Appendix H). Each inference model is trained on 1024 realizations of three environments, whereas the fourth one is held back for DG testing. The tasks are to identify the parents of the current target variable Y, and to train a transferable regression model based on this parent hypothesis. We measure performance by the accuracy of the detected parent sets and by the L2 regression errors relative to the regression function using the ground-truth parents. We evaluate four models derived from our theory: two normalizing flows as in Eq. 7 with and without gating mechanisms (FlowG, Flow) and two additive noise models, again with and without gating mechanism (ANMG, ANM), using a feed-forward network with the objective in Eq. 9 (ANMG) and Eq. 8 (ANM).

For comparison, we train three baselines: ICP (a causal discovery algorithm also exploiting ICM, but restricted to linear regression, [30]), a variant of the PC-Algorithm (PC-Alg, see Appendix H.4) and standard empirical-risk-minimization ERM, a feed-forward network minimizing the L2-loss, which ignores the causal structure by regressing Y on all other variables. We normalize our results with a ground truth model (CERM), which is identical to ERM, but restricted to the true causal parents of the respective Y. The accuracy of parent detection is shown in Fig. 1a. The score indicates the fraction of the experiments where the exact set of all causal parents was found and all non-parents were excluded. We see that the PC algorithm performs unsatisfactorily, whereas ICP exhibits the expected behavior: it works well for variables without parents and for linear SCMs, i.e. exactly within its specification. Among our models, only the gating ones explicitly identify the parents. They clearly outperform the baselines, with a slight edge for ANMG, as long as its assumption of additive noise is fulfilled. Figure 1b and Table 1 report regression errors for seen and unseen environments, with CERM indicating the theoretical lower bound. The PC algorithm is excluded from this experiment due to its poor detection of the direct causes. ICP wins for linear SCMs, but otherwise has largest errors, since it cannot accurately account for non-linear mechanisms. ERM gives reasonable test errors (while overfitting the training data), but generalizes poorly to unseen environments, as expected. Our models perform quite similarly to CERM. We again find a slight edge for ANMG, except under multiplicative noise, where ANMG's additive noise assumption is violated and Flow is superior. All methods (including CERM) occasionally fail in the domain generalization task, indicating that some DG problems are more difficult than others, e.g. when the differences between seen environments are too small to reliably identify the invariant mechanism or the unseen environment requires extrapolation beyond the training data boundaries. Models without gating (Flow, ANM) seem to be slightly more robust in this respect. A detailed analysis of our experiments can be found in Appendix H.

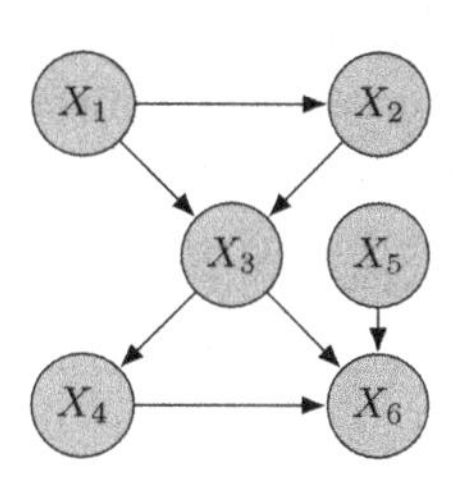

Fig. 2. Directed graph of our SCM. Target variable Y is chosen among $X_1, \ldots, X_6$ in turn.

5.2 Colored MNIST

To demonstrate that our model is able to perform DG in the classification case, we use the same data generating process as in the colored variant of the MNIST-dataset established by [2], but create training instances online rather than upfront. The response is reduced to two labels – 0 for all images with digit $\{0, \ldots, 4\}$ and 1 for digits $\{5, \ldots 9\}$ – with deliberate label noise that limits the achievable shape-based classification accuracy to 75%. To confuse the classifier, digits are additionally colored such that colors are spuriously associated with the true labels at accuracies of 90% resp. 80% in the first two environments, whereas the association is only 10% correct in the third environment. A classifier naively trained on the first two environments will identify color as the best predictor,

Table 1. Medians and upper 95% quantiles for domain generalization L2 errors (i.e. on unseen environments) for different model types and data-generating mechanisms (lower is better).

Models	Linear	Tanhshrink	Softplus	ReLU	Mult. Noise
FlowG (ours)	1.05...4.2	1.08...4.8	1.09...5.52	1.08...5.7	1.55...8.64
ANMG (ours)	1.02...1.56	**1.03**...2.23	**1.04**...4.66	**1.03**...4.32	1.46...4.22
Flow (ours)	1.08...1.61	1.14...1.57	1.14...1.55	1.14...1.54	**1.35**...4.07
ANM (ours)	1.05...1.52	1.15...1.47	1.14...1.47	1.15...1.54	1.48...4.19
ICP (Peters et al. 2016)	**0.99**...25.7	1.44...20.39	3.9...23.77	4.37...23.49	8.94...33.49
ERM	1.79...3.84	1.89...3.89	1.99...3.71	2.01...3.62	2.08...5.86
CERM (true parents)	1.06...1.89	1.06...1.84	1.06...2.11	1.07...2.15	1.37...5.1

but will perform terribly when tested on the third environment. In contrast, a robust model will ignore the unstable relation between colors and labels and use the invariant relation, namely the one between digit shapes and labels, for prediction. We supplement the HSIC loss with a Wasserstein term to explicitly enforce $R \perp E$, i.e. $\mathcal{L}_I = \text{HSIC} + \text{L2}(\text{sort}(R^{e_1}), \text{sort}(R^{e_2}))$ (see Appendix D). This gives a better training signal as the HSIC alone, since the difference in label-color association between environments 1 and 2 (90% vs. 80%) is deliberately chosen very small to make the task hard to learn. Experimental details can be found in Appendix I. Figure 3a shows the results for our model: Naive training ($\lambda_I = 0$, i.e. invariance of residuals is not enforced) gives accuracies corresponding to the association between colors and labels and thus completely fails in test environment 3. In contrast, our model performs close to the best possible rate for invariant classifiers in environments 1 and 2 and still achieves 68.5% in environment 3. This is essentially on par with preexisting methods. For instance, IRM achieves 71% on the third environment for this particular dataset, although the dataset itself is not particularly suitable for meaningful quantitative comparisons. Figure 3b demonstrates the trade-off between goodness of fit in the training environments 1 and 2 and the robustness of the resulting classifier: the model's ability to perform DG to the unseen environment 3 improves as λ_I increases. If λ_I is too large, it dominates the classification training signal and performance breaks down in all environments. However, the choice of λ_I is not critical, as good results are obtained over a wide range of settings.

6 Discussion

In this paper, we have introduced a new method to find invariant and causal models by exploiting the principle of ICM. Our method works by gradient descent in contrast to combinatorial optimization procedures. This circumvents scalability issues and allows us to extract invariant features even when the raw data representation is not in itself meaningful (e.g. we only observe pixel values). In comparison to alternative approaches, our use of normalizing flows places fewer restrictions on the underlying true generative process. We have also shown under

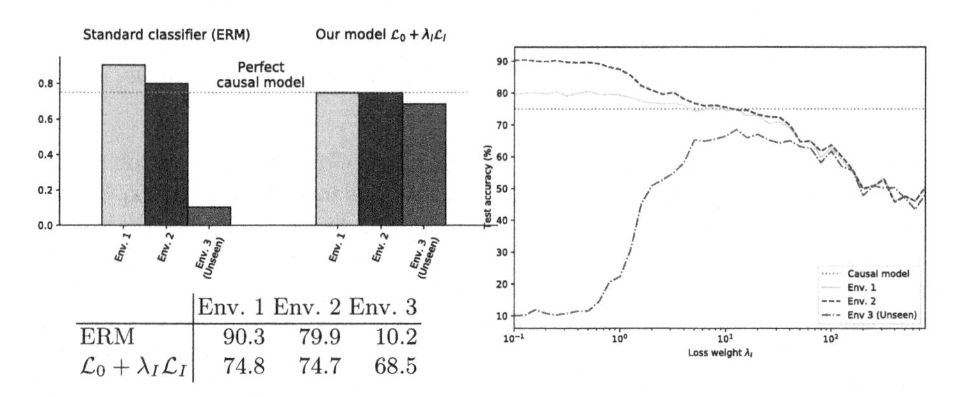

	Env. 1	Env. 2	Env. 3
ERM	90.3	79.9	10.2
$\mathcal{L}_0 + \lambda_I \mathcal{L}_I$	74.8	74.7	68.5

Fig. 3. (a) Accuracy of a standard classifier and our model (b) Performance of the model in the three environments, depending on the hyperparameter λ_I.

which circumstances our method guarantees to find the underlying causal model. Moreover, we demonstrated theoretically and empirically that our method is able to learn robust models w.r.t. distribution shift. Future work includes ablations studies in order to improve the understanding of the influence of single components, e.g. the choice of the maxmin objective over the average mutual information or the Wasserstein loss and the HSIC loss. As a next step, we will examine our approach in more complex scenarios where, for instance, the invariance assumption may only hold approximately.

Acknowledgments. JM received funding by the Heidelberg Collaboratory for Image Processing (HCI). LA received funding by the Federal Ministry of Education and Research of Germany project High Performance Deep Learning Framework (No 01IH 17002). CR and UK received financial support from the European Research Council (ERC) under the European Unions Horizon2020 research and innovation program (grant agreement No647769).

Furthermore, we thank our colleagues Felix Draxler, Jakob Kruse and Michael Aichmüller for their help, support and fruitful discussions.

Appendix

A Causality: Basics

Structural Causal Models. (SCM) allow us to express causal relations on a functional level. Following [31] we define a SCM in the following way:

Definition 1. *A Structural Causal Model (SCM) $\mathcal{S} = (S, P_{\mathbf{N}})$ consists of a collection S of D (structural) assignments*

$$X_j := f_j(\widetilde{\mathbf{X}}_{pa(j)}, N_j), \quad j = 1, \ldots, D \tag{11}$$

where $pa(j) \subset \{1, \dots, j-1\}$ *are called parents of* X_j. $P_{\boldsymbol{N}}$ *denotes the distribution over the noise variables* $\boldsymbol{N} = (N_1, \dots, N_D)$ *which are assumed to be jointly independent.*

An SCM defined as above produces an acyclic graph G and induces a probability distribution over $P_{\widetilde{\mathbf{X}}}$ which allows for the *causal factorization* as in Eq. 3 [31]. Children of X_i in G are denoted as $ch(i)$ or $ch(X_i)$. An SCM satisfies the *causal sufficiency* assumption if all the noise variables in Definition 1 are indeed jointly independent. A random variable H in the SCM is called *confounder* between two variables X_i, X_j if it causes both of them. If a confounder is not observed, we call it hidden confounder. If there exists a hidden confounder, the causal sufficiency assumption is violated.

The random variables in an SCM correspond to vertices in a graph and the structural assignments S define the edges of this graph. Two sets of vertices $\boldsymbol{A}, \boldsymbol{B}$ are said to be d-separated if there exists a set of vertices $\boldsymbol{C}$ such that every path between $\boldsymbol{A}$ and $\boldsymbol{B}$ is blocked. For details see e.g. [31]. The subscript $\perp_d$ denotes d-separability which in this case is denoted by $\boldsymbol{A} \perp_d \boldsymbol{B}$. An SCM generates a probability distribution $P_{\widetilde{\mathbf{X}}}$ which satisfies the *Causal Markov Condition*, that is $\boldsymbol{A} \perp_d \boldsymbol{B} \mid \boldsymbol{C}$ results in $\boldsymbol{A} \perp \boldsymbol{B} \mid \boldsymbol{C}$ for sets or random variables $\boldsymbol{A}, \boldsymbol{B}, \boldsymbol{C} \subset \widetilde{\mathbf{X}}$. The Causal Markov Condition can be seen as an inherent property of a causal system which leaves marks in the data distribution.

A distribution $P_{\widetilde{\mathbf{X}}}$ is said to be *faithful* to the graph G if $\boldsymbol{A} \perp \boldsymbol{B} \mid \boldsymbol{C}$ results in $\mathbf{A} \perp_d \boldsymbol{B} \mid \boldsymbol{C}$ for all $\mathbf{A}, \mathbf{B}, \mathbf{C} \subset \widetilde{\mathbf{X}}$. This means from the distribution $P_{\widetilde{\mathbf{X}}}$ statements about the underlying graph G can be made.

Assuming both, faithfulness and the Causal Markov condition, we obtain that the d-separation statements in G are equivalent to the conditional independence statements in $P_{\widetilde{\mathbf{X}}}$. These two assumptions allow for a whole class of causal discovery algorithms like the PC- or IC-algorithm [29,36].

The smallest set $\boldsymbol{M}$ such that $Y \perp_d \mathbf{X} \setminus (\{Y\} \cup \boldsymbol{M})$ is called *Markov Blanket*. It is given by $\boldsymbol{M} = \mathbf{X}_{pa(Y)} \cup \mathbf{X}_{ch(Y)} \cup \mathbf{X}_{pa(ch(Y))} \setminus \{Y\}$. The *Markov Blanket* of Y is the only set of vertices which are necessary to predict Y.

B Discussion and Illustration of Assumptions

B.1 Causal Sufficiency and Omitted Variables

Assumption 2 implies that Y must not directly depend on E. In addition, it has important consequences when there exist omitted variables $\mathbf{W}$, which influence Y but have not been measured. Specifically, if the omitted variables depend on the environment (hence $\mathbf{W} \not\perp E$) or $\mathbf{W}$ contains a hidden confounder of $\mathbf{X}_{pa(Y)}$ and Y while $\mathbf{X}_{pa(Y)} \not\perp E$ (the system is not causally sufficient and $\mathbf{X}_{pa(Y)}$ becomes a "collider", hence $\mathbf{W} \not\perp E \mid \mathbf{X}_{pa(Y)}$), then Y and E are no longer d-separated by $\mathbf{X}_{pa(Y)}$ and Assumption 2 is unsatisfiable. Then our method will be unable to find an invariant mechanism.

B.2 Using Causal Effects for Prediction

Our estimator might use predictor variables beyond $\mathbf{X}_{pa(Y)}$ as well, e.g. children of Y or parents of children, provided their relationships to Y do not depend on the environment. The case of children is especially interesting: Suppose X_j is a noisy measurement of Y, described by the causal mechanism $P(X_j \mid Y)$. As long as the measurement device works identically in all environments, including X_j as a predictor of Y is desirable, despite it being a child.

B.3 Examples

Domain generalization is in general impossible without strong assumptions (in contrast to classical supervised learning). In our view, the interesting question is "Which strong assumptions are the most useful in a given setting?". For instance, [14] use Assumption 2 to identify causes for birth rates in different countries. If all variables mediating the influence of continent/country (environment variable) on birth rates (target variable) are included in the model (e.g. GDP, Education), this assumption is reasonable. The same may hold for other epidemiological investigations as well. [33] suppose Assumption 2 in the field of finance.

Another reasonable example are data augmentations in computer vision. Deliberate image rotations, shifts and distortions can be considered as environment interventions that preserve the relation between semantic image features and object classes (see e.g. [25]), i.e. verify assumption 2. In general, assumption 2 may be justified when one studies a fundamental mechanism that can reasonably be assumed to remain invariant across environments, but is obscured by unstable relationships between observable variables.

B.4 Robustness Example

To illustrate the impact of causality on robustness, consider the following example: Suppose we would like to estimate the gas consumption of a car. In a sufficiently narrow setting, the total amount of money spent on gas might be a simple and accurate predictor. However, gas prices vary dramatically between countries and over time, so statistical models relying on it will not be robust, even if they fit the training data very well. Gas costs are an *effect* of gas consumption, and this relationship is unstable due to external influences. In contrast, predictions on the basis of the *causes* of gas consumption (e.g. car model, local speed limits and geography, owner's driving habits) tend to be much more robust, because these causal relations are intrinsic to the system and not subjected to external influences. Note that there is a trade-off here: Including gas costs in the model will improve estimation accuracy when gas prices remain sufficiently stable, but will impair results otherwise. By considering the same phenomenon in several environments simultaneously, we hope to gain enough information to adjust this trade-off properly.

In the gas example, countries can be considered as environments that "intervene" on the relation between consumed gas and money spent, e.g. by applying different tax policies. In contrast, interventions changing the impact of motor properties or geography on gas consumption are much less plausible - powerful motors and steep roads will always lead to higher consumption. From a causal standpoint, finding robust models is therefore a causal discovery task [24].

C Normalizing Flows

Normalizing flows are a specific type of neural network architecture which are by construction invertible and have a tractable Jacobian. They are used for density estimation and sampling of a target density (for an overview see [28]). This in turn allows optimizing information theoretic objectives in a convenient and mathematically sound way.

Similarly as in the paper, we denote with $\mathcal{H}$ the set of feature extractors $h\colon \mathbb{R}^D \to \mathbb{R}^M$ where M is chosen a priori. The set of all one-dimensional (conditional) normalizing flows is denoted by $\mathcal{T}$. Together with a reference distribution p_{ref}, a normalizing flow T defines a new distribution $\nu_T = (T(\cdot; h(\mathbf{x})))^{-1}_{\#} p_{ref}$ which is called the push-forward of the reference distribution p_{ref} [23]. By drawing samples from p_{ref} and applying T on these samples we obtain samples from this new distribution. The density of this so-obtained distribution p_{ν_T} can be derived from the change of variables formula:

$$p_{\nu_T}(y \mid h(\mathbf{x})) = p_{ref}(T(y; h(\mathbf{x})))|\nabla_y T(y; h(\mathbf{x}))| \tag{12}$$

The KL-divergence between the target distribution $p_{Y|h(\mathbf{X})}$ and the flow-based model p_{ν_T} can be written as follows:

$$\begin{aligned}
&\mathbb{E}_{h(\mathbf{X})}[D_{\mathrm{KL}}(p_{Y|h(\mathbf{X})}\|p_{\nu_T})] \\
=&\mathbb{E}_{h(\mathbf{X})}\left[\mathbb{E}_{Y|h(\mathbf{X})}\left[\log\left(\frac{p_{Y|h(\mathbf{X})}}{p_{\nu_T}}\right)\right]\right] \\
=&-H(Y \mid h(\mathbf{X})) - \mathbb{E}_{h(\mathbf{X}),Y}[\log p_{\nu_T}(Y \mid h(\mathbf{X}))] \\
=&-H(Y \mid h(\mathbf{X})) + \mathbb{E}_{h(\mathbf{X}),Y}[-\log p_{ref}(T(y; h(\mathbf{x})) \\
&-\log|\nabla_y T(y; h(\mathbf{x}))|]
\end{aligned} \tag{13}$$

The last two terms in Eq. 13 correspond to the negative log-likelihood (NLL) for conditional flows with distribution p_{ref} in latent space. If the reference distribution is assumed to be standard normal, the NLL is given as in Sect. 3.

We restate Lemma 1 with a more general notation. Note that the argmax or argmin is a set.

Lemma 1. *Let* $\mathbf{X}, Y$ *be random variables. We furthermore assume that for each* $h \in \mathcal{H}$ *there exists one* $T \in \mathcal{T}$ *with* $\mathbb{E}_{h(\mathbf{X})}[D_{KL}(p_{Y|h(\mathbf{X})}\|p_{\nu_T})] = 0$. *Then, the following two statements are true*

(a) Let

$$h^\star, T^\star = \arg\min_{h\in\mathcal{H}, T\in\mathcal{T}} -\mathbb{E}_{h(\mathbf{X}),Y}[\log p_{\nu_T}(Y \mid h(\mathbf{X}))]$$

then it holds $h^\star = g^\star$ *where* $g^\star = \arg\max_{g\in\mathcal{H}} I(g(\mathbf{X}); Y)$

(b) Let

$$T^\star = \arg\min_{T \in \mathcal{T}} \mathbb{E}_{h(\mathbf{X})}[D_{KL}(p_{Y|h(\mathbf{X})} \| p_{\nu_T})]$$

then it holds $h(\mathbf{X}) \perp T^\star(Y; h(\mathbf{X}))$

Proof. (a) From Eq. 13, we obtain $-\mathbb{E}_{h(\mathbf{X}),Y}[\log p_{\nu_T}(Y \mid h(\mathbf{X}))] \geq H(Y \mid h(\mathbf{X}))$ for all $h \in \mathcal{H}, T \in \mathcal{T}$. We furthermore have $\min_{T \in \mathcal{T}} -\mathbb{E}_{h(\mathbf{X}),Y}[\log p_{\nu_T}(Y \mid h(\mathbf{X}))] = H(Y \mid h(\mathbf{X}))$ due to our assumptions on $\mathcal{T}$.

Therefore, $\min_{h \in \mathcal{H}, T \in \mathcal{T}} -\mathbb{E}_{h(\mathbf{X}),Y}[\log p_{\nu_T}(Y \mid h(\mathbf{X}))] = \min_{h \in \mathcal{H}} H(Y \mid h(\mathbf{X}))$. Since we have $I(Y; h(\mathbf{X})) = H(Y) - H(Y \mid h(\mathbf{X}))$ and only the second term depends on h, statement (a) holds true.

(b) For convenience, we denote $T(Y; h(\mathbf{X})) = R$ and $h(\mathbf{X}) = Z$. We have $\mathbb{E}_Z[D_{\mathrm{KL}}(p_{Y|Z} \| p_{\nu_{T^\star}})] = 0$ and therefore $p_{Y|Z}(y \mid z) = p_{ref}(T(y; z)) |\nabla_y T^{-1}(y; z)|$.

Then it holds

$$\begin{aligned} p_{R|Z}(r \mid z) &= p_{Y|Z}(T^{-1}(r; z)|z) \cdot |\nabla_y T^{-1}(r; z)| \\ &= p_{ref}(T(T^{-1}(r; z); z)) \cdot |\nabla_y T(y; z)| \\ &\quad \cdot |\nabla_y T^{-1}(r; z)| \\ &= p_{ref}(r) \cdot 1 \end{aligned}$$

Since the density p_{ref} is independent of Z, we obtain $R \perp Z$ which concludes the proof of (b)

Statement (a) describes an optimization problem that allows to find features which share maximal information with the target variable Y. Due to statement (b) it is possible to draw samples from the conditional distribution $P(Y \mid h(\mathbf{X}))$ via the reference distribution.

Let $\mathcal{H}_\perp$ the set of features which satisfy the invariance property, i.e. $Y \perp E \mid h(\mathbf{X})$ for all $h \in \mathcal{H}_\perp$. In the following, we sketch why $\arg\min_{h \in \mathcal{H}_\perp} \max_{e \in \mathcal{E}} \min_{T \in \mathcal{T}} \mathcal{L}^e_{\mathrm{NLL}}(T; h) = \arg\max_{h \in \mathcal{H}_\perp} \min_{e \in \mathcal{E}} I(Y^e; h(\mathbf{X}^e))$ follows from Lemma 1.

Let $h \in \mathcal{H}_\perp$. Then, it is easily seen that there exists a $T^\star \in \mathcal{T}$ with (1) $\mathcal{L}_{\mathrm{NLL}}(T^\star; h) = \min_{T \in \mathcal{T}} \mathcal{L}_{\mathrm{NLL}}(T, h)$ and (2) $\mathcal{L}^e_{\mathrm{NLL}}(T^\star, h) = \min_{T \in \mathcal{T}} \mathcal{L}^e_{\mathrm{NLL}}(T, h)$ for all $e \in \mathcal{E}$ since the conditional densities $p(y \mid h(\mathbf{X}))$ are invariant across all environments. Hence we have $H(Y^e \mid h(\mathbf{X}^e)) = \mathcal{L}^e_{\mathrm{NLL}}(T^\star; h)$ for all $e \in \mathcal{E}$. Therefore, $\arg\min_{h \in \mathcal{H}_\perp} \max_{e \in \mathcal{E}} \min_{T \in \mathcal{T}} \mathcal{L}^e_{\mathrm{NLL}}(T; h) = \arg\max_{h \in \mathcal{H}_\perp} \min_{e \in \mathcal{E}} I(Y^e; h(\mathbf{X}^e))$ due to $I(Y^e; h(\mathbf{X}^e)) = H(Y^e) - H(Y^e \mid h(\mathbf{X}^e))$.

C.1 Normalizing Flows and Additive Noise Models

In our case, we represent the conditional distribution $P(Y \mid h(\mathbf{X}))$ using a *conditional* normalizing flow (see e.g. [1]). In our work, we seek a mapping $R = T(Y; h(\mathbf{X}))$ that is diffeomorphic in Y such that $R \sim \mathcal{N}(0, 1) \perp h(\mathbf{X})$ when

$Y \sim P(Y \mid h(\mathbf{X}))$. This is a generalization of the well-studied additive Gaussian noise model $R = Y - f(h(\mathbf{X}))$, see Appendix E. The inverse $Y = F(R; h(\mathbf{X}))$ takes the role of a structural equation for the mechanism $p(Y \mid h(\mathbf{X}))$ with R being the corresponding noise variable.[1]

D HSIC and Wasserstein Loss

The Hilbert-Schmidt Independence Criterion (HSIC) is a kernel based measure for independence which is in expectation 0 if and only if the compared random variables are independent [12]. An empirical estimate of HSIC(A, B) for two random variables A, B is given by

$$\widehat{\text{HSIC}}(\{a_j\}_{j=1}^n, \{b_j\}_{j=1}^n) = \frac{1}{(n-1)^2}\text{tr}(KHK'H) \tag{14}$$

where tr is the trace operator. $K_{ij} = k(a^i, a^j)$ and $K'_{ij} = k'(b^i, b^j)$ are kernel matrices for given kernels k and k'. The matrix H is a centering matrix $H_{i,j} = \delta_{i,j} - 1/n$.

The one dimensional Wasserstein loss compares the similarity of two distributions [18]. This loss has expectation 0 if both distributions are equal. An empirical estimate of the one dimensional Wasserstein loss for two random variables A, B is given by

$$\mathcal{L}_W = \|\text{sort}(\{a_j\}_{j=1}^n) - \text{sort}(\{b_j\}_{j=1}^n)\|_2$$

Here, the two batches are sorted in ascending order and then compared in the L2-Norm. We assume that both batches have the same size.

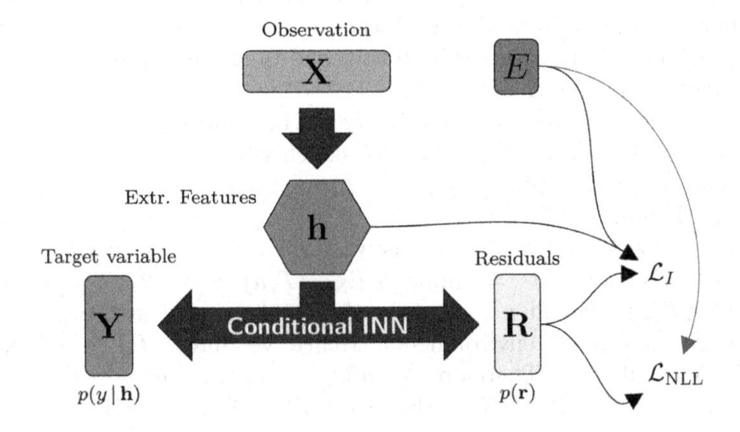

Fig. 4. Illustration of Architecture of Conditional Invertible Neural Network (Conditional INN) which implements Eq. 7. h is a feature extractor implemented as feed forward neural network. $\mathcal{L}_I$ is the invariance loss that measures the dependence between residuals R and $(E, h(\mathbf{X}))$ and $\mathcal{L}_{\text{NLL}}$ is the negative log-likelihood as in Eq. 5.

[1] F is the concatenation of the normal CDF with theinverse CDF of $P(Y \mid h(\mathbf{X}))$, see [32].

E Additive Noise Models and Robust Prediction

Let f_θ be a regression function. Solving for the noise term gives $R = Y - f_\theta(\mathbf{X})$ which corresponds to a diffeomorphism in Y, namely $T_\theta(Y; X) = Y - f_\theta(\mathbf{X})$. If we make two simplified assumptions: (i) the noise is Gaussian with zero mean and (ii) $R \perp f_\theta(\mathbf{X})$, then we obtain

$$\begin{aligned} I(Y; f_\theta(\mathbf{X})) &= H(Y) - H(Y \mid f_\theta(\mathbf{X})) = H(Y) - H(R \mid f_\theta(\mathbf{X})) \\ &\overset{(ii)}{=} H(Y) - H(R) \overset{(i)}{=} H(Y) - 1/2 \log(2\pi e \sigma^2) \end{aligned}$$

where $\sigma^2 = \mathbb{E}[(Y - f_\theta(\mathbf{X}))^2]$. In this case maximizing the mutual information $I(Y; f_\theta(\mathbf{X}))$ amounts to minimizing $\mathbb{E}[(Y - f_\theta(\mathbf{X}))^2]$ w.r.t. θ, i.e. the standard L2-loss for regression problems. From this, we obtain an approximation of Eq. 7 via

$$\arg\min_\theta \Big(\max_{e \in \mathcal{E}_{\text{seen}}} \Big\{ \mathbb{E}\big[(Y^e - f_\theta(\mathbf{X}^e))^2\big] \Big\} + \lambda_I \mathcal{L}_I(P_R, P_{f_\theta(\mathbf{X}),E}) \Big) \qquad (15)$$

where $R^e = Y - f_\theta(\mathbf{X}^e)$ and $\lambda_I > 0$. Under the conditions stated above, the objective achieves the mentioned trade-off between information and invariance.

Alternatively we can view the problem as to find features $h_\phi \colon \mathbb{R}^D \to \mathbb{R}^m$ such that $I(h_\phi(\mathbf{X}), Y)$ gets maximized under the assumption that there exists a model $f_\theta(h_\phi(\mathbf{X})) + R = Y$ where R is independent of $h_\phi(\mathbf{X})$ and R is gaussian. In this case we obtain similarly as above the learning objective

$$\arg\min_{\theta,\phi} \Big(\max_{e \in \mathcal{E}_{\text{seen}}} \Big\{ \mathbb{E}\big[(Y^e - f_\theta(h_\phi(\mathbf{X}^e)))^2\big] \Big\} + \lambda_I \mathcal{L}_I(P_R, P_{h_\phi(\mathbf{X}),E}) \Big) \qquad (16)$$

F Algorithm

In order to optimize the DG problem in Eq. 4, we optimize a normalizing flow T_θ and a feed forward neural network h_ϕ as described in Algorithm 1. There is an inherent trade-off between robustness and goodness-of-fit. The hyperparameter λ_I describes this trade-off and is chosen a priori.

If we choose a gating mechanisms h_ϕ as feature extractor similar to [17], then a complexity loss is added to the loss in the gradient update step. The architecture is illustrated in Fig. 4. Figure 5 shows the architecture with gating function.

In case we assume that the underlying mechanisms elaborates the noise in an additive manner, we could replace the normalizing flow T_θ with a feed forward neural network f_θ and execute Algorithm 2.

If we choose a gating mechanism, minor adjustments have to be made to Algorithm 2 such that we optimize Eq. 9. The classification case can be obtained similarly as described in Sect. 4.

Data: Samples from $P_{\mathbf{X}^e,Y^e}$ in different environments $e \in \mathcal{E}_{\text{seen}}$.
Initialize: Parameters θ, ϕ;
for *number of training iterations* **do**
  for $e \in \mathcal{E}_{seen}$ **do**
    Sample minibatch $\{(y_1^e, \mathbf{x}_1^e), \ldots, (y_m^e, \mathbf{x}_m^e)\}$ from $P_{Y,\mathbf{X}|E=e}$ for $e \in \mathcal{E}_{\text{seen}}$;;
    Compute $r_j^e = T_\theta(y_j^e; h_\phi(\mathbf{x}_j^e))$;;
  end
  Update θ, ϕ by descending alongside the stochastic gradient

$$\nabla_{\theta,\phi}\Big(\max_{e\in\mathcal{E}_{\text{seen}}}\Big\{\sum_{i=1}^{m}\big[\tfrac{1}{2}\|T_\theta(y_i^e;h_\phi(\mathbf{x}_i^e))\|^2 - \log\nabla_y T_\theta(y_i^e;h_\phi(\mathbf{x}_i^e))\big]\Big\} + \lambda_I \mathcal{L}_I(\{r_j^e\}_{j,e},\{h_\phi(\mathbf{x}_j^e),e\}_{j,e})\Big);$$

end
Result: In case of convergence, we obtain $T_{\theta^\star}, h_{\phi^\star}$ with

$$\theta^\star,\phi^\star = \arg\min_{\theta,\phi}\Big(\max_{e\in\mathcal{E}_{\text{seen}}}\Big\{\mathbb{E}_{\mathbf{X}^e,Y^e}\big[\tfrac{1}{2}\|T_\theta(Y^e;h_\phi(\mathbf{X}^e))\|^2 - \log\nabla_y T_\theta(Y^e;h_\phi(\mathbf{X}^e))\big]\Big\} + \lambda_I \mathcal{L}_I(P_R, P_{h_\phi(\mathbf{X}),E})\Big)$$

Algorithm 1: DG training with normalizing flows

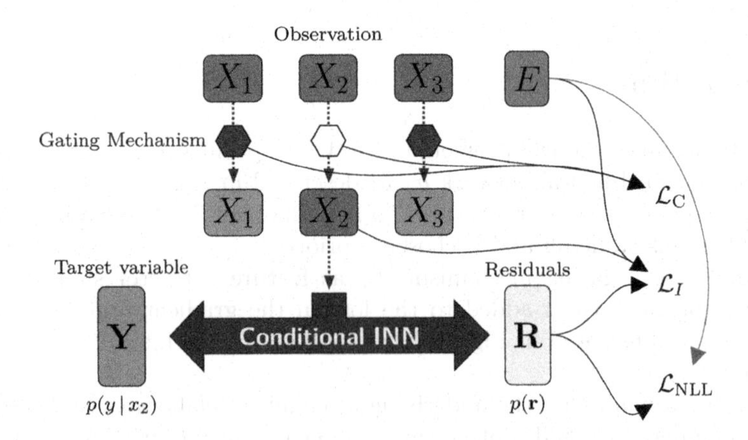

Fig. 5. Illustration of Architecture of Conditional Invertible Neural Network (Conditional INN) model which implements Eq. 7 where the feature extractor h is a gating mechanism. Architecture is depicted for three input variables. $\mathcal{L}_I$ is the invariance loss that measures the dependence between residuals R and $(E, h(\mathbf{X}))$, $\mathcal{L}_{\text{NLL}}$ is the negative log-likelihood as in Eq. 5 and $\mathcal{L}_{\text{C}}$ is the complexity loss as described in Sect. 4.3.

Data: Samples from $P_{\mathbf{X}^e,Y^e}$ in different environments $e \in \mathcal{E}_{\text{seen}}$.
Initialize: Parameters θ, ϕ;
for *number of training iterations* **do**
  for $e \in \mathcal{E}_{seen}$ **do**
    Sample minibatch $\{(y_1^e, \mathbf{x}_1^e), \ldots, (y_m^e, \mathbf{x}_m^e)\}$ from $P_{Y,\mathbf{X}|E=e}$ for $e \in \mathcal{E}_{\text{seen}}$;;
    Compute $r_j^e = y_j^e - f_\theta(\mathbf{x}_j^e)$;;
  end
  Update θ by descending alongside the stochastic gradient

$$\nabla_\theta \Big(\max_{e \in \mathcal{E}_{\text{seen}}} \Big\{ \sum_{i=1}^{m} |r_j^e|^2 \Big\} + \lambda_I \mathcal{L}_I(\{r_j^e\}_{j,e}, \{f_\theta(\mathbf{x}_j^e), e\}_{j,e}) \Big);$$

end
Result: In case of convergence, we obtain $f_{\theta^\star}$ with

$$\theta^\star = \arg\min_\theta \Big(\max_{e \in \mathcal{E}_{\text{seen}}} \Big\{ \mathbb{E}_{\mathbf{X}^e,Y^e} \big[|Y^e - f_\theta(\mathbf{X}^e)|^2 \big] \Big\} + \lambda_I \mathcal{L}_I(P_R, P_{f_\theta(\mathbf{X}),E}) \Big)$$

Algorithm 2: DG training under the assumption of additive noise

G Identifiability Result

Under certain conditions on the environment and the underlying causal graph, the direct causes of Y become identifiable:

Proposition 1. *We assume that the underlying causal graph G is faithful with respect to $P_{\tilde{\mathbf{X}},E}$. We further assume that every child of Y in G is also a child of E in G. A variable selection $h(\mathbf{X}) = \mathbf{X}_S$ corresponds to the direct causes if the following conditions are met: (i) $T(Y; h(\mathbf{X})) \perp E, h(\mathbf{X})$ are satisfied for a diffeomorphism $T(\cdot; h(\mathbf{X}))$, (ii) $h(\mathbf{X})$ is maximally informative about Y and (iii) $h(\mathbf{X})$ contains only variables from the Markov blanket of Y.*

Proof. Let $S(\mathcal{E}_{\text{seen}})$ denote a subset of $\mathbf{X}$ which corresponds to the variable selection due to h. Without loss of generality, we assume $S(\mathcal{E}_{\text{seen}}) \subset \mathbf{M}$ where $\mathbf{M}$ is the Markov Blanket. This assumption is reasonable since we have $Y \perp \mathbf{X} \setminus \mathbf{M} \mid \mathbf{M}$ in the asymptotic limit.

Since $pa(Y)$ cannot contain colliders between Y and E, we obtain that $Y \perp E \mid S(\mathcal{E}_{\text{seen}})$ implies $Y \perp E \mid (S(\mathcal{E}_{\text{seen}}) \cup pa(Y))$. This means using $pa(Y)$ as predictors does not harm the constraint in the optimization problem. Due to faithfulness and since the parents of Y are directly connected to Y, we obtain that $pa(Y) \subset S(\mathcal{E}_{\text{seen}})$.

For each subset $\mathbf{X}_S \subset \mathbf{X}$ for which there exists an $X_i \in \mathbf{X}_S \cap \mathbf{X}_{ch(Y)}$, we have $\mathbf{X}_S \not\perp Y \mid E$. This follows from the fact that X_i is a collider, in particular $E \rightarrow X_i \leftarrow Y$. Conditioning on X_i leads to the result that Y and E are not

d-separated anymore. Hence, we obtain $Y \not\perp \mathbf{X}_S \mid E$ due to the faithfulness assumption. Hence, for each $\mathbf{X}_S$ with $Y \perp E \mid \mathbf{X}_S$ we have $\mathbf{X}_S \cap \mathbf{X}_{ch(Y)} = \emptyset$ and therefore $\mathbf{X}_{ch(Y)} \cap S(\mathcal{E}_{\text{seen}}) = \emptyset$.

Since $\mathbf{X}_{pa(Y)} \subset S(\mathcal{E}_{\text{seen}})$, we obtain that $Y \perp \mathbf{X}_{pa(ch(Y))} \mid \mathbf{X}_{pa(Y)}$ and therefore the parents of $ch(Y)$ are not in $S(\mathcal{E}_{\text{seen}})$ except when they are parents of Y.

Therefore, we obtain that $S(\mathcal{E}_{\text{seen}}) = \mathbf{X}_{pa(Y)}$.

One might argue that the conditions are very strict in order to obtain the true direct causes. But the conditions set in Proposition 1 are necessary if we do not impose additional constraints on the true underlying causal mechanisms, e.g. linearity as done by [30]. For instance if $E \rightarrow X_1 \rightarrow Y \rightarrow X_2$, a model including X_1 and X_2 as predictor might be a better predictor than the one using only X_1. From the Causal Markov Condition we obtain $E \perp Y \mid X_1, X_2$ which results in $X_1, X_2 \in S(\mathcal{E}_{\text{seen}})$. Under certain conditions however, the relation $Y \rightarrow X_2$ might be invariant across $\mathcal{E}$. This is for instance the case when X_2 is a measurement of Y. In this cases it might be useful to use X_2 for a good prediction.

G.1 Gating Architecture

We employ the same gating architecture as in [17] which was first proposed in [21] as a Bernoulli reparameterization trick. They use this reparameterization trick in their original work in order to train neural networks with L0-Regularization in a gradient based manner. [17] apply the L0-Regularization on the input to learn a gating mechanism. Similarly we use the L0-Regularization to learn a gating mechanism.

The gating architecture h_ϕ is parameterized via $\phi = (\boldsymbol{\alpha}, \boldsymbol{\beta})$ where $\boldsymbol{\alpha} = (\alpha_1, \ldots, \alpha_D)$ and $\boldsymbol{\beta} = (\beta_1, \ldots, \beta_D)$. Let $\gamma < 0$ and $\zeta > 0$ be fixed. Then we map $\boldsymbol{u} \sim \mathcal{U}[0,1]^D$ via $\boldsymbol{s}(\boldsymbol{u}) = \text{Sigmoid}((\log \boldsymbol{u} - \log(1 - \boldsymbol{u}) + \boldsymbol{\alpha})/\boldsymbol{\beta})$, to $\boldsymbol{z} = \min(1, \max(0, \boldsymbol{s}(\boldsymbol{u})(\zeta - \gamma) + \gamma))$. This is how we sample the gates for each batch during training. The gates are then multiplied element-wise with the input $\boldsymbol{z} \odot \mathbf{X}$. In principle we could sample many $u \sim \mathcal{U}[0,1]$, but we observe that one sample of $u \sim \mathcal{U}[0,1]$ per batch suffices for our examples. At test time we use the following estimator for the gates:

$$\hat{\boldsymbol{z}} = \min(1, \max(0, \text{Sigmoid}(\boldsymbol{\alpha})(\zeta - \gamma) + \gamma))$$

Similarly as during training time, we multiply $\hat{\boldsymbol{z}}$ with the input. After sufficient training $\hat{\boldsymbol{z}}$ is a hard 0-1 mask. The complexity loss is defined via

$$\mathcal{L}(h_\theta) = \sum_{j=1}^{D} \text{Sigmoid}\Big(\alpha_j - \beta_j \log \frac{-\gamma}{\zeta}\Big). \tag{17}$$

For a detailed derivation of the reparameterization and complexity loss, see [21].

H Experimental Setting for Synthetic Dataset

H.1 Data Generation

In Sect. 5 we described how we choose different Structural Causal Models (SCM). In the following we describe details of this process.

We simulate the datasets in a way that the conditions in Proposition 1 are met. We choose different variables in the graph shown in Fig. 2 as target variable. Hence, we consider different "topological" scenarios. We assume the data is generated by some underlying SCM. We define the structural assignments in the SCM as follows

(a) $f_i^{(1)}(\mathbf{X}_{pa(i)}, N_i) = \sum_{j\in pa(i)} a_j X_j + N_i$ [Linear]

(b) $f_i^{(2)}(\mathbf{X}_{pa(i)}, N_i) = \sum_{j\in pa(i)} a_j X_j - \tanh(a_j X_j) + N_i$

[Tanhshrink]

(c) $f_i^{(3)}(\mathbf{X}_{pa(i)}, N_i) = \sum_{j\in pa(i)} \log(1 + \exp(a_j X_j)) + N_i$

[Softplus]

(d) $f_i^{(4)}(\mathbf{X}_{pa(i)}, N_i) = \sum_{j\in pa(i)} \max\{0, a_j X_j)\} + N_i$

[ReLU]

(e) $f_i^{(5)}(\mathbf{X}_{pa(i)}, N_i) = \Big(\sum_{j\in pa(i)} a_j X_j\Big) \cdot (1 + \frac{1}{4} N_i) + N_i$

[Mult. Noise]

with $N_i \sim \mathcal{N}(0, c_i^2)$ where $c_i \sim \mathcal{U}[0.8, 1.2]$, $i \in \{0, \ldots, 5\}$ and $a_i \in \{-1, 1\}$ according to Fig. 6. Note that the mechanisms in (b), (c) and (d) are non-linear with additive noise and (e) elaborates the noise in a non-linear manner.

We consider hard- and soft-interventions on the assignments f_i. We either intervene on all variables except the target variable at once *or* on all parents and children of the target variable (Intervention Location). We consider three types of interventions:

- *Hard-Intervention* on X_i: Force $X_i \sim e_1 + e_2\mathcal{N}(0,1)$ where we sample for each environment $e_2 \sim \mathcal{U}([1.5, 2.5])$ and $e_1 \sim \mathcal{U}([0.5, 1.5] \cup [-1.5, -0.5])$
- *Soft-Intervention* I on X_i: Add $e_1 + e_2\mathcal{N}(0,1)$ to X_i where we sample for each environment $e_2 \sim \mathcal{U}([1.5, 2.5])$ and $e_1 \sim \mathcal{U}([0.5, 1.5] \cup [-1.5, -0.5])$
- *Soft-Intervention* II on X_i: Set the noise distribution N_i to $\mathcal{N}(0, 2^2)$ for $E = 2$ and to $\mathcal{N}(0, 0.2^2)$ for $E = 3$

Per run, we consider one environment without intervention ($E = 1$) and two environments with either both soft- or hard-interventions ($E = 2, 3$). We also create a fourth environment to measure a models' ability for out-of-distribution generalization:

- *Hard-Intervention*: Force $X_i \sim e + \mathcal{N}(0, 4^2)$ where $e = e_1 \pm 1$ with e_1 from environment $E = 1$. The sign $\{+, -\}$ is chosen once for each i with equal probability.
- *Soft-Intervention* I: Add $e + \mathcal{N}(0, 4^2)$ to X_i where $e = e_1 \pm 1$ with e_1 from environment $E = 1$. The sign $\{+, -\}$ is chosen once for each i with equal probability as for the *do-intervention* case.
- *Soft-Intervention* II: Half of the samples have noise N_i distributed due to $\mathcal{N}(0, 1.2^2)$ and the other half of the samples have noise distributed as $\mathcal{N}(0, 3^2)$

We randomly sample causal graphs as described above. Per environment, we consider 1024 samples.

H.2 Training Details

All used feed forward neural networks have two internal layers of size 256. For the normalizing flows we use a 2 layer *MTA-Flow* described in Appendix H.3 with K=32. As optimizer we use Adam with a learning rate of 10^{-3} and a L2-Regularizer weighted by 10^{-5} for all models. Each model is trained with a batch size of 256. We train each model for 1000 epochs and decay the learning rate every 400 epochs by 0.5. For each model we use $\lambda_I = 256$ and the HSIC $\mathcal{L}_I$ employs a Gaussian kernel with $\sigma = 1$. The gating architecture was trained without the complexity loss for 200 epochs and then with complexity loss weighted by 5. For the Flow model without gating architecture we use a feed forward neural network h_ϕ with two internal layers of size 256 mapping to an one dimensional vector. In total, we evaluated our models on 1365 created datasets as described in H.1.

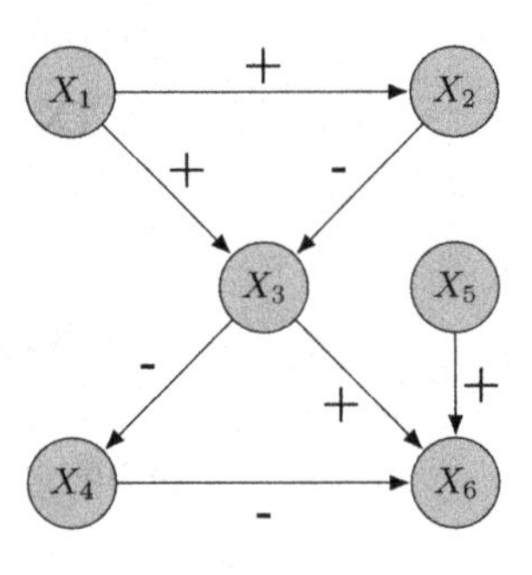

Fig. 6. The signs of the coefficients a_j for the mechanisms of the different SCMs

Once the normalizing flow T is learned, we predict y given features $h(\mathbf{x})$ using 512 normally distributed samples u_i which are mapped to samples from $p(y|h(\mathbf{x}))$ by the trained normalizing flow $T(u_i; h(\mathbf{x}))$. As prediction we use the mean of these samples.

H.3 One-Dimensional Normalizing Flow

We use as one-dimension normalizing flow the *More-Than-Affine-Flow (MTA-Flow)*, which was developed by us. An overview of different architectures for one-dimensional normalizing flows can be found in [28]. For each layer of the flow, a conditioner network C maps the conditional data $h(\mathbf{X})$ to a set of parameters $a, b \in \mathbb{R}$ and $\mathbf{w}, \mathbf{v}, \mathbf{r} \in \mathbb{R}^K$ for a chosen $K \in \mathbb{N}$. It builds the transformer τ for each layer as

$$\begin{aligned} z &= \tau(y \mid h(\mathbf{X})) \\ &:= a\left(y + \frac{1}{N(\mathbf{w}, \mathbf{v})} \sum_{i=1}^{K} w_i f(v_i y + r_i)\right) + b, \end{aligned} \tag{18}$$

where f is any almost everywhere smooth function with a derivative bounded by 1. In this work we used a gaussian function with normalized derivative for f. The division by

$$N(\mathbf{w}, \mathbf{v}) := \varepsilon^{-1}\left(\sum_{i=1}^{K} |w_i v_i| + \delta\right), \tag{19}$$

with numeric stabilizers $\varepsilon < 1$ and $\delta > 0$, assures the strict monotonicity of τ and thus its invertibility $\forall x \in \mathbb{R}$. We also used a slightly different version of the *MTA-Flow* which uses the ELU activation function and – because of its monotonicity – can use a relaxed normalizing expression $N(\mathbf{w}, \mathbf{v})$.

H.4 PC-Variant

Since we are interested in the direct causes of Y, the widely applied PC-Algorithm gives not the complete answer to the query for the parents of Y. This is due to the fact that it is not able to orient all edges. To compare the PC-Algorithm we include the environment as system-intern variable and use a conservative assignment scheme where non-oriented edges are thrown away. This assignment scheme corresponds to the conservative nature of the ICP.

For further interest going beyond this work, we consider diverse variants of the PC-Algorithm. We consider two orientation schemes: A *conservative* one, where non-oriented edges are thrown away and a *non-conservative* one where non-oriented edges from a node X_i to Y are considered parents of Y.

We furthermore consider three scenarios: (1) the samples across all environments are pooled, (2) only the observational data (from the first environment) is given, and (3) the environment variable is considered as system-intern variable and is seen by the PC-Algorithm (similar as in [26]). Results are shown in Fig. 7. In order to obtain these results, we sampled 1500 graphs as described above and applied on each of these datasets a PC-Variant. Best accuracies are achieved if we consider the environment variable as system-intern variable and use the non-conservative orientation scheme (EnvIn).

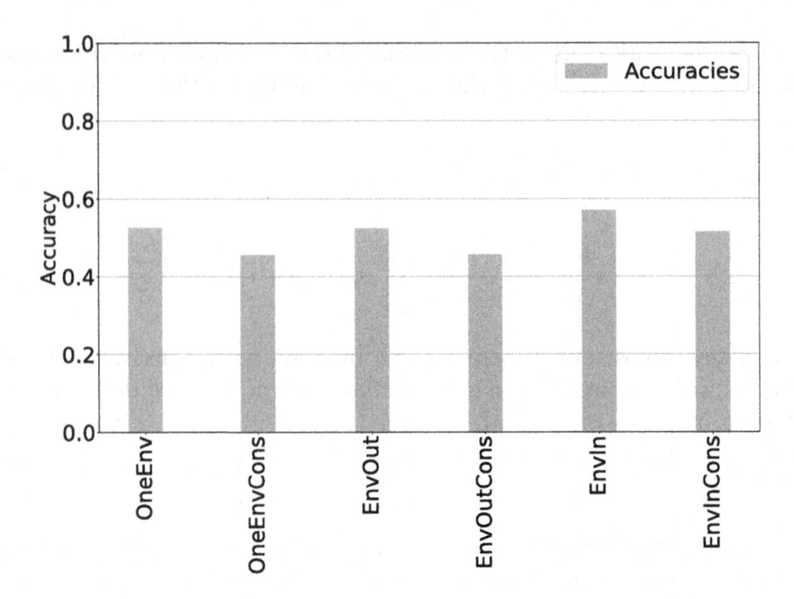

Fig. 7. Detection accuracies of direct causes for different variants of the PC-Algorithm. EnvOut means we pool over all environments and EnvIn means the environment is treated as system intern variable E. The suffix Cons means we us the conservative assignment scheme. OneEnv means we only consider the observational environment for inference.

H.5 Variable Selection

We consider the task of finding the direct causes of a target variable Y. Our models based on the gating mechanism perform a variable selection and are therefore compared to the PC-Algorithm and ICP. In the following we show the accuracies of this variable selection according to different scenarios.

Figure 8 shows the accuracies of ICP, the PC-Algorithm and our models pooled over all scenarios. Our models perform comparably well and better than the baseline in the causal discovery task.

In the following we show results due to different mechanisms, target variables, intervention types and intervention locations. Figure 9a shows the accuracies of all models across different target variables. Parentless target variables, i.e. $Y = X_4$ or $Y = X_0$ are easy to solve for ICP due to its conservative nature. All our models solve the parentless case quite well. Performance of the PC-variant depends strongly on the position of the target variable in the SCM indicating that its conservative assignment scheme has a strong influence on its performance. As expected, the PC-variant deals well with with $Y = X_6$ which is a childless collider. The causal discovery task seems to be particularly hard for variable $Y = X_6$ for all other models. This is the variable which has the most parents.

The type of intervention and its location seem to play a minor role as shown in Fig. 9a and Fig. 9a.

Figure 9b shows that ICP performs well if the underlying causal model is linear, but degrades if the mechanism become non-linear. The PC-Algorithm

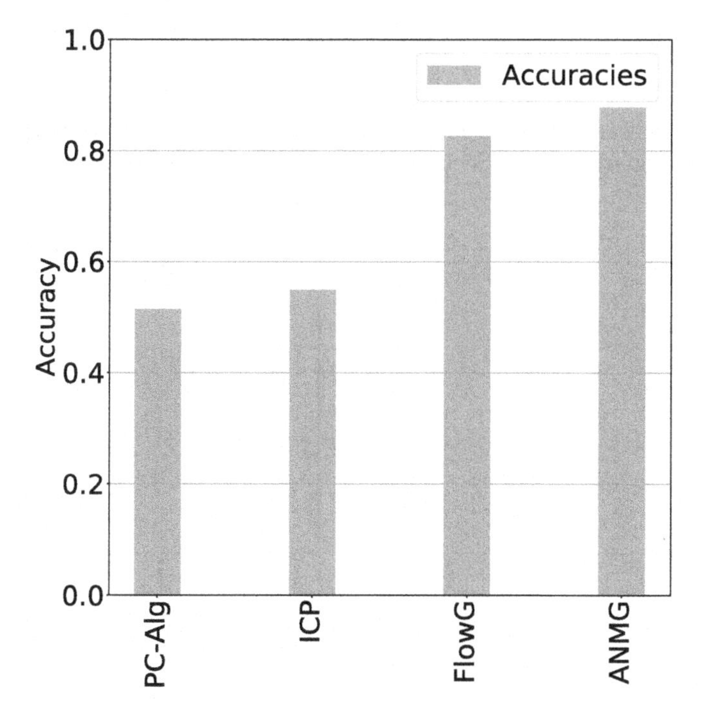

Fig. 8. Accuracies for different models across all scenarios. FlowG and ANMG are our models.

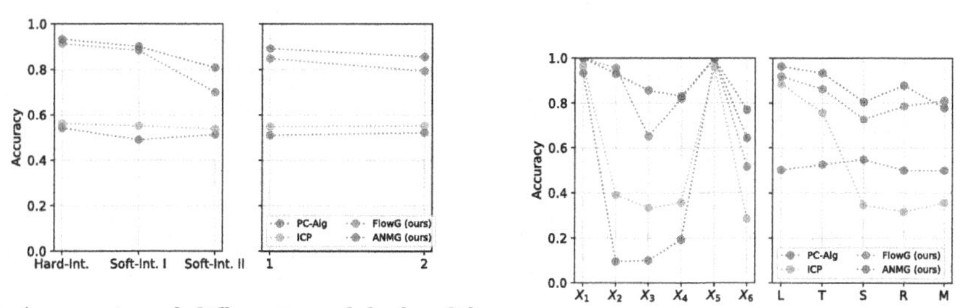

(a) Accuracies of different models for different intervention types and locations. 1 stands for intervention on all variables except Y and 2 stands for interventions on the parents and children only.

(b) Accuracies of different models according to target variables and mechanisms of the underlying SCM.

Fig. 9. Comparison of models across different scenarios in the causal discovery task

performs under all mechanisms comparably, but not well. ANMG performs quite well in all cases and even slightly better than FlowG in the cases of additive noise. However in the case of non-additive noise FlowG performs quite well whereas ANMG perform slightly worse – arguably because their requirements (additive noise) on the underlying mechanisms are not met.

H.6 Transfer Study

In the following we show the performance of different models on the training set, a test set of the same distribution and a set drawn from an unseen environment for different scenarios. As in Sect. 5, we use the L2-Loss on samples of an unseen

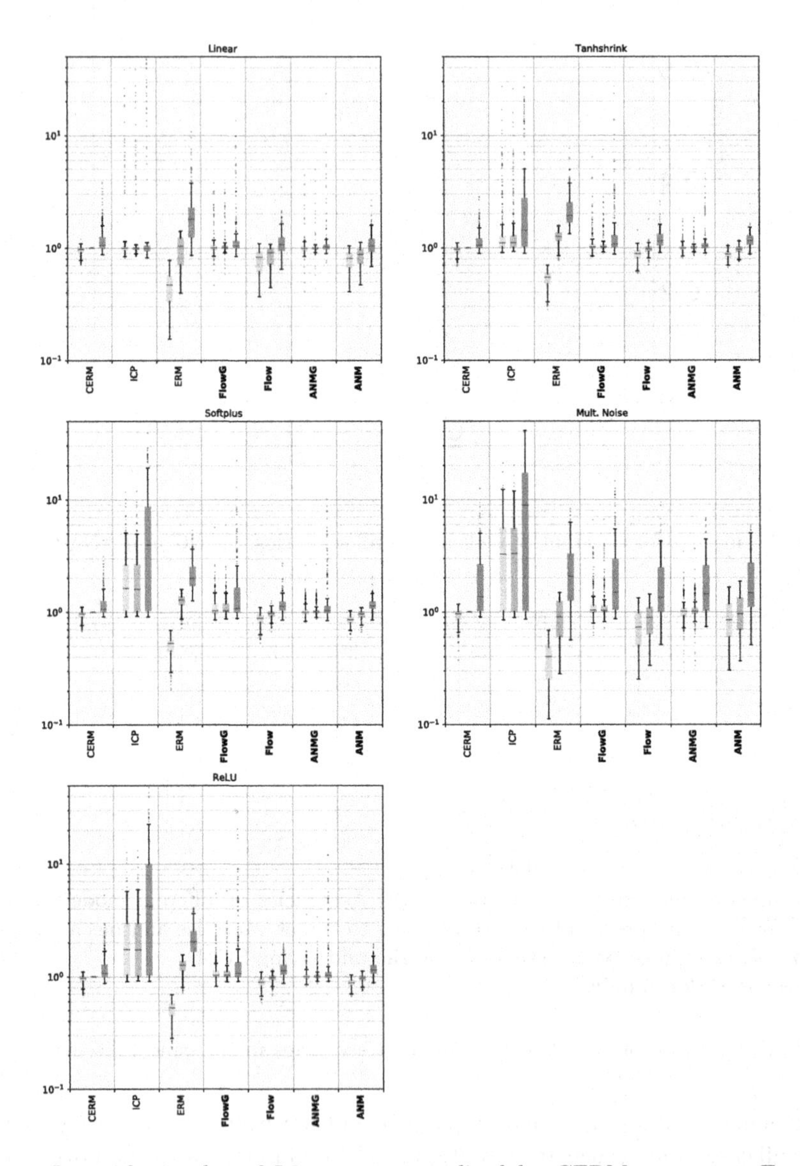

Fig. 10. Logarithmic plot of L2 errors, normalized by CERM test error. For each method (ours in bold) from left to right: training error, test error on seen environments, domain generalization error on unseen environments. Scenarios for different mechanisms are shown.

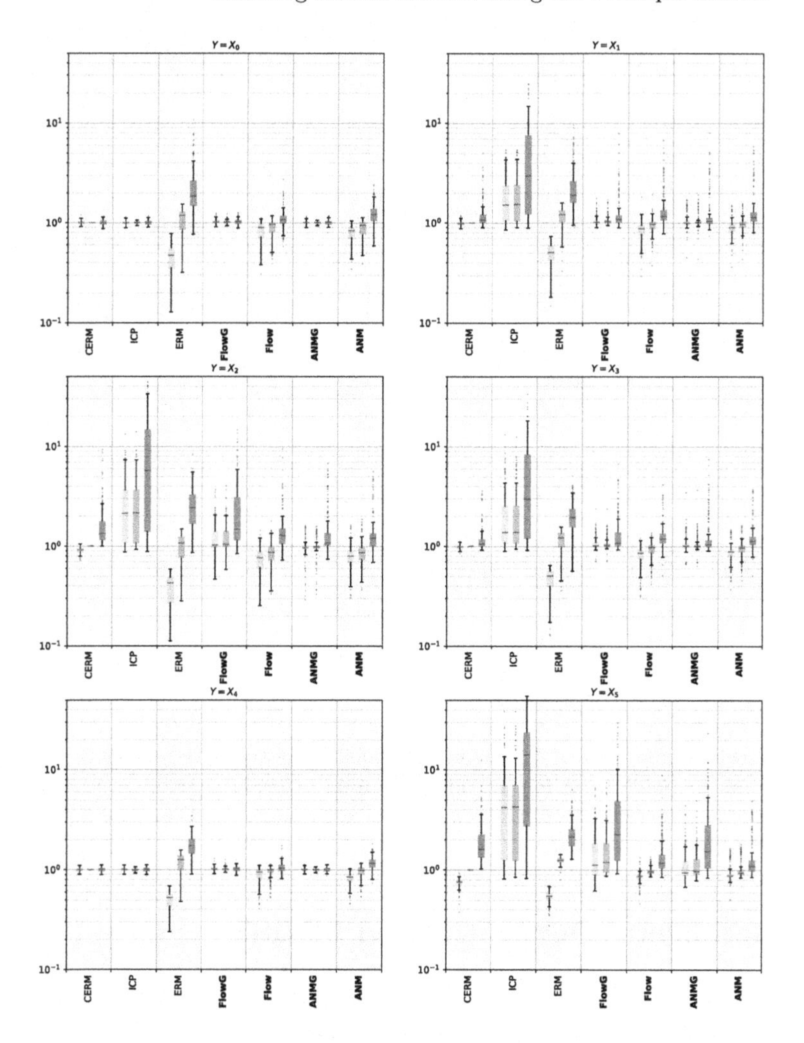

Fig. 11. Logarithmic plot of L2 errors, normalized by CERM test error. For each method (ours in bold) from left to right: training error, test error on seen environments, domain generalization error on unseen environments. Scenarios for different target variables are shown.

environment to measure out-of-distribution generalization. Figure 10, 11 and 12 show results according to the underlying mechanisms, target variable or type of intervention respectively. The boxes show the quartiles and the upper whiskers ranges from third quartile to $1.5 \cdot IQR$ where IQR is the interquartile range. Similar for the lower whisker.

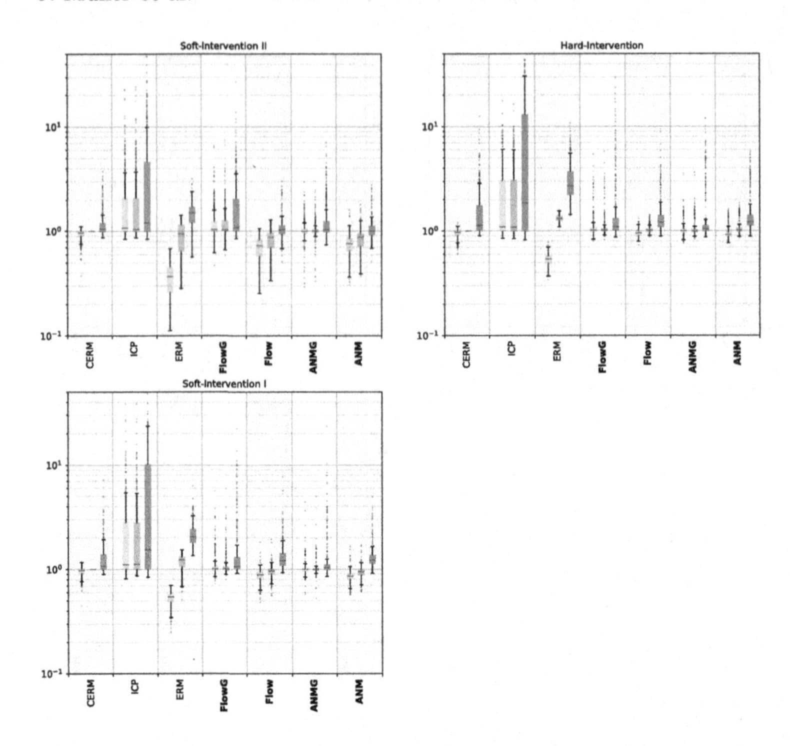

Fig. 12. Logarithmic plot of L2 errors, normalized by CERM test error. For each method (ours in bold) from left to right: training error, test error on seen environments, domain generalization error on unseen environments. Scenarios for different intervention types are shown.

I Experimental Details Colored MNIST

For the training, we use a feed forward neural network consisting of a feature selector followed by a classificator. The feature selector consists of two convolutional layers with a kernel size of 3 with 16 respectively 32 channels followed by a max pooling layer with kernel size 2, one dropout layer ($p = 0.2$) and a fully connected layer mapping to 16 feature dimensions. After the first convolutional layer and after the pooling layer a PReLU activation function is applied. For the classification we use a PReLU activation function followed by a Dropout layer ($p = 0.2$) and a linear layer which maps the 16 features onto the two classes corresponding to the labels.

We use the data generating process from [2]. 50 000 samples are used for training and 10 000 samples as test set. For training, we choose a batch size of 1000 and train our models for 60 epochs. We choose a starting learning rate of $6 \cdot 10^{-3}$. The learning rate is decayed by 0.33 after 20 epochs. We use an L2-Regularization loss weighted by 10^{-5}. After each epoch we randomly reassign the colors and the labels with the corresponding probabilities. The one-dimensional Wasserstein loss is applied dimension-wise and the maximum over dimensions is computed in order to compare residuals. For the HSIC we use a cauchy kernel with $\sigma = 1$. The invariance loss $\mathcal{L}_I$ is simply the sum of the HSIC and Wasserstein term. For Fig. 3a we trained our model with $\lambda_I \approx 13$. This hyperparameter is chosen from the best run in Fig. 3b. For stability in the case of large λ_I, we divide the total loss by λ_I during training to produce the results in Fig. 3b. For the reported accuracy of IRM, we train with the same network architecture on the dataset where we created training instances online.

References

1. Ardizzone, L., Lüth, C., Kruse, J., Rother, C., Köthe, U.: Guided image generation with conditional invertible neural networks. arXiv preprint arXiv:1907.02392 (2019)
2. Arjovsky, M., Bottou, L., Gulrajani, I., Lopez-Paz, D.: Invariant risk minimization. arXiv preprint arXiv:1907.02893 (2019)
3. Barber, D., Agakov, F.V.: The IM algorithm: a variational approach to information maximization. In: Advances in Neural Information Processing Systems (2003)
4. Bareinboim, E., Pearl, J.: Causal inference and the data-fusion problem. Proc. Natl. Acad. Sci. **113**(27), 7345–7352 (2016)
5. Ben-David, S., Blitzer, J., Crammer, K., Pereira, F.: Analysis of representations for domain adaptation. In: Advances in Neural Information Processing Systems, pp. 137–144 (2007)
6. Chickering, D.M.: Optimal structure identification with greedy search. J. Mach. Learn. Res. **3**(Nov), 507–554 (2002)
7. Frisch, R.: Statistical versus theoretical relations in economic macrodynamics. In: Hendry, D.F., Morgan, M.S. (eds.) Paper given at League of Nations (1995). The Foundations of Econometric Analysis (1938)
8. Ganin, Y., et al.: Domain-adversarial training of neural networks. J. Mach. Learn. Res. **17**(1), 1–35 (2016). 2096–2030
9. Ghassami, A., Kiyavash, N., Huang, B., Zhang, K.: Multi-domain causal structure learning in linear systems. In: Advances in Neural Information Processing Systems, pp. 6266–6276 (2018)
10. Ghassami, A., Salehkaleybar, S., Kiyavash, N., Zhang, K.: Learning causal structures using regression invariance. In: Advances in Neural Information Processing Systems, pp. 3011–3021 (2017)
11. Greenfeld, D., Shalit, U.: Robust learning with the Hilbert-Schmidt independence criterion. arXiv preprint arXiv:1910.00270 (2019)
12. Gretton, A., Bousquet, O., Smola, A., Schölkopf, B.: Measuring statistical dependence with Hilbert-Schmidt norms. In: Jain, S., Simon, H.U., Tomita, E. (eds.) ALT 2005. LNCS (LNAI), vol. 3734, pp. 63–77. Springer, Heidelberg (2005). https://doi.org/10.1007/11564089_7
13. Heckman, J.J., Pinto, R.: Causal analysis after haavelmo. Technical report, National Bureau of Economic Research (2013)
14. Heinze-Deml, C., Peters, J., Meinshausen, N.: Invariant causal prediction for nonlinear models. J. Causal Inference **6**(2) (2018)
15. Hoover, K.D.: The logic of causal inference: econometrics and the conditional analysis of causation. Econ. Philos. **6**(2), 207–234 (1990)
16. Huang, B., et al.: Causal discovery from heterogeneous/nonstationary data. J. Mach. Learn. Res. **21**(89), 1–53 (2020)
17. Kalainathan, D., Goudet, O., Guyon, I., Lopez-Paz, D., Sebag, M.: Sam: structural agnostic model, causal discovery and penalized adversarial learning. arXiv preprint arXiv:1803.04929 (2018)

18. Kolouri, S., Pope, P.E., Martin, C.E., Rohde, G.K.: Sliced-Wasserstein autoencoder: an embarrassingly simple generative model. arXiv preprint arXiv:1804.01947 (2018)
19. Krueger, D., et al.: Out-of-distribution generalization via risk extrapolation (REx). arXiv preprint arXiv:2003.00688 (2020)
20. Lake, B.M., Ullman, T.D., Tenenbaum, J.B., Gershman, S.J.: Building machines that learn and think like people. Behav. Brain Sciences, **40** (2017)
21. Louizos, C., Welling, M., Kingma, D.P.: Learning sparse neural networks through l_0 regularization. arXiv preprint arXiv:1712.01312 (2017)
22. Magliacane, S., van Ommen, T., Claassen, T., Bongers, S., Versteeg, P., Mooij, J.M.: Domain adaptation by using causal inference to predict invariant conditional distributions. In: Advances in Neural Information Processing Systems, pp. 10846–10856 (2018)
23. Marzouk, Y., Moselhy, T., Parno, M., Spantini, A.: Sampling via measure transport: an introduction. In: Ghanem, R., Higdon, D., Owhadi, H. (eds.) Handbook of Uncertainty Quantification, pp. 1–41. Springer, Heidelberg (2016). https://doi.org/10.1007/978-3-319-11259-6_23-1
24. Meinshausen, N.: Causality from a distributional robustness point of view. In: 2018 IEEE Data Science Workshop (DSW), pp. 6–10. IEEE (2018)
25. Mitrovic, J., McWilliams, B., Walker, J., Buesing, L., Blundell, C.: Representation learning via invariant causal mechanisms. arXiv preprint arXiv:2010.07922 (2020)
26. Mooij, J.M., Magliacane, S., Claassen, T.: Joint causal inference from multiple contexts. arXiv preprint arXiv:1611.10351 (2016)
27. Pan, S.J., Tsang, I.W., Kwok, J.T., Yang, Q.: Domain adaptation via transfer component analysis. IEEE Trans. Neural Netw. **22**(2), 199–210 (2010)
28. Papamakarios, G., Nalisnick, E., Rezende, D.J., Mohamed, S., Lakshminarayanan, B.: Normalizing flows for probabilistic modeling and inference. arXiv preprint arXiv:1912.02762 (2019)
29. Pearl, J.: Causality. Cambridge University Press, Cambridge (2009)
30. Peters, J., Bühlmann, P., Meinshausen, N.: Causal inference by using invariant prediction: identification and confidence intervals. J. R. Stat. Soc.: Ser. B (Stat. Methodol.) **78**(5), 947–1012 (2016)
31. Peters, J., Janzing, D., Schölkopf, B.: Elements of Causal Inference: Foundations and Learning Algorithms. MIT Press, Cambridge (2017)
32. Peters, J., Mooij, J.M., Janzing, D., Schölkopf, B.: Causal discovery with continuous additive noise models. J. Mach. Learn. Res. **15**(1), 2009–2053 (2014)
33. Pfister, N., Bühlmann, P., Peters, J.: Invariant causal prediction for sequential data. J. Am. Stat. Assoc. **114**(527), 1264–1276 (2019)
34. Qin, Z., Kim, D.: Rethinking softmax with cross-entropy: neural network classifier as mutual information estimator. arXiv preprint arXiv:1911.10688 (2019)
35. Rojas-Carulla, M., Schölkopf, B., Turner, R., Peters, J.: Invariant models for causal transfer learning. J. Mach. Learn. Res. **19**(1), 1309–1342 (2018)
36. Spirtes, P., Glymour, C.: An algorithm for fast recovery of sparse causal graphs. Soc. Sci. Comput. Rev. **9**(1), 62–72 (1991)
37. Tian, J., Pearl, J.: Causal discovery from changes. In: Uncertainty in Artificial Intelligence (UAI), pp. 512–521 (2001)
38. Xie, C., Chen, F., Liu, Y., Li, Z.: Risk variance penalization: from distributional robustness to causality. arXiv preprint arXiv:2006.07544 (2020)
39. Zhang, K., Gong, M., Schölkopf, B.: Multi-source domain adaptation: a causal view. In: Twenty-Ninth AAAI Conference on Artificial Intelligence (2015)

Reintroducing Straight-Through Estimators as Principled Methods for Stochastic Binary Networks

Alexander Shekhovtsov[1(✉)] and Viktor Yanush[2]

[1] Czech Technical University in Prague, Prague, Czech Republic
shekhole@fel.cvut.cz
[2] Samsung-HSE Laboratory National Research University
Higher School of Economics, Moscow, Russia

Abstract. Training neural networks with binary weights and activations is a challenging problem due to the lack of gradients and difficulty of optimization over discrete weights. Many successful experimental results have been achieved with empirical straight-through (ST) approaches, proposing a variety of ad-hoc rules for propagating gradients through non-differentiable activations and updating discrete weights. At the same time, ST methods can be truly derived as estimators in the stochastic binary network (SBN) model with Bernoulli weights. We advance these derivations to a more complete and systematic study. We analyze properties, estimation accuracy, obtain different forms of correct ST estimators for activations and weights, explain existing empirical approaches and their shortcomings, explain how latent weights arise from the mirror descent method when optimizing over probabilities. This allows to reintroduce ST methods, long known empirically, as sound approximations, apply them with clarity and develop further improvements.

1 Introduction

Neural networks with binary weights and activations have much lower computation costs and memory consumption than their real-valued counterparts [18, 26, 45]. They are therefore very attractive for applications in mobile devices, robotics and other resource-limited settings, in particular for solving vision and speech recognition problems [8, 56].

The seminal works that showed feasibility of training networks with binary weights [15] and binary weights and activations [27] used the empirical straight-through gradient estimation approach. In this approach the derivative of a step

We gratefully acknowledge support by Czech OP VVV project "Research Center for Informatics (CZ.02.1.01/0.0/0.0/16019/0000765)".

Supplementary Information The online version contains supplementary material available at https://doi.org/10.1007/978-3-030-92659-5_7.

C. Bauckhage et al. (Eds.): DAGM GCPR 2021, LNCS 13024, pp. 111–126, 2021.
https://doi.org/10.1007/978-3-030-92659-5_7

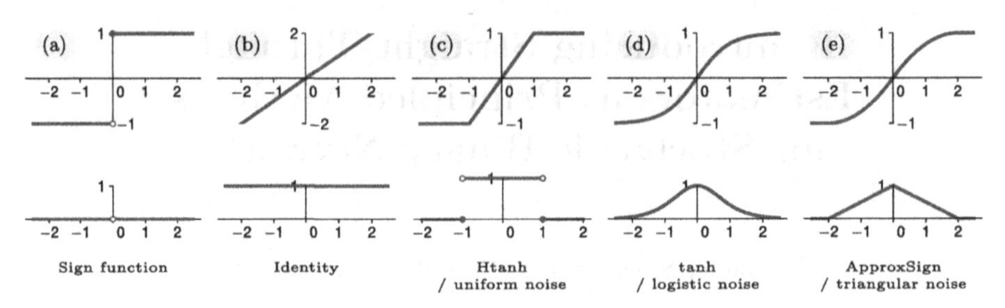

Fig. 1. The sign function and different proxy functions for derivatives used in empirical ST estimators. Variants (c-e) can be obtained by choosing the noise distribution in our framework. Specifically for a real-valued noise z with cdf F, in the upper plots we show $\mathbb{E}_z[\text{sign}(a - z)] = 2F - 1$ and, respectively, twice the density, $2F'$ in the lower plots. Choosing *uniform* distribution for z gives the density $p(z) = \frac{1}{2}\mathbb{1}_{z\in[-1,1]}$ and recovers the common Htanh proxy in (c). The *logistic* noise has cdf $F(z) = \sigma(2z)$, which recovers tanh proxy in (d). The *triangular* noise has density $p(z) = \max(0, |(2 - x)/4|)$, which recovers a scaled version of ApproxSign [34] in (e). The scaling (standard deviation) of the noise in each case is chosen so that $2F'(0) = 1$. The identity ST form in (b) we recover as latent weight updates with mirror descent.

function like sign, which is zero, is substituted with the derivative of some other function, hereafter called a *proxy* function, on the backward pass. One possible choice is to use *identity* proxy, *i.e.*, to completely bypass sign on the backward pass, hence the name *straight-through* [5]. This ad-hoc solution appears to work surprisingly well and the later mainstream research on binary neural networks heavily relies on it [2,6,9,11,18,34,36,45,52,60].

The *de-facto standard* straight-through approach in the above mentioned works is to use deterministic binarization and the clipped identity proxy as proposed by Hubara et al. [27]. However, other proxy functions were experimentally tried, including tanh and piece-wise quadratic ApproxSign [18,34], illustrated in Fig. 1. This gives rise to a diversity of empirical ST methods, where various choices are studied purely experimentally [2,6,52]. Since binary weights can be also represented as a sign mapping of some real-valued *latent weights*, the same type of methods is applied to weights. However, often a different proxy is used for the weights, producing additional unclear choices. The dynamics and interpretation of latent weights are also studied purely empirically [51]. With such obscurity of latent weights, Helwegen et al. [24] argues that "latent weights do not exist" meaning that discrete optimization over binary weights needs to be considered. The existing partial justifications of deterministic straight-through approaches are limited to one-layer networks with Gaussian data [58] or binarization of weights only [1] and do not lead to practical recommendations.

In contrast to the deterministic variant used by the mainstream SOTA, straight-through methods were originally proposed (also empirically) for stochastic autoencoders [25] and studied in models with stochastic binary neurons [5,44]. In the *stochastic binary network* (SBN) model which we consider, all hidden units and/or weights are Bernoulli random variables. The expected loss is a truly differentiable function of parameters (*i.e.*, weight probabilities) and its gradient can be estimated. This framework allows to pose questions such as: "What is the true expected gradient?" and "How far from it is the estimate computed by ST?" Towards computing the true gradient, unbiased gradient estimators were developed [20,55,57], which however have not been applied to networks with deep binary dependencies due to increased variance in deep layers and complexity that grows quadratically with the number of layers [48]. Towards explaining ST methods in SBNs, Tokui & Sato [54] and Shekhovtsov et al. [48] showed how to *derive* ST under linearizing approximations in SBNs. These results however were secondary in these works, obtained from more complex methods. They remained unnoticed in the works applying ST in practice and recent works on its analysis [13,58]. They are not properly related to existing empirical ST variants for activations and weights and did not propose analysis.

The goal of this work is to reintroduce straight-through estimators in a principled way in SBNs, to formalize and systematize empirical ST approaches for activation and weights in shallow and deep models. Towards this goal we review the derivation and formalize many empirical variants and algorithms using the derived method and sound optimization frameworks: we show how different kinds of ST estimators can occur as valid modeling choices or valid optimization choices. We further study properties of ST estimator and its utility for optimization: we theoretically predict and experimentally verify the improvement of accuracy with network width and show that popular modifications such as deterministic ST decrease this accuracy. For deep SBNs with binary weights we demonstrate that several estimators lead to equivalent results, as long as they are applied consistently with the model and the optimization algorithm.

More details on the related work, including alternative approaches for SBNs we discuss in Appendix A.

2 Derivation and Analysis

Notation. We model random states $\boldsymbol{x} \in \{-1, 1\}^n$ using the noisy sign mapping:

$$x_i = \text{sign}(a_i - z_i), \tag{1}$$

where z_i are real-valued independent noises with a fixed cdf F and a_i are (input-dependent) parameters. Equivalently to (1), we can say that x_i follows $\{-1, 1\}$ valued Bernoulli distribution with probability $p(x_i{=}1) = \mathbb{P}(a_i{-}z_i \geq 0) = \mathbb{P}(z_i \leq a_i) = F(a_i)$. The noise cdf F will play an important role in understanding different schemes. For logistic noise, its cdf F is the logistic sigmoid function σ.

Derivation. Straight-through method was first proposed empirically [25,32] in the context of stochastic autoencoders, highly relevant to date [*e.g.* 16]. In contrast to more recent works applying variants of deterministic ST methods, these earlier works considered stochastic networks. It turns out that in this context it is possible to derive ST estimators exactly in the same form as originally proposed by Hinton. This is why we will first derive, using observations of [48,54], analyze and verify it for stochastic autoencoders.

Algorithm 1: Straight-Through-Activations

```
/* a: preactivation           */
/* F: injected noise cdf      */
/* x ∈ {−1,1}^n               */
1 Forward( a )
2   p = F(a);
3   return
      x ∼ 2Bernoulli(p) − 1;
4 Backward( dL/dx )
5   return
      dL/da ≡ 2diag(F'(a)) dL/dx;
```

Algorithm 2: Straight-Through-Weights

```
/* η: latent weights          */
/* F: weight noise cdf        */
/* w ∈ {−1,1}^d               */
1 Forward( η )
2   p = F(η);
3   return
      w ∼ 2Bernoulli(p) − 1;
4 Backward( dL/dw )
5   return dL/dη ≡ 2 dL/dw;
```

Let $\boldsymbol{y}$ denote observed variables. The *encoder network*, parametrized by $\boldsymbol{\phi}$, computes logits $\boldsymbol{a}(\boldsymbol{y};\boldsymbol{\phi})$ and samples a binary latent state $\boldsymbol{x}$ via (1). As noises $\boldsymbol{z}$ are independent, the conditional distribution of hidden states given observations $p(\boldsymbol{x}|\boldsymbol{y};\boldsymbol{\phi})$ factors as $\prod_i p(x_i|\boldsymbol{y};\boldsymbol{\phi})$. The *decoder* reconstructs observations with $p^{\text{dec}}(\boldsymbol{y}|\boldsymbol{x};\boldsymbol{\theta})$—another neural network parametrized by $\boldsymbol{\theta}$. The autoencoder reconstruction loss is defined as

$$\mathbb{E}_{\boldsymbol{y}\sim\text{data}}\big[\mathbb{E}_{\boldsymbol{x}\sim p(\boldsymbol{x}|\boldsymbol{y};\boldsymbol{\phi})}[-\log p^{\text{dec}}(\boldsymbol{y}|\boldsymbol{x};\boldsymbol{\theta})]\big]. \tag{2}$$

The main challenge is in estimating the gradient w.r.t. the encoder parameters $\boldsymbol{\phi}$ (differentiation in $\boldsymbol{\theta}$ can be simply taken under the expectation). The problem for a fixed observation $\boldsymbol{y}$ takes the form

$$\tfrac{\partial}{\partial\boldsymbol{\phi}}\mathbb{E}_{\boldsymbol{x}\sim p(\boldsymbol{x};\boldsymbol{\phi})}[\mathcal{L}(\boldsymbol{x})] = \tfrac{\partial}{\partial\boldsymbol{\phi}}\mathbb{E}_{\boldsymbol{z}}[\mathcal{L}(\text{sign}(\boldsymbol{a}-\boldsymbol{z}))], \tag{3}$$

where $p(\boldsymbol{x};\boldsymbol{\phi})$ is a shorthand for $p(\boldsymbol{x}|\boldsymbol{y};\boldsymbol{\phi})$ and $\mathcal{L}(\boldsymbol{x}) = -\log p^{\text{dec}}(\boldsymbol{y}|\boldsymbol{x};\boldsymbol{\theta})$. The reparametrization trick, *i.e.*, to draw one sample of $\boldsymbol{z}$ in (3) and differentiate $\mathcal{L}(\text{sign}(\boldsymbol{a}-\boldsymbol{z}))$ fails: since the loss as a function of $\boldsymbol{a}$ and $\boldsymbol{z}$ is not *continuously differentiable we cannot interchange the gradient and the expectation in* $\boldsymbol{z}$[1]. If we nevertheless attempt the interchange, we obtain that the gradient of $\text{sign}(\boldsymbol{a}-\boldsymbol{z})$ is zero as well as its expectation. Instead, the following steps lead to an unbiased low-variance estimator. From the LHS of (3) we express the derivative as

$$\tfrac{\partial}{\partial\boldsymbol{\phi}}\textstyle\sum_{\boldsymbol{x}}\big(\prod_i p(x_i;\boldsymbol{\phi})\big)\mathcal{L}(\boldsymbol{x}) = \sum_{\boldsymbol{x}}\sum_i\big(\prod_{i'\neq i} p(x_{i'};\boldsymbol{\phi})\big)\big(\tfrac{\partial}{\partial\boldsymbol{\phi}}p(x_i;\boldsymbol{\phi})\big)\mathcal{L}(\boldsymbol{x}). \tag{4}$$

Then we apply *derandomization* [40, ch.8.7], which performs summation over x_i holding the rest of the state $\boldsymbol{x}$ fixed. Because x_i takes only two values, we have

$$\begin{aligned}\textstyle\sum_{x_i}\tfrac{\partial p(x_i;\boldsymbol{\phi})}{\partial\boldsymbol{\phi}}\mathcal{L}(\boldsymbol{x}) &= \tfrac{\partial p(x_i;\boldsymbol{\phi})}{\partial\boldsymbol{\phi}}\mathcal{L}(\boldsymbol{x}) + \tfrac{\partial(1-p(x_i;\boldsymbol{\phi}))}{\partial\boldsymbol{\phi}}\mathcal{L}(\boldsymbol{x}_{\downarrow i})\\ &= \tfrac{\partial}{\partial\boldsymbol{\phi}}p(x_i;\boldsymbol{\phi})\big(\mathcal{L}(\boldsymbol{x})-\mathcal{L}(\boldsymbol{x}_{\downarrow i})\big),\end{aligned} \tag{5}$$

[1] The conditions allow to apply Leibniz integral rule to exchange derivative and integral. Other conditions may suffice, *e.g.*, when using weak derivatives [17].

where $\boldsymbol{x}_{\downarrow i}$ denotes the full state vector $\boldsymbol{x}$ with the sign of x_i flipped. Since this expression is now invariant of x_i, we can multiply it with $1 = \sum_{x_i} p(x_i; \boldsymbol{\phi})$ and express the gradient (4) in the form:

$$\begin{aligned}\sum_i \sum_{\boldsymbol{x}_{\neg i}} \big(\prod_{i' \neq i} p(x_{i'}; \boldsymbol{\phi})\big) \sum_{x_i} p(x_i; \boldsymbol{\phi}) \tfrac{\partial p(x_i;\boldsymbol{\phi})}{\partial \boldsymbol{\phi}} \big(\mathcal{L}(\boldsymbol{x}) - \mathcal{L}(\boldsymbol{x}_{\downarrow i})\big) \\ \sum_{\boldsymbol{x}} \big(\prod_{i'} p(x_{i'}; \boldsymbol{\phi})\big) \sum_i \tfrac{\partial p(x_i;\boldsymbol{\phi})}{\partial \boldsymbol{\phi}} \big(\mathcal{L}(\boldsymbol{x}) - \mathcal{L}(\boldsymbol{x}_{\downarrow i})\big) \\ = \mathbb{E}_{\boldsymbol{x} \sim p(\boldsymbol{x};\boldsymbol{\phi})} \sum_i \tfrac{\partial p(x_i,\boldsymbol{\phi})}{\partial \boldsymbol{\phi}} \big(\mathcal{L}(\boldsymbol{x}) - \mathcal{L}(\boldsymbol{x}_{\downarrow i})\big),\end{aligned} \tag{6}$$

where $\boldsymbol{x}_{\neg i}$ denotes all states excluding x_i. To obtain an unbiased estimate, it suffices to take one sample $\boldsymbol{x} \sim p(\boldsymbol{x}; \boldsymbol{\phi})$ and compute the sum in i in (6). This estimator is known as *local expectations* [53] and coincides in this case with GO-gradient [14], RAM [54] and PSA [48].

However, evaluating $\mathcal{L}(\boldsymbol{x}_{\downarrow i})$ for all i may be impractical. A huge simplification is obtained if we assume that the change of the loss $\mathcal{L}$ when only a single latent bit x_i is changed can be approximated via linearization. Assuming that $\mathcal{L}$ is defined as a differentiable mapping $\mathbb{R}^n \to \mathbb{R}$ (*i.e.*, that the loss is built up of arithmetic operations and differentiable functions), we can approximate

$$\mathcal{L}(\boldsymbol{x}) - \mathcal{L}(\boldsymbol{x}_{\downarrow i}) \approx 2x_i \tfrac{\partial \mathcal{L}(\boldsymbol{x})}{\partial x_i}, \tag{7}$$

where we used the identity $x_i - (-x_i) = 2x_i$. Expanding the derivative of conditional density $\frac{\partial}{\partial \boldsymbol{\phi}} p(x_i; \boldsymbol{\phi}) = x_i F'(a_i(\boldsymbol{\phi})) \frac{\partial}{\partial \boldsymbol{\phi}} a_i(\boldsymbol{\phi})$, we obtain

$$\tfrac{\partial p(x_i,\boldsymbol{\phi})}{\partial \boldsymbol{\phi}} \big(\mathcal{L}(\boldsymbol{x}) - \mathcal{L}(\boldsymbol{x}_{\downarrow i})\big) \approx 2F'(a_i(\boldsymbol{\phi})) \tfrac{\partial a_i(\boldsymbol{\phi})}{\partial \boldsymbol{\phi}} \tfrac{\partial \mathcal{L}(\boldsymbol{x})}{\partial x_i}. \tag{8}$$

If we now define that $\frac{\partial x_i}{\partial a_i} \equiv 2F'(a_i)$, the summation over i in (6) with the approximation (8) can be written in the form of a chain rule:

$$\sum_i 2F'(a_i(\boldsymbol{\phi})) \tfrac{\partial a_i(\boldsymbol{\phi})}{\partial \boldsymbol{\phi}} \tfrac{\partial \mathcal{L}(\boldsymbol{x})}{\partial x_i} = \sum_i \tfrac{\partial \mathcal{L}(\boldsymbol{x})}{\partial x_i} \tfrac{\partial x_i}{\partial a_i} \tfrac{\partial a_i(\boldsymbol{\phi})}{\partial \boldsymbol{\phi}}. \tag{9}$$

To clarify, the estimator is already defined by the LHS of (9). We simply want to compute this expression by (ab)using the standard tools, and this is the sole purpose of introducing $\frac{\partial x_i}{\partial a_i}$. Indeed the RHS of (9) is a product of matrices that would occur in standard backpropagation. We thus obtained ST algorithm Algorithm 1. We can observe that *it matches exactly to the one described by Hinton* [25]*: to sample on the forward pass and use the derivative of the noise cdf on the backward pass*, up to the multiplier 2 which occurred due to the use of ± 1 encoding for $\boldsymbol{x}$.

2.1 Analysis

Next we study properties of the derived ST algorithm and its relation to empirical variants. We will denote a modification of Algorithm 1 that does not use sampling in Line 3, but instead computes $x = \text{sign}(a)$, a *deterministic ST*; and a modification that uses derivative of some other function G instead of F in Line 5 as *using a proxy G*.

Invariances. Observe that binary activations (and hence the forward pass) stay invariant under transformations: $\text{sign}(a_i - z_i) = \text{sign}(T(a_i) - T(z_i))$ for any strictly monotone mapping T. Consistently, *the ST gradient by Algorithm 1 is also invariant to* T. In contrast, empirical straight-through approaches, in which the derivative proxy is hand-designed, fail to maintain this property. In particular, rescaling the proxy leads to different estimators.

Furthermore, when applying transform $T = F$ (the noise cdf), the backpropagation rule in line 5 of Algorithm 1 becomes equivalent to using the identity proxy. Hence we see that a common description of ST in the literature as "to back-propagate through the hard threshold function as if it had been the identity function" is also correct, *but only for the case of uniform noise* in $[-1, 1]$. Otherwise, and especially so for deterministic ST, this description is ambiguous because the resulting gradient estimator crucially depends on what transformations were applied under the hard threshold.

ST Variants. Using the invariance property, many works applying randomized ST estimators are easily seen to be equivalent to Algorithm 1: [16,44,49]. Furthermore, using different noise distributions for $\boldsymbol{z}$, we can obtain correct ST analogues for common choices of sign proxies used in empirical ST works as shown in Fig. 1 (c–e). In our framework they correspond to the choice of parametrization of the conditional Bernoulli distribution, which should be understood similarly to how a neural network can be parametrized in different ways.

If the "straight-through" idea is applied informally, however, this may lead to confusion and poor performance. The most cited reference for the ST estimator is Bengio et al. [5]. However, [5, Eq. 13] defines in fact the identity ST variant, incorrectly attributing it to Hinton (see Appendix A). We will show this variant to be less accurate for hidden units, both theoretically and experimentally. Pervez et al. [42] use ± 1 binary encoding but apply ST estimator without coefficient 2. When such estimator is used in VAE, where the gradient of the prior KL divergence is computed analytically, it leads to a significant bias of the total gradient towards the prior. In Fig. 2 we illustrate that the difference in performance may be substantial. We analyze other techniques introduced in FouST in more detail in [47]. An inappropriate scaling by a factor of 2 can be as well detrimental in deep models, where the factor would be applied multiple times (in each layer).

Bias Analysis. Given a rather crude linearization involved, it is indeed hard to obtain fine theoretical guarantees about the ST method. We propose an analysis targeting understanding the effect of common empirical variants and understanding conditions under which the estimator becomes more accurate. The respective formal theorems are given in Appendix B.

I) When ST is unbiased? As we used linearization as the only biased approximation, it follows that *Algorithm 1 is unbiased if the objective function* $\mathcal{L}$ *is multilinear in* $\boldsymbol{x}$. A simple counter-example, where ST is biased, is $\mathcal{L}(x) = x^2$. In this case the expected value of the loss is 1, independently of a that determines x; and the true gradient is zero. However the expected ST gradient is $\mathbb{E}[2F'(a)2x] = 4F'(a)(2F(a) - 1)$, which may be positive or negative depend-

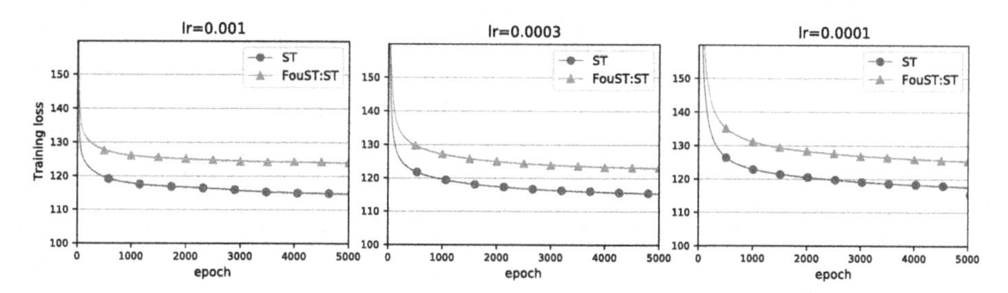

Fig. 2. Training VAE on MNIST, closely following experimental setup [42]. The plots show training loss (negative ELBO) during epochs for different learning rates. The variant of ST algorithm used [42] is misspecified because of the scaling factor and performs substantially worse at for all learning rates. Full experiment specification is given in Appendix D.1.

ing on a. On the other hand, any function of binary variables has an equivalent multilinear expression. In particular, if we consider $\mathcal{L}(\boldsymbol{x}) = \|\boldsymbol{W}\boldsymbol{x}-\boldsymbol{y}\|^2$, analyzed by Yin et al. [58], then $\tilde{\mathcal{L}}(\boldsymbol{x}) = \|\boldsymbol{W}\boldsymbol{x}-\boldsymbol{y}\|^2 - \sum_i x_i^2\|\boldsymbol{W}_{:,i}\|^2 + \sum_i \|\boldsymbol{W}_{:,i}\|^2$ *coincides with* $\mathcal{L}$ *on all binary configurations and is multilinear.* It follows that ST applied to $\tilde{\mathcal{L}}$ gives an unbiased gradient estimate of $\mathbb{E}[\mathcal{L}]$, an immediate improvement compared to [58]. In the special case when $\mathcal{L}$ is linear in $\boldsymbol{x}$, the ST estimator is not only unbiased but has a zero variance, *i.e.*, it is exact.

II) How does using a mismatched proxy in Line 5 of Algorithm 1 affect the gradient in $\boldsymbol{\phi}$? Since $\operatorname{diag}(F')$ occurs in the backward chain, we call estimators that use some matrix $\boldsymbol{\Lambda}$ instead of $\operatorname{diag}(F')$ as *internally rescaled.* We show that *for any* $\boldsymbol{\Lambda} \succcurlyeq 0$, *the expected rescaled estimator has non-negative scalar product with the expected original estimator.* Note that this is not completely obvious as the claim is about the final gradient in the model parameters $\boldsymbol{\phi}$ (*e.g.*, weights of the encoder network in the case of autoencoders). However, if the ST gradient by Algorithm 1 is biased (when $\mathcal{L}$ is not multi-linear) but is nevertheless an ascent direction in expectation, the expected rescaled estimator may fail to be an ascent direction, *i.e.*, to have a positive scalar product with the true gradient.

III) When does ST gradient provide a valid ascent direction? Assuming that all partial derivatives $g_i(\boldsymbol{x}) = \frac{\partial \mathcal{L}(\boldsymbol{x})}{\partial x_i}$ are L-Lipschitz continuous for some L, we can show that *the expected ST gradient is an ascent direction for any network if and only if* $\left|\mathbb{E}_{\boldsymbol{x}}[g_i(\boldsymbol{x})]\right| > L$ *for all* i.

IV) Can we decrease the bias? Assume that the loss function is applied to a linear transform of Bernoulli variables, *i.e.*, takes the form $\mathcal{L}(\boldsymbol{x}) = \ell(\boldsymbol{W}\boldsymbol{x})$. A typical initialization uses random $\boldsymbol{W}$ normalized by the size of the fan-in, *i.e.*, such that $\|\boldsymbol{W}_{k,:}\|_2 = 1\ \forall k$. In this case *the Lipschitz constant of gradients of* $\mathcal{L}$ *scales as* $O(1/\sqrt{n})$, *where* n *is the number of binary variables.* Therefore, *using more binary variables decreases the bias, at least at initialization.*

V) Does deterministic ST give an ascent direction? Let $\boldsymbol{g}^*$ be the deterministic ST gradient for the state $\boldsymbol{x}^* = \operatorname{sign}(\boldsymbol{a})$ and $p^* = p(\boldsymbol{x}^*|\boldsymbol{a})$ be its probability. *We show that deterministic ST gradient forms a positive scalar product* *with the expected ST gradient if* $|g^*_i| \geq 2L(1 - p^*)$ *and with the true gradient if* $|g^*_i| \geq 2L(1-p^*)+L$. From this we conclude that deterministic ST positively correlates with the true gradient when $\mathcal{L}$ is multilinear, improves with the number of hidden units in the case described by IV and approaches expected stochastic ST as units learn to be deterministic so that the factor $(1 - p^*)$ decreases.

Deep ST. So far we derived and analyzed ST for a single layer model. It turns out that simply applying Algorithm 1 in each layer of a deep model with conditional Bernoulli units gives the correct extension for this case. We will not focus on deriving deep ST here, but remark that it can be derived rigorously by chaining derandomization and linearization steps, discussed above, for each layer [48]. In particular, [48] show that ST can be obtained by making additional linearizations in their (more accurate) PSA method. The insights from the derivation are twofold. First, since derandomization is performed recurrently, the variance for deep layers is significantly reduced. Second, we know which approximations contribute to the bias, they are indeed the linearizations of all conditional Bernoulli probabilities in all layers and of the loss function as a function of the last Bernoulli layer. We may expect that using more units, similarly to property IV, would improve linearizing approximations of intermediate layers increasing the accuracy of deep ST gradient.

3 Latent Weights Do Exist!

Responding to the work "Latent weights do not exist: Rethinking binarized neural network optimization" [24] and the lack of formal basis to introduce latent weights in the literature (*e.g.*, [27]), we show that such weights can be formally defined in SBNs and that several empirical update rules do in fact correspond to sound optimization schemes: projected gradient descent, mirror descent, variational Bayesian learning.

Let $\boldsymbol{w}$ be ±1-Bernoulli weights with $p(w_i{=}1) = \theta_i$, let $\mathcal{L}(\boldsymbol{w})$ be the loss function for a fixed training input. Consistently with the model for activations (1), we can define $w_i = \text{sign}(\eta_i - z_i)$ in order to model weights w_i using parameters $\eta_i \in \mathbb{R}$ which we will call *latent weights*. It follows that $\theta_i = F_z(\eta_i)$. We need to tackle two problems in order to optimize $\mathbb{E}_{\boldsymbol{w}\sim p(\boldsymbol{w}\,|\,\boldsymbol{\theta})}[\mathcal{L}(\boldsymbol{w})]$ in probabilities $\boldsymbol{\theta}$: i) how to estimate the gradient and ii) how to handle constraints $\boldsymbol{\theta} \in [0, 1]^m$.

Projected Gradient. A basic approach to handle constraints is the *projected gradient descent*:

$$\boldsymbol{\theta}^{t+1} := \text{clip}(\boldsymbol{\theta}^t - \varepsilon \boldsymbol{g}^t, 0, 1), \tag{10}$$

where $\boldsymbol{g}^t = \nabla_{\boldsymbol{\theta}} \mathbb{E}_{\boldsymbol{w}\sim p(\boldsymbol{w}\,|\,\boldsymbol{\theta}^t)}[\mathcal{L}(\boldsymbol{w})]$ and $\text{clip}(\boldsymbol{x}, a, b) := \max(\min(\boldsymbol{x}, b), a)$ is the projection. Observe that for the uniform noise distribution on $[-1, 1]$ with $F(z) = \text{clip}(\frac{z+1}{2}, 0, 1)$, we have $\theta_i = p(w_i{=}1) = F(\eta_i) = \text{clip}(\frac{\eta_i+1}{2}, 0, 1)$. Because this F is linear on $[-1, 1]$, the update (10) can be equivalently reparametrized in $\boldsymbol{\eta}$ as

$$\boldsymbol{\eta}^{t+1} := \text{clip}(\boldsymbol{\eta}^t - \varepsilon' \boldsymbol{h}^t, -1, 1), \tag{11}$$

where $\boldsymbol{h}^t = \nabla_{\boldsymbol{\eta}} \mathbb{E}_{\boldsymbol{w} \sim p(\boldsymbol{w}|F(\boldsymbol{\eta}))}[\mathcal{L}(\boldsymbol{w})]$ and $\varepsilon' = 4\varepsilon$. The gradient in the latent weights, $\boldsymbol{h}^t$, can be estimated by Algorithm 1 and simplifies by expanding $2F' = 1$. We obtained that *the emperically proposed method of Hubara et al.* [27, Alg.1] *with stochastic rounding and with real-valued weights identified with* $\boldsymbol{\eta}$ *is equivalent to PGD on* $\boldsymbol{\eta}$ *with constraints* $\eta \in [-1, 1]^m$ *and ST gradient by Algorithm* 1.

Mirror Descent. As an alternative approach to handle constraints $\boldsymbol{\theta} \in [0, 1]^m$, we study the application of mirror descent (MD) and connect it with the identity ST update variants. A step of MD is found by solving the following proximal problem:

$$\boldsymbol{\theta}^{t+1} = \min_{\boldsymbol{\theta}} \left[\langle \boldsymbol{g}^t, \boldsymbol{\theta} - \boldsymbol{\theta}^t \rangle + \tfrac{1}{\varepsilon} D(\boldsymbol{\theta}, \boldsymbol{\theta}^t) \right]. \tag{12}$$

The divergence term $\frac{1}{\varepsilon} D(\boldsymbol{\theta}, \boldsymbol{\theta}^t)$ weights how much we trust the linear approximation $\langle \boldsymbol{g}^t, \boldsymbol{\theta} - \boldsymbol{\theta}^t \rangle$ when considering a step from $\boldsymbol{\theta}^t$ to $\boldsymbol{\theta}$. When the gradient is stochastic we speak of *stochastic mirror descent* (SMD) [3,59]. A common choice of divergence to handle probability constraints is the KL-divergence $D(\theta_i, \theta_i^t) = \mathrm{KL}(\mathrm{Ber}(\theta_i), \mathrm{Ber}(\theta_i^t)) = \theta_i \log(\frac{\theta_i}{\theta_i^t}) + (1 - \theta_i) \log(\frac{1-\theta_i}{1-\theta_i^t})$. Solving for a stationary point of (12) gives

$$0 = g_i^t + \tfrac{1}{\varepsilon} \big(\log(\tfrac{\theta_i}{1-\theta_i}) - \log(\tfrac{\theta_i^t}{1-\theta_i^t}) \big). \tag{13}$$

Observe that when $F = \sigma$ we have $\log(\frac{\theta_i}{1-\theta_i}) = \eta_i$. Then the MD step can be written in the well-known convenient form using the latent weights $\boldsymbol{\eta}$ (natural parameters of Bernoulli distribution):

$$\boldsymbol{\theta}^t := \sigma(\boldsymbol{\eta}^t); \qquad \boldsymbol{\eta}^{t+1} := \boldsymbol{\eta}^t - \varepsilon \nabla_{\boldsymbol{\theta}} \mathcal{L}(\boldsymbol{\theta}^t). \tag{14}$$

We thus have obtained the rule where on the forward pass $\boldsymbol{\theta} = \sigma(\boldsymbol{\eta})$ defines the sampling probability of $\boldsymbol{w}$ and on the backward pass the derivative of σ, that otherwise occurs in Line 5 of Algorithm 1, *is bypassed exactly as if the identity proxy was used.* We define such ST rule for optimization in weights as Algorithm 2. Its correctness is not limited to logistic noise. We show that for any strictly monotone noise distribution F there is a corresponding divergence function D:

Proposition 1. *Common SGD in latent weights* $\boldsymbol{\eta}$ *using the* identity straight-through-weights *Algorithm 2 implements SMD in the weight probabilities* $\boldsymbol{\theta}$ *with the divergence corresponding to* F.

Proof in Appendix C. Proposition 1 reveals that although Bernoulli weights can be modeled the same way as activations using the injected noise model $\boldsymbol{w} = \mathrm{sign}(\boldsymbol{\eta} - \boldsymbol{z})$, *the noise distribution* F *for weights correspond to the choice of the optimization proximity scheme.*

Despite generality of Proposition 1, we view the KL divergence as a more reliable choice in practice. Azizan et al. [3] have shown that the optimization with SMD has an inductive bias to find the closest solution to the initialization point as measured by the divergence used in MD, which has a strong impact on generalization. This suggests that MD with KL divergence will prefer higher entropy solutions, making more diverse predictions. It follows that SGD on latent weights with logistic noise and identity straight-through Algorithm 2 enjoys the same properties.

Variational Bayesian Learning. Extending the results above, we study the variational Bayesian learning formulation and show the following:

Proposition 2. *Common SGD in latent weights* $\boldsymbol{\eta}$ with a weight decay and identity straight-through-weights *Algorithm 2 is equivalent to optimizing a factorized variational approximation to the weight posterior* $p(\boldsymbol{w}|data)$ *using a composite SMD method.*

Proof in Appendix C.2. As we can see, powerful and sound learning techniques can be obtained in a form of simple update rules using identity straight-through estimators. Therefore, identity-ST is fully rehabilitated in this context.

4 Experiments

Stochastic Autoencoders. Previous work has demonstrated that Gumbel-Softmax (biased) and ARM (unbiased) estimators give better results than ST on training variational autoencoders with Bernoulli latents [16,29,57]. However, only the test performance was revealed to readers. We investigate in more detail what happens during training. Except of studying the training loss under the same training setup, we measure the gradient approximation accuracy using ARM with 1000 samples as the reference.

We train a simple yet realistic variant of stochastic autoencoder for the task of text retrieval with binary representation on *20newsgroups* dataset. The autoencoder is trained by minimizing the reconstruction loss (2). Please refer to Appendix D.2 for full specification of the model and experimental setup.

For each estimator we perform the following protocol. First, we train the model with this estimator using Adam with $lr = 0.001$ for 1000 epochs. We then switch the estimator to ARM with 10 samples and continue training for 500 more epochs (denoted as ARM-10 correction phase). Figure 3 top shows the training performance for different number of latent bits n. It is seen (esp. for 8 and 64 bits) that some estimators (esp. ST and det_ST) appear to make no visible progress, and even increase the loss, while switching them to ARM makes a rapid improvement. Does it mean that these estimators are bad and ARM is very good? An explanation of this phenomenon is offered in Fig. 5. The rapid improvement by ARM is possible because these estimators have accumulated a significant bias due to a systematic error component, which nevertheless can be easily corrected by an unbiased estimator.

To measure the bias and alignment of directions, as theoretically analyzed in Sect. 2.1, we evaluate different estimators at the same parameter points located

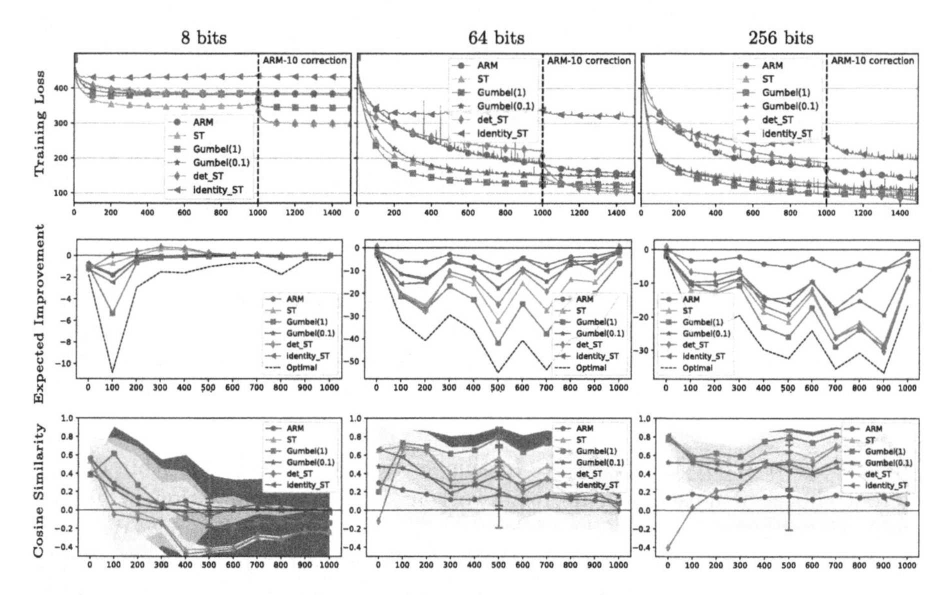

Fig. 3. Comparison of the training performance and gradient estimation accuracy for a stochastic autoencoder with different number of latent Bernoulli units (bits). *Training Loss:* each estimator is applied for 1000 epochs and then switched to ARM-10 in order to correct the accumulated bias. *Expected improvement*: lower is better (measures expected change of the loss), the dashed line shows the maximal possible improvement knowing the true gradient. *Cosine similarity*: higher is better, close to 1 means that the direction is accurate while below 0 means the estimated gradient is not an ascent direction; error bars indicate empirical 70% confidence intervals using 100 trials.

along the learning trajectory of the reference ARM estimator. At each such point we estimate the true gradient $\boldsymbol{g}$ by ARM-1000. To measure the quality of a candidate 1-sample estimator $\tilde{\boldsymbol{g}}$ we compute the *expected cosine similarity* and the *expected improvement*, defined respectively as:

$$\mathrm{ECS} = \mathbb{E}\Big[\langle \boldsymbol{g}, \tilde{\boldsymbol{g}}\rangle / (\|\boldsymbol{g}\|\|\tilde{\boldsymbol{g}}\|)\Big], \qquad \mathrm{EI} = -\mathbb{E}[\langle \boldsymbol{g}, \tilde{\boldsymbol{g}}\rangle] / \sqrt{\mathbb{E}[\|\tilde{\boldsymbol{g}}\|^2]}, \tag{15}$$

The expectations are taken over 100 trials and all batches. A detailed explanation of these metrics is given in Appendix D.2. These measurements, displayed in Fig. 3 for different bit length, clearly show that with a small bit length biased estimators consistently run into producing wrong directions. *Identity ST and deterministic ST clearly introduce an extra bias to ST.* However, when we increase the number of latent bits, the accuracy of all biased estimators improves, confirming our analysis **IV**, **V**.

The practical takeaways are as follows: 1) biased estimators may perform significantly better than unbiased but might require a correction of the systematically accumulated bias; 2) with more units the ST approximation clearly improves and the bias has a less detrimental effect, requiring less correction; 3) Algorithm 1 is more accurate than other ST variants in estimating the true gradient.

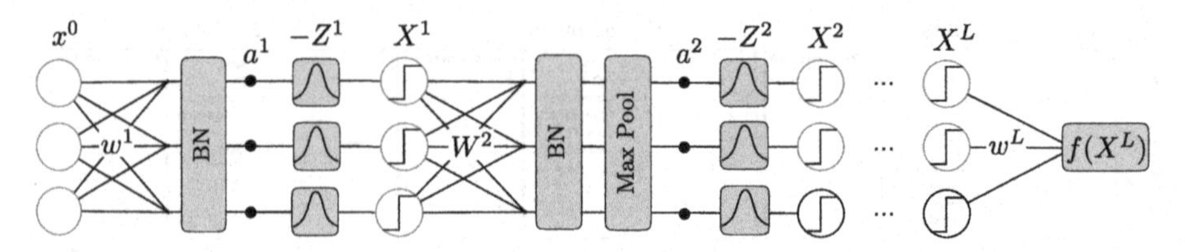

Fig. 4. Stochastic Binary Network: first and last layer have real-valued weights. BN layers have real-valued scale and bias parameters that can adjust scaling of activations relative to noise. Z are independent injected noises with a chosen distribution. Binary weights W_{ij} are random ± 1 Bernoulli(θ_{ij}) with learnable probabilities θ_{ij}. In experiments we consider SBN with a convolutional architecture same as [15,27]: (2×128C3) – MP2 – (2×256C3) – MP2 – (2×512C3) – MP2 – (2×1024FC) – 10FC – softmax.

Classification with Deep SBN. In this section we verify Algorithm 1 with different choice of noises in a deep network and verify optimization in binary weight probabilities using SGD on latent weights with Algorithm 2. We consider CIFAR-10 dataset and use the SBN model illustrated in Fig. 4. The SBN model, its initialization and the full learning setup is detailed in Appendix D.3. We trained this SBN with three choices of noise distributions corresponding to proxies used by prior work as in Fig. 1 (c–e). Table 1 shows the test results in comparison with baselines.

We see that training with different choices of noise distributions, corresponding to different ST rules, all achieves similar results. This is in contrast to empirical studies advocating specific proxies and is allowed by the consistency of the model, initialization and training. The identity ST applied to weights, implementing SMD updates, works well. Comparing to empirical ST baselines (all except Peters & Welling), we see that there is no significant difference in the 'det' column indicating that our derived ST method is on par with the well-guessed baselines. If the same networks are tested in the stochastic mode ('10-sample' column), there is a clear boost of performance, indicating an advantage of SBN models. Out of the two experiments of Hubara et al., randomized training (rand.) also appears better confirming advantage of stochastic ST. In the stochastic mode, there is a small gap to Peters & Welling, who use a different estimation method and pretraining. Pretraining a real valued network also seem important, *e.g.*, [19] report 91.7% accuracy with VGG-Small using pretraining and a smooth transition from continuous to binarized model. When our method is applied with an initialization from a pretrained model, improved results (92.6% 10-sample test accuracy) can be obtained with even a smaller network [35]. There are however even more superior results in the literature, *e.g.*, using neural architecture search with residual real connections, advanced data augmentation techniques and model distillation [10] achieve 96.1%.

The takeaway message here is that ST can be considered in the context of deep SBN models as a simple and robust method if the estimator matches the model and is applied correctly. Since we achieve experimentally near 100% training accuracy in all cases, the optimization fully succeeds and thus the bias of ST is tolerable.

Table 1. Test accuracy for different methods on CIFAR-10 with the same/similar architecture. SBN can be tested either with zero noises (*det*) or using an ensemble of several samples (we use *10-sample*). Standard deviations are given w.r.t. to 4 trials with random initialization. The two quotations for Hubara et al. [27] refer to their result with Torch7 implementation using randomized Htanh and Theano implementation using deterministic Htanh, respectively.

Method	det	10-sample
Stochastic training		
Our SBN, logistic noise	89.6 ± 0.1	90.6 ± 0.2
Our SBN, uniform noise	89.7 ± 0.2	90.5 ± 0.2
Our SBN, triangular noise	89.5 ± 0.2	90.0 ± 0.3
Hubara et al. [27] (rand.)	89.85	-
Peters & Welling [43]	88.61	16-sample: 91.2
Deterministic training		
Rastegari et al. [45]	89.83	-
Hubara et al. [27] (det.)	88.60	-

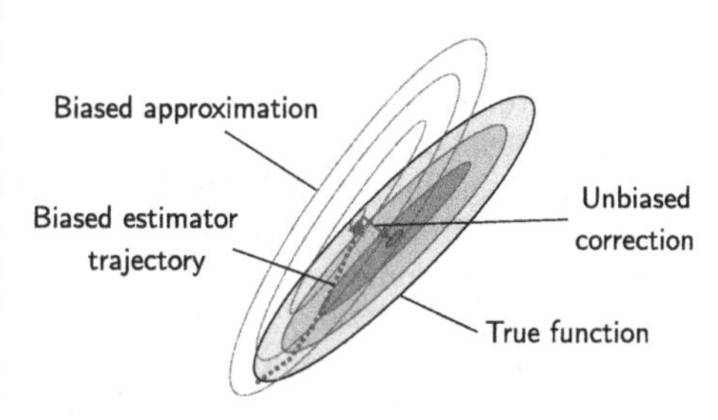

Fig. 5. Schematic explanation of the optimization process using a biased estimator followed by a correction with an unbiased estimator. Initially, the biased estimator makes good progress, but then the value of the true loss function may start growing while the optimization steps nevertheless come closer to the optimal location in the parameter space.

5 Conclusion

We have put many ST methods on a solid basis by deriving and explaining them from the first principles in one framework. It is well-defined what they estimate and what the bias means. We obtained two different main estimators for propagating activations and weights, bringing the understanding which function they have, what approximations they involve and what are the limitations imposed by these approximations. The resulting methods in all cases are strikingly simple, no wonder they have been first discovered empirically long ago. We showed how our theory leads to a useful understanding of bias properties and to reasonable choices that allow for a more reliable application of these methods. We hope that researchers will continue to use these simple techniques, now with less guesswork and obscurity, as well as develop improvements to them.

References

1. Ajanthan, T., Gupta, K., Torr, P.H., Hartley, R., Dokania, P.K.: Mirror descent view for neural network quantization. arXiv preprint arXiv:1910.08237 (2019)
2. Alizadeh, M., Fernandez-Marques, J., Lane, N.D., Gal, Y.: An empirical study of binary neural networks' optimisation. In: ICLR (2019)
3. Azizan, N., Lale, S., Hassibi, B.: A study of generalization of stochastic mirror descent algorithms on overparameterized nonlinear models. In: ICASSP, pp. 3132–3136 (2020)

4. Bai, Y., Wang, Y.-X., Liberty, E.: ProxQuant: quantized neural networks via proximal operators. In: ICLR (2019)
5. Bengio, Y., Léonard, N., Courville, A.: Estimating or propagating gradients through stochastic neurons for conditional computation. arXiv preprint arXiv:1308.3432 (2013)
6. Bethge, J., Yang, H., Bornstein, M., Meinel, C.: Back to simplicity: how to train accurate BNNs from scratch? CoRR, abs/1906.08637 (2019)
7. Boros, E., Hammer, P.: Pseudo-Boolean optimization. Discret. Appl. Math. **1–3**(123), 155–225 (2002)
8. Bulat, A., Tzimiropoulos, G.: Binarized convolutional landmark localizers for human pose estimation and face alignment with limited resources. In: ICCV, October 2017
9. Bulat, A., Tzimiropoulos, G., Kossaifi, J., Pantic, M.: Improved training of binary networks for human pose estimation and image recognition. arXiv (2019)
10. Bulat, A., Martinez, B., Tzimiropoulos, G.: BATS: binary architecture search (2020)
11. Bulat, A., Martinez, B., Tzimiropoulos, G.: High-capacity expert binary networks. In: ICLR (2021)
12. Chaidaroon, S., Fang, Y.: Variational deep semantic hashing for text documents. In: SIGIR Conference on Research and Development in Information Retrieval, pp. 75–84 (2017)
13. Cheng, P., Liu, C., Li, C., Shen, D., Henao, R., Carin, L.: Straight-through estimator as projected Wasserstein gradient flow. arXiv preprint arXiv:1910.02176 (2019)
14. Cong, Y., Zhao, M., Bai, K., Carin, L.: GO gradient for expectation-based objectives. In: ICLR (2019)
15. Courbariaux, M., Bengio, Y., David, J.-P.: BinaryConnect: training deep neural networks with binary weights during propagations. In: NeurIPS, pp. 3123–3131 (2015)
16. Dadaneh, S.Z., Boluki, S., Yin, M., Zhou, M., Qian, X.: Pairwise supervised hashing with Bernoulli variational auto-encoder and self-control gradient estimator. arXiv, abs/2005.10477 (2020)
17. Dai, B., Guo, R., Kumar, S., He, N., Song, L.: Stochastic generative hashing. In: ICML 2017, pp. 913–922 (2017)
18. Esser, S.K., et al.: Convolutional networks for fast, energy-efficient neuromorphic computing. Proc. Natl. Acad. Sci. **113**(41), 11441–11446 (2016)
19. Gong, R., et al.: Differentiable soft quantization: bridging full-precision and low-bit neural networks. In: Proceedings of the IEEE/CVF International Conference on Computer Vision (ICCV), October 2019
20. Grathwohl, W., Choi, D., Wu, Y., Roeder, G., Duvenaud, D.: Backpropagation through the void: optimizing control variates for black-box gradient estimation. In: ICLR (2018)
21. Graves, A.: Practical variational inference for neural networks. In: NeurIPS, pp. 2348–2356 (2011)
22. Gregor, K., Danihelka, I., Mnih, A., Blundell, C., Wierstra, D.: Deep autoregressive networks. In: ICML (2014)
23. He, K., Zhang, X., Ren, S., Sun, J.: Delving deep into rectifiers: surpassing human-level performance on ImageNet classification. In: ICCV, pp. 1026–1034 (2015)
24. Helwegen, K., Widdicombe, J., Geiger, L., Liu, Z., Cheng, K.-T., Nusselder, R.: Latent weights do not exist: rethinking binarized neural network optimization. In: NeurIPS, pp. 7531–7542 (2019)

25. Hinton, G.: Lecture 15D - Semantic hashing: 3:05–3:35 (2012). https://www.cs.toronto.edu/~hinton/coursera/lecture15/lec15d.mp4
26. Horowitz, M.: Computing's energy problem (and what we can do about it). In: International Solid-State Circuits Conference Digest of Technical Papers (ISSCC), pp. 10–14 (2014)
27. Hubara, I., Courbariaux, M., Soudry, D., El-Yaniv, R., Bengio, Y.: Binarized neural networks. In: NeurIPS, pp. 4107–4115 (2016)
28. Ioffe, S., Szegedy, C.: Batch normalization: accelerating deep network training by reducing internal covariate shift. In: ICML, vol. 37, pp. 448–456 (2015)
29. Jang, E., Gu, S., Poole, B.: Categorical reparameterization with Gumbel-Softmax. In: ICLR (2017)
30. Khan, E., Rue, H.: Learning algorithms from Bayesian principles. Draft v. 0.7, August 2020
31. Kingma, D.P., Ba, J.: Adam: a method for stochastic optimization. In: ICLR (2015)
32. Krizhevsky, A., Hinton, G.E.: Using very deep autoencoders for content-based image retrieval. In: ESANN (2011)
33. Lin, W., Khan, M.E., Schmidt, M.: Fast and simple natural-gradient variational inference with mixture of exponential-family approximations. In: ICML, vol. 97, June 2019
34. Liu, Z., Wu, B., Luo, W., Yang, X., Liu, W., Cheng, K.-T.: Bi-real net: enhancing the performance of 1-bit CNNs with improved representational capability and advanced training algorithm. In: ECCV, pp. 722–737 (2018)
35. Livochka, A., Shekhovtsov, A.: Initialization and transfer learning of stochastic binary networks from real-valued ones. In: Proceedings of the IEEE/CVF Conference on Computer Vision and Pattern Recognition (CVPR) Workshops (2021)
36. Martínez, B., Yang, J., Bulat, A., Tzimiropoulos, G.: Training binary neural networks with real-to-binary convolutions. In: ICLR (2020)
37. Meng, X., Bachmann, R., Khan, M.E.: Training binary neural networks using the Bayesian learning rule. In: ICML (2020)
38. Nanculef, R., Mena, F.A., Macaluso, A., Lodi, S., Sartori, C.: Self-supervised Bernoulli autoencoders for semi-supervised hashing. CoRR, abs/2007.08799 (2020)
39. Nemirovsky, A.S., Yudin, D.B.: Problem complexity and method efficiency in optimization (1983)
40. Owen, A.B.: Monte Carlo theory, methods and examples (2013)
41. Paszke, A., et al.: Pytorch: an imperative style, high-performance deep learning library. In: NeurIPS, pp. 8024–8035 (2019)
42. Pervez, A., Cohen, T., Gavves, E.: Low bias low variance gradient estimates for Boolean stochastic networks. In: ICML, vol. 119, pp. 7632–7640, 13–18 July 2020
43. Peters, J.W., Welling, M.: Probabilistic binary neural networks. arXiv preprint arXiv:1809.03368 (2018)
44. Raiko, T., Berglund, M., Alain, G., Dinh, L.: Techniques for learning binary stochastic feedforward neural networks. In: ICLR (2015)
45. Rastegari, M., Ordonez, V., Redmon, J., Farhadi, A.: XNOR-Net: ImageNet classification using binary convolutional neural networks. In: Leibe, B., Matas, J., Sebe, N., Welling, M. (eds.) ECCV 2016. LNCS, vol. 9908, pp. 525–542. Springer, Cham (2016). https://doi.org/10.1007/978-3-319-46493-0_32
46. Roth, W., Schindler, G., Fröning, H., Pernkopf, F.: Training discrete-valued neural networks with sign activations using weight distributions. In: European Conference on Machine Learning (ECML) (2019)
47. Shekhovtsov, A.: Bias-variance tradeoffs in single-sample binary gradient estimators. In: GCPR (2021)

48. Shekhovtsov, A., Yanush, V., Flach, B.: Path sample-analytic gradient estimators for stochastic binary networks. In: NeurIPS (2020)
49. Shen, D., et al.: NASH: toward end-to-end neural architecture for generative semantic hashing. In: Proceedings of the 56th Annual Meeting of the Association for Computational Linguistics, ACL 2018, Melbourne, Australia, July 15–20 2018, Volume 1: Long Papers, pp. 2041–2050 (2018)
50. Srivastava, N., Hinton, G., Krizhevsky, A., Sutskever, I., Salakhutdinov, R.: Dropout: a simple way to prevent neural networks from overfitting. JMLR **15**, 1929–1958 (2014)
51. Sun, Z., Yao, A.: Weights having stable signs are important: finding primary subnetworks and kernels to compress binary weight networks (2021)
52. Tang, W., Hua, G., Wang, L.: How to train a compact binary neural network with high accuracy? In: AAAI (2017)
53. Titsias, M.K., Lázaro-Gredilla, M.: Local expectation gradients for black box variational inference. In: NeurIPS, pp. 2638–2646 (2015)
54. Tokui, S., Sato, I.: Evaluating the variance of likelihood-ratio gradient estimators. In: ICML, pp. 3414–3423 (2017)
55. Tucker, G., Mnih, A., Maddison, C.J., Lawson, J., Sohl-Dickstein, J.: REBAR: low-variance, unbiased gradient estimates for discrete latent variable models. In: NeurIPS (2017)
56. Xiang, X., Qian, Y., Yu, K.: Binary deep neural networks for speech recognition. In: INTERSPEECH (2017)
57. Yin, M., Zhou, M.: ARM: augment-REINFORCE-merge gradient for stochastic binary networks. In: ICLR (2019)
58. Yin, P., Lyu, J., Zhang, S., Osher, S., Qi, Y., Xin, J.: Understanding straight-through estimator in training activation quantized neural nets. arXiv preprint arXiv:1903.05662 (2019)
59. Zhang, S., He, N.: On the convergence rate of stochastic mirror descent for nonsmooth nonconvex optimization. arXiv, Optimization and Control (2018)
60. Zhou, S., Wu, Y., Ni, Z., Zhou, X., Wen, H., Zou, Y.: DoReFa-Net: training low bitwidth convolutional neural networks with low bitwidth gradients. arXiv preprint arXiv:1606.06160 (2016)

Bias-Variance Tradeoffs in Single-Sample Binary Gradient Estimators

Alexander Shekhovtsov(✉)

Czech Technical University in Prague, Prague, Czech Republic
shekhole@fel.cvut.cz

Abstract. Discrete and especially binary random variables occur in many machine learning models, notably in variational autoencoders with binary latent states and in stochastic binary networks. When learning such models, a key tool is an estimator of the gradient of the expected loss with respect to the probabilities of binary variables. The straight-through (ST) estimator gained popularity due to its simplicity and efficiency, in particular in deep networks where unbiased estimators are impractical. Several techniques were proposed to improve over ST while keeping the same low computational complexity: Gumbel-Softmax, ST-Gumbel-Softmax, BayesBiNN, FouST. We conduct a theoretical analysis of bias and variance of these methods in order to understand tradeoffs and verify the originally claimed properties. The presented theoretical results allow for better understanding of these methods and in some cases reveal serious issues.

1 Introduction

Binary variables occur in many models of interest. Variational autoencoders (VAE) with binary latent states are used to learn generative models with compressed representations [10,11,22,33] and to learn binary hash codes for text and image retrieval [6,7,20,30]. Neural networks with binary activations and weights are extremely computationally efficient and attractive for embedded applications, in particular pushed forward in the vision research [1–5,8,12,15,17,25,31,34,36]. Training these discrete models is possible via the stochastic relaxation, equivalent to training a Stochastic Binary Networks (SBN) [23,24,26,27,29]. In this relaxation, each binary weight is replaced with a Bernoulli random variable and each binary activation is replaced with a conditional Bernoulli variable. The gradient of the expected loss in the weight probabilities is well defined and SGD optimization can be applied.

The author gratefully acknowledges support by Czech OP VVV project "Research Center for Informatics (CZ.02.1.01/0.0/0.0/16019/0000765)".

Supplementary Information The online version contains supplementary material available at https://doi.org/10.1007/978-3-030-92659-5_8.

C. Bauckhage et al. (Eds.): DAGM GCPR 2021, LNCS 13024, pp. 127–141, 2021.
https://doi.org/10.1007/978-3-030-92659-5_8

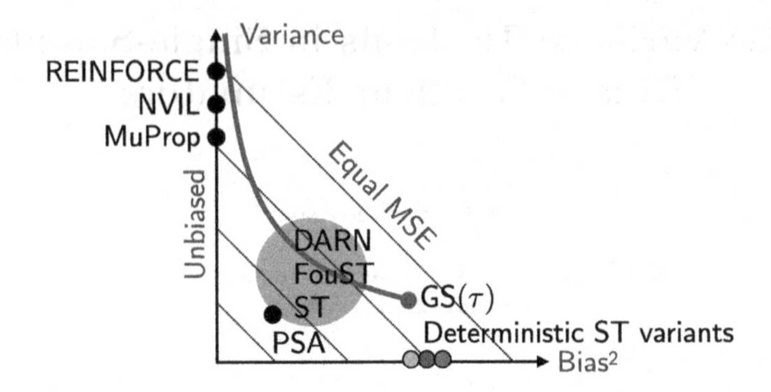

Fig. 1. Schematic illustration of bias-variance tradeoffs (we do not pretend on exactness, but see experimental evaluations in [28,29]; notice that the Mean Squared Error (MSE) is the sum of variance and squared bias). Unbiased methods have a prohibitively high variance for deep models. PSA achieves a significant reduction in variance at a price of a small bias, but has a limited applicability. According to [29], ST estimator can be as accurate as PSA in wide deep models. We analytically study methods in the gray area: GS, DARN and FouST in order to find out whether they can offer a sound improvement over ST. In particular, for GS estimator the figure illustrates its possible tradeoffs when varying the temperature parameter according to the asymptotes we prove.

For the problem of estimating gradient of expectation in probabilities of (conditional) Bernoulli variables, several unbiased estimators were proposed [9,11,19,32,35]. However, in the context of deep SBNs these methods become impractical: MuProp [11] and REINFORCE with baselines [19] have a prohibitively high variance in deep layers [28, Figs. C6, C7] while other methods' complexity grows quadratically with the number of Bernoulli layers. In these cases, biased estimators were more successful in practice: straight-through (ST) [28], Gumbel-Softmax (GS) [13,16] and their variants. In order to approximate the gradient of the expectation these methods use a single sample of all random entities and the derivative of the objective function extended to the real-valued domain. A more accurate PSA method was presented in [29], which has low computation complexity, but applies only to SBNs of classical structure[1] and requires specialized convolutions. Notably, it was experimentally reported [29, Fig. 4] that the baseline ST performs nearly identically to PSA in moderate size SBNs. Figure 1 schematically illustrates the bias-variance tradeoff with different approaches.

Contribution. In this work we analyze theoretical properties of several recent single-sample gradient based methods: GS, ST-GS [13], BayesBiNN [18] and FouST [22]. We focus on clarifying these techniques, studying their limitations and identifying incorrect and over-claimed results. We give a detailed analysis of bias and variance of GS and ST-GS estimators. Next we analyze the application

[1] Feed-forward, with no residual connections and only linear layers between Bernoulli activations.

of GS in BayesBiNN. We show that a correct implementation would result in an extremely high variance. However due to a hidden issue, the estimator in effect reduces to a deterministic straight-through (with zero variance). A long-range effect of this swap is that BayesBiNN fails to solve the variational Bayesian learning problem as claimed. FouST [22] proposed several techniques for *lowering* bias and variance of the baseline ST estimator. We show that the baseline ST estimator was applied incorrectly and that some of the proposed improvements may increase bias and or variance.

We believe these results are valuable for researchers interested in applying these methods, working on improved gradient estimators or developing Bayesian learning methods. Incorrect results with hidden issues in the area could mislead many researchers and slow down development of new methods.

Outline. The paper is organized as follows. In Sect. 2 we briefly review the baseline ST estimator. In the subsequent sections we analyze Gumbel-Softmax estimator (Sect. 3), BayesBiNN (Sect. 4) and FouST estimator (Sect. 5). Proofs are provided in the respective Appendices A to C. As most of our results are theoretical, simplifying derivation or identifying limitations and misspecifications of the preceding work, we do not propose extensive experiments. Instead, we refer to the literature for the experimental evidence that already exists and only conduct specific experimental tests as necessary. In Sect. 6 we summarize our findings and discuss how they can facilitate future research.

2 Background

We define a stochastic binary unit $x \sim \text{Bernoulli}(p)$ as $x = 1$ with probability p and $x = 0$ with probability $1 - p$. Let $f(x)$ be a loss function, which in general may depend on other parameters and may be stochastic aside from the dependence on x. This is particularly the case when f is a function of multiple binary stochastic variables and we study its dependence on one of them explicitly. The goal of binary gradient estimators is to estimate

$$g = \frac{\mathrm{d}}{\mathrm{d}p}\mathbb{E}[f(x)], \tag{1}$$

where $\mathbb{E}$ is the total expectation. Gradient estimators which we consider make a stochastic estimate of the total expectation by taking a single joint sample. We will study their properties with respect to x only given the rest of the sample fixed. In particular, we will confine the notion of bias and variance to the conditional expectation $\mathbb{E}_x$ and the conditional variance $\mathbb{V}_x$. We will assume that the function $f(x)$ is defined on the interval $[0, 1]$ and is differentiable on this interval. This is typically the case when f is defined as a composition of simple functions, such as in neural networks. While for discrete inputs x, the continuous definition of f is irrelevant, it will be utilized by approximations exploiting its derivatives.

The expectation $\mathbb{E}_x[f(x)]$ can be written as

$$(1 - p)f(0) + pf(1), \tag{2}$$

Its gradient in p is respectively

$$g = \frac{\mathrm{d}}{\mathrm{d}p}\mathbb{E}_x[f(x)] = f(1) - f(0). \tag{3}$$

While this is simple for one random variable x, it requires evaluating f at two points. With n binary units in the network, in order to estimate all gradients stochastically, we would need to evaluate the loss $2n$ times, which is prohibitive.

Of high practical interest are stochastic estimators that evaluate f only at a single joint sample (perform a single forward pass). Arguably, the most simple such estimator is the straight-through (ST) estimator:

$$\hat{g}_{\mathrm{ST}} = f'(x). \tag{4}$$

For an in-depth introduction and more detained study of its properties we refer to [28]. The mean and variance of this ST estimator are given by

$$\mathbb{E}_x[\hat{g}_{\mathrm{ST}}] = (1-p)f'(0) + pf'(1), \tag{5a}$$

$$\mathbb{V}_x[\hat{g}_{\mathrm{ST}}] = \mathbb{E}_x[\hat{g}^2_{\mathrm{ST}}] - (\mathbb{E}_x[\hat{g}_{\mathrm{ST}}])^2 = p(1-p)(f'(1) - f'(0))^2. \tag{5b}$$

If $f(x)$ is linear in x, *i.e.*, $f(x) = hx + c$, where h and c may depend on other variables, then $f'(0) = f'(1) = h$ and $f(1) - f(0) = h$. In this linear case we obtain

$$\mathbb{E}_x[\hat{g}_{\mathrm{ST}}] = h, \tag{6a}$$

$$\mathbb{V}_x[\hat{g}_{\mathrm{ST}}] = 0. \tag{6b}$$

From the first expression we see that the estimator is unbiased and from the second one we see that its variance (due to x) is zero. It is therefore a reasonable baseline: if f is close to linear, we may expect the estimator to behave well. Indeed, there is a theoretical and experimental evidence [28] that in typical neural networks the more units are used per layer, the closer we are to the linear regime (at least initially) and the better the utility of the estimate for optimization. Furthermore, [29] show that in SBNs of moderate size, the accuracy of ST estimator is on par with a more accurate PSA estimator.

We will study alternative single-sample approaches and improvements proposed to the basic ST. In order to analyze BayesBiNN and FouST we will switch to the ± 1 encoding. We will write $y \sim \mathrm{Bin}(p)$ to denote a random variable with values $\{-1, 1\}$ parametrized by $p = \mathbb{P}_y(y{=}1)$. Alternatively, we will parametrize the same distribution using the expectation $\mu = 2p - 1$ and denote this distribution as $\mathrm{Bin}(\mu)$ (the naming convention and the context should make it unambiguous). Note that the mean of Bernoulli(p) is p. The ST estimator of the gradient in the mean parameter μ in both $\{0, 1\}$ and $\{-1, 1\}$ valued cases is conveniently given by the same equation (4).

Proof. Indeed, $\mathbb{E}_y[f(y)]$ with $y \sim \mathrm{Bin}(\mu)$ can be equivalently expressed as $\mathbb{E}_x[\tilde{f}(x)]$ with $x \sim \mathrm{Bernoulli}(p)$, where $p = \frac{\mu+1}{2}$ and $\tilde{f}(x) = f(2x - 1)$. The

ST estimator of gradient in the Bernoulli probability p for a sample x can then be written as

$$\hat{g}_{\mathrm{ST}} = \tilde{f}'(x) = 2f'(y), \tag{7}$$

where $y = 2x - 1$ is a sample from $\mathrm{Bin}(\mu)$. The gradient estimate in μ becomes $2f'(y)\frac{\partial p}{\partial \mu} = f'(y)$. □

3 Gumbel Softmax and ST Gumbel-Softmax

Gumbel Softmax [13] and Concrete relaxation [16] enable differentiability through discrete variables by relaxing them to real-valued variables that follow a distribution closely approximating the original discrete distribution. The two works [13,16] have contemporaneously introduced the same relaxation, but the name Gumbel Softmax (GS) became more popular in the literature.

A categorical discrete random variable x with K category probabilities π_k can be sampled as

$$x = \arg\max_k(\log \pi_k - \Gamma_k), \tag{8}$$

where Γ_k are independent Gumbel noises. This is known as Gumbel reparametrization. In the binary case with categories $k \in \{1, 0\}$ we can express it as

$$x = [\![\log \pi_1 - \Gamma_1 \geq \log \pi_0 - \Gamma_0]\!], \tag{9}$$

where $[\![\cdot]\!]$ is the Iverson bracket. More compactly, denoting $p = \pi_1$,

$$x = [\![\log \frac{p}{1-p} - (\Gamma_1 - \Gamma_0) \geq 0]\!]. \tag{10}$$

The difference of two Gumbel variables $z = \Gamma_1 - \Gamma_0$ follows the logistic distribution. Its cdf is $\sigma(z) = \frac{1}{1+e^{-z}}$. Denoting $\eta = \mathrm{logit}(p)$, we obtain the well-known noisy step function representation:

$$x = [\![\eta - z \geq 0]\!]. \tag{11}$$

This reparametrization of binary variables is exact but does not yet allow for differentiation of a single sample because we cannot take the derivative under the expectation of this function in (1). The relaxation [13,16] replaces the threshold function by a continuously differentiable approximation $\sigma_\tau(\eta) := \sigma(\eta/\tau) = \frac{1}{1+e^{-\eta/\tau}}$. As the temperature parameter $\tau > 0$ decreases towards 0, the function $\sigma_\tau(\eta)$ approaches the step function. The GS estimator of the derivative in η is then defined as the total derivative of f at a random relaxed sample:

$$z \sim \mathrm{Logistic}, \tag{12a}$$

$$\tilde{x} = \sigma_\tau(\eta - z), \tag{12b}$$

$$\frac{\hat{d}f}{d\eta} := \frac{\mathrm{d}f(\tilde{x})}{\mathrm{d}\eta} = f'(\tilde{x})\frac{\partial \tilde{x}}{\partial \eta}. \tag{12c}$$

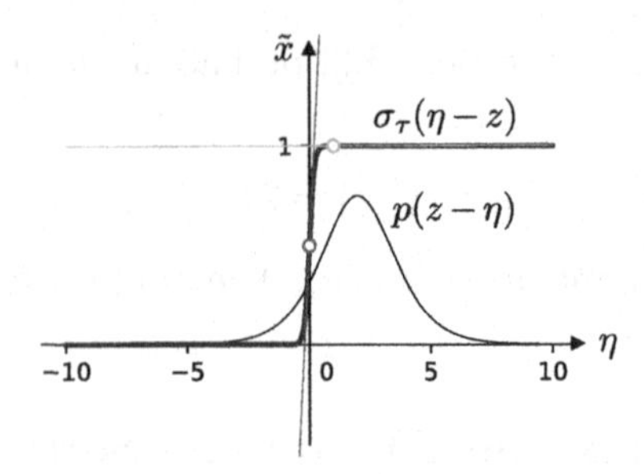

Fig. 2. GS Estimator: relaxed samples $\tilde{x}$ are obtained and differentiated as follows. Noisy inputs, following a shifted logistic distribution (black density), are passed through a smoothed step function σ_τ (blue). Observe that for a small τ, the derivative is often (in probability of $\eta - z$) close to zero (green) and, very rarely, when $|\eta - z|$ is small, it becomes $O(1/\tau)$ large (red). (Color figure online)

A possible delusion about GS gradient estimator is that it can be made arbitrary accurate by using a sufficiently small temperature τ. This is however not so simple and we will clarify theoretical reasons for why it is so. An intuitive explanation is proposed in Fig. 2. Formally, we show the following properties.

Proposition 1. *GS estimator is asymptotically unbiased as $\tau \to 0$ and the bias decreases at the rate $O(\tau)$ in general and at the rate $O(\tau^2)$ for linear functions.*

Proof in Appendix A. The decrease of the bias with $\tau \to 0$ is a desirable property, but this advantage is practically nullified by the fast increase of the variance:

Proposition 2. *The variance of GS estimator grows at the rate $O(\frac{1}{\tau})$.*

Proof in Appendix A. This fast growth of the variance prohibits the use of small temperatures in practice. In more detail the behavior of the gradient estimator is described by the following two propositions.

Proposition 3. *For any given realization $z \neq \eta$ the norm of GS estimator asymptotically vanishes at the exponential rate $O(\frac{1}{\tau}c^{1/\tau})$ with $c = e^{-|x|} < 1$.*

Proof in Appendix A. For small x, where c is close to one, the term $1/\tau$ dominates at first. In particular for $z = \eta$, the asymptote is $O(1/\tau)$. So while for the most of noise realizations the gradient magnitude vanishes exponentially quickly, it is compensated by a significant grows at rate $1/\tau$ around $z = \eta$. In practice it means that most of the time a value of gradient close to zero is measured and occasionally, very rarely, a value of $O(1/\tau)$ is obtained.

Proposition 4. *The probability to observe GS gradient of norm at least ε is asymptotically $O(\tau \log(\frac{1}{\varepsilon}))$, where the asymptote is $\tau \to 0$, $\varepsilon \to 0$.*

Proof in Appendix A.

Unlike ST, GS estimator with $\tau > 0$ is biased even for linear objectives. Even for a single neuron and a linear objective it has a non-zero variance. Propositions

3 and 4 apply also to the case of a layer with multiple units since they just analyze the factor $\frac{\partial}{\partial \eta}\sigma_\tau(\eta - z)$, which is present independently at all units. Proposition 4 can be extended to deep networks with L layers of Bernoulli variables, in which case the chain derivative will encounter L such factors and we obtain that the probability to observe a gradient with norm at least ε will vanish at the rate $O(\tau^L)$.

These facts should convince the reader of the following: it is not possible to use a very small τ, not even with an annealing schedule starting from $\tau = 1$. For a very small τ the most likely consequence would be to never encounter a numerically non-zero gradient during the whole training. For moderately small τ the variance would be prohibitively high. Indeed, Jang et al. [13] anneal τ only down to 0.5 in their experiments.

A major issue with this and other relaxation techniques (*i.e.* techniques using relaxed samples $\tilde{x} \in \mathbb{R}$) is that the relaxation biases all the expectations. There is only one forward pass and hence the relaxed samples $\tilde{x}$ are used for all purposes, not only for the purpose of estimating the gradient with respect to the given neuron. It biases all expectations for all other units in the same layer as well as in preceding and subsequent layers (in SBN). Let for example f depend on additional parameters θ in a differentiable way. More concretely, θ could be parameters of the decoder in VAE. With a Bernoulli sample x, an unbiased estimate of gradient in θ can be obtained simply as $\frac{\partial}{\partial \theta} f(x; \theta)$. However, if we replace the sample with a relaxed sample $\tilde{x}$, the estimate $\frac{\partial}{\partial \theta} f(\tilde{x}; \theta)$ becomes biased because the distribution of $\tilde{x}$ only approximates the distribution of x. If y were other binary variables relaxed in a similar way, the gradient estimate for x will become more biased because $E_{\tilde{y}}[\nabla_{\tilde{x}} f(\tilde{x}, \tilde{y})]$ is a biased estimate of $E_y[\nabla_{\tilde{x}} f(\tilde{x}, y)]$ desired. Similarly, in a deep SBN, the relaxation applied in one layer of the model additionally biases all expectations for units in layers below and above. In practice the accuracy for VAEs is relatively good [13], [28, Fig. 3] while for deep SBNs a bias higher than of ST is observed for $\tau = 1$ in a synthetic model with 2 or more layers and with $\tau = 0.1$ for a model with 7 (or more) layers [29, Fig. C.6]. When training moderate size SBNs on real data, it performs worse than ST [29, Fig. 4].

ST Gumbel-Softmax. Addressing the issue that relaxed variables deviate from binary samples on the forward pass, Jang et al. [13] proposed the following empirical modification. *ST Gumbel-Softmax* estimator [13] keeps the relaxed sample for the gradient but uses the correct Bernoulli sample on the forward pass:

$$z \sim \text{Logistic}, \tag{13a}$$

$$\tilde{x} = \sigma_\tau(\eta - z), \tag{13b}$$

$$x = [\![\eta - z \geq 0]\!], \tag{13c}$$

$$\hat{g}_{\text{ST-GS}(\tau)} = f'(x)\frac{\partial \tilde{x}}{\partial \eta}. \tag{13d}$$

Note that x is now distributed as Bernoulli(p) with $p = \sigma(\eta)$ so the forward pass is fixed. We show the following asymptotic properties.

Proposition 5. *ST Gumbel-Softmax estimator* [13] *is asymptotically unbiased for quadratic functions and the variance grows as* $O(1/\tau)$ *for* $\tau \to 0$.

Proof in Appendix A.

To summarize, ST-GS is more expensive than ST as it involves sampling from logistic distribution (and keeping samples), it is biased for $\tau > 0$. It becomes unbiased for quadratic functions as $\tau \to 0$, which would be an improvement over ST, but the variance grows as $\frac{1}{\tau}$.

4 BayesBiNN

Meng et al. [18], motivated by the need to reduce the variance of REINFORCE, apply GS estimator. However, in their large-scale experiments they use temperature $\tau = 10^{-10}$. According to the previous section, the variance of GS estimator should go through the roof as it grows as $O(\frac{1}{\tau})$. It is practically prohibitive as the learning would require an extremely small learning rate and a very long training time as well as high numerical accuracy. Nevertheless, good experimental results are demonstrated [18]. We identify a hidden implementation issue which completely changes the gradient estimator and enables learning.

First, we explain the issue. Meng et al. [18] model stochastic binary weights as $w \sim \mathrm{Bin}(\mu)$ and express GS estimator as follows.

Proposition 6. (Meng et al. [18] Lemma 1). *Let* $w \sim Bin(\mu)$ *and let* $f\colon \{-1,1\} \to \mathbb{R}$ *be a loss function. Using parametrization* $\mu = \tanh(\lambda)$, $\lambda \in \mathbb{R}$, *GS estimator of gradient* $\frac{\mathrm{d}\mathbb{E}_w[f]}{\mathrm{d}\mu}$ *can be expressed as*

$$\delta \sim \frac{1}{2}\text{Logistic}, \tag{14a}$$

$$\tilde{w} = \tanh_\tau(\lambda - \delta) \equiv \tanh(\frac{\lambda - \delta}{\tau}), \tag{14b}$$

$$J = \frac{1 - \tilde{w}^2}{\tau(1 - \mu^2)}, \tag{14c}$$

$$\hat{g} = Jf'(\tilde{w}), \tag{14d}$$

which we verify in Appendix B. However, the actual implementation of the scaling factor J used in the experiments [18] according to the published code[2] introduces a technical $\epsilon = 10^{-10}$ as follows:

$$J := \frac{1 - \tilde{w}^2 + \epsilon}{\tau(1 - \mu^2 + \epsilon)}. \tag{15}$$

It turns out this changes the nature of the gradient estimator and of the learning algorithm. The BayesBiNN algorithm [18, Table 1middle] performs the update:

$$\lambda := (1 - \alpha)\lambda - \alpha s f'(\tilde{w}), \tag{16}$$

where $s = NJ$, N is the number of training samples and α is the learning rate.

[2] https://github.com/team-approx-bayes/BayesBiNN.

Proposition 7. *With the setting of the hyper-parameters $\tau = O(10^{-10})$ [18, Table 7] and $\epsilon = 10^{-10}$ (author's implementation) in large-scale experiments (MNIST, CIFAR10, CIFAR100), the BayesBiNN algorithm is practically equivalent to the following deterministic algorithm:*

$$w := sign(\bar{\lambda}); \tag{17a}$$
$$\bar{\lambda} := (1 - \alpha)\bar{\lambda} - \alpha f'(w). \tag{17b}$$

In particular, it does not depend on the values of τ and N.

Proof in Appendix B. Experimentally, we have verified, using authors implementation, that indeed parameters λ (16) grow to the order 10^{10} during the first iterations, as predicted by our calculations in the proof.

Notice that the step made in (17b) consists of a decay term $-\alpha\bar{\lambda}$ and the gradient descent term $-\alpha f'(w)$, where the gradient $f'(w)$ is a straight-through estimate for the *deterministic* forward pass $w = \text{sign}(\bar{\lambda})$. Therefore the deterministic ST is effectively used. It is seen that the decay term is the only remaining difference to the deterministic STE algorithm [18, Table 1left], the method is contrasted to. From the point of view of our study, we should remark that the deterministic ST estimator used in effect indeed decreases the variance (down to zero) however it increases the bias compared to the baseline stochastic ST [28].

The issue has also downstream consequences for the intended Bayesian learning. The claim of Proposition 7 that the method does not depend on τ and N is perhaps somewhat unexpected, but it makes sense indeed. The initial BayesBiNN algorithm of course depends on τ and N. However due to the issue with the implementation of Gumbel Softmax estimator, for a sufficiently small value of τ it falls into a regime which is significantly different from the initial Bayesian learning rule and is instead more accurately described by (17). In this regime, the result it produces does not dependent on the particular values of τ and N. While we do not know what problem it is solving in the end, it is certainly not solving the intended variational Bayesian learning problem. This is so because the variational Bayesian learning problem and its solution do depend on N in a critical way. The algorithm (17) indeed does not solve any variational problem as there is no variational distribution involved (nothing sampled). Yet, the decay term $-\alpha\lambda$ stays effective: if the data gradient becomes small, the decay term implements some small "forgetting" of the learned information and may be responsible for an improved generalization observed in the experiments [18].

5 FouST

Pervez et al. [22] introduced several methods to improve ST estimators using Fourier analyzes of Boolean functions [21] and Taylor series. The proposed methods are guided by this analysis but lack formal guarantees. We study the effect of the proposed improvements analytically.

One issue with the experimental evaluation [22] is that the baseline ST estimator [22, Eq. 7] is *misspecified*: it is adopted from the works considering $\{0, 1\}$ Bernoulli variables without correcting for $\{-1, 1\}$ case as in (7), differing by a coefficient 2. The reason for this misspecifications is that ST is know rather as a folklore, vaguely defined, method (see [28]). While in learning with a simple expected loss this coefficient can be compensated by the tuned learning rate, it can lead to a more serious issues, in particular in VAEs with Bernoulli latents and deep SBNs. VAE training objective [14] has the data evidence part, where binary gradient estimator is required and the prior KL divergence part, which is typically computed analytically and differentiated exactly. Rescaling the gradient of the evidence part only introduces a bias which cannot be compensated by tuning the learning rate. Indeed, it is equivalent to optimizing the objective with the evidence part rescaled. In [28, Fig. 2] we show that this effect is significant. In the reminder of the section we will assume that the correct ST estimator (7) is used as the starting point.

5.1 Lowering Bias by Importance Sampling

The method [22, Sec. 4.1] "Lowering Bias by Importance Sampling", as noted by authors, obtains DARN gradient estimator [10, Appendix A] who derived it by applying a (biased) control variate estimate in the REINFORCE method. Transformed to the encoding with ± 1 variables, it expresses as

$$\hat{g}_{\text{DARN}} = f'(x)/p(x). \tag{18}$$

By design [10], this method is unbiased for quadratic functions, which is straightforward to verify by inspecting its expectation

$$\mathbb{E}[\hat{g}_{\text{DARN}}] = f'(1) + f'(0). \tag{19}$$

While, this is in general an improvement over ST—we may expect that functions close to linear will have a lower bias, it is not difficult to construct an example when it can increase the bias compared to ST.

Example 1. The method [22, Sec. 4.1] "Lowering Bias by Importance Sampling", also denoted as Importance Reweighing (IR), can increase bias.

Let $p \in [0, 1]$ and $x \sim \text{Bin}(p)$. Let $f(x) = |x + a|$. The derivative of $\mathbb{E}[f(x)]$ in p is

$$\frac{\mathrm{d}}{\mathrm{d}p}((1-p)f(-1) + pf(1)) = f(1) - f(-1) = f(1) = 2a. \tag{20}$$

The expectation of $\hat{g}_{\text{ST}}$ is given by

$$(1-p)2f'(-1) + p2f'(1) = 2(2p-1). \tag{21}$$

The expectation of $\hat{g}_{\text{DARN}}$ is given by

$$f'(-1) + f'(1) = 0. \tag{22}$$

The bias of DARN is $2|a|$ while the bias of ST is $2|a+1-2p|$. Therefore for $a > 0$ and $p > 0.5$, the bias of DARN estimator is higher. In particular for $a = 0.9$ and $p = 0.95$ the bias of ST estimator equals 0 while the bias of DARN estimator equals 1.8.

Furthermore, we can straightforwardly express its variance.

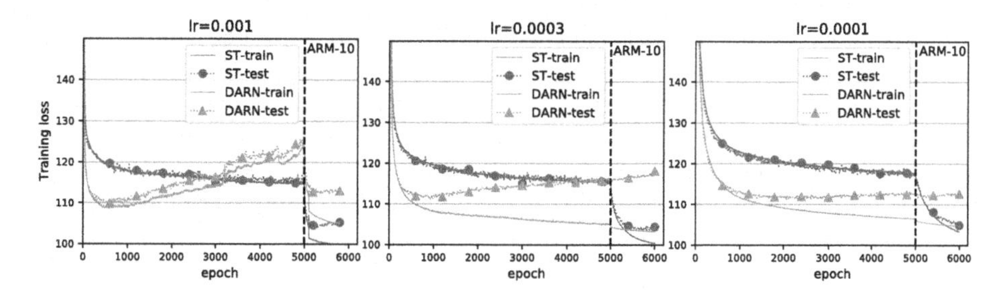

Fig. 3. Experimental comparison of DARN and ST estimators on MNIST VAE. The plots show training and test loss (negative ELBO) during training for different learning rates. After 5000 epochs, an unbiased ARM-10 estimator is applied in order to measure (and correct) the accumulated bias. At the smaller learning rates, where DARN does not diverge, it clearly has a much smaller accumulated bias but manages to overfit significantly.

Proposition 8. *The variance of* $\hat{g}_{\text{DARN}}$ *is expressed as*

$$\mathbb{V}_z[\hat{g}_{\text{DARN}}] = \frac{(f'(1) - p(f'(1) + f'(-1)))^2}{p(1-p)}. \tag{23}$$

It has asymptotes $O(\frac{f'(-1)^2}{1-p})$ *for* $p \to 1$ *and* $O(\frac{f'(1)^2}{p})$ *for* $p \to 0$.

The asymptotes indicate that the variance can grow unbounded for units approaching deterministic mode. If applied in a deep network with L layers, L expressions (18) are multiplied and the variance can grow respectively. Interestingly though, if the probability p is defined using the sigmoid function as $p = \sigma(\eta)$, then the gradient in η additionally multiplies by the Jacobian $\sigma'(\eta) = p(1-p)$, and the variance of the gradient in η becomes bounded. Moreover, a numerically stable implementation can simplify $p(1-p)/p(x)$ for both outcomes of x. We conjecture that this estimator can be particularly useful with this parametrization of the probability (which is commonly used in VAEs and SBNs).

Experimental evidence [11, Fig. 2.a], where DARN estimator is denoted as "$\frac{1}{2}$" shows that the plain ST performs similar for the structural output prediction problem. However, [11, Fig. 3.a] gives a stronger evidence in favor of DARN for VAE. In Fig. 3 we show experiment for the MNIST VAE problem, reproducing the experiment [11,22] (up to data binarization and implementation details). The exact specification is given in [28, Appendix D.1]. It is seen that DARN improves the training performance but needs an earlier stopping and or more regularization. Interestingly, with a correction of accumulated bias using unbiased ARM [35] method with 10 samples, ST leads to better final training and test performance.

5.2 Reducing Variance via the Fourier Noise Operator

The Fourier noise operator [22, Sec. 2] is defined as follows. For $\rho \in [0, 1]$, let $x' \sim N_\rho(x)$ denote that x' is set equal to x with probability ρ and chosen as an independent sample from Bin(p) with probability $1-\rho$. The Fourier noise operator smooths the loss function and is defined as $T_\rho[f](x) = \mathbb{E}_{x' \sim N_\rho(x)}[f(x')]$. When applied to f before taking the gradient, it can indeed reduce both bias and variance, ultimately down to zero when $\rho = 0$. Indeed, in this case x' is independent of x and $T_\rho[f](x) = \mathbb{E}[f(x)]$, which is a constant function of x. However, the exact expectation in x' is intractable. The computational method proposed in [22, Sec. 4.2] approximates the gradient of this expectation using S samples $x^{(s)} \sim N_\rho(x)$ as

$$\hat{g}_\rho = \frac{1}{S}\sum\nolimits_s \hat{g}(x^{(s)}), \tag{24}$$

where $\hat{g}$ is the base ST or DARN estimator. We show the following.

Proposition 9. *The method* [22, Sec. 4.2] *"Reducing Variance via the Fourier Noise operator" does not reduce the bias (unlike T_ρ) and increases variance in comparison to the trivial baseline that averages independent samples.*

Proof in Appendix C.

This result is found in a sharp contradiction with the experiments [22, Figure 4], where independent samples perform worse than correlated. We do not have a satisfactory explanation for this discrepancy except for the misspecified ST. Since the author's implementation is not public, it is infeasible to reproduce this experiment in order to verify whether a similar improvement can be observed with the well-specified ST. Lastly, note, that unlike correlated sampling, uncorrelated sampling can be naturally applied with multiple stochastic layers.

5.3 Lowering Bias by Discounting Taylor Coefficients

For the technique [22, Sec. 4.3.1] "Lowering Bias by Discounting Taylor Coefficients" we present an alternative view, not requiring Taylor series expansion of f, thus simplifying the construction. Following [22, Sec. 4.3.1] we assume that the importance reweighing was applied. Since the technique samples f' at non-binary points, we refer to it as a *relaxed* DARN estimator. It can be defined as

$$\tilde{g}_{\text{DARN}}(x,u) = \frac{f'(xu)}{p(x)}, \text{where} \quad u \sim \mathcal{U}[0,1]. \tag{25}$$

In the total expectation, when we draw x and u multiple times, the gradient estimates are averaged out. The expectation over u alone effectively integrates the derivative to obtain:

$$\mathbb{E}_u\left[\tilde{g}_{\text{DARN}}(x,u)\right] = \begin{cases} \frac{1}{p}\int_0^1 f'(u)\mathrm{d}u = \frac{1}{p}(f(1)-f(0)), & \text{if} \quad x=1, \\ \frac{1}{1-p}\int_0^1 f'(-u)\mathrm{d}u = \frac{1}{1-p}(f(-1)-f(0)), & \text{if} \quad x=-1. \end{cases} \tag{26}$$

In the expectation over x we therefore obtain

$$\mathbb{E}_{x,u}[\tilde{g}_{\text{DARN}}(x,u)] = f(1) - f(-1), \tag{27}$$

which is the correct derivative. One issue, discussed by [22] is that variance increases (as there is more noise in the system). However, a major issue similar to GS estimator Sect. 3, reoccurs here, that all related expectations become biased. In particular (25) becomes biased in the presence of other variables Pervez et al. [22, Sec. 4.3.1] propose to use $u \in \mathcal{U}[a,1]$ with $a > 0$, corresponding to shorter integration intervals around ± 1 states, in order to find an optimal tradeoff.

5.4 Lowering Bias by Representation Rescaling

Consider the estimator $\hat{g}$ of the gradient of function $\mathbb{E}_x[f(x)]$ where $x \sim \text{Bin}(p)$. Representation rescaling is defined in [22, Algorithm 1] as drawing $\tilde{x} \sim \frac{1}{\tau}\text{Bin}(p)$ instead of x and then using FouST estimator based on the derivative $f'(\tilde{x})$. It is claimed that using a scaled representation can decrease the bias of the gradient estimate. However, the following issue occurs.

Proposition 10. *The method* [22, Sec. 4.3.2] *"Lowering Bias by Representation Rescaling" compares biases of gradient estimators of different functions.*

Proof. Sampling $\tilde{x}$ can be equivalently defined as $\tilde{x} = x/\tau$. Bypassing the analysis of Taylor coefficients [22], it is easy to see that for a smooth function f, as $\tau \to \infty$, $f(x/\tau)$ approaches a linear function of x and therefore the bias of the ST estimator of $\mathbb{E}_x[f(x/\tau)]$ approaches zero. However, clearly $\mathbb{E}_x[f(x/\tau)]$ is a different function from $\mathbb{E}_x[f(x)]$ which we wish to optimize. □

We explain, why this method nevertheless has effect. Choosing and fixing the scaling hyper-parameter τ is equivalent to staring from a different initial point, where (initially random) weights are scaled by $1/\tau$. At this initial point, the network is found to be closer to a linear regime, where the ST estimator is more accurate and possibly the vanishing gradient issue is mitigated. Thus the method can have a positive effect on the learning as observed in [22, Appendix Table 3].

6 Conclusion

We theoretically analyzed properties of several methods for estimation of binary gradients and gained interesting new insights.

- For GS and ST-GS estimator we proposed a simplified presentation for the binary case and explained detrimental effects of low and high temperatures. We showed that bias of ST-GS estimator approaches that of DARN, connecting these two techniques.
- For BayesBiNN we identified a hidden issue that completely changes the behavior of the method from the intended variational Bayesian learning with Gumbel-Softmax estimator, theoretically impossible due to the used temperature $\tau = 10^{-10}$, to non-Bayesian learning with deterministic ST estimator and latent weight decay. As this learning method shows improved experimental results, it becomes an open problem to clearly understand and advance the mechanism which facilitates this.

- In our analysis of techniques comprising FouST estimator, we provided additional insights and showed that some of these techniques are not well justified. It remains open, whether they are nevertheless efficient in practice in some cases for other unknown reasons, not taken into account in this analysis.

Overall we believe our analysis clarifies the surveyed methods and uncovers several issues which limit their applicability in practice. It provides tools and clears the ground for any future research which may propose new improvements and would need to compare with existing methods both theoretically and experimentally. We hope that this study will additionally motivate such research.

References

1. Alizadeh, M., Fernandez-Marques, J., Lane, N.D., Gal, Y.: An empirical study of binary neural networks' optimisation. In: ICLR (2019)
2. Bethge, J., Yang, H., Bornstein, M., Meinel, C.: Back to simplicity: how to train accurate BNNs from scratch? CoRR, abs/1906.08637 (2019)
3. Bulat, A., Tzimiropoulos, G.: Binarized convolutional landmark localizers for human pose estimation and face alignment with limited resources. In: ICCV (2017)
4. Bulat, A., Tzimiropoulos, G., Kossaifi, J., Pantic, M.: Improved training of binary networks for human pose estimation and image recognition. arXiv (2019)
5. Bulat, A., Martinez, B., Tzimiropoulos, G.: High-capacity expert binary networks. In: ICLR (2021)
6. Chaidaroon, S., Fang, Y.: Variational deep semantic hashing for text documents. In: SIGIR Conference on Research and Development in Information Retrieval, pp. 75–84 (2017)
7. Dadaneh, S. Z., Boluki, S., Yin, M., Zhou, M., Qian, X.: Pairwise supervised hashing with Bernoulli variational auto-encoder and self-control gradient estimator. ArXiv, abs/2005.10477 (2020)
8. Esser, S.K., et al.: Convolutional networks for fast, energy-efficient neuromorphic computing. Proc. Natl. Acad. Sci. **113**(41), 11441–11446 (2016)
9. Grathwohl, W., Choi, D., Wu, Y., Roeder, G., Duvenaud, D.: Backpropagation through the void: optimizing control variates for black-box gradient estimation. In: ICLR (2018)
10. Gregor, K., Danihelka, I., Mnih, A., Blundell, C., Wierstra, D.: Deep autoregressive networks. In: ICML (2014)
11. Gu, S., Levine, S., Sutskever, I., Mnih, A.: MuProp: unbiased backpropagation for stochastic neural networks. In: 4th International Conference on Learning Representations (ICLR), May 2016
12. Horowitz, M.: Computing's energy problem (and what we can do about it). In: International Solid-State Circuits Conference Digest of Technical Papers (ISSCC), pp. 10–14 (2014)
13. Jang, E., Gu, S., Poole, B.: Categorical reparameterization with gumbel-softmax. In: ICLR (2017)
14. Kingma, D.P., Welling, M.: Auto-encoding variational Bayes. CoRR, abs/1312.6114 (2013)
15. Liu, Z., Wu, B., Luo, W., Yang, X., Liu, W., Cheng, K.-T.: Bi-real net: enhancing the performance of 1-Bit CNNs with improved representational capability and advanced training algorithm. In: Ferrari, V., Hebert, M., Sminchisescu, C., Weiss,

Y. (eds.) ECCV 2018. LNCS, vol. 11219, pp. 747–763. Springer, Cham (2018). https://doi.org/10.1007/978-3-030-01267-0_44
16. Maddison, C.J., Mnih, A., Teh, Y.W.: The concrete distribution: a continuous relaxation of discrete random variables. In: ICLR (2017)
17. Martínez, B., Yang, J., Bulat, A., Tzimiropoulos, G.: Training binary neural networks with real-to-binary convolutions. In: ICLR (2020)
18. Meng, X., Bachmann, R., Khan, M.E.: Training binary neural networks using the Bayesian learning rule. In: ICML (2020)
19. Mnih, A., Gregor, K.: Neural variational inference and learning in belief networks. In: ICML of JMLR Proceedings, vol. 32, pp. 1791–1799 (2014)
20. Ñanculef, R., Mena, F.A., Macaluso, A., Lodi, S., Sartori, C.: Self-supervised Bernoulli autoencoders for semi-supervised hashing. CoRR, abs/2007.08799 (2020)
21. O'Donnell, R.: Analysis of Boolean Functions. Cambridge University Press, Cambridge (2014). ISBN 1107038324
22. Pervez, A., Cohen, T., Gavves, E.: Low bias low variance gradient estimates for Boolean stochastic networks. In: ICML, vol. 119, pp. 7632–7640 (2020)
23. Peters, J.W., Welling, M.: Probabilistic binary neural networks. arXiv preprint arXiv:1809.03368 (2018)
24. Raiko, T., Berglund, M., Alain, G., Dinh, L.: Techniques for learning binary stochastic feedforward neural networks. In: ICLR (2015)
25. Rastegari, M., Ordonez, V., Redmon, J., Farhadi, A.: XNOR-Net: ImageNet classification using binary convolutional neural networks. In: Leibe, B., Matas, J., Sebe, N., Welling, M. (eds.) ECCV 2016. LNCS, vol. 9908, pp. 525–542. Springer, Cham (2016). https://doi.org/10.1007/978-3-319-46493-0_32
26. Roth, W., Schindler, G., Fröning, H., Pernkopf, F.: Training discrete-valued neural networks with sign activations using weight distributions. In: Brefeld, U., Fromont, E., Hotho, A., Knobbe, A., Maathuis, M., Robardet, C. (eds.) ECML PKDD 2019. LNCS (LNAI), vol. 11907, pp. 382–398. Springer, Cham (2020). https://doi.org/10.1007/978-3-030-46147-8_23
27. Shayer, O., Levi, D., Fetaya, E.: Learning discrete weights using the local reparameterization trick. In: ICLR (2018)
28. Shekhovtsov, A., Yanush, V.: Reintroducing straight-through estimators as principled methods for stochastic binary networks. In: GCPR (2021)
29. Shekhovtsov, A., Yanush, V., Flach, B.: Path sample-analytic gradient estimators for stochastic binary networks. In: NeurIPS (2020)
30. Shen, D., et al.: NASH: toward end-to-end neural architecture for generative semantic hashing. In: Annual Meeting of the Association for Computational Linguistics (2018)
31. Tang, W., Hua, G., Wang, L.: How to train a compact binary neural network with high accuracy? In: AAAI (2017)
32. Tucker, G., Mnih, A., Maddison, C.J., Lawson, J., Sohl-Dickstein, J.: REBAR: low-variance, unbiased gradient estimates for discrete latent variable models. In: NeurIPS (2017)
33. Vahdat, A., Andriyash, E., Macready, W.: Undirected graphical models as approximate posteriors. In: ICML, vol. 119, pp. 9680–9689 (2020)
34. Xiang, X., Qian, Y., Yu, K.: Binary deep neural networks for speech recognition. In: INTERSPEECH (2017)
35. Yin, M., Zhou, M.: ARM: augment-REINFORCE-merge gradient for stochastic binary networks. In: ICLR (2019)
36. Zhou, S., Wu, Y., Ni, Z., Zhou, X., Wen, H., Zou, Y.: DoReFa-Net: training low bitwidth convolutional neural networks with low bitwidth gradients. arXiv preprint arXiv:1606.06160 (2016)

End-to-End Learning of Fisher Vector Encodings for Part Features in Fine-Grained Recognition

Dimitri Korsch[1(✉)], Paul Bodesheim[1], and Joachim Denzler[1,2,3]

[1] Computer Vision Group, Friedrich-Schiller-University Jena, Jena, Germany
dimitri.korsch@uni-jena.de

[2] Michael Stifel Center Jena for Data-Driven and Simulation Science, Jena, Germany

[3] German Aerospace Center (DLR), Institute for Data Science, Jena, Germany

Abstract. Part-based approaches for fine-grained recognition do not show the expected performance gain over global methods, although explicitly focusing on small details that are relevant for distinguishing highly similar classes. We assume that part-based methods suffer from a missing representation of local features, which is invariant to the order of parts and can handle a varying number of visible parts appropriately. The order of parts is artificial and often only given by ground-truth annotations, whereas viewpoint variations and occlusions result in not observable parts. Therefore, we propose integrating a Fisher vector encoding of part features into convolutional neural networks. The parameters for this encoding are estimated by an online EM algorithm jointly with those of the neural network and are more precise than the estimates of previous works. Our approach improves state-of-the-art accuracies for three bird species classification datasets.

Keywords: End-to-end learning · Fisher vector encoding · Part-based fine-grained recognition · Online EM algorithm

1 Introduction

Part- or attention-based approaches [9,10,15,48,49,51] are common choices for fine-grained visual categorization because they explicitly focus on small details that are relevant for distinguishing highly similar classes, e.g., different bird species. Quite surprisingly, methods that perform the categorization with global image features [7,27,33,34,41,52] also achieve excellent results. It is hard to tell from the empirical results reported in the literature which general approach (global or part-based) is superior, given that all of them show comparable results in terms of recognition performance. We hypothesize that part-based algorithms cannot exploit their full potential due to the problems that arise from

Supplementary Information The online version contains supplementary material available at https://doi.org/10.1007/978-3-030-92659-5_9.

C. Bauckhage et al. (Eds.): DAGM GCPR 2021, LNCS 13024, pp. 142–158, 2021.
https://doi.org/10.1007/978-3-030-92659-5_9

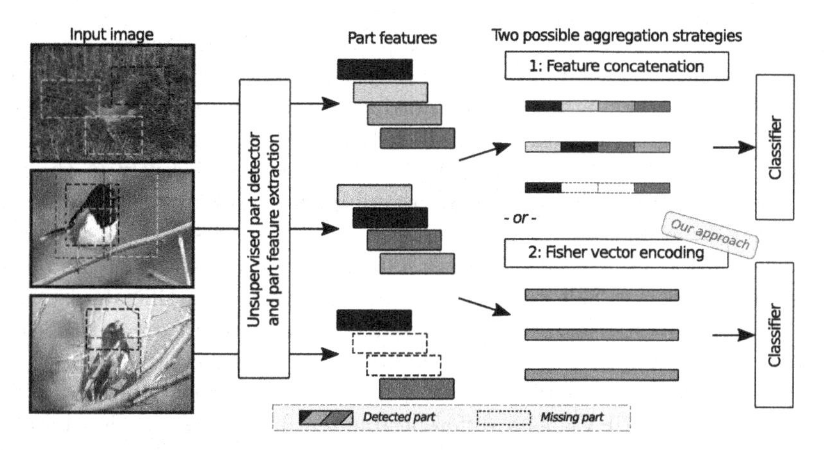

Fig. 1. When some parts can not be detected because they are not visible, the resulting gaps of missing features need to be filled when features are concatenated. Furthermore, the semantic meaning of extracted part features is not clear when applying unsupervised part detection algorithms that do not preserve a consistent order of part features. With our approach, we compute a Fisher vector encoding as a unified representation of fixed length for an arbitrary number of unordered part features, which can be used by any type of classifier, including simple linear classifiers and fully-connected layers in a deep neural network.

the initial detection of parts, especially regarding a unified representation of individual part features after the detection. Since learning individual part detectors requires part annotations and thus additional, time-consuming efforts by domain experts, methods for unsupervised part detection have been developed that already obtain remarkable classification results [10,24,49]. However, unsupervised part detection faces various challenges, such as missing parts caused by different types of occlusions and parts with ambiguous semantic meaning, as shown in Fig. 1. Hence, it remains unclear whether detected parts are reasonable and semantically consistent. Furthermore, part-based classifiers usually require a fixed number of parts to be determined for each image and a pre-defined order of the extracted part features. These are strong restrictions for the application, especially when considering varying poses and viewpoints that lead to hidden parts. We believe that a common way for representing a *varying number of unordered part features* obtained from every single image is rarely used in the context of fine-grained classification.

Fisher vector encoding (FVE) [30,31] is a well-known feature transformation method typically applied to local descriptors of key points. When applied to part features, it allows for computing a unified representation of fixed length for each image, independent of the number of detected parts (Fig. 1). Note that the problem of missing parts can also be observed for ground-truth part annotations of fine-grained bird species datasets [3,42,44]. Due to various poses and occlusions, some bird parts are not visible and cannot be annotated. These datasets are often used to evaluate the performance of classifiers without taking the part detection into account, and this requires a proper gap-filling strategy for missing parts. In contrast, no gap-filling is needed when using the FVE. Furthermore, FVE allows for neglecting an artificial order of parts since there is usually no natural order of parts, and each part should be treated equally.

Applying FVE together with CNNs has been done differently in the past: either GMM and FVE are computed after training the CNN model [6,36,50] such that only FVE parameters are adapted to the CNN features and not vice versa, or the FVE has been realized as a trainable layer [1,11,40,45]. Although the latter allows for end-to-end training of GMM parameters, artificial constraints are required to obtain reasonable values (e.g., positive variances). Furthermore, only the classification loss influences the mixture estimations, and there is no objective involved for modeling the data distribution correctly. In contrast, we propose to realize the FVE as a differentiable feature transformation based on a GMM estimated with an iterative EM algorithm using mini-batch updates of the parameters. First, the differentiable transformation allows end-to-end training of the feature extraction and classification weights. Next, the FVE directly influences the feature extraction within the CNN such that the features adapt to the encoding. Finally, since the parameters of the GMM are estimated with an EM algorithm instead of gradient descent, the resulting GMM describes the input data more precisely.

In our experiments, we show that the FVE outperforms other feature aggregation methods and that our approach estimates the GMM parameters more precisely than gradient descent methods. Furthermore, our approach described in Sect. 3 improves state-of-the-art accuracies for three fine-grained bird species categorization datasets: CUB-200-2011 [44] (from 90.3% to 91.1%), NA-Birds [42] (from 89.2% to 90.3%), and Birdsnap [3] (from 84.3% to 84.9%).

2 Related Work

First, we discuss related work on FVE in the context of deep neural networks. Some of these approaches either do not allow for learning the GMM parameters end-to-end but rather estimate parameters separately after neural network training. Others treat GMM parameters as conventional network parameters learned without any clustering objective by artificially enforcing reasonable values for the mixture parameters. Second, we review existing algorithms for iterative EM algorithms since we borrow ideas from these approaches. Surprisingly, none of these iterative approaches has been integrated in a CNN yet. Third, we list current state-of-the-art techniques for fine-grained categorization, mainly focusing on part-based methods. Some approaches [6,50] use FVE in fine-grained approaches, but all of them deploy the encoding in an offline manner, i.e., the FVE is used *after* learning CNN parameters for feature extraction.

2.1 Variants of Deep Fisher Vector Encoding (Deep FVE)

Simonyan *et al.* [35] presented a Fisher vector layer for building deep networks. They encode an input image or pre-extracted SIFT features in multiple layers with an FVE, but the entire network is trained greedily layer-by-layer and not end-to-end due to some restrictions. Sydorov *et al.* [37] suggested another deep architecture that learns an FVE by updating GMM parameters based on gradients that are backpropagated from a Hinge loss of an SVM classifier. Cimpoi *et al.* [6] proposed a CNN together with an FVE for local CNN features, and the resulting feature representation is used by a one-vs-rest SVM. Song *et al.* [36] improved this approach, but still without end-to-end learning.

In contrast to the methods above, Wieshollek *et al.* [45] and Tang *et al.* [40] deploy the FVE directly in a neural network. As a result, features are learned jointly with the classification and mixture model parameters. However, although GMM parameters are estimated jointly with other network parameters using gradient descent, the training procedure has some drawbacks. First, artificial constraints have to be applied to obtain reasonable mixture parameters, e.g., positive variances. Next, due to the formulas for the FVE, the resulting gradients cause numerically unstable computations. Finally, these approaches require a proper initialization of the mixture parameters, which implies the computation of the features of the entire dataset. Performing such an initialization on large dataset results in an unreasonable computation and storage overhead. Arandjelovic *et al.* [1] employ a simplified version of the FVE, called Vector of Locally Aggregated Descriptors (VLAD) [20], and rewrite its computation in terms of trainable network parameters that are optimized end-to-end via gradient descent. However, for this NetVLAD-Layer [1] as well as for [40,45], it is arguable whether estimated GMM parameters are reasonable due to gradient-based optimization with back-propagation. In contrast, our proposed method does not require any preliminary initialization since it uses an iterative EM algorithm to estimate the mixture parameters end-to-end, leading to meaningful cluster representations.

2.2 EM Algorithms and GMM Estimation

The standard technique for estimating GMM parameters is the EM algorithm. It is an iterative process of alternating between maximum likelihood estimation of mixture parameters and computing soft assignments of samples to the mixture components. In the default setting, all samples are used in both steps, but this leads to an increased runtime for large-scale datasets. To reduce the computational costs, one can only use a subset of samples for the parameter estimation or rely on existing online versions of the EM algorithm [4,29]. For example, Cappé and Moulines [4] approximate the expectation over the entire dataset with an exponential moving average over batches of the data. Based on this work, Chen *et al.* [5] propose a variance reduction of the estimates in each step, which results in faster and more stable convergence. A comparison of these algorithms is carried out by Karimi *et al.* [21], who also introduce a new estimation algorithm. Further approaches [2,13,26] propose similar solutions with different applications and motivations for the iterative parameter update.

In our work, we employ the ideas of iterative parameter update coupled with a bias correction. Furthermore, we demonstrate how to integrate these ideas in a neural network and estimate the parameters jointly with the network weights.

2.3 Fine-grained Visual Categorization

In the literature, two main directions can be observed for fine-grained recognition: global and part-based methods. Global methods use the input image as a whole and employ clever strategies for pre-training [7], augmentation [25,41], or pooling [27,33,34,52]. In contrast, part- or attention-based approaches apply sophisticated detection techniques to determine interesting image regions and to extract detailed local features from these patches. It results in part features as an additional source of information for boosting the classification performance.

He *et al.* [15] propose a reinforcement learning method for estimating how many and which image regions are helpful for the classification. They use multi-scale image representations for localizing the object and afterward estimate discriminative part regions. Ge *et al.* [10] present the current state-of-the-art approach on the CUB-200-2011 dataset. Based on weakly supervised instance detection and segmentations, part proposals are generated and constrained by a part model. The final classification is performed with a stacked LSTM classifier and context encoding. The method of Zhang *et al.* [49] also yields good results on the CUB-200-2011 and the NA-Birds dataset. Expert models arranged in multiple stages predict class assignments and attention maps that the final expert uses to crop the image and refine the observed data. Finally, a gating network is used to weigh the decisions of the individual experts.

Compared to the previous approaches, we use a different part detection method described at the beginning of the next section before presenting the details of our proposed FVE for part features.

3 Fisher Vector Encoding (FVE) of Part Features

In this section, we present our approach for an FVE of part features, which allows for joint end-to-end learning of all parameters, i.e., the parameters of the underlying GMM and the parameters of the CNN that computes the part features. It can be applied to any set of extracted parts from an image. Hence it is possible to combine it with different part detection algorithms. In this paper, we use the code[1] for a part detection method provided by Korsch *et al.* [24]. The authors use an initial classification of the entire input image to identify features used for this classification. Then, the pixels in the receptive field of these features are clustered and divided into candidate regions. Bounding boxes are estimated around these regions and used as parts in the final part-based classification.

As shown in Fig. 2, given a set of parts specified by their corresponding image regions, we propose the computation of a set of local features for each part with a CNN. We denote the output of a CNN as a *ConvMap* $C \in \mathbb{R}^{H \times W \times D}$, that consists of $N = H \cdot W$ local D-dimensional features $\mathcal{X} = \{\vec{x}_1, \ldots, \vec{x}_N\}$. Usually, CNNs contain *global average pooling (GAP)* to reduce $\mathcal{X}$ to a single feature representation: $GAP(\mathcal{X}) = \frac{1}{N}\sum_{n=1}^{N} \vec{x}_n$. Common part-based approaches [10, 24,46,47] extract a set of *ConvMaps* $\mathcal{C} = \{C_1, \ldots, C_T\}$ from a single image

[1] https://github.com/cvjena/l1_parts.

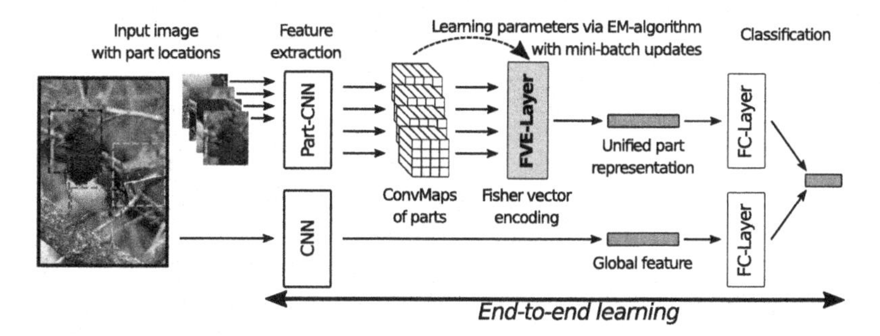

Fig. 2. Overview of our proposed method. During training, we estimate the parameters of the GMM that leads to the FVE using an EM-algorithm with mini-batch updates described in Sect. 3.2. The resulting FVE-Layer, which is explained in Sect. 3.3, can be integrated in any deep network architecture. We use this new layer for computing a unified part representation that aggregates local features extracted from *ConvMaps* of a CNN. Our approach enables joint end-to-end learning of both CNN parameters and GMM parameters for the FVE.

I by processing T image regions and use GAP for each *ConvMap* followed by concatenation of the resulting T part features. Since each part t is represented by N local features of the corresponding *ConvMap* C_t, an image I can be described by a set of $N \cdot T$ local *ConvMap* features $\mathcal{X}_I = \{\vec{x}_{1,1}, \ldots, \vec{x}_{n,t}, \ldots, \vec{x}_{N,T}\}$. We then use FVE to transform this set into a single feature: $FVE(\mathcal{X}_I) = \vec{f} \in \mathbb{R}^{\hat{D}}$. Note that the set $\mathcal{X}_I$ is just another representation for the set of *ConvMaps* $\mathcal{C}$, and we can also directly write $FVE(\mathcal{C})$ instead of $FVE(\mathcal{X}_I)$ to indicate that the FVE is a transformation of the *ConvMaps* $\mathcal{C}$.

3.1 Fisher Vector Encoding

First, we assume that the CNN computes local features *i.i.d.* from an input image. We further assume that all local descriptors $\mathcal{X}_I$ of the extracted *ConvMaps* $\mathcal{C}$ come from the same distribution with density function $p(\vec{x}_{n,t})$, represented by a finite mixture model with K components: $p(\vec{x}_{n,t}|\Theta) = \sum_{k=1}^{K} \alpha_k p_k(\vec{x}_{n,t}|\theta_k)$ with mixture weights α_k that add up to 1 ($\sum_{k=1}^{K} \alpha_k = 1$), and model parameters $\Theta = \{\alpha_1, \theta_1, \ldots, \alpha_K, \theta_K\}$. Without any prior knowledge, let the density function of each component be a Gaussian distribution with mean vector $\vec{\mu}_k$ and diagonal covariance matrix $\vec{\sigma}_k$: $p_k(\vec{x}_{n,t}|\theta_k) = \mathcal{N}(\vec{x}_{n,t}|\vec{\mu}_k, \vec{\sigma}_k)$ leading to a Gaussian Mixture Model (GMM) with parameters $\Theta = \{\alpha_1, \vec{\mu}_1, \vec{\sigma}_1 \ldots \alpha_K, \vec{\mu}_K, \vec{\sigma}_K\}$.

Following Jaakola and Haussler [19], and Perronnin *et al.* [30,31], the FVE is derived by considering the gradients of the log-likelihood with respect to the GMM parameters Θ and assuming independence of the part features: $\mathcal{F}_\Theta(\mathcal{C}) = \sum_{n,t=1}^{N,T} \nabla_\Theta \log p(\vec{x}_{n,t}|\Theta)$. These gradients, also called Fisher scores, describe how parameters contribute to the process of generating a particular feature. We use the approximated normalized Fisher scores introduced by [30,31]:

$$\mathcal{F}_{\vec{\mu}_k}(\mathcal{C}) = \frac{1}{\sqrt{NT\alpha_k}} \sum_{n,t=1}^{N,T} w_{n,t,k} \left(\frac{\vec{x}_{n,t} - \vec{\mu}_k}{\vec{\sigma}_k} \right), \tag{1}$$

$$\mathcal{F}_{\vec{\sigma}_k}(\mathcal{C}) = \frac{1}{\sqrt{2\,NT\alpha_k}} \sum_{n,t=1}^{N,T} w_{n,t,k} \left(\frac{(\vec{x}_{n,t} - \vec{\mu}_k)^2}{\vec{\sigma}_k^2} - 1 \right) \tag{2}$$

as feature encoding with $w_{n,t,k} = \frac{\alpha_k p(\vec{x}_{n,t}|\theta_k)}{\sum_{l=1}^{K} \alpha_l p(\vec{x}_{n,t}|\theta_l)}$ denoting the soft assignment of the feature $\vec{x}_{n,t}$ to a component k.

Finally, these scores can be computed for all parameters $\vec{\mu}_k$ and $\vec{\sigma}_k$ of the estimated GMM. We use the concatenation of the scores as FVE of the part features, which results in a unified representation of dimension $\hat{D} = 2KD$ that is independent of the number of part features T.

3.2 Estimation of the Mixture Model Parameters

The computations of the FVE require a GMM, and we illustrate two ways for estimating its parameters from data jointly with the CNN parameters. In our experiments, we use both methods and compare them by the classification accuracy and quality of the estimated GMM parameters.

Gradient-Based Estimation. This idea is covered in different works [1,40,45]: since all of the operations in Eqs. (1) and (2) are differentiable, it is straightforward to implement the FVE as a differentiable FVE-Layer such that the parameters Θ are estimated via gradient descent. However, the GMM constraints (positive variances and prior weights adding up to one) need to be enforced. Wieschollek *et al.* [45] propose to model the variances $\sigma_k^2 = \epsilon + \exp(s_k)$ and the mixture weights $\alpha_k = \frac{\mathrm{sigm}(a_k)}{\sum_j \mathrm{sigm}(a_j)}$ by estimating s_k and a_k instead of σ_k^2 and α_k.

Online-EM Estimation. Different variants for an online EM algorithm can be found in the literature [2,4,5,21,26,29]. The main idea is to approximate the expectations over the entire dataset with exponential moving averages (EMAs) over batches of the data: $\Theta[t] = \lambda \cdot \Theta[t-1] + (1-\lambda) \cdot \Theta^{\mathrm{new}}$ with $\lambda \in (0,1)$ and $[t]$ indicating the training step t. We follow this approach and propose an FVE-Layer with online parameter estimation via EMAs. It is worth mentioning that the parameters of the widely used batch-normalization layer [18] are also estimated with EMAs. However, our FVE-Layer differs from a batch-normalization layer in two ways. First, we estimate a mixture of Gaussians and not only the mean and variance of the inputs. Second, we encode the input according to Eqs. (1) and (2) instead of whitening the inputs. Additionally, we perform bias correction via $\hat{\Theta}[t] = \frac{\Theta[t]}{1-\lambda^t}$ since plain EMAs are biased estimators. Similar bias correction is also done in the Adam optimizer [23].

Finally, we have observed that some of these D-dimensional local feature vectors ($H{\cdot}W$ vectors for each of the T parts) have low L^2-norm, especially if the corresponding receptive field mainly contains background pixels. However, since we are only interested in using local features that exceed a certain activation level, i.e., that carry important information, we include an additional filtering step for the local feature vectors before both the estimation of the GMM parameters and the computation of the FVE. We only use local features with an L^2-norm greater than the mean L^2-norm of all local features obtained from the same image. We found that this filtering leads to more stable and balanced estimates for the GMM parameters during our experiments.

3.3 Training with the FVE-Layer

We implement the proposed FVE-Layer utilizing the calculations from Eqs. (1) and (2) as well as the online parameter estimation introduced in the previous section. For end-to-end training of the CNN layers preceding the FVE-Layer, we estimate the gradients of the encoding w.r.t. the inputs similar to [45]:

$$\frac{\partial \mathcal{F}_{\mu_{k,d}}(\mathcal{C})}{\partial x_{n,t,d_*}} = \frac{1}{\sqrt{NT\alpha_k}} \left[\frac{\partial w_{n,t,k}}{\partial x_{n,t,d_*}} \left(\frac{x_{n,t,d} - \mu_{k,d}}{\sigma_{k,d}} \right) + \delta_{d,d_*} \frac{w_{n,t,k}}{\sigma_{k,d_*}} \right], \tag{3}$$

$$\begin{aligned} \frac{\partial \mathcal{F}_{\sigma_{k,d}}(\mathcal{C})}{\partial x_{n,t,d_*}} &= \frac{1}{\sqrt{2\,NT\alpha_k}} \left[\frac{\partial w_{n,t,k}}{\partial x_{n,t,d_*}} \left(\frac{(x_{n,t,d} - \mu_{k,d})^2}{(\sigma_{k,d})^2} - 1 \right) \right. \\ &== \frac{1}{\sqrt{2\,NT\alpha_k}} \left. + \delta_{d,d_*} \frac{2 w_{n,t,k}(x_{n,t,d_*} - \mu_{k,d_*})}{(\sigma_{k,d_*})^2} \right]. \end{aligned} \tag{4}$$

In both equations, we use δ_{d,d_*} to denote the Kronecker delta being 1 if $d = d_*$ and 0 else, as well as the derivative of $w_{n,t,k}$ w.r.t. x_{n,t,d_*} that is given by $\frac{\partial w_{n,t,k}}{\partial x_{n,t,d_*}} = w_{n,t,k} \left(-\frac{(x_{n,t,d_*} - \mu_{k,d_*})}{(\sigma_{k,d_*})^2} + \sum_{\ell=1}^{K} w_{n,t,\ell} \frac{(x_{n,t,d_*} - \mu_{\ell,d_*})}{(\sigma_{\ell,d_*})^2} \right)$. Further details for the derivation of these gradients can be found in the supplementary material.

Though these gradients are computed within the deep learning framework by the autograd functionality, it is important to mention that we observed some numerical instabilities during the training, especially with high dimensional mixture components. To circumvent this issue, we perform an auxiliary classification on the inputs of the FVE-Layer, similar to Szegedy *et al.* [38]. The auxiliary classification branch consists of a global average pooling and a linear layer. Finally, we combine the resulting auxiliary loss with the loss computed from the prediction on the encoded part features: $\mathcal{L}_{parts} = \beta \cdot \mathcal{L}_{aux} + (1 - \beta) \cdot \mathcal{L}_{FVE}$. We set β to 0.5 and multiply it by 0.5 after 20 epochs, such that the effect of the auxiliary classification decreases over time. Our motivation behind the initial value of β is to give both losses equal impact at the beginning and to increase the impact of the encoded part features as the training goes on. We tested other initial values, but they had minor effects on the classification performance, except that disabling the auxiliary loss resulted in degraded training stability, as mentioned before. Hence, we have chosen the β value that matches best the arithmetic mean of the losses. Furthermore, we omit the auxiliary branch for the final classification and perform the part classification entirely on the features encoded by the FVE.

The final loss, consisting of the losses computed from the global and the part predictions, is computed in a similar way: $\mathcal{L}_{final} = \frac{1}{2}\left(\mathcal{L}_{parts} + \mathcal{L}_{global}\right)$. For cross-entropy, this combination is equivalent to computing the final prediction as a geometric mean of the class probabilities or the arithmetic mean of the normalized log-likelihoods (see Sect. S2 in the supplementary material for more details).

4 Experimental Results

4.1 Datasets

We evaluate our method on widely used datasets for fine-grained categorization. First, we use three datasets for bird species recognition: *CUB-200-2011* [44], *NA-Birds* [42], and *Birdsnap* [3], since this is the most challenging domain when considering current state-of-the-art results with accuracies of around 90 % or less (see Table 1). For other fine-grained domains like aircraft, cars, or flowers, the methods already achieve accuracies above 95 %. The three bird datasets contain between 200 and 555 different species. CUB-200-2011 is the most popular fine-grained dataset for benchmarking because of its balanced sample distribution, but it is also the smallest one with only 5994 training and 5794 test images. The other two datasets are more imbalanced but contain much more training images: 23929 and 40871, respectively. Besides class labels, bounding boxes and part annotations are available for all three datasets.

Additionally, we evaluate our method on datasets for dogs and moths species to show the applicability of our approach for other domains. *Stanford Dogs* [22] consists of 120 classes and 20580 images with class labels and bounding box annotations. Since the entire dataset is part of the ImageNet dataset [8], we only use neural networks pre-trained on the iNaturalist 2017 dataset [43] to avoid pre-training on the test images. The *EU-Moths* dataset[2] contains 200 moth species common in Central Europe. Each of the species is represented by approximately 11 images. The insects are photographed manually and mainly on a relatively homogeneous background. We manually annotated bounding boxes for each specimen and used the cropped images for training. We trained the CNN on a random balanced split of 8 training and 3 test images per class.

4.2 Implementation Details

As a primary backbone of the presented method, we take the InceptionV3 CNN architecture [39]. We use the pre-trained weights proposed by Cui *et al.* [7]. They have pre-trained the network on the iNaturalist 2017 dataset [43] and could show that this is more beneficial for animal datasets than pre-training on ImageNet. For some experiments (Sect. 4.4 and 4.5), we also use a ResNet-50 CNN architecture [14] pre-trained on the ImageNet dataset [32].

[2] https://www.inf-cv.uni-jena.de/eu_moths_dataset.

Table 1. Comparison of our proposed FVE for part features with various state-of-the-art methods on three bird datasets (**bold** = best per dataset).

METHOD	CUB-200-2011	NA-BIRDS	BIRDSNAP
Cui *et al.* [7]	89.3	87.9	–
Stacked LSTM [10]	90.3	–	–
FixSENet-154 [41]	88.7	89.2	84.3
CS-Parts [24]	89.5	88.5	–
MGE-CNN [49]	89.4	88.6	–
WS-DAN [16]	89.4	–	–
PAIRS [12]	89.2	87.9	–
API-Net [53]	90.0	88.1	–
No Parts (baseline)	89.5 $\pm$0.2	86.9 $\pm$0.1	81.9 $\pm$0.5
GAP (parts of [24])	90.9 $\pm$0.1	89.9 $\pm$0.1	84.0 $\pm$0.2
Gradient-based FVE (parts of [24])	**91.2** $\pm$0.3	**90.4** $\pm$0.1	**85.3** $\pm$0.2
EM-based FVE (parts of [24])	**91.1** $\pm$0.2	**90.3** $\pm$0.1	**84.9** $\pm$0.2

For a fair comparison, we use fixed hyperparameters for every experiment. We train each model for 60 epochs with an AdamW optimizer [28], setting the learning rate to 2e-3 and α to 0.01. Due to limited GPU memory, we apply the gradient accumulation technique. We use a batch size of 12 and accumulate the gradients over four training iterations before we perform a weight update, which results in an effective batch size of 48. Furthermore, we repeat each experiment at least 5 times to observe the significance and robustness of the presented approach. The source code for our approach is publicly available on GitHub[3].

4.3 Fine-grained Classification

In our first experiment, we test our proposed FVE-Layer together with an unsupervised part detector that provides classification-specific parts (CS-Parts) [24]. However, in contrast to Korsch *et al.* [24], we use a separate CNN for calculating part features. This part-CNN is fine-tuned on the detected parts, and the extracted features are adapted to the FVE. Besides the part-CNN, we also extract features from the global image with another CNN. For the part-CNN, the prediction is performed based on the FVE of the part features, whereas the prediction on the global image is made based on standard CNN features. Both predictions are then weighted equally and summed up to the final prediction. The GMM parameters are either estimated via gradient descent or by our proposed online EM algorithm. After investigating the effect of both the number and dimension of the mixture components (see Sect. S3.2 in supplementary material), we use one component with a dimension of 2048. We select these hyperparameters since (1) this setup introduces the least number of additional

[3] https://github.com/cvjena/fve_experiments.

Table 2. Results on the Stanford Dogs and EU-Moths datasets. For Stanford Dogs, we only compare to methods that do not use ImageNet pre-training. Similar to our work, they utilize a pre-training on the iNaturalist 2017 dataset. This kind of pre-training results in a more fair comparison, since the training set of ImageNet contains the test set of Stanford Dogs.

Method	Stanford Dogs	EU-Moths
Cui *et al.* [7]	78.5	–
DATL [17] (with [16])	79.1	–
No Parts (baseline)	77.5 $\pm$0.5	90.5 $\pm$0.5
GAP (parts of [24])	77.8 $\pm$0.4	91.0 $\pm$0.5
Gradient-based FVE (parts of [24])	**79.1** $\pm$0.3	**93.0** $\pm$1.2
EM-based FVE (parts of [24])	**79.2** $\pm$0.1	**92.0** $\pm$0.9

parameters and (2) favors both GMM parameter estimation methods equally. We also compare the FVE method with results based on concatenated part features and GAP and with a baseline using only the prediction obtained from the global image (no parts). In our preliminary experiments (see Sect. S3.1 in supplementary material), we also showed the superiority of the joint training of the GMM parameters and the CNN weights against the conventional pipeline that is for example used in [6,36].

For bird species classification, Table 1 contains our results as well as accuracies reported in previous work. Our proposed FVE for part features performs best on all three datasets. The results for NA-Birds stand out, since our approach is the only one reaching an accuracy greater than 90 % on this challenging dataset Accuracies for dogs and moth species are shown in Table 2. Again, our FVE approach performs best, showing its suitability beyond the bird species domain.

Nevertheless, it is worth mentioning that the GMM parameter estimation method has a minor effect on classification accuracy. We observed in our preliminary experiments that the GMM parameters have little effect on the expressiveness of the final encoded feature. Even if the GMM parameters are initialized randomly and are not further adapted to the data, the classification performance remains equally high compared to the reported values in Table 1. This also explains why the Gradient-based estimation of the parameters performs on a par with the EM-based estimation, even though the GMM parameters do not change much from their initialization values (see Sect. S5 in supplementary material). Finally, we think the observation of why the randomly initialized GMM parameters perform as well as the trained parameters is out of the scope of this paper, and we would investigate this in the future.

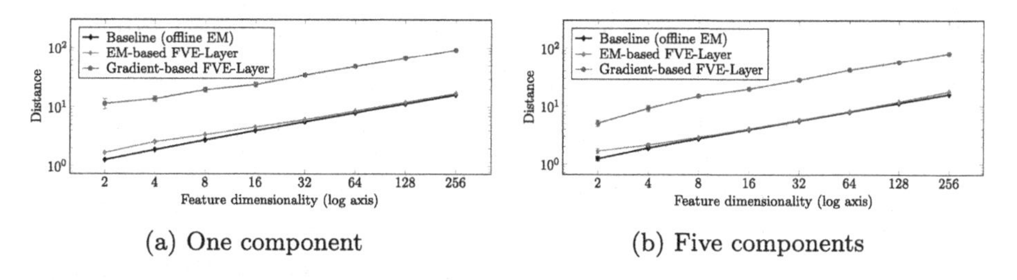

(a) One component (b) Five components

Fig. 3. Comparison of the weighted normalized Euclidean distance (see Eq. (5)) on a generated dataset. The dataset consist of five classes and the experiment was performed with one and five GMM components.

4.4 Quality of the Estimated GMM Parameters

We now investigate the quality of the estimated GMM parameters with respect to the two proposed approaches (gradient-based vs. EM-based). For this purpose, we first compare them on a generated dataset consisting of feature vectors for five classes sampled from different normal distributions. The corresponding mean vectors (class centroids) are arranged on the unit sphere, and we set the variance of the class distributions to $\frac{1}{5}$ such that the features do not overlap. We trained a simple neural network consisting of the FVE-Layer and a linear layer as a classifier. The data dimension has been varied, using powers of two in the range 2 to 256, which also defines the dimensions of the mixture components. Each setup was repeated five times. As a baseline, we also estimated the GMM parameters with a standard EM algorithm independent of neural network training.

In Fig. 3, we visualize the normalized Euclidean distance of the estimated parameters averaged over the entire dataset. The normalized Euclidean distance for a single feature vector can be computed via:

$$D(\vec{x}_{n,t}|\theta) = \sum_k w_{n,t,k} \cdot \sqrt{\sum_d \frac{\left(x_{n,t,d} - \mu_{k,d}\right)^2}{\left(\sigma_{k,d}\right)^2}} \quad . \tag{5}$$

It is the sum of distances to the mean vectors of the mixture components, normalized by the corresponding variances and weighted by the soft-assignments $w_{n,t,k}$. Furthermore, we estimate the same distances for the CUB-200-2011 dataset, shown in Table 3. Both evaluations on synthetic and real data show that GMMs estimated by our proposed online EM algorithm fit the data much better, resulting in more precise clusters due to lower normalized Euclidean distances. Moreover, we show in the supplementary material that the gradient-based method changes the GMM parameters only slightly for two-dimensional data, whereas the online EM algorithm estimates parameters as good as the offline EM algorithm such that the mixtures match the data distributions well.

Table 3. Comparison of weighted normalized Euclidean distances (see Eq. (5)) for the three birds datasets, evaluated for two CNN architectures.

	GMM Estimation for ResNet50		GMM Estimation for InceptionV3	
	EM-based	Gradient-based	EM-based	Gradient-based
CUB-200-2011	21.49 (±0.68)	46.85 (±0.18)	33.21 (±0.13)	37.15 (±0.18)
NA-Birds	18.36 (±1.00)	45.35 (±0.18)	36.62 (±0.38)	35.50 (±0.15)
Birdsnap	11.08 (±2.01)	46.75 (±0.23)	32.75 (±0.31)	35.62 (±0.30)

Table 4. Ablation study on the CUB-200-2011 dataset with two different CNNs.

Method	ResNet50	InceptionV3
Baseline CNN	84.4 ± 0.3	89.5 ± 0.2
Parts [24] + GAP	79.7 ± 0.3	88.8 ± 0.2
Parts [24] + FVE	82.6 ± 0.4	89.1 ± 0.2
Baseline CNN + Parts [24] + GAP	85.9 ± 0.2	90.9 ± 0.1
Baseline CNN + Parts [24] + FVE	86.4 ± 0.2	91.1 ± 0.2

4.5 Ablation Study

In an ablation study, we investigate the impact of the FVE in the proposed approach. As seen in Fig. 2, our method consists of two branches: classification of the global image and classification based on estimated parts. In Table 4, we show accuracies achieved by the individual branches as well as by the combined classification for two CNN architectures. We see that part features with FVE result in better classification accuracies compared to GAP. This effect propagates to the final accuracy, resulting in improved classification performance.

5 Conclusions

In this paper, we have proposed a new FVE-Layer for aggregating part features of a CNN in the context of fine-grained categorization, which uses an online EM algorithm for estimating the underlying GMM jointly with other network parameters. With this layer, we are able to compute a unified fixed-length representation for a varying number of local part features, which allows a deep neural network to cope with missing parts as well as with an arbitrary order of part features, e.g., given by an unsupervised part detector. In our experiments, we have achieved state-of-the-art recognition accuracies on three fine-grained datasets for bird species classification: CUB-200-2011 (91.1 %), NA-Birds (90.3 %), and Birdsnap (84.9 %). Furthermore, we have shown that compared to existing deep FVE implementations, our online EM-based approach results in more accurate estimates of the mixture model.

References

1. Arandjelovic, R., Gronat, P., Torii, A., Pajdla, T., Sivic, J.: NetVLAD: CNN architecture for weakly supervised place recognition. In: Proceedings of the IEEE Conference on Computer Vision and Pattern Recognition, pp. 5297–5307 (2016)
2. Awwad Shiekh Hasan, B., Gan, J.Q.: Sequential EM for unsupervised adaptive gaussian mixture model based classifier. In: Perner, P. (ed.) MLDM 2009. LNCS (LNAI), vol. 5632, pp. 96–106. Springer, Heidelberg (2009). https://doi.org/10.1007/978-3-642-03070-3_8
3. Berg, T., Liu, J., Woo Lee, S., Alexander, M.L., Jacobs, D.W., Belhumeur, P.N.: Birdsnap: large-scale fine-grained visual categorization of birds. In: Proceedings of the IEEE Conference on Computer Vision and Pattern Recognition, pp. 2011–2018 (2014)
4. Cappe, O., Moulines, E.: On-line expectation-maximization algorithm for latent data models. J. R. Stat. Soc. Ser. B (Stat. Methodol.) **71**(3), 593–613 (2009)
5. Chen, J., Zhu, J., Teh, Y.W., Zhang, T.: Stochastic expectation maximization with variance reduction. In: Advances in Neural Information Processing Systems, pp. 7967–7977 (2018)
6. Cimpoi, M., Maji, S., Vedaldi, A.: Deep filter banks for texture recognition and segmentation. In: Proceedings of the IEEE Conference on Computer Vision and Pattern Recognition, pp. 3828–3836 (2015)
7. Cui, Y., Song, Y., Sun, C., Howard, A., Belongie, S.: Large scale fine-grained categorization and domain-specific transfer learning. In: Proceedings of the IEEE Conference on Computer Vision and Pattern Recognition (2018)
8. Deng, J., Dong, W., Socher, R., Li, L.J., Li, K., Fei-Fei, L.: ImageNet: a large-scale hierarchical image database. In: 2009 IEEE Conference on Computer Vision and Pattern Recognition, pp. 248–255. IEEE (2009)
9. Fu, J., Zheng, H., Mei, T.: Look closer to see better: recurrent attention convolutional neural network for fine-grained image recognition. In: Proceedings of the IEEE Conference on Computer Vision and Pattern Recognition (2017)
10. Ge, W., Lin, X., Yu, Y.: Weakly supervised complementary parts models for fine-grained image classification from the bottom up. In: Proceedings of the IEEE Conference on Computer Vision and Pattern Recognition, pp. 3034–3043 (2019)
11. Gong, Y., Wang, L., Guo, R., Lazebnik, S.: Multi-scale orderless pooling of deep convolutional activation features. In: Fleet, D., Pajdla, T., Schiele, B., Tuytelaars, T. (eds.) ECCV 2014. LNCS, vol. 8695, pp. 392–407. Springer, Cham (2014). https://doi.org/10.1007/978-3-319-10584-0_26
12. Guo, P., Farrell, R.: Aligned to the object, not to the image: a unified pose-aligned representation for fine-grained recognition. In: IEEE Winter Conference on Applications of Computer Vision, pp. 1876–1885. IEEE (2019)
13. Hasan, B.A.S., Gan, J.Q.: Unsupervised adaptive GMM for bci. In: 2009 4th International IEEE/EMBS Conference on Neural Engineering, pp. 295–298 (2009)
14. He, K., Zhang, X., Ren, S., Sun, J.: Deep residual learning for image recognition. In: IEEE Conference on Computer Vision and Pattern Recognition, pp. 770–778 (2016)
15. He, X., Peng, Y., Zhao, J.: Which and how many regions to gaze: focus discriminative regions for fine-grained visual categorization. Int. J. Comput. Vis. **127**, 1–21 (2019). https://doi.org/10.1007/s11263-019-01176-2
16. Hu, T., Qi, H., Huang, Q., Lu, Y.: See better before looking closer: weakly supervised data augmentation network for fine-grained visual classification. arXiv preprint arXiv:1901.09891 (2019)

17. Imran, A., Athitsos, V.: Domain adaptive transfer learning on visual attention aware data augmentation for fine-grained visual categorization. In: Bebis, G., et al. (eds.) ISVC 2020. LNCS, vol. 12510, pp. 53–65. Springer, Cham (2020). https://doi.org/10.1007/978-3-030-64559-5_5
18. Ioffe, S., Szegedy, C.: Batch normalization: accelerating deep network training by reducing internal covariate shift. arXiv preprint arXiv:1502.03167 (2015)
19. Jaakkola, T., Haussler, D.: Exploiting generative models in discriminative classifiers. In: Advances in Neural Information Processing Systems, pp. 487–493 (1999)
20. Jégou, H., Douze, M., Schmid, C., Pérez, P.: Aggregating local descriptors into a compact image representation. In: 2010 IEEE Computer Society Conference on Computer Vision and Pattern Recognition, pp. 3304–3311. IEEE (2010)
21. Karimi, B., Wai, H.T., Moulines, É., Lavielle, M.: On the global convergence of (fast) incremental expectation maximization methods. In: Advances in Neural Information Processing Systems, pp. 2833–2843 (2019)
22. Khosla, A., Jayadevaprakash, N., Yao, B., Li, F.F.: Novel dataset for fine-grained image categorization: Stanford dogs. In: Proceedings of CVPR Workshop on Fine-Grained Visual Categorization (FGVC), vol. 2. Citeseer (2011)
23. Kingma, D.P., Ba, J.: Adam: a method for stochastic optimization. arXiv preprint arXiv:1412.6980 (2014)
24. Korsch, D., Bodesheim, P., Denzler, J.: Classification-specific parts for improving fine-grained visual categorization. In: Fink, G.A., Frintrop, S., Jiang, X. (eds.) DAGM GCPR 2019. LNCS, vol. 11824, pp. 62–75. Springer, Cham (2019). https://doi.org/10.1007/978-3-030-33676-9_5
25. Krause, J., et al.: The unreasonable effectiveness of noisy data for fine-grained recognition. In: Leibe, B., Matas, J., Sebe, N., Welling, M. (eds.) ECCV 2016. LNCS, vol. 9907, pp. 301–320. Springer, Cham (2016). https://doi.org/10.1007/978-3-319-46487-9_19
26. Liang, P., Klein, D.: Online EM for unsupervised models. In: Proceedings of Human Language Technologies: The 2009 Annual Conference of the North American Chapter of the Association for Computational Linguistics, pp. 611–619 (2009)
27. Lin, T.Y., RoyChowdhury, A., Maji, S.: Bilinear CNN models for fine-grained visual recognition. In: Proceedings of the IEEE International Conference on Computer Vision, pp. 1449–1457 (2015)
28. Loshchilov, I., Hutter, F.: Decoupled weight decay regularization. arXiv preprint arXiv:1711.05101 (2017)
29. Neal, R.M., Hinton, G.E.: A view of the EM algorithm that justifies incremental, sparse, and other variants. In: Jordan, M.I. (ed.) Learning in Graphical Models. ASID, vol. 89, pp. 355–368. Springer, Heidelberg (1998). https://doi.org/10.1007/978-94-011-5014-9_12
30. Perronnin, F., Dance, C.: Fisher kernels on visual vocabularies for image categorization. In: Proceedings of the IEEE Conference on Computer Vision and Pattern Recognition, pp. 1–8. IEEE (2007)
31. Perronnin, F., Sánchez, J., Mensink, T.: Improving the fisher kernel for large-scale image classification. In: Daniilidis, K., Maragos, P., Paragios, N. (eds.) ECCV 2010. LNCS, vol. 6314, pp. 143–156. Springer, Heidelberg (2010). https://doi.org/10.1007/978-3-642-15561-1_11
32. Russakovsky, O., et al.: ImageNet large scale visual recognition challenge. Int. J. Comput. Vision **115**(3), 211–252 (2015). https://doi.org/10.1007/s11263-015-0816-y

33. Simon, M., Gao, Y., Darrell, T., Denzler, J., Rodner, E.: Generalized orderless pooling performs implicit salient matching. In: Proceedings of the IEEE International Conference on Computer Vision, pp. 4970–4979 (2017)
34. Simon, M., Rodner, E., Darell, T., Denzler, J.: The whole is more than its parts? From explicit to implicit pose normalization. IEEE Trans. Pattern Anal. Mach. Intell. **42**, 1–13 (2018)
35. Simonyan, K., Vedaldi, A., Zisserman, A.: Deep fisher networks for large-scale image classification. In: Advances in Neural Information Processing Systems, pp. 163–171 (2013)
36. Song, Y., Zhang, F., Li, Q., Huang, H., O'Donnell, L.J., Cai, W.: Locally-transferred fisher vectors for texture classification. In: Proceedings of the IEEE International Conference on Computer Vision, pp. 4912–4920 (2017)
37. Sydorov, V., Sakurada, M., Lampert, C.H.: Deep fisher kernels-end to end learning of the fisher kernel GMM parameters. In: Proceedings of the IEEE Conference on Computer Vision and Pattern Recognition, pp. 1402–1409 (2014)
38. Szegedy, C., et al.: Going deeper with convolutions. In: Proceedings of the IEEE Conference on Computer Vision and Pattern Recognition, pp. 1–9 (2015)
39. Szegedy, C., Vanhoucke, V., Ioffe, S., Shlens, J., Wojna, Z.: Rethinking the inception architecture for computer vision. In: Proceedings of the IEEE Conference on Computer Vision and Pattern Recognition (2016)
40. Tang, P., Wang, X., Shi, B., Bai, X., Liu, W., Tu, Z.: Deep FisherNet for image classification. IEEE Trans. Neural Netw. Learn. Syst. **30**(7), 2244–2250 (2018)
41. Touvron, H., Vedaldi, A., Douze, M., Jégou, H.: Fixing the train-test resolution discrepancy. In: Advances in Neural Information Processing Systems, pp. 8250–8260 (2019)
42. Van Horn, G., et al.: Building a bird recognition app and large scale dataset with citizen scientists: the fine print in fine-grained dataset collection. In: Proceedings of the IEEE Conference on Computer Vision and Pattern Recognition, pp. 595–604 (2015)
43. Van Horn, G., et al.: The iNaturalist species classification and detection dataset. In: Proceedings of the IEEE Conference on Computer Vision and Pattern Recognition, pp. 8769–8778 (2018)
44. Wah, C., Branson, S., Welinder, P., Perona, P., Belongie, S.: The CALTECH-UCSD birds-200-2011 dataset. Technical report CNS-TR-2011-001, California Institute of Technology (2011)
45. Wiescholłek, P., Groh, F., Lensch, H.: Backpropagation training for fisher vectors within neural networks. arXiv preprint arXiv:1702.02549 (2017)
46. Yang, S., Liu, S., Yang, C., Wang, C.: Re-rank coarse classification with local region enhanced features for fine-grained image recognition. arXiv preprint arXiv:2102.09875 (2021)
47. Yang, Z., Luo, T., Wang, D., Hu, Z., Gao, J., Wang, L.: Learning to navigate for fine-grained classification. In: Ferrari, V., Hebert, M., Sminchisescu, C., Weiss, Y. (eds.) Computer Vision – ECCV 2018. LNCS, vol. 11218, pp. 438–454. Springer, Cham (2018). https://doi.org/10.1007/978-3-030-01264-9_26
48. Zhang, J., Zhang, R., Huang, Y., Zou, Q.: Unsupervised part mining for fine-grained image classification. arXiv preprint arXiv:1902.09941 (2019)
49. Zhang, L., Huang, S., Liu, W., Tao, D.: Learning a mixture of granularity-specific experts for fine-grained categorization. In: Proceedings of the IEEE International Conference on Computer Vision, pp. 8331–8340 (2019)

50. Zhang, X., Xiong, H., Zhou, W., Lin, W., Tian, Q.: Picking deep filter responses for fine-grained image recognition. In: Proceedings of the IEEE Conference on Computer Vision and Pattern Recognition, pp. 1134–1142 (2016)
51. Zheng, H., Fu, J., Mei, T., Luo, J.: Learning multi-attention convolutional neural network for fine-grained image recognition. In: Proceedings of the IEEE International Conference on Computer Vision (2017)
52. Zheng, H., Fu, J., Zha, Z.J., Luo, J.: Learning deep bilinear transformation for fine-grained image representation. In: Advances in Neural Information Processing Systems, pp. 4279–4288 (2019)
53. Zhuang, P., Wang, Y., Qiao, Y.: Learning attentive pairwise interaction for fine-grained classification. In: Proceedings of the AAAI Conference on Artificial Intelligence, vol. 34, pp. 13130–13137 (2020)

Investigating the Consistency of Uncertainty Sampling in Deep Active Learning

Niklas Penzel[1(✉)], Christian Reimers[1], Clemens-Alexander Brust[1], and Joachim Denzler[1,2]

[1] Computer Vision Group, Friedrich Schiller University Jena, Ernst-Abbe-Platz 2, 07743 Jena, Germany
{niklas.penzel,christian.reimers,clemens-alexander.brust, joachim.denzler}@uni-jena.de
[2] Institute of Data Science, German Aerospace Center (DLR), Mälzerstraße 3, 07745 Jena, Germany

Abstract. Uncertainty sampling is a widely used active learning strategy to select unlabeled examples for annotation. However, previous work hints at weaknesses of uncertainty sampling when combined with deep learning, where the amount of data is even more significant. To investigate these problems, we analyze the properties of the latent statistical estimators of uncertainty sampling in simple scenarios. We prove that uncertainty sampling converges towards some decision boundary. Additionally, we show that it can be inconsistent, leading to incorrect estimates of the optimal latent boundary. The inconsistency depends on the latent class distribution, more specifically on the class overlap. Further, we empirically analyze the variance of the decision boundary and find that the performance of uncertainty sampling is also connected to the class regions overlap. We argue that our findings could be the first step towards explaining the poor performance of uncertainty sampling combined with deep models.

Keywords: Uncertainty sampling · Consistency · Active learning

1 Introduction

Annotating data points is a laborious and often expensive task, especially if highly trained experts are necessary, *e.g.*, in medical areas. Hence, intelligently selecting data points out of a large collection of unlabeled examples to most efficiently use human resources is a critical problem in machine learning. Active learning (AL) is one approach that tackles this problem by iteratively using a selection strategy based on a classifier trained on some initially labeled data.

Supplementary Information The online version contains supplementary material available at https://doi.org/10.1007/978-3-030-92659-5_10.

C. Bauckhage et al. (Eds.): DAGM GCPR 2021, LNCS 13024, pp. 159–173, 2021.
https://doi.org/10.1007/978-3-030-92659-5_10

There are many different selection strategies possible in the AL framework [9, 11,17,24,26,27,29].

One popular strategy is uncertainty sampling, which was developed to train statistical text classifiers [17]. Uncertainty sampling uses a metric to estimate the uncertainty of the model prediction and queries the examples where the classifier is most uncertain. The original metric is the confidence of the classifier's prediction, but other popular metrics include the distance to the decision boundary, *i.e.*, margin sampling [23], and the entropy of the posterior distribution [28]. Uncertainty sampling performs well when combined with classical machine learning models, *e.g.* conditional random fields [25], support vector machines [18], or decision trees [19].

However, current state-of-the-art classification models are deep architectures like convolutional neural networks (CNNs). Combining sophisticated AL with CNNs can result in only marginal improvements [4]. One possible explanation for this behavior could be the bias introduced by the sampling strategy [21].

Towards understanding this unexpected behavior, we analyze uncertainty sampling in one-dimensional scenarios and derive the usually latent estimators of the AL system. The scenarios we investigate are closely related to binary logistic regression. Further, we can interpret the softmax activation of the output layer of a CNN as multinomial logistic regression [14, p. 266] in the extracted feature space. Hence, we argue that our approach relates to the stated goal.

Our main contribution is proving that a simple active learning system using uncertainty sampling converges against some decision boundary. We do this by analyzing the statistical estimators introduced by uncertainty sampling. To the best of our knowledge, we are the first to investigate uncertainty sampling on the level of the resulting statistical estimators. We find that the consistency depends on the latent class distribution. Furthermore, our empirical analysis reveals that the performance depends highly on the overlap of the latent class regions.

After introducing the problem in Sect. 3, we state our main findings in Sect. 4, including the proof that uncertainty sampling possibly converges towards undesired decision boundaries. Furthermore, we empirically validate and extend our findings in Sect. 5.

2 Related Work

Multiple authors report poor performances when combining different AL strategies with deep models, *i.e.*, CNNs [4,21]. Chan *et al.* conduct an ablation study with self-supervised initialization, semi-supervised learning, and AL [4]. They find that the inclusion of AL only leads to marginal benefits and speculate that the pretraining and semi-supervision already subsume the advantages.

Mittal *et al.* focus their critique on the evaluation scheme generally used to assess deep AL methods [21]. They find that changes in the training procedure significantly impact the behavior of the selection strategies. Furthermore, employing data augmentations makes the AL strategies hard to distinguish but increases overall performance. They speculate that AL may introduce a bias into the distribution of the labeled examples resulting in undesired behavior.

One bias of uncertainty sampling is visualized in the work of Sener and Savarese [24]. They use t-SNE [20] to visualize the coverage of the CIFAR-10 dataset [16] when selecting examples using uncertainty sampling. A bias towards certain areas of the feature space and a lack of selected points in other regions is visible. In contrast, the approach of Sener and Savarese leads by design to a more even coverage indicating that t-SNE truly uncovers a bias of uncertainty sampling large batches. Such a bias could be advantageous and necessary to outperform randomly sampling new points, but their empirical evaluation shows that uncertainty sampling does not perform better than random sampling.

In contrast to these previous works, we are focussing on identifying the issues of uncertainty sampling by performing a detailed theoretic analysis of an AL system in simple scenarios. We theoretically derive and investigate the usually latent parameter estimators of the decision boundary.

A significant theoretic result is provided in the work of Dasgupta [6]. They prove there is no AL strategy that can outperform random sampling in all cases. In other words, there are datasets or data distributions where randomly selecting new examples is the optimal strategy.

Additionally, Dasgupta *et al.* [7] theoretically analyze the rate of convergence of a perceptron algorithm in an AL system, but they assume linear separable classes. Similarly, the analysis of the query by committee strategy by Freund *et al.* [8] also assumes that a perfect solution exists. We do not need this assumption in our analysis. Balcan *et al.* theoretically investigate the rate of convergence of the agnostic AL strategy without the assumption of an ideal solution [2]. Also related to our theoretical approach is the work of Mussmann and Liang [22]. They focus on uncertainty sampling and logistic regression and theoretically derive bounds for the data efficiency given the inverse error rate. The authors also note that the data efficiency decreases if the means of two generated normally distributed classes are moved closer together. In contrast to these works, we focus on the consistency of the decision boundary estimators instead of the data efficiency or rate of convergence.

3 Problem Setting

Here, we describe the classifier and estimators we want to analyze in Sect. 4. We start by describing a simple binary one-dimensional problem. Let us assume there is some latent mixture consisting of two components. These components define the class distributions and can be described by the density functions p_1 and p_2. We want to determine the class of a point, *i.e.*, whether it was drawn from the distribution of class 1 or class 2. Towards this goal, we want to estimate the optimal decision boundary M. Given such a decision boundary, the classifier is a simple threshold operation.

Let $\mathcal{X}_1$ and $\mathcal{X}_2$ be sets of sample points drawn from classes 1 and 2, respectively. We estimate the decision boundary without knowledge of the latent distribution by assuming that both classes are normally distributed. To further simplify the problem, we also assume that both classes are equally likely, *i.e.*, the mixture weights are $\frac{1}{2}$, and that both classes share the same variance σ^2. This approach is closely related to linear discriminant analysis (LDA) [14, p. 242]. LDA also assumes normal distributions but results in logistic functions describing the probability that an example is of a certain class. Another related approach is logistic regression [14, p. 250] where a logistic function is estimated using the maximum likelihood principle. Though in our simple scenario, both LDA and logistic regression result in the same decision boundary. In contrast, we directly estimate the decision boundary. Our approach enables us to study the statistical properties of the related estimators.

Under the described assumptions, the important estimators are

$$\hat{\mu}_i = \frac{1}{|\mathcal{X}_i|} \sum_{X \in \mathcal{X}_i} X, \quad (1) \qquad \hat{M} = \frac{\hat{\mu}_1 + \hat{\mu}_2}{2}, \quad (2)$$

where $i \in \{1, 2\}$ denotes the class. The mean estimators $\hat{\mu}_i$ are the maximum likelihood estimators for Gaussians [3, p. 93]. A derivation for the decision boundary estimator $\hat{M}$ can be found in Appendix A. Note that the shared variance σ^2 is not needed to estimate M. Hence, we can ignore the variance of the class distributions and do not need to estimate it.

The AL system estimates the means and decision boundaries for multiple time steps $t \geq 0$, resulting in a sequence of estimators. In the beginning we start with m_0 examples per class, *i.e.*, $|\mathcal{X}_1^{(0)}| = |\mathcal{X}_2^{(0)}| = m_0$ holds. During each time step t, we select one example X_{t+1} that we annotate before estimating the next set of parameters. Therefore, in our AL system, we get the estimators.

$$\hat{\mu}_{t,i} = \frac{1}{|\mathcal{X}_i^{(t)}|} \sum_{X \in \mathcal{X}_i^{(t)}} X, \quad (3) \qquad \hat{M}_t = \frac{\hat{\mu}_{t,1} + \hat{\mu}_{t,2}}{2}, \quad (4)$$

in all time steps $t \geq 0$. To analyze the convergence of such a uncertainty sampling using system in Sect. 4 we additionally introduce a concrete update formula for the mean estimators:

$$\hat{\mu}_{t+1,i} = \frac{|\mathcal{X}_i^{(t)}|}{|\mathcal{X}_i^{(t)}| + 1} \hat{\mu}_{t,i} + \frac{X_{t+1}}{|\mathcal{X}_i^{(t)}| + 1}. \quad (5)$$

A complete derivation of Eq. (5) can be found in Appendix B.

To select examples X_{t+1}, let us assume we can generate a sample according to the latent distribution, *i.e.*, we do not have a pool of finitely many unlabeled examples. This does not assume knowledge of the latent distribution but merely that we can run the process that generates examples. Sampling according to the latent distribution is known as the random baseline in AL. After querying an example, we employ experts to annotate and add it to the corresponding class set $\mathcal{X}_i^{(t+1)}$. In the following time step, we estimate updated parameters that potentially better fit the latent distribution.

To perform uncertainty sampling instead, we use an uncertainty metric to assess different examples and query the example where the classifier is most uncertain. In our scenario, we use these metrics to calculate the example where our classifier is most uncertain. Afterward, experts label this example according to the latent class distribution at this specific point. There are three common uncertainty metrics and we show in the following that they are equivalent in our scenario.

Claim. Given the described classifier, the uncertainty metrics (i) least confidence, (ii) margin, and (iii) entropy are equivalent and generate the same point.

Proof. To validate the claim, it is enough to show that the example generated by all three metrics is the same in any given time step t.

(i) Least confidence sampling queries the sample where the prediction confidence of the classifier is minimal. In a binary problem, this confidence is at least $\frac{1}{2}$, or else we predict the other class. The point where the probability for both classes is equal is exactly the intersection of the latest estimations of our Gaussian mixture components. Hence, the point X_{t+1} where the classifier is least confident is precisely the last decision boundary estimate $\hat{M}_t$.
(ii) Margin sampling selects the point closest to a decision boundary of the classifier. Here, margin sampling chooses the best, *i.e.*, the latest estimation of the decision boundary. Hence, the new point X_{t+1} is exactly $\hat{M}_t$.
(iii) When using entropy as an uncertainty metric, we query the example that maximizes the posterior distribution's entropy. The categorical distribution that maximizes the entropy is the uniform distribution over both classes [5]. Given our classifier, this is precisely the decision boundary where both classes are equally likely. Hence, X_{t+1} is our latest decision boundary estimate $\hat{M}_t$.

□

In this one-dimensional scenario, three commonly used uncertainty metrics select the same point which allows us to simplify our investigation. Instead of looking at different metrics, we will analyze the system that queries and annotates the latest decision boundary estimate in each step, *i.e.*, $X_{t+1} = \hat{M}_t$. Note that adding the last decision boundary estimate to either of the two classes leads to future estimates of the means being skewed towards our estimations of M.

The theoretical part of our analysis focuses mainly on AL, where we have no unlabeled pool but can generate examples directly instead. Before we start to analyze the consistency and convergence of such an AL system, we want to briefly discuss the differences to a finite pool of unlabeled data points. In real-world problems, we often do not have access to the example generating system but instead a fixed number T of unlabeled examples. Let the set $\mathfrak{U}$ be the unlabeled pool with $|\mathfrak{U}| = T$. Each datapoint X from $\mathfrak{U}$ belongs either to class 1 or 2.

Let us assume that $\mathfrak{U}$ is sampled from the undistorted latent distribution defining the problem. Given the sequence $(\mathfrak{U}_t)_{t\in\{0,\ldots,T\}}$ of examples still unknown in time step t, we observe that $\mathfrak{U} = \mathfrak{U}_0 \supset \mathfrak{U}_1 \supset \ldots \supset \mathfrak{U}_{T-1} \supset \mathfrak{U}_T = \emptyset$ applies.

Given such a finite pool, random sampling selects an element from $\mathfrak{U}_t$ uniformly, which is equivalent to sampling from the latent distribution. In contrast, uncertainty sampling selects the example

$$X_{t+1} = \underset{X \in \mathfrak{U}_t}{\arg\min} |\hat{M}_t - X|, \tag{6}$$

closest to the latest decision boundary estimate, because the actual $\hat{M}_t$ is likely not included in $\mathfrak{U}_t$.

4 Theoretical Investigation of Uncertainty Sampling

Towards the goal of analyzing the convergence of an uncertainty sampling system, we look at Eq. (4). Given this definition of $\hat{M}_t$ we know that exactly one of the statements $\hat{\mu}_{t,1} < \hat{M}_t < \hat{\mu}_{t,2}$, $\hat{\mu}_{t,1} > \hat{M}_t > \hat{\mu}_{t,2}$, or $\hat{\mu}_{t,1} = \hat{M}_t = \hat{\mu}_{t,2}$ applies. We now analyze these cases to show that uncertainty sampling converges.

We start with the case $\hat{\mu}_{t,1} = \hat{M}_t = \hat{\mu}_{t,2}$. In this case, the AL system already converged. No matter to which class we add $\hat{M}_t$, both $\hat{\mu}_{t,1}$ and $\hat{\mu}_{t,2}$ will not change for $t \to \infty$ which follows directly from Eq. (5).

Let us now look at the other two cases. We note that in a time step t the variables $\hat{\mu}_{t,1}$ and $\hat{\mu}_{t,2}$ define a random interval. Without loss of generality let us assume $\hat{\mu}_{t,1} < \hat{M}_t < \hat{\mu}_{t,2}$ applies. In step t, $\hat{M}_t$ is annotated and used to calculate $\hat{\mu}_{t+1,1}$ and $\hat{\mu}_{t+1,2}$. Let us assume the label turns out to be one. Then $\hat{\mu}_{t+1,2}$ is equal to $\hat{\mu}_{t,2}$. In contrast, we know that $\hat{\mu}_{t+1,1} > \hat{\mu}_{t,1}$ because of Eq. (5) and $\hat{M}_t > \hat{\mu}_{t,1}$. Further, we use Eq. (4) to see that $\hat{\mu}_{t,1} < \hat{\mu}_{t+1,1} < \hat{M}_{t+1} < \hat{\mu}_{t+1,2} = \hat{\mu}_{t,2}$ holds. The consequence is $[\hat{\mu}_{t+1,1}, \hat{\mu}_{t+1,2}] \subset [\hat{\mu}_{t,1}, \hat{\mu}_{t,2}]$. We can derive the same result if $\hat{M}_t$ is added to class 2 because then $\hat{\mu}_{t+1,1} = \hat{\mu}_{t,1}$ and $\hat{\mu}_{t+1,2} < \hat{\mu}_{t,2}$.

To show that uncertainty sampling converges, we now analyze these nested intervals. Given the sequences of estimators $(\hat{\mu}_{t,1})_{t \in \mathbb{N}_0}$, $(\hat{\mu}_{t,2})_{t \in \mathbb{N}_0}$ and $(\hat{M}_t)_{t \in \mathbb{N}_0}$, we already know that $\hat{M}_t \in [\hat{\mu}_{t,1}, \hat{\mu}_{t,2}]$ and $[\hat{\mu}_{t+1,1}, \hat{\mu}_{t+1,2}] \subset [\hat{\mu}_{t,1}, \hat{\mu}_{t,2}]$. If the length of nested intervals becomes arbitrarily small, then the intersection of these nested intervals is a single number [10, p. 29]. In other words, if the length converges towards zero then $\hat{\mu}_{t,1} \leq \hat{M}_\infty \leq \hat{\mu}_{t,2}$ is true for all $t \geq 0$ and for some value $\hat{M}_\infty$. It is enough to show that the nested intervals $[\hat{\mu}_{t,1}, \hat{\mu}_{t,2}]$ become arbitrarily small to prove that uncertainty sampling converges in our scenario.

Theorem 1. *The nested intervals $[\hat{\mu}_{t,1}, \hat{\mu}_{t,2}]$ in our one-dimensional scenario become arbitrarily small for $t \to \infty$.*

Proof. Let ϵ_t be the length of the interval $I_t = [\hat{\mu}_{t,1}, \hat{\mu}_{t,2}]$ in time step t. It is given by $\epsilon_t = \hat{\mu}_{t,2} - \hat{\mu}_{t,1}$. To prove that these interval lengths become arbitrarily small, we must show

$$\forall \delta > 0 : \lim_{t \to \infty} \epsilon_t < \delta. \tag{7}$$

The decision boundary in step t is exactly the middle of the interval $[\hat{\mu}_{t,1}, \hat{\mu}_{t,2}]$. This fact leads to the observations

$$\hat{M}_t = \hat{\mu}_{t,1} + \frac{\epsilon_t}{2} = \hat{\mu}_{t,2} - \frac{\epsilon_t}{2}. \tag{8}$$

In time step t let there be k_t examples in class 1 additionally to the m_0 initial samples, *i.e.*, $|\mathcal{X}_1^t| = m_0 + k_t$. Consequently, we know class 2 contains $|\mathcal{X}_2^t| = m_0 + t - k_t$ examples. Furthermore, there are two possible labels for $\hat{M}_t$.

Case 1: Let $\hat{M}_t$ be labeled as class 1. By using Eq. (5), Eq. (8) and some algebraic manipulations, we get the interval boundaries of I_{t+1}:

$$\hat{\mu}_{t+1,2} = \hat{\mu}_{t,2}, \text{ and} \tag{9}$$

$$\hat{\mu}_{t+1,1} = \frac{(m_0 + k_t)\hat{\mu}_{t,1} + \hat{M}_t}{m_0 + k_t + 1} = \frac{(m_0 + k_t)\hat{\mu}_{t,1} + \hat{\mu}_{t,1} + \frac{\epsilon_t}{2}}{m_0 + k_t + 1} \tag{10}$$

$$= \hat{\mu}_{t,1} + \frac{\epsilon_t}{2(m_0 + k_t + 1)}. \tag{11}$$

Therefore, the length ϵ_{t+1} of the interval I_{t+1} is

$$\epsilon_{t+1} = \hat{\mu}_{t,2} - \hat{\mu}_{t,1} - \frac{\epsilon_t}{2(m_0 + k_t + 1)} \tag{12}$$

$$= \epsilon_t - \frac{\epsilon_t}{2(m_0 + k_t + 1)} \leq \epsilon_t - \frac{\epsilon_t}{2(m_0 + t + 1)}. \tag{13}$$

Case 2: Let $\hat{M}_t$ be labeled as class 2. By using Eq. (5), Eq. (8) and some algebraic manipulations, we get the interval boundaries of I_{t+1}:

$$\hat{\mu}_{t+1,1} = \hat{\mu}_{t,1}, \text{ and} \tag{14}$$

$$\hat{\mu}_{t+1,2} = \frac{(m_0 + t - k_t)\hat{\mu}_{t,2} + \hat{M}_t}{m_0 + t - k_t + 1} = \frac{(m_0 + t - k_t)\hat{\mu}_{t,2} + \hat{\mu}_{t,2} - \frac{\epsilon_t}{2}}{m_0 + t - k_t + 1} \tag{15}$$

$$= \hat{\mu}_{t,2} - \frac{\epsilon_t}{2(m_0 + t - k_t + 1)}. \tag{16}$$

Therefore, the length ϵ_{t+1} of the interval I_{t+1} is

$$\epsilon_{t+1} = \hat{\mu}_{t,2} - \frac{\epsilon_t}{2(m_0 + t - k_t + 1)} - \hat{\mu}_{t,1} \tag{17}$$

$$= \epsilon_t - \frac{\epsilon_t}{2(m_0 + t - k_t + 1)} \leq \epsilon_t - \frac{\epsilon_t}{2(m_0 + t + 1)}. \tag{18}$$

In both cases we derive an upper bound for the length of the interval by increasing the denominator of a negative fraction and get

$$\epsilon_{t+1} \leq \epsilon_t - \frac{\epsilon_t}{2(m_0 + t + 1)} = \underbrace{\left(1 - \frac{1}{2(m_0 + t + 1)}\right)}_{=:l_t} \epsilon_t. \tag{19}$$

We now use this property recursively until we reach time step 0 resulting in

$$\epsilon_{t+1} \leq \epsilon_0 \prod_{j=0}^{t} l_j = \epsilon_0 \exp\left(\sum_{j=0}^{t} \log(l_j)\right). \tag{20}$$

Using this bound (Eq. (20)) and recalling Eq. (7), we must now show

$$\forall \delta > 0 : \lim_{t \to \infty} \epsilon_0 \exp\left(\sum_{j=0}^{t-1} \log(l_j)\right) < \delta. \tag{21}$$

Dividing by ϵ_0, we see it is enough to show that $\sum_{j=0}^{\infty} \log(l_j)$ diverges towards $-\infty$. Towards this goal, we use a known bound [1] of the natural logarithm and derive an upper bound for $\log(l_t)$

$$\log(l_t) \leq -\frac{1}{2m_0 + 2}\left(\frac{1}{t+1}\right). \tag{22}$$

The complete derivation can be found in Appendix C.

We can use this bound for $\log(l_t)$ and some algebraic manipulations to derive an upper bound for the limit we are interested in

$$\lim_{t \to \infty} \sum_{j=0}^{t-1} \log(l_j) \leq \lim_{t \to \infty} \sum_{j=0}^{t-1} -\frac{1}{2m_0 + 2}\left(\frac{1}{j+1}\right) = \lim_{t \to \infty} -\frac{1}{2m_0 + 2} \sum_{j=1}^{t} \frac{1}{j} \tag{23}$$

$$= -\frac{1}{2m_0 + 2} \sum_{j=1}^{\infty} \frac{1}{j}. \tag{24}$$

The harmonic series multiplied by a negative constant is an upper bound for the limit of the series of $\log(l_t)$. This upper bound diverges towards $-\infty$ because the harmonic series itself diverges towards ∞ [15]. Consequently, we know

$$\lim_{t \to \infty} \sum_{j=0}^{t-1} \log(l_j) = -\infty. \tag{25}$$

Using this limit as well as the facts $\delta > 0$ and $\epsilon_0 > 0$, we see

$$\forall \delta > 0 : \lim_{t \to \infty} \exp\left(\sum_{j=0}^{t-1} \log(l_j)\right) = 0 < \frac{\delta}{\epsilon_0}. \tag{26}$$

Hence, the interval lengths ϵ_t converge towards zero for $t \to \infty$. □

Until now, our analysis is independent of the specific latent class distribution. We have shown that in the one-dimensional case estimating the decision boundary M of a mixture of two classes for a growing number of examples using uncertainty sampling as a selection strategy converges towards some value $\hat{M}_\infty$. The question about the consistency of uncertainty sampling now reduces to: Is $\hat{M}_\infty$ equal to the optimal decision boundary M?

Let us assume, for example, that the latent distribution is a mixture of Gaussians. As already observed, $\hat{\mu}_{t,1} < \hat{M}_t < \hat{\mu}_{t,2}$ and $\hat{\mu}_{t+1,1} < \hat{M}_{t+1} < \hat{\mu}_{t+1,2}$ apply. Also either $\hat{\mu}_{t+1,1} > \hat{\mu}_{t,1}$ or $\hat{\mu}_{t+1,2} < \hat{\mu}_{t,2}$ holds true. if for any time step t,

$\hat{\mu}_{t,1}$ becomes larger than M, then the decision boundary M is not reachable anymore. In this case, $\hat{M}_\infty > M$ applies. The decision boundary M is likewise unreachable, if for any time step t, $\hat{\mu}_{t,2} < M$ applies. This behavior is possible because the probability density function of Gaussians is greater than zero for all possible examples $\hat{M}_t$ selected by uncertainty sampling. Hence, it is possible that $\hat{M}_t$ converges to a value $\hat{M}_\infty \neq M$. The stochastic process $\{\hat{M}_t\}_{t \in \mathbb{N}_0}$ defined by uncertainty sampling is not consistent given a mixture of Gaussians.

For further analysis, we look at the overlap ξ of the latent mixture [13]. Let the latent distribution be a mixture consisting of two components with the densities p_1 and p_2. The overlap of such a mixture is defined as

$$\xi = \int_{\mathbb{R}} \min(p_1(x), p_2(x))dx. \tag{27}$$

The behavior we determined for a mixture of two Gaussians occurs because both densities are greater than zero for all possible values. Hence, the overlap is greater than zero. More examples could lead to a wrong and unfixable decision boundary if the latent class regions overlap. We use the example of two Gaussians later on in Sect. 5.1 to empirically estimate the likelihood of such an event.

Let us now look at distributions without overlap. Without loss of generality let $\mu_1 < \mu_2$. If the latent distribution has separate class regions, *e.g.*, a uniform mixture, where the class regions are next to each other, $\hat{M}_t > M$ cannot be added to class 1. Equivalently a $\hat{M}_t < M$ can never be added to class 2 because the density of class 2 at such a point is zero. Assuming the density of the latent mixture is greater than zero for all $\hat{M}_t$, then these estimators of the decision boundary are consistent because $\forall t \in \mathbb{N}_0 : M \in [\hat{\mu}_{t,1}, \hat{\mu}_{t,2}]$.

In contrast, let us look at the case of a pool of T unlabeled examples $\mathfrak{U}$. Then the resulting sequence of intervals $[\hat{\mu}_{t,1}, \hat{\mu}_{t,2}]$ are not necessarily nested. There is the possibility that a later interval can be larger than the interval in step t if an example outside the interval is closest to the decision boundary estimate and therefore selected. Another way to think about it is to observe that for $t \to T$, uncertainty sampling tends towards random sampling because $\mathfrak{U}$ is an unbiased random sample from the latent distribution.

Let us assume our annotation budget is quite limited. We can only label T' out of a vast pool of T unlabeled examples. Under these circumstances, we claim that the undesirable properties of the infinite case of uncertainty sampling derived in this section approximately apply. In Sect. 5.2, we give empirical evidence towards this claim.

5 Empirical Investigation of Uncertainty Sampling

To corroborate our theoretical analysis in Sect. 4, we further investigate the consistency, convergence and performance of uncertainty sampling. In Sect. 5.1 we investigate the likelihood that the optimal decision boundary becomes unachievable. To analyze the performance we look at the empirical variance of the decision boundary estimators in Sect. 5.2.

5.1 Inconsistency Given Overlapping Class Regions

Section 4 shows that uncertainty sampling can lead to inconsistent estimators depending on the overlap of the latent mixture. Further, there are sequences of estimates where the optimal decision boundary can never be achieved. In this section, we empirically evaluate how likely such a scenario occurs. Towards this goal we want to estimate $P(M \in [\hat{\mu}_{t,1}, \hat{\mu}_{t,2}])$, *i.e.* the probability that the latent decision boundary is still achievable in step t.

To approximate these probabilities, we train an AL system 1,000 times and estimate the parameters μ_1, μ_2, and M in 500 consecutive time steps. In a given step t, the label of the estimate $\hat{M}_t$ depends on the latent mixture at this point. As a latent distribution, we select a mixture of two Gaussians with the same variance. This setup is a perfect fit for our classifier and contains an optimal latent decision boundary, *i.e.*, the intersection of both densities. To now analyze different overlapping scenarios, we repeat the experiment with different combinations of mixture parameters. We set the means either to -1 and 1, or -3 and 3, respectively. Regarding the variance, we evaluate values between 0.1 and 10.

Furthermore, to investigate if the probabilities $P(M \in [\hat{\mu}_{t,1}, \hat{\mu}_{t,2}])$ only depend on the overlap of the latent mixture, we also run the experiment for multiple parameter configurations with approximately the same overlap ξ. These configurations can be found in Table 1 in Appendix D.

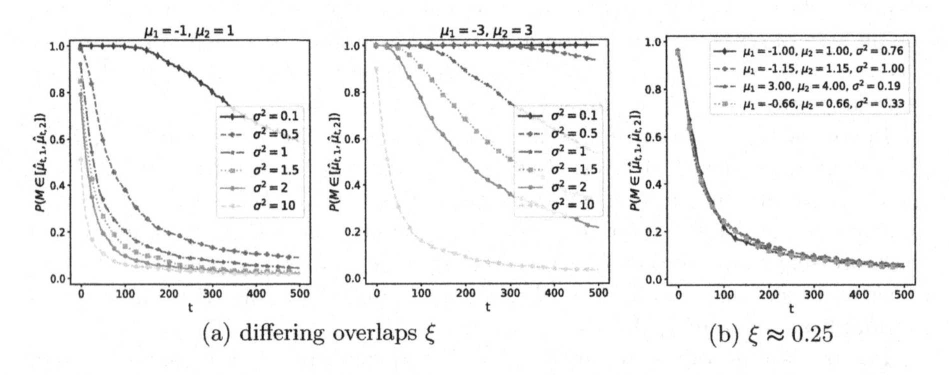

(a) differing overlaps ξ (b) $\xi \approx 0.25$

Fig. 1. Estimations of the probabilities $P(M \in [\hat{\mu}_{t,1}, \hat{\mu}_{t,2}])$ for multiple parameter configurations of Gaussian mixtures. a. contains parameter compositions with differing overlap ξ. In contrast, b. displays the course of $P(M \in [\hat{\mu}_{t,1}, \hat{\mu}_{t,2}])$ for different parameter combinations defining mixtures with an overlap $\xi \approx 0.25$.

Results: Figure 1a displays the empirical estimations of the probabilities $P(M \in [\hat{\mu}_{t,1}, \hat{\mu}_{t,2}])$. We can see that the possibility that the optimal decision boundary M is achievable declines for later time steps. A smaller σ^2 and larger distances between both means, *i.e.*, more separate class regions, correlate with an increased likelihood that M is still learnable.

The influence of the overlap ξ of the latent mixture on the course of $P(M \in [\hat{\mu}_{t,1}, \hat{\mu}_{t,2}])$ can be seen in Fig. 1b. All different parameter configurations with approximately the same overlap result in roughly the same curve.

We do not know if the probabilities $P(M \in [\hat{\mu}_{t,1}, \hat{\mu}_{t,2}])$ converge towards zero for $t \to \infty$ and $\xi > 0$ but our results point in that direction. Whether this conjecture holds true or not, we find that given $\xi > 0$, uncertainty sampling is not only inconsistent, but for later time steps t, it is also unlikely that the optimal decision boundary is still achievable.

5.2 Empirical Variance of the Decision Boundary

To further analyze uncertainty sampling especially compared to random sampling, we investigate the estimators $\hat{M}_t$ using both strategies. To compare two estimators for the same parameter, we look at their respective variances. A smaller variance leads, on average, to better estimations. We approximate the variances of $\hat{M}_t$ by simulating AL systems employing uncertainty sampling or random sampling, 10,000 times with $m_0 = 3$ initial examples per class.

First, we use multiple pairs of Gaussians with approximately the same overlap ξ. Table 1 in Appendix D lists all parameter configurations.

Second, we look at two latent mixtures sharing the same class means and variance to compare separate class regions and overlapping class regions. The first one is the mixture of uniform distributions $\mathcal{U}(-2, 0)$ and $\mathcal{U}(0, 2)$. The second mixture is the mixture of Gaussians $\mathcal{N}(-1, \frac{1}{3}^2)$ and $\mathcal{N}(1, \frac{1}{3}^2)$. Derivation of these parameter configurations can be found in Appendix E.

We analyze these scenarios by first running the experiment 10,000 times, starting with $m_0 = 3$ examples per class and then estimating the empirical variance. Finally, to additionally investigate the case of a large unlabeled pool of examples, we draw $T = 25{,}006$ examples balanced from both classes in each experiment round. After three initial examples per class in $t = 0$, a pool of exactly 25,000 unlabeled examples remains. From this pool, we use Eq. (6) to select the unlabeled example closest to the estimated decision boundary in each time step. We expect similar results as in the case of AL systems directly sampling from the latent mixture.

Results. Figure 2 displays the variances of $\hat{M}_t$ for multiple time steps given latent mixtures with $\xi \approx 0.25$. We can see for all four parameter configurations that the variances converge approximately equally fast. However, they converge towards different values. The exact value seems to depend on the variance σ^2.

Figure 3 displays the results for the other empirical variance experiments. These results include the overlapping Gaussian mixtures and the separate uniform mixtures in both the finite unlabeled pool and generating examples cases. The first observation is that the results given a finite pool of unlabeled examples are nearly identical to our other results. Consequently, we conclude that our results for the infinite case of generating samples to be labeled are approximately valid for querying from a large pool of unlabeled examples.

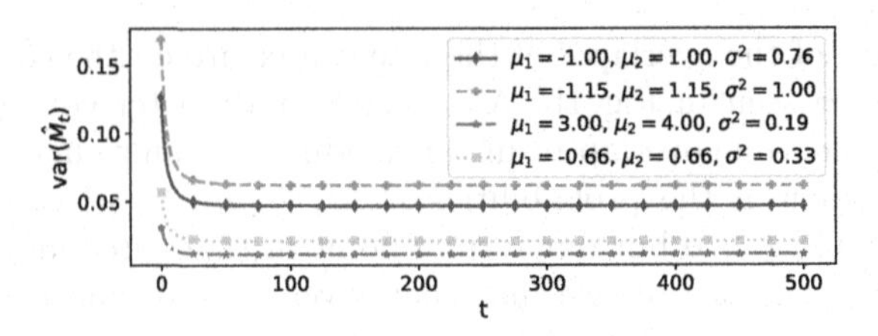

Fig. 2. The estimated empirical variance of the decision boundary estimates inbetween time steps $t = 0$ and $t = 500$. The parameter configurations define mixtures with approximately the same overlap $\xi \approx 0.25$.

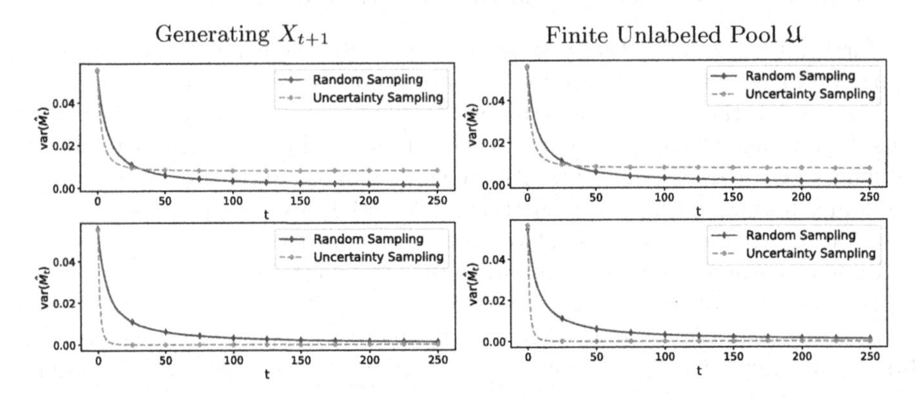

Fig. 3. The first row of plots shows the time evolution of the empirical variances of the decision boundaries in an AL scenario with overlapping class regions represented by a Gaussian mixture. The second row shows the equivalent results for a scenario with separate class regions represented by a mixture of uniform distributions. In contrast, the two columns denote generating unlabeled examples and selecting points from a finite pool, respectively.

Uncertainty sampling outperforms random sampling in the case of separate class regions. However, if the class regions overlap, uncertainty sampling performs marginally better for the first few steps but converges towards a higher, *i.e.*, worse, variance. This behavior occurs because, given overlapping class regions, the AL system converges towards different values in different experiment rounds (Sect. 5.1). Hence, given overlapping class regions, the variance of $\hat{M}_t$ is greater than zero for all time steps. In contrast, random sampling results in consistent mean estimators of the Gaussian mixture components [12, p. 217]. Hence, they stochastically converge towards μ_1 and μ_2, respectively. Consequently, the estimates of the decision boundary converge stochastically towards the optimal boundary resulting in zero variance for $t \to \infty$.

Summarizing this experiment: uncertainty sampling can outperform random sampling, but it highly depends on the latent class distribution. One reason could be the inaccurate intuition that selecting points close to the decision boundary reduces the uncertainty in this area. This intuition only holds if the class regions of the latent distribution are separate. Given overlapping class regions, some uncertainty always remains, leading to the observed performance decrease of uncertainty sampling.

6 Conclusions

We analyzed the latent estimators of an AL system performing uncertainty sampling in one-dimensional scenarios. We find that uncertainty sampling converges, but performance and consistency depend on the overlap of the latent distribution. Our results are congruent with the work of Mussmann and Liang [22], who also find worse properties for more overlapping class regions.

Towards understanding the problems uncertainty sampling causes for deep AL systems, we reduce the classifier to the backbone feature extractor and the classification output layer. Given a CNN, we can interpret the softmax activation of the output layer as a multinomial logistic regression [14, p. 266] in the extracted feature space. Our analysis is closely related to the binary case logistic regression. Hence we argue, that this work is a first step towards understanding the underwhelming performance of uncertainty sampling combined with deep classifiers.

References

1. Upper bound of natural logarithm. https://proofwiki.org/wiki/Upper_Bound_of_Natural_Logarithm. Accessed 19 May 2021
2. Balcan, M.F., Beygelzimer, A., Langford, J.: Agnostic active learning. In: Proceedings of the 23rd International Conference on Machine Learning, ICML '06, pp. 65–72. Association for Computing Machinery, New York (2006). https://doi.org/10.1145/1143844.1143853
3. Bishop, C.M.: Pattern Recognition and Machine Learning. Information Science and Statistics, Springer, Heidelberg (2006)
4. Chan, Y.C., Li, M., Oymak, S.: On the marginal benefit of active learning: does self-supervision eat its cake? In: ICASSP 2021–2021 IEEE International Conference on Acoustics, Speech and Signal Processing (ICASSP), pp. 3455–3459. IEEE (2021)
5. Conrad, K.: Probability distributions and maximum entropy (2005)
6. Dasgupta, S.: Analysis of a greedy active learning strategy. In: Saul, L., Weiss, Y., Bottou, L. (eds.) Advances in Neural Information Processing Systems, vol. 17. MIT Press (2005). https://proceedings.neurips.cc/paper/2004/file/c61fbef63df5ff317aecdc3670094472-Paper.pdf
7. Dasgupta, S., Kalai, A.T., Monteleoni, C.: Analysis of perceptron-based active learning. In: Auer, P., Meir, R. (eds.) COLT 2005. LNCS (LNAI), vol. 3559, pp. 249–263. Springer, Heidelberg (2005). https://doi.org/10.1007/11503415_17
8. Freund, Y., Seung, H.S., Shamir, E., Tishby, N.: Information, prediction, and query by committee. In: Hanson, S., Cowan, J., Giles, C. (eds.) Advances in Neural Information Processing Systems. vol. 5. Morgan-Kaufmann (1993). https://proceedings.neurips.cc/paper/1992/file/3871bd64012152bfb53fdf04b401193f-Paper.pdf

9. Freytag, A., Rodner, E., Denzler, J.: Selecting influential examples: active learning with expected model output changes. In: Fleet, D., Pajdla, T., Schiele, B., Tuytelaars, T. (eds.) ECCV 2014. LNCS, vol. 8692, pp. 562–577. Springer, Cham (2014). https://doi.org/10.1007/978-3-319-10593-2_37
10. Fridy, J.A.: Introductory Analysis: The Theory of Calculus. Gulf Professional Publishing, Houston (2000)
11. Gal, Y., Islam, R., Ghahramani, Z.: Deep Bayesian active learning with image data. In: International Conference on Machine Learning, pp. 1183–1192. PMLR (2017)
12. Georgii, H.O.: Stochastik: Einführung in die Wahrscheinlichkeitstheorie und Statistik. De Gruyter (2009). publication Title: Stochastik
13. Inman, H.F., Bradley, E.L., Jr.: The overlapping coefficient as a measure of agreement between probability distributions and point estimation of the overlap of two normal densities. Commun. Stat. Theory Methods **18**(10), 3851–3874 (1989)
14. Izenman, A.J.: Modern Multivariate Statistical Techniques: Regression, Classification, and Manifold Learning. Springer Texts in Statistics, Springer, New York (2008). https://doi.org/10.1007/978-0-387-78189-1
15. Kifowit, S.J., Stamps, T.A.: The harmonic series diverges again and again. The AMATYC Review, Spring, p. 13 (2006)
16. Krizhevsky, A.: Learning multiple layers of features from tiny images, p. 60 (2009)
17. Lewis, D.D., Gale, W.A.: A sequential algorithm for training text classifiers. In: Croft, B.W., van Rijsbergen, C.J. (eds.) SIGIR'94, pp. 3–12. Springer, London (1994). https://doi.org/10.1007/978-1-4471-2099-5_1
18. Loosli, G., Canu, S., Bottou, L.: Training invariant support vector machines using selective sampling. Large Scale Kernel Mach. 2 (2007)
19. Ma, L., Destercke, S., Wang, Y.: Online active learning of decision trees with evidential data. Pattern Recognit. **52**, 33–45 (2016). https://doi.org/10.1016/j.patcog.2015.10.014. https://www.sciencedirect.com/science/article/pii/S0031320315003933
20. Maaten, L.V.d., Hinton, G.: Visualizing data using t-SNE. J. Mach. Learn. Res. **9**(86), 2579–2605 (2008). http://jmlr.org/papers/v9/vandermaaten08a.html
21. Mittal, S., Tatarchenko, M., Çiçek, Ö., Brox, T.: Parting with illusions about deep active learning. arXiv preprint arXiv:1912.05361 (2019)
22. Mussmann, S., Liang, P.: On the relationship between data efficiency and error for uncertainty sampling. In: International Conference on Machine Learning, pp. 3674–3682. PMLR (2018)
23. Scheffer, T., Decomain, C., Wrobel, S.: Active hidden Markov models for information extraction. In: Hoffmann, F., Hand, D.J., Adams, N., Fisher, D., Guimaraes, G. (eds.) IDA 2001. LNCS, vol. 2189, pp. 309–318. Springer, Heidelberg (2001). https://doi.org/10.1007/3-540-44816-0_31
24. Sener, O., Savarese, S.: Active learning for convolutional neural networks: a core-set approach. arXiv preprint arXiv:1708.00489 (2017)
25. Settles, B., Craven, M.: An analysis of active learning strategies for sequence labeling tasks. In: Proceedings of the Conference on Empirical Methods in Natural Language Processing - EMNLP '08, p. 1070. Association for Computational Linguistics (2008). https://doi.org/10.3115/1613715.1613855. http://portal.acm.org/citation.cfm?doid=1613715.1613855

26. Settles, B., Craven, M., Ray, S.: Multiple-instance active learning. In: Platt, J., Koller, D., Singer, Y., Roweis, S. (eds.) Advances in Neural Information Processing Systems, vol. 20, pp. 1289–1296. Curran Associates, Inc. (2008). https://proceedings.neurips.cc/paper/2007/file/a1519de5b5d44b31a01de013b9b51a80-Paper.pdf
27. Seung, H.S., Opper, M., Sompolinsky, H.: Query by committee. In: Proceedings of the Fifth Annual Workshop on Computational Learning Theory, COLT '92, pp. 287–294. Association for Computing Machinery, New York (1992). https://doi.org/10.1145/130385.130417
28. Shannon, C.E.: A mathematical theory of communication. Bell Syst. Tech. J. **27**(3), 379–423 (1948)
29. Sinha, S., Ebrahimi, S., Darrell, T.: Variational adversarial active learning. In: Proceedings of the IEEE/CVF International Conference on Computer Vision, pp. 5972–5981 (2019)

ScaleNet: An Unsupervised Representation Learning Method for Limited Information

Huili Huang and M. Mahdi Roozbahani(✉)

School of Computational Science and Engineering, Georgia Institute of Technology, 756 W Peachtree St NW, Atlanta, GA 30308, USA
{hhuang413,mahdir}@gatech.edu

Abstract. Although large-scale labeled data are essential for deep convolutional neural networks (ConvNets) to learn high-level semantic visual representations, it is time-consuming and impractical to collect and annotate large-scale datasets. A simple and efficient unsupervised representation learning method named ScaleNet based on multi-scale images is proposed in this study to enhance the performance of ConvNets when limited information is available. The input images are first resized to a smaller size and fed to the ConvNet to recognize the rotation degree. Next, the ConvNet learns the rotation-prediction task for the original size images based on the parameters transferred from the previous model. The CIFAR-10 and ImageNet datasets are examined on different architectures such as AlexNet and ResNet50 in this study. The current study demonstrates that specific image features, such as Harris corner information, play a critical role in the efficiency of the rotation-prediction task. The ScaleNet supersedes the RotNet by $\approx 7\%$ in the limited CIFAR-10 dataset. The transferred parameters from a ScaleNet model with limited data improve the ImageNet Classification task by about 6% compared to the RotNet model. This study shows the capability of the ScaleNet method to improve other cutting-edge models such as SimCLR by learning effective features for classification tasks.

Keywords: Self-supervised learning · Representation learning · Computer vision

1 Introduction

Deep convolutional neural networks [32] (ConvNets) are widely used for Computer Vision tasks such as object recognition [10,16,41] and image classification [50]. ConvNets generally perform better when they are trained by a massive amount of manually labeled data. A large-scale dataset allows ConvNets to capture more higher-level representations and avoid over-fitting. The prior studies show that these models produce excellent results when implemented for vision tasks, such as object detection [15] and image captioning [27]. However, collecting and labeling the large-scale training dataset is a very time-consuming and

C. Bauckhage et al. (Eds.): DAGM GCPR 2021, LNCS 13024, pp. 174–188, 2021.
https://doi.org/10.1007/978-3-030-92659-5_11

expensive task in fields such as neuroscience [42], medical diagnosis [40], material science [9], and chemistry application [17].

Researchers have investigated different approaches to learn effective visual representations on limited labeled data. Multiple studies employed data augmentation techniques such as scaling, rotating, cropping, and generating synthetic samples to produce more training samples [22,23,45]. Transfer learning methods are implemented to learn the high-level ConvNet-based representations in limited information content [47,51].

Self-supervised learning is a novel machine learning paradigm that has been employed in different fields, such as representation learning [3] and natural language processing [31]. Self-supervised learning trains models to solve the pretext task to learn the intrinsic visual representations that are semantically meaningful for the target task without human annotation. Zhang and et al. [49] introduced an image colorization method to train a model that colors photographs automatically. Doersch and et al. [11] presented a pretext task that predicts the relative location of image patches. Noroozi and et al. [37] developed their semi-supervised model to learn the visual features by solving the jigsaw puzzle. MoCo [20], SimCLR [6], BYOL [18], SIMSIAM [8] as contrastive learning methods, modify a two-branch network architectures to generate two different augmentations of an image and maximize the similarity outputs from two branches. Image clustering-based methods, such as DeepCluster [4], SwAV [5], and SeLa [2], generate labels by an unsupervised learning method initially and then use the subsequent assignments in a supervised way to update the weights of the network. RotNet is one of the simplest self-supervised learning approaches that capture visual representation via rotation-prediction task [28]. The 2D rotational transformations (0°, 90°, 180°, 270°) of an image is recognized by training a ConvNet model during the pretext task. This method simplifies the self-supervised learning implementation and learns desirable visual features for downstream tasks. Since RotNet only affects the input images of the ConvNet, it can be combined with other architectures such as GAN [7] or be used for 3D structured datasets [26].

The main milestone of the current research is to enhance the performance of self-supervised learning in the presence of limited information for current existing self-supervised learning models, such as the RotNet and SimCLR, as opposed to re-introducing a new architecture for these models. Recent studies propose several new architectures based on RotNet architectures. For example, Feng et al. [12] improved the rotation-prediction task by learning the split representation that contains rotation-related and unrelated features. Jenni et al. [24] claimed a new architecture to discriminate the transformations such as rotations, warping, and LCI to learn visual features. The current study focuses on improving self-supervised learning methods with limited information, including limited training samples, missing corner information and lacking color information. A multi-scale self-supervised learning model named ScaleNet is proposed to improve the quality of learned representation for limited data. This simple and efficient framework comprises the following three components: resizing the original input images to a smaller size, feeding resized dataset to a ConvNet for the rotation recognition task, and training the CovNet using the larger size (e.g. original size) input dataset based on the parameters learned from the previous ConvNet model. The ScaleNet method is trained using different architectures such as AlexNet [30] and ResNet50 [21]. Results show that the ScaleNet outperforms the RotNet by 1.23% in the absence of image corner information using the CIFAR-10 dataset and 7.03% with a limited CIFAR-10 dataset. The performance of the SimCLR with a limited dataset and small batch size is improved by ~4% using a multi-scale SimCLR model. The experiments outlined in this study demonstrate that the performance of the RotNet model for a larger dataset is enhanced by using the parameters learned from the ScaleNet model trained with a smaller dataset.

2 ScaleNet

2.1 Self-supervised Learning Based on Geometrical Transformation

Assume $y_i \in Y$ is the human-annotated label for image $x_i \in X$, and $y_i^* \in Y^*$ is the pseudo label generated automatically by a self-supervised ConvNet. Instead of training the ConvNet model $F(.)$ by human-labeled data (X and Y) using a supervised learning method, self-supervised learning trains $F(.)$ with the images X and pseudo labels Y^*.

The length and width of images in the dataset X are R and C, respectively, and the parameter α is provided as the resize operator ($\alpha \leq 1$). The length and width of the new dataset, X_α, are αR and $\alpha C(X_\alpha$ represents the rescaled dataset from X and $x_{\alpha i}$ is one sample from dataset X_α).

A geometrical transformation operator is defined as $G = \{g(.|y)\}_{y=1}^K$, where K denotes different geometrical transformations such as resizing or rotation. $g(X|y)$ is the geometric transformation that applies to a dataset X. For example, $X_\alpha^{y^*} = g(X_\alpha|y^*)$ is the resized dataset X_α, transformed by the geometric transformation and labeled as y^* by ConvNet $F_\alpha(.)$. The ConvNet model $F_\alpha(.)$ is trained on the pseudo-label dataset $X_\alpha^{y^*}$. The probability distribution over all possible geometric transformations is:

$$F_\alpha(X_\alpha^{y^*}|\theta) = \{F_\alpha^y(X_\alpha^{y^*}|\theta)\}_{y=1}^K \tag{1}$$

where $F_\alpha(.)$ gets input from X_α, $F_\alpha^y(X_\alpha^{y^*}|\theta)$ is the predicted probability of the geometric transformation with label y, and θ is the parameter that learned by model $F_\alpha(.)$.

A cross-entropy loss is implemented in this study. Given a set of N training images $D = \{x_i\}_{i=0}^N$ and the ConvNet $F_1(.)$ as an example, the overall training loss is defined as:

$$Loss(D) = \min_\theta \frac{1}{N} \sum_{i=1}^N loss(x_i, \theta) \tag{2}$$

where the $loss(x_i, \theta)$ is the loss between the predicted probability distribution over K and the rotation y:

$$loss(x_i, \theta) = -\frac{1}{K} \sum_{y=1}^K \log(F_1^y(g(x_i|y)|\theta)) \tag{3}$$

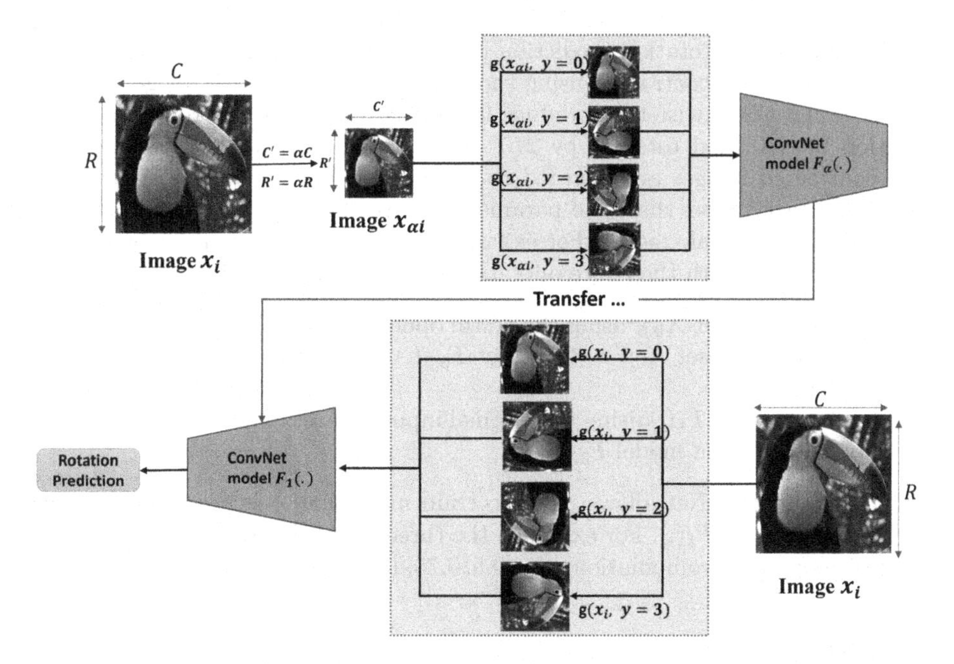

Fig. 1. The Input image x_i is first scaled to a smaller image $x_{\alpha i}$ and rotated 0, 90, 180, 270°. The ConvNet model $F_\alpha(.)$ is trained to predict the rotation degree of an image $x_{\alpha i}$. The ConvNet model $F_\alpha(.)$ is then transferred to the ConvNet model $F_1(.)$ to train the rotation-prediction task.

2.2 Multi-Scale Rotation-Prediction Based on Different Scales

A geometrical transformation G is defined to rotate images by a multiplier of 90° [14]. $G = \{g(X|y)\}_{y=1}^{4}$ represents the rotation and $y = 0, 1, 2, 3$ is the multiplier to generate 0, 90, 180, 270 rotated images, respectively. The horizontal and vertical flipping operations are implemented to rotate images. The rotation task is performed in three ways: transposing and then vertical flipping to rotate an image by 90°, vertical flipping and then transposing to rotate by 270°, and vertical flipping and then horizontal flipping to rotate by 180°.

The main goal of the rotation-prediction pretext task is to enforce the ConvNet to recognize high-level objects such as the horse's leg or bird's eyes [14,25]. Gidaris et al. [14] demonstrated that the RotNet model focused on almost the same image regions as the supervised model by comparing the attention maps. Similar to other data augmentation methods such as flipping and cropping [43], rotating images hinder the model from learning impractical low-level visual features that are trivial information for downstream tasks [14]. The RotNet learns these features by unlocking different underlying structures of images such as texture, shape, and high-level semantic features by rotating images.

The multi-scale rotation prediction is considered a pretext task (Fig. 1). The original images are contracted using the resize operator α. Each image is transformed counterclockwise to reproduce four images: no rotation, rotation by 90, rotation by 180, and rotation by 270° and labeled as 0, 1, 2, 3, respectively. The contracted images are then fed to a ConvNet model $F_\alpha(.)$ for a rotation-prediction task. After that, the parameters learned from $F_\alpha(.)$ are transferred to another larger scale model. For example, The training steps for a two-model ScaleNet trained with the resize operator combination $\alpha = [0.5, 1]$ are:

- Obtaining dataset $X_{0.5}$ using the resize operator $\alpha = 0.5$.
- Feeding the dataset $X_{0.5}$ to ConvNet $F_{0.5}(.)$ to learn the rotation-prediction task.
- Training CovNet $F_1(.)$ using the original input dataset X based on the parameters learned from model $F_{0.5}(.)$.

Note that the ScaleNet allows users to train more than one pre-trained model before training the $F_1(.)$. For example, the three-model ScaleNet is trained with the resize operator combination $\alpha = [0.5, 0.75, 1]$.

3 Experiments

The rotation-prediction task is trained on different scales of images. Note that the current study focuses on enhancing the self-supervised learning method with limited information. First, the value of the resize operators α and the learning rate are analyzed to find the best model for the ScaleNet. Second, the effects of limited information are evaluated. For example, missing corner information and lacking color information on the pretext task is studied. Third, a multi-scale self-supervised learning technique is implemented in SimCLR with a small batch size and limited information to confirm the effectiveness of the ScaleNet further. Finally, the relationship between the quality of the feature maps and the depth of the model is analyzed. All experiments are trained on the CIFAR-10 [29] or ImageNet [10] datasets in this study. Input data are standardized (zero mean and unit standard deviation). A Stochastic gradient descent (SGD) is applied to optimize the rotation-prediction network, and Adaptive Moment Estimation (Adam) is applied to train the SimCLR network. NVIDIA K80 is used for ImageNet and SimCLR models and RTX 2070 super for other models.

3.1 CIFAR-10 Experiment

Determining the Resize Operator: The resize operator α is evaluated quantitatively using different multi-model ScaleNets. Three kinds of models are trained in this Section: RotNet model, two-model ScaleNet, three-model ScaleNet. The RotNet model is similar to the $\alpha = 1$ model, where the rotation recognition task is performed on original images. The two-model ScaleNet is initially trained on a smaller size of images using one of the resize operators of $\alpha = 0.25$, $\alpha = 0.5$, or $\alpha = 0.75$. The pre-trained model's parameters are then transferred to train a new model with an α value of 1, which is the pre-trained model for the downstream task. For the three-model ScaleNet, model's parameters with a smaller α value (e.g., $\alpha = 0.5$) are used to train the model with a larger α value (e.g., $\alpha = 0.75$). The weights of the conv. layers for the downstream task model are transferred from the last pre-trained ScaleNet model (e.g., $\alpha = 0.75$). Since the same dataset is trained, there is no need to freeze the weights of the conv. layers [44]. The initial learning rate of the CIFAR-10 classification task (downstream task) is 0.001, which decays by a factor of 5 when the epoch reaches 80, 160, 200. All experiments are constructed using ResNet50 architecture with the batch size 128, momentum value 0.9, and the weight decay value $5e-4$. Each ConvNet model is trained with 100 epochs.

A smaller learning rate is used to train a larger α value model in the ScaleNet to prevents the significant distortion of the ConvNet weights and avoids overfitting [36]. The different learning rate combinations are investigated to confirm the study by Ng et al. [36]. For example, $[0.1, 0.1]$, $[0.1, 0.05]$, and $[0.1, 0.01]$ are selected for the two-model ScaleNet. The initial learning rate decays by a factor of 5 when the epoch reaches 30, 60, and 80. Since there are ten categories in the CIFAR-10 dataset, the classification layer output is adjusted from four (four rotation degrees) to ten for the downstream task. Table 1 shows that setting a lower learning rate improves the performance of the downstream task and the learning rate combinations of $[0.1, 0.05]$, and $[0.1, 0.05, 0.01]$ achieve highest accuracy in the classification task. The learning rate combination $[0.1, 0.05]$ for two-model ScaleNet and $[0.1, 0.05, 0.01]$ for three-model ScaleNet are then used in the evaluation of the resize operator α in Sect. 3.1.

Table 2 shows that the ScaleNet model with an α value of 0.5 outperforms the RotNet by about 1%. Although the improvement is minimal, it manifests itself when limited information is available, explained in Sect. 3.1 latter. As shown in Table 2, the $\alpha = 0.5$ operator performs better than other operators. Suppose the pixel is identified by a pair of coordinates (x, y), only pixels in even coordinates like $(2, 2)$ are retained during the bi-linear interpolation with $\alpha = 0.5$. Other resize operators compute the resized pixel value by averaging over four surrounding pixels. The $\alpha = 0.25$ scales down images to a tiny size (8×8)

Table 1. Evaluation of different learning rates for the ScaleNet model. All models are trained on ResNet50. Different Learning rate combinations are tested based on the two-model and three-model ScaleNet. $\alpha = 0.5$ corresponds to the 16×16 input images. $\alpha = 0.75$ corresponds to 24×24 input images.

Method	α	Learning rate	Pretext task	Classification
Two-model ScaleNet	0.5, 1	0.1, 0.1	84.73	86.81
Two-model ScaleNet	0.5, 1	0.1, 0.05	84.65	**88.19**
Two-model ScaleNet	0.5, 1	0.1, 0.01	83.0	88.00
Three-model ScaleNet	0.5, 0.75, 1	0.1, 0.05,0.05	85.77	86.16
Three-model ScaleNet	0.5, 0.75, 1	0.1, 0.05, 0.01	85.08	**88.04**

Table 2. Evaluation of the resize operator α. All models are trained on ResNet50. $\alpha = 0.25$ corresponds to the 8×8 input images. $\alpha = 0.5$ corresponds to 16×16 input images. $\alpha = 0.75$ corresponds to 24×24 input images. Each experiment was run three times. The same learning rate 0.1 is implemented for the RotNet [14]

Method	α	Learning rate	Pretext task	Classification
RotNet	1	0.1	83.64	87.17 ± 0.13
Two-model ScaleNet	0.25, 1	0.1, 0.05	82.71	86.19 ± 0.16
Two-model ScaleNet	0.5, 1	0.1, 0.05	84.65	88.19 ± 0.04
Two-model ScaleNet	0.75, 1	0.1, 0.05	85.13	87.63 ± 0.21
Three-model ScaleNet	0.25, 0.5,1	0.1, 0.05,0.01	81.49	86.31 ± 0.17
Three-model ScaleNet	0.25, 0.75,1	0.1, 0.05,0.01	83.14	86.59 ± 0.35
Three-model ScaleNet	0.5, 0.75, 1	0.1, 0.05, 0.01	85.08	88.00 ± 0.05

that obscures essential information for ConvNets tasks. Considering Table 1 and Table 2 results, the two-model ScaleNet with an α value of $[0.5, 1]$, and a learning rate with the combination of $[0.1, 0.05]$ are employed for later experiments in this study on the CIFAR-10 dataset.

Table 3. Experiments are performed using the RotNet and ScaleNet architecture. The input of RotNet/ScaleNet, Harris RotNet/ScaleNet, Hybrid Rot/ScaleNet, Gray-scale RotNet/ScaleNet, Gray-scale-harris RotNet/ScaleNet are the original images, the images without corner information, a random combination of original and Harris images, pseudo-gray-scale images, and pseudo-gray-scale images without corner information, respectively. Each experiment runs three times to get the average results

Method	Pretext task	Classification(4K)	Classification
RotNet	82.45	71.98	87.17
Harris RotNet	50.38	62.80	86.04
Hybrid RotNet	81.76	69.32	87.17
Gray-scale RotNet	80.33	70.63	86.96
Gray-scale-harris RotNet	65.34	61.63	85.77
ScaleNet	84.65	73.69(+1.71%)	88.19(+1.02%)
Harris ScaleNet	45.83	69.83(+7.03%)	87.27(+1.23%)
Hybrid ScaleNet	81.79	73.16(+3.84%)	88.12(+0.95%)
Gray-scale ScaleNet	80.71	72.48(+1.85%)	87.2(+0.24%)
Gray-scale-harris ScaleNet	69.65	68.09(+6.46%)	86.04(+0.27%)

ScaleNet Performance Using Limited Information: The effects of corner features, the color of images, and limited data are examined for the transformational model in Sect. 3.1. The limited information is exercised in this study using a combination of the limited CIFAR-10 data (randomly selecting 4000 out of 50000), missing corner information, and gray-vs-color images. A two-model ScaleNet is trained using resize operators of [0.5, 1] as explained in Table 1 and Table 2. Three models are constructed in this study: the model using the original/pseudo-gray-scale image, the Harris model that is trained on the original/pseudo-gray-scale images with missing corner information, and the hybrid model where the input images are a random combination of original/pseudo-gray-scale images with and without corner information. The pseudo-gray-scale images are trained based on the RotNet and ScaleNet to study the color effects of rotation-prediction results. A pseudo-gray-scale image is generated by replacing each channel with its gray-scale image to adjust the three-channel input for the pretext and downstream task. Although corners maintain only a small percentage of an image, it restores crucial information such as invariant features. The corner information of images is extracted by the Harris corner detector [19]. The detected corner pixels by Harris are replaced with 255 (white dot), and the new images (Harris images) with missing corner information are fed to ConvNet models (Harris models).

Table 3 shows the pretext and the classification task accuracy reduce due to the absence of the corner information for both the RotNet and ScaleNet models. The ScaleNet noticeably outperforms the RotNet by about 7% for the classification task with limited information (missing corners and a limited number of data). This experiment clearly shows the adjustment of the ScaleNet to understand and detect influential features with limited information, that will be crucial for many fields with a lack of data [13,33,42]. The same behavior is observed to a lesser extent when the gray-scale images are used instead of color images. According to Table 3 results, a gray-scale channel would be sufficient to achieve a comparable accuracy similar to color images.

Harris corner features are invariant under rotation, but they are not invariant under scaling. A rotation-prediction task (Pretext task) is relatively dependent on corner features according to Table 3 results. It is shown that pretext accuracy significantly reduces by removing corner features for both the ScaleNet and RotNet. The classification task for 4K data shows the light on the Harris ScaleNet improves the accuracy noticeably comparing to the RotNet. It sheds the fact that the resultant parameters by the ScaleNet are not only rotation-invariant but also scale-invariant. However, the RotNet parameters are trained to be rotation-invariant features. The ScaleNet improves the ConvNets parameters by unlocking underlying features that are crucial in classification tasks.

Multi-SimCLR Performance Using Limited Information: The linear evaluation based on SimCLR [6] is introduced to further confirm the ScaleNet effectiveness in the presence of limited information. Given two different augmented data $\tilde{x_i}$ and $\tilde{x_j}$ from the same sample $x \in X$, the ConvNet $F(.)$ maximizes the agreement between the augmented samples. The ConvNet $F(.)$ is a symmetric architecture including the data augmentations from a family of augmentations $t \sim T$, the base encoder networks $f(.)$, and the projection heads $g(.)$.

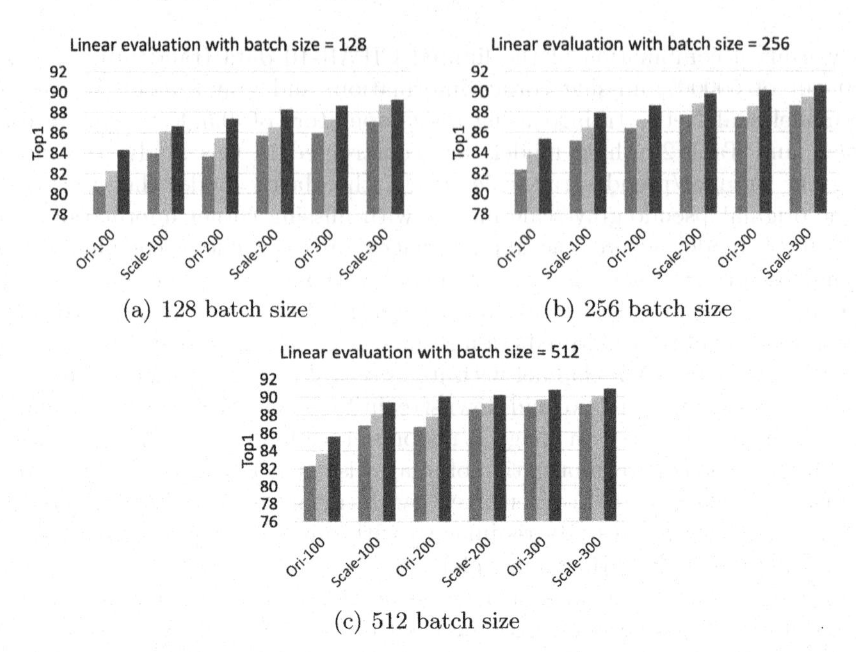

(a) 128 batch size

(b) 256 batch size

(c) 512 batch size

Fig. 2. Linear evaluation of multi-scale SimCLR trained with different epochs and batch sizes on the CIFAR-10 dataset. The Ori-* and Scale-* represent the original SimCLR model [6] and the multi-scale SimCLR, respectively. The Ori/Scale-100, Ori/Scale-200, Ori/Scale-300 show the original/multi-scale SimCLR method results trained with 100, 200, 300 epochs, respectively. The blue, yellow, and red bar show the classification task trained with 4K, 10K, 50K data, respectively. All experiments run two times to compute the average results (Color figure online)

The NT-Xent [6,39,46] is implemented as the loss function. Given the results from the projection head $z_i = g(f(\tilde{x_i})), z_j = g(f(\tilde{x_j}))$, the loss function is:

$$Loss(i,j) = -\log \frac{exp(sim(z_i, z_j))/\tau}{\sum_{k=1}^{2N} \mathbb{1}_{[k \neq i]} exp(sim(z_i, z_k)/\tau)} \tag{4}$$

where $sim(u,v) = u^{\mathsf{T}} v / \|u\| \|v\|$, τ is a temperature parameter, and $\mathbb{1}_{[k \neq i]} \in \{0,1\}$ is an indicator function that equals 1 if $k \neq i$. The multi-scale SimCLR with an α value of $[0.5, 1]$ and a learning rate of $[0.001, 0.001]$ is selected as the best candidate model according to Table 2 and Table 1 experiments. Chen's study[6] demonstrated the contrastive learning benefits from larger batch size. However, training a small-scale dataset with large batch size such as 1024 and 2048 leads to a decrease in performance, even if tuning learning rate to a heuristic [35]. The experiment in this section use small batch size to enhance the performance of SimCLR for limited data. The batch size in {128, 256, 512} is used in this experiment because the minimum training dataset only contains 4K samples. 0.5 and 0.001 are chosen for the temperature of the pretext and the learning rate of the downstream task, respectively. The $F_{0.5}(.)$ model is trained with epoch size 300 for faster convergence. The trained parameters are transferred to $F_1(.)$ with {128, 256, 512} batch size. The Adam optimizer is applied instead of the LARS because of its effective capability to train models with small batch sizes [48]. The inception crop and color distortion are used as the data augmentation methods, and ResNet50 architecture is implemented for this experiment, similar to the SimCLR study [6]. The classification results trained for 4K, 10K, and 50K (whole dataset) data with different batch sizes are shown in Fig. 2. Note that all experiment results for the original SimCLR method are in the acceptable range of the SimCLR study [6]. Figure 2 demonstrates the multi-scale SimCLR clearly improves the classification results with limited information. The best multi-scale SimCLR model considerably outperforms the SimCLR by 4.55%, 4.47%, and 3.79% when trained on 4K, 10K, 50K data, respectively. The results further assert the ScaleNet process is an effective technique for self-supervised learning algorithm with limited information, and multi-scale training is able to capture necessary high-level representation that cannot be learned in the existing SimCLR model [6].

3.2 ImageNet Experiment

The ScaleNet is examined by training on the ImageNet dataset similar to other self-supervised learning methods [14,37,38]. The datasets that contain 65 and 240 samples per category are generated from ImageNet, respectively. For the ScaleNet model trained on 65000 samples, the combination of the resize operator is $[0.5, 1]$. The ScaleNet and RotNet are trained on AlexNet with batch size 192, momentum 0.9, and weight decay $5e-4$, respectively. The learning rate combination is [0.01, 0.001] for the ScaleNet to avoid over-fitting [36]. The pretext model runs for 30 epochs with a learning rate decay factor of 10, which is applied after 10 and 20 epochs. A logistic regression model is added to the last layer for the classification task. All weights of conv. layers are frozen during the downstream task. The initial learning rate of the ImageNet classification task (downstream task) is 0.01, which decays by a factor of 5 when the epoch reaches 5, 15, 25.

Table 4 shows that the generated feature maps by the ScaleNet perform better in all conv. blocks experiments comparing to the RotNet. The feature map generated by the 4th conv. layer achieves the highest classification accuracy in the ScaleNet experiment. The classification accuracy decreases after the 4th conv. layer as the ConvNet model starts learning the specific features of the rotation-prediction task. The accuracy of the ScaleNet model trained on 65,000 images is 0.93% higher than the RotNet model.

Note that the trained ScaleNet model on a limited dataset is capable of improving the RotNet model accuracy for a larger dataset when it uses the ScaleNet pre-trained parameters. This experiment is conducted by training the ScaleNet and RotNet model with 65,000 samples and transferring their parameters to train the RotNet with 240,000 samples using RotNet, respectively. Table 4

Table 4. The ImageNet classification with linear layers, where the logistic regression is used in the classification layer for the downstream task. The ScaleNet and RotNet model parameters trained on 65,000 samples are used in the RotNet model with 240,000 samples as a pre-trained model and called RotNet (from RotNet) and RotNet (from ScaleNet), respectively. The rate 0.01 is implemented for the RotNet [14]

Model	Samples	ConvB1	ConvB2	ConvB3	ConvB4	ConvB5
RotNet	65,000	5.62	9.24	11.30	11.71	11.75
ScaleNet	65,000	**5.93**	**9.91**	**12.05**	**12.64**	**12.25**
RotNet (from RotNet)	240,000	11.30	17.60	21.26	22.15	21.90
RotNet (from ScaleNet)	240,000	**14.55**	**23.18**	**27.75**	**27.50**	**26.61**

shows the RotNet (from ScaleNet) clearly outperforms the RotNet (from RotNet) in the classification task by 6.49%. This result shows the capability of the ScaleNet to detect crucial features for a limited dataset and provide effective parameters from a limited dataset for a transfer learning task.

Attention Map Analysis of ResNet50: Gidaris's study [14] explains the critical role of the first conv. layer edge filter in the rotation-prediction task. The study proves that the RotNet catches more information in various directions than the supervised learning method. The attention maps of the ScaleNet and RotNet are generated in this section. The ScaleNet model with $\alpha = [0.5, 1]$ is used for the 2-model ScaleNet. The ResNet50 architecture contains four stages after the first conv. layer, where each stage includes a convolution and identity blocks. Each conv. block and identity block consist of 3 conv. layers. Attention maps based on Grad-CAM are produced according to the first three stages of the ResNet50 to explore the results in more detail. Figure 3 indicates that both models learn the high-level semantic representations of the image to fulfill the rotation-prediction tasks. The attention maps of the first stage of the ScaleNet and RotNet indicate the ScaleNet catches more edge information than the RotNet, which confirms that the edge information is of vital importance for the rotation-prediction task. When the model goes more in-depth, the ScaleNet model focuses on more specific objects while the areas of interest in RotNet are distributed uniformly. Based on the attention maps in Stage-3, the ScaleNet already focuses on the head of the cat and the dog collar while the RotNet model is trying to understand the shape of the animals. The visualization result implies that the ScaleNet focuses more on the edge information and specific semantic features, which efficiently improves image classification tasks [1,34]. The current study results indicate that the ScaleNet learns more high-level semantic visual representations than the RotNet, which are crucial in many computer vision tasks such as object detection and segmentation [14].

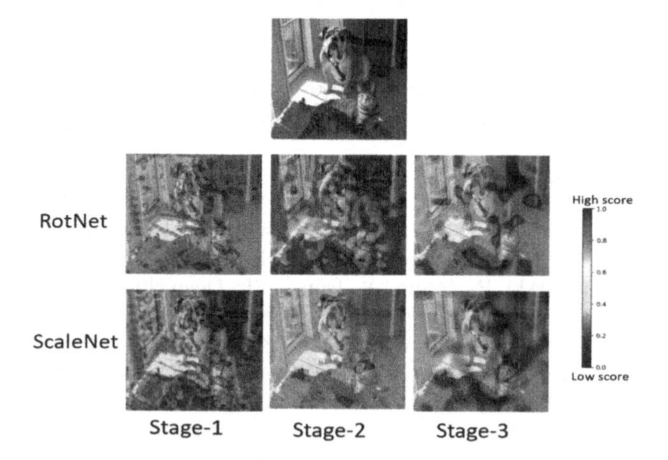

Fig. 3. The attention maps generated by the ResNet50 model. The heat map is implemented to show the area of attention by ConvNet. Red regions correspond to a high score for the class. The first row is the original image. The second row and third row show the attention map generated by the RotNet and ScaleNet, respectively. Stage-1, Stage-2, Stage-3 are the first three stages of ResNet50, respectively.

4 Conclusion

An unsupervised representation learning method named ScaleNet is proposed in this study. The ScaleNet method trains multi-scale images to extract high-quality representations. The results demonstrate that ScaleNet outperforms the RotNet with different architectures such as ResNet50 and AlexNet, especially in limited information. The ScaleNet model not only learns detailed features, but the experiments outlined in this study also show that training the ScaleNet on small datasets effectively improves the performance of the RotNet on larger datasets. Furthermore, the ScaleNet process is proved to enhance the performance of other self-supervised learning models such as SimCLR. The ScaleNet is an excellent step towards improving the performance of self-supervised learning in the presence of limited information. Future studies are categorized in three steps: (i) investigating the influence of other affine transformations apart from scales, such as stretching the image. (ii) investigating the impact of additional image information apart from color and corner. (iii) implementing the ScaleNet approach in other self-supervised learning models.

Acknowledgment. This study is supported by Google Cloud Platform (GCP) Research by providing credit supports to implement all deep learning algorithms related to SimCLR and ImageNet using virtual machines. The author would like to thank J. David Frost, Kevin Tynes, and Russell Strauss for their feedback on the draft.

References

1. Abdillah, B., Jati, G., Jatmiko, W.: Improvement CNN performance by edge detection preprocessing for vehicle classification problem. In: MHS, pp. 1–7. IEEE Press, Nagoya (2018)
2. Asano, Y.M., Rupprecht, C., Vedaldi, A.: Self-labelling via simultaneous clustering and representation learning. In: ICLR. OpenReview.net, Addis Ababa (2019)
3. Bengio, Y., Courville, A., Vincent, P.: Representation learning: a review and new perspectives. IEEE Trans. Pattern Anal. Mach. Intell. **35**(8), 1798–1828 (2013)
4. Caron, M., Bojanowski, P., Joulin, A., Douze, M.: Deep clustering for unsupervised learning of visual features. In: Ferrari, V., Hebert, M., Sminchisescu, C., Weiss, Y. (eds.) Computer Vision – ECCV 2018. LNCS, vol. 11218, pp. 139–156. Springer, Cham (2018). https://doi.org/10.1007/978-3-030-01264-9_9
5. Caron, M., Misra, I., Mairal, J., Goyal, P., Bojanowski, P., Joulin, A.: Unsupervised learning of visual features by contrasting cluster assignments. In: NIPS, pp. 9912–9924. MIT Press, Cambridge (2020)
6. Chen, T., Kornblith, S., Norouzi, M., Hinton, G.: A simple framework for contrastive learning of visual representations. In: ICML, pp. 1597–1607. PMLR, California (2020)
7. Chen, T., Zhai, X., Ritter, M., Lucic, M., Houlsby, N.: Self-supervised gans via auxiliary rotation loss. In: CVPR, pp. 12154–12163. IEEE Press, California (2019)
8. Chen, X., He, K.: Exploring simple Siamese representation learning. In: CVPR, pp. 15750–15758. IEEE Press (2021)
9. Cubuk, E.D., Sendek, A.D., Reed, E.J.: Screening billions of candidates for solid lithium-ion conductors: a transfer learning approach for small data. J. Chem. Phys. **150**(21), 214701 (2019)
10. Deng, J., Dong, W., Socher, R., Li, L.J., Li, K., Fei-Fei, L.: ImageNet: a large-scale hierarchical image database. In: CVPR, pp. 248–255. IEEE Press, Georgia (2009)
11. Doersch, C., Gupta, A., Efros, A.A.: Unsupervised visual representation learning by context prediction. In: ICCV, pp. 1422–1430. IEEE Press, Santiago (2015)
12. Feng, Z., Xu, C., Tao, D.: Self-supervised representation learning by rotation feature decoupling. In: CVPR, pp. 10364–10374. IEEE Press, California (2019)
13. Frid-Adar, M., Diamant, I., Klang, E., Amitai, M., Goldberger, J., Greenspan, H.: GAN-based synthetic medical image augmentation for increased CNN performance in liver lesion classification. Neurocomputing **321**, 321–331 (2018)
14. Gidaris, S., Singh, P., Komodakis, N.: Unsupervised representation learning by predicting image rotations. In: ICLR. OpenReview.net, British Columbia (2018)
15. Girshick, R.: Fast R-CNN. In: ICCV, pp. 1440–1448. IEEE Press, Santiago (2015)
16. Girshick, R., Donahue, J., Darrell, T., Malik, J.: Rich feature hierarchies for accurate object detection and semantic segmentation. In: CVPR, pp. 580–587. IEEE Press, Ohio (2014)
17. Grambow, C.A., Li, Y.P., Green, W.H.: Accurate thermochemistry with small data sets: a bond additivity correction and transfer learning approach. J. Phys. Chem. A **123**(27), 5826–5835 (2019)
18. Grill, J.B., et al.: Bootstrap your own latent - a new approach to self-supervised learning. In: NIPS, pp. 21271–21284. MIT Press, Cambridge (2020)
19. Harris, C.G., Stephens, M., et al.: A combined corner and edge detector. In: AVC, pp. 1–6. Alvey Vision Club, Manchester (1988)
20. He, K., Fan, H., Wu, Y., Xie, S., Girshick, R.: Momentum contrast for unsupervised visual representation learning. In: CVPR, pp. 9729–9738. IEEE Press, Seattle (2020)

21. He, K., Zhang, X., Ren, S., Sun, J.: Deep residual learning for image recognition. In: CVPR, pp. 770–778. IEEE Press, Nevada (2016)
22. Hu, G., Peng, X., Yang, Y., Hospedales, T.M., Verbeek, J.: Frankenstein: learning deep face representations using small data. IEEE Trans. Image Process. **27**(1), 293–303 (2017)
23. Inoue, H.: Data augmentation by pairing samples for images classification (2018)
24. Jenni, S., Jin, H., Favaro, P.: Steering self-supervised feature learning beyond local pixel statistics. In: CVPR, pp. 6408–6417. IEEE Press, California (2020)
25. Jing, L., Yang, X., Liu, J., Tian, Y.: Self-supervised spatiotemporal feature learning via video rotation prediction. arXiv preprint arXiv:1811.11387 (2018)
26. Kanezaki, A., Matsushita, Y., Nishida, Y.: RotationNet: joint object categorization and pose estimation using multiviews from unsupervised viewpoints. In: CVPR, pp. 5010–5019. IEEE Press, Utah (2018)
27. Karpathy, A., Fei-Fei, L.: Deep visual-semantic alignments for generating image descriptions. In: CVPR, pp. 3128–3137. IEEE Press, Massachusetts (2015)
28. Kolesnikov, A., Zhai, X., Beyer, L.: Revisiting self-supervised visual representation learning. In: CVPR, pp. 1920–1929. IEEE Press, California (2019)
29. Krizhevsky, A., Hinton, G., et al.: Learning multiple layers of features from tiny images (2009)
30. Krizhevsky, A., Sutskever, I., Hinton, G.E.: ImageNet classification with deep convolutional neural networks. In: NIPS, pp. 1097–1105. MIT Press, Cambridge (2012)
31. Lan, Z., Chen, M., Goodman, S., Gimpel, K., Sharma, P., Soricut, R.: ALBERT: A lite BERT for self-supervised learning of language representations. In: ICLR. OpenReview.net, Addis Ababa (2020)
32. Lecun, Y., Bottou, L., Bengio, Y., Haffner, P.: Gradient-based learning applied to document recognition. Proc. IEEE **86**(11), 2278–2324 (1998)
33. Lee, H., Kwon, H.: Going deeper with contextual CNN for hyperspectral image classification. IEEE Trans. Image Process. **26**(10), 4843–4855 (2017)
34. Marmanis, D., Schindler, K., Wegner, J.D., Galliani, S., Datcu, M., Stilla, U.: Classification with an edge: improving semantic image segmentation with boundary detection. ISPRS J. Photogramm. Remote. Sens. **135**, 158–172 (2018)
35. Mishkin, D., Sergievskiy, N., Matas, J.: Systematic evaluation of convolution neural network advances on the ImageNet. Comput. Vis. Image Underst. **161**, 11–19 (2017)
36. Ng, H.W., Nguyen, V.D., Vonikakis, V., Winkler, S.: Deep learning for emotion recognition on small datasets using transfer learning. In: ICMI, pp. 443–449. ACM, Seattle (2015)
37. Noroozi, M., Favaro, P.: Unsupervised learning of visual representations by solving jigsaw puzzles. In: Leibe, B., Matas, J., Sebe, N., Welling, M. (eds.) ECCV 2016. LNCS, vol. 9910, pp. 69–84. Springer, Cham (2016). https://doi.org/10.1007/978-3-319-46466-4_5
38. Noroozi, M., Pirsiavash, H., Favaro, P.: Representation learning by learning to count. In: ICCV, pp. 5898–5906. IEEE Press, Venice (2017)
39. Oord, A.V.D., Li, Y., Vinyals, O.: Representation learning with contrastive predictive coding. arXiv preprint arXiv:1807.03748 (2018)
40. Rajpurkar, P., et al.: AppendiXNet: deep learning for diagnosis of appendicitis from a small dataset of CT exams using video pretraining. Sci. Rep. **10**(1), 1–7 (2020)
41. Ren, S., He, K., Girshick, R., Sun, J.: Faster R-CNN: towards real-time object detection with region proposal networks. In: NIPS, pp. 91–99. MIT Press, Cambridge (2015)

42. Ronneberger, O., Fischer, P., Brox, T.: U-Net: convolutional networks for biomedical image segmentation. In: Navab, N., Hornegger, J., Wells, W.M., Frangi, A.F. (eds.) MICCAI 2015. LNCS, vol. 9351, pp. 234–241. Springer, Cham (2015). https://doi.org/10.1007/978-3-319-24574-4_28
43. Shijie, J., Ping, W., Peiyi, J., Siping, H.: Research on data augmentation for image classification based on convolution neural networks. In: CAC, pp. 4165–4170. IEEE Press, Jinan (2017)
44. Soekhoe, D., van der Putten, P., Plaat, A.: On the impact of data set size in transfer learning using deep neural networks. In: Boström, H., Knobbe, A., Soares, C., Papapetrou, P. (eds.) IDA 2016. LNCS, vol. 9897, pp. 50–60. Springer, Cham (2016). https://doi.org/10.1007/978-3-319-46349-0_5
45. Wu, R., Yan, S., Shan, Y., Dang, Q., Sun, G.: Deep image: scaling up image recognition. arXiv preprint arXiv:1501.02876, **7**(8) (2015)
46. Wu, Z., Xiong, Y., Yu, S.X., Lin, D.: Unsupervised feature learning via non-parametric instance discrimination. In: CVPR, pp. 3733–3742. IEEE Press, Utah (2018)
47. Yosinski, J., Clune, J., Bengio, Y., Lipson, H.: How transferable are features in deep neural networks. In: NIPS, pp. 3320–3328. MIT Press, Cambridge (2014)
48. You, Y., Gitman, I., Ginsburg, B.: Large batch training of convolutional networks. arXiv preprint arXiv:1708.03888 (2017)
49. Zhang, R., Isola, P., Efros, A.A.: Colorful image colorization. In: Leibe, B., Matas, J., Sebe, N., Welling, M. (eds.) ECCV 2016. LNCS, vol. 9907, pp. 649–666. Springer, Cham (2016). https://doi.org/10.1007/978-3-319-46487-9_40
50. Zhou, B., Lapedriza, A., Xiao, J., Torralba, A., Oliva, A.: Learning deep features for scene recognition using places database. In: NIPS, pp. 487–495. MIT Press, Cambridge (2014)
51. Zhuang, F., et al.: A comprehensive survey on transfer learning. Proc. IEEE **109**(1), 43–76 (2021)

Actions, Events, and Segmentation

A New Split for Evaluating True Zero-Shot Action Recognition

Shreyank N. Gowda[1(✉)], Laura Sevilla-Lara[1], Kiyoon Kim[1], Frank Keller[1], and Marcus Rohrbach[2]

[1] University of Edinburgh, Edinburgh, UK
[2] Facebook AI Research, Menlo Park, USA

Abstract. Zero-shot action recognition is the task of classifying action categories that are not available in the training set. In this setting, the standard evaluation protocol is to use existing action recognition datasets (e.g. UCF101) and *randomly* split the classes into seen and unseen. However, most recent work builds on representations pre-trained on the Kinetics dataset, where classes largely overlap with classes in the zero-shot evaluation datasets. As a result, classes which are supposed to be unseen, are present during supervised pre-training, invalidating the condition of the zero-shot setting. A similar concern was previously noted several years ago for image based zero-shot recognition, but has not been considered by the zero-shot *action* recognition community. In this paper, we propose a new split for *true* zero-shot action recognition with no overlap between unseen test classes and training or pre-training classes. We benchmark several recent approaches on the proposed True Zero-Shot (**TruZe**) Split for UCF101 and HMDB51, with zero-shot and generalized zero-shot evaluation. In our extensive analysis we find that our TruZe splits are significantly harder than comparable random splits as nothing is leaking from pre-training, i.e. unseen performance is consistently lower, up to 8.9% for zero-shot action recognition. In an additional evaluation we also find that similar issues exist in the splits used in few-shot action recognition, here we see differences of up to 17.1%. We publish our splits (Splits can be found at https://github.com/kini5gowda/TruZe) and hope that our benchmark analysis will change how the field is evaluating zero- and few-shot action recognition moving forward.

1 Introduction

Much of the recent progress in action recognition is due to the availability of large annotated datasets. Given how impractical it is to obtain thousands of videos in order to recognize a single class label, researchers have turned to the problem of zero-shot learning (ZSL). Each class label has semantic embeddings that are either manually annotated or inferred through semantic knowledge using

Supplementary Information The online version contains supplementary material available at https://doi.org/10.1007/978-3-030-92659-5_12.

C. Bauckhage et al. (Eds.): DAGM GCPR 2021, LNCS 13024, pp. 191–205, 2021.
https://doi.org/10.1007/978-3-030-92659-5_12

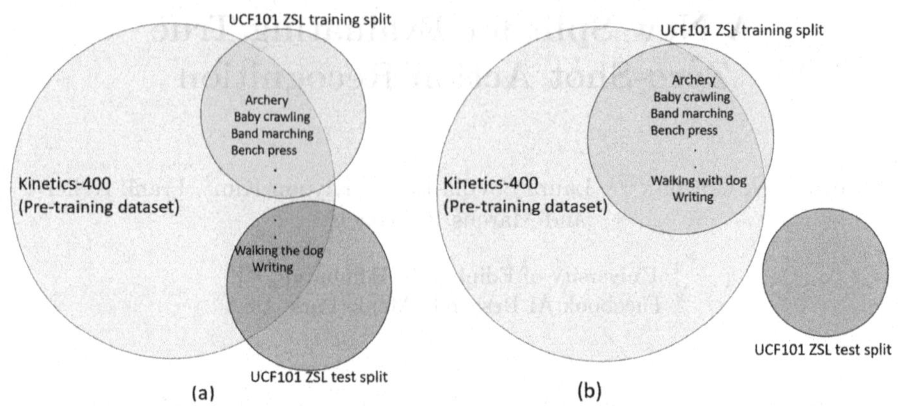

Fig. 1. An illustration of the overlap in *classes* of the pretraining dataset (grey), training split (yellow) and zero-shot test split (green). Current evaluation protocol (a) picks classes at random with 51 training classes and 50 test classes. There is always some overlap and also chances of an extremely high overlap. We propose a stricter evaluation protocol (b) where there is no overlap at test time, maintaining the ZSL premise. (Color figure online)

word embeddings. These embeddings help obtain relationships between training classes (that have many samples) and test classes (that have zero samples). Typically, the model predicts the semantic embedding of the input video and matches it to a test class using the nearest neighbor's search.

However, work in video ZSL [1,7,10] often uses a pre-trained model to represent videos. While pre-trained models help obtaining good visual representations, overlap with test classes can invalidate the premise of zero-shot learning, making it difficult to compare approaches fairly.

In the image domain [19,21,24,25], this problem has also been observed. Typically image models are pre-trained on ImageNet [5]. Xian et al. [25] showed that, in image ZSL, if the pre-training dataset has overlapping classes with the test set, the accuracy is inflated at test time. Hence, the authors propose a new split that avoids that problem, and it is now widely used. Similarly, most video models are pre-trained on Kinetics-400 [3], which has a large overlap with the typical ZSL action recognition benchmarks (UCF101, HMDB51 and Olympics). This pre-training gives leads to inflated accuracies, creating the need for a new split. Figure 1 shows an illustration of these overlap issues.

Contributions: First, we show the significant difference in performance caused by pre-training on classes that are included in the test set, across all networks and all datasets. Second, we measure the extent of the overlap between Kinetics-400 and the datasets typically used for ZSL testing: UCF101, HMDB51 and Olympics datasets. We do this by computing both visual and semantic similarity between classes. Finally, we propose a fair split of the classes that takes this class overlap into account, and does not break the premise of ZSL. We show that current models do indeed perform more poorly in this split, which is further proof of the significance of the problem. We hope that this split will be useful to the community, will avoid the need of random splits, and help an actually fair comparison among methods.

2 Related Work

Previous work [25] has studied the effect of pre-training on classes that overlap with test classes in the image domain. The authors compute the extent of overlap between testing datasets and Imagenet [5], where models are typically pre-trained. The overlapping classes correspond to the training classes, while the non-overlapping classes correspond to the test classes. Figure 1 shows an illustration of the proposed evaluation protocol. Unlike the traditional evaluation protocol that chooses classes at random from the list of classes, without typically looking at the list of overlapping classes from the pre-trained dataset, we strictly remove all classes that have a high threshold of visual or semantic similarity (see Sect. 5).

Roitberg et al. [15] proposed to look at the overlapping classes in videos by using a corrective method that would automatically remove categories that are similar. This was done by utilizing a pairwise similarity of labels within the same dataset. While they showed that using pre-trained models resulted in improved accuracy due to class overlap, the evaluation included only one dataset, and only looked at the semantic similarity of labels. Adding visual similarity, helps discovering overlapping classes like "typing" in UCF101 and "using computer" in Kinetics. Therefore, in our proposed split we use both semantic and visual similarity across classes.

Busto et al. [2] provide a mapping of shared classes between UCF101 and Kinetics as part of a domain adaptation problem. They manually find semantic matches based on class names. However, their mapping was not based on visual and semantic similarity as they have classes such as typing and writing on board as part of UCF101 classes not similar to any Kinetics class. Also, they use "floor gymnastics" as an unknown class in UCF101, however, Kinetics has "gymnastics tumbling" which is the same action. We see that samples from the "typing" class consist of a large proportion of people using their computers and this maps directly to the "using computers" class in Kinetics. Based on our visual and semantic similarity approach, we obtain a slightly different set of classes to those proposed by Busto et al.

Recently, end-to-end training [1] has been proposed for ZSL in video classification. As part of the evaluation protocol, to uphold the ZSL premise, the authors propose to train a model on Kinetics by removing the set of overlapping classes (using semantic matching) and using this as a pre-trained model. While this is a promising way to ensure the following of the premise of ZSL, it is very computationally expensive. We also show that having a better backbone (see Sect. 6.4) results in better accuracy, and as such, training end-to-end is expensive. As a result, using a proposed split instead whilst having the opportunity to use any backbone seems an easier approach.

3 ZSL Preliminaries

Consider S to be the training set of seen classes, composed of tuples $(x, y, a(y))$, where x represents the visual features of each sample in S (spatio-temporal features in the case of video), y corresponds to the class label in the set of seen class labels Y_s, and $a(y)$ represents the semantic embedding of class y. These semantic embeddings are either annotated manually or computed using a language-based embedding, e.g. word2vec [11] or sen2vec [13].

Let U be the set of unseen classes, composed of tuples $(u, a(u))$, where u is a class in the label set Y_u, and $a(u)$ are the corresponding semantic representations. Y_s and Y_u do not overlap, i.e.

$$Y_s \bigcap Y_u = \emptyset \tag{1}$$

In ZSL for video classification, given an input video, the task is to predict a class label in the unseen set of classes, $f_{ZSL} : X \rightarrow Y_u$. An extension of the problem is the related generalized zero-shot learning (GZSL) setting, where given a video, the task is to predict a class label in the union of the seen and unseen classes, as $f_{GZSL} : X \rightarrow Y_s \cup Y_u$.

When relying on a pre-trained model to obtain visual features, we denote the pre-trained classes as the set Y_p. For the ZSL premise to be maintained, there must be no overlap with the unseen classes:

$$Y_p \bigcap Y_u = \emptyset. \tag{2}$$

The core problem we address in this paper is that while prior work generally adheres to Eq. 1, recent use of pre-trained models does not adhere to Eq. 2. Instead, we propose the TruZe split in Sect. 5.2, which adheres to both Eq. (1) *and* Eq. (2).

3.1 Visual and Semantic Embeddings

Early work computed **visual embeddings** (or representations) using hand-crafted features such as Improved Dense Trajectories (IDT) [20], which include tracked trajectories of detected interest points within a video, and four descriptors. More recent work often uses deep features such as those from 3D convolutional networks (e.g., I3D [3] or C3D [18]). These 3D CNNs are used to learn spatio-temporal representation of the video. In our experiments, we will use both types of visual representations.

To obtain **semantic embeddings**, previous work [9] uses manual attribute annotations for each class. For example, the action of kicking would have motion of the leg and motion of twisting the upper body. However, such attributes are not available for all datasets. An alternative approach is to use word embeddings such as word2vec [11] for each class label. This gets rid of the requirement of manual attributes. More recently, Gowda et al. [7] showed that using sen2vec [13] instead of word2vec yields better results as action labels are typically multi-worded and averaging them using word2vec makes it lose context. Based on this, in our experiments we use sen2vec.

4 Evaluated Methods

We consider early approaches that use IDT features such as ZSL by bi-directional latent embedding learning (BiDiLEL), ZSL by single latent embedding (Latem) [23] and synthesized classifiers (SYNC) [4]. Using features that are not learned, allows us to control for the effect of pre-training when using random splits, and when using the proposed split (PS). We then evaluate recent state-of-the-art approaches such as feature generating networks (WGAN) [24], out-of-distribution detection networks (OD) [10] and end-to-end learning for ZSL (E2E) [1] as well. Let us briefly have a look at these methods.

Latem [23] uses piece-wise linear compatibility to understand the visual-semantic embedding relationship with the help of latent variables. Here, each latent variable is encoded to account for the various visual properties of the input data. The authors project the visual embedding to the semantic embedding space.

BiDiLEL [21] projects both the visual and semantic embeddings into a common latent space (instead of projecting to the semantic space) so that the intrinsic relationship between them is maintained.

SYNC [4] uses a weighted bipartite graph in order to learn a projection between the semantic embeddings and the classifier model space. They generate the graph by using a set of "phantom" classes synthesized in order to ensure aligned semantic embedding space and classifier model space and minimize the distortion error.

WGAN [24] uses a Wasserstein GAN to synthesize the unseen features of classes, with additional losses in the form of cosine and cycle-consistency losses. These losses help enhancing the feature generation process.

OD [10] trains an out-of-distribution detector to distinguish the generated features from those of the seen class features and in turn to help with classification in the generalized zero-shot learning setting.

E2E [1] is a recent approach that leverages end-to-end training to alleviate the problem of overlapping classes. This is done by removing all overlapping classes in the pre-training dataset and then using a CNN trained on the remaining classes to generate the visual features for the ZSL videos.

CLASTER [7] uses clustering of visual-semantic embeddings optimised by reinforcement learning.

5 Evaluation Protocol

5.1 Datasets

The three most popular benchmarks for ZSL in videos are UCF101 [17], HMDB51 [8] and Olympics [12]. The typical evaluation protocol in video ZSL is to use a 50-50 split of each dataset, where 50 % of the labels are used as the

Table 1. Datasets and their splits used for ZSL in action recognition. Traditionally, 'Random Split' was followed where the seen and unseen classes were randomly selected. However, we can see the extent of overlap in the 'overlapping classes' column. Using the extent of overlap we define our 'TruZe split'. For the full list of seen and unseen classes, please look at the supplementary. *Note that for the all experiments in this paper we use a random split which matches the number of classes of our TruZe, e.g. 29/22 for HMDB51.

Dataset	Videos	Classes	Random Split* (Seen/Unseen)	Overlapping classes with kinetics	TruZe Split (Seen/Unseen)
Olympics	783	16	8/8	13	–
HMDB51	6766	51	26/25	29	29/22
UCF101	13320	101	51/50	70	70/31

training set and 50 % as the test set. In order to provide comparisons to prior work [1,4,7,10,21,23] and for the purpose of communicating replicable research results, we study UCF101, HMDB51, and Olympics, as well as the the relationship to the pre-training dataset Kinetics-400.

In our experiments (see Sect. 6), we find overlapping classes between Kinetics-400 and each of the ZSL datasets, and move them to the training split. Thus, instead of using 50-50, we need to use 70-31 (number of labels for train and test) for UCF101 and 29-22 (number of labels for train and test) for HMDB51. We see that the number of overlapping classes in the case of Olympics is 13 out of 16, and hence we choose not to proceed further with it. More details can be found in Table 1. For a fair comparison between the TruZe and random split, we use the same proportions (i.e., 70-31 in UCF101 and so on) in the experiments with random splits. We create ten such random splits and use these same splits for all models.

5.2 TruZe Split

We now describe the process of creating the proposed TruZe split, to avoid the coincidental influence of pre-training on ZSL. First, we identify overlapping classes between the pre-training Kinetics-400 dataset and each ZSL dataset. To do this, we compute visual and semantic similarities, and discard those classes that are too similar.

To calculate visual similarity, we use an I3D model pre-trained on Kinetics-400 and evaluate all video samples in UCF101, HMDB51 and Olympics using the Kinetics labels. This helps us to detect similarities that are often not recognized in terms of semantic similarities. Some examples include typing (class in UCF101) that the model detects as using computer (class in Kinetics), applying eye makeup (class in UCF101) that the model detects as filling eyebrows (class in Kinetics).

To calculate semantic similarity, we use a sen2vec model pre-trained on Wikipedia that helps us compare action phrases and outputs a similarity score. We combine the visual and semantic similarity to obtain a list of extremely

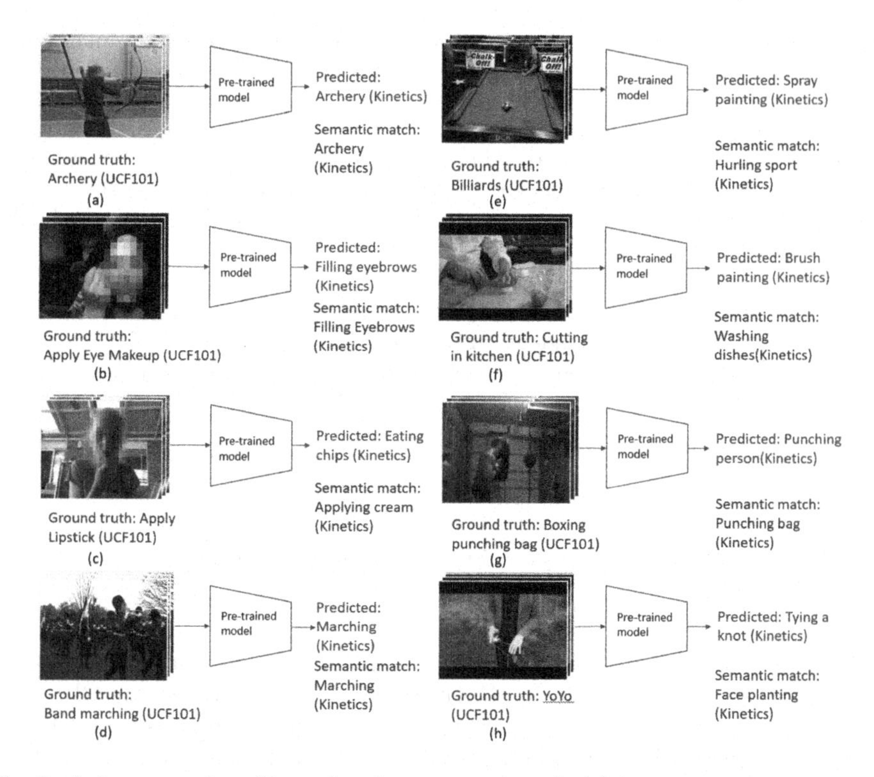

Fig. 2. A few examples of how the classes are selected. (a) is an example of an exact match between the testing dataset (in this case UCF101) and the pre-trained dataset (Kinetics). (b), (d) and (g) are examples of visual-semantic similar matches where the output and semantically closest classes are the same. (c), (e), (f) and (h) are examples of classes without overlap in terms of both visual and semantic similarity.

similar classes to the ones present in Kinetics. This list of classes is present in the supplementary. The classes that even have a slight overlap or are a subset of a class in Kinetics are all chosen as part of the seen set (for example, cricket bowling and cricket shot in UCF101 are part of the seen set due to the superclass playing cricket in Kinetics). A few examples of the selection of classes is show in Fig. 2.

We discard classes from the test set based on the following rules:

- Discard exact matches. For example, archery in UCF101 is also present in Kinetics.
- Discard matches that can be either superset or subset. For example, UCF101 has classes such as cricket shot and cricket bowling while Kinetics has playing cricket (superset). We manually do this based on the output of the closest semantic match.
- Discard matches that predict the same visual and semantic match. For example, *apply eye makeup* (UCF101 label) predicts *filling eyebrows* as the visual match using Kinetics labels and the closest semantic match to classes in Kinetics is also *filling eyebrows*. We also manually confirm this.

Table 2. Results with different splits for **Zero-Shot Learning (ZSL)**. Column 'Random' corresponds to the accuracy using splits in the traditional fashion (random selection of train and test classes, but with the same number of classes in train/test as in TruZe), 'TruZe' corresponds to the accuracy using our proposed split and 'Diff' corresponds to the difference in accuracy between using random splits and our proposed split. We run 10 independent runs for different random splits and report the average accuracy. We see positive differences in the 'Diff' column which we believe is due to the overlapping classes in Kinetics.

Method	UCF101			HMDB51		
	Random	TruZe	Diff	Random	TruZe	Diff
Latem [23]	21.4	15.5	5.9	17.8	9.4	8.4
SYNC [4]	22.1	15.3	6.8	18.1	11.6	6.5
BiDiLEL [21]	21.3	15.7	5.6	18.4	10.5	7.9
OD [10]	28.4	22.9	5.5	30.6	21.7	8.9
E2E [1]	46.6	45.5	1.1	33.2	31.5	1.7
CLASTER [7]	47.1	45.2	1.9	36.6	33.2	3.4

We move all the discarded classes to the training set. This leaves a 70-31 split on UCF101 and a 29-22 split on HMDB51. We also see that in the Olympics dataset, there are 13 directly overlapping classes out of 16 classes and hence dropped the dataset from further analysis. One particular interesting scenario is the "pizza tossing" class in UCF101. In Kinetics, there is a class called "making pizza", however, the action of tossing is not performed in them and hence we use "pizza tossing" as an unseen class.

6 Experimental Results

6.1 Results on ZSL and Generalized ZSL

We first consider the results on ZSL. Here, as explained before, only samples from the unseen class are passed as input to the model at test time. Since TruZe separates the overlapping classes from the pre-training dataset, we expect a lower accuracy on this split compared to the traditionally used random splits. We compare BiDiLEL [21], Latem [23], SYNC [4], OD [10], E2E [1] and CLASTER [7] and report the results in Table 2. As expected, we see in the 'Diff' column for both UCF101 and HMDB51 a positive difference, indicating that the accuracy is lower for the TruZe split.

Generalized ZSL (GZSL) looks at a more realistic scenario, wherein the samples at test time belong to both seen and unseen classes. The reported accuracy is then the harmonic mean of the seen and unseen class accuracies. Since we separate out the overlapping classes, we expect to see an increase in the seen class accuracy and a decrease in the unseen class accuracy. We report GZSL results on OD, WGAN and CLASTER in Table 3. The semantic embedding used for

Table 3. Results with different splits for **Generalized Zero-Shot Learning (GZSL)**. 'Rand' corresponds to the splits using random classes over 10 independent runs, 'TruZe' corresponds to the proposed split. Acc_U and Acc_S correspond to unseen class accuracy and seen class accuracy respectively. The semantic embedding used is sen2vec. 'diff' corresponds to the difference between 'Rand' and 'TruZe'. We see consistent positive difference in performance on the unseen classes and negative difference in the performance of the seen classes while using the 'TruZe'.

Method	Acc_U			Acc_S			Harmonic mean			Dataset
	Rand	TruZe	Diff	Rand	TruZe	Diff	Rand	TruZe	Diff	
WGAN [24]	27.9	21.3	6.6	58.2	63.2	−5.0	37.7	31.8	5.9	HMDB51
WGAN [24]	28.2	23.9	4.3	74.9	75.6	−0.7	41.0	36.3	4.7	UCF101
OD [10]	34.1	24.7	9.4	58.5	62.8	−4.3	43.1	35.5	7.6	HMDB51
OD [10]	32.6	29.1	3.5	76.1	78.4	−2.3	45.6	42.4	3.2	UCF101
CLASTER [7]	41.8	38.4	3.4	52.3	53.1	−0.8	46.4	44.5	1.9	HMDB51
CLASTER [7]	37.5	35.6	1.9	68.8	70.6	−1.8	48.5	47.3	1.2	UCF101

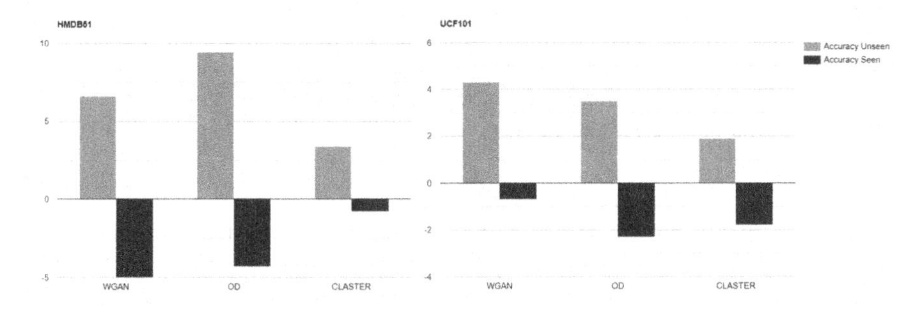

Fig. 3. Graphical representation of the difference in performances of different models on GZSL. We see consistent positive difference in performance on the unseen classes and negative difference in the performance of the seen classes while using the 'TruZe'. The x-axis corresponds to difference in accuracy (Random splits accuracy - TruZe split accuracy) and the y-axis to different methods.

all models is sen2vec. We use 70 classes for training chosen at random along with 31 test classes (also chosen at random) for UCF101 and 29 training with 22 testing for HMDB51. As expected, the average unseen class accuracy drops in the proposed split and the average seen class accuracy increases. We expect this as the unseen classes are more disjoint in the proposed split than using random splits. For easier understanding, we convert the differences in Table 3 to a graph and this can be seen in Fig. 3.

6.2 Extension to Few-shot Learning

Few-shot learning (FSL) is another scenario we consider. Since the premise is the same as ZSL, except that we have a few samples instead of zero. Again,

Table 4. Few Shot Learning (FSL) with different splits on UCF101. Accuracies are reported for 5-way, 1, 2, 3, 4, 5-shot classification. 'SS' corresponds to the split used in [14,26] and 'TruZe' corresponds to the proposed split. We can see that using our proposed split results in a drop in performance of up to 6.2 % for UCF101. This shows TruZe is much harder even in the FSL scenario.

Method	SS					TruZe					Diff				
	1	2	3	4	5	1	2	3	4	5	1	2	3	4	5
C3D-PN [16]	57.1	66.4	71.7	75.5	78.2	50.9	61.9	67.5	72.9	75.4	6.2	4.5	4.2	2.6	2.8
ARN [26]	66.3	73.1	77.9	80.4	83.1	61.2	70.7	75.2	78.8	80.2	5.1	2.4	2.7	1.6	2.9
TRX [14]	77.5	88.8	92.8	94.7	96.1	75.2	88.1	91.5	93.1	93.5	2.5	0.7	1.3	1.6	2.6

Table 5. Few Shot Learning (FSL) with different splits on HMDB51. Accuracies are reported for 5-way, 1, 2, 3, 4, 5-shot classification. 'SS' corresponds to the split used in [14,26] and 'TruZe' corresponds to the proposed split. We can see that using our proposed split results in a drop in performance of up to 17.1 % for HMDB51. This shows TruZe is much harder even in the FSL scenario.

Method	SS					TruZe					Diff				
	1	2	3	4	5	1	2	3	4	5	1	2	3	4	5
C3D-PN [16]	38.1	47.5	50.3	55.6	57.4	28.8	38.5	43.4	46.7	49.1	9.3	9.0	6.9	8.9	8.3
ARN [26]	45.5	50.1	54.2	58.7	60.6	31.9	42.3	46.5	49.8	53.2	12.6	7.8	7.7	8.9	7.4
TRX [14]	50.5	62.7	66.9	73.5	75.6	33.5	46.7	49.8	57.9	61.5	17.0	16.0	17.1	15.6	14.1

usually, the splits used are random, and as such, the pre-trained model has seen hundreds of samples of classes that are supposed to belong to the test set. We report results on the 5-way, 1,2,3,4,5-shot case for temporal relational cross-transformers (TRX) [14], action relation network (ARN) [26], and C3D prototypical net (C3D-PN) [16]. Results are reported in Table 4 and Table 5. The standard split (SS) used here is taken from the one proposed in ARN [26]. Similar to the SS, we divide the classes in UCF101 and HMDB51 to (70,10,21) and (31,10,10), respectively, where the order corresponds to the number of training classes, validation classes and test classes. We see that the proposed split is much harder than SS. Consistent drops in performance can be seen on every split and for every model. Performance drops of upto 6.2% on UCF101 and 17.1% on HMDB51 can be seen. Our proposed splits are available in the supplementary material.

6.3 Is Overlap the Reason for Performance Difference Between Random and Our TruZe Split?

In order to understand the difference in model performance due to the overlapping classes, we compare the performance of each model for the random split (with five runs) vs the proposed split by using visual features represented by IDT and I3D. We depict the difference in performance in the form of a bar graph for better visual understanding. This is seen in Fig. 4. The higher the difference, the

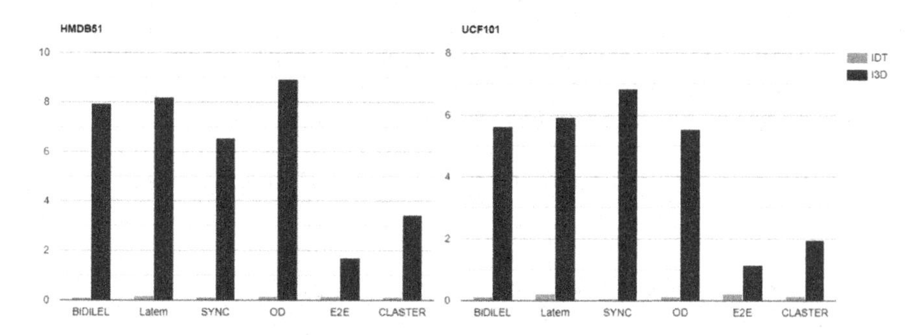

Fig. 4. The difference of accuracy for different models using IDT and I3D using manual annotations as the semantic embedding. The larger the bar, the more significant the difference. We can see a clear difference when using I3D and this difference is due to the presence of overlapping classes in the test set. The y-axis corresponds to the difference in performance in percentage and the x-axis corresponds to various models.

bigger the impact of performance. We can see that there is a big difference when using I3D features compared to using IDT features (where there is a minimal difference). Since IDT features are independent of any pre-training model, the difference in performance is negligible. The difference while using I3D features can be attributed to the presence of overlapping classes in the random splits compared to the proposed split.

6.4 Use of Different Backbone Networks

An end-to-end approach was proposed in [1] where a 3D CNN was trained in an end-to-end manner on classes in Kinetics not semantically similar to UCF101 and HMDB51 to overcome the overlapping classes conundrum. While this approach is useful, training more complex models end-to-end is not feasible for everyone due to the high computational cost involved. We show that using more recent state-of-the-art approaches as the backbone, there is a slight improvement in model performance and hence believe that having a proposed split instead of training end-to-end would be more easily affordable for the general public. Table 6 shows the results of using different backbones for extracting visual features on some of the recent state-of-the-art ZSL approaches. We use Non-Local networks [22] that build on I3D by adding long-term spatio-temporal dependencies in the form of non-local connections (referred as NL-I3D in Table 6). We also use slow-fast networks [6] that is a recent state-of-the-art approach that uses two pathways, a slow and a fast, to capture motion and fine temporal information. We can see minor but consistent improvements using stronger backbones, and this suggests that having a proposed split is an economical way of maximising the use of state-of-the-art models as backbone networks. We see gains of up to 0.6% in UCF101 and 0.8% in HMDB51.

Table 6. Results comparison using different backbones to extract visual features for the ZSL models. We evaluate OD, E2E and CLASTER using I3D, NL-I3D and SlowFast networks as backbones. All results are on the proposed split. We see that stronger backbones result in improved performance of the ZSL model.

Method	Backbone	UCF101 Accuracy	HMDB51 Accuracy
WGAN [24]	I3D	22.5	21.1
WGAN [24]	NL-I3D	22.7	21.3
WGAN [24]	SlowFast	**23.1**	**21.5**
OD [10]	I3D	22.9	21.7
OD [10]	NL-I3D	23.2	22.0
OD [10]	SlowFast	**23.4**	**22.5**
CLASTER [7]	I3D	45.2	33.2
CLASTER [7]	NL-I3D	45.3	33.6
CLASTER [7]	SlowFast	**45.5**	**33.9**

7 Implementation Details

7.1 Visual Features

We use either IDT [20] or I3D [3] for the visual features. Using the fisher vector obtained from a 256 component Gaussian mixture model, we generate visual feature representations using IDT (contains four different descriptors). To reduce this, PCA is used to obtain a 3000-dimensional vector for each descriptor. Concatenating these (all four descriptors), we obtain a 12000-dimensional vector for each video. In the case of I3D features, we use RGB and flow features taken from the *mixed 5c* layer from a pre-trained I3D (pre-trained on Kinetics-400). The output of the flow network is averaged across the temporal dimension and pooled by four in the spatial dimension, and then flattened to a vector of size 4096. We then concatenate the two.

7.2 Semantic Embedding

While manual annotations are available for UCF101 in the form of a vector of size 40, there is no such annotation available for HMDB51. Hence, we use sen2vec embeddings of the action classes where the sen2vec model is pre-trained on Wikipedia. While most approaches use word2vec and average embeddings for each word in the label, we use sen2vec which obtains an embedding for the entire label.

7.3 Hyperparameters for Evaluated Methods

We use the optimal parameters reported in BiDiLEL [21]. The values for α and $k_G$values are set to 10 and d_y is set to 150. SYNC [4] has a parameter σ that models correlation between a real class and a phantom class and this is set to 1, while the balance coefficient is set to 2^{-10}. For Latem [23] the learning rate, number of epochs, and number of embeddings are 0.1, 200 and 10 respectively. For OD [10], WGAN [24], E2E [1] and CLASTER [7] we follow the settings provided by the authors. For few-shot learning, we use the hyperparameters defined in the papers [14,26]. We compare against the standard split proposed in [26]. For the proposed split, we change the classes slightly for fair comparison to the standard split. Now the splits for HMDB51 and UCF101 are (31,10,10) and (70,10,21) where the order corresponds to (train,val,test).

8 Discussion and Conclusion

As we see in Fig. 4 using IDT features which do not require pre-training on Kinetics resulted in a negligible change in performance comparing the TruZe split vs the random splits. However, using I3D features saw a stark difference due to the overlapping classes in the pre-trained dataset.

We see that the proposed split is harder in all scenarios (ZSL, GZSL, and FSL) whilst maintaining the premise of the problem. The differences are significant in most cases: between 2.2–2.8 % for UCF101 and 7.4–14.1 % for HMDB51 in FSL, an increase of 1.0–4.1 % for UCF101 and 1.9–7.6 % for HMDB51 (with respect to the harmonic mean of seen and unseen classes) in GZSL and an increase of 1.2–6.1 % for UCF101 and 1.7–8.1 % for HMDB51 in ZSL. It is also important to note that different methods are differently affected, suggesting that some method in the past have claimed improvements due to not adhering to the zero-shot premise, which is highly concerning.

We also see that changing the backbone network increases the performance slightly for each model, and as a result, the end-to-end pre-training [1] can prove very expensive. As such, having a proposed split makes things easier as we can directly use pre-trained models off the shelf. We see gains of up to 1.3% in UCF101 and 1.1% in HMDB51.

Details to our TruZe splits can be found in supplemental material and we will release them publicly so the research community can fairly compare zero-shot and few-shot action recognition approaches and compare to the benchmark results provided in this paper.

References

1. Brattoli, B., Tighe, J., Zhdanov, F., Perona, P., Chalupka, K.: Rethinking zero-shot video classification: end-to-end training for realistic applications. In: Proceedings of the IEEE/CVF Conference on Computer Vision and Pattern Recognition, pp. 4613–4623 (2020)
2. Busto, P.P., Iqbal, A., Gall, J.: Open set domain adaptation for image and action recognition. IEEE Trans. Pattern Anal. Mach. Intell. **42**(2), 413–429 (2018)
3. Carreira, J., Zisserman, A.: Quo vadis, action recognition? A new model and the kinetics dataset. In: Proceedings of the IEEE Conference on Computer Vision and Pattern Recognition, pp. 6299–6308 (2017)

4. Changpinyo, S., Chao, W.L., Gong, B., Sha, F.: Synthesized classifiers for zero-shot learning. In: Proceedings of the IEEE Conference on Computer Vision and Pattern Recognition, pp. 5327–5336 (2016)
5. Deng, J., Dong, W., Socher, R., Li, L.J., Li, K., Fei-Fei, L.: ImageNet: a large-scale hierarchical image database. In: 2009 IEEE Conference on Computer Vision and Pattern Recognition, pp. 248–255. IEEE (2009)
6. Feichtenhofer, C., Fan, H., Malik, J., He, K.: Slowfast networks for video recognition. In: Proceedings of the IEEE/CVF International Conference on Computer Vision, pp. 6202–6211 (2019)
7. Gowda, S.N., Sevilla-Lara, L., Keller, F., Rohrbach, M.: Claster: clustering with reinforcement learning for zero-shot action recognition. arXiv preprint arXiv:2101.07042 (2021)
8. Kuehne, H., Jhuang, H., Garrote, E., Poggio, T., Serre, T.: HMDB: a large video database for human motion recognition. In: 2011 International Conference on Computer Vision, pp. 2556–2563. IEEE (2011)
9. Liu, J., Kuipers, B., Savarese, S.: Recognizing human actions by attributes. In: CVPR 2011, pp. 3337–3344. IEEE (2011)
10. Mandal, D., et al.: Out-of-distribution detection for generalized zero-shot action recognition. In: Proceedings of the IEEE/CVF Conference on Computer Vision and Pattern Recognition, pp. 9985–9993 (2019)
11. Mikolov, T., Sutskever, I., Chen, K., Corrado, G., Dean, J.: Distributed representations of words and phrases and their compositionality. arXiv preprint arXiv:1310.4546 (2013)
12. Niebles, J.C., Chen, C.-W., Fei-Fei, L.: Modeling temporal structure of decomposable motion segments for activity classification. In: Daniilidis, K., Maragos, P., Paragios, N. (eds.) ECCV 2010. LNCS, vol. 6312, pp. 392–405. Springer, Heidelberg (2010). https://doi.org/10.1007/978-3-642-15552-9_29
13. Pagliardini, M., Gupta, P., Jaggi, M.: Unsupervised learning of sentence embeddings using compositional n-gram features. In: Proceedings of the 2018 Conference of the North American Chapter of the Association for Computational Linguistics: Human Language Technologies, vol. 1 (Long Papers), pp. 528–540 (2018)
14. Perrett, T., Masullo, A., Burghardt, T., Mirmehdi, M., Damen, D.: Temporal-relational crosstransformers for few-shot action recognition. arXiv preprint arXiv:2101.06184 (2021)
15. Roitberg, A., Martinez, M., Haurilet, M., Stiefelhagen, R.: Towards a fair evaluation of zero-shot action recognition using external data. In: Leal-Taixé, L., Roth, S. (eds.) ECCV 2018. LNCS, vol. 11132, pp. 97–105. Springer, Cham (2019). https://doi.org/10.1007/978-3-030-11018-5_8
16. Snell, J., Swersky, K., Zemel, R.S.: Prototypical networks for few-shot learning. arXiv preprint arXiv:1703.05175 (2017)
17. Soomro, K., Zamir, A.R., Shah, M.: UCF101: a dataset of 101 human actions classes from videos in the wild. arXiv preprint arXiv:1212.0402 (2012)
18. Tran, D., Bourdev, L., Fergus, R., Torresani, L., Paluri, M.: Learning spatiotemporal features with 3d convolutional networks. In: Proceedings of the IEEE International Conference on Computer Vision, pp. 4489–4497 (2015)
19. Verma, V.K., Arora, G., Mishra, A., Rai, P.: Generalized zero-shot learning via synthesized examples. In: Proceedings of the IEEE Conference on Computer Vision and Pattern Recognition, pp. 4281–4289 (2018)
20. Wang, H., Schmid, C.: Action recognition with improved trajectories. In: Proceedings of the IEEE International Conference on Computer Vision, pp. 3551–3558 (2013)

21. Wang, Q., Chen, K.: Zero-shot visual recognition via bidirectional latent embedding. Int. J. Comput. Vision **124**(3), 356–383 (2017). https://doi.org/10.1007/s11263-017-1027-5
22. Wang, X., Girshick, R., Gupta, A., He, K.: Non-local neural networks. In: Proceedings of the IEEE Conference on Computer Vision and Pattern Recognition, pp. 7794–7803 (2018)
23. Xian, Y., Akata, Z., Sharma, G., Nguyen, Q., Hein, M., Schiele, B.: Latent embeddings for zero-shot classification. In: Proceedings of the IEEE Conference on Computer Vision and Pattern Recognition, pp. 69–77 (2016)
24. Xian, Y., Lorenz, T., Schiele, B., Akata, Z.: Feature generating networks for zero-shot learning. In: Proceedings of the IEEE Conference on Computer Vision and Pattern Recognition, pp. 5542–5551 (2018)
25. Xian, Y., Schiele, B., Akata, Z.: Zero-shot learning-the good, the bad and the ugly. In: Proceedings of the IEEE Conference on Computer Vision and Pattern Recognition, pp. 4582–4591 (2017)
26. Zhang, H., Zhang, L., Qi, X., Li, H., Torr, P.H.S., Koniusz, P.: Few-shot action recognition with permutation-invariant attention. In: Vedaldi, A., Bischof, H., Brox, T., Frahm, J.-M. (eds.) ECCV 2020. LNCS, vol. 12350, pp. 525–542. Springer, Cham (2020). https://doi.org/10.1007/978-3-030-58558-7_31

Video Instance Segmentation with Recurrent Graph Neural Networks

Joakim Johnander[1,2(✉)], Emil Brissman[1,3], Martin Danelljan[4], and Michael Felsberg[1,5]

[1] Computer Vision Laboratory, Department of Electrical Engineering, Linköping University, Linköping, Sweden
joakim.johnander@liu.se
[2] Zenseact, Gothenburg, Sweden
[3] Saab, Bröderna Ugglas gata, 582 54 Linköping, Sweden
[4] Computer Vision Lab, ETH Zürich, Zürich, Switzerland
[5] School of Engineering, University of KwaZulu-Natal, Durban, South Africa

Abstract. Video instance segmentation is one of the core problems in computer vision. Formulating a purely learning-based method, which models the generic track management required to solve the video instance segmentation task, is a highly challenging problem. In this work, we propose a novel learning framework where the entire video instance segmentation problem is modeled jointly. To this end, we design a graph neural network that in each frame jointly processes all detections and a memory of previously seen tracks. Past information is considered and processed via a recurrent connection. We demonstrate the effectiveness of the proposed approach in comprehensive experiments. Our approach, operating at over 25 FPS, outperforms previous video real-time methods. We further conduct detailed ablative experiments that validate the different aspects of our approach.

1 Introduction

Video instance segmentation (VIS) is the task of simultaneously detecting, segmenting, and tracking object instances from a set of predefined classes. This task has a wide range of applications in autonomous driving [14,32], data annotation [4,19], and biology [10,26,33]. In contrast to image instance segmentation, the temporal aspect of its video counterpart poses several additional challenges. Preserving correct instance identities in each frame is made difficult by the presence of other, similar instances. Objects may be subject to occlusions, fast motion, or major appearance changes. Moreover, the videos can include wild camera motion and severe background clutter.

Supplementary Information The online version contains supplementary material available at https://doi.org/10.1007/978-3-030-92659-5_13.

C. Bauckhage et al. (Eds.): DAGM GCPR 2021, LNCS 13024, pp. 206–221, 2021.
https://doi.org/10.1007/978-3-030-92659-5_13

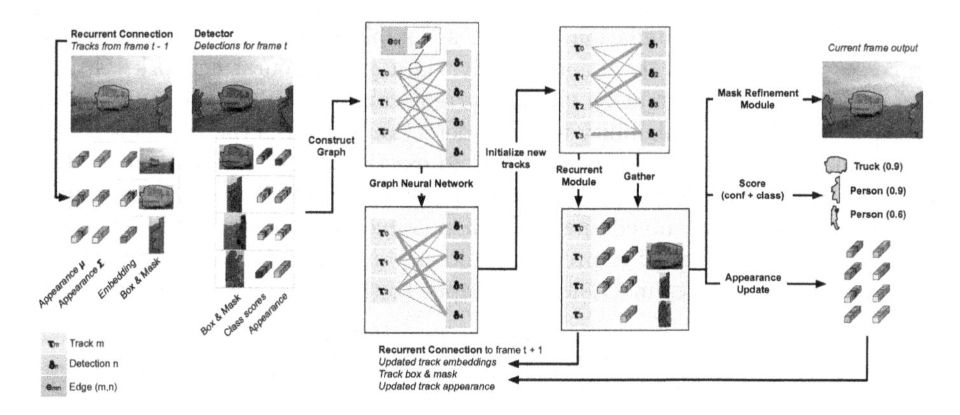

Fig. 1. Illustration of the proposed approach. An instance segmentation method is applied to each frame. The set of detections is, together with a maintained memory of tracks, used to construct a graph. Each node and each edge is represented with an embedding. These are processed by a graph neural network and directly used to predict assignment and track initialization (see Sect. 3.1). The track embeddings are further processed by a recurrent module, producing final track embeddings (Sect. 3.3). These are used to predict track confidences and class memberships (Sect. 3.4), masks (Sect. 3.4), and updated appearance descriptors (Sect. 3.2). Last, the final track embeddings are propagated to the next frame via the recurrent connection.

Prior works have taken inspiration from the related areas of multiple object tracking, video object detection, instance segmentation, and video object segmentation [1,6,30]. Most methods adopt the tracking-by-detection paradigm popular in multiple object tracking [9]. In this paradigm, an instance segmentation method provides detections in each frame, reducing the task to the formation of *tracks* from these detections. Given a set of already initialized tracks, one must determine for each detection whether it belongs to one of the tracks, is a false positive, or if it should initialize a new track. Most approaches [6,11,21,30] learn to match pairs of detections and then rely on heuristics to form the final output, e.g., initializing new tracks, predicting confidences, removing tracks, and predicting class memberships.

The aforementioned pipelines suffer from two major drawbacks. (i) The learnt models lack flexibility, and are for instance unable to reason globally over all detections or access information temporally [11,30]. (ii) The model learning stage does not closely model the inference, for instance by utilizing only pairs of frames or ignoring subsequent detection merging stages [6,11,21,30]. This means that the method never gets the chance to learn many of the aspects of the VIS problem – such as dealing with false positives in the employed instance segmentation method or handling uncertain detections.

We address these two drawbacks by proposing a novel spatiotemporal learning framework for video instance segmentation that closely models the inference stage during training. Our network proceeds frame by frame, and is in each frame asked to create tracks, associate detections to tracks, and score existing tracks. We use this formulation to train a flexible model, that in each frame processes all tracks and detections jointly via a graph neural network (GNN), and considers past information via a recurrent connection. The model predicts, for each detection, a probability that the detection should initialize a new track. The model also predicts, for each pair of an existing track and a detection, the probability that the track and the detection correspond to the same instance. Finally, it predicts an embedding for each existing track. The embedding serves two purposes: (i) it is used to predict confidence and class for the track; and (ii) it is via the recurrent connection fed as input to the GNN in the next frame.

Contributions: Our main contributions are as follows. **(i)** We propose a new framework for training video instance segmentation methods. The methods proceed frame-by-frame and are in each, given detections from an instance segmentation network, trained to match detections to tracks, initialize new tracks, predict segmentations, and score tracks. **(ii)** We present a suitable and flexible model based on Graph Neural Networks and Recurrent Neural Networks. **(iii)** We show that the GNN successfully learns to propagate information between different tracks and detections in order to predict matches, initialize new tracks, and predict track confidence and class. **(iv)** A recurrent connection permits us to feed information about the tracks to the next time step. We show that while a naïve implementation of such a connection leads to highly unstable training, an adaption of the long short-term memory effectively solves this issue. **(v)** We model the instance appearance as a Gaussian distribution and introduce a learnable update formulation. **(vi)** We analyze the effectiveness of our approach in comprehensive experiments. Our method outperforms previous near real-time approaches with a relative mAP gain of 9.0% on the YouTubeVIS dataset [30].

2 Related Work

The video instance segmentation (VIS) problem was introduced by Yang *et al.* [30]. With it, they proposed several simple and straightforward approaches to tackle the task. They follow the tracking-by-detection paradigm and first apply an instance segmentation method to provide detections in each frame, and then form tracks based on these detections. They experiment with several approaches to matching different detections, such as mask propagation with a video object segmentation method [27]; application of a multiple object tracking method [29] in which the image-plane bounding boxes are Kalman filtered and targets are re-detected with a learnt re-identification mechanism; and similarity learning of instance-specific appearance descriptors [30]. Additionally, they experiment with the offline temporal filtering proposed in [16].

Cao *et al.* [11] propose to improve the underlying instance segmentation method, obtaining both better performance and computational efficiency. Luiten *et al.* [21] propose (i) to improve the instance segmentation method by applying different networks for classification, segmentation, and proposal generation; and (ii) to form tracks with the offline algorithm proposed in [22]. Bertasius *et al.* [6] also utilize a more powerful instance segmentation method [7], and propose a novel mask propagation method based on deformable convolutions. Both [21] and [6] achieve strong performance, but at a very high computational cost.

All of these approaches follow the tracking-by-detection paradigm and try various ways to improve the underlying instance segmentation method or the association of detections. The latter relies mostly on heuristics [21,30] and is often not end-to-end trainable. Furthermore, the track scoring step, where the class and confidence is predicted, has received little attention and is in existing approaches calculated with a majority vote and an averaging operation, as pointed out in the introduction. The work of Athar *et al.* [1] instead proposes an end-to-end trainable approach that is trained to predict instance center heatmaps and an embedding for each pixel. A track is constructed from strong responses in the heatmap. The embedding at that location is matched with the embeddings of all other pixels, and sufficiently similar pixels are assigned to that track.

Our approach is closely related to two works on multiple object tracking (MOT) [9,28] and a work on feature matching [25]. These works associate detections or feature points by forming a bipartite graph and applying a Graph Neural Network. The strength of this approach is that the neural network simultaneously reasons about all available information. However, the setting of these works differs significantly from video instance segmentation. MOT is typically restricted to a specific type of scene, such as automotive, and usually with only one or two classes. Furthermore, for both MOT and feature matching, no classification or confidence is to be provided for the tracks. This is reflected in the way [9,25,28] utilizes their GNNs, where only either nodes or edges are of interest, not both. The other part exists solely for the purpose of passing messages. As we explain in Sect. 3, we will instead utilize both edges and nodes: the edges to predict association and the nodes to predict class membership and confidence.

3 Method

We propose an approach for video instance segmentation, consisting of a single neural network. Our model proceeds frame by frame, and performs the following steps: (i) predict tentative single-image instance segmentations, (ii) associate detections to existing tracks, (iii) initialize new tracks, (iv) score existing tracks, (v) update the states of each track.

The instance segmentations together with the existing tracks are fed into a graph neural network (GNN). The GNN processes all tracks and detections jointly to produce output embeddings that are used for association and scoring. These output embeddings are furthermore fed as input to the GNN in the next time step, permitting the GNN to process both present and previous information. An overview of the approach is provided in Fig. 1.

3.1 Track-Detection Association

We maintain a memory of previously seen objects, or *tracks*, which is updated over time. In each frame, an instance segmentation method produces tentative

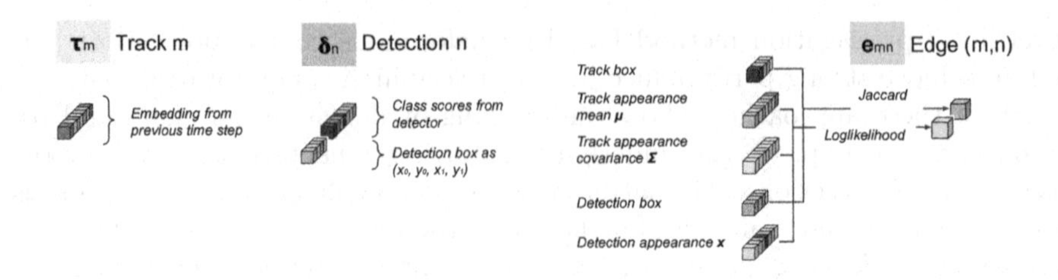

Fig. 2. Overview of the node and edge initialization during graph construction.

detections. The aim of our model is to associate detections with tracks, determining whether or not track m corresponds to detection n. In addition, the model needs to decide, for each detection n, whether it should initialize a new track.

Motivation. Most existing methods [11,21,30] associate tracks to detections by training a network to extract appearance descriptors. The descriptors are trained to be similar if they correspond to the same object, and dissimilar if they correspond to different objects. The issue with such an approach is that appearance descriptors corresponding to visually and semantically similar, but different instances, will be trained to be different. In such scenarios it might be better to let the appearance descriptors be similar, and instead rely on for instance spatial information. The network should therefore assess all available information before making its decision.

Further information is obtained from track-detection pairs other than the one considered. It may be difficult to determine whether a track and a detection match in isolation, for instance with cluttered scenes or when visibility is poor. In such scenarios, the instance segmentation method might provide multiple detections that all overlap the same object to some extent. Another difficult scenario is when there is sudden and severe camera motion, in which case we might need global reasoning in order to either disregard spatial similarity or treat it differently. We therefore hypothesize that it is important for the network to reason about all tracks and detections simultaneously.

The same is true when determining whether a detection should initialize a new track. How well a detection matches existing tracks must influence this decision. Previous works [11,30] achieve this observation with a hard decision. In these, a new track will be initialized for each detection that does not match an existing track. We avoid this heuristic and instead let the network process all tracks and detections simultaneously and jointly predict track-detection assignment and track initialization. It should be noted, however, that the detections are noisy in general. Making the correct decision may be outright impossible. In such scenarios we would expect the model to create a track, and over time as more information is accumulated, re-evaluate whether the track is novel, previously seen, or from a false positive in the detector.

Graph Construction. For each detection n we construct an embedding δ_n. It is initialized as the concatenation of the bounding box and classification scores output by the detector. Each track in memory has an embedding τ_m, that was produced by our model in the previous time step via the *recurrent connection.* We represent the relationship between each track-detection pair with an embedding e_{mn}. This embedding will later be used to predict the probability that track m matches detection n. It is initialized as the concatenation of (i) the spatial similarity between them, based on the Jaccard index between their bounding boxes (see [30]); and (ii) the appearance similarity between them, as described in Sect. 3.2. This construction is illustrated in Fig. 2. Further, we let the relationship between each detection and a corresponding potential new track be represented with an embedding e_{0n}, and let τ_0 represent an empty track embedding. We treat e_{0n} and τ_0 the way we treat other edges and tracks, but they are processed with their own set of weights. The initialization of the edges e_{0n} is done without the spatial similarity and only with the appearance similarity. We maintain a separate appearance model for the empty track, based on the appearance of the entire scene. The elements τ_m, δ_n, and e_{mn} constitute a bipartite graph, as illustrated in Fig. 1.

GNN-Based Association. The idea is to propagate information between the different embeddings in a learnable way, providing us with updated embeddings that we can directly use to predict the quantities needed for video instance segmentation. To this end we use layers that perform updates of the form

$$e_{mn}^{i+1} = f_i^e([e_{mn}^i, \tau_m^i, \delta_n^i]) \ , \tag{1a}$$

$$\tau_m^{i+1} = f_i^\tau([\tau_m^i, \sum_j g_i^\tau(e_{mj}^i)e_{mj}^i]) \ , \tag{1b}$$

$$\delta_n^{i+1} = f_i^\delta([\delta_n^i, \sum_i g_i^\delta(e_{in}^i)e_{in}^i]) \ . \tag{1c}$$

Here, i enumerates the network layers. The functions f_i^e, f_i^τ, and f_i^δ are linear layers followed by a ReLU activation. The gating functions g_i^τ and g_i^δ are multi-layer perceptrons ending with the logistic sigmoid. $[\cdot,\cdot]$ denotes concatenation.

The aforementioned formulation has the structure of a Graph Neural Network (GNN) block [2], with both τ_m and δ_n as nodes, and e_{mn} as edges. These layers permit information exchange between the embeddings. The layer deviates slightly from the literature. First, we have two types of nodes and use two different updates for them. This is similar to the work of Brasó *et al.* [9] where message passing forward and backward in time uses two different neural networks. Second, the accumulation in the nodes in (1b) and (1c) uses an additional gate, permitting the nodes to dynamically select from which message information should be accumulated. This is sensible in our setting, as for instance class information should be passed from detection to track if and only if the track and detection match well.

We construct our graph neural network by stacking GNN blocks. For added expressivity at small computational cost, we interleave them with residual blocks [17] in which there is no information exchange between different graph elements. That is, for these blocks the f_i rely only on their first arguments. Note that these blocks use fully connected layers instead of 2D convolutions. The GNN will provide us with updated edge embeddings which we use for association of detections to tracks, and updated node embeddings which will be used to score tracks and as input to the GNN in the next frame.

Association Prediction. We predict the probability that the track m matches the detection n by feeding the edge embeddings e_{mn} through a logistic model

$$\Pr(m \, \text{matches} \, n) = \text{sigmoid}(w \cdot e_{mn} + b). \tag{2}$$

If the probability is high, they are considered to match and the track will obtain the segmentation of that detection. New tracks are initialized in a similar fashion. The edge embeddings e_{0n} are fed through another logistic model to predict the probability that the detection n should initialize a new track. If the probability is beyond a threshold, we initialize a new track with the embedding of that detection δ_n. This threshold is intentionally selected to be quite low. This leads to additional false positives, but our model can mark them as such by giving them low class scores and not assigning any segmentation pixels to them.

Note that we treat the track-detection association as multiple binary classification problems. This may lead to a single detection being assigned to multiple tracks. An alternative would be to instead consider the classification of a single detection as a multiclass classification problem. We observed, however, that this led to slightly inferior results and that it was uncommon for a single detection to be assigned to more than one track.

3.2 Modelling Appearance

In order to accurately match tracks and detections, we create instance-specific appearance models for each tracked object. To this end, we add an appearance network, comprising a few convolutional layers, and apply it to feature maps of the backbone ResNet [17]. The output of the appearance network is pooled with the masks provided by the detections, resulting in an appearance descriptor for each detection. The tracks gather appearance from the detections and over time construct an appearance model of that track. The similarity in appearance between a track and a detection will serve as an important additional cue during matching. The aim for the appearance network is to learn a rich representation that allows us to discriminate between visually or semantically similar instances.

Our initial experiments of integrating appearance information directly into the GNN, similar to [9,25,28], did not lead to noticeable improvement. This is likely due to differences between the problems. The video instance segmentation problem is fairly unconstrained: there is significant variation in scenes and objects considered, and compared to its variation, there are quite few labelled training sequences available. In contrast, multiple object tracking typically works with a single type of scene or a single category of objects and feature matching is learnt with magnitudes more training examples than what is available for video instance segmentation.

In order to sidestep this issue, we treat appearance separately and allow the GNN to observe *only* the appearance similarity, and *not* the actual appearance. Each track models its appearance as a multidimensional Gaussian distribution with diagonal covariance. When the track is initialized, we use the appearance vector of the initializing detection as mean μ and a fixed covariance Σ. We feed appearance information into the GNN via the track-detection edges. The edge between track m and detection n is initialized with the loglikelihood of the detection appearance given the track distribution. The GNN is able to utilize this information when calculating the matching probability of each track-detection pair. Afterwards, the appearance (μ, Σ) of each track is updated with the appearance x of the best matching detection. The update is based on the Bayesian update of a Gaussian under a conjugate prior. We use a normal-inverse-chi-square prior [23],

$$\mu^+ = \kappa x + (1-\kappa)\mu \ , \tag{3a}$$

$$\Sigma^+ = \nu\tilde{\Sigma} + (1-\nu)\Sigma + \frac{\kappa(1-\nu)}{\kappa+\nu}(x-\mu)^2 \ . \tag{3b}$$

The term $\tilde{\Sigma}$ corresponds to the sample variance and the update rates κ and ν would usually be the number of samples in the update relative the strength of the prior. For added flexibility we predict these values based on the track embedding, permitting the network to learn a good update strategy.

3.3 Recurrent Connection

In order to process object tracks, it is crucial to propagate information over time. We achieve this with a recurrent connection, which brings the benefit of end-to-end training. However, naïvely adding recurrent connections leads to highly unstable training and in extension, poor video instance segmentation results. Even with careful weight initialization and low learning rate, both activation and gradient spikes arise. This is a well-known problem when training recurrent neural networks and is usually tackled with the Long Short-Term Memory (LSTM) [18] or Gated Recurrent Unit [13]. These modules use a system of multiplicative sigmoid-activated *gates*, and have been repeatedly shown to be able to well model sequential data while avoiding aforementioned issues [13,15,18].

We adapt the LSTM to our scenario. Typically, the output of the LSTM is fed as its input in the next time step. We instead feed the output of the LSTM as input to the GNN in the next time step, and the output of the GNN as input to the LSTM. First, denote the output of the GNN as

$$\{\tilde{\tau}_m^t\}, \{\tilde{\delta}_n^t\}, \{\tilde{e}_{mn}^t\} = \mathrm{GNN}(\{\tau_m^{t-1}\}, \{\delta_n^t\}, \{e_{mn}^t\}) \ , \tag{4}$$

where superscript t denotes time. Next, we feed each track embedding $\tilde{\tau}_m^t$ through the LSTM system of gates

$$\alpha_m^{\text{forget}} = \sigma(h^{\text{forget}}(\tilde{\tau}_m^t)) \ , \tag{5a}$$

$$\alpha_m^{\text{input}} = \sigma(h^{\text{input}}(\tilde{\tau}_m^t)) \ , \tag{5b}$$

$$\alpha_m^{\text{output}} = \sigma(h^{\text{output}}(\tilde{\tau}_m^t)) \ , \tag{5c}$$

$$\tilde{c}_m^t = \tanh(h^{\text{cell}}(\tilde{\tau}_m^t)) \ , \tag{5d}$$

$$c_m^t = \alpha_m^{\text{forget}} \odot c_i^{t-1} + \alpha_m^{\text{input}} \odot \tilde{c}_m^t \ , \tag{5e}$$

$$\tau_m^t = \alpha_m^{\text{output}} \odot \tanh(c_m^t) \ . \tag{5f}$$

The functions $h^{\text{forget}}, h^{\text{input}}, h^{\text{output}}, h^{\text{cell}}$ are linear neural network layers. $\odot$ is the element-wise product, tanh the hyperbolic tangent, and σ the logistic sigmoid.

3.4 VIS Output Prediction

Track scoring. For the VIS task, we need to constantly assess the validity and class membership of each active track τ_m. To this end, we predict a confidence value and the class of existing tracks in each frame. The confidence reflects our trust about whether or not the track is a true positive. It is updated over time together with the class prediction as more information becomes available. This provides the model with the option of effectively removing tracks by reducing their scores. Existing approaches [6,11,21,30] score tracks by averaging the detection confidence of the detections deemed to correspond to the track. Class predictions are made with a majority vote. The drawback is that other available information, such as how certain we are that each detection indeed belongs to the track or the consistency of the detections, is not taken into account.

We address the problem of track scoring and classification using the GNN introduced in Sect. 3.1 together with a recurrent connection (Sect. 3.3). The track embeddings $\{\tau_m\}_m$ gather information from all detections via the GNN, and accumulate this information over time via the recurrent connection. We then predict the confidence and class for each track based on its embedding. This is achieved via linear layer followed by softmax.

Segmentation. In each frame, we report a segmentation. This segmentation is based on both the track embeddings and the masks provided with the detections. Each track that matches sufficiently well with a detection claims the mask of that detection. This mask together with the track embedding are then fed through a small CNN that reweights and refines the mask. This permits our model to not assign pixels to tracks that it believes are false positives.

3.5 Training

We train the network by feeding a sequence of T frames through it as we would during inference at test-time. In each frame t, the neural network predicts track-detection match probabilities, track initialization probabilities, track class probabilities, and track segmentation probabilities

$$\mathbf{y}_t^{\text{match}} \in [0,1]^{M_t \times N_t} \;, \tag{6a}$$
$$\mathbf{y}_t^{\text{init}} \in [0,1]^{N_t} \;, \tag{6b}$$
$$\mathbf{y}_t^{\text{score}} \in [0,1]^{M_{t+1} \times C} \;, \tag{6c}$$
$$\mathbf{y}_t^{\text{seg}} \in [0,1]^{M_{t+1} \times H \times W} \;. \tag{6d}$$

Here, M_t denotes the number of tracks in frame t prior to initializing new tracks; N_t the number of detections obtained from the detector in frame t; C the number of object categories, including background; and $H \times W$ the image size. The four components in (6) permits the model to conduct video instance segmentation. We penalize each with a corresponding loss component

$$\mathcal{L} = \lambda^1 \mathcal{L}^{\text{score}} + \lambda^2 \mathcal{L}^{\text{seg}} + \lambda^3 \mathcal{L}^{\text{match}} + \lambda^4 \mathcal{L}^{\text{init}} \;. \tag{7}$$

The component $\mathcal{L}^{\text{score}}$ rewards the network for correct prediction of the class scores; $\mathcal{L}^{\text{seg}}$ for segmentation refinement; $\mathcal{L}^{\text{match}}$ for assignment of detections to tracks; and $\mathcal{L}^{\text{init}}$ for initialization of new tracks. We weight the components with constants $(\lambda^1, \lambda^2, \lambda^3, \lambda^4)$.

In order to compute the loss, we determine the identity of each track and each detection. The identity is either one of the annotated objects or background. First, for each frame, the detections are matched to the annotated objects in that frame. Detections can claim the identity of an annotated object if their bounding boxes overlap by at least 50%. If multiple detections overlap with the same object, only the best matching detection claims its identity. Detections that do not claim the identity of an annotated object are marked as background. Thus, each annotated object will correspond to a maximum of one detection in each frame. Next, the tracks are assigned identities. Each track was initialized by a single detection at some frame and the track can claim the identity of that detection. However, if multiple tracks try to claim the identity of a single annotated object, only the first initialized of those tracks gets that identity. The others are assigned as background. Thus, each annotated object will correspond to a maximum of one track.

Using the track and detection identities we compute the loss components. Each component is normalized with the batchsize and video length, but not with the number of tracks or detections. Detections or tracks that are false positives will therefore not reduce the loss for other tracks or detections, as they otherwise would.

$\mathcal{L}^{\textbf{match}}$ is the binary cross-entropy loss. The target for $\mathbf{y}_{t,m,n}^{\text{match}}$ is 1 if track m and detection n has the same identity and that identity corresponds to an annotated object. If their identities differ or if the identity is background, the target is 0.

$\mathcal{L}^{\textbf{init}}$ is the binary cross-entropy loss. The target for $\mathbf{y}_{t,n}^{\text{init}}$ is 1 if detection n initializes a track with the identity of an annotated object. Otherwise, the target is 0.

$\mathcal{L}^{\text{score}}$ is the cross-entropy loss. If track m corresponds to an annotated object, the target for $\mathbf{y}_{t,m}^{\text{score}}$ is the category of that object. Otherwise the target is the background class. We found that it was difficult to score tracks early on in some scenarios and therefore we weight the loss over the sequence, giving higher weight to later frames.
$\mathcal{L}^{\text{seg}}$ is the Lovasz loss [5]. The target for $\mathbf{y}_t^{\text{seg}}$ is obtained by mapping the annotated object identities in the ground-truth segmentation to the track identities. In scenarios where a single annotated object gives rise to multiple tracks, the network is rewarded for assigning pixels only to the track that claimed the identity of that object.

4 Experiments

We evaluate the proposed approach for video instance segmentation on YouTube-VIS [30] (2019), a benchmark comprising 40 object categories in 2k training videos and 300 validation videos. Performance is measured in terms of video mean average precision (mAP). We first provide qualitative results, showing that the proposed neural network learns to tackle the video instance segmentation problem. Next, we quantitatively compare to the state-of-the-art. Last, we analyze the different components and aspects of our approach in an ablation study.

4.1 Implementation Details

We implement the proposed approach in PyTorch [24] and will make code available upon publication. We aim for real-time performance and therefore select YOLACT [8] as base instance segmentation method. We use the implementation publicly provided by the authors. The detector and our ResNet50 [17] backbone are initialized with weights provided with the YOLACT implementation. We fine-tune the detector on images from YouTubeVIS and OpenImages [3,20] for 120 epochs à 933 iterations, with a batch size of 8. Next, we freeze the backbone and the detector, and train all other modules: the appearance network, the GNN, and the recurrent module. We train for 150 epochs à 633 iterations with a batch of 4 video clips, each 10 frames sampled randomly from YouTubeVIS. During training, 200 sequences of YouTubeVIS are held-out for hyperparameter selection. For additional model and training details, see Supplementary material.

4.2 Qualitative Results

In Fig. 3 we show the output of the detector and the tracks predicted by our approach. The detector may provide noisy class predictions. Our model learns to filter these predictions and accurately predict the correct class. When the detector fails to detect an object, our approach pauses the corresponding track until the detector again finds the object. If the detector provides a false positive, our approach initializes a track that is later marked as background and rendered inactive. The proposed model has learnt to deal with mistakes made by the detector. For additional qualitative results, see the Supplementary material.

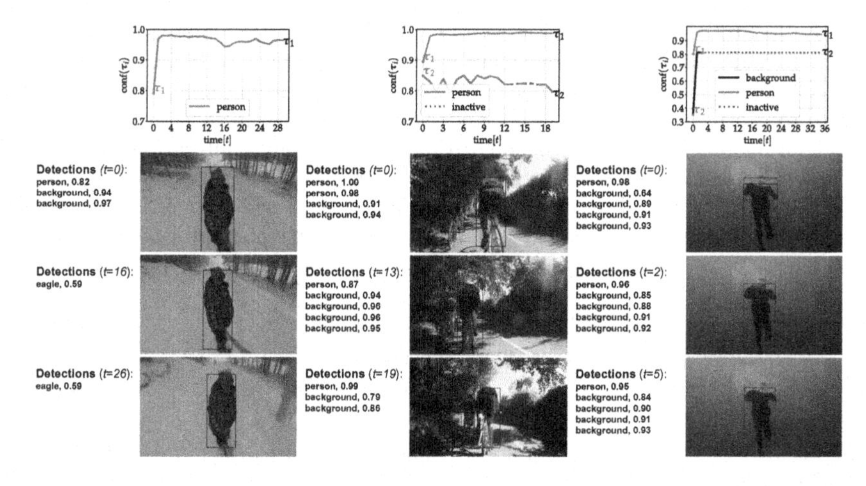

Fig. 3. Track score plots (top) and detections (3 bottom rows) for three videos. The plot colour is the ground-truth class for that track and the value is the confidence for that class, ideally 1.00. In the left video, the detector makes noisy class predictions, but our approach learns to filter this noise. In the center, there is a missed detection. Our method renders the track inactive and resumes it in subsequent frames where the detector finds both objects. To the right, a false positive in the detector leads to a false track. This track is, however, quickly marked as background with high confidence.

4.3 Quantitative Comparison

Next, we compare our approach to the state-of-the-art, including the baselines proposed in [30]. The results are shown in Table 1a. Our approach, running at 30 fps, outperforms all near real-time methods. DeepSORT [29], which relies on Kalman-filtering the bounding boxes and a learnt appearance descriptor used for re-identification, obtains an mAP score of 26.1. MaskTrack R-CNN [30] gets a score of 30.3. SipMask [11] improves MaskTrack R-CNN by changing its detector and reach a score of 33.7. Using a ResNet50 backbone, we run at similar speed and outperform all three methods with an absolute gain of 9.2, 5.0, and 1.6 respectively.

While [6,21] obtain higher mAP, those methods are more than a magnitude slower, and thus infeasible for real-time applications or for processing large amounts of data. STEm-Seg [1] reports results using both a ResNet50 and a ResNet101 backbone. We show a gain of 4.7 mAP with ResNet50. We also try with a ResNet101 backbone, retraining our base detector and approach. This leads to a performance of 37.7 mAP, an absolute gain of 3.1.

4.4 Ablation Study

Last, we analyze the different aspects of the proposed approach, with results provided in Table 1b. For additional experiments, see Supplementary material.

Table 1. (a) State-of-the-art comparison on the YouTubeVIS validation dataset [30]. The proposed approach outperforms all near real-time approaches. †: No speed reported in [21] or [6], but each utilize components ([22] and [12]) with a reported speed of 1 fps and 2 fps respectively. (b) Performance under different configurations on the YouTube-VIS validation set. Each experiment corresponds to a single alteration to the final approach. The first set of experiments seeks to simplify the different modules in the final approach. The second set of experiments tackles the association and scoring tasks of the video instance segmentation problem using the mechanism proposed by Yang et al. [11,30].

Method	fps	mAP
OSMN MaskProp [31]		23.4
FEELVOS [27]		26.9
IoUTracker+ [30]		23.6
OSMN [31]		27.5
DeepSORT [29]		26.1
SeqTracker [30]		27.5
MaskTrack R-CNN [30]	20	30.3
SipMask [11]	30	32.5
SipMask ms-train [11]	30	33.7
STEm-Seg ResNet50 [1]		30.6
STEm-Seg ResNet101 [1]	7	34.6
VIS2019 Winner [21]	$< 1^{\dagger}$	44.8
MaskProp [6]	$< 2^{\dagger}$	46.6
Ours (ResNet50)	30	*35.3*
Ours (ResNet101)	25	*37.7*

(a)

Configuration	mAP
Our final approach	35.3
No GNN	28.6
No LSTM-like gating	Diverges
Simple recurrent gate	31.5
No appearance	34.5
Appearance baked into embedding	34.4
Association from [11,30]	29.2
Scoring from [11,30]	31.5

(b)

No GNN. We first analyze the benefit of processing tracks and detections jointly using our GNN. This is done by restricting the GNN module. First, a neural network predicts the probability that each track-detection pair matches, based only on the appearance and spatial similarities. Next, new tracks are initialized from detections that are not assigned to any track. Last, each track embedding is updated with the best matching edge and detection. This leads to a substantial 6.7 drop in mAP, demonstrating the importance of our GNN.

Simpler Recurrent Module. We experiment with the LSTM-like gating mechanism. We first try to remove it, directly feeding the track embeddings output from the GNN as input in the subsequent frame. We found that this configuration leads to unstable training and in all attempts diverge. We therefore also try a simpler mechanism, adding only a single gate and a tanh activation. This setting leads to more stable training, but provides deteriorated performance.

Simpler Appearance. We measure the impact of the appearance by removing it. We also experiment with removing its separate treatment. The appearance is instead baked into the detection node embeddings. Both of these configurations lead to performance drops.

Association or Scoring from [30]**.** The proposed model is trained to (i) associate detections to tracks and (ii) score tracks. We try to let each of these two tasks instead be performed by the simpler mechanisms used in [11,30]. This leads to performance drops of 6.1 and 3.8 mAP respectively.

5 Conclusion

We introduced a novel learning formulation together with an intuitive and flexible model for video instance segmentation. The model proceeds frame by frame, uses as input the detections produced by an instance segmentation method, and incrementally forms tracks. It assigns detections to existing tracks, initializes new tracks, and updates class and confidence in existing tracks. We demonstrate via qualitative and quantitative experiments that the model learns to create accurate tracks, and provide an analysis of its various aspects via ablation experiments.

Acknowledgement. This work was partially supported by the Wallenberg Artificial Intelligence, Autonomous Systems and Software Program (WASP) funded by Knut and Alice Wallenberg Foundation; and the Excellence Center at Linköping-Lund in Information Technology (ELLIT); and the Swedish National Infrastructure for Computing (SNIC), partially funded by the Swedish Research Council through grant agreement no. 2018-05973.

References

1. Athar, A., Mahadevan, S., Ošep, A., Leal-Taixé, L., Leibe, B.: STEm-Seg: spatio-temporal embeddings for instance segmentation in videos. In: Vedaldi, A., Bischof, H., Brox, T., Frahm, J.-M. (eds.) ECCV 2020. LNCS, vol. 12356, pp. 158–177. Springer, Cham (2020). https://doi.org/10.1007/978-3-030-58621-8_10
2. Battaglia, P.W., et al.: Relational inductive biases, deep learning, and graph networks. CoRR abs/1806.01261 (2018)
3. Benenson, R., Popov, S., Ferrari, V.: Large-scale interactive object segmentation with human annotators. In: CVPR (2019)
4. Berg, A., Johnander, J., Durand de Gevigney, F., Ahlberg, J., Felsberg, M.: Semi-automatic annotation of objects in visual-thermal video. In: Proceedings of the IEEE International Conference on Computer Vision Workshops (2019)
5. Berman, M., Rannen Triki, A., Blaschko, M.B.: The lovász-softmax loss: a tractable surrogate for the optimization of the intersection-over-union measure in neural networks. In: Proceedings of the IEEE Conference on Computer Vision and Pattern Recognition, pp. 4413–4421 (2018)
6. Bertasius, G., Torresani, L.: Classifying, segmenting, and tracking object instances in video with mask propagation. In: Proceedings of the IEEE/CVF Conference on Computer Vision and Pattern Recognition, pp. 9739–9748 (2020)
7. Bertasius, G., Torresani, L., Shi, J.: Object detection in video with spatiotemporal sampling networks. In: Ferrari, V., Hebert, M., Sminchisescu, C., Weiss, Y. (eds.) ECCV 2018. LNCS, vol. 11216, pp. 342–357. Springer, Cham (2018). https://doi.org/10.1007/978-3-030-01258-8_21

8. Bolya, D., Zhou, C., Xiao, F., Lee, Y.J.: YOLACT: real-time instance segmentation. In: Proceedings of the IEEE International Conference on Computer Vision, pp. 9157–9166 (2019)
9. Brasó, G., Leal-Taixé, L.: Learning a neural solver for multiple object tracking. In: Proceedings of the IEEE/CVF Conference on Computer Vision and Pattern Recognition, pp. 6247–6257 (2020)
10. Burghardt, T., Ćalić, J.: Analysing animal behaviour in wildlife videos using face detection and tracking. IEE Proc.-Vis. Image Signal Process. **153**(3), 305–312 (2006)
11. Cao, J., Anwer, R.M., Cholakkal, H., Khan, F.S., Pang, Y., Shao, L.: SipMask: spatial information preservation for fast image and video instance segmentation. In: Vedaldi, A., Bischof, H., Brox, T., Frahm, J.-M. (eds.) ECCV 2020. LNCS, vol. 12359, pp. 1–18. Springer, Cham (2020). https://doi.org/10.1007/978-3-030-58568-6_1
12. Chen, K., et al.: Hybrid task cascade for instance segmentation. In: Proceedings of the IEEE Conference on Computer Vision and Pattern Recognition, pp. 4974–4983 (2019)
13. Cho, K., van Merriënboer, B., Bahdanau, D., Bengio, Y.: On the properties of neural machine translation: encoder-decoder approaches. Syntax, Semantics and Structure in Statistical Translation, p. 103 (2014)
14. Cordts, M., et al.: The cityscapes dataset for semantic urban scene understanding. In: Proceedings of the IEEE Conference on Computer Vision and Pattern Recognition, pp. 3213–3223 (2016)
15. Greff, K., Srivastava, R.K., Koutník, J., Steunebrink, B.R., Schmidhuber, J.: LSTM: a search space odyssey. IEEE Trans. Neural Netw. Learn. Syst. **28**(10), 2222–2232 (2016)
16. Han, W., et al.: SEQ-NMS for video object detection. arXiv preprint arXiv:1602.08465 (2016)
17. He, K., Zhang, X., Ren, S., Sun, J.: Deep residual learning for image recognition. In: Proceedings of the IEEE Conference on Computer Vision and Pattern Recognition, pp. 770–778 (2016)
18. Hochreiter, S., Schmidhuber, J.: Long short-term memory. Neural Comput. **9**(8), 1735–1780 (1997)
19. Izquierdo, R., Quintanar, A., Parra, I., Fernández-Llorca, D., Sotelo, M.: The prevention dataset: a novel benchmark for prediction of vehicles intentions. In: 2019 IEEE Intelligent Transportation Systems Conference (ITSC), pp. 3114–3121. IEEE (2019)
20. Kuznetsova, A., et al.: The open images dataset v4: unified image classification, object detection, and visual relationship detection at scale. IJCV **128**, 1956–1981 (2020)
21. Luiten, J., Torr, P., Leibe, B.: Video instance segmentation 2019: a winning approach for combined detection, segmentation, classification and tracking. In: Proceedings of the IEEE International Conference on Computer Vision Workshops (2019)
22. Luiten, J., Zulfikar, I.E., Leibe, B.: UnOVOST: unsupervised offline video object segmentation and tracking. In: 2020 IEEE Winter Conference on Applications of Computer Vision (WACV), pp. 1989–1998. IEEE (2020)
23. Murphy, K.P.: Conjugate Bayesian analysis of the Gaussian distribution. def $1(2\sigma 2)$, 16 (2007)
24. Paszke, A., et al.: Automatic differentiation in PyTorch (2017)

25. Sarlin, P.E., DeTone, D., Malisiewicz, T., Rabinovich, A.: SuperGlue: learning feature matching with graph neural networks. In: Proceedings of the IEEE/CVF Conference on Computer Vision and Pattern Recognition, pp. 4938–4947 (2020)
26. T'Jampens, R., Hernandez, F., Vandecasteele, F., Verstockt, S.: Automatic detection, tracking and counting of birds in marine video content. In: 2016 Sixth International Conference on Image Processing Theory, Tools and Applications (IPTA), pp. 1–6. IEEE (2016)
27. Voigtlaender, P., Chai, Y., Schroff, F., Adam, H., Leibe, B., Chen, L.C.: FEELVOS: Fast end-to-end embedding learning for video object segmentation. In: Proceedings of the IEEE Conference on Computer Vision and Pattern Recognition, pp. 9481–9490 (2019)
28. Weng, X., Wang, Y., Man, Y., Kitani, K.M.: GNN3DMOT: graph neural network for 3d multi-object tracking with 2d–3d multi-feature learning. In: 2020 IEEE/CVF Conference on Computer Vision and Pattern Recognition, CVPR 2020, Seattle, WA, USA, June 13–19, 2020, pp. 6498–6507. IEEE (2020). https://doi.org/10.1109/CVPR42600.2020.00653
29. Wojke, N., Bewley, A., Paulus, D.: Simple online and realtime tracking with a deep association metric. In: 2017 IEEE International Conference on Image Processing (ICIP), pp. 3645–3649. IEEE (2017)
30. Yang, L., Fan, Y., Xu, N.: Video instance segmentation. In: Proceedings of the IEEE International Conference on Computer Vision, pp. 5188–5197 (2019)
31. Yang, L., Wang, Y., Xiong, X., Yang, J., Katsaggelos, A.K.: Efficient video object segmentation via network modulation. In: The IEEE Conference on Computer Vision and Pattern Recognition (CVPR) (2018)
32. Yu, F., et al.: Bdd100k: A diverse driving dataset for heterogeneous multitask learning. In: Proceedings of the IEEE/CVF Conference on Computer Vision and Pattern Recognition, pp. 2636–2645 (2020)
33. Zhang, X.Y., Wu, X.J., Zhou, X., Wang, X.G., Zhang, Y.Y.: Automatic detection and tracking of maneuverable birds in videos. In: 2008 International Conference on Computational Intelligence and Security, vol. 1, pp. 185–189. IEEE (2008)

Distractor-Aware Video Object Segmentation

Andreas Robinson(✉), Abdelrahman Eldesokey, and Michael Felsberg

Linköping University, Linköping, Sweden
{andreas.robinson,abdelrahman.eldesokey,michael.felsberg}@liu.se

Abstract. Semi-supervised video object segmentation is a challenging task that aims to segment a target throughout a video sequence given an initial mask at the first frame. Discriminative approaches have demonstrated competitive performance on this task at a sensible complexity. These approaches typically formulate the problem as a one-versus-one classification between the target and the background. However, in reality, a video sequence usually encompasses a target, background, and possibly other distracting objects. Those objects increase the risk of introducing false positives, especially if they share visual similarities with the target. Therefore, it is more effective to separate distractors from the background, and handle them independently.

We propose a one-versus-many scheme to address this situation by separating distractors into their own class. This separation allows imposing special attention to challenging regions that are most likely to degrade the performance. We demonstrate the prominence of this formulation by modifying the learning-what-to-learn [3] method to be distractor-aware. Our proposed approach sets a new state-of-the-art on the DAVIS 2017 validation dataset, and improves over the baseline on the DAVIS 2017 test-dev benchmark by 4.6% points.

1 Introduction

Semi-supervised video object segmentation (VOS) aims to segment a target throughout a video sequence, given an initial segmentation mask in the first frame. This task can be very challenging due to camera motion, occlusion, and background clutter. Several deep learning based methods have been proposed recently to address these challenges [3,8,9,11,12,15]. Among those, discriminative methods [3,11] have shown competitive performance at a reasonable computational cost, making them suitable for real-time applications, e.g., enhancing visual object tracking in crowded scenes, removing or replacing the background in video sequences or live conference calls, for privacy masking in surveillance videos, or as an attention mechanism in downstream vision tasks such as action recognition.

The majority of the discriminative approaches formulate the problem as a one-versus-one classification between the target and the background. Based on this, they attempt to construct a robust representation of the target that is as distinct as possible from the background. However, the background usually includes other objects that

Supplementary Information The online version contains supplementary material available at https://doi.org/10.1007/978-3-030-92659-5_14.

C. Bauckhage et al. (Eds.): DAGM GCPR 2021, LNCS 13024, pp. 222–234, 2021.
https://doi.org/10.1007/978-3-030-92659-5_14

Fig. 1. The impact of incorporating distractor-awareness into the baseline (LWL [3]). Our distractor-aware approach produces more accurate predictions than the baseline in highly ambiguous regions, where objects share visual similarities.

could be visually similar to the target. In this case, it might be challenging to find a good representation that discriminates the target from those distracting objects. This can produce false positives as the classifier is likely to fail given the underlying coarse representation between the target and the background. Figure 1 shows an example where a top-performing discriminative approach fails to discriminate between the target and other objects of the same type.

In this paper, we address this aforementioned challenge by reformulating the problem as a one-versus-many classification. We propose to separate the distracting objects from the background, and handle them as a distinct class. As a result, making the network aware of these distractors during training, promotes the learning of a robust representation of the target that is more discriminative against both the background *and* distractors.

We demonstrate the effectiveness of our approach by modifying the learning-what-to-learn (LWL) approach [3] to become distractor-aware. First, we modify the learning pipeline to incorporate information about the distractors. In case of videos with multiple objects, we initialize other objects in the scene as distractors.

Second, to enhance the discriminative power of the network, we integrate high-resolution features to the target model. Finally, we introduce the use of adaptive refinement and upsampling [13] as a regularization to enforce local consistency at uncertain regions such as edges, and object-to-object boundaries.

Experiments show that our proposed framework sets a new state-of-the-art result on the DAVIS 2017 *val* dataset [10]. Moreover, we improve the results over the baseline on the DAVIS 2017 *test-dev* benchmark by a large margin. On the YouTube-VOS 2018 dataset [14], which is characterized by limited annotation accuracy at object-to-distractor boundaries, our method still shows significant improvement over the baseline.

The remainder of the paper is organized as follows: we start by providing an overview of existing discriminative VOS approaches in the literature, distractor-awareness in other vision tasks, and existing upsampling and refinement methods in

VOS. Next, we briefly describe the baseline approach, Learning-what-to-learn [3], followed by our proposed distractors modelling. Finally, we provide quantitative and qualitative results for our proposed approach in comparison with the baseline, and existing state-of-the-art methods as well as an ablation study.

2 Related Work

The video object segmentation (VOS) task can be tackled in a semi-supervised or an unsupervised manner, but we only consider the former in this paper. Semi-supervised approaches from the literature can usually be categorized as either generative or discriminative. Generative approaches such as [6], focus on constructing a robust model of the target of interest ignoring other objects in the scene. In contrast, discriminative approaches [11, 12] attempt to solve the task as a classification problem between the target and the background. A more recent method [3] follows an embedding approach to learn features that are as discriminative as possible. With the emergence of deep learning, the robustness of feature representations has significantly improved, boosting the performance of most variants of VOS approaches. However, the current top-performing semi-supervised VOS methods are mainly discriminative, taking into account both the target and the background when solving the task. Therefore, we focus on discriminative approaches in this paper.

Discriminative Video Object Segmentation. Discriminative VOS methods were introduced quite recently. Yang *et al.* [15] proposed to build a target model from separate target and background feature embeddings, extracted from past images and target masks. For test frames, extracted features are matched against the two pretrained models of the target and the background. STM [9] incorporates a feature memory, encoding past images and segmentation masks. Similarly, Lu *et al.* [7] introduce an episodic graph memory unit to mine newly available features, and update the model accordingly. In contrast to Yang *et al.*, both methods [7,9] produce the final target mask using a dedicated decoder network, from concatenated memory features and new image features. Seong *et al.* [12] extended STM [9] further with a soft Gaussian-shaped attention mask to limit confusion with distant objects. Robinson *et al.* [11] introduced the use of discriminative correlation filters to construct a target model that produces a coarse segmentation mask. This coarse mask is then enhanced and refined through a decoder network. Learning-what-to-learn [3] improved it further by learning to produce target embeddings that are more reliable for training the target model. All of these aforementioned papers adopt a one-vs-one classification between the target and the background, where other objects in the sequence are considered as background. In contrast, our approach reduces the likelihood of predicting false positives when some background objects share visual similarities with the target.

Distractor-Aware Modelling. Distractor-aware modeling can be realized as a kind of hard-example mining when training a model. The general concept is to identify inputs that are more likely to confuse a given model and emphasis them during training, One of the earliest uses of this concept is found in the classical human detection algorithm, histograms of oriented gradients [4]. A more recent example [5] proposed marking flickering object detections in a video as hard-negatives, assigning them higher priority during training. A recent visual object tracking method [2] ranks target proposals based on their signal strength, where the strongest is assumed to be the target and the rest are distractors. The method tracks the complete scene state in a dense vector field over all spatial locations, using it to classify regions as either target, distractor or background. A similar Siamese-based approach [16] also classifies a target and any distractors through ranking of detection strengths. In both approaches, distractors are provided as hard examples during online training their respective tracker target models. Zhu *et al.* [16] also introduced distractor awareness to Siamese visual tracking networks as they noticed that the standard trackers posses imbalanced distribution of training data leading to less discriminative features. They propose an effective sampling strategy to make the model more discriminative towards semantic distractors. In this paper, we follow the strategy adopted by these approaches, and we introduce distractor-awareness to the task of video object segmentation.

Segmentation Mask Upsampling and Refinement. Existing VOS CNNs employ different upsampling and refinement approaches to provide the final segmentation mask at the full resolution. RGMP and STM [8,9] employ two residual blocks with a bilinear interpolation in between for refinement and upsampling. Seong *et al.* [12] used a residual block followed by two refinement modules with skip connections. In contrast, Robinson *et al.* and Bhat *et al.* [3,11] replace the residual blocks with standard convolution, and employ bicubic rather than bilinear interpolation. All these approaches provide spatially independent prediction with no regularization to enforce local consistency, especially at uncertain regions such as edges and object boundaries. We employ the convex upsampler [13] to jointly upsample the final mask while enforcing spatial consistency.

3 Method

Ideally, in video object segmentation, it is desired to produce pixel-wise predictions y either as *target* $\mathcal{T}$ or *background* $\mathcal{B}$. In a probabilistic sense, we are interested in maximizing the posterior probability for the target given an input embedding X:

$$\begin{aligned} P(Y=\mathcal{T}|X) &= \frac{P(X,Y=\mathcal{T})}{P(X)} = \frac{P(X,Y=\mathcal{T})}{P(X,Y=\mathcal{T})+P(X,Y=\mathcal{B})} \\ &= \frac{1}{1+\dfrac{P(X,Y=\mathcal{B})}{P(X,Y=\mathcal{T})}} = \frac{1}{1+\dfrac{P(X|Y=\mathcal{B})P(Y=\mathcal{B})}{P(X|Y=\mathcal{T})P(Y=\mathcal{T})}}, \end{aligned} \tag{1}$$

The ratio in the denominator determines the posterior probability. If a pixel belongs to the target, the ratio becomes small and the posterior probability tends to 1. Contrarily, if a pixel belongs to the background and is quite distinct from the target, the ratio becomes large and the posterior probability goes to 0. However, if the target prior is large, and the background and target likelihoods are similar because X contains features of a distractor, the posterior can easily be larger than 0.5 and produce false positives. We propose splitting the non-target into two classes, background $\mathcal{B}$ and distractor $\mathcal{D}$:

$$\begin{aligned} P(X,Y\neq\mathcal{T}) &= P(X,Y=\mathcal{B})+P(X,Y=\mathcal{D}) \\ &= P(X|Y=\mathcal{B})P(Y=\mathcal{B})+P(X|Y=\mathcal{D})P(Y=\mathcal{D}). \end{aligned} \tag{2}$$

Consequently, the ratio in (1) will have two terms in the numerator as denoted by (2). This modification limits the occurrences of false positives as it models ambiguous pixels from the first frame and propagates them. As an example, if the likelihood of a certain pixel is similar between the target and the distractor at an intermediate frame, the propagated prior from previous frames will cause the ratio to be large, and the probability to drop. In the following sections, we will describe how to modify an existing baseline to be distractor-aware.

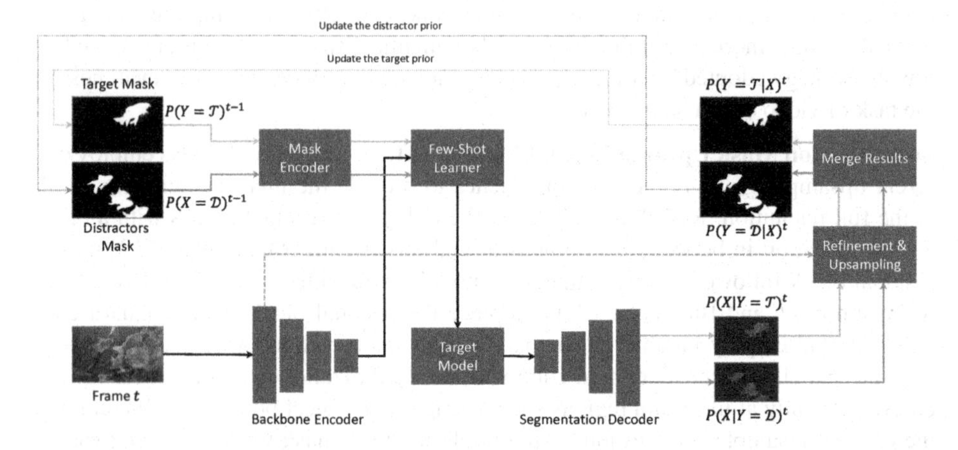

Fig. 2. An overview of our distractor-aware approach. We extend the baseline LWL to incorporate information about distractors throughout its mask encoder, target model, and segmentation decoder stages, and replace its upsampler with a joint refinement and upsampling approach, acting as local regularization.

3.1 Baseline Approach

We base our approach on the recently published method learning-what-to-learn [3] (LWL). In the LWL baseline, a target segmentation mask is encoded by a mask encoder into an multi-channel embedding at one sixteenth of the original image resolution. At the same time, a standard backbone encoder network (ResNet-50) is used to extract features from the whole image (see Fig. 2). At the first frame, both the mask embeddings and the image features are utilized as training data for a target model $T_\theta(x) = x * \theta$ that is trained with a few-shot learner. In subsequent video frames, this target model generates new multi-channel embeddings from the corresponding image features. A decoder network finally recovers segmentation masks at full resolution from these embeddings. Eventually, the newly predicted masks and deep features are added to the training data, so that the target model can be tuned, adapting it to the changing target appearance. Note that multiple objects are tracked independently and individual target models do not interact with each other. In other words, the target model is not aware of any objects in the scene since everything other than the target is treated as background.

3.2 Introducing Distractor-Awareness

As describe in the related work section, there are several ways to detect distractors. Here, we consider other objects in the scene as potential distractors. More specifically, under the assumption that segmentation masks are binary, given a target masks $\mathbf{1}_{t_i}$, we generate a distractor mask as:

$$\mathbf{1}_{d_i} = \bigcup_{j \neq i} \mathbf{1}_{t_j} \; \forall j \in \mathbb{I}, \tag{3}$$

where $\mathbb{I} = \{1...N\}$ is the set of target IDs in the current video sequence. Both the target and the distractor masks are set to the ground truth in the first frame, and updated with the previous predictions in later frames. To accommodate the distractor, we add a second input channel to the mask encoder and a second output channel to the segmentation decoder (see Fig. 2).

As the segmentation of frames progresses, we merge the decoded and upsampled target masks to form new distractors. However, in this case, the propagated masks are no longer binary and (3) needs to be replaced. For this, we develop a per-pixel winner-take-all (WTA) function.

Let $p_{t_i}(x) \in \mathbb{R}^{H \times W}$ be the target segment probability map, i.e. the network decoder output after a sigmoid activation, of the target with index i. Now let

$$p_{\max}(x) = \sup_j p_{t_j}(x) \; \forall j \in \mathbb{I}, \tag{4}$$

$$p_{\min}(x) = \inf_j p_{t_j}(x) \; \forall j \in \mathbb{I}, \tag{5}$$

merging the highest and lowest probabilities (per pixel), into $p_{\max}$ and $p_{\min}$.

Now let, $L(x) \in (\mathbb{I} \cup \{0\})^{H \times W}$ be the map of merged segmentation labels (after softmax-aggregation, as introduced in [8]), with 0 the background label.

Also let

$$\mathbf{1}_f(x) = \begin{cases} 1 & \text{if } L(x) > 0, \\ 0 & \text{if } L(x) = 0. \end{cases} \tag{6}$$

indicate regions with any foreground pixel, and

$$\mathbf{1}_{d_i}(x) = \begin{cases} \mathbf{1}_f(x) & \text{if } L(x) \neq i, \\ 0 & \text{otherwise.} \end{cases} \tag{7}$$

indicate distractors of target i. A new probability map of distractor i is generated as

$$p_{d_i}(x) = \mathbf{1}_{d_i}(x) p_{\max}(x) + (1 - \mathbf{1}_f(x)) p_{\min}(x). \tag{8}$$

In less formal terms, we let the the most certain predictions of target and background "win" in every pixel. p_{d_i} takes information from every p_{t_j} except that of target i.

In the ablation study, we compare the performance of this function to that of feeding back the decoded distractor to the few-shot learner without modification.

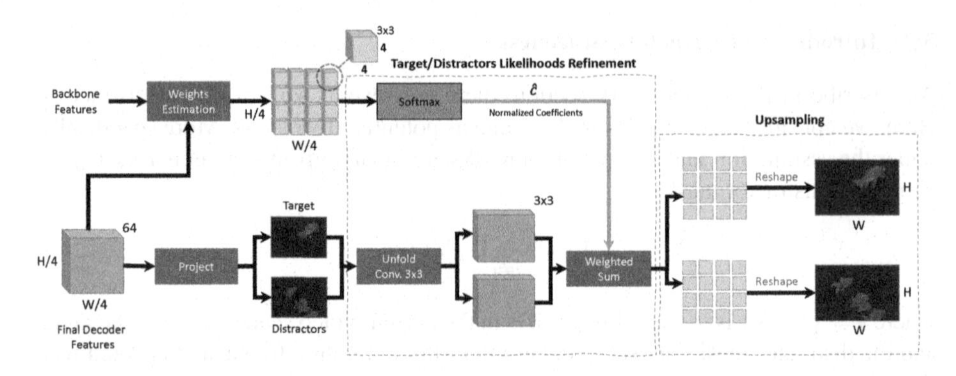

Fig. 3. An overview of the refinement and upsampling module. The final decoder features are first projected into target and distractor logit maps. A weights estimation network then predicts a 5D tensor of weights in 3×3 windows around each pixel in the logit maps (shown in green), as well as interpolation data (shown in yellow). The weights are mapped to normalized probability vectors with the softmax function and then used refine the target and the distractor logits with weighted summation. The refined logits are finally upsampled to full resolution with the interpolation data. (Color figure online)

3.3 Joint Refinement and Upsampling for Local Consistency

The majority of existing VOS methods adopt a binary classification scheme between the foreground and the background. It is typically challenging to produce accurate predictions at uncertain regions such as target boundaries due to multi-modality along edges. This problem is aggravated when we introduce a new class for distractors, as the decision is now made between three classes instead of two. We tackle this problem by introducing two modifications to the baseline encoder and decoder, respectively.

First, we provide high-resolution feature maps from the backbone when training the target model in the few-shot learner. Unlike the baseline, we employ backbone features with a 1/8th of the full resolution instead of 1/16th. These higher-resolution features provide finer details to the target model, especially along edges, to help resolving ambiguities in uncertain regions. Second, we replace the baseline upsampler on the decoder output, with a joint refinement and upsampling unit based on the convex combination module of [13]. This is originally employed to upsample flow fields, while we adopt it to produce consistent selections between the target and distractors in logit space.

The proposed joint refinement/upsampling approach is illustrated in Fig. 3. First, the output from the decoder is projected using two convolution layers to two channels resembling the likelihoods of the target and the distractor. Then, the likelihoods are unfolded into 3×3 patches around each pixel for computational efficiency. In parallel, we employ a weights estimation network that jointly perform likelihoods refinement, and mask upsampling. The former allows modifying the likelihoods of the target and the distractor based on their neighbors to enforce some local consistency, while the latter produces pixel values needed for upsampling.

For the refinement, the weights estimation network predicts a local coefficients vector $\mathbf{c}_x$ for a 3×3 window centered around each pixel $x \in X$:

$$\mathbf{c}_x = [c_{-4}, \ldots, c_{-1} c_{-1}, c_0, c_1, \ldots, c_4], \tag{9}$$

where c_0 is the coefficient on top of x. However, these coefficients are not normalized, and to preserve likelihood ratios we first map $\mathbf{c}_x$ to a normalized vector $\hat{\mathbf{c}}_x$ using the softmax function. To modify the likelihoods, we subsequently apply the coefficients to the likelihoods using a weighted sum resembling convolution:

$$P(X|Y)' * \hat{\mathbf{c}}_x[x] = \sum_{m \in [-4:4]} P(X|Y)[x - m]\, \hat{\mathbf{c}}_x[m]. \tag{10}$$

For the upsampling, we need to predict pixel values for 4×4 patches around each pixel for a scaling of 4. Those values are also predicted using the weights-estimation network and are needed to produce the full resolution prediction. This combined refinement-and-upsampling feature volume takes the shape of a local 3D tensor $\mathbf{A}_x \in \mathbb{R}^{4 \times 4 \times 9}$ for each pixel $x \in X$.

4 Experiments

In this section, we provide implementation details including the training procedure and loss function. We compare against state-of-the-art approaches on DAVIS 2017 [10], and YouTube-VOS 2018 [14]. In addition, we provide an extensive ablation analysis.

4.1 Implementation Details

Training Procedure. Similar to the baseline, our training setup imitates the segmentation processing during inference. Each training sample is a mini-video of four frames with one main target object to segment. The few-shot learner is provided the first frame to train a target model. The target model and the decoder then predicts segmentation masks from the subsequent three frames. The predictions in turn, are both used to update the target model and compute the network training loss.

To extend this procedure for distractors, we replicate the segmentation process for all other objects that are present in the ground-truth label map. These additional segments are then merged and provided as distractors to the main target. However, we set the network into evaluation mode when processing these targets to conserve memory.

We employ the same data augmentation and first two training phases as the baseline [3]. The proposed refinement and upsampling module is trained in a third phase for 6,000 iterations on DAVIS 2017, with learning rate 10^{-4}, all other weights frozen.

Loss Function. LWL was trained with the Lovász-softmax loss [1], a differentiable relaxation of the Jaccard similarity measure, and we continue to do so here. However, in the SOTA experiments, we did initially not see any improvements with our method on YouTube-VOS. We hypothesize that this is due to large size-differences between objects in the same sequence in YouTube-VOS. To counter this, we split the loss into two terms, or a balanced loss:

$$L = \text{Lovasz}(T) + w(\hat{T}, \hat{D})\text{Lovasz}(D) \tag{11}$$

where T and D are the output batch from the decoder network, separated into target and distractor channels. $w(\hat{T}, \hat{D})$ reduces the influence of large objects as a function of the number of ground-truth target pixels $|\hat{T}|$ and ground-truth distractor pixels $|\hat{D}|$ across the training batch:

$$w(\hat{T}, \hat{D}) = \min(|\hat{T}|/|\hat{D}|, 1.0) \tag{12}$$

In other words, when the distractors jointly occupy a larger region than the target, their influence on the loss is reduced.

Relaxed Distractor Loss. Some sequences have only one target. This implies that no distractors exist, since we derive them from every other target. To not unnecessarily over-constrain the training, we partially disable the computation of the loss in training samples in these cases. Specifically, we require the distractor output to be zero in the area under the target, but allow it to take any value elsewhere.

We also train a variant with a "hard" loss, requiring that no distractor is output when no distractor is given, and compare them in the experiments.

4.2 State-of-the-Art Comparisons

Table 1. Results on DAVIS 2017 and YouTube-VOS 2018, comparing our method to the LWL baseline and the state-of-the-art. The LWL scores are from our own run with the official code.

Method	DAVIS'17 val			DAVIS'17 test-dev			YouTube-VOS'18 valid				
Name	$\mathcal{G}$	$\mathcal{J}$	$\mathcal{F}$	$\mathcal{G}$	$\mathcal{J}$	$\mathcal{F}$	$\mathcal{G}$	$\mathcal{J}_s$	$\mathcal{J}_u$	$\mathcal{F}_s$	$\mathcal{F}_u$
CFBI [15]	81.9	79.1	84.6	74.8	71.1	78.5	81.4	81.1	75.3	85.8	83.4
CFBI-MS [15]	**83.3**	**80.5**	**86.0**	**77.5**	73.8	**81.1**	**82.7**	**82.2**	**76.9**	**86.8**	**85.0**
KMN [12]	82.8	80.0	85.6	77.2	**74.1**	80.3	81.4	81.4	75.3	85.6	83.3
STM [9]	81.8	79.2	84.3	72.2	69.3	75.2	79.4	79.7	72.8	84.2	80.9
GMVOS [7]	82.8	80.2	85.2	–	–	–	80.2	80.7	74.0	85.1	80.9
LWL [3]	80.1	77.4	82.8	70.8	68.0	73.7	80.7	79.5	75.6	84.0	83.5
Ours	**83.7**	**81.1**	**86.2**	74.1	71.2	77.1	79.8	79.8	74.0	84.1	81.3
Ours (Balanced loss)	82.6	80.8	85.3	**75.4**	**72.5**	**78.2**	**81.5**	**80.4**	**76.0**	**85.1**	**84.5**

We compare against the most recent state-of-the-art approaches for video object segmentation: CFBI [15], STM [9], KMN [12], GMVOS [7], and the baseline LWL [3]. CFBI-MS is a multi-scale variant operating on three different scales. Our method in contrast, is single scale and thus more comparable to the plain CFBI. Running the official LWL implementation we noticed that our results do not coincide exactly with those reported in [3]. Therefore, we use these newly obtained scores to accurately compare the baseline to our method.

Table 1 summarizes the quantiative results on DAVIS 2017 and the YouTube-VOS 2018 validation split. On DAVIS val, our proposed approach without the balanced loss sets a new state-of-the-art on DAVIS val, while improving over the baseline on test-dev by 3.6% points. With the balanced loss however, test-dev surprisingly improves 4.6% points, while reducing the gain on val to 2.5% points. On the YouTube-VOS 2018 validation split, our approach improves the baseline by 0.8% points when trained *with* the balanced loss, while scoring similarly to other state-of-the-art approaches.

Some qualitative results are found in the supplement.

4.3 Ablation Study

Table 2. Ablation study on DAVIS 2017. See Sect. 4.3 for details. Interesting scores are printed in bold type.

Parameters				DAVIS'17 val			DAVIS'17 test-dev		
L2	D	U	B	$\mathcal{G}$	$\mathcal{J}$	$\mathcal{F}$	$\mathcal{G}$	$\mathcal{J}$	$\mathcal{F}$
				80.1	77.4	82.8	70.8	68.0	73.7
	✓			80.7	77.9	83.4	71.0	67.7	74.2
	✓	✓		80.7	78.2	83.3	71.7	68.7	74.8
✓				81.9	78.9	84.9	72.7	69.5	75.9
✓		✓		81.6	78.9	84.3	71.6	68.5	74.7
✓	✓			81.3	78.8	83.9	73.2	70.0	76.4
✓	✓	✓		**83.7**	81.1	86.2	**74.1**	71.2	77.1
Hard loss				**80.2**	77.5	82.8	**74.9**	72.4	77.4
No Distractors				81.1	78.7	83.4	71.2	68.0	74.4
No WTA				83.0	80.5	85.5	71.8	68.4	75.2
✓	✓	✓	✓	**82.6**	80.8	85.3	**75.4**	72.5	78.2

In this section, we analyze the impact of different components of our method on the overall performance. We use the DAVIS 2017 dataset [10] for this purpose, where we include both *val* and *test-dev* splits for diversity. The results are shown in Table 2. Since LWL [3] is our baseline method, we reevaluate their official pretrained model. As mentioned above, there is a discrepancy between our obtained scores and the published ones. This could be due to several factors, e.g. hardware/software differences and driver variations. However, we use the scores that we obtained for a valid comparison.

We first enable or disable various combinations of these components: the high-resolution features from ResNet block/layer 2 (L2), the distractor-awareness (D), the refinement/upsampling module (U). Enabling D alone causes a slight improvement. Adding U to this improves the test-dev split further. We believe that the lack of high-resolution features prevents further gains. Similarly, introducing L2 alone, causes a marginal improvement on both splits and enabling U at the same time, shows no improvement on the valdiation set, and hurts the performance on the test-set. This can be explained by our argument in Sect. 3.3: When the decision is made between two classes (target and background), there is no need for the upsampler to resolve ambiguities. As to why this combinatios hurts test-dev performance, is unclear. However, enabling all three (L2+D+U) at the same time, yields significantly better results over the baseline.

With L2, D and U included, we then separately replace the relaxed distractor loss with the "hard" loss (Hard loss). The results then drops back to the baseline for DAVIS validation split, but oddly *improves* on the test-dev split.

We also test a pair of inference-time modifications. First, we zero out the distractor masks so that the few shot learner do not use them. Second, we replace the WTA distractor-merging function with a simple pass-through; Distractors are directly routed from the decoder output back to the few-shot learner, bypassing the merging step. As one would expect, the results drop in both cases, proving that WTA is important. Surprisingly though, disabling the WTA function hits DAVIS test-dev split much harder than the validation split.

Finally, we add the balanced loss (B) to the L2+D+U variant. This clearly damages the DAVIS validation scores but improves the test-dev results. However, at the same time it also improves the YouTube-VOS results in Table 1.

These results suggest that the method works, but also reveal differences between the DAVIS dataset splits that we cannot explain at this time.

4.4 A Note on WTA vs. Softmax Aggregation

The softmax aggregation introduced in [8], and used to merge estimations of multiple targets, was introduced as a superior alternative to a winner-take-all approach. In our case however, we found that to merge distractors, WTA performs better. We hypothesize that lacking a dedicated background channel is beneficial to the refinement and upsampling module, as the decoder may output low activations on both target and distractors when the classification is uncertain.

4.5 Emergent Distractors

An essential question is how our framework would behave when there is only one labeled object in the scene. Interestingly, the model learns to identify ambiguous regions. Figure 4 shows an example where our approach learns to identify the camel in the background without any explicit supervision (if columns 2+3 from the left). We attribute this to the relaxed distractor loss described in Sect. 4.1. To test this, we modify the loss to be "hard" (H). As suspected, this suppresses the behaviour greatly, while increasing the decoders' certainty in both the target and background (columns 4+5).

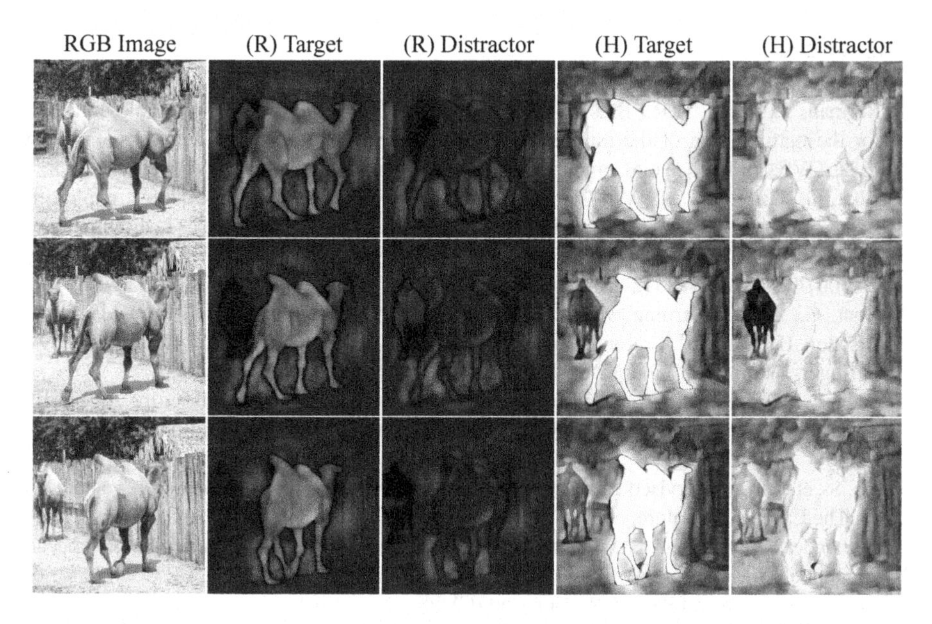

Fig. 4. Frames 40, 50 60 (from top to bottom) from the 'camel' sequence, with target/distractor score maps. Our model can learn to identify distractors in case they were not explicitly provided, provided it is trained with the relaxed distractor loss. Red/yellow indicates positive log-likelihoods, blue/cyan negative. The colors represent the same value range in all columns. (R) and (H) indicate results from models trained with the relaxed and hard distractor losses, respectively. (Color figure online)

5 Conclusion

We have proposed a distractor-aware, discriminative video object segmentation approach. In contrast to existing methods, our proposed method encodes distractors into a separate class, to exploit information about other objects in the scene that are likely to be confused with the target. Moreover, we propose the use of joint refinement and upsampling to regularize the likelihoods for highly uncertain regions with their neighborhoods. We demonstrated the effectiveness of our approach by modifying an existing state-of-the-art approach to be distractor-aware. Our modification sets a new state-of-the-art on the DAVIS 2017 *val* dataset, while improving over the baseline with a remarkable margin on the DAVIS 2017 *test-dev* dataset. These results clearly indicate the efficacy of explicitly modelling distractors when solving video object segmentation.

Acknowledgements. This project was partially supported by the Wallenberg AI, Autonomous Systems and Software Program (WASP) funded by the Knut and Alice Wallenberg Foundation, the Excellence Center at Linköping-Lund in Information Technology (ELLIIT), the Swedish Research Council grant no. 2018-04673, and the Swedish Foundation for Strategic Research (SSF) project Symbicloud.

References

1. Berman, M., Triki, A.R., Blaschko, M.B.: The lovász-softmax loss: a tractable surrogate for the optimization of the intersection-over-union measure in neural networks. In: 2018 IEEE/CVF Conference on Computer Vision and Pattern Recognition (CVPR), June 2018
2. Bhat, G., Danelljan, M., Van Gool, L., Timofte, R.: Know your surroundings: exploiting scene information for object tracking. In: Vedaldi, A., Bischof, H., Brox, T., Frahm, J.-M. (eds.) ECCV 2020. LNCS, vol. 12368, pp. 205–221. Springer, Cham (2020). https://doi.org/10.1007/978-3-030-58592-1_13
3. Bhat, G., et al.: Learning what to learn for video object segmentation. In: Vedaldi, A., Bischof, H., Brox, T., Frahm, J.-M. (eds.) ECCV 2020. LNCS, vol. 12347, pp. 777–794. Springer, Cham (2020). https://doi.org/10.1007/978-3-030-58536-5_46
4. Dalal, N., Triggs, B.: Histograms of oriented gradients for human detection. In: 2005 IEEE Computer Society Conference on Computer Vision and Pattern Recognition (CVPR 2005) (2005)
5. Jin, S., et al.: Unsupervised hard example mining from videos for improved object detection. In: The European Conference on Computer Vision (ECCV), September 2018
6. Johnander, J., Danelljan, M., Brissman, E., Khan, F.S., Felsberg, M.: A generative appearance model for end-to-end video object segmentation. In: 2019 IEEE/CVF Conference on Computer Vision and Pattern Recognition (CVPR), June 2019
7. Lu, X., Wang, W., Danelljan, M., Zhou, T., Shen, J., Van Gool, L.: Video object segmentation with episodic graph memory networks. In: Vedaldi, A., Bischof, H., Brox, T., Frahm, J.-M. (eds.) ECCV 2020. LNCS, vol. 12348, pp. 661–679. Springer, Cham (2020). https://doi.org/10.1007/978-3-030-58580-8_39
8. Oh, S.W., Lee, J.Y., Sunkavalli, K., Kim, S.J.: Fast video object segmentation by reference-guided mask propagation. In: 2018 IEEE/CVF Conference on Computer Vision and Pattern Recognition (CVPR), June 2018
9. Oh, S.W., Lee, J.Y., Xu, N., Kim, S.J.: Video object segmentation using space-time memory networks. In: 2019 IEEE/CVF Conference on Computer Vision and Pattern Recognition (CVPR), June 2019
10. Pont-Tuset, J., Perazzi, F., Caelles, S., Arbeláez, P., Sorkine-Hornung, A., Van Gool, L.: The 2017 Davis challenge on video object segmentation. arXiv:1704.00675 (2017)
11. Robinson, A., Lawin, F.J., Danelljan, M., Khan, F.S., Felsberg, M.: Learning fast and robust target models for video object segmentation. In: 2020 IEEE/CVF Conference on Computer Vision and Pattern Recognition (CVPR), June 2020
12. Seong, H., Hyun, J., Kim, E.: Kernelized memory network for video object segmentation. In: Vedaldi, A., Bischof, H., Brox, T., Frahm, J.-M. (eds.) ECCV 2020. LNCS, vol. 12367, pp. 629–645. Springer, Cham (2020). https://doi.org/10.1007/978-3-030-58542-6_38
13. Teed, Z., Deng, J.: RAFT: recurrent all-pairs field transforms for optical flow. In: Vedaldi, A., Bischof, H., Brox, T., Frahm, J.-M. (eds.) ECCV 2020. LNCS, vol. 12347, pp. 402–419. Springer, Cham (2020). https://doi.org/10.1007/978-3-030-58536-5_24
14. Xu, N., et al.: YouTube-VOS: a large-scale video object segmentation benchmark. arXiv preprint arXiv:1809.03327 (2018)
15. Yang, Z., Wei, Y., Yang, Y.: Collaborative video object segmentation by foreground-background integration. In: Vedaldi, A., Bischof, H., Brox, T., Frahm, J.-M. (eds.) ECCV 2020. LNCS, vol. 12350, pp. 332–348. Springer, Cham (2020). https://doi.org/10.1007/978-3-030-58558-7_20
16. Zhu, Z., Wang, Q., Li, B., Wu, W., Yan, J., Hu, W.: Distractor-aware Siamese networks for visual object tracking. In: The European Conference on Computer Vision (ECCV), September 2018

$(SP)^2$Net for Generalized Zero-Label Semantic Segmentation

Anurag Das[1(✉)], Yongqin Xian[4], Yang He[5], Bernt Schiele[1], and Zeynep Akata[2,3]

[1] MPI for Informatics, Saarland Informatics Campus, Saarbrücken, Germany
andas@mpi-inf.mpg.de

[2] MPI for Intelligent Systems, Tubingen, Germany

[3] University of Tübingen, Tübingen, Germany

[4] ETH Zurich, Zürich, Switzerland

[5] Amazon, Bellevue, USA

Abstract. Generalized zero-label semantic segmentation aims to make pixel-level predictions for both seen and unseen classes in an image. Prior works approach this task by leveraging semantic word embeddings to learn a semantic projection layer or generate features of unseen classes. However, those methods rely on standard segmentation networks that may not generalize well to unseen classes. To address this issue, we propose to leverage a class-agnostic segmentation prior provided by superpixels and introduce a superpixel pooling (SP-pooling) module as an intermediate layer of a segmentation network. Also, while prior works ignore the pixels of unseen classes that appear in training images, we propose to minimize the log probability of seen classes alleviating biased predictions in those ignore regions. We show that our $(SP)^2$Net significantly outperforms the state-of-the-art on different data splits of PASCAL VOC 2012 and PASCAL-Context benchmarks.

Keywords: Scene understanding · Zero-shot semantic segmentation

1 Introduction

Training a deep CNN for semantic segmentation often requires highly costly pixel-level annotations. This problem was tackled first by weakly supervised learning approaches that utilize weaker forms of annotations, e.g., bounding boxes [16], key points [4], image-level labels [17], and scribble-level annotations [20]. Recently, few-shot semantic segmentation methods have taken a different route by segmenting novel classes with only a few labeled training examples.

Y. Xian and Y. He—The majority of the work was done when Yongqin Xian and Yang He were with MPI for Informatics.

Supplementary Information The online version contains supplementary material available at https://doi.org/10.1007/978-3-030-92659-5_15.

C. Bauckhage et al. (Eds.): DAGM GCPR 2021, LNCS 13024, pp. 235–249, 2021.
https://doi.org/10.1007/978-3-030-92659-5_15

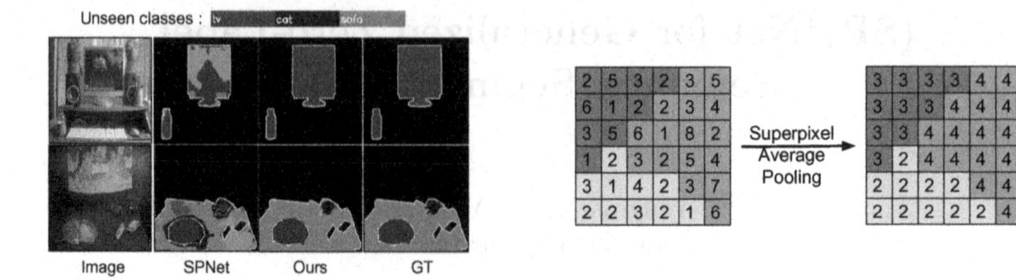

Fig. 1. Left: In this example, our model can predict unseen classes (tv in row 1, and sofa + cat in row 2) correctly compared to the baseline SPNet model. We integrate a superpixel pooling module in the segmentation network along with a bias reduction loss resulting in better generalization on the unseen classes as seen from the example, right: Illustration of our superpixel pooling. The color indicates individual superpixel region. (Color figure online)

However, those methods are not capable of making pixel-wise "zero-shot" predictions for the classes without a single label, which is an important real-world scenario. Therefore, successful methods in this task have the potential to significantly reduce the labeling efforts.

In the "zero-label" [31] setting, semantic segmentation models may have access to the pixels from novel classes but they do not have access to their pixel labels. In generalized zero-label semantic segmentation (GZLSS) [5,11,19,31], the goal is to make pixel-level predictions for both seen classes with abundant labels and novel classes without any label. Prior methods mainly focus on learning feature generators [5,11,19] or a semantic projection layer [31]. Moreover, those works rely on standard segmentation networks, i.e., DeepLab-v3+ [7] trained with large quantities of annotated data and do not generalize well to unseen classes without any training example.

Our main focus is to enable the segmentation networks to achieve better generalization for unseen classes. At a high-level, we would like to explore superpixels [3,24], i.e., groups of pixels that share similar visual characteristics, to learn more generic image features. While superpixels are intuitively beneficial for segmentation tasks, how and where to incorporate them into a convolutional neural network is not obvious. We believe that aggregating features from superpixel regions provides a generic class-agnostic segmentation prior for segmentation networks such as DeepLab-v3+ [7] and PSPNet [33]. To this end, we propose a superpixel pooling module as an intermediate layer of segmentation networks. The resulting architecture lends itself better for generalization to seen and unseen classes.

Furthermore, GZLSS suffers from the severe data imbalance issue, biasing the predictions towards seen classes. Therefore, we devise a simple solution to resolve this issue. Our main assumption is that the ignore regions of training images do not contain pixels from seen classes. Note that this assumption holds true according to the definition of GZLSS [31] where the ignore regions include only pixels from novel classes and background. Based on this, we propose a bias reduction loss that minimizes the log-likelihood of seen class predictions in the ignore regions. The insight is to treat the unlabeled unseen classes as negatives for the seen classes and thus reduce their confidence in the pixels that definitely do not belong to them. Compared to the previous balancing strategies, i.e., feature generation [5,19], our bias reduction loss is highly efficient and allows us to train the network in a single-stage.

Our (SP)^{2}Net augments the semantic projection network (SPNet) [31] with the proposed superpixel pooling (SP) module and bias reduction loss. On PASCAL VOC 2012, our (SP)^{2}Net improves the averaged harmonic mean mIoU (on 5 splits) of the previous state of the art by 9.8%, while on the challenging PASCAL-Context dataset, we achieve a remarkable improvement of 9.4%. We further provide an extensive model analysis and qualitative results to give insights and show the effectiveness of our approach.

2 Related Work

Zero-Label Semantic Segmentation. aims to segment novel classes without having their annotated masks during training. We focus on generalized zero-label semantic segmentation (GZLSS) where the model is required to segment both seen and novel classes. Prior works [5,19,31] tackle this task by adapting technics from zero-shot image classification into segmentation networks, e.g., learning semantic projection layer [31] and generating pixel-wise features [5,19] of novel classes, ignoring the challenges that are specific to semantic segmentation. Our proposed method improves the SPNet [31] by incorporating superpixels [24] into the network for learning dense features that generalize better to novel classes. Moreover, we leverage ignore regions in training images to alleviate the imbalance issue, which is more efficient than previous feature generation methods [5,19] and can be trained in a single stage.

Zero-Shot Learning. aims to predict the novel classes that are not observed, by using semantic embeddings, e.g., attributes [18], and word2vec [26], which encode the similarities between seen and unseen classes. Early works tackle this task by learning attribute classifiers [14,18] or learning a compatibility function [2,9,28, 30,32] between image and semantic embeddings.

In contrast, we are interested in the GZLSS problem which requires making pixel-wise predictions. We focus on addressing the challenges that are specific to the semantic segmentation problem, i.e., leveraging superpixels for better context modeling and alleviating imbalanced issues using ignore regions.

Superpixel and Semantic Segmentation. Superpixel [1,3,24,29] and semantic segmentation [6,7,13,23,33] have a long history in computer vision, which provides a pre-segmentation and understanding of images. The convolutional encoder-decoder architectures represent images into structural feature maps and predict the class labels in the end. Generally, it requires expensive dense annotations to train the structural output models. To address this issue, superpixel has been combined with modern deep neural networks to boost semantic segmentation models trained on a variety of supervisions [10,13,16,17,21]. In particular, [17] shows surprisingly promising results in a weakly supervised learning

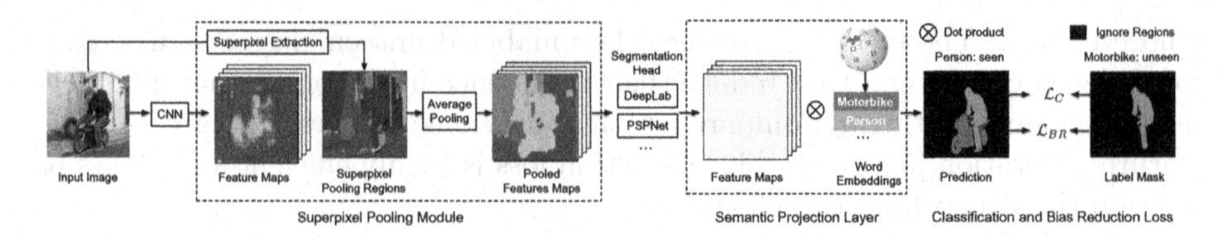

Fig. 2. Overview of our (SP)^{2}Net for generalized zero-label semantic segmentation. We propose to capture class-agnostic segmentation priors by introducing a Superpixel Pooling Module, to be integrated into standard segmentation networks, followed by a semantic projection layer [31] to embed pixel features into a semantic space. Finally, we devise a bias reduction loss ($\mathcal{L}_{BR}$) to alleviate biased predictions in ignore regions that belong to unseen classes (e.g. motorcycle).

setup, owing to its clustering capability on similar pixels. [17] applies the SP-pooling before the GAP layer in a standard FCN. In contrast, we integrate the SPpooling into the ASPP module of the semantic projection network. Different from previous works, we aim to improve GZLSS by incorporating superpixels into segmentation networks. To the best of our knowledge, we are the first to leverage superpixels in GZLSS.

3 Background: Semantic Projection Network

Problem Formulation. We denote the set of seen classes as $\mathcal{S}$, a disjoint set of novel classes as $\mathcal{U}$ and the union of them as $\mathcal{Y} = \mathcal{S} \cup \mathcal{U}$. Let $\mathcal{T} = \{(x, y)|x \in \mathcal{X}, y_i \in \{I, \mathcal{S}\}\}$ be the training set where x is an image of spatial size $H \times W$ in the RGB image space $\mathcal{X}$, y is its label mask with the same size, and y_i is the class label at pixel x_i belonging to either one of the seen classes $\mathcal{S}$ or ignore region labeled as I. Moreover, each class label is represented by the word embedding (e.g., word2vec [25]) associated with its class name. We denote the word embedding matrices of seen and novel classes with $A^s \in \mathbb{R}^{D \times |\mathcal{S}|}$ and $A^u \in \mathbb{R}^{D \times |\mathcal{U}|}$ where D is the dimension of the word embedding. Given $\mathcal{T}$, A^s and A^u, the goal of generalized zero-label semantic segmentation (GZLSS) is to learn to make pixel-wise predictions among both seen and novel classes.

Semantic Projection. We follow SPNet [31] to segment novel classes via mapping pixel features into a semantic embedding space. Specifically, SPNet consists of a visual-semantic embedding module and a semantic projection layer. The former (denoted as ϕ) is based on a standard segmentation network (e.g., DeepLab-v3+ [7]), encoding each pixel x_i as a D-dimensional feature embedding $\phi(x)_i$ in the semantic embedding space. The latter computes the compatibility scores between the pixel and word embeddings followed by applying softmax that maps scores into a probability distribution, $P_c(x_i) = \frac{\exp s_c(x_i)}{\sum_{c' \in \mathcal{Y}} \exp s_{c'}(x_i)}$,

where $s_c(x_i) = \phi(x)_i^T a_c$ and a_c denotes the word embedding of class c. The scoring function is capable to compute the compatibility score of a given pixel to any class using its word embedding, thus enabling zero-shot prediction.

For a particular labeled training pixel (x_i, y_i) from seen classes $\mathcal{S}$, the following cross-entropy loss is optimized, $\mathcal{L}_C(x_i, y_i) = -\log P_{y_i}(x_i)$.

4 Method: (SP)^{2}Net

We propose to incorporate a class-agnostic segmentation prior to the network to achieve better generalization for novel classes and leverage the ignore regions to address the biased prediction issue.

Figure 2 shows an overview of our approach (SP)^{2}Net, consisting of three main components: (1) a superpixel pooling module to capture class-agnostic segmentation priors, (2) a semantic projection layer [31] for segmenting novel classes (introduced in Sect. 3), (3) a bias reduction loss in ignore regions. In the following, we will describe our major contributions, i.e., superpixel pooling and the bias reduction loss.

4.1 Superpixel Pooling Module

Unlike SPNet [31] and other prior works [5,19], we argue that relying on a standard segmentation network may not generalize well to unseen classes. Therefore, we develop a superpixel pooling (SP-pooling) module which aims to facilitate feature learning for GZLSS.

Integrating Superpixels. As labeled images from unseen classes are not available in GZLSS, learning generic features or introducing prior information becomes important for achieving good generalization on unseen classes. We believe that superpixels could be particularly helpful for GZLSS because they can provide precise and generic object boundaries. We incorporate the superpixels as a pooling layer in our segmentation network. The simplest idea is to apply the superpixel pooling as a post processing step on the output probability scores of the network, i.e., after the semantic projection layer (see Fig. 2). Although this indeed yields smooth predictions, the superpixels do not benefit feature learning. An alternative is to pool the final feature map $\phi(x)$ after the segmentation head i.e., atrous spatial pyramid pooling (ASPP) module of DeepLab-v3+ [7] or PSPNet [33]. However, we found that applying SP-pooling before the ASPP yields best performance as superpixels can guide the ASPP to learn more generic class-agnostic information.

Superpixel Pooling. Pooling operation plays an important role in semantic segmentation to extract global features [22] or pyramid features [34].

For efficiency, we apply simple average pooling to each superpixel region (see Fig. 1, right), which is parameter-free, and not sensitive to the spatial size of the region. More specifically, given an input feature map $F_{in} \in \mathbb{R}^{K \times H \times W}$ and corresponding superpixels $\{s_n | n = 1, ..., N\}$ of the input image, we compute

$$F_{out}(k, i, j) = \frac{1}{|s_n|} \sum_{(p,q) \in s_n} F_{in}(k, p, q), \tag{1}$$

where (i, j) is a pixel in superpixel s_n and k is the index of feature channels. After repeating the above computation for every pixel, we obtain a pooled feature map $F_{out} \in \mathbb{R}^{K \times H \times W}$ of the same size as the input F_{in}. Note that individual input images in a mini-batch may have a different number of superpixels, but the output size of the pooling operation does not depend on that number as Eq. 1 assigns each pixel within the same superpixel with the same feature embedding. Our superpixel pooling module not only provides boundary information, but also context information as prior for learning better representations to segment novel classes.

4.2 Bias Reduction Loss

In this section, we first explain ignore regions followed by introducing our bias reduction loss to alleviate the imbalance issue in GZLSS.

Ignore Regions. In zero-shot image classification, the training set must exclude any image from novel classes to satisfy the "zero-shot" assumption. In contrast, in semantic segmentation, where an image consists of dense labels from multiple object classes, it is not realistic to build a training set that contains no pixels from novel classes due to a large number of class co-occurrences. Therefore, a common practice in GZLSS is to follow the training set of supervised learning but ignoring the pixels from novel classes. Note that prior work [31] allows the models to process those pixels i.e., during the forward pass, but do not apply any loss on them as their labels are not accessible.

Bias Reduction Loss. We argue that ignoring certain regions in the image severely biases the predictions. Indeed, a DNN trained with only seen classes will be overconfident even on regions of novel classes at test time. Although their labels are not available, it is certain that the ignore regions do not belong to the seen classes, which can serve as strong prior information for alleviating the bias issue. Formally, for a particular unlabeled pixel x_i from the ignore regions i.e., $y_i \notin \mathcal{S}$, we propose the following bias reduction loss,

$$\mathcal{L}_{BR}(x_i) = \sum_{c \in \mathcal{S}} -\log(1 - P_c(x_i)) \quad (2)$$

where P_c denotes the probability of pixel x_i being predicted as class c. This loss essentially treats seen classes as negative in the ignore regions and reduces the probability of those pixels being classified as any of the seen classes.

Discussion. This loss enjoys several advantages over existing balancing strategies for GZLSS. SPNet [31] adopts a post hoc calibration technique that reduces the scores of seen classes by a constant factor at the test time. However, tuning a global calibration factor that works for all pixels is extremely hard because the optimal factors for different pixels can be completely different. In contrast, we learn to alleviate the bias issue from the training data, resulting in a model that does not require any calibration at test time. Compared to feature generation approaches [5,19], our bias reduction loss is conceptually simpler and can be optimized end-to-end. Specifically, those approaches require an additional training stage to learn the feature generator and novel class classifiers. Despite the simplicity, we outperform them [5,19] significantly as shown experimentally.

4.3 Training and Inference

As a preprocessing step, we first compute superpixels for each image with an off-the-shelf superpixel method (more details in Sect. 5). Our full model (SP)^{2}Net then minimizes the following objective:

$$\sum_{i=0}^{H\times W} \mathbb{1}[y_i \in \mathcal{S}]\mathcal{L}_C(x_i, y_i) + \mathbb{1}[y_i \notin \mathcal{S}]\lambda\mathcal{L}_{BR}(x_i) \tag{3}$$

where $\mathbb{1}[y_i \in \mathcal{S}]$ denotes an indicator function ($= 1$ if $y_i \in \mathcal{S}$ otherwise 0). $\mathcal{L}_C$ is the classification loss that learns the semantic projection layer on pixels of seen classes. $\mathcal{L}_{BR}$ is the bias reduction loss defined in Eq. 2 which handles the biased prediction issue in ignore regions. λ is hyperparameter to tune, controlling the trade-off between learning semantic projection and bias reduction. Our proposed superpixel pooling layer and bias reduction loss are both differentiable, which allows us to train the model end-to-end.

Once trained, we make a prediction by searching for the class with the highest probability among both seen and unseen classes, i.e. $\arg\max_{c\in\mathcal{Y}} P_c(x_i)$.

5 Experiments

In this section, we first describe our experimental setting, then we present (1) our results comparing with the state-of-the-art for the GZLSS task on 10 different data splits from two benchmark datasets, (2) model analysis on each model component and impact of hyperparameters, (3) our qualitative results comparing with SPNet.

Datasets. We use PASCAL VOC 2012 [8] (10582 train/1449 val images from 20 classes) and PASCAL-Context [27] (4998 train/5105 val images from 59 classes) datasets. We adopt the same data splits (having 2, 4, 6, 8, and 10 unseen classes in an incremental manner respectively) used by ZS3Net [5] and CSRL [19] for a fair comparison.

Implementation Details. Unless otherwise stated, we follow ZS3Net [5] and CSRL [19] to use DeepLab-v3+ [7] with the ImageNet-pretrained ResNet-101 [12] as the backbone for a fair comparison. For the semantic embeddings, we use the concatenations of fasttext [15] and word2vec [25] embeddings (each with dimension 300) because of its superior performance as shown in SPNet [31]. We adopt the SGD optimizer with initial learning rate 2.5×10^{-4} and use "poly" learning rate decay [6] with power = 0.9. We set momentum and weight decay rate to 0.9 and 0.0005 respectively. Unless otherwise stated, we apply our superpixel pooling module before the ASPP layer in DeepLab-v3+ as it performs better.

Table 1. Comparing with the state-of-the-art methods on the generalised zero-label semantic segmentation task on 10 different splits (U-k: split with k unseen classes) of PASCAL VOC 2012 and PASCAL-Context datasets. We report mIoU (in %) on seen classes (S), unseen classes (U) and harmonic mean of them (HM). (SP)^{2}Net (SLIC): Our method with SLIC based superpixels, (SP)^{2}Net (COB): Our method with COB based superpixels.

Method	PASCAL VOC 2012 splits														
	U2			U4			U6			U8			U10		
	S	U	HM	S	U	HM	S	U	HM	S	U	HM	S	U	HM
ZS3Net [5]	72.0	35.4	47.5	66.4	23.2	34.4	47.3	24.2	32.0	29.2	22.9	25.7	33.9	18.9	23.6
CSRL [19]	73.4	45.7	56.3	69.8	31.7	43.6	66.2	29.4	40.7	62.4	26.4	37.6	59.2	**21.0**	31.0
SPNet [31]	72.0	24.2	36.3	70.3	32.1	44.1	70.3	26.6	38.6	66.2	22.7	33.9	68.8	17.9	28.4
SPNet + $\mathcal{L}_{BR}$	69.1	40.4	51.0	61.3	**44.8**	**51.8**	71.5	36.5	48.4	65.6	23.1	34.2	**75.1**	16.3	26.8
(SP)^{2}Net (SLIC)	**75.8**	70.1	**72.8**	70.2	35.6	47.2	71.7	47.2	56.9	**74.2**	26.9	39.5	72.1	16.9	27.4
(SP)^{2}Net (COB)	73.4	**71.3**	72.4	**81.7**	37.9	**51.8**	**78.6**	**52.8**	**63.1**	72.8	**27.6**	**40.1**	73.2	19.7	**31.0**
Method	PASCAL Context splits														
	U2			U4			U6			U8			U10		
	S	U	HM	S	U	HM	S	U	HM	S	U	HM	S	U	HM
ZS3Net [5]	41.6	21.6	28.4	37.2	24.9	29.8	32.1	20.7	25.2	20.9	16.0	18.1	20.8	12.7	15.8
CSRL [19]	41.9	27.8	33.4	**39.8**	23.9	29.9	35.5	22.0	27.2	31.7	18.1	23.0	29.4	14.6	19.5
SPNet [31]	**42.9**	4.6	8.3	36.0	11.1	17.0	36.3	11.5	17.4	**33.2**	12.6	18.3	30.3	10.0	15.0
SPNet + $\mathcal{L}_{BR}$	34.4	16.2	22.1	38.0	23.8	29.3	**39.5**	16.7	23.4	26.3	17.1	20.7	33.5	12.8	18.6
(SP)^{2}Net (SLIC)	37.7	**54.5**	**44.5**	31.6	**48.9**	38.4	28.6	27.7	28.2	31.1	24.9	27.7	30.2	15.0	20.1
(SP)^{2}Net (COB)	38.1	52.9	44.3	34.6	47.0	**39.9**	35.6	**45.1**	**39.8**	33.1	**26.8**	**29.6**	**36.6**	**20.7**	**26.4**

Superpixel Extraction and Correction. We employ the pretrained convolutional oriented boundaries (COB) [24] provided by the authors to compute the superpixels because it is computationally efficient and generalizes well to unseen categories. Please note that COB is trained with only boundary supervision on PASCAL datasets and it does not access any semantic class label. We compute the boundary probability with COB followed by applying a threshold (K) to obtain the superpixels of different scales. A higher value of K implies larger superpixels that are noisy, while a lower value means smaller superpixels. Unless otherwise stated, we use K = 0.2 for our experiments. We present the efficiency of superpixel pooling method with another simple superpixel method, i.e., Simple Linear Iterative Clustering (SLIC) [1] for our experiments. The SLIC method uses clustering based method to generate superpixels and does not require any pretraining. For SLIC, we use OpenCV's SLIC method for generating the superpixels. Specifically, we set the region size as 60, unless stated otherwise, for our work. Larger region size generates noisy superpixels which can hurt performance, while smaller region size results in oversegmentation.

We observe that superpixels can be noisy and propose to fix the issue by a simple heuristic: ignoring the superpixels that do not intersect with the ground truth label mask sufficiently (see supplement for details). Note that such correction is only applied during the training phase and not in inference as the ground truth is not available for test images.

Evaluation Metric. We compute mean Intersection of Union (mIoU) since it is widely used for semantic segmentation [6,7]. We follow [5,19,31] to report mIoU for seen classes (S), unseen classes (U) and harmonic mean (HM) of both, which is computed as $HM = \frac{2*mIoU_{seen}*mIoU_{unseen}}{mIoU_{seen}+mIoU_{unseen}}$. The HM measures how well the model balances seen and unseen mIoU, i.e., the HM would be high if both are high.

5.1 Comparing with State-of-the-Art

We compare our (SP)^{2}Net[1] to the following methods.

- **SPNet** [31] embeds pixel features into a semantic space and produces a probability distribution with softmax.
- **ZS3Net** [5] learns a feature generator using word embeddings, that generates novel class features and apply a classifier on these learnt features.
- **CSRL** [19] augments ZS3Net by a structural feature generator that relates seen and novel classes. It is currently the state of the art in GZLSS on both our benchmarks.

We evaluate SPNet ourselves as it is not evaluated on the same benchmark, while results of ZS3Net and CSRL are directly taken from the papers. In addition, we combine SPNet and our bias reduction loss, i.e., SPNet + $\mathcal{L}_{BR}$.

We report the results of generalized zero-label semantic segmentation under 5 different data splits of PASCAL VOC and Context dataset in Table 1. First, our (SP)^{2}Net with either SLIC or COB[2] based superpixels outperforms feature generation approaches (i.e., CSRL and ZS3Net) significantly in almost all cases in terms of harmonic mean mIoU (HM) establishing a new state-of-the-art on PASCAL VOC and PASCAL Context datasets. For PASCAL VOC, our (SP)^{2}Net (COB) achieves remarkable performance gains compared to closest baseline CSRL, i.e., +16.1% on U2 split, +6.7% on U4 split, +22.4% on U6 split and +2.5% on U8 split. For PASCAL Context, which is more challenging than PASCAL VOC, our (SP)^{2}Net (COB) still achieves large performance gains over the closest baseline CSRL, i.e.,+10.9% on U-2 split, +10% on U-4 split, +12.6% on U-6 split, +6.6% on U-8 split, and +6.9% on U-10 split.

Moreover, we observe that both (SP)^{2}Net (COB) and (SP)^{2}Net (SLIC) outperform SPNet + $\mathcal{L}_{BR}$ significantly in 4 out of 5 cases on PASCAL VOC dataset in terms of HM, e.g., 72.8% of (SP)^{2}Net (COB) vs 51.0% of SPNet + $\mathcal{L}_{BR}$ on Unseen-2 split, 63.1% of (SP)^{2}Net (COB) vs 48.4% of SPNet + $\mathcal{L}_{BR}$ on Unseen-6 split, and similarly for 5 out of 5 splits for Context dataset. These compelling results clearly show the importance of our superpixel pooling module. Indeed, by integrating the class-agnostic segmentation prior from superpixels, our (SP)^{2}Net learns dense image features that are more suitable for GZLSS. Another observation is that SPNet + $\mathcal{L}_{BR}$ improves the HM of SPNet in almost all cases and even surprisingly outperforms CSRL in some cases, confirming that our bias

[1] (SP)^{2}Net refers to (SP)^{2}Net with COB superpixels,unless otherwise stated.

[2] COB based superpixels are pretrained on object boundaries on PASCAL Dataset.

reduction loss $\mathcal{L}_{BR}$ is able to alleviates the strong bias towards seen classes. It is worth noting that CSRL, ZS3Net and SPNet indeed process the ignore regions but they fail to apply any loss on those pixels. However, our (SP)^{2}Net employs the bias reduction loss which makes full use of the ignore regions for balancing the model.

Finally, we would like to emphasize that our (SP)^{2}Net (SLIC) outperforms the state-of-the-art method, i.e., CSRL in 9 out of 10 splits, confirming the importance of our superpixel pooling and bias reduction loss. In fact, (SP)^{2}Net (SLIC) is able to achieve on par results to (SP)^{2}Net (COB) under U2 and U8 splits of PASCAL VOC and U2, U4 and U8 splits of PASCAL-Context, showing that our method is not limited to COB based superpixels. This is encouraging because SLIC is a simple clustering-based approach and does not require any pretraining.

5.2 Model Analysis

In this section, we conduct extensive ablation experiments to show the effectiveness of our network design and impact of hyperparameters. Here, we report results for the U-6 split of PASCAL VOC 2012 and provide results for all splits for both datasets in the supplementary showing that our observations hold across most data splits.

Table 2. Ablation results on U-6 split of PASCAL VOC 2012 dataset. Left: Applying our superpixel pooling (SP-pooling) module at different layers of DeepLab-v3+. Right: Ablation results for model components i.e. superpixel pooling module and bias reduction loss. SPNet: our base model [31] based on DeepLab-v3+ [7], SP-pooling: superpixel pooling module, $\mathcal{L}_{BR}$: our bias reduction loss.

Location of SP-pooling	S	U	HM
w/o SP-pooling	71.5	36.5	48.4
Output layer	72.4	37.8	49.7
After ASPP	71.4	37.3	49.1
Before ASPP	**78.6**	**52.8**	**63.1**

Method	S	U	HM
SPNet [31]	70.3	26.6	38.6
+ SP-pooling	70.0	28.2	40.2
+ $\mathcal{L}_{BR}$	71.5	36.5	48.4
+ SP-pooling + $\mathcal{L}_{BR}$	**78.6**	**52.8**	**63.1**

Location of Superpixel Pooling. We show the results of applying superpixel pooling at different layers of the network in Table 2 (left). We tried three different locations for DeepLab-v3+ [7], i.e., on the output probability scores of the output layer (a post-processing step on the predictions), on the output feature maps of its Atrous Spatial Pyramid Pooling (ASPP) module, and on the input feature maps of the ASPP module. We have the following observations.

First, all three variants of our superpixel pooling module improve the network compared to without using SP-pooling, confirming again the advantage of using superpixels. Second, while SP-pooling on the output layer and after ASPP only slightly improve the results, we observe that applying the SP-pooling before ASPP significantly boosts the HM of the baseline without SP-pooling, e.g., from 48.4% to 63.1% on PASCAL VOC 2012. This is understandable as pooling the outputs or last layer simply smooths the predictions or has limited effects on improving the feature learning. On the other hand, pooling before ASPP provides the ASPP module the class-agnostic segmentation prior to learn better features.

Ablations on Superpixels and Bias Reduction Loss. We perform ablation studies with respect to our superpixel pooling and bias reduction loss and report the results in Table 2 (right). We start from our baseline SPNet and gradually add components to it. First, we observe that our superpixel module, i.e., SPNet + SP, improves the seen, unseen and harmonic mean mIoU of SPNet significantly on both datasets. This again confirms our claims that the superpixel provide a strong class-agnostic segmentation prior and is able to facilitate the network to learn better dense image features for the seen as well novel classes. Second, adding the bias reduction loss $\mathcal{L}_{BR}$ to SPNet immediately leads to a huge boost on the harmonic mean on both datasets, implying that the bias reduction loss is an effective way to alleviate the imbalance issue. Finally, putting all components together, i.e., SPNet + SP + $\mathcal{L}_{BR}$, yields our full model (SP)^{2}Net which outperforms all other baselines consistently on all the metrics and datasets and indicates the complements of the superpixel pooling and bias reduction loss.

Table 3. Ablation results under the U-6 split of PASCAL VOC 2012. Left: superpixel pooling module (SP-pooling) based on two segmentation networks, i.e., DeepLab-v3+ [7] and PSPNet [33]. Right: Effect of using different λ

Method	Segmentation Networks	PASCAL VOC 2012 S	U	HM
w/o SP-pooling	DeepLab-v3+	71.5	36.5	48.4
w/ SP-pooling	DeepLab-v3+	**78.6**	**52.8**	**63.1**
w/o SP-pooling	PSPNet	69.7	22.3	33.8
w/ SP-pooling	PSPNet	**70.5**	**25.5**	**37.4**

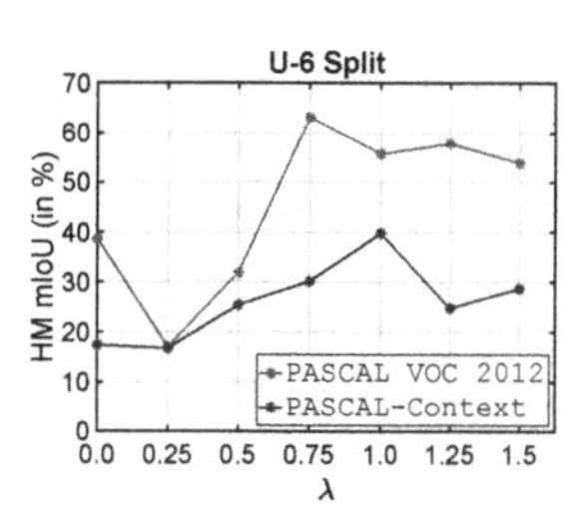

Effect of Segmentation Networks. The goal of this study is to determine the effect of using different segmentation networks that provide critical segmentation heads on top of the CNN backbone. We investigate two popular segmentation networks, i.e., DeepLab-v3+ [7] and PSPNet [33] (both are with ResNet101 as the backbone). In Table 3 (left), we first observe that the results with PSP-Net are lower than the ones obtained with DeepLab-v3+. This indicates that DeepLab-v3+ is a stronger architecture compared to PSPNet in GZLSS, which is reasonable because DeepLab-v3+ indeed outperforms PSPNet on various segmentation benchmarks. Besides, most importantly, with both segmentation networks, we observe that our SP-pooling significantly improves the results without SP-pooling. In particular, on PASCAL VOC 2012, the HM increases from 48.4%

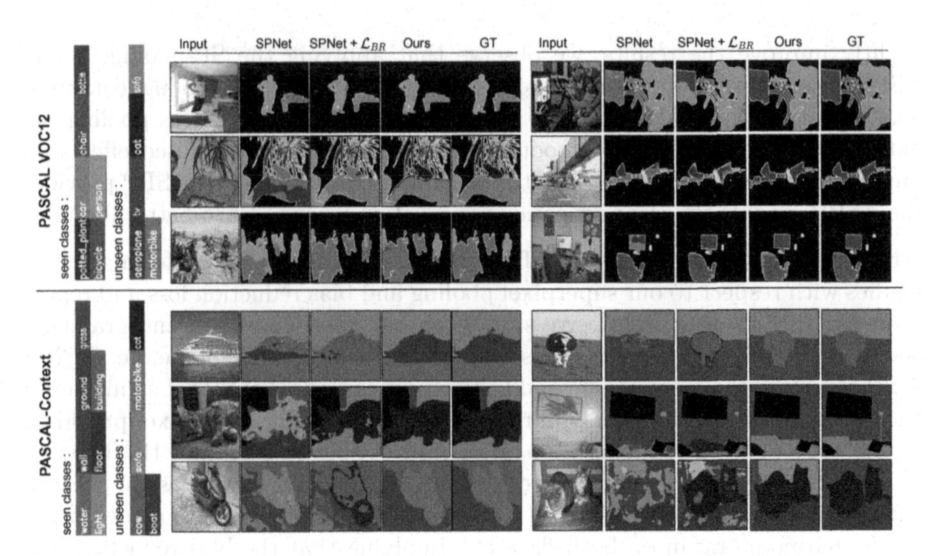

Fig. 3. Qualitative comparisons with SPNet [31] and SPNet + $\mathcal{L}_{BR}$ on PASCAL VOC 2012 and PASCAL-Context datasets under the U-6 setting. Black color indicates ignore background regions.

to 63.1% for DeepLab-v3+ and from 33.8% to 37.4% for PSPNet. These results show that our superpixel pooling module is not limited to DeepLab-v3+, but also a simple yet effective architecture change for other segmentation networks such as PSPNet.

Impact of the Hyperparameter λ. In Eq. 3, λ plays a trade-off between the classification loss $\mathcal{L}_C$ and bias reduction loss $\mathcal{L}_{BR}$. We show the results of using different λ in table 3 (right) and observe the following. (1) Without the bias reduction loss ($\lambda = 0$), the harmonic mean is rather low, indicating that the model fails to handle the imbalance issue well. (2) using a lambda in the range of $[0.75, 1.5]$ significantly boosts the harmonic mean, confirming the effectiveness of our bias reduction loss. (3) it seems to have a trend of performance drop with a large λ because the model would fail to learn a precise semantic projection layer when putting too much weights on the bias reduction loss.

5.3 Qualitative Results

We show our qualitative results in Fig. 3 for the unseen-6 split of PASCAL VOC 2012 and PASCAL-Context datasets. For PASCAL VOC 2012, we highlight the following unseen classes that our model segment much better than the SPNet: sofa (row 1, row 2), cat (row 2), and motorbike (row 3). Similarly, for PASCAL-Context dataset, where models also predict fine-grained classes for the background including water, wall, grass etc., our model segments the unseen classes better than the baseline: cat (row2, row3), motorbike (row3) and sofa (row2). We point out several key observations: (1) By introducing bias reduction loss with SPNet (SPNet + $\mathcal{L}_{BR}$), we are able to overcome the biasness to the seen classes and then successfully recognize some unseen objects such as tv, sofa, cat, cow etc., although the output masks remain to improve further. (2) Superpixel plays an important role for stronger performance w.r.t more smooth prediction over objects (e.g., the areoplane in the second row) as well as learning better representations for zero-label segmentation (e.g., the cat in the second row). (3) Because of the challenges of zero-label setup, objects from seen classes are also imperfectly predicted, for example, the bottle in the first row is partially predicted as the unseen class motorbike. In contrast, our full model with superpixels segments the entire outline and outputs more favorable masks than SPNet and SPNet + $\mathcal{L}_{BR}$. (4) (3) Our (SP)^{2}Net segments the background classes more smoothly on the inner regions and precisely on the boundaries.

We conclude that, our (SP)^{2}Net obtains significant improvements over the baseline SPNet [31] and ranks at today's state of the art performance for the task of generalized zero-label semantic segmentation according to a series of qualitative and quantitative comparison to the baseline method [31] and other recent approaches [5,19].

6 Conclusion

We present a novel approach (SP)^{2}Net for the challenging generalized zero-label semantic segmentation task. The main novelties lie in the superpixel pooling module that aggregates features from adaptive superpixel regions and an efficient bias reduction loss that minimizes the confidence of seen classes in the ignore regions. We empirically show that our superpixel pooling module significantly improves the generalization to novel classes and our bias reduction loss effectively alleviates the data imbalance issue. We benchmark our (SP)^{2}Net against various baselines on 10 different splits from two datasets and establish a new state of the art on all data splits in GZLSS.

Acknowledgements. This work has been partially funded by the ERC (853489 - DEXIM) and by the DFG (2064/1 - Project number 390727645).

References

1. Achanta, R., Shaji, A., Smith, K., Lucchi, A., Fua, P., Süsstrunk, S.: SLIC superpixels compared to state-of-the-art superpixel methods. TPAMI (2012)
2. Akata, Z., Perronnin, F., Harchaoui, Z., Schmid, C.: Label-embedding for image classification. TPAMI (2016)
3. Arbeláez, P., Pont-Tuset, J., Barron, J.T., Marques, F., Malik, J.: Multiscale combinatorial grouping. In: Proceedings of the IEEE conference on computer vision and pattern recognition, pp. 328–335 (2014)
4. Bearman, A., Russakovsky, O., Ferrari, V., Fei-Fei, L.: What's the point: semantic segmentation with point supervision. In: Leibe, B., Matas, J., Sebe, N., Welling, M. (eds.) ECCV 2016. LNCS, vol. 9911, pp. 549–565. Springer, Cham (2016). https://doi.org/10.1007/978-3-319-46478-7_34

5. Bucher, M., Vu, T.H., Cord, M., Pérez, P.: Zero-shot semantic segmentation. In: NeurIPS (2019)
6. Chen, L.C., Papandreou, G., Kokkinos, I., Murphy, K., Yuille, A.L.: DeepLab: semantic image segmentation with deep convolutional nets, atrous convolution, and fully connected CRFs. TPAMI (2017)
7. Chen, L.C., Zhu, Y., Papandreou, G., Schroff, F., Adam, H.: Encoder-decoder with atrous separable convolution for semantic image segmentation. In: ECCV (2018)
8. Everingham, M., Van Gool, L., Williams, C.K.I., Winn, J., Zisserman, A.: The pascal Visual Object Classes (VOC) challenge. Int. J. Comput. Vis. **88**, 303–338 (2010). https://doi.org/10.1007/s11263-009-0275-4
9. Frome, A., et al.: Devise: a deep visual-semantic embedding model. In: NeurIPS (2013)
10. Gadde, R., Jampani, V., Kiefel, M., Kappler, D., Gehler, P.V.: Superpixel convolutional networks using bilateral inceptions. In: Leibe, B., Matas, J., Sebe, N., Welling, M. (eds.) ECCV 2016. LNCS, vol. 9905, pp. 597–613. Springer, Cham (2016). https://doi.org/10.1007/978-3-319-46448-0_36
11. Gu, Z., Zhou, S., Niu, L., Zhao, Z., Zhang, L.: Context-aware feature generation for zero-shot semantic segmentation. In: ACM Multimedia (2020)
12. He, K., Zhang, X., Ren, S., Sun, J.: Deep residual learning for image recognition. In: CVPR (2016)
13. He, Y., Chiu, W.C., Keuper, M., Fritz, M.: Std2p: RGBD semantic segmentation using spatio-temporal data-driven pooling. In: CVPR (2017)
14. Jayaraman, D., Grauman, K.: Zero-shot recognition with unreliable attributes. In: NIPS (2014)
15. Joulin, A., Grave, E., Bojanowski, P., Douze, M., Jégou, H., Mikolov, T.: Fasttext.zip: compressing text classification models. arXiv preprint arXiv:1612.03651 (2016)
16. Khoreva, A., Benenson, R., Hosang, J., Hein, M., Schiele, B.: Simple does it: weakly supervised instance and semantic segmentation. In: CVPR (2017)
17. Kwak, S., Hong, S., Han, B.: Weakly supervised semantic segmentation using superpixel pooling network. In: AAAI (2017)
18. Lampert, C.H., Nickisch, H., Harmeling, S.: Learning to detect unseen object classes by between-class attribute transfer. In: CVPR (2009)
19. Li, P., Wei, Y., Yang, Y.: Consistent structural relation learning for zero-shot segmentation. In: NeurIPS (2020)
20. Lin, D., Dai, J., Jia, J., He, K., Sun, J.: Scribblesup: Scribble-supervised convolutional networks for semantic segmentation. In: CVPR (2016)
21. Lin, D., Ji, Y., Lischinski, D., Cohen-Or, D., Huang, H.: Multi-scale context intertwining for semantic segmentation. In: ECCV (2018)
22. Liu, W., Rabinovich, A., Berg, A.C.: Parsenet: looking wider to see better. arXiv preprint arXiv:1506.04579 (2015)
23. Long, J., Shelhamer, E., Darrell, T.: Fully convolutional networks for semantic segmentation. In: CVPR (2015)
24. Maninis, K.K., Pont-Tuset, J., Arbeláez, P., Van Gool, L.: Convolutional oriented boundaries: From image segmentation to high-level tasks. IEEE TPAMI (2017)
25. Mikolov, T., Chen, K., Corrado, G., Dean, J.: Efficient estimation of word representations in vector space. arXiv preprint arXiv:1301.3781 (2013)
26. Mikolov, T., Sutskever, I., Chen, K., Corrado, G.S., Dean, J.: Distributed representations of words and phrases and their compositionality. In: NIPS (2013)

27. Mottaghi, R., et al.: The role of context for object detection and semantic segmentation in the wild. In: Proceedings of the IEEE Conference on Computer Vision and Pattern Recognition, pp. 891–898 (2014)
28. Romera-Paredes, B., OX, E., Torr, P.H.: An embarrassingly simple approach to zero-shot learning. In: ICML (2015)
29. Stutz, D., Hermans, A., Leibe, B.: Superpixels: an evaluation of the state-of-the-art. In: CVIU (2018)
30. Xian, Y., Akata, Z., Sharma, G., Nguyen, Q., Hein, M., Schiele, B.: Latent embeddings for zero-shot classification. In: CVPR (2016)
31. Xian, Y., Choudhury, S., He, Y., Schiele, B., Akata, Z.: Semantic projection network for zero-and few-label semantic segmentation. In: CVPR, pp. 8256–8265 (2019)
32. Zhang, L., Xiang, T., Gong, S.: Learning a deep embedding model for zero-shot learning. In: CVPR (2017)
33. Zhao, H., Shi, J., Qi, X., Wang, X., Jia, J.: Pyramid scene parsing network. In: Proceedings of the IEEE Conference on Computer Vision and Pattern Recognition, pp. 2881–2890 (2017)
34. Zhao, H., Shi, J., Qi, X., Wang, X., Jia, J.: Pyramid scene parsing network. In: CVPR (2017)

Contrastive Representation Learning for Hand Shape Estimation

Christian Zimmermann, Max Argus(✉), and Thomas Brox

University of Freiburg, Freiburg im Breisgau, Germany
argus@cs.uni-freiburg.de
https://lmb.informatik.uni-freiburg.de/projects/contra-hand/

Abstract. This work presents improvements in monocular hand shape estimation by building on top of recent advances in unsupervised learning. We extend momentum contrastive learning and contribute a structured collection of hand images, well suited for visual representation learning, which we call *HanCo*. We find that the representation learned by established contrastive learning methods can be improved significantly by exploiting advanced background removal techniques and multi-view information. These allow us to generate more diverse instance pairs than those obtained by augmentations commonly used in exemplar based approaches. Our method leads to a more suitable representation for the hand shape estimation task and shows a 4.7% reduction in mesh error and a 3.6% improvement in F-score compared to an ImageNet pretrained baseline. We make our benchmark dataset publicly available, to encourage further research into this direction.

Keywords: Hand shape estimation · Self-supervised learning · Contrastive learning · Dataset

1 Introduction

Leveraging unlabeled data for training machine learning is a long standing goal in research and its importance has increased dramatically with the advances made by data-driven deep learning methods. Using unlabeled data is an attractive proposition, because more training data usually leads to improved results. On the other hand, label acquisition for supervised training is difficult, time-consuming, and cost intensive.

While the use of unsupervised learning is conceptually desirable the research community struggled long to compete with simple transfer learning approaches using large image classification benchmark datasets. For a long time, there has been a substantial gap between the performance of these methods and the results of supervised pretraining on ImageNet. However, recent work [14] succeeded to surpass ImageNet based pretraining for multiple downstream tasks. The core innovation was to use a consistent, dynamic and large dictionary of embeddings in combination with a contrastive loss, which is a practice we are following in this work as well. Similarly, we make use

C. Zimmermann and M. Argus—These authors contributed equally.

C. Bauckhage et al. (Eds.): DAGM GCPR 2021, LNCS 13024, pp. 250–264, 2021.
https://doi.org/10.1007/978-3-030-92659-5_16

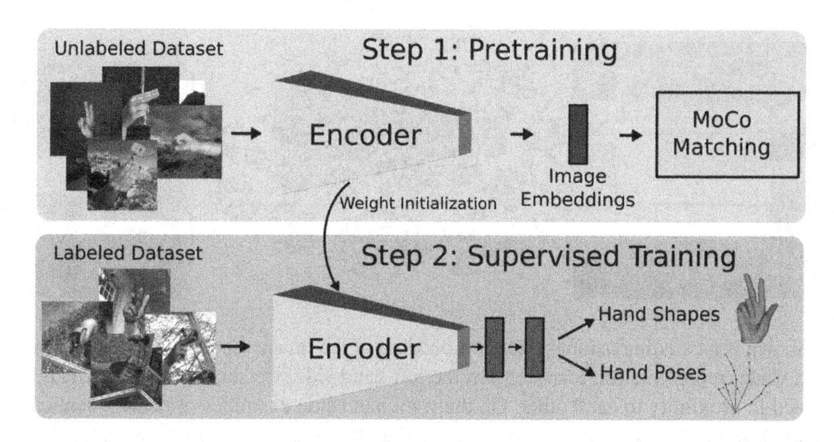

Fig. 1. We train a self-supervised feature representation on our proposed large dataset of unlabeled hand images. The resulting encoder weights are then used to initialize supervised training based on a smaller labeled dataset. This pretraining scheme yields useful image embeddings that can be used to query the dataset, as well as increasing the performance of hand shape estimation.

of strong geometric and color space augmentations, like flipping, cropping, as well as modification of hue and brightness, to generate positive pairs at training time.

Additionally, we find that large improvements lie in the use of strategies that extend the simple exemplar strategy of using a single image and heavy augmentation to generate a positive pair of samples. More precisely, we explore sampling strategies that exploit the structure in *HanCo* that is available without additional labeling cost. The data was captured in an controlled multi-view setting as video sequences, which allows us to easily extract foreground segmentation masks and sample correlated hand poses by selecting simultaneously recorded images of neighboring cameras. This allows us to generate more expressive positive pairs during self-supervised learning.

Hand shape estimation is a task where it is inherently hard to acquire diverse training data at a large scale. This stems from frequent ambiguities and its high dimensionality, which further raises the value of self-supervision techniques. Its need for large amounts of training data also makes hand shape estimation an ideal testbed for developing self-supervision techniques. Most concurrent work in the field of hand shape estimation follows the strategy of weak-supervision, where other modalities are used to supervise hand shape training indirectly. Instead, we explore an orthogonal approach: pretraining the network on data of the source domain which eliminates both the need for hand shape labels as well as additional modalities.

In our work, we find that using momentum contrastive learning yields a valuable starting point for hand shape estimation. It can find a meaningful visual representation from pure self-supervision that allows us to surpass the ImageNet pretrained baseline significantly. We provide a comprehensive analysis on the learned representation and show how the procedure can be used for identifying clusters of hand poses within the data or perform retrieval of similar poses.

For the purpose of further research into self-supervised pretraining we release a dataset that is well structured for this purpose it includes a) a larger number of images, b) temporal structure, and c) multi-view camera calibration.

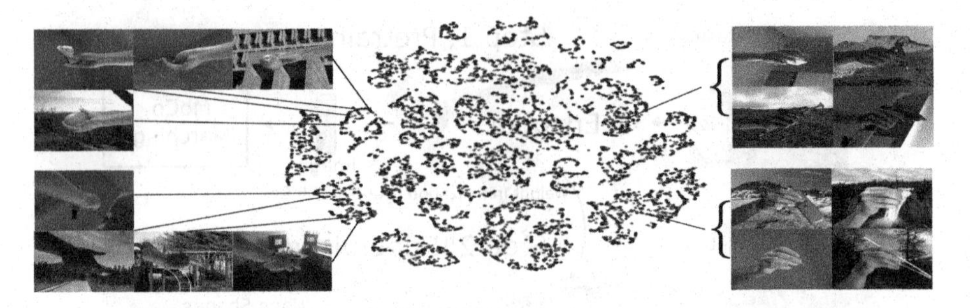

Fig. 2. Shown is a two dimensional *t-SNE* embedding of the image representation found by unsupervised visual representation learning. On the left hand side, we show that similar hand poses are located in proximity to each other. On the right hand side examples of nearly identical hand poses are shown which are approximately mapped to the same point. The color of the dots indicates which pre-processing method has been applied to the sample, a good mixing shows that the embedding focuses on the hand pose instead. Blue represents the unaltered camera-recorded images, red are images that follow the simple cut and paste strategy, while pink and bright blue correspond to images processed with the methods by Tsai *et al.* [33] and Zhang *et al.* [40] respectively.

2 Related Work

Visual Representation Learning. Approaches that aim to learn representations from collection of images without any labels can be roughly categorized into generative and discriminative approaches. While earlier work was targetting generative approaches the focus shifted towards discriminative methods that either leverage contrastive learning or formulate auxiliary tasks using pseudo labels.

Generative approaches aim to recover the input, subject to a certain type of pertubation or regularization. Examples are DCGAN by Radford *et al.* [26], image colorization [39], denoising autoencoders [34] or image in-painting [24].

Popular auxiliary tasks include solving Jigsaw Puzzles [22], predicting image rotations [11], or clustering features during training [5].

In contrast to auxiliary tasks the scheme of contrastive loss functions [12], doesn't define pseudo-labels, but uses a dynamic dictionary of keys that are sampled from the data and resemble the current representation of samples via the encoding network. For training one matching pair of keys is generated and the objective drives the matching keys to be similar, while being dissimilar to the other entries in the dictionary. Most popular downstream task is image classification, where contrastive learning approaches have yielded impressive results [4,6,7]. In this setting pretrained features are evaluated by a common protocol in which these features are frozen and only a supervised linear classifier is trained on the global average pooling of these features. In this work we follow the paradigm of contrastive learning, hence the recent successes, and build on the work by Chen *et al.* [7]. However, we find retraining of the complete convolutional backbone is necessary for hand shape estimation, which is following the transfer learning by fine-tuning [8,37] idea. Furthermore, we extend sampling of matching keys beyond exemplar and augmentation-based strategies.

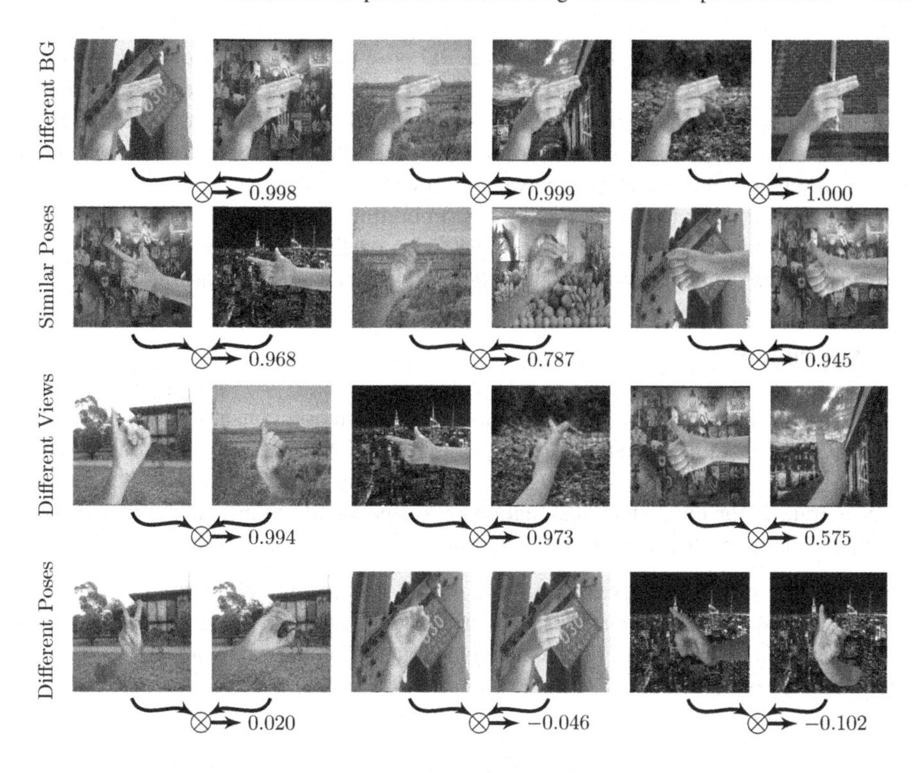

Fig. 3. Examples showing the cosine-similarity scores for the embeddings produced by pretraining. The first row shows embeddings for the same hand pose with different backgrounds, the embedding learns to focus on the hand and ignore the background. In the second row we show images for similar hand poses, with different background, these also score highly. In the third row different hand poses, with the same background produce low scores.

Hand Shape Estimation. Most recent works in the realm of hand shape estimation rely on a deep neural network estimating parameters of a hand shape model. By far, the most popular choice is MANO presented by Romero *et al.* [28]. There are little approaches using another surface topology; examples are Moon *et al.* [21] and Ge *et al.* [10], but MANO was used by the majority of works [2,13,18,42] and is also used in our work.

Commonly, these approaches are trained with a combination of losses, frequently incorporating additional supervision coming from shape derived modalities like depth [3,21,35], silhouette [1,2,20], or keypoints [2,13,18,21,42], which is referred to as weak-supervision. This allows to incorporate datasets into training that don't contain shape annotation, significantly reducing the effort for label acquisition. Sometimes, these approaches incorporate adversarial losses [17], which can help to avoid degenerate solutions on the weakly labeled datasets. Other works focusing on hand pose estimation propose weak supervision by incorporating biomechanical constraints [30] or sharing a latent embedding space between multiple modalities [31,32].

One specific way of supervision between modalities is leveraging geometric constraints of multiple camera views, which was explored in human pose estimation before:

Rhodin *et al.* [27] proposed to run separate 3D pose detectors per view and constrain the estimated poses with respect to each other. Simon *et al.* [29] presents an iterative bootstrapping procedure of self-labeling with using 3D consistency as a quality criterion. Yao *et al.* [36] and He *et al.* [16] supervise 2D keypoint detectors by explicitly incorporating epipolar constraints between two views.

In our work the focus is not on incorporating constraints by adding weak-supervision to the models' training, but we focus on finding a good initialization for supervised hand shape training using methods from unsupervised learning.

3 Approach

Our approach to improve monocular hand shape estimation consists of two steps and is summarized in Fig. 1: First, we are pretraining the CNN encoder backbone on large amounts of unlabeled data using unsupervised learning on a pretext task. Second, the CNN is trained for hand shape estimation in a supervised manner, using the network weights from the pretext task as initialization.

Momentum Contrastive Learning. *MoCo* [14] is a recent self-supervised learning method that performs contrastive learning as a dictionary look-up. *MoCo* uses two encoder networks to encode different augmentations of the same image instance as query and key pairs. Given two images $\boldsymbol{I}_i \in \mathbb{R}^{H \times W}$ and $\boldsymbol{I}_j \in \mathbb{R}^{H \times W}$ the embeddings are calculated as

$$q = f(\theta,\ \boldsymbol{I}_i) \quad \text{and} \tag{1}$$

$$k = f(\tilde{\theta},\ \boldsymbol{I}_j). \tag{2}$$

which yields a query q and key k. The same function f is used in both cases, but parameterized differently. The query function uses θ which is directly updated by the optimization, while the key function $\tilde{\theta}$ is updated indirectly. At a given optimizations step n it is calculated as

$$\tilde{\theta}_n = m \cdot \tilde{\theta}_{n-1} + (1 - m) \cdot \theta_n \tag{3}$$

using the momentum factor m, which is chosen close to 1.0 to ensure a slowly adapting encoding of the key values k.

During training a large queue of dictionary keys k is accumulated over iterations which allows for efficient training as a large set of negative samples has been found to be critical for contrastive training [14]. Following this methodology *MoCo* produces feature representations that transfer well to a variety of downstream tasks. As training objective the InfoNCE loss [23] is used, which relates the inner product of the matching key-query-pair to the inner products of all negative pairs in a softmax cross-entropy kind of fashion.

At test time the similarity of a key-value-pair can be computed using cosine similarity

$$\text{cossim}(q,\ k) = \frac{q \cdot k}{\|q\|_2 \cdot \|k\|_2} \tag{4}$$

which can return values ranging from 1.0 in the similar case to -1.0 in the dissimilar case.

MoCo relies entirely on standard image space augmentations to generate positive pairs of image during representation learning. A function $g(.)$, subject to a random vector ζ, is applied to the same image instance $\boldsymbol{I}$ two times to generate different augmentations

$$\boldsymbol{I}_i = g(\boldsymbol{I},\, \zeta_1) \tag{5}$$

$$\boldsymbol{I}_j = g(\boldsymbol{I},\, \zeta_2) \tag{6}$$

that are considered as the matching pair. The function $g(.)$ performs randomized: crops of the image, color jitter, grayscale, and conversion and Gaussian blur. We omit randomized image flipping as this augmentation changes the semantic information of the hand pose.

The structured nature of *HanCo* allows us going beyond these augmentation-based strategies. Here we are looking for strategies that preserve the hand pose, but change other aspects of the image. We investigate three different configurations a) background randomization, b) temporal sampling and c) multi-view sampling.

HanCo consists of short video clips that are recorded by multiple cameras simultaneously at 5 Hz. The clips are up to 40 s long and have an average length of 14 s. For simplicity and without loss of generality we describe the sampling methods for a single sequence only. Extending the approach towards multiple sequences is straight forward, by first sampling a sequence and then applying the described procedure. Formally, we can sample from a pool of images $\boldsymbol{I}_t^c$ recorded at a timestep t from camera c.

During background randomization we use a single image $\boldsymbol{I}_t^c$ with its foreground segmentation as source to cut the hand from the source image and paste it into a randomly sampled background image. Example outputs of background randomization are shown in the first row of Fig. 3.

For temporal sampling we exploit the fact, that our unlabeled dataset stems from a video stream which naturally constraints subsequent hand poses to be highly correlated. A positive pair of samples is generated by sampling two neighboring frames $\boldsymbol{I}_t^c$ and $\boldsymbol{I}_{t+1}^c$ for a given camera c. Due to hand movement the hand posture is likely to change slightly from t to $t+1$, which naturally captures the fact that similar poses should be encoded with similar embeddings.

As the data is recorded using a calibrated multi-camera setup and frame capturing is temporally synchronized, views from different cameras at a particular point in time show the same hand pose. This can be used as a powerful method of "augmentation" as different views change many aspects of the image but not the hand pose. Consequently, we generate a positive sample pair $\boldsymbol{I}_t^{c_1}$ and $\boldsymbol{I}_t^{c_2}$ in the multi-view case by sampling neighboring cameras c_1 and c_2 at a certain timestep t. The dataset contains an 8 camera setup, with cameras mounted on each of the corners of a cube, in order to simplify the task of instance recognition, we chose to sample neighboring cameras, meaning those connected by no more than one edge of the cubical fixture.

Hand Shape Estimation. Compared to unsupervised pretraining the network architecture used for shape estimation is modified by changing the number of neurons of the last fully-connected layer from 128 to 61 neurons in order to estimate the *MANO* parameter

Fig. 4. Given a query image our learned embedding allows to identify images showing similar poses. This enables identifying clusters in the data, without the need of pose annotations. The nearest neighbors are queried from a random subset of 25,000 samples of *HanCo*.

vector. Consequently, the approach is identical to the one presented by Zimmermann *et al.* [42] with the difference being only that the weights of the convolutional backbone are initialized through unsupervised contrastive learning and not ImageNet pretraining. The network is being trained to estimate the *MANO* parameter vector $\tilde{\boldsymbol{\theta}} \in \mathbb{R}^{61}$ using the following loss:

$$\begin{aligned} \mathcal{L} = w_{\mathrm{3D}} & \left\| \boldsymbol{P} - \tilde{\boldsymbol{P}} \right\| + \\ w_{\mathrm{2D}} & \left\| \Pi(\boldsymbol{P}) - \Pi(\tilde{\boldsymbol{P}}) \right\| + \\ w_{\mathrm{p}} & \left\| \boldsymbol{\theta} - \tilde{\boldsymbol{\theta}} \right\| . \end{aligned} \tag{7}$$

We deploy L_2 losses for all components and weight with $w_{\mathrm{3D}} = 1000$, $w_{\mathrm{2D}} = 10$, and $w_{\mathrm{p}} = 1$ respectively. To derive the predicted keypoints $\tilde{\boldsymbol{P}}$ from the estimated shape $\tilde{\boldsymbol{\theta}}$ in a differentiable way the MANO *et al.* [28] model implementation in *PyTorch* by Hasson *et al.* [13] is used. $\boldsymbol{P} \in \mathbb{R}^{21}$ is the ground truth 3D location of the hand joints and $\boldsymbol{\theta}$ the ground truth set of MANO parameter, both of which are provided by the training dataset. Denoted by $\Pi(.)$ is the projection operator mapping from 3D space to image pixel coordinates.

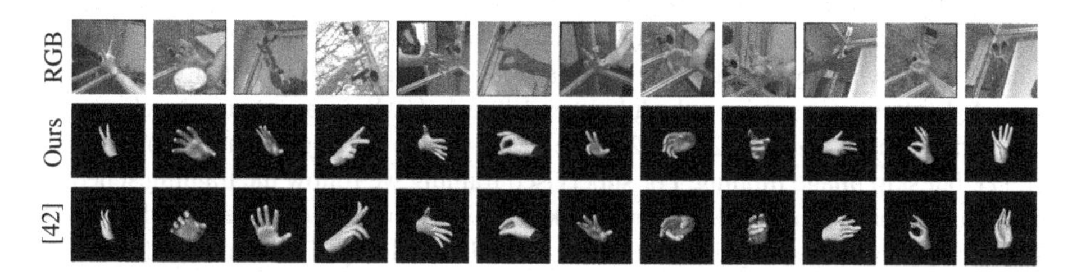

Fig. 5. Qualitative comparison of MANO predictions between *Ours-Multi view* and the approach by Zimmermann *et al.* [42] showing improvements in hand mesh predictions yielded by our self-supervised pretraining on the evaluation split of *FreiHAND* [42]. Generally, our predictions look seem to capture the global pose and grasp of the hand more accurately, which results into a lower mesh error and higher F@5 score.

4 Experiments

Dataset. Our experiments are conducted on data recorded by Zimmermann *et al.* [42], which the authors kindly made available to us. The images show 32 subjects which are recorded by 8 cameras simultaneously that are mounted on a cubical aluminum fixture. The cameras face towards the center of the cube and are calibrated. One part of the data was recorded against a green background, which allows extracting of the foreground segmentation automatically and to perform background randomization without any additional effort. Another part of the data was intended for evaluation and was recorded without green backgrounds. For a subset of both parts there are hand shape labels, which were created by the annotation method [42]. The set of annotated frames was published before as *FreiHAND* [42], which we use for evaluation of the supervised hand shape estimation approaches.

For compositing the hand foreground with random backgrounds we have collected 2193 background images from Flickr, that are showing various landscapes, city scenes and indoor shots. We manually inspect these images, to ensure they don't contain shots targeted at humans. There are three different methods to augment the cut and paste version: The colorization approach by Zhang *et al.* [40] is used in both its automatic and sampling-based mode. Also we use the deep harmonization approach proposed by Tsai *et al.* [33] that can remove color bleeding at the foreground boundaries. The background post-processing methods are also reflected by the point colors in the *t-SNE* embedding plot (Fig. 2).

The complete dataset is used for visual representation learning and we refer to it as *HanCo*, which provides $107,538$ recorded time instances or poses, each recorded by 8 cameras. All available frames are used for unsupervised training, which results into $860,304$ frames, while a subset of $63,864$ frames contains hand shape annotations and are recorded against green screen, which we use for supervised training of the monocular hand shape estimation network.

Training Details. For training of neural networks the *PyTorch* framework is used and we rely on ResNet-50 as convolutional backbone [15,25].

During unsupervised training we follow the procedure by Chen *et al.* [7] and train with the following hyper-parameters: a base learning rate of 0.015 is used, which is annealed following a cosine schedule over 100 epochs. We train with a batch size of 128 and an image size of 224×224 pixels. We follow the augmentations of Chen *et al.* [7], but skip image flipping. For supervised training we follow the training schedule by Zimmermann *et al.* [42]. The network is trained for $500,000$ steps with a batch size of 16. A base learning rate of 0.0001 is used, and decayed by a factor of 0.5 after $220,000$ and $340,000$ steps.

Evaluation of Embeddings. First, we perform a qualitative evaluation of the learned embedding produced by pretraining. For this purpose, we sample a random subset of $25,000$ images from the unlabeled dataset. The query network is used to compute a 128 dimensional feature vector for each image. Using the feature vectors we find a *t-SNE* [19] representation that is shown in Fig. 2. It is apparent, that similar poses cluster closely together, while different poses are clearly separated.

Pairs of images, together with their cosine similarity scores (4) are shown in Fig. 5, the cosine similarity scores reveal many desirable properties of the embedding: The representation is invariant to the picked background (first row), small changes in hand pose only result in negligible drop in similarity (second row). Viewing the same hand pose from different directions results in high similarity scores, though this can be subject to occlusion (third row). Large changes in hand pose induce a significant drop of the score (last row). This opens up the possibility to use the learned embedding for retrieval tasks, which is shown in Fig. 4. Given a query image it is possible to identify images showing similar and different hand poses, without an explicit hand pose annotation.

Hand Shape Estimation. For comparison we follow [42] and rely on the established metrics *mesh error* in cm and F-score evaluated at thresholds of 5 mm and 15 mm. All of them being reported for the Procrustes aligned estimates as calculated by the online Codalab evaluation service [41].

The results of our approach are compared to literature in Table 1. The results reported share a similar architecture that is based on a ResNet-50 and the differences can be attributed to the training objective and data used. The approach presented by Boukhayma *et al.* [2] performs pretraining on a large synthetic dataset and subsequent training on a combination of datasets containing real images with 2D, 3D and hand segmentation annotation. The datasets used are *MPII+NZSL* [29], *Panoptic* [29] and *Stereo* [38]. Another rendered dataset is proposed and used for training by Hasson *et al.* [13], which is combined with real images from Garcia *et al.* [9]. Zimmermann *et al.* [42] use only the real images from the *FreiHAND* dataset for training, which is the setting we are also using for the results reported.

Table 1 summarizes the results of the quantitative evaluation on *FreiHAND*. It shows that training the network from random initialization leads to unsatisfactory results reported by *Ours-Scratch*, which indicates that a proper network initialization is important. *Ours-Fixed* is training only the fully-connected layers starting from the weights found by *MoCo* while keeping the convolutional part fixed. This achieves results that fall far behind in comparison. Additionally, training the parameters associated with batch normalization gives a significant boost in accuracy as reported by *Ours-Fixed-BN*. The entry named *Ours-Augmentation* does not make use of the advanced back-

Table 1. Pretraining the convolutional backbone using momentum contrastive learning improves over previous results by -4.7% in terms of mesh error, 3.6% in terms of F@5 mm and 0.9% in terms of F@15 mm (comparing Zimmermann *et al.* [42] and *Ours-Multi View*). This table shows that fixing the *MoCo* learned convolutional backbone and only training the fully-connected part during hand shape estimation (see *Ours-Fixed*) can not compete with state-of-the-art approaches. *Ours-Fixed-BN* shows that additionally training the batch normalization parameters leads to substantial improvements. Consequently, leaving all parameters open for optimization (*Ours-Augmentation*) leads to further improvements. In *Ours-Scratch* all parameters are trained from random initialization, which performs much better than fixing the convolutional layers, but is still behind the reported results in literature, which illustrates the importance of a good network initialization. Applying, our proposed sampling strategies *Ours-Background*, *Ours-Temporal* or *Ours-Multi View* does improve results over using an augmentation based sampling strategy like Chen *et al.* [7], denoted by *Ours-Augmentation*.

Method	Mesh error in cm ↓	F@5 mm ↑	F@15 mm ↑
Boukhayma *et al.* [2]	1.30	0.435	0.898
Hasson *et al.* [13]	1.32	0.436	0.908
Zimmermann *et al.* [42]	1.07	0.529	0.935
Ours-fixed	2.13	0.299	0.803
Ours-fixed-BN	1.25	0.463	0.914
Ours-scratch	1.24	0.475	0.915
Ours-augmentation	1.09	0.521	0.934
Ours-background	1.04	0.538	0.940
Ours-temporal	1.04	0.538	0.939
Ours-multi view	**1.02**	**0.548**	**0.943**

ground randomization methods, and is the direct application of the proposed *MoCo* approach [7] to our data. In this case all network weights are being trained. It performs significantly better than the fixed approaches and training from scratch, but lacks behind ImageNet based initialization used by Zimmermann *et al.* [42]. Finally, the rows *Ours-Background*, *Ours-Temporal* and *Ours-Multi view* report results for the proposed sampling strategies for positive pairs while training all network parameters. All methods are able to outperform the ImageNet trained baseline by Zimmermann *et al.* [42]. We find the differences between *Ours-Background* and *Ours-Temporal* to be negligible, while *Ours-Multi view* shows an significant improvement over Zimmermann *et al.* [42]. This shows the influence and importance of the proposed sampling strategies.

To quantify the effect of our proposed multi-view pretraining strategy we do an ablation study in which the quantity of training samples for supervised training is reduced and we evaluate how well our proposed multi-view sampling strategy compares to an ImageNet pretrained baseline. This is shown in Fig. 6. In this experiment, the amount of labeled data used during supervised training is varied between 20% and 100% of the full *FreiHAND* training dataset, which corresponds to between 12,772 and 63,864 training samples. Except for varying the training data we follow the approach by Zimmermann *et al.* [42]. The figure shows curves for the minimal, maximal, and mean F@5 scores achieved over three runs per setting for Procrustes aligned predictions.

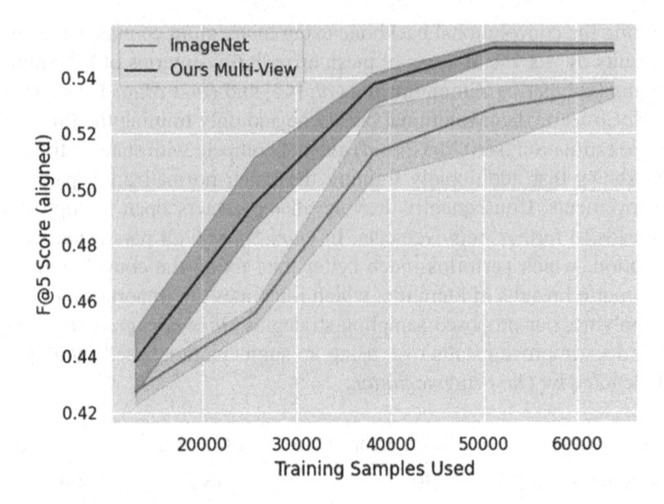

Fig. 6. Comparison between our proposed pretraining method (blue) and an ImageNet baseline (red) for varying fractions of the training dataset used during supervised learning of hand shape estimation. The lines represent the mean result over 3 runs per setting, while the shaded area indicates the range of obtained results. Our proposed *Multi-View* sampling approach consistently outperforms the baseline, with the largest differences occurring when using approximately 40% of the full training dataset, indicating a sweet-spot where pretraining is most beneficial. Learning from very large or small datasets reduces the gains from pretraining, for small datasets we hypothesize that there is not sufficient variation in the training dataset to properly learn the task of hand shape estimation anymore. (Color figure online)

We observe that *Our Multi-View* consistently outperforms the ImageNet pretrained baseline and that more data improves the performance of both methods. The differences between both methods are largest when using 40% of the full training dataset, for a very large or small supervised datasets the differences between the two methods become smaller. We hypothesize, that there is a sweet spot between enough training data, to diminish the value of pretraining, and a minimal amount of training data needed to learn the task of hand shape estimation reasonably well from the supervised data.

A qualitative comparison between our method and results by [42] is presented in Fig. 5. It shows that the differences between both methods are visually subtle, but in general hand articulation is captured more accurately by our methods which results into lower *mesh error* and higher F-scores.

5 Conclusion

Our work shows that unsupervised visual representation learning is beneficial for hand shape estimation and that sampling meaningful positive pairs is crucial. We only scratch on the surface of possible sampling strategies and find that sampling needs to be inline with the downstream task at hand.

By making the data available publicly, we encourage the community strongly to explore some of the open directions. These include to extend sampling towards other sources of consistency that remain untouched by our work, *e.g.*, temporal consistency of hand pose within one recorded sequence leaves opportunities for future exploration.

Another direction points into combining recently proposed weak-supervision methods with the presented pretraining methods and leverage the given calibration information at training time.

References

1. Baek, S., Kim, K.I., Kim, T.: Pushing the envelope for rgb-based dense 3d hand pose estimation via neural rendering. In: IEEE Conference on Computer Vision and Pattern Recognition, CVPR 2019, Long Beach, CA, USA, 16–20 June 2019, pp. 1067–1076. Computer Vision Foundation/IEEE (2019). https://doi.org/10.1109/CVPR.2019.00116, http://openaccess.thecvf.com/content_CVPR_2019/html/Baek_Pushing_the_Envelope_for_RGB-Based_Dense_3D_Hand_Pose_Estimation_CVPR_2019_paper.html
2. Boukhayma, A., de Bem, R., Torr, P.H.S.: 3D hand shape and pose from images in the wild. In: IEEE Conference on Computer Vision and Pattern Recognition, CVPR 2019, Long Beach, CA, USA, 16–20 June 2019, pp. 10843–10852. Computer Vision Foundation/IEEE (2019). https://doi.org/10.1109/CVPR.2019.01110, http://openaccess.thecvf.com/content_CVPR_2019/html/Boukhayma_3D_Hand_Shape_and_Pose_From_Images_in_the_Wild_CVPR_2019_paper.html
3. Cai, Y., Ge, L., Cai, J., Yuan, J.: Weakly-supervised 3D hand pose estimation from monocular RGB images. In: Ferrari, V., Hebert, M., Sminchisescu, C., Weiss, Y. (eds.) ECCV 2018. LNCS, vol. 11210, pp. 678–694. Springer, Cham (2018). https://doi.org/10.1007/978-3-030-01231-1_41
4. Caron, M., Misra, I., Mairal, J., Goyal, P., Bojanowski, P., Joulin, A.: Unsupervised learning of visual features by contrasting cluster assignments. ArXiv abs/2006.09882 (2020)
5. Caron, M., Bojanowski, P., Joulin, A., Douze, M.: Deep clustering for unsupervised learning of visual features. In: Ferrari, V., Hebert, M., Sminchisescu, C., Weiss, Y. (eds.) Computer Vision – ECCV 2018. LNCS, vol. 11218, pp. 139–156. Springer, Cham (2018). https://doi.org/10.1007/978-3-030-01264-9_9
6. Chen, T., Kornblith, S., Swersky, K., Norouzi, M., Hinton, G.: Big self-supervised models are strong semi-supervised learners. arXiv preprint arXiv:2006.10029 (2020)
7. Chen, X., Fan, H., Girshick, R.B., He, K.: Improved baselines with momentum contrastive learning. CoRR abs/2003.04297 (2020). https://arxiv.org/abs/2003.04297
8. Donahue, J., et al.: Decaf: a deep convolutional activation feature for generic visual recognition. In: ICML (2014)
9. Garcia-Hernando, G., Yuan, S., Baek, S., Kim, T.: First-person hand action benchmark with RGB-D videos and 3d hand pose annotations. In: 2018 IEEE Conference on Computer Vision and Pattern Recognition, CVPR 2018, Salt Lake City, UT, USA, 18–22 June 2018, pp. 409–419. IEEE Computer Society (2018). https://doi.org/10.1109/CVPR.2018.00050, http://openaccess.thecvf.com/content_cvpr_2018/html/Garcia-Hernando_First-Person_Hand_Action_CVPR_2018_paper.html
10. Ge, L., et al.: 3D hand shape and pose estimation from a single RGB image. In: IEEE Conference on Computer Vision and Pattern Recognition, CVPR 2019, Long Beach, CA, USA, 16–20 June 2019, pp. 10833–10842. Computer Vision Foundation/IEEE (2019). https://doi.org/10.1109/CVPR.2019.01109, http://openaccess.thecvf.com/content_CVPR_2019/html/Ge_3D_Hand_Shape_and_Pose_Estimation_From_a_Single_RGB_CVPR_2019_paper.html

11. Gidaris, S., Singh, P., Komodakis, N.: Unsupervised representation learning by predicting image rotations. In: 6th International Conference on Learning Representations, ICLR 2018, Vancouver, BC, Canada, April 30 - May 3, 2018, Conference Track Proceedings. OpenReview.net (2018). https://openreview.net/forum?id=S1v4N2l0-
12. Hadsell, R., Chopra, S., LeCun, Y.: Dimensionality reduction by learning an invariant mapping. In: 2006 IEEE Computer Society Conference on Computer Vision and Pattern Recognition (CVPR 2006), 17–22 June 2006, New York, NY, USA, pp. 1735–1742. IEEE Computer Society (2006). https://doi.org/10.1109/CVPR.2006.100
13. Hasson, Y., et al.: Learning joint reconstruction of hands and manipulated objects. In: IEEE Conference on Computer Vision and Pattern Recognition, CVPR 2019, Long Beach, CA, USA, 16–20 June 2019, pp. 11807–11816. Computer Vision Foundation/IEEE (2019). https://doi.org/10.1109/CVPR.2019.01208, http://openaccess.thecvf.com/content_CVPR_2019/html/Hasson_Learning_Joint_Reconstruction_of_Hands_and_Manipulated_Objects_CVPR_2019_paper.html
14. He, K., Fan, H., Wu, Y., Xie, S., Girshick, R.B.: Momentum contrast for unsupervised visual representation learning. In: 2020 IEEE/CVF Conference on Computer Vision and Pattern Recognition, CVPR 2020, Seattle, WA, USA, 13–19 June 2020, pp. 9726–9735. IEEE (2020). https://doi.org/10.1109/CVPR42600.2020.00975
15. He, K., Zhang, X., Ren, S., Sun, J.: Deep residual learning for image recognition. In: 2016 IEEE Conference on Computer Vision and Pattern Recognition, CVPR 2016, Las Vegas, NV, USA, 27–30 June 2016, pp. 770–778. IEEE Computer Society (2016). https://doi.org/10.1109/CVPR.2016.90
16. He, Y., Yan, R., Fragkiadaki, K., Yu, S.: Epipolar transformers. In: 2020 IEEE/CVF Conference on Computer Vision and Pattern Recognition, CVPR 2020, Seattle, WA, USA, 13–19 June 2020, pp. 7776–7785. IEEE (2020). https://doi.org/10.1109/CVPR42600.2020.00780
17. Kanazawa, A., Black, M.J., Jacobs, D.W., Malik, J.: End-to-end recovery of human shape and pose. In: 2018 IEEE Conference on Computer Vision and Pattern Recognition, CVPR 2018, Salt Lake City, UT, USA, 18–22 June 2018, pp. 7122–7131. IEEE Computer Society (2018). https://doi.org/10.1109/CVPR.2018.00744, http://openaccess.thecvf.com/content_cvpr_2018/html/Kanazawa_End-to-End_Recovery_of_CVPR_2018_paper.html
18. Kulon, D., Güler, R.A., Kokkinos, I., Bronstein, M.M., Zafeiriou, S.: Weakly-supervised mesh-convolutional hand reconstruction in the wild. In: 2020 IEEE/CVF Conference on Computer Vision and Pattern Recognition, CVPR 2020, Seattle, WA, USA, 13–19 June 2020, pp. 4989–4999. IEEE (2020). https://doi.org/10.1109/CVPR42600.2020.00504
19. van der Maaten, L., Hinton, G.: Visualizing high-dimensional data using t-SNE. J. Mach. Learn. Res. **9**, 2579–2605 (2008)
20. Malik, J., Elhayek, A., Stricker, D.: WHSP-Net: a weakly-supervised approach for 3D hand shape and pose recovery from a single depth image. Sensors **19**(17), 3784 (2019)
21. Moon, G., Shiratori, T., Lee, K.M.: DeepHandMesh: a weakly-supervised deep encoder-decoder framework for high-fidelity hand mesh modeling. In: Vedaldi, A., Bischof, H., Brox, T., Frahm, J.-M. (eds.) ECCV 2020. LNCS, vol. 12347, pp. 440–455. Springer, Cham (2020). https://doi.org/10.1007/978-3-030-58536-5_26
22. Noroozi, M., Favaro, P.: Unsupervised learning of visual representations by solving jigsaw puzzles. In: Leibe, B., Matas, J., Sebe, N., Welling, M. (eds.) ECCV 2016. LNCS, vol. 9910, pp. 69–84. Springer, Cham (2016). https://doi.org/10.1007/978-3-319-46466-4_5
23. van den Oord, A., Li, Y., Vinyals, O.: Representation learning with contrastive predictive coding. CoRR abs/1807.03748 (2018). http://arxiv.org/abs/1807.03748
24. Pathak, D., Krähenbühl, P., Donahue, J., Darrell, T., Efros, A.A.: Context encoders: feature learning by inpainting. In: 2016 IEEE Conference on Computer Vision and Pattern Recognition, CVPR 2016, Las Vegas, NV, USA, 27–30 June 2016, pp. 2536–2544. IEEE Computer Society (2016). https://doi.org/10.1109/CVPR.2016.278

25. Qiao, S., Wang, H., Liu, C., Shen, W., Yuille, A.L.: Weight standardization. CoRR abs/1903.10520 (2019). http://arxiv.org/abs/1903.10520
26. Radford, A., Metz, L., Chintala, S.: Unsupervised representation learning with deep convolutional generative adversarial networks. In: Bengio, Y., LeCun, Y. (eds.) 4th International Conference on Learning Representations, ICLR 2016, San Juan, Puerto Rico, 2–4 May 2016, Conference Track Proceedings (2016). http://arxiv.org/abs/1511.06434
27. Rhodin, H., et al.: Learning monocular 3D human pose estimation from multi-view images. In: 2018 IEEE Conference on Computer Vision and Pattern Recognition, CVPR 2018, Salt Lake City, UT, USA, 18–22 June 2018, pp. 8437–8446. IEEE Computer Society (2018). https://doi.org/10.1109/CVPR.2018.00880, http://openaccess.thecvf.com/content_cvpr_2018/html/Rhodin_Learning_Monocular_3D_CVPR_2018_paper.html
28. Romero, J., Tzionas, D., Black, M.J.: Embodied hands: modeling and capturing hands and bodies together. ACM Trans. Graph. **36**(6), 245:1–245:17 (2017). https://doi.org/10.1145/3130800.3130883
29. Simon, T., Joo, H., Matthews, I.A., Sheikh, Y.: Hand keypoint detection in single images using multiview bootstrapping. In: 2017 IEEE Conference on Computer Vision and Pattern Recognition, CVPR 2017, Honolulu, HI, USA, 21–26 July 2017, pp. 4645–4653. IEEE Computer Society (2017). https://doi.org/10.1109/CVPR.2017.494
30. Spurr, A., Iqbal, U., Molchanov, P., Hilliges, O., Kautz, J.: Weakly supervised 3D hand pose estimation via biomechanical constraints. In: Vedaldi, A., Bischof, H., Brox, T., Frahm, J.-M. (eds.) ECCV 2020. LNCS, vol. 12362, pp. 211–228. Springer, Cham (2020). https://doi.org/10.1007/978-3-030-58520-4_13
31. Spurr, A., Song, J., Park, S., Hilliges, O.: Cross-modal deep variational hand pose estimation. In: 2018 IEEE Conference on Computer Vision and Pattern Recognition, CVPR 2018, Salt Lake City, UT, USA, 18–22 June 2018, pp. 89–98. IEEE Computer Society (2018). https://doi.org/10.1109/CVPR.2018.00017, http://openaccess.thecvf.com/content_cvpr_2018/html/Spurr_Cross-Modal_Deep_Variational_CVPR_2018_paper.html
32. Theodoridis, T., Chatzis, T., Solachidis, V., Dimitropoulos, K., Daras, P.: Cross-modal variational alignment of latent spaces. In: 2020 IEEE/CVF Conference on Computer Vision and Pattern Recognition, CVPR Workshops 2020, Seattle, WA, USA, 14–19 June 2020, pp. 4127–4136. IEEE (2020). https://doi.org/10.1109/CVPRW50498.2020.00488
33. Tsai, Y., Shen, X., Lin, Z., Sunkavalli, K., Lu, X., Yang, M.: Deep image harmonization. In: 2017 IEEE Conference on Computer Vision and Pattern Recognition, CVPR 2017, Honolulu, HI, USA, 21–26 July 2017, pp. 2799–2807. IEEE Computer Society (2017). https://doi.org/10.1109/CVPR.2017.299
34. Vincent, P., Larochelle, H., Bengio, Y., Manzagol, P.: Extracting and composing robust features with denoising autoencoders. In: Cohen, W.W., McCallum, A., Roweis, S.T. (eds.) Machine Learning, Proceedings of the Twenty-Fifth International Conference (ICML 2008), Helsinki, Finland, June 5–9, 2008. ACM International Conference Proceeding Series, pp. 1096–1103. ACM, New York (2008). https://doi.org/10.1145/1390156.1390294
35. Wan, C., Probst, T., Gool, L.V., Yao, A.: Self-supervised 3D hand pose estimation through training by fitting. In: IEEE Conference on Computer Vision and Pattern Recognition, CVPR 2019, Long Beach, CA, USA, 16–20 June 2019, pp. 10853–10862. Computer Vision Foundation/IEEE (2019). https://doi.org/10.1109/CVPR.2019.01111, http://openaccess.thecvf.com/content_CVPR_2019/html/Wan_Self-Supervised_3D_Hand_Pose_Estimation_Through_Training_by_Fitting_CVPR_2019_paper.html
36. Yao, Y., Jafarian, Y., Park, H.S.: MONET: multiview semi-supervised keypoint detection via epipolar divergence. In: 2019 IEEE/CVF International Conference on Computer Vision, ICCV 2019, Seoul, Korea (South), October 27 - November 2, 2019, pp. 753–762. IEEE (2019). https://doi.org/10.1109/ICCV.2019.00084

37. Yosinski, J., Clune, J., Bengio, Y., Lipson, H.: How transferable are features in deep neural networks? In: NIPS (2014)
38. Zhang, J., Jiao, J., Chen, M., Qu, L., Xu, X., Yang, Q.: 3D hand pose tracking and estimation using stereo matching. CoRR abs/1610.07214 (2016). http://arxiv.org/abs/1610.07214
39. Zhang, R., Isola, P., Efros, A.A.: Colorful image colorization. In: Leibe, B., Matas, J., Sebe, N., Welling, M. (eds.) ECCV 2016. LNCS, vol. 9907, pp. 649–666. Springer, Cham (2016). https://doi.org/10.1007/978-3-319-46487-9_40
40. Zhang, R., et al.: Real-time user-guided image colorization with learned deep priors. ACM Trans. Graph. **36**(4), 119:1-119:11 (2017). https://doi.org/10.1145/3072959.3073703. Kindly provide year of the publication for the Ref. [41]
41. Zimmermann, C.: Freihand competition. https://competitions.codalab.org/competitions/21238
42. Zimmermann, C., Ceylan, D., Yang, J., Russell, B.C., Argus, M.J., Brox, T.: Freihand: a dataset for markerless capture of hand pose and shape from single RGB images. In: 2019 IEEE/CVF International Conference on Computer Vision, ICCV 2019, Seoul, Korea (South), October 27 - November 2, 2019, pp. 813–822. IEEE (2019). https://doi.org/10.1109/ICCV.2019.00090

Fusion-GCN: Multimodal Action Recognition Using Graph Convolutional Networks

Michael Duhme, Raphael Memmesheimer(✉), and Dietrich Paulus

Arbeitsgruppe Aktives, University of Koblenz-Landau, Koblenz, Germany
{mduhme,raphael,paulus}@uni-koblenz.de.de

Abstract. In this paper we present Fusion-GCN, an approach for multimodal action recognition using Graph Convolutional Network (GCNs). Action recognition methods based around Graph Convolutional Network (GCNs) recently yielded state-of-the-art performance for skeleton-based action recognition. With Fusion-GCN, we propose to integrate various sensor data modalities into a graph that is trained using a GCN model for multi-modal action recognition. Additional sensor measurements are incorporated into the graph representation either on a channel dimension (introducing additional node attributes) or spatial dimension (introducing new nodes). Fusion-GCN was evaluated on two publicly available datasets, the UTD-MHAD- and MMACT datasets, and demonstrates flexible fusion of RGB sequences, inertial measurements and skeleton sequences. Our approach gets comparable results on the UTD-MHAD dataset and improves the baseline on the large-scale MMACT dataset by a significant margin of up to 12.37% (F1-Measure) with the fusion of skeleton estimates and accelerometer measurements.

1 Introduction

Automatic Human Action Recognition (HAR) is a research area that is utilized in various fields of application where human monitoring is infeasible due to the amount of data and scenarios where quick reaction times are vital, such as surveillance and real-time monitoring of suspicious and abnormal behavior in public areas [12,33,34,49] or intelligent hospitals and healthcare sectors [8,9] with scenarios such as fall detection [36,45], monitoring of medication intake [13] and detection of other potentially life-threatening situations [8]. Additional areas of applications include video retrieval [40], robotics [41], smart home automation [21], autonomous vehicles [52]. In recent years, approaches based on neural networks, especially GCNs, like ST-GCN [51] or 2s-AGCN [43], have achieved state-of-the-art results in classifying human actions from time series data.

GCNs can be seen as an extension to Convolutional Neural Network (CNNs) that work on graph-structured data [19]. Its network layers operate by including a binary or weighted adjacency matrix, that describes the connections between each of the individual graph nodes. As of now, due to their graph-structured representation in the form of joints (graph nodes) and bones (graph edges), research for HAR using GCNs is mostly limited to skeleton-based recognition. However, the fusion of additional modalities into

C. Bauckhage et al. (Eds.): DAGM GCPR 2021, LNCS 13024, pp. 265–281, 2021.
https://doi.org/10.1007/978-3-030-92659-5_17

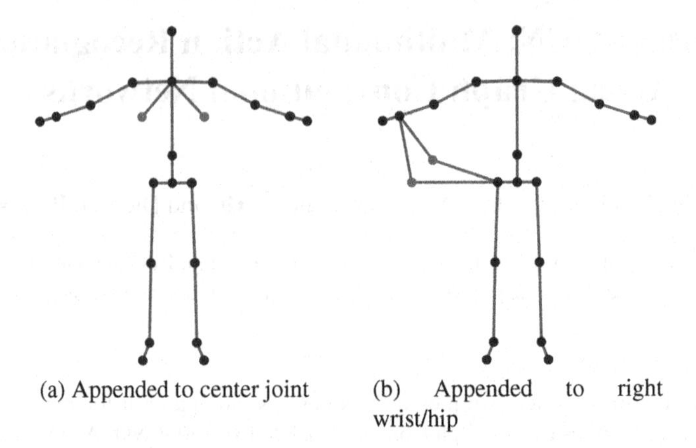

(a) Appended to center joint
(b) Appended to right wrist/hip

Fig. 1. Showing the skeleton as included in UTD-MHAD. IMU nodes are either appended to the central node (neck) or to both the right wrist and right hip. Two additional representations arise when all newly added nodes are themselves connected by edges.

GCNs models are currently neglected. For that reason, taking skeleton-based action recognition as the foundation, our objective is to research possibilities of incorporating other vision-based modalities and modalities from worn sensors into existing GCN models for skeleton-based action recognition through data fusion and augmentation of skeleton sequences. Figure 1 gives an example of two suggestions on how inertial measurements can be incorporated into a skeleton graph. To the best of our knowledge, Fusion-GCN is the first approach proposing to flexibly incorporate additional sensor modalities into the skeleton graph for HAR. We evaluated our approach on two multimodal datasets, UTD-MHAD [5] and MMACT [20].

The contributions of this paper can be summarized as: (1) We propose the fusion of multiple modalities by incorporating sensor measurements or extracted features into a graph representation. The proposed approach significantly lifts the state-of-the-art on the large-scale MMACT dataset. (2) We propose modality fusion for GCNs on two dimensionality levels: (a) the fusion at a channel dimension to incorporate additional modalities directly into the already existing skeleton nodes, (b) the fusion at a spatial dimension, to incorporate additional modalities as new nodes spatially connected to existing graph nodes. (3) We demonstrate applicability of the flexible fusion of various modalities like skeleton, inertial, RGB data in an early fusion approach.

The code for Fusion-GCN to reproduce and verify our results is publicly available on https://github.com/mduhme/fusion-gcn.

2 Related Work

In this section, we present related work from the skeleton-based action recognition domain that is based on GCN and further present recent work on multimodal action recognition.

Skeleton-based Action Recognition. Approaches based on GCNs have recently shown great applicability on non-Eucliean data [38] like naturally graph-structure represented skeletons and have recently defined the state-of-the-art. Skeletons, as provided by large-scale datasets [42], commonly are extracted from depth cameras [44]. RGB images can be transformed into human pose feature that yield a similar skeleton-graph in 2D [4,22,29] and in 3D [14,29,31]. All of those approaches output skeleton-graphs that are suitable as input for our fusion approach as a base structure for the incorporation of additional modalities. The Spatial-Temporal Graph Convolutional Network (ST-GCN) [51] is one of the first proposed models for skeleton-based HAR that utilizes GCNs based on the propagation rule introduced by Kipf and Welling [19]. The Adaptive Graph Convolutional Network (AGCN) [43] builds on these fundamental ideas with the proposal of learning the graph topology in an end-to-end-manner. Peng et al. [38] propose a Neural Architecture Search (NAS) approach for finding neural architectures to overcome the limitations of GCN caused by fixed graph structures. Cai et al. [1] proposes to add flow patches to handle subtle movements into a GCN. Approaches based on GCN [6,23,37,47] have been constantly improving the state-of-the-art on skeleton-based action recognition recently.

Multimodal Action Recognition. Cheron et al. [7] design CNN input features based on the positions of individual skeleton joints. Here, human poses are applied to RGB images and optical flow images. The pixel coordinates that represent skeleton joints are then grouped hierarchically starting from smaller body parts, such as arms, and upper body to full body. For each group, an RGB image and optical flow patch is cropped and passed to a 2D-CNN. The resulting feature vectors are then processed and concatenated to form a single vector, which is used to predict the corresponding action label. Similarly, Cao et al. [2] propose to fuse pose-guided features from RGB-Videos. Cao et al. [3] further, refine this method by using different aggregation techniques and an attention model. Islam and Iqbal [15] propose to fuse data of RGB, skeleton and inertial sensor modalities by using a separate encoder for each modality to create a similar shaped vector representation. The different streams are fused using either summation or vector concatenation. With Multi-GAT [16] an additional message-passing graphical attention mechanism was introduced. Li et al. [24] propose another architecture that entails skeleton-guided RGB features. For this, they employ ST-GCN to extract a skeleton feature vector and R(2+1)D [48] to encode the RGB video. Both output features are fused either by concatenation or by compact bilinear correlation.

The above-mentioned multimodal action recognition approaches follow a late-fusion method, that fuse various models for each modality. This allows a flexible per modality model-design, but comes at the computational cost of the multiple streams that need to be trained. For early fusion approaches, multiple modalities are fused on a representation level [32], reducing the training process to a single model but potentially loosing the more descriptive features from per-modality models. Kong et al. [20] presented a multi modality distillation model. Teacher models are trained separately using a 1D-CNN. The semantic embeddings from the teaching models are weighted with an attention mechanism and are ensembled with a soft target distillation loss into the student network. Similarly, Liu et al. [27] utilize distilled sensor information to improve the vision modality. Luo et al. [30] propose a graph distillation method to incorporate rich privileged information from a large-scale multi-modal dataset in the source domain, and improves the learning in the target domain More fundamentally, multimodality in neural networks is recently also tackled by the multimodal neurons that respond to photos, conceptual drawings and images of text [10]. Joze et al. [17] propose a novel intermediate fusion scheme in addition to early and late-fusion, they share intermediate layer features between different modalities in CNN streams. Perez-Rua et al. [39] presented an approach for finding neural architecture search for the fusion of multiple modalities. To the best of our knowledge, our Fusion-GCNapproach is the first that proposes to incorporate additional modalities directly into the skeleton-graphs as an early fusion scheme.

3 Approach

In the context of multimodal action recognition, early and late fusion methods have been established to either fuse on a representation or feature level. We present approaches for fusion of multiple modalities at representation level to create a single graph which is passed to a GCN.

3.1 Incorporating Additional Modalities into a Graph Model

Early fusion denotes the combination of structurally equivalent streams of data before sending them to a larger (GCN) model, whereas late fusion combines resulting outputs of multiple neural network models. For early fusion, one network handles multiple data sources which are required to have near identical shape to achieve fusion. As done by Song et al. [46], each modality may be processed by some form of an encoder to attain a common structure before being fused and passed on to further networks. Following a skeleton-based approach, for example, by employing a well established GCN model like ST-GCN or AGCN as the main component, RGB and inertial measurements are remodeled and factored into the skeleton structure. With Fusion-GCN we suggest the flexible integration of additional sensor modalities into a skeleton graph by either adding additional node attributes *(fusion on a channel dimension)* or introducing additional nodes *(fusion at a spatial dimension)*. In detail, the exact possible fusion approach is as follows.

Let $\boldsymbol{X}_{SK} \in \mathbb{R}^{(M, C_{SK}, T_{SK}, N_{SK})}$ be a skeleton sequence input, where M is the number of actors that are involved in an action, C_{SK} is the initial channel dimension (2D or 3D joint coordinates) and sizes T_{SK} and N_{SK} are sequence length and number of skeleton graph nodes. An input of shape $\mathbb{R}^{(M,C,T,N)}$ is required when passing data to a spatial-temporal GCN model, such as ST-GCN. Furthermore, let $\boldsymbol{X}_{RGB} \in \mathbb{R}^{(C_{RGB}, T_{RGB}, H_{RGB}, W_{RGB})}$ be the shape of an RGB video with channels C_{RGB}, frames T_{RGB} and image size $H_{RGB} \times W_{RGB}$. For sensor data, the input is defined as $\boldsymbol{X}_{IMU} \in \mathbb{R}^{(M, C_{IMU}, S_{IMU}, T_{IMU})}$, where T_{IMU} is the sequence length, S_{IMU} is the number of sensors and C_{IMU} is the channel dimension. For example, given gyroscope and accelerometer with x-, y- and z-values each, the structure would be $S_{IMU} = 2$ and $C_{IMU} = 3$. Similar to skeleton data, M denotes the person wearing the sensor and its value is equivalent to that of skeleton, that is, $M_{SK} = M_{IMU}$.

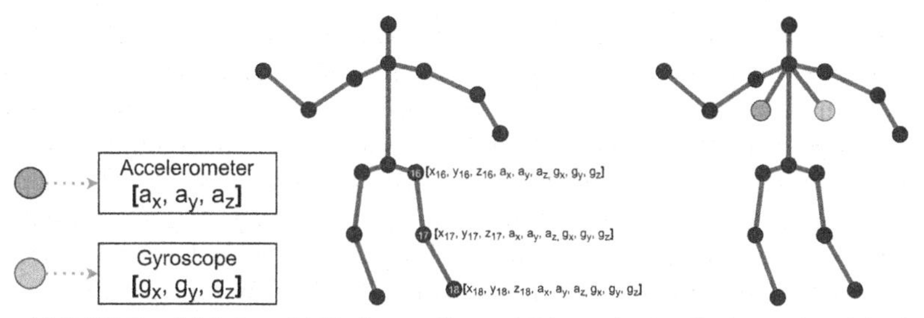

(a) IMU Graph Nodes (b) Fusion at Channel Dimension(c) Fusion at Spatial Dimension

Fig. 2. Options for fusion of skeleton graph and IMU signal values, viewed as skeleton nodes. If both skeleton joint coordinates and wearable sensor signals share a common channel dimension, the skeleton graph can be augmented by simply appending signal nodes at some predefined location.

Considering a multimodal model using a skeleton-based GCN approach, early fusion can now be seen as a task of restructuring non-skeleton modalities to be similar to skeleton sequences by finding a mapping $\mathbb{R}^{(C_{RGB},T_{RGB},H_{RGB},W_{RGB})} \rightarrow \mathbb{R}^{(M,C,T,N)}$ or $\mathbb{R}^{(M,C_{IMU},S_{IMU},T_{IMU})} \rightarrow \mathbb{R}^{(M,C,T,N)}$ with some C, T and N. This problem can be reduced: If the sequence length of some modalities is different, $T_{SK} \neq T_{RGB} \neq T_{IMU}$, a common T can be achieved by resampling T_{RGB} and T_{IMU} to be of the same length as the target modality T_{SK}. Early fusion is then characterized by two variants of feature concatenation to fuse data:

1. Given $\boldsymbol{X}_{SK}$ and an embedding $\boldsymbol{X}_E \in \mathbb{R}^{(M,C_E,T,N_E)}$ with sizes C_E and N_E where $N = N_{SK} = N_E$, fusion at the channel dimension means creating a fused feature $\boldsymbol{X} \in \mathbb{R}^{(M,C_{SK}+C_E,T,N)}$. An example is shown in Fig. 2b.
2. Given an embedding where $C = C_{SK} = C_E$ instead, a second possibility is fusion at the spatial dimension, that is, creating a feature $\boldsymbol{X} \in \mathbb{R}^{(M,C,T,N_{SK}+N_E)}$. Effectively, this amounts to producing $M \cdot T \cdot N_E$ additional graph nodes and distributing them to the existing skeleton graph at each time step by resizing its adjacency matrix and including new connections. In other words, the already existing skeleton graph is extended by multiple new nodes with an identical number of channels. An example is shown in Fig. 2c.

The following sections introduce multiple approaches for techniques about the early fusion of RGB video and IMU sensor modalities together with skeleton sequences by outlining the neural network architecture.

3.2 Fusion of Skeleton Sequences and RGB Video

This section explores possibilities for fusion of skeleton sequences and 2D data modalities. Descriptions and the following experiments are limited to RGB video, but all introduced approaches are in the same way applicable to depth sequences. As previously established, early fusion of RGB video and skeleton sequences in preparation for a skeleton-based GCN model is a problem of finding a mapping $\mathbb{R}^{(C_{RGB},H_{RGB},W_{RGB})} \rightarrow$

$\mathbb{R}^{(M,C,N)}$. An initial approach uses a CNN to compute vector representations of $N \cdot M \cdot T$ skeleton-guided RGB patches that are cropped around projected skeleton joint positions. Inspired by the work of Wang et al. [50] and Norcliffe-Brown et al. [35], a similar approach involves using an encoder network to extract relevant features from each image of the RGB video. This way, instead of analyzing $N \cdot M \cdot T$ cropped images, the T images of each video are utilized in their entirety. A CNN is used to extract features for every frame and fuse the resulting features with the corresponding skeleton graph, before the fused data is forwarded to a GCN. By running this procedure as part of the training process and performing fusion with skeleton sequences, the intention is to let the encoder network extract those RGB features that are relevant to the skeleton modality. For example, an action involving an object cannot be fully represented by merely the skeleton modality because an object is never part of the extracted skeleton. Objects are only visible in RGB video.

3.3 Fusion of Skeleton Sequences and IMU Signals

Fusion of skeleton and data from wearable sensors, such as IMUs, is applicable in the same way as described in the fusion scheme from the previous section. In preparation to fuse both modalities, they again need to be adjusted to have an equal sequence length first. Then, assuming both the skeleton joint coordinates and the signal values have a common channel dimension $C = 3$ and because $M_{SK} = M_{IMU}$, since all people wear a sensor, the only differing sizes between skeleton modality and IMU modality are N, the number of skeleton graph nodes, and S, the number of sensor signals. Leaving aside its structure, the skeleton graph is a collection of N nodes. A similar understanding can be applied to the S different sensors. They can be understood as a collection of S graph nodes (see Fig. 2a). The fusion of sensor signals with the skeleton graph is therefore trivial because the shape is almost identical. According to channel dimension fusion as described in the previous section, the channels of all S signals can be broadcasted to the x-, y- and z-values of all N skeleton nodes to create the GCN input feature $\boldsymbol{X} \in \mathbb{R}^{(M,(1+S)\cdot C,T,N)}$, as presented in Fig. 2b. The alternative is to append all S signal nodes onto the skeleton graph at some predefined location to create the GCN input feature $\boldsymbol{X} \in \mathbb{R}^{(M,C,T,N+S)}$, as illustrated in Fig. 2c. Similar to the RGB fusion approaches, channel dimension fusion does not necessarily require both modalities to have the same dimension C if vector concatenation is used. In contrast, the additional nodes are required to have the same dimension as all existing nodes if spatial dimension fusion is intended.

3.4 Combining Multiple Fusion Approaches

All the introduced fusion approaches can be combined into a single model, as illustrated by Fig. 3. First, the RGB modality needs to be processed using one of the variants discussed in Sect. 3.2. Ideally, this component runs as part of the supervised training process to allow the network to adjust the RGB feature extraction process based on the interrelation of its output with the skeleton graph. Similarly, sensor signals need to be processed using one of the variants discussed previously for that modality. Assuming all sequences are identical in length, to combine the different representations, let

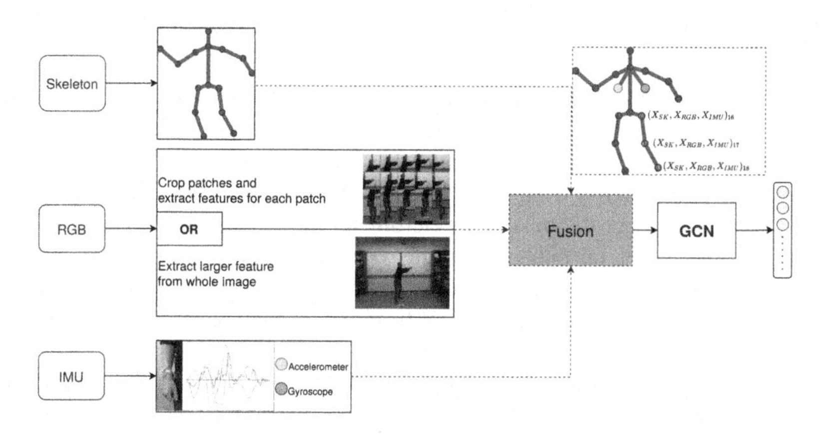

Fig. 3. All described approaches can be flexible fused together for early fusion and passed to a GCN. Fusion can be realized independent of a channel or spatial fusion dimension. Here we give an example of a mixed (channel and spatial) fusion.

$\boldsymbol{X}_{SK} \in \mathbb{R}^{(M,C_{SK},T,N_{SK})}$ be the sequence of skeleton graphs. For RGB, let $\boldsymbol{X}_{RGB1} \in \mathbb{R}^{(M,C_E,T,N)}$ be the C_E-sized channel features obtained from computing individual patch features or feature extraction for the whole image or $\boldsymbol{X}_{RGB2} \in \mathbb{R}^{(M,C,T,N_E)}$ be the RGB feature representing additional graph nodes. Respectively, the two variants of generated IMU features are $\boldsymbol{X}_{IMU1} \in \mathbb{R}^{(M,S\cdot C_{IMU},T,N)}$ or $\boldsymbol{X}_{IMU2} \in \mathbb{R}^{(M,C_{IMU},T,S)}$. The following possibilities to fuse different combinations of these representations arise.

- $(\boldsymbol{X}_{SK}, \boldsymbol{X}_{RGB1}, \boldsymbol{X}_{IMU1}) \rightarrow \boldsymbol{X}_{FUSED} \in \mathbb{R}^{(M,C_{SK}+C_E+S\cdot C_{IMU},T,N_{SK})}$ is the feature when combining modalities at channel dimension by vector concatenation.
- $(\boldsymbol{X}_{SK}, \boldsymbol{X}_{RGB1}, \boldsymbol{X}_{IMU2}) \rightarrow \boldsymbol{X}_{FUSED} \in \mathbb{R}^{(M,C_{SK}+C_E,T,N_{SK}+S)}$ combines skeleton with computed RGB features at channel dimension and expands the skeleton graph by including additional signal nodes. Since $C_{IMU} = C_{SK}$, the newly added nodes also need to be extended to have $C_{SK} + C_E$ channels. In contrast to skeleton nodes, there exists no associated cropped patch or RGB value. Therefore, the remaining C_E values can be filled with zeros. Conversely, the same applies when replacing $\boldsymbol{X}_{RGB1}$ with $\boldsymbol{X}_{RGB2}$ and $\boldsymbol{X}_{IMU2}$ with $\boldsymbol{X}_{IMU1}$.
- $(\boldsymbol{X}_{SK}, \boldsymbol{X}_{RGB2}, \boldsymbol{X}_{IMU2}) \rightarrow \boldsymbol{X}_{FUSED} \in \mathbb{R}^{(M,C,T,N_{SK}+N_E+S)}$ introduces new nodes for both RGB and signal modalities. This is accomplished by appending them to a specific location in the graph.

4 Experiments

We conducted experiments on two public available datasets and various modality fusion experiments. If not stated otherwise we use the top-1 accuracy as reporting metric for the final epoch of the trained model.

4.1 Datasets

UTD-MHAD. UTD-MHAD [5] is a relatively small dataset containing 861 samples and 27 action classes, which thereby results in shorter training durations for neural networks. Eight individuals (four females and four males) perform each action a total of four times, captured from a front-view perspective by a single Kinect camera. UTD-MHAD also includes gyroscope and accelerometer modalities by letting each subject wear the inertial sensor on either the right wrist or on the right hip, depending on whether an action is primarily performed using the hands or the legs. For the following experiments using this dataset, the protocol from the original paper [5] is used.

MMACT. The MMACT dataset [20] contains more than 35k data samples and 35 available action classes. With 20 subjects and four scenes with four currently available different camera perspectives each, the dataset offers a larger variation of scenarios. RGB videos are captured with a resolution of 1920×1080 pixels at a frame rate of 30 frames per second. For inertial sensors, acceleration, gyroscope and orientation data is obtained from a smartphone carried inside the pocket of a subject's pants. Another source for acceleration data is a smartwatch, resulting in data from four sensors in total. For the following experiments using this dataset, the protocol from the original paper [20] is used which proposes a cross-subject and a cross-view split. Since skeleton sequences are missing in the dataset, we create them from RGB data using OpenPose [4].

4.2 Implementation

Models are implemented using PyTorch 1.6 and trained on a Nvidia RTX 2080 GPU with 8 GB of video memory. To guarantee a deterministic and reproducible behavior, all training procedures are initialized with a fixed random seed. Unless stated otherwise, experiments regarding UTD-MHAD use a cosine annealing learning rate scheduler [28] with a total of 60 epochs, warm restarts after 20 and 40 epochs, an initial learning rate of $1\text{e}{-3}$ and ADAM [18] optimization. Experiments using RGB data instead run for 50 epochs without warm restarts. Training for MMAct adopts the hyperparameters used by Shi et al. [43]. For the MMACT, skeleton and RGB features were extracted for every third frame for more efficient pre-processing and training. The base GCN model is a single-stream AGCN for all experiments.

4.3 Comparison to the State-of-the-Art

UTD-MHAD. Table 1a shows a ranking of all conducted experiments in comparison with other recent state-of-the-art techniques that implement multimodal HAR on UTD-MHAD with the proposed cross-subject protocol. Without GCNs and all perform better than the default skeleton-only approach using a single-stream AGCN. Additionally, another benchmark using GCNs on UTD-MHAD does not exist, thus, making a direct comparison of different approaches difficult. From the listing in Table 1a, it is clear that all fusion approaches skeleton and IMU modalities achieve the highest classification performance out of all methods introduced in this work. In comparison to the best performing fusion approach of skeleton with IMU nodes appended at its central node. MCRL [26] uses a fusion of skeleton, depth and RGB to reach 93.02% (−1.4%)

Table 1. Comparison to the state-of-the-art

Approach	Acc
Skeleton	92.32
RGB Patch Features R-18	27.67
RGB Encoder R-18	27.21
R(2+1)D	61.63
Skeleton + RGB Encoder R(2+1)D	91.62
Skeleton + RGB Encoder R-18	89.83
Skeleton + RGB Patch Features R-18	73.49
Skeleton + RGB Patch Features R-18 (no MLP)	44.60
Skeleton + IMU (Center)	94.42
Skeleton + IMU (Wrist/Hip)	94.07
Skeleton + IMU (Center + Add. Edges)	93.26
Skeleton + IMU (Wrist/Hip + Add. Edges)	93.26
Skeleton + IMU (Channel Fusion)	90.29
Skeleton + IMU + RGB Patch Features R-18	78.90
Skeleton + IMU + RGB Encoder R-18	92.33
Skeleton + IMU + RGB Encoder R(2+1)D	92.85
PoseMap [25]	**94.50**
Gimme Signals [32]	93.33
MCRL [26]	93.02

(a) UTD-MHAD

Approach	Acc	F1-measure
Skl	87.85	88.65
Skl+Acc(W+P)+Gyo+Ori	84.85	85.50
Skl+Acc(W+P)+Gyo+Ori (Add. Edges)	84.40	84.78
Skl+Acc(W)	**89.32**	89.55
Skl+Acc(P)	87.70	88.72
Skl+Gyo	86.35	87.41
Skl+Ori	87.65	88.64
Skl+Acc(W+P)	89.30	**89.60**
SMD [11] (Acc+RGB)	-	63.89
MMD [20] (Acc+Gyo+Ori+RGB)	-	64.33
MMAD [20] (Acc+Gyo+Ori+RGB)	-	66.45
Multi-GAT [16]	-	75.24
SAKDN [27]		77.23

(b) MMAct

validation accuracy on UTD-MHAD. Gimme Signals [32] reach 93.33% (−1.09%) using a CNN and augmented image representations of skeleton sequences. PoseMap [25] achieves 94.5% (+0.08%) accuracy using pose heatmaps generated from RGB videos. This method slightly outperforms the proposed fusion approach.

MMACT. To show better generalization, we also conducted experiments on the large-scale MMACT dataset which contains more modalities, classes and samples as the UTD-MHAD dataset. Note we only use the cross-subject protocol, the signal modalities can not be seperated by view. A comparison of approaches regarding the MMAct dataset is given in Table 1b. Kong et al. [20] propose along with the MMACT dataset the MMAD approach, a multimodal distillation method utiliziting an attention mechanism that incorporates acceleration, gyroscope, orientation and RGB. For evaluation, they use the F1-measure and reach an average of 66.45%. Without the attention mechanism, the approach (MMD) yields 64.33%. An approach utilizing the standard distillation approach Single Modality Distillation (SMD) yields 63.89%. The current baseline is set by SAKDN [27] which distills sensor information to enhance action recognition for the vision modality. Experiments show that the skeleton-based approach can be further improved by fusion with just the acceleration data to reach a recognition F1-measure of 89.60% (+12.37%). The MMACT dataset contains two accelerometers, where only the one from the smartwatch yields a mention-able improvement. The most significant improvement of our proposed approach is yielded by introducing the skeleton graph. In contrast, while the fusion approaches of skeleton and all four sensors do not improve the purely skeleton-based approach of 88.65% (+13.41%), with 85.5% (+10.26%) without additional edges and 84.78% (+9.54%) with additional edges, both reach a higher F1-measure than the baseline but also impact the pure skeleton-based recognition negatively.

4.4 Ablation Study

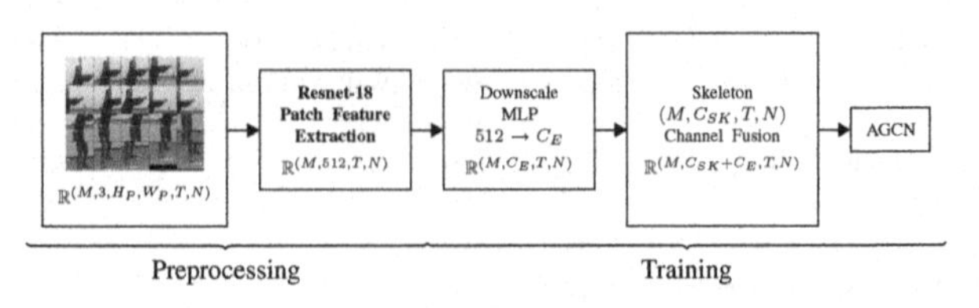

(a) Skeleton + RGB Patch Feature Fusion

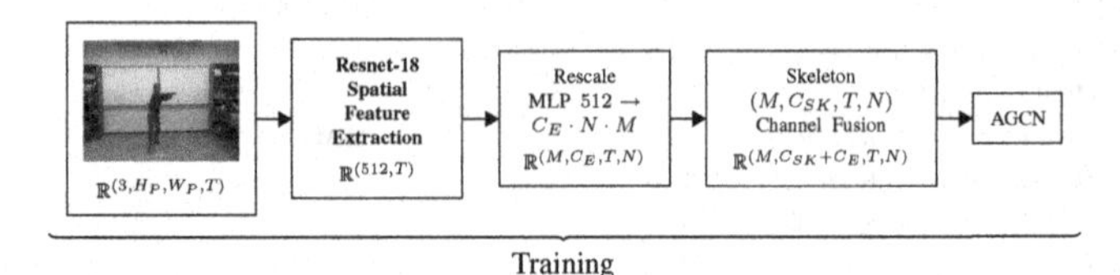

(b) Skeleton + Resnet-18 Generated Feature Fusion

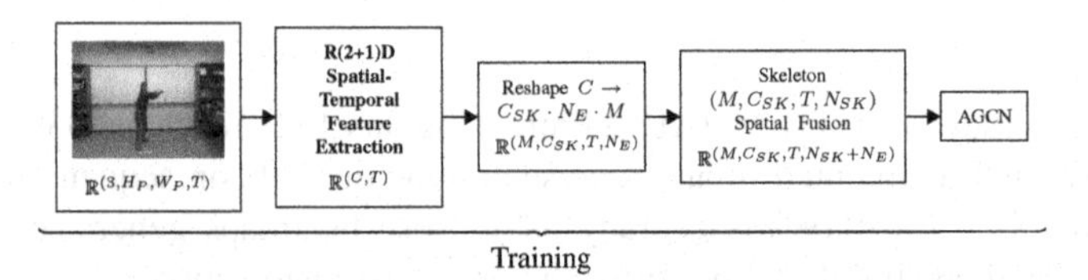

(c) Skeleton + Modified R(2+1)D Generated Feature Fusion

Fig. 4. The three different skeleton + RGB Fusion models with reference of an image from UTD-MHAD. The first model generates a feature for each node, while the last two generate a feature for the entire image that is distributed to the nodes and adjusted as part of the supervised training.

Fusion of Skeleton and RGB. Skeletons and RGB videos are combined using the three approaches depicted in Fig. 4. Figure 4a shows an approach using RGB patches that are cropped around each skeleton node and passed to a Resnet-18 to compute a feature vector $\boldsymbol{X}_{RGB} \in \mathbb{R}^{(M,512,T,N)}$ as part of preprocessing. The second approach, shown in Fig. 4b, uses Resnet-18 to compute a feature vector for each image. The resulting feature vector is rescaled to the size $C_E \cdot N \cdot M$ and reshaped to be able to be fused with skeleton data. Similarly, in Fig. 4c, the third approach uses R(2+1)D. In terms of parameters, the basis Skeleton model has 3.454.099 parameters, only 2.532 parameters are added for incorporation of inertial measurements into the model Skeleton+IMU(Center) 3.456.631 for a 2.2% accuracy improvement. Fusion with an RGB encoder adds five times more parameters (Skeleton+RGB Encoder Resnet-18 with 17.868.514) and a massive training overhead.

Table 1a shows that the RGB approaches viewed individually (without fusion) do not reach the performance of R(2+1)D pre-trained for action recognition. Results regarding the fusion models show a low accuracy of 73.49% for RGB patch features that have been created outside the training process and 44.6% for the same procedure without a downscaling Multilayer Perceptron (MLP). A similar conclusion can be drawn from the remaining two fusion models. Using R(2+1)D to produce features shows a slightly increased effectiveness of +1.79% (91.62%) over Resnet-18 (89.83%) but -0.7% in comparison to the solely skeleton-based approach.

Fusion of Skeleton and IMU. Fusion of skeletal and inertial sensor data is done according to Fig. 2. Figure 1 shows the skeleton structure of UTD-MHAD and illustrates two possibilities for fusing the red IMU graph nodes with the skeleton by connecting them to different skeleton joint nodes. In Fig. 1a, nodes are appended at the central skeleton joint as it is defined in ST-GCN and AGCN papers. The configuration depicted in Fig. 1b is attributed to the way sensors are worn by subjects of UTD-MHAD. This configuration is therefore not used for MMAct. Additional configurations arise when additional edges are drawn between the newly added nodes. According to Fig. 2b, another experiment involves broadcasting the $\mathbb{R}^6$-sized IMU feature vector to each skeleton joint and fuse them at channel dimension.

From the results in Table 1a, it is observable that all skeleton graphs with additional associated IMU nodes at each point in time improve the classification performance by at least one percent. In comparison to a skeleton-only approach, variants with additional edges between the newly added nodes perform generally worse than their not-connected counterparts and are both at 93.26% (+0.94%). The average classification accuracy of both other variants reaches 94.42% (+2.1%) and 94.07% (+1.75%). Despite having a slightly increased accuracy for appending new nodes to the existing central node, both variants almost reach equal performance and the location where nodes are appended seemingly does not matter much. While all experiments with fusion at spatial dimension show increased accuracies, the only experiment that does not surpass the skeleton-based approach is about fusion of both modalities at channel dimension, reaching 90.29% (−2.03%) accuracy.

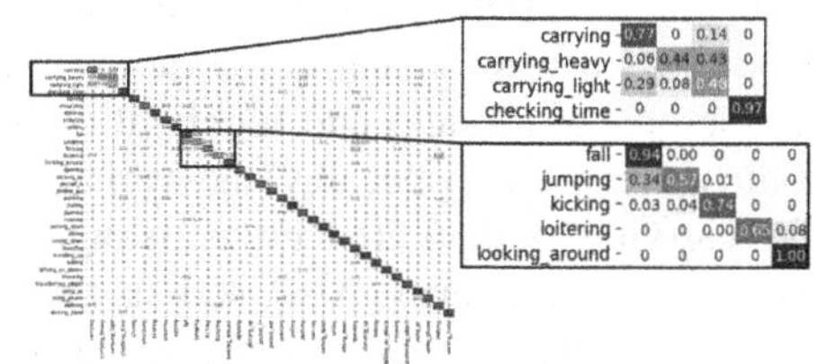

Fig. 5. Confusion matrix for the results on MMACT with the fusion of skeleton and accelerometer measurements from the smartwatch with highlighted high-confused actions.

For MMAct, all experiments are conducted using only the configuration in Fig. 1a and its variation with interconnected nodes. Table 1b shows that the skeleton-based approach reaches 87.85% accuracy for a cross-subject split, fusion approaches including all four sensors perform worse and reach only 84.85% (−3%) and 84.4% (−3.45%). Mixed results are achieved when individual sensors are not part of the fusion model. Fusion using only one of the phone's individual sensors, acceleration, gyroscope or orientation, reaches comparable results with 87.70% (−0.15%), 86.35% (−1.5%) and 87.65% (−0.2%) accuracy, respectively. On the contrary, performing a fusion of skeleton and acceleration data obtained by the smartwatch or with the fusion of both acceleration sensors shows an improved accuracy of 89.32% (1.47%) and 89.30% (1.45%).

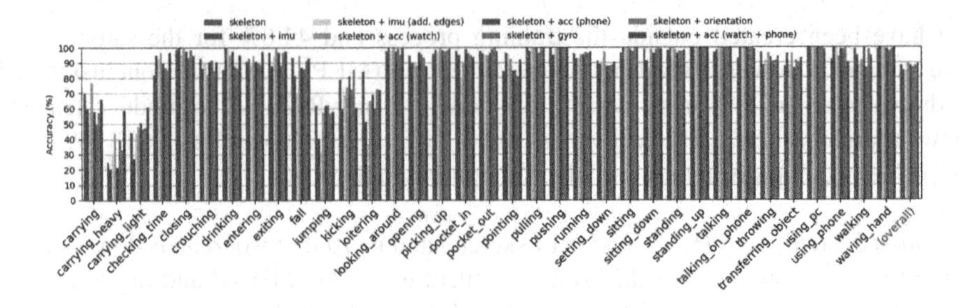

Fig. 6. Class specific accuracy for all MMACT classes for the fusion of various data modalities with Fusion-GCN.

Table 2 shows the top-5 improved classes by the fusion with the accelerometer measurements of a smartwatch. All the top-5 improved actions have a high arm movement in common. In Fig. 5 we give a confusion matrix for the Skeleton + Accelerometer (Watch) and highlight the most confused classes. Especially the variations of the "carrying" actions are hard to distinguish by their obvious similarity. Also, actions that contain sudden movements with high acceleration peaks are often confused ("jumping" is often considered as "falling"). In general, most of the activities can be recognized quite well. Figure 6 gives a general comparison of all class-specific results on different fusion experiments. Especially the fusion from skeleton-sequences with the accelerometer measurements (skeleton + acc (watch)) suggest a high improvement on many classes, especially the similar "carrying" classes.

Table 2. Top-5 most improved classes by the fusion of skeleton (Skl) and additional accelerometer (Acc) data from the smartwatch.

Class	Skl	Skl + Acc
carrying_heavy	24.69	43.83
checking_time	81.93	96.58
Drinking	85.00	95.00
transferring_object	87.23	96.10
Pointing	84.52	92.34

Fusion of Skeleton, RGB and IMU. One experiment is conducted using skeleton, RGB and IMU with IMU nodes appended to the skeleton central node without additional edges in combination with and all three RGB early fusion approaches. The results in Table 1a show that, like previously except for the RGB patch feature model, all models achieve an accuracy over 90%, albeit not reaching the same values as the skeleton and IMU fusion approach.

4.5 Limitations and Discussion

Comparing skeleton and skeleton + IMU, the fused approach generally has less misclassifications in all areas. Especially similar actions, such as "throw", "catch", "knock" or "tennis swing", are able to be classified more confidently. The only action with decreased recognition accuracy using the fused approach is "jog" which is misclassified more often as "walk", two similar actions and some of the few with sparse involvement of arm movement. Common problems for all RGB approaches regarding UTD-MHAD are a small number of training samples, resulting in overfitting in some cases that can not be lifted by either weight decay or dropout. Another fact is the absence of object interactions in UTD-MHAD. With the exception of "sit2stand" and "stand2sit", actions such as "throwing", "catching", "pickup_throw" or sports activities never include any objects. As pointed out previously, skeleton is focused purely on human movements and, by that, omits all other objects inside of a scene. RGB still contains such visual information, making it supposedly more efficient in recognizing object interactions. In contrast, many of MMACT's actions, like "transferring_object", "using_pc", "using_phone" or "carrying", make use of real objects. While fusion with RGB modality achieves similar accuracies as other approaches, incorporating the data into the network increases the training time by up to a magnitude of ten; hence, the RGB fusion models do not provide a viable alternative to skeleton and IMU regarding the current preprocessing and training configurations. Therefore, due to timely constraints, experiments for fusion of skeleton and RGB modalities on the larger dataset MMACT are omitted.

5 Conclusion

With Fusion-GCN, we presented an approach for multimodal action recognition using GCNs. To incorporate additional modalities we suggest two different fusion dimensions, either on a channel- or spatial dimension. Further integration into early- and late fusion approaches have been presented. In our experiments we considered the flexible fusion of skeleton sequences, with inertial measurements, accelerometer-, gyro-, orientation- measurements separately, as well as RGB features. Our presented fusion approach successfully improved the previous baselines on the large-scale MMACT dataset by a significant margin. Further, it was showcased that additional modalities can further improve recognition from skeleton-sequences. However, the addition of too many modalities decreased the performance. We believe that Fusion-GCN demonstrated successfully that GCNs serve as good basis for multimodal action recognition and could potentially guide future research in this domain.

References

1. Cai, J., Jiang, N., Han, X., Jia, K., Lu, J.: JOLO-GCN: mining joint-centered light-weight information for skeleton-based action recognition. In: IEEE Winter Conference on Applications of Computer Vision, WACV 2021, Waikoloa, HI, USA, 3–8 January 2021, pp. 2734–2743. IEEE (2021). https://doi.org/10.1109/WACV48630.2021.00278
2. Cao, C., Zhang, Y., Zhang, C., Lu, H.: Action recognition with joints-pooled 3d deep convolutional descriptors. In: Kambhampati, S. (ed.) Proceedings of the Twenty-Fifth International Joint Conference on Artificial Intelligence, IJCAI 2016, New York, NY, USA, 9–15 July 2016, pp. 3324–3330. IJCAI/AAAI Press (2016). http://www.ijcai.org/Abstract/16/470
3. Cao, C., Zhang, Y., Zhang, C., Lu, H.: Body joint guided 3-D deep convolutional descriptors for action recognition. IEEE Trans. Cybern. **48**(3), 1095–1108 (2018)
4. Cao, Z., Hidalgo, G., Simon, T., Wei, S., Sheikh, Y.: OpenPose: realtime multi-person 2D pose estimation using part affinity fields. IEEE Trans. Pattern Anal. Mach. Intell. **43**(1), 172–186 (2021)

5. Chen, C., Jafari, R., Kehtarnavaz, N.: UTD-MHAD: a multimodal dataset for human action recognition utilizing a depth camera and a wearable inertial sensor. In: 2015 IEEE International Conference on Image Processing, ICIP 2015, Quebec City, QC, Canada, 27–30 September 2015, pp. 168–172. IEEE (2015). https://doi.org/10.1109/ICIP.2015.7350781
6. Cheng, K., Zhang, Y., He, X., Chen, W., Cheng, J., Lu, H.: Skeleton-based action recognition with shift graph convolutional network. In: 2020 IEEE/CVF Conference on Computer Vision and Pattern Recognition, CVPR 2020, Seattle, WA, USA, 13–19 June 2020, pp. 180–189. IEEE (2020). https://doi.org/10.1109/CVPR42600.2020.00026
7. Chéron, G., Laptev, I., Schmid, C.: P-CNN: pose-based CNN features for action recognition. In: 2015 IEEE International Conference on Computer Vision, ICCV 2015, Santiago, Chile, 7–13 December 2015, pp. 3218–3226. IEEE Computer Society (2015). https://doi.org/10.1109/ICCV.2015.368
8. Duong, T.V., Bui, H.H., Phung, D.Q., Venkatesh, S.: Activity recognition and abnormality detection with the switching hidden semi-Markov model. In: 2005 IEEE Computer Society Conference on Computer Vision and Pattern Recognition (CVPR 2005), 20–26 June 2005, San Diego, CA, USA, pp. 838–845. IEEE Computer Society (2005). https://doi.org/10.1109/CVPR.2005.61
9. Gao, Y., et al.: Human action monitoring for healthcare based on deep learning. IEEE Access **6**, 52277–52285 (2018)
10. Goh, G., et al.: Multimodal neurons in artificial neural networks. Distill **6**(3), e30 (2021)
11. Hinton, G.E., Vinyals, O., Dean, J.: Distilling the knowledge in a neural network. CoRR abs/1503.02531 (2015). http://arxiv.org/abs/1503.02531
12. Hu, W., Xie, D., Fu, Z., Zeng, W., Maybank, S.J.: Semantic-based surveillance video retrieval. IEEE Trans. Image Process. **16**(4), 1168–1181 (2007)
13. Huynh, H.H., Meunier, J., Sequeira, J., Daniel, M.: Real time detection, tracking and recognition of medication intake. Int. J. Comput. Inf. Eng. **3**(12), 2801–2808 (2009). https://publications.waset.org/vol/36
14. Iqbal, U., Doering, A., Yasin, H., Krüger, B., Weber, A., Gall, J.: A dual-source approach for 3D human pose estimation from single images. Comput. Vis. Image Underst. **172**, 37–49 (2018)
15. Islam, M.M., Iqbal, T.: HAMLET: a hierarchical multimodal attention-based human activity recognition algorithm. CoRR abs/2008.01148 (2020). https://arxiv.org/abs/2008.01148
16. Islam, M.M., Iqbal, T.: Multi-GAT: a graphical attention-based hierarchical multimodal representation learning approach for human activity recognition. IEEE Robot. Autom. Lett. **6**(2), 1729–1736 (2021). https://doi.org/10.1109/LRA.2021.3059624
17. Joze, H.R.V., Shaban, A., Iuzzolino, M.L., Koishida, K.: MMTM: multimodal transfer module for CNN fusion. In: 2020 IEEE/CVF Conference on Computer Vision and Pattern Recognition, CVPR 2020, Seattle, WA, USA, 13–19 June 2020, pp. 13286–13296. IEEE (2020). https://doi.org/10.1109/CVPR42600.2020.01330
18. Kingma, D.P., Ba, J.: Adam: a method for stochastic optimization. In: Bengio, Y., LeCun, Y. (eds.) 3rd International Conference on Learning Representations, ICLR 2015, San Diego, CA, USA, 7–9 May 2015, Conference Track Proceedings (2015). http://arxiv.org/abs/1412.6980
19. Kipf, T.N., Welling, M.: Semi-supervised classification with graph convolutional networks. In: 5th International Conference on Learning Representations, ICLR 2017, Toulon, France, 24–26 April 2017, Conference Track Proceedings. OpenReview.net (2017). https://openreview.net/forum?id=SJU4ayYgl
20. Kong, Q., Wu, Z., Deng, Z., Klinkigt, M., Tong, B., Murakami, T.: MMAct: a large-scale dataset for cross modal human action understanding. In: 2019 IEEE/CVF International Conference on Computer Vision, ICCV 2019, Seoul, Korea (South), 27 October–2 November 2019, pp. 8657–8666. IEEE (2019). https://doi.org/10.1109/ICCV.2019.00875

21. Kotyan, S., Kumar, N., Sahu, P.K., Udutalapally, V.: HAUAR: home automation using action recognition. CoRR abs/1904.10354 (2019). http://arxiv.org/abs/1904.10354
22. Kreiss, S., Bertoni, L., Alahi, A.: PifPaf: composite fields for human pose estimation. In: IEEE Conference on Computer Vision and Pattern Recognition, CVPR 2019, Long Beach, CA, USA, 16–20 June 2019, pp. 11977–11986. Computer Vision Foundation/IEEE (2019). https://doi.org/10.1109/CVPR.2019.01225. http://openaccess.thecvf.com/content_CVPR_2019/html/Kreiss_PifPaf_Composite_Fields_for_Human_Pose_Estimation_CVPR_2019_paper.html
23. Li, B., Li, X., Zhang, Z., Wu, F.: Spatio-temporal graph routing for skeleton-based action recognition. In: The Thirty-Third AAAI Conference on Artificial Intelligence, AAAI 2019, The Thirty-First Innovative Applications of Artificial Intelligence Conference, IAAI 2019, The Ninth AAAI Symposium on Educational Advances in Artificial Intelligence, EAAI 2019, Honolulu, Hawaii, USA, 27 January–1 February 2019, pp. 8561–8568. AAAI Press (2019). https://doi.org/10.1609/aaai.v33i01.33018561
24. Li, J., Xie, X., Pan, Q., Cao, Y., Zhao, Z., Shi, G.: SGM-net: skeleton-guided multimodal network for action recognition. Pattern Recognit. **104**, 107356 (2020)
25. Liu, M., Yuan, J.: Recognizing human actions as the evolution of pose estimation maps. In: 2018 IEEE Conference on Computer Vision and Pattern Recognition, CVPR 2018, Salt Lake City, UT, USA, 18–22 June 2018, pp. 1159–1168. IEEE Computer Society (2018). https://doi.org/10.1109/CVPR.2018.00127. http://openaccess.thecvf.com/content_cvpr_2018/html/Liu_Recognizing_Human_Actions_CVPR_2018_paper.html
26. Liu, T., Kong, J., Jiang, M.: RGB-D action recognition using multimodal correlative representation learning model. IEEE Sens. J. **19**(5), 1862–1872 (2019). https://doi.org/10.1109/JSEN.2018.2884443
27. Liu, Y., Wang, K., Li, G., Lin, L.: Semantics-aware adaptive knowledge distillation for sensor-to-vision action recognition. IEEE Trans. Image Process. **30**, 5573–5588 (2021)
28. Loshchilov, I., Hutter, F.: SGDR: stochastic gradient descent with warm restarts. In: 5th International Conference on Learning Representations, ICLR 2017, Toulon, France, 24–26 April 2017, Conference Track Proceedings. OpenReview.net (2017). https://openreview.net/forum?id=Skq89Scxx
29. Lugaresi, C., et al.: MediaPipe: a framework for building perception pipelines. CoRR abs/1906.08172 (2019). http://arxiv.org/abs/1906.08172
30. Luo, Z., Hsieh, J.-T., Jiang, L., Niebles, J.C., Fei-Fei, L.: Graph distillation for action detection with privileged modalities. In: Ferrari, V., Hebert, M., Sminchisescu, C., Weiss, Y. (eds.) ECCV 2018, Part XIV. LNCS, vol. 11218, pp. 174–192. Springer, Cham (2018). https://doi.org/10.1007/978-3-030-01264-9_11
31. Mehta, D., et al.: XNect: real-time multi-person 3D motion capture with a single RGB camera. ACM Trans. Graph. **39**(4), 82 (2020)
32. Memmesheimer, R., Theisen, N., Paulus, D.: Gimme signals: discriminative signal encoding for multimodal activity recognition. In: IEEE/RSJ International Conference on Intelligent Robots and Systems, IROS 2020, Las Vegas, NV, USA, 24 October 2020–24 January 2021, pp. 10394–10401. IEEE (2020). https://doi.org/10.1109/IROS45743.2020.9341699
33. Ni, B., Yan, S., Kassim, A.A.: Recognizing human group activities with localized causalities. In: 2009 IEEE Computer Society Conference on Computer Vision and Pattern Recognition (CVPR 2009), 20–25 June 2009, Miami, Florida, USA, pp. 1470–1477. IEEE Computer Society (2009). https://doi.org/10.1109/CVPR.2009.5206853
34. Niu, W., Long, J., Han, D., Wang, Y.F.: Human activity detection and recognition for video surveillance. In: Proceedings of the 2004 IEEE International Conference on Multimedia and Expo, ICME 2004, 27–30 June 2004, Taipei, Taiwan, pp. 719–722. IEEE Computer Society (2004)

35. Norcliffe-Brown, W., Vafeias, S., Parisot, S.: Learning conditioned graph structures for interpretable visual question answering. In: Bengio, S., Wallach, H.M., Larochelle, H., Grauman, K., Cesa-Bianchi, N., Garnett, R. (eds.) Advances in Neural Information Processing Systems 31: Annual Conference on Neural Information Processing Systems 2018, NeurIPS 2018, 3–8 December 2018, Montréal, Canada, pp. 8344–8353 (2018). https://proceedings.neurips.cc/paper/2018/hash/4aeae10ea1c6433c926cdfa558d31134-Abstract.html
36. Noury, N., et al.: Fall detection-principles and methods. In: 2007 29th Annual International Conference of the IEEE Engineering in Medicine and Biology Society, pp. 1663–1666. IEEE (2007)
37. Papadopoulos, K., Ghorbel, E., Aouada, D., Ottersten, B.E.: Vertex feature encoding and hierarchical temporal modeling in a spatial-temporal graph convolutional network for action recognition. CoRR abs/1912.09745 (2019). http://arxiv.org/abs/1912.09745
38. Peng, W., Hong, X., Chen, H., Zhao, G.: Learning graph convolutional network for skeleton-based human action recognition by neural searching. In: The Thirty-Fourth AAAI Conference on Artificial Intelligence, AAAI 2020, The Thirty-Second Innovative Applications of Artificial Intelligence Conference, IAAI 2020, The Tenth AAAI Symposium on Educational Advances in Artificial Intelligence, EAAI 2020, New York, NY, USA, 7–12 February 2020, pp. 2669–2676. AAAI Press (2020). https://aaai.org/ojs/index.php/AAAI/article/view/5652
39. Perez-Rua, J., Vielzeuf, V., Pateux, S., Baccouche, M., Jurie, F.: MFAS: multimodal fusion architecture search. In: IEEE Conference on Computer Vision and Pattern Recognition, CVPR 2019, Long Beach, CA, USA, 16–20 June 2019, pp. 6966–6975. Computer Vision Foundation/IEEE (2019). https://doi.org/10.1109/CVPR.2019.00713. http://openaccess.thecvf.com/content_CVPR_2019/html/Perez-Rua_MFAS_Multimodal_Fusion_Architecture_Search_CVPR_2019_paper.html
40. Ramezani, M., Yaghmaee, F.: A review on human action analysis in videos for retrieval applications. Artif. Intell. Rev. **46**(4), 485–514 (2016)
41. Ryoo, M.S., Fuchs, T.J., Xia, L., Aggarwal, J.K., Matthies, L.H.: Robot-centric activity prediction from first-person videos: what will they do to me? In: Adams, J.A., Smart, W.D., Mutlu, B., Takayama, L. (eds.) Proceedings of the Tenth Annual ACM/IEEE International Conference on Human-Robot Interaction, HRI 2015, Portland, OR, USA, 2–5 March 2015, pp. 295–302. ACM (2015). https://doi.org/10.1145/2696454.2696462
42. Shahroudy, A., Liu, J., Ng, T.T., Wang, G.: NTU RGB+D: a large scale dataset for 3D human activity analysis. CoRR abs/1604.02808 (2016). http://arxiv.org/abs/1604.02808
43. Shi, L., Zhang, Y., Cheng, J., Lu, H.: Two-stream adaptive graph convolutional networks for skeleton-based action recognition. In: IEEE Conference on Computer Vision and Pattern Recognition, CVPR 2019, Long Beach, CA, USA, 16–20 June 2019, pp. 12026–12035. Computer Vision Foundation/IEEE (2019). https://doi.org/10.1109/CVPR.2019.01230. http://openaccess.thecvf.com/content_CVPR_2019/html/Shi_Two-Stream_Adaptive_Graph_Convolutional_Networks_for_Skeleton-Based_Action_Recognition_CVPR_2019_paper.html
44. Shotton, J., et al.: Real-time human pose recognition in parts from single depth images. In: Cipolla, R., Battiato, S., Farinella, G.M. (eds.) Machine Learning for Computer Vision. Studies in Computational Intelligence, vol. 411, pp. 119–135. Springer, Heidelberg (2013). https://doi.org/10.1007/978-3-642-28661-2_5
45. Solbach, M.D., Tsotsos, J.K.: Vision-based fallen person detection for the elderly. In: 2017 IEEE International Conference on Computer Vision Workshops, ICCV Workshops 2017, Venice, Italy, 22–29 October 2017, pp. 1433–1442. IEEE Computer Society (2017). https://doi.org/10.1109/ICCVW.2017.170
46. Song, S., Lan, C., Xing, J., Zeng, W., Liu, J.: Skeleton-indexed deep multi-modal feature learning for high performance human action recognition. In: 2018 IEEE International Conference on Multimedia and Expo, ICME 2018, San Diego, CA, USA, 23–27 July 2018, pp. 1–6. IEEE Computer Society (2018). https://doi.org/10.1109/ICME.2018.8486486

47. Song, Y., Zhang, Z., Shan, C., Wang, L.: Stronger, faster and more explainable: a graph convolutional baseline for skeleton-based action recognition. In: Chen, C.W., et al. (eds.) MM 2020: The 28th ACM International Conference on Multimedia, Virtual Event/Seattle, WA, USA, 12–16 October 2020, pp. 1625–1633. ACM (2020). https://doi.org/10.1145/3394171.3413802
48. Tran, D., Wang, H., Torresani, L., Ray, J., LeCun, Y., Paluri, M.: A closer look at spatiotemporal convolutions for action recognition. In: 2018 IEEE Conference on Computer Vision and Pattern Recognition, CVPR 2018, Salt Lake City, UT, USA, 18–22 June 2018, pp. 6450–6459. IEEE Computer Society (2018). https://doi.org/10.1109/CVPR.2018.00675. http://openaccess.thecvf.com/content_cvpr_2018/html/Tran_A_Closer_Look_CVPR_2018_paper.html
49. Tripathi, R.K., Jalal, A.S., Agrawal, S.C.: Suspicious human activity recognition: a review. Artif. Intell. Rev. **50**(2), 283–339 (2018)
50. Wang, X., Gupta, A.: Videos as space-time region graphs. In: Ferrari, V., Hebert, M., Sminchisescu, C., Weiss, Y. (eds.) ECCV 2018, Part V. LNCS, vol. 11209, pp. 413–431. Springer, Cham (2018). https://doi.org/10.1007/978-3-030-01228-1_25
51. Yan, S., Xiong, Y., Lin, D.: Spatial temporal graph convolutional networks for skeleton-based action recognition. In: McIlraith, S.A., Weinberger, K.Q. (eds.) Proceedings of the Thirty-Second AAAI Conference on Artificial Intelligence, (AAAI-18), the 30th Innovative Applications of Artificial Intelligence (IAAI-18), and the 8th AAAI Symposium on Educational Advances in Artificial Intelligence (EAAI-18), New Orleans, Louisiana, USA, 2–7 February 2018, pp. 7444–7452. AAAI Press (2018). https://www.aaai.org/ocs/index.php/AAAI/AAAI18/paper/view/17135
52. Zheng, Y., Bao, H., Xu, C.: A method for improved pedestrian gesture recognition in self-driving cars. Aust. J. Mech. Eng. **16**(sup1), 78–85 (2018)

FIFA: Fast Inference Approximation for Action Segmentation

Yaser Souri[1(✉)], Yazan Abu Farha[1], Fabien Despinoy[2], Gianpiero Francesca[2], and Juergen Gall[1]

[1] University of Bonn, Bonn, Germany
{souri,abufarha,gall}@iai.uni-bonn.de
[2] Toyota Motor Europe, Brussels, Belgium
{fabien.despinoy,gianpiero.francesca}@toyota-motor.com

Abstract. We introduce FIFA, a fast approximate inference method for action segmentation and alignment. Unlike previous approaches, FIFA does not rely on expensive dynamic programming for inference. Instead, it uses an approximate differentiable energy function that can be minimized using gradient-descent. FIFA is a general approach that can replace exact inference, improving its speed by more than 5 times while maintaining its performance. FIFA is an anytime inference algorithm that provides a better speed vs. accuracy trade-off compared to exact inference. We apply FIFA on top of state-of-the-art approaches for weakly supervised action segmentation and alignment as well as fully supervised action segmentation. FIFA achieves state-of-the-art results for most metrics on two action segmentation datasets.

Keywords: Video understanding · Action segmentation · Approximate inference

1 Introduction

Action segmentation is the task of predicting the action label for each frame in the input video. Action segmentation is usually studied in the context of activities performed by a single person, where temporal smoothness of actions are assumed. Fully supervised approaches for action segmentation [1,21,26,34] already achieve good performance on this task. Most approaches for fully supervised action segmentation make frame-wise predictions [1,21,26] while trying to model the temporal relationship between the action labels. These approaches usually suffer from over-segmentation. Recent works [13,34] try to overcome the over-segmentation problem by finding the action boundaries and temporally smoothing the predictions inside each action segment. But these post-processing approaches still can not guarantee temporal smoothness.

Supplementary Information The online version contains supplementary material available at https://doi.org/10.1007/978-3-030-92659-5_18.

C. Bauckhage et al. (Eds.): DAGM GCPR 2021, LNCS 13024, pp. 282–296, 2021.
https://doi.org/10.1007/978-3-030-92659-5_18

Action segmentation inference is the problem of making segment-wise smooth predictions from frame-wise probabilities given a known grammar of the actions and their average lengths [30]. The typical inference in action segmentation involves solving an expensive Viterbi-like dynamic programming problem that finds the best action sequence and its corresponding lengths. In the literature, weakly supervised action segmentation approaches [18,25,29,30,32] usually use inference at test time. Despite being very useful for action segmentation, the inference problem remains the main computational bottleneck in the action segmentation pipeline [32].

In this paper, we propose FIFA, a fast anytime approximate inference procedure that achieves comparable performance with respect to the dynamic programming based Viterbi decoding inference at a fraction of the computational time. Instead of relying on dynamic programming, we formulate the energy function as an approximate differentiable function of segment lengths parameters and use gradient-descent-based methods to search for a configuration that minimizes the approximate energy function. Given a transcript of actions and the corresponding initial lengths configuration, we define the energy function as a sum over segment level energies. The segment level energy consists of two terms: a length energy term that penalizes the deviations from a global length model and an observation energy term that measures the compatibility between the current configuration and the predicted frame-wise probabilities. A naive approach to model the observation energy would be to sum up the negative log probabilities of the action labels that are defined based on the length configuration. Nevertheless, such an approach is not differentiable with respect to the segment lengths. In order to optimize the energy using gradient descent-based methods, the observation energy has to be differentiable with respect to the segment lengths. To this end, we construct a plateau-shaped mask for each segment which temporally locates the segment within the video. This mask is parameterized by the segment lengths, the position in the video, and a sharpness parameter. The observation energy is then defined as a product of a segment mask and the predicted frame-wise negative log probabilities, followed by a sum pooling operation. Finally, a gradient descent-based method is used to find a configuration for the segment lengths that minimizes the total energy.

FIFA is a general inference approach and can be applied at test time on top of different action segmentation approaches for fast inference. We evaluate our approach on top of the state-of-the-art methods for weakly supervised temporal action segmentation, weakly supervised action alignment, and fully supervised action segmentation. Results on the Breakfast [16] and Hollywood extended [4] datasets show that FIFA achieves state-of-the-art results on most metrics. Compared to the exact inference using the Viterbi decoding, FIFA is at least 5 times faster. Furthermore, FIFA is an anytime algorithm which can be stopped after each step of the gradient-based optimization, therefore it provides a better speed vs. accuracy trade-off compared to exact inference.

2 Related Work

In this section we highlight relevant works addressing fully and weakly supervised action segmentation that have been recently proposed.

Fully Supervised Action Segmentation. In fully supervised action segmentation, frame-level labels are used for training. Initial attempts for action segmentation applied action classifiers on a sliding window over the video frames [15,31]. However, these approaches did not capture the dependencies between the action segments. With the objective of capturing the context over long video sequences, context free grammars [28,33] or hidden Markov models (HMMs) [17,19,22] are typically combined with frame-wise classifiers. Recently, temporal convolutional networks showed good performance for the temporal action segmentation task using encoder-decoder architectures [21,24] or even multi-stage architectures [1,26]. Many approaches further improve the multi-stage architectures by applying post-processing based on boundary-aware pooling operation [13,34] or graph-based reasoning [12]. Without any inference most of the fully-supervised approaches therefore suffer from oversegmentation at test time.

Weakly Supervised Action Segmentation. To reduce the annotation cost, many approaches that rely on a weaker form of supervision have been proposed. Earlier approaches apply discriminative clustering to align video frames to movie scripts [7]. Bojanowski *et al.* [4] proposed to use as supervision the transcripts in the form of ordered lists of actions. Indeed, many approaches rely on this form of supervision to train a segmentation model using connectionist temporal classification [11], dynamic time warping [5] or energy-based learning [25]. In [6], an iterative training procedure is used to refine the transcript. A soft labeling mechanism is further applied at the boundaries between action segments. Kuehne *et al.* [18] applied a speech recognition system based on a HMM and Gaussian mixture model (GMM) to align video frames to transcripts. The approach generates pseudo ground truth labels for the training videos and iteratively refines them. A similar idea has been recently used in [19,29]. Richard *et al.* [30] combined the frame-wise loss function with the Viterbi algorithm to generate the target labels. At inference time, these approaches iterate over the training transcripts and select the one that matches the test video best. By contrast, Souri *et al.* [32] predict the transcript besides the frame-wise scores at inference time. State-of-the-art weakly supervised action segmentation approaches require time consuming dynamic programming based inference at test time.

Energy-Based Inference. In energy-based inference methods, gradient descent is used at inference time as described in [23]. The goal is to minimize an energy function that measures the compatibility between the input variables and the predicted variables. This idea has been exploited for many structured prediction tasks such as image generation [8,14], machine translation [10] and structured prediction energy networks [3]. Belanger and McCallum [2] relaxed the discrete output space for multi-label classification tasks to a continuous space and used gradient descent to approximate the solution. Gradient-based methods have also been used for other applications such as generating adversarial examples [9] and learning text embeddings [20].

3 Background

The following sections introduce all the concepts and notations required to understand the proposed FIFA methodology.

3.1 Action Segmentation

In action segmentation, we want to temporally localize all the action segments occurring in a video. In this paper, we consider the case where the actions are from a predefined set of M classes (a background class is used to cover uninteresting parts of a video). The input video of length T is usually represented as a set of d dimensional features vectors $x_{1:T} = (x_1, \ldots, x_T)$. These features are extracted offline and are assumed to be the input to the action segmentation model. The output of action segmentation can be represented in two ways:

- Frame-wise representation $y_{1:T} = (y_1, \ldots, y_T)$ where y_t represents the action label at time t.
- Segment-wise representation $s_{1:N} = (s_1, \ldots, s_N)$ where segment s_n is represented by both the action label of the segment c_n and its corresponding length ℓ_n, i.e., $s_n = (c_n, \ell_n)$. The ordered list of actions $c_{1:N}$ is usually referred to as the *transcript*.

These two representations are equal and redundant, i.e., it is possible to compute one from the other. In order to transfer from the segment-wise to the frame-wise representation, we introduce a mapping $\alpha(t; c_{1:N}, \ell_{1:N})$ which outputs the action label at frame t given the segment-wise labeling.

The target labels to train a segmentation model, depend on the level of supervision. In fully supervised action segmentation [1,26,34], the target label for each frame is provided. However, in weakly supervised approaches [25,30,32] only the ordered list of action labels are provided during training while their lengths are unknown.

Recent fully supervised approaches for action segmentation like MSTCN [1] and its variants directly predict the frame-wise representation $y_{1:T}$ by choosing the action label with the highest probability for each frame independently. This results in predictions that are sometimes oversegmented.

Conversely, recent weakly supervised action segmentation approaches like NNV [30] and follow-up work include an inference stage during testing where they explicitly predict the segment-wise representation. This inference stage involves a dynamic programming algorithm for solving an optimization problem which is a computational bottleneck for these approaches.

3.2 Inference in Action Segmentation

During testing, the inference stage involves an optimization problem to find the most likely segmentation for the input video, i.e.,

$$c_{1:N}, \ell_{1:N} = \underset{\hat{c}_{1:N}, \hat{\ell}_{1:N}}{\operatorname{argmax}} \Big\{ p(\hat{c}_{1:N}, \hat{\ell}_{1:N} | x_{1:T}) \Big\}. \tag{1}$$

Given the transcript $c_{1:N}$, the inference stage boils down to finding the segment lengths $\ell_{1:N}$ by aligning the transcript to the input video, i.e.,

$$\ell_{1:N} = \underset{\hat{\ell}_{1:N}}{\operatorname{argmax}} \Big\{ p(\hat{\ell}_{1:N} | x_{1:T}, c_{1:N}) \Big\}. \tag{2}$$

In approaches like NNV [30] and CDFL [25], the transcript is found by iterating over the transcripts seen during training and selecting the transcript that achieves the most likely alignment by optimizing (2). In MuCon [32], the transcript is predicted by a sequence to sequence network.

The probability defined in (2) is broken down by making independences assumption between frames

$$p(\hat{\ell}_{1:N} | x_{1:T}, c_{1:N}) = \prod_{t=1}^{T} p(\alpha(t; c_{1:N}, \hat{\ell}_{1:N}) | x_t) \cdot \prod_{n=1}^{N} p(\hat{\ell}_n | c_n) \tag{3}$$

where $p(\alpha(t)|x_t)$ is referred to as the observation model and $p(\ell_n|c_n)$ as the length model. Here, $\alpha(t)$ is the mapping from time t to the action label given the segment-wise labeling. The observation model estimates the frame-wise action probabilities and is implemented using a neural network. The length model is used to constrain the inference defined in (2) with the assumption that the lengths of segments for the same action follow a particular probability distribution. The segment length is usually modelled by a Poisson distribution with a class dependent mean parameter λ_{c_n}, i.e.,

$$p(\ell_n | c_n) = \frac{\lambda_{c_n}^{\ell_n} \exp(-\lambda_{c_n})}{\ell_n!}. \tag{4}$$

This optimization is solved using an expensive dynamic programming based Viterbi decoding [30]. For details on how to solve this optimization problem using Viterbi decoding please refer to the supplementary material.

4 FIFA: Fast Inference Approximation

Our goal is to introduce a fast inference algorithm for action segmentation. We want the fast inference to be applicable in both weakly supervised and fully supervised action segmentation. We also want the fast inference to be flexible enough to work with different action segmentation methods. To this end, we introduce FIFA, a novel approach for fast inference for action segmentation.

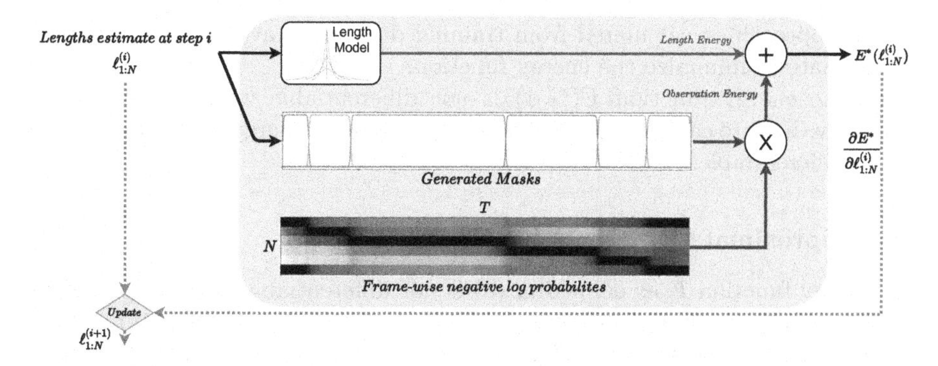

Fig. 1. Overview of the FIFA optimization process. At each step in the optimization, a set of masks are generated using the current length estimates. Using the generated masks and the frame-wise negative log probabilities, the observation energy is calculated in an approximate but differentiable manner. The length energy is calculated from the current length estimate and added to the observation energy to calculate the total energy value. Taking the gradient of the total energy with respect to the length estimates we can update it using a gradient step.

In the following, for brevity, we write the mapping $\alpha(t; c_{1:N}, \ell_{1:N})$ simply as $\alpha(t)$. Maximizing probability (2) can be rewritten as minimizing the negative logarithm of that probability

$$\text{argmax}\Big\{p(\hat{\ell}_{1:N}|x_{1:T}, c_{1:N})\Big\} = \text{argmin}\Big\{-\log\big(p(\hat{\ell}_{1:N}|x_{1:T}, c_{1:N})\big)\Big\} \tag{5}$$

which we refer to as the energy $E(\ell_{1:N})$. Using (3) the energy is rewritten as

$$\begin{aligned} E(\ell_{1:N}) &= -\log\Big(p(\ell_{1:N}|x_{1:T}, c_{1:N})\Big) \\ &= -\log\Big(\prod_{t=1}^{T} p\big(\alpha(t)|x_t\big) \cdot \prod_{n=1}^{N} p\big(\ell_n|c_n\big)\Big) \\ &= \underbrace{\sum_{t=1}^{T} -\log p\big(\alpha(t)|x_t\big)}_{E_o} + \underbrace{\sum_{n=1}^{N} -\log p\big(\ell_n|c_n\big)}_{E_\ell}. \end{aligned} \tag{6}$$

The first term in (6), E_o is referred to as the observation energy. This term calculates the cost of assigning the labels for each frame and is calculated from the frame-wise probability estimates. The second term E_ℓ is referred to as the length energy. This term is the cost of each segment having a length given that we assume an average length for actions of a specific class.

We propose to optimize the energy defined in (6) using gradient based optimization in order to avoid the need for time-consuming dynamic programming. We start with an initial estimate of the lengths (obtained from the length model of each approach or calculated from training data when available) and update our estimate to minimize the energy function.

As the energy function $E(\ell_{1:N})$ is not differentiable with respect to the lengths, we have to calculate a relaxed and approximate energy function $E^*(\ell_{1:N})$ that is differentiable.

4.1 Approximate Differentiable Energy E^*

The energy function E as defined in (6) is not differentiable in two parts. First the observation energy term E_o is not differentiable because of the $\alpha(t)$ function. Second, the length energy term E_ℓ is not differentiable because it expects natural numbers as input and cannot be computed on real values which are dealt with in gradient-based optimization. Below we describe how we approximate and make each of the terms differentiable.

Approximate Differentiable Observation Energy. Consider a $N \times T$ matrix P containing negative log probabilities, i.e.,

$$P[n,t] = -\log p(c_n|x_t). \tag{7}$$

Furthermore, we define a mask matrix M with the same size $N \times T$ where

$$M[n,t] = \begin{cases} 0 & if\ \alpha(t) \neq c_n \\ 1 & if\ \alpha(t) = c_n \end{cases}. \tag{8}$$

Using the mask matrix we can rewrite the observation energy term as

$$E_o = \sum_{t=1}^{T} \sum_{n=1}^{N} M[n,t] \cdot P[n,t]. \tag{9}$$

In order to make the observation energy term differentiable with respect to the length, we propose to construct an approximate differentiable mask matrix M^*. We use the following smooth and parametric plateau function

$$f(t|\lambda^c, \lambda^w, \lambda^s) = \frac{1}{(e^{\lambda^s(t-\lambda^c-\lambda^w)}+1)(e^{\lambda^s(-t+\lambda^c-\lambda^w)}+1)} \tag{10}$$

from [27]. This plateau function has three parameters and it is differentiable with respect to them: λ^c controls the center of the plateau, λ^w is the width and λ^s is the sharpness of the plateau function.

While the sharpness of the plateau functions λ^s is fixed as a hyper-parameter of our approach, the center λ^c and the width λ^w are computed from the lengths $\ell_{1:N}$. First we calculate the starting position of each plateau function b_n as

$$b_1 = 0, b_n = \sum_{n'=1}^{n-1} \ell_{n'}. \tag{11}$$

We can then define both the center and the width parameters of each plateau function as

$$\begin{aligned} \lambda^c_n &= b_n + \ell_n/2, \\ \lambda^w_n &= \ell_n/2 \end{aligned} \tag{12}$$

and define each row of the approximate mask as

$$M^*[n,t] = f(t|\lambda^c_n, \lambda^w_n, \lambda^s). \tag{13}$$

Now we can calculate a differentiable approximate observation energy similar to (9) as

$$E^*_o = \sum_{t=1}^{T} \sum_{n=1}^{N} M^*[n,t] \cdot P[n,t]. \tag{14}$$

Approximate Differentiable Length Energy. For the gradient-based optimization, we must relax the length values to be positive real values instead of natural numbers. As the Poisson distribution (4) is only defined on natural numbers, we propose to use a substitute distribution defined on real numbers. As a replacement, we experiment with a Laplace distribution and a Gaussian distribution. In both cases, the scale or the width parameter of the distribution is assumed to be fixed.

We can rewrite the length energy E_ℓ as the approximate length energy

$$E^*_\ell(\ell_{1:N}) = \sum_{n=1}^{N} -\log p(\ell_n | \lambda^\ell_{c_n}), \tag{15}$$

where $\lambda^\ell_{c_n}$ is the expected value for the length of a segment from the action c_n. In case of the Laplace distribution this length energy will be

$$E^*_\ell(\ell_{1:N}) = \frac{1}{Z} \sum_{n=1}^{N} |\ell_n - \lambda^\ell_{c_n}|, \tag{16}$$

where Z is the constant normalization factor. This means that the length energy will penalize any deviation from the expected average length linearly. Similarly, for the Gaussian distribution, the length energy will be

$$E^*_\ell(\ell_{1:N}) = \frac{1}{Z} \sum_{n=1}^{N} |\ell_n - \lambda^\ell_{c_n}|^2, \tag{17}$$

which means that the Gaussian length energy will penalize any deviation from the expected average length quadratically.

With the objective to maintain a positive value for the length during the optimization process, we estimate the length in log space and convert it to absolute space only in order to compute both the approximate mask matrix M^* and the approximate length energy E^*_ℓ.

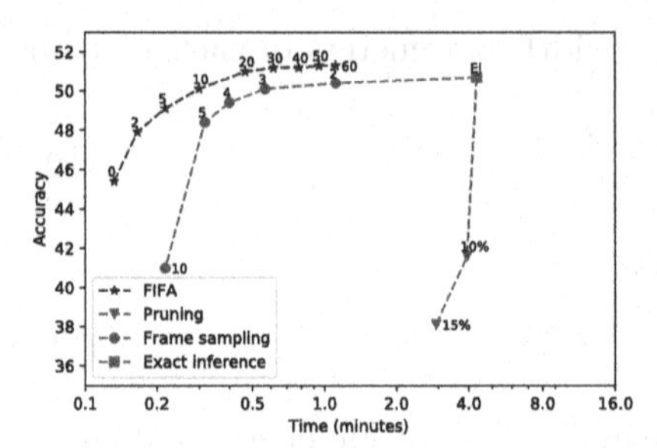

Fig. 2. Speed vs. accuracy trade-off of different inference approaches applied to the MuCon method. Using FIFA we can achieve a better speed vs. accuracy trade-off compared to frame sampling or hypothesis pruning in exact inference.

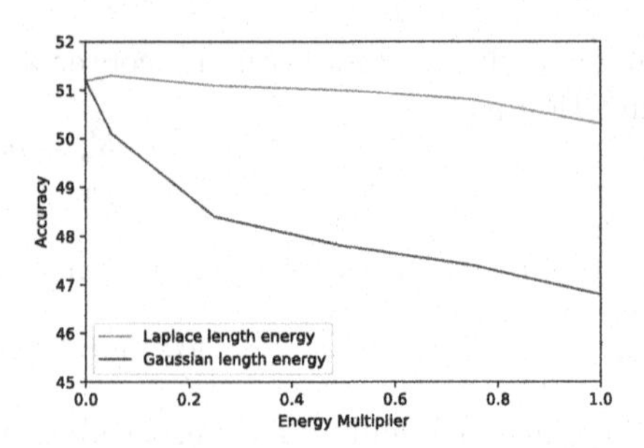

Fig. 3. Effect of the length energy multiplier for Laplace and Gaussian length energy. Accuracy is calculated on the Breakfast dataset using FIFA applied to the MuCon approach trained in the weakly supervised action segmentation setting.

Approximate Energy Optimization. The total approximate energy function is defined as a weighted sum of both the approximate observation and the approximate length energy functions

$$E^*(\ell_{1:N}) = E_o^*(\ell_{1:N}, Y) + \beta E_\ell^*(\ell_{1:N}) \tag{18}$$

where β is the multiplier for the length energy.

Given an initial length estimate $\ell^0_{1:N}$, we iteratively update this estimate to minimize the total energy. Figure 1 illustrates the optimization step for our approach. During each optimization step, we first calculate the energy E^* and then calculate the gradients of the energy with respect to the length values. Using the calculated gradients, we update the length estimate using a gradient descent update rule such as SGD or Adam. After a certain number of gradient steps (50 steps in our experiments) we will finally predict the segment length.

If the transcript for a test video is provided then it is used, e.g., using the MuCon [32] approach or in a weakly supervised action alignment setting. However, if the latter is not known, e.g., in a fully supervised approach or CDFL [25] for weakly supervised action segmentation, we perform the optimization for each of the transcripts seen during training and select the most likely one based on the final energy value at the end of the optimization.

The initial length estimates are calculated from the length model of each approach in case of a weakly supervised setting whereas in a fully supervised setting the average length of each action class is calculated from the training data and used as the initial length estimates. The initial length estimates are also used as the expected length parameters for the length energy calculations.

The hyper-parameters like the choice of the optimizer, number of steps, learning rate, and the mask sharpness, remain as the hyper-parameters of our approach.

5 Experiments

5.1 Evaluation Protocols and Datasets

We evaluate FIFA on 3 different tasks: weakly supervised action segmentation, fully supervised action segmentation, and weakly supervised action alignment. Results for action alignment are included in the supplementary material. We obtain the source code for the state-of-the-art approaches on each of these tasks and train a model using the standard training configuration of each model. Then we apply FIFA as a replacement for an existing inference stage or as an additional inference stage.

We evaluate our model using the Breakfast [16] and Hollywood extended [4] datasets on the 3 different tasks. Details of the datasets are included in the supplementary material.

5.2 Results and Discussions

In this section, we study the speed-accuracy trade-off and the impact of the length model. Additional ablation experiments are included in the supplementary material.

Speed vs. Accuracy Trade-off. One of the major benefits of FIFA is that it is anytime. It provides the flexibility of choosing the number of optimization steps. The number of steps of the optimization can be a tool to trade-off speed vs. accuracy. In exact inference, we can use frame-sampling, i.e., lowering the resolution of the input features, or hypothesis pruning, i.e., beam search for speed vs. accuracy trade-off.

Figure 2 plots the speed vs. accuracy trade-off of exact inference compared to FIFA. We observe that FIFA provides a much better speed-accuracy trade-off as compared to frame-sampling for exact inference. The best performance is achieved after 50 steps with 5.9% improvement on the MoF accuracy compared to not performing any optimization (0 steps).

Impact of the Length Energy Multiplier. For the length energy, we assume that the segment lengths follow a Laplace distribution. Figure 3 shows the impact of the length energy multiplier on the weakly supervised action segmentation performance on the Breakfast dataset. The choice of this parameter depends on the dataset. While the best accuracy is achieved with a multiplier of 0.05, our approach is robust to the choice of these hyper-parameters on this dataset. We further experimented with a Gaussian length energy. However, as shown in the figure, the performance is much worse compared to the Laplace energy. This is due to the quadratic penalty that dominates the total energy, which makes the optimization biased towards the initial estimate and ignores the observation energy.

Impact of the Length Initialization. Since FIFA starts with an initial estimate for the lengths, the choice of initialization might have an impact on the performance. Table 1 shows the effect of initializing the lengths with equal values compared to using the length model of MuCon [32] for the weakly supervised action segmentation on the Breakfast dataset. As shown in the table, FIFA is more robust to initialization compared to the exact inference as the drop in performance is approximately half of the exact inference.

Table 1. Impact of the length initialization for MuCon using exact inference and FIFA for weakly supervised action segmentation on the Breakfast dataset.

Inference Method	Initialization	MoF
Exact	MuCon [32]	**50.7**
	Equal	48.8 (−1.9)
FIFA	MuCon [32]	**51.3**
	Equal	50.2 (−1.1)

5.3 Comparison to State of the Art

In this section, we compare FIFA to other state-of-the-art approaches.

Weakly Supervised Action Segmentation. We apply FIFA on top of two state-of-the-art approaches for weakly supervised action segmentation namely MuCon [32] and CDFL [25] on the Breakfast dataset [16] and report the results in Table 2. FIFA applied on CDFL achieves a 12 times faster inference speed while obtaining results comparable to exact inference. FIFA applied to MuCon achieves a 5 times faster inference speed and obtains a new state-of-the-art performance on the Breakfast dataset on most of the metrics.

We also reported the inference speed of ISBA [6] and NNV [30] (since the source code of D3TW [5] is not available we could not measure its inference speed) and reported them for the sake of completeness in Table 2. ISBA has the fastest inference time as during testing it does not perform any optimization. ISBA makes framewise predictions which results in over-segmentation and low performance across all metrics.

Table 2. Results for weakly supervised action segmentation on the Breakfast dataset. * indicates results obtained by running the code on our machine.

Method	MoF	MoF-BG	IoU	IoD	Time (min)
ISBA [6]	38.4	38.4	24.2	40.6	0.01
NNV [30]	43.0	–	–	–	234
D3TW [5]	45.7	–	–	–	–
CDFL [25]	50.2	48.0	33.7	45.4	–
CDFL*	49.4	47.5	35.2	46.4	260
FIFA + CDFL*	47.9	46.3	34.7	48.0	20.4 (×12.8)
MuCon [32]	47.1	–	–	–	–
MuCon*	<u>50.7</u>	<u>50.3</u>	<u>40.9</u>	**54.0**	4.1
FIFA + MuCon*	**51.3**	**50.7**	**41.1**	<u>53.3</u>	0.8 (×5.1)

Table 3. Results for fully supervised action segmentation setup on the Breakfast dataset. * indicates results obtained by running the code on our machine.

Method	F1@{10, 25, 50}			Edit	MoF
BCN [34]	68.7	65.5	<u>55.0</u>	66.2	**70.4**
ASRF [13]	<u>74.3</u>	68.9	**56.1**	72.4	67.6
MS-TCN++ [26]	64.1	58.6	45.9	65.6	67.6
FIFA + MS-TCN++*	<u>74.3</u>	<u>69.0</u>	54.3	<u>77.3</u>	67.9
MS-TCN [1]	52.6	48.1	37.9	61.7	66.3
FIFA + MS-TCN*	**75.5**	**70.2**	54.8	**78.5**	<u>68.6</u>

Similarly for the Hollywood extended dataset [4], we apply FIFA to MuCon [32] and report the results in Table 4. FIFA applied on MuCon achieves a 4 times faster inference speed while obtaining results comparable to MuCon with exact inference.

Fully Supervised Action Segmentation. In the fully supervised action segmentation setting, we apply FIFA on top of MS-TCN [1] and its variant MS-TCN++ [26] on the Breakfast dataset [16] and report the results in Table 3. MS-TCN and MS-TCN++ do not perform any inference at test time. This usually results in over-segmentation and low F1 and Edit scores. Applying FIFA on top of these approaches improves the F1 and Edit scores significantly. FIFA applied on top of MS-TCN achieves state-of-the-art performance on most metrics.

For the Hollywood extended dataset [4], we train MS-TCN [1] and report results comparing exact inference (EI) compared to FIFA in Table 5. We observe that MS-TCN using an inference algorithm achieves new state-of-the-art results on this dataset. FIFA is comparable or better than exact inference on this dataset.

Table 4. Results for weakly supervised action segmentation on the Hollywood extended dataset. Time is reported in seconds. * indicates results obtained by running the code on our machine.

Method	MoF-BG	IoU	Time (speedup)
ISBA [6]	34.5	12.6	–
D3TW [25]	33.6		–
CDFL [25]	40.6	**19.5**	–
MuCon [32]	**41.6**		–
MuCon*	40.1	<u>13.9</u>	53
MuCon + FIFA*	<u>41.2</u>	13.7	13 (×4.1)

Table 5. Results for fully supervised action segmentation on the Hollywood extended dataset. * indicates results obtained by running the code on our machine. EI stands for Exact Inference.

Method	MoF	MoF-BG	IoU	IoD
HTK [18]	39.5		8.4	
ED-TCN [6]	36.7	27.3	10.9	13.1
ISBA [6]	54.8	33.1	20.4	28.8
MSTCN [32] (+ EI)*	<u>64.9</u>	**35.0**	<u>22.6</u>	<u>33.2</u>
MSTCN + FIFA*	**66.2**	<u>34.8</u>	**23.9**	**35.8**

5.4 Qualitative Example

A qualitative example of the FIFA optimization process is depicted in Fig. 4. For further qualitative examples, failure cases, and details please refer to the supplementary material.

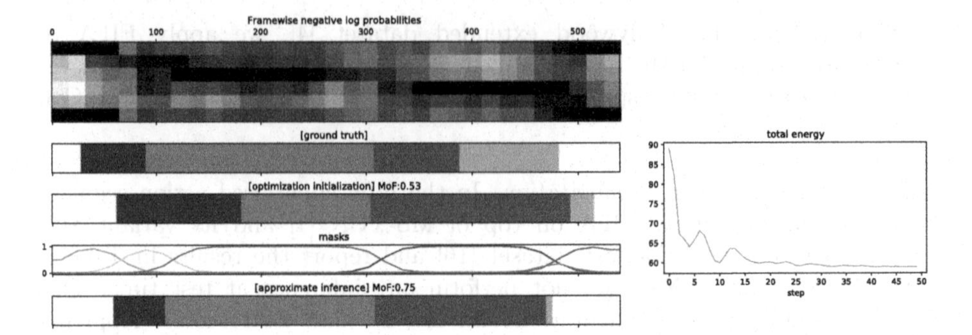

Fig. 4. Visualization of the FIFA optimization process. On the right the values of the total approximate energy is plotted. On the left, negative log probability values, ground truth segmentation, optimization initialization, the masks and the segmentation after inference are plotted.

6 Conclusion

In this paper, we proposed FIFA a fast approximate inference procedure for action segmentation and alignment. Unlike previous methods, our approach does not rely on dynamic programming-based Viterbi decoding for inference. Instead, FIFA optimizes a differentiable energy function that can be minimized using gradient-descent which allows for a fast and accurate inference during testing. We evaluated FIFA on top of fully and weakly supervised methods trained on the Breakfast and Hollywood extended datasets. The results show that FIFA is able to achieve comparable or better performance, while being at least 5 times faster than exact inference.

Acknowledgement. The work has been partially funded by the Deutsche Forschungsgemeinschaft (DFG, German Research Foundation) - GA 1927/4-2 (FOR 2535 Anticipating Human Behavior).

References

1. Abu Farha, Y., Gall, J.: MS-TCN: multi-stage temporal convolutional network for action segmentation. In: CVPR (2019)
2. Belanger, D., McCallum, A.: Structured prediction energy networks. In: ICML (2016)
3. Belanger, D., Yang, B., McCallum, A.: End-to-end learning for structured prediction energy networks. In: ICML (2017)
4. Bojanowski, P., et al.: Weakly supervised action labeling in videos under ordering constraints. In: Fleet, D., Pajdla, T., Schiele, B., Tuytelaars, T. (eds.) ECCV 2014. LNCS, vol. 8693, pp. 628–643. Springer, Cham (2014). https://doi.org/10.1007/978-3-319-10602-1_41
5. Chang, C., Huang, D., Sui, Y., Fei-Fei, L., Niebles, J.C.: D^{3}TW: discriminative differentiable dynamic time warping for weakly supervised action alignment and segmentation. In: CVPR (2019)

6. Ding, L., Xu, C.: Weakly-supervised action segmentation with iterative soft boundary assignment. In: CVPR (2018)
7. Duchenne, O., Laptev, I., Sivic, J., Bach, F., Ponce, J.: Automatic annotation of human actions in video. In: ICCV (2009)
8. Gatys, L.A., Ecker, A.S., Bethge, M.: A neural algorithm of artistic style. arXiv preprint arXiv:1508.06576 (2015)
9. Goodfellow, I.J., Shlens, J., Szegedy, C.: Explaining and harnessing adversarial examples. In: ICLR (2015)
10. Hoang, C.D.V., Haffari, G., Cohn, T.: Towards decoding as continuous optimization in neural machine translation. In: EMNLP (2017)
11. Huang, D.-A., Fei-Fei, L., Niebles, J.C.: Connectionist temporal modeling for weakly supervised action labeling. In: Leibe, B., Matas, J., Sebe, N., Welling, M. (eds.) ECCV 2016. LNCS, vol. 9908, pp. 137–153. Springer, Cham (2016). https://doi.org/10.1007/978-3-319-46493-0_9
12. Huang, Y., Sugano, Y., Sato, Y.: Improving action segmentation via graph-based temporal reasoning. In: CVPR (2020)
13. Ishikawa, Y., Kasai, S., Aoki, Y., Kataoka, H.: Alleviating over-segmentation errors by detecting action boundaries. In: WACV (2021)
14. Johnson, J., Alahi, A., Fei-Fei, L.: Perceptual losses for real-time style transfer and super-resolution. In: Leibe, B., Matas, J., Sebe, N., Welling, M. (eds.) ECCV 2016. LNCS, vol. 9906, pp. 694–711. Springer, Cham (2016). https://doi.org/10.1007/978-3-319-46475-6_43
15. Karaman, S., Seidenari, L., Del Bimbo, A.: Fast saliency based pooling of fisher encoded dense trajectories. In: ECCV THUMOS Workshop (2014)
16. Kuehne, H., Arslan, A., Serre, T.: The language of actions: recovering the syntax and semantics of goal-directed human activities. In: CVPR (2014)
17. Kuehne, H., Gall, J., Serre, T.: An end-to-end generative framework for video segmentation and recognition. In: WACV (2016)
18. Kuehne, H., Richard, A., Gall, J.: Weakly supervised learning of actions from transcripts. CVIU **163**, 78–89 (2017)
19. Kuehne, H., Richard, A., Gall, J.: A Hybrid RNN-HMM approach for weakly supervised temporal action segmentation. PAMI **42**(04), 765–779 (2020)
20. Le, Q., Mikolov, T.: Distributed representations of sentences and documents. In: ICML (2014)
21. Lea, C., Flynn, M.D., Vidal, R., Reiter, A., Hager, G.D.: Temporal convolutional networks for action segmentation and detection. In: CVPR (2017)
22. Lea, C., Reiter, A., Vidal, R., Hager, G.D.: Segmental spatiotemporal CNNs for fine-grained action segmentation. In: Leibe, B., Matas, J., Sebe, N., Welling, M. (eds.) ECCV 2016. LNCS, vol. 9907, pp. 36–52. Springer, Cham (2016). https://doi.org/10.1007/978-3-319-46487-9_3
23. LeCun, Y., Chopra, S., Hadsell, R., Ranzato, M., Huang, F.: A tutorial on energy-based learning. In: Predicting Structured Data, no. 1 (2006)
24. Lei, P., Todorovic, S.: Temporal deformable residual networks for action segmentation in videos. In: CVPR (2018)
25. Li, J., Lei, P., Todorovic, S.: Weakly supervised energy-based learning for action segmentation. In: ICCV (2019)
26. Li, S., Abu Farha, Y., Liu, Y., Cheng, M.M., Gall, J.: MS-TCN++: multi-stage temporal convolutional network for action segmentation. PAMI (2020)
27. Moltisanti, D., Fidler, S., Damen, D.: Action recognition from single timestamp supervision in untrimmed videos. In: CVPR (2019)

28. Pirsiavash, H., Ramanan, D.: Parsing videos of actions with segmental grammars. In: CVPR (2014)
29. Richard, A., Kuehne, H., Gall, J.: Weakly supervised action learning with RNN based fine-to-coarse modeling. In: CVPR (2017)
30. Richard, A., Kuehne, H., Iqbal, A., Gall, J.: Neuralnetwork-viterbi: a framework for weakly supervised video learning. In: CVPR (2018)
31. Rohrbach, M., Amin, S., Andriluka, M., Schiele, B.: A database for fine grained activity detection of cooking activities. In: CVPR (2012)
32. Souri, Y., Fayyaz, M., Minciullo, L., Francesca, G., Gall, J.: Fast weakly supervised action segmentation using mutual consistency. PAMI (2021)
33. Vo, N.N., Bobick, A.F.: From stochastic grammar to Bayes network: probabilistic parsing of complex activity. In: CVPR (2014)
34. Wang, Z., Gao, Z., Wang, L., Li, Z., Wu, G.: Boundary-aware cascade networks for temporal action segmentation. In: Vedaldi, A., Bischof, H., Brox, T., Frahm, J.-M. (eds.) ECCV 2020. LNCS, vol. 12370, pp. 34–51. Springer, Cham (2020). https://doi.org/10.1007/978-3-030-58595-2_3

Hybrid SNN-ANN: Energy-Efficient Classification and Object Detection for Event-Based Vision

Alexander Kugele[1,2(✉)], Thomas Pfeil[1], Michael Pfeiffer[1], and Elisabetta Chicca[2,3]

[1] Bosch Center for Artificial Intelligence, 71272 Renningen, Germany
{alexander.kugele,thomas.pfeil,michael.pfeiffer3}@de.bosch.com
[2] Bio-Inspired Circuits and Systems (BICS) Lab, Zernike Inst Adv Mat, University of Groningen, Nijenborgh 4, 9747 Groningen, AG, The Netherlands
e.chicca@rug.nl
[3] Groningen Cognitive Systems and Materials Center (CogniGron), University of Groningen, Nijenborgh 4, 9747 Groningen, AG, The Netherlands

Abstract. Event-based vision sensors encode local pixel-wise brightness changes in streams of events rather than full image frames and yield sparse, energy-efficient encodings of scenes, in addition to low latency, high dynamic range, and lack of motion blur. Recent progress in object recognition from event-based sensors has come from conversions of successful deep neural network architectures, which are trained with backpropagation. However, using these approaches for event streams requires a transformation to a synchronous paradigm, which not only loses computational efficiency, but also misses opportunities to extract spatio-temporal features. In this article we propose a hybrid architecture for end-to-end training of deep neural networks for event-based pattern recognition and object detection, combining a spiking neural network (SNN) backbone for efficient event-based feature extraction, and a subsequent classical analog neural network (ANN) head to solve synchronous classification and detection tasks. This is achieved by combining standard backpropagation with surrogate gradient training to propagate gradients inside the SNN layers. Hybrid SNN-ANNs can be trained without additional conversion steps, and result in highly accurate networks that are substantially more computationally efficient than their ANN counterparts. We demonstrate results on event-based classification and object detection datasets, in which only the architecture of the ANN heads need to be adapted to the tasks, and no conversion of the event-based input is necessary. Since ANNs and SNNs require different hardware paradigms to maximize their efficiency, we envision that SNN backbone and ANN head can be executed on different processing units, and thus analyze the necessary bandwidth to communicate between the two parts. Hybrid networks are promising architectures to further advance machine learning approaches for event-based vision, without having to compromise on efficiency.

Supplementary Information The online version contains supplementary material available at https://doi.org/10.1007/978-3-030-92659-5_19.

C. Bauckhage et al. (Eds.): DAGM GCPR 2021, LNCS 13024, pp. 297–312, 2021.
https://doi.org/10.1007/978-3-030-92659-5_19

Keywords: Object detection · Event-based vision · Spiking neural networks

1 Introduction

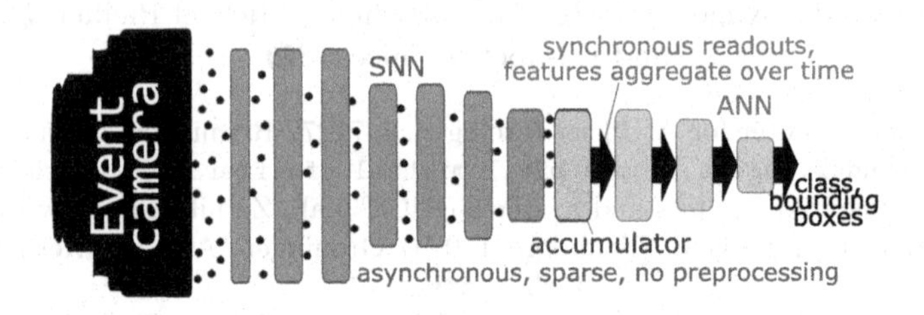

Fig. 1. Hybrid SNN-ANN models consist of an SNN backbone to compute features directly from event camera outputs which are accumulated and processed by an ANN head for classification and object detection. Using sparse, binary communication (dots) instead of dense tensors (arrows) to process events enables efficient inference. SNN and ANN can be on completely different devices.

Event-based vision sensors address the increasing need for fast and energy-efficient visual perception [10,22,24,33,41], and have enabled new use cases such as high-speed navigation [7], gesture recognition [26], visual odometry and SLAM [28,46,48]. These sensors excel at very low latency and high dynamic range, while their event-based encoding creates a sparse spatio-temporal representation of dynamic scenes. Every event indicates the position, precise time, and polarity of a local brightness change.

In conventional frame-based computer vision deep learning-based methods have led to vastly improved performance in object classification and detection, so it is natural to expect a boost in performance also from applying deep neural networks to event-based vision. However, such an approach has to overcome the incompatibility of machine learning algorithms developed for a synchronous processing paradigm, and the sparse, asynchronous nature of event-based inputs. Recent successful approaches for processing event data have therefore relied on early conversions of events into filtered representations that are more suitable to apply standard machine learning methods [1,32,44]. Biologically inspired spiking neural networks (SNNs) in principle do not require any conversion of event data and can process data from event-based sensors with minimal preprocessing.

However, high performing SNNs rely on conversion from standard deep networks [38,40], thereby losing the opportunity to work directly with precisely timed events. Other approaches like evolutionary methods [30] or local learning rules like STDP [2,16] are not yet competitive in performance.

Here we describe a hybrid approach that allows end-to-end training of neural networks for event-based object recognition and detection. It combines sparse spike-based processing of events in early layers with off-the-shelf ANN layers to process the sparse, abstract features (see Fig. 1 for an illustration). This is made possible by combining standard backpropagation with recently developed surrogate gradient methods to train deep SNNs [4,21,29,35,42,47]. The advantage of the hybrid approach is that early layers can operate in the extremely efficient event-based computing paradigm, which can run on special purpose hardware implementations [3,5,9,27,34,39]. The hybrid approach is also optimized for running SNN parts and conventional neural networks on separate pieces of hardware by minimizing the necessary bandwidth for communication. In our experiments we demonstrate that the hybrid SNN-ANN approach yields very competitive accuracy at significantly reduced computational costs, and is thus ideally suited for embedded perception applications that can exploit the advantages of the event-based sensor frontend.

Our main contributions are as follows:

- We propose a novel hybrid architecture that efficiently integrates information over time without needing to transform input events into other representations.
- We propose the first truly end-to-end training scheme for hybrid SNN-ANN architectures on event camera datasets for classification and object detection.
- We investigate how to reduce communication bandwidths for efficient hardware implementations that run SNNs and ANNs on separate chips.
- We analyze how transfer learning of SNN layers increases the accuracy of our proposed architecture.

2 Related Work

A variety of **low level representations** for event-based vision data have been explored: The HOTS method [19] defines a time surface, *i.e.*, a two-dimensional representation of an event stream by convolving a kernel over the event stream. This method was improved in [44] by choosing an exponentially decaying kernel and adding local memory for increased efficiency. In [12], a more general take on event stream processing is proposed, utilizing general kernel functions that can be learned and project the event stream to different representations. Notably, using a kernel on event camera data and aggregating the result is the same as using a spiking neural network layer. Our approach allows learning a more general low level representation by using a deep SNN with an exponentially decaying kernel, compared to only one layer in [12] and learnable weights compared to [19,44].

Conversion approaches such as [38,40] transform trained ANNs into SNNs for inference, and so far have set accuracy benchmarks for deep SNNs. Conversion methods train on image frames and do not utilize the membrane potential or delays to integrate information over time. In [18] networks are unrolled over multiple time steps, which allows training on event camera datasets. However, temporal integration is only encoded in the structure of the underlying ANN, but not in the dynamics of spiking neurons. In their formulation, the efficiency of the SNN is limited by the rate coding of the neurons, which is potentially

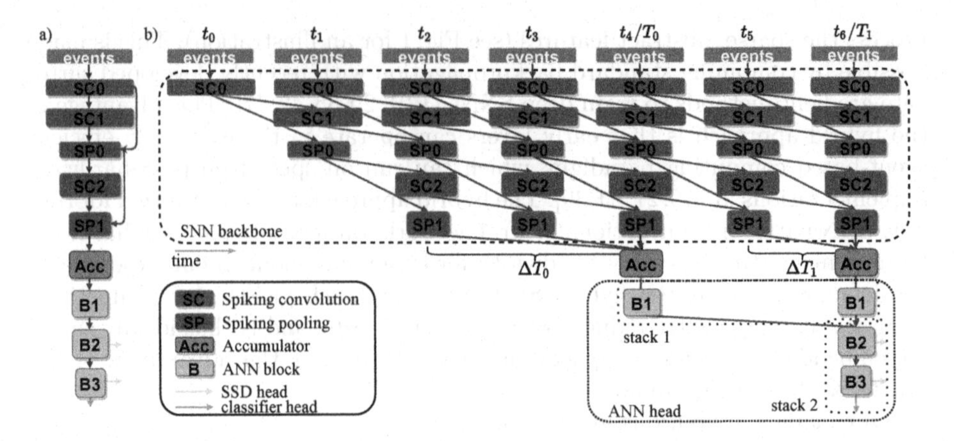

Fig. 2. Training a hybrid network with a DenseNet backbone. a) Compact representation. b) Network rolled out over time steps. The SNN backbone computes sparse features from event camera data that are accumulated at time intervals ΔT_i. The ANN head integrates features from multiple outputs (2, in this example) for a prediction (classification or object detection). We use a time step of 1 ms to integrate information over small time scales. During inference, the SNN backbone runs asynchronously without a time-stepped rollout, enabling potential savings in computation. Layers with the same name share weights.

more inefficient than encodings learned via end-to-end training in our hybrid SNN-ANN approach. In addition, conversion methods typically do not optimize for energy-efficiency.

SNN training with variants of backpropagation has been demonstrated in [21,42,47], albeit on simpler architectures (*e.g.*, only one fully-connected layer in [21]) and, in general, without delays during simulation. A mixed approach is used in [35], first training and converting an ANN and then training the converted SNN. Our approach uses synaptic delays and skip connections, exploring how complex ANN architectures translate to SNN architectures. The closest architecture to ours is from [20], which is trained from scratch to predict optical flow. Their U-Net with an SNN encoder transmits information from all SNN layers, leading to a high bandwidth from SNN to ANN. We improve on this by using only a single layer to transmit information from SNN to ANN, and extend to applications in classification and object detection tasks.

Conventional ANN architectures are used in [36] to solve the task of image reconstruction from events with a recurrent U-Net, and they subsequently show that classification and object detection are possible on the reconstructed images. No SNN layers are used in this case, resulting in a computationally expensive network and the need to preprocess the event camera data. Faster and more efficient training and inference for object detection is presented in [32], who propose a recurrent convolutional architecture that does not need to explicitly reconstruct frames. As the sparsity of the event stream is not utilized, it is expected that gains in energy-efficiency are possible with SNN approaches.

3 Methods

This section introduces the proposed hybrid SNN-ANN architecture and describes training, inference and metrics we used to evaluate our method. Our hybrid network consists of two parts: An SNN backbone that computes features from raw event camera inputs, and an ANN head that infers labels or bounding boxes at predefined times (see Fig. 2). The overall task is to find an efficient mapping from a sequence of event camera data E in a time interval T to a prediction P, which can be a label l or a set of bounding boxes B. Our approach consists of three stages: First, continuously in time, an intermediate representation $I = S(E)$ is generated using the SNN backbone. Second, this intermediate representation is accumulated at predefined points in time. Third, when all accumulators are filled, the accumulated intermediate representations are mapped via the ANN head A to the final prediction $P = A(\mathrm{Acc}(I)) = A(\mathrm{Acc}(S(E)))$. The following sections describe all three parts in more detail.

3.1 SNN Backbone

Spiking neural networks (SNNs) are biologically inspired neural networks, where each neuron has an internal state. Neurons communicate via spikes, i.e. binary events in time, to signal the crossing of their firing thresholds. Upon receiving a spike i, the synaptic current I changes proportionally to the synaptic weight w, which in turn leads to a change of the neuron's internal state V. Because of the binary and sparse communication, these networks can be significantly more energy-efficient than dense matrix multiplications in conventional ANNs (see also [37]).

The task of the SNN backbone S is to map a sequence of raw event camera inputs E into a compressed, sparse, and abstract intermediate representation $I = S(E)$ in an energy-efficient way. More concretely, the input is a stream of events $e_i = (t_i, x_i, y_i, p_i)$, representing the time t_i and polarity p_i of an input event at location (x_i, y_i). Polarity is a binary $\{-1, 1\}$ signal that encodes if the brightness change is positive (brighter) or negative (darker). This stream is processed by the SNN without further preprocessing. In our implementation, the first layer has two input channels representing the two polarities. In contrast to previous work such as [20] and [18], the input events are never transformed into voxel grids or rate averages, but directly processed by the SNN to compute the intermediate representation $S(E)$. The spiking neuron model we use is the leaky integrate-and-fire model [13], simulated using the forward Euler method,

$$I_i = \alpha I_{i-1} + \textstyle\sum_j w_j S_j \tag{1}$$

$$V_i = \beta V_{i-1} + I_i \tag{2}$$

$$S_i = \Theta(V_i - V_{\mathrm{th}}) \tag{3}$$

where V is the membrane potential, I is the presynaptic potential, $S_{j,i}$ are the binary input and output spikes, V_{th} is the threshold voltage, and $t_d = t_i - t_{i-1}$ is an update step. The membrane leakage β and the synaptic leakage α are given as $\exp(-t_d/\tau_{\text{mem}})$ and $\exp(-t_d/\tau_{\text{syn}})$, respectively. To simulate the membrane reset, we add a term to Eq. (2) which implements the reset-by-subtraction [38] mechanism,

$$V_i = V_i - V_{\text{th}} S_{i-1}. \tag{4}$$

The threshold is always initialized as $V_{\text{th}} = 1$ and trained jointly with the weights (see appendix for details).

We simulate our network by unrolling it in time with a fixed time step t_d. Figure 2b shows the training graph for a rollout of 7 time steps. Our simulation allows choosing arbitrary delays for the connections of different layers, which determines how information is processed in time. Inspired by recent advances, we choose to implement streaming rollouts [8] in all our simulations. This means that each connection has a delay of one time step, in accordance with the minimum delay of large-scale simulators for neuroscience [45]. This allows integrating temporal information via delayed connections, in addition to the internal state.

The SNN backbone $S(E)$ outputs a set of sequences of spikes in predefined time intervals $T_{\text{out},i}$: $I = (e_j)_{t_j \in T_{\text{out},i}}$. An example is shown in the dashed box of Fig. 2. Details about the backbones used can be found in Sect. 3.8. The figure shows an unrolled DenseNet backbone with two blocks (SC0 to SP0 and SP0 to SP1, respectively), where two output intervals are defined as $\Delta T_0 = [t_2, T_0]$ and $\Delta T_1 = [t_5, T_1]$. This structure is used during training, where a time-stepped simulator approximates the continuous-time SNN. During inference, the SNN backbone runs asynchronously, enabling savings in computation.

3.2 Accumulator

Our model connects the sparse, continuous representation of the SNN with the dense, time-stepped input of the ANN with an accumulator layer for each output interval ΔT_i (Acc in Fig. 2). The task of this layer is to transform the sparse data to a dense tensor. We choose the simple approach of summing all spikes in each time interval ΔT_i to get a dense tensor with the feature map shape $(c_{\text{f}}, h_{\text{f}}, w_{\text{f}})$.

3.3 ANN Head

The ANN head processes the accumulated representations from the accumulators to predict classes or bounding boxes. The general structure of the ANN head can be described with three parameters: the number of SNN outputs n_{out}, the number of stacks n_{s} and the number of blocks per stack n_{l}. The exemplary graph in Fig. 2 has $n_{\text{out}} = 2$, $n_{\text{s}} = 2$ and $n_{\text{l}} = 2$. Having multiple outputs and stacks allows to increase the temporal receptive field, *i.e.*, the time interval the ANN uses for its predictions (for details see appendix). All blocks with the same name share their weights to reduce the number of parameters. The number of blocks can be different for each stack. In most experiments we use two stacks with 1 and 3 blocks, respectively. The dense representation for each ΔT_i is then further processed by each stack, where results are summed at the end of a stack and used as input for the next stack. Each block in a stack consists of batch normalization BN, convolution C, dropout Drop, ReLU and pooling P, *e.g.*, $\text{B0}(x) = \text{P}(\text{ReLU}(\text{Drop}(\text{C}(\text{BN}(x)))))$. We use a convolutional layer with kernel size 2, stride 2, learnable weights and ReLU activation as pooling layer. Whenever we use dropout in conjunction with weight sharing, we also share the dropout mask. In the case of classification, a linear layer is attached to the last block in the last stack (see Sect. 3.6). For object detection, an SSD head is attached to selected blocks in the last stack (see Sect. 3.7).

3.4 Energy-efficiency

We use the same metric as [38], *i.e.*, we count the number of operations of our network. Due to the sparse processing and binary activations in the SNN, the number of operations is given by the synaptic operations, *i.e.*, the sum of all post-synaptic current updates. For ANN layers, we count the number of multiply-add operations, as the information is processed with dense arrays. For this metric, it is assumed that both SNN and ANN are run on dedicated hardware. Then the total energy is proportional to the number of operations plus a constant offset. We discuss benefits, drawbacks and alternatives to this metric in the appendix. To regularize the number of spikes we utilize an L_1 loss on all activations of the SNN backbone

$$L_{\text{s}} = \sum_{l,b,i,x,y} \frac{\lambda_s}{LBTW_l H_l} |S_{l,i,b,x,y}|. \tag{5}$$

with a scaling factor λ_s, the number of layers L, the batch size B, total simulation time T, and width W_l and height H_l of the respective layer. This also reduces the bandwidth, as discussed in Sect. 3.5.

3.5 Bandwidth

As we expect that special hardware could be used to execute at least the SNN backbone during inference, we want to minimize the bandwidth between SNN and ANN to avoid latency issues. We design our architectures such that only the last layer is connected to the ANN and use L_1 loss on the activations to regularize the number of spikes. This is in contrast to [20], where each layer has to be propagated to the ANN. We report all bandwidths in MegaEvents per second (MEv/s) and MegaBytes per second (MB/s). One event equals 6.125 Bytes, because we assume it consists of 32 bit time +16 bit spatial coordinates +1 bit polarity.

3.6 Classification

To classify, we attach a single linear layer to the last block in the last stack of our hybrid network. We use the negative log-likelihood loss of the output of

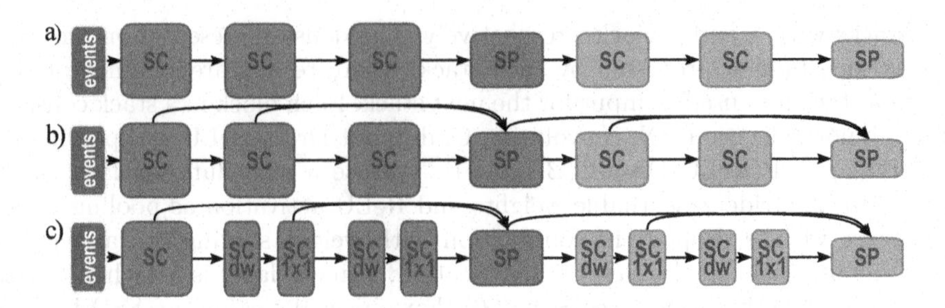

Fig. 3. The different backbones in compact representation with spiking convolutional (**SC**) and pooling (**SP**) layers. a) VGG. b) DenseNet. c) DenseSep (DenseNet with depthwise separable convolutions). Depthwise separable convolutions consist of a depthwise convolution (**dw**) and a convolution with kernel size **1** × **1**. Shades of blue mark different blocks. All depicted networks have 2 blocks and 2 layers per block. Multiple inputs are concatenated.

this layer. Additionally, we use L_2-regularization (weight decay) with a factor of 0.001. We use the Adam optimizer [17] with a learning rate of $\eta = 0.01$ and a learning rate schedule to divide it by 5 at 20%, 80% and 90% of the total number of epochs $n_e = 100$.

3.7 Object Detection

We use the SSD (single shot detection) architecture [25], which consists of a backbone feature extractor and multiple predictor heads. Features are extracted from the backbone at different scales and for each set of features, bounding boxes and associated classes are predicted, relative to predefined prior boxes. In the appendix, we present a novel, general way to tune the default prior boxes in a fast and efficient way. During inference, non-maximum suppression is used on the output of all blocks to select non-overlapping bounding boxes with a high confidence. The performance of the network is measured with the VOC mean average precision (mAP) [6]. If not otherwise denoted, we use the same learning hyperparameters as in Sect. 3.6 but a smaller learning rate of 0.001.

3.8 Backbone Architectures

Three different architecture types are used in our experiments: VGG [43], DenseNet [15] and DenseSep. A VGG network with N_b blocks and N_l layers per block is a feed-forward neural network with $N_b \cdot N_l$ layers, where each output channel is given by $g \cdot l$ with l the layer index (starting from 1). The DenseNet structure consists of N_b blocks, where each block consists of N_l connected layers per block. DenseSep is a combination of the depthwise separable block of MobileNet [14] and the DenseNet structure (see Fig. 3).

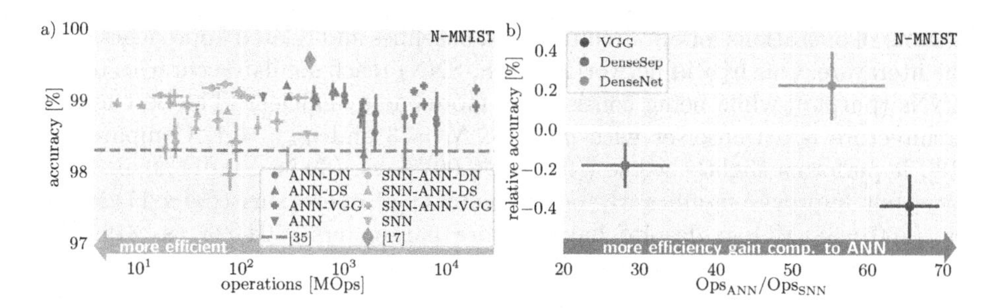

Fig. 4. a) Accuracy on the N-MNIST test set vs. number of operations for different architectures. Hybrid SNN-ANN architectures (green) overcome the efficiency limit of conversion-based architectures (blue diamond) with only a minor drop in accuracy. SNN-ANN architectures are more efficient than almost all ANNs (purple). Compared to the SNN and ANN baseline, our hybrid networks increase accuracy and energy-efficiency. **b)** Relative accuracy on the N-MNIST test set vs. relative number of operations. The VGG backbone has the highest gain in energy-efficiency, while losing significantly in accuracy. Other backbones show a minor decrease with a significant gain in energy-efficiency. We report the mean and error of the mean over 6 repetitions. (Color figure online)

4 Results

4.1 Classification on N-MNIST

We train a hybrid SNN-ANN network for classification on the N-MNIST dataset [31]. In N-MNIST, each MNIST digit is displayed on an LCD monitor while an event camera performs three saccades within roughly 300 ms. It contains the same 60 000 train and 10 000 test samples as MNIST at a resolution of 34×34 pixels. Here we compare the performance of our hybrid network to networks where the SNN backbone is replaced with ANN layers (ANN-ANN), two baselines and two approaches from the literature: A conversion approach [18], where a trained ANN is converted to a rate-coded SNN and an ANN that reconstructs frames from events and classifies the frames [36]. The first baseline is a feed-forward ANN with the same structure as our VGG SNN-ANNs, but where all time steps are concatenated and presented to the network as different channels. The second baseline is an SNN of the same structure, where we accumulate the spikes of the last linear layer and treat this sum as logits for classification. The SNN is unrolled over the same number of time steps than the SNN-ANNs. We report results for multiple network sizes and configurations to show that hybrid SNN-ANNs generally perform better in terms of energy-efficiency and bandwidth (detailed architecture parameters can be found in the appendix). Mean values of accuracy and number of operations are reported together with the error of the mean over 6 repetitions, using different initial seeds for the weights. For one DenseSep network only 5 iterations are reported, because training of one trial did not converge, resulting in a significant outlier compared to the performance of all other networks. In Fig. 4a, we show the accuracy on the N-MNIST test set vs. the number of operations for our architectures, baselines and related approaches from the literature. Our hybrid networks (green, SNN) reach similar accuracies to the ANNs (purple), while being consistently more energy-efficient. The best hybrid architecture is a DenseNet with $g = 16$, $N_\mathrm{l} = 3$ and $n_\mathrm{out} = 1$. Compared to [18], it performs slightly worse ($(99.10 \pm 0.09)\,\%$ vs. $(99.56 \pm 0.01)\,\%$) in accuracy, but improves significantly on the number of operations ((94 ± 17) MOps vs. (460 ± 38) MOps) despite having more parameters (504 025 vs. 319 890). The average bandwidth between SNN and ANN is (0.250 ± 0.044) MEv/s for this architecture, or approximately 1.53 MB/s. Our smallest DenseNet with 64 875 parameters ($g = 8$, $N_\mathrm{l} = 2$, $n_\mathrm{out} = 1$) is even more efficient, needing only (15.9 ± 2.5) MOps at a bandwidth of (0.144 ± 0.034) MEv/s to reach $(99.06 \pm 0.03)\,\%$ accuracy. Our results are mostly above [36] in terms of accuracy, although our networks are much smaller (their networks have over 10 M parameters). Due to the large parameter size, we also estimate that we should be more energy-efficient, although the authors do not provide any numbers in their publication.

In Fig. 4b, we plot the average per backbone over all architectures in Fig. 4a, relative to the averages of the ANN-ANN architectures. All hybrid SNN-ANNs improve the number of operations significantly over ANN-ANN implementations with at most a minor loss in accuracy. Hybrid DenseNets provide the best accuracy-efficiency trade-off with an average improvement of about a factor of 56 in energy-efficiency while also increasing accuracy by approximately 0.2%. For VGG architectures, the energy-efficiency is increased by a factor of roughly 65 at a loss in accuracy of 0.4 between hybrid and ANN-ANN architectures. Hybrid DenseSep architectures lose approximately 0.2 accuracy points, but gain a factor of about 28 in energy-efficiency. We assume that the DenseSep architectures perform the worst in comparison for two reasons: First, they are already the most efficient ANN architectures, so improving is harder than for less optimized architectures. Second, as the effective number of layers is higher in this architecture compared to the other two, optimizing becomes more difficult and gradient deficiencies (vanishing, exploding, errors of surrogate gradient) accumulate more. Hyperparameter optimization (learning rate, dropout rates, regularization factors for weights and activations) was not performed for each network separately, but only once for a DenseNet with $N_\mathrm{l} = 2, g = 16, n_\mathrm{out} = 1$ on a validation set that consisted of 10% of the training data, *i.e.*, 6000 samples.

4.2 Classification on N-CARS

N-CARS is a binary car vs. background classification task, where each sample has a length of 100 ms. We train the same networks as in Sect. 4.1 with a growth factor $g = 16$ and compare to the same baselines and two results from the literature. We see the same trend in energy-efficiency improvements over ANNs, with factors ranging from 15 (DenseSep) to 110 (VGG). Relative accuracy decreases significantly for DenseSeps by 1.5 points, but is 0.75 points higher for VGG. These results, together with the results in Fig. 4b suggest, that the more efficient an architecture already is, the less can be gained from training an equiva-

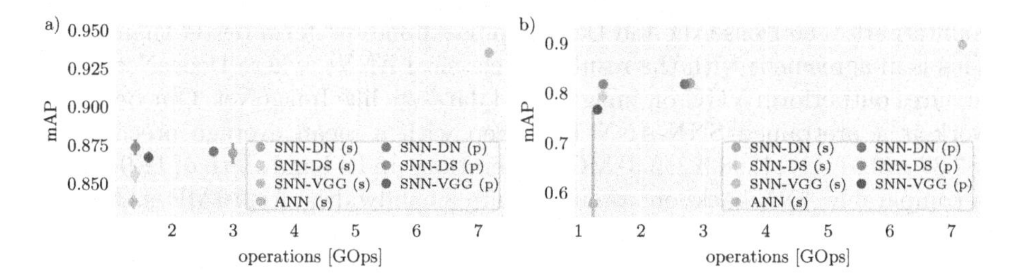

Fig. 5. Mean average precision (mAP) over number of operations for `shapes_translation-30/90` (**a**/**b**, mean over 4 trials). The architectures with the best results on the `N-MNIST` training are either trained from scratch (s) or initialized with the weights from the training (pretrained, p). Pretrained architectures improve over random initialization (VGG, DenseSep) or are on par with it (DenseNet).

lent hybrid SNN, although the gains of at least a factor of 15 in energy-efficiency is still significant. In comparison to our ANN and SNN baselines, our SNN-ANNs perform better in terms of accuracy and number of operations. Two out of four SNN runs could not go beyond chance level and were therefore excluded from the validation. In conclusion, using hybrid SNN-ANNs saves energy compared to SNNs and ANNs of similar structure. In this experiment, our architecture is not competitive to state-of-the-art [18] in accuracy, but has a similar energy demand with approximately double the number of parameters. Detailed results and figures can be found in the appendix.

4.3 Object Detection on `shapes_translation`

The `shapes_translation` dataset [28] contains frames, events, IMU measurements and calibration information of a scene of different shapes pinned to a wall. As training data for event-based vision object detection is scarce [11], we labelled bounding boxes for each of the 1356 images and 10 shapes, resulting in 9707 ground truth bounding boxes. A detailed description and an example of the dataset can be found in the appendix. We provide two train/test splits: `shapes_translation-90` where 90% of the data is randomly assigned to the train set and a more difficult split `shapes_translation-30`, where only 30% of the data is used in the train set. In section Sect. 4.4 and Sect. 4.5, we present the results on `shapes_translation-30/90`.

4.4 Results on `shapes_translation-90`

For all backbones, we take the architecture parameters from the best network in the `N-MNIST` task. See the appendix for details. We want to compare hybrid networks trained from scratch with networks where the SNN backbone is initialized with network weights trained on `N-MNIST`. This allows investigating the effect of transfer learning during training. Results are shown in Fig. 5. The networks with pretrained weights always converge to higher mAP than their non-pretrained counterpart. The DenseNet and DenseSep backbones perform better than VGG. This is in agreement with the results for classical ANNs, where DenseNet architectures outperform VGG on image-based datasets like ImageNet. Our best network is a pretrained SNN-ANN DenseSep with a mean average precision of (87.37 ± 0.51) %, (1398.9 ± 2.3) MOps operations and a bandwidth of 11.0 MB/s. A comparable ANN backbone would require a bandwidth of 1210 MB/s. A regular SSD architecture with the same backbone as our VGG network, where all time steps are concatenated over the channels outperforms our networks in terms of mAP, but also needs significantly more operations. We report the detailed results in the appendix.

4.5 Results on `shapes_translation-30`

We do the same evaluation as in Sect. 4.4, but with a 30/70% training/test data split (Fig. 5). The mAP is higher for networks with pretrained weights for DenseSep and VGG and on par with DenseNet. As with `shapes_translation-90`, the backbones with skip connections perform better than the VGG backbone. One of the four VGG experiments did not fully converge, explaining the high standard deviation. Our best network is an SNN-ANN DenseNet trained from scratch with (2790 ± 50) MOps operations, a mean average precision of (82.0 ± 1.0) %, and a bandwidth of 2.68 MB/s. A comparable ANN backbone would have a bandwidth of 864 MB/s. As in Sect. 4.4 the regular SSD architecture is better in mAP but worse in the number of operations. We report the detailed results in the appendix.

5 Conclusion

In this paper, we introduced a novel hybrid SNN-ANN architecture for efficient classification and object detection on event data that can be trained end-to-end with backpropagation. Hybrid networks can overcome the energy-efficiency limit of rate-coded converted SNN architectures, improving by up to a factor of 10. In comparison to similar ANN architectures, they improve by a factor of 15 to 110 on energy-efficiency with only a minor loss in accuracy. Their flexible design allows efficient custom hardware implementations for both the SNN and ANN part, while minimizing the required communication bandwidth. Our SNN-ANN networks learn general features of event camera data, that can be utilized to boost the object detection performance on a transfer learning task.

We expect that the generality of the features can be improved when learning on larger and more diverse datasets. Our work is particularly suited for datasets where temporal integration happens on a short time interval, but struggles for longer time intervals, *e.g.*, multiple seconds due to the immense number of rollout steps needed. Recent advances in deep learning, particularly C-RBP [23] can potentially help to overcome this by using recurrent backpropagation that has a constant memory complexity with number of steps (compared to backpropagation, where the memory-complexity is linear). This also would ensure that our networks converge to a fixed point over time, potentially making predictions more stable. More work on surrogate gradients and methods to stabilize training can further help to increase both energy-efficiency and performance of hybrid networks. Our work is a first step towards ever more powerful event-based perception architectures that are going to challenge the performance of image-based deep learning methods.

Acknowledgments. The authors would like to acknowledge the financial support of the CogniGron research center and the Ubbo Emmius Funds (Univ. of Groningen). Furthermore, this publication has received funding from the European Union's Horizon 2020 research innovation programme under grant agreement 732642 (ULPEC project).

References

1. Amir, A., et al.: A low power, fully event-based gesture recognition system. In: Proceedings of the IEEE Conference on Computer Vision and Pattern Recognition, pp. 7243–7252 (2017)
2. Barbier, T., Teulière, C., Triesch, J.: Unsupervised learning of spatio-temporal receptive fields from an event-based vision sensor. In: Farkaš, I., Masulli, P., Wermter, S. (eds.) ICANN 2020. LNCS, vol. 12397, pp. 622–633. Springer, Cham (2020). https://doi.org/10.1007/978-3-030-61616-8_50
3. Billaudelle, S., et al.: Versatile emulation of spiking neural networks on an accelerated neuromorphic substrate. In: 2020 IEEE International Symposium on Circuits and Systems (ISCAS), pp. 1–5 (2020). https://doi.org/10.1109/ISCAS45731.2020.9180741
4. Cramer, B., et al.: Surrogate gradients for analog neuromorphic computing. arXiv 2006.07239 (2021)
5. Davies, M., et al.: Loihi: a neuromorphic manycore processor with on-chip learning. IEEE Micro **38**(1), 82–99 (2018). https://doi.org/10.1109/MM.2018.112130359
6. Everingham, M., Gool, L.V., Williams, C.K.I., Winn, J., Zisserman, A.: The pascal visual object classes (VOC) challenge. Int. J. Comput. Vis. **88**, 303–308 (2009). https://www.microsoft.com/en-us/research/publication/the-pascal-visual-object-classes-voc-challenge/, printed version publication date: June 2010
7. Falanga, D., Kleber, K., Scaramuzza, D.: Dynamic obstacle avoidance for quadrotors with event cameras. Sci. Robot. **5**(40) (2020). https://doi.org/10.1126/scirobotics.aaz9712
8. Fischer, V., Koehler, J., Pfeil, T.: The streaming rollout of deep networks - towards fully model-parallel execution. In: Bengio, S., Wallach, H., Larochelle, H., Grauman, K., Cesa-Bianchi, N., Garnett, R. (eds.) Advances in Neural Information Processing Systems, vol. 31, pp. 4039–4050. Curran Associates, Inc. (2018). http://papers.nips.cc/paper/7659-the-streaming-rollout-of-deep-networks-towards-fully-model-parallel-execution.pdf
9. Furber, S.B., et al.: Overview of the SpiNNaker system architecture. IEEE Trans. Comput. **62**(12), 2454–2467 (2013). https://doi.org/10.1109/TC.2012.142
10. Gallego, G., et al.: Event-based vision: a survey. IEEE Trans. Pattern Anal. Mach. Intell. 1 (2020). https://doi.org/10.1109/tpami.2020.3008413, http://dx.doi.org/10.1109/TPAMI.2020.3008413

11. Gehrig, D., Gehrig, M., Hidalgo-Carrio, J., Scaramuzza, D.: Video to events: recycling video datasets for event cameras. In: IEEE/CVF Conference on Computer Vision and Pattern Recognition (CVPR) (June 2020)
12. Gehrig, D., Loquercio, A., Derpanis, K.G., Scaramuzza, D.: End-to-end learning of representations for asynchronous event-based data. In: Proceedings of the IEEE/CVF International Conference on Computer Vision (ICCV) (October 2019)
13. Gerstner, W., Kistler, W.M., Naud, R., Paninski, L.: Neuronal dynamics: from single neurons to networks and models of cognition (2014)
14. Howard, A.G., et al.: MobileNets: Efficient convolutional neural networks for mobile vision applications. arXiv 1704.04861 (2017)
15. Huang, G., Liu, Z., Weinberger, K.Q.: Densely connected convolutional networks. In: The IEEE Conference on Computer Vision and Pattern Recognition (CVPR) (2017)
16. Kheradpisheh, S.R., Ganjtabesh, M., Thorpe, S.J., Masquelier, T.: STDP-based spiking deep convolutional neural networks for object recognition. Neural Netw. **99**, 56–67 (2018)
17. Kingma, D.P., Ba, J.: Adam: A method for stochastic optimization. CoRR abs/1412.6980 (2015)
18. Kugele, A., Pfeil, T., Pfeiffer, M., Chicca, E.: Efficient processing of spatio-temporal data streams with spiking neural networks. Front. Neurosci. **14**, 439 (2020). https://doi.org/10.3389/fnins.2020.00439
19. Lagorce, X., Orchard, G., Galluppi, F., Shi, B.E., Benosman, R.B.: HOTS: a hierarchy of event-based time-surfaces for pattern recognition. IEEE Trans. Pattern Anal. Mach. Intell. **39**(7), 1346–1359 (2017). https://doi.org/10.1109/TPAMI.2016.2574707
20. Lee, C., Kosta, A.K., Zhu, A.Z., Chaney, K., Daniilidis, K., Roy, K.: Spike-FlowNet: event-based optical flow estimation with energy-efficient hybrid neural networks. In: Vedaldi, A., Bischof, H., Brox, T., Frahm, J.-M. (eds.) ECCV 2020. LNCS, vol. 12374, pp. 366–382. Springer, Cham (2020). https://doi.org/10.1007/978-3-030-58526-6_22
21. Lee, J.H., Delbruck, T., Pfeiffer, M.: Training deep spiking neural networks using backpropagation. Front. Neurosci. **10**, 508 (2016). https://doi.org/10.3389/fnins.2016.00508
22. Lichtsteiner, P., Posch, C., Delbruck, T.: A 128×128 120 dB 15μs latency asynchronous temporal contrast vision sensor. IEEE J. Solid-State Circuits **43**(2), 566–576 (2008). https://doi.org/10.1109/JSSC.2007.914337
23. Linsley, D., Karkada Ashok, A., Govindarajan, L.N., Liu, R., Serre, T.: Stable and expressive recurrent vision models. In: Larochelle, H., Ranzato, M., Hadsell, R., Balcan, M.F., Lin, H. (eds.) Advances in Neural Information Processing Systems, vol. 33, pp. 10456–10467. Curran Associates, Inc. (2020). https://proceedings.neurips.cc/paper/2020/file/766d856ef1a6b02f93d894415e6bfa0e-Paper.pdf
24. Liu, S.C., Delbruck, T.: Neuromorphic sensory systems. Curr. Opin. Neurobiol. **20**(3), 288–295 (2010)
25. Liu, W., et al.: SSD: single shot multibox detector. In: Leibe, B., Matas, J., Sebe, N., Welling, M. (eds.) ECCV 2016. LNCS, vol. 9905, pp. 21–37. Springer, Cham (2016). https://doi.org/10.1007/978-3-319-46448-0_2
26. Maro, J.M., Ieng, S.H., Benosman, R.: Event-based gesture recognition with dynamic background suppression using smartphone computational capabilities. Front. Neurosci. **14**, 275 (2020)
27. Merolla, P.A., et al.: A million spiking-neuron integrated circuit with a scalable communication network and interface. Science **345**(6197), 668–673 (2014)

28. Mueggler, E., Rebecq, H., Gallego, G., Delbruck, T., Scaramuzza, D.: The event-camera dataset and simulator: Event-based data for pose estimation, visual odometry, and SLAM **36**, 142–149 (2017)
29. Neftci, E.O., Mostafa, H., Zenke, F.: Surrogate gradient learning in spiking neural networks: bringing the power of gradient-based optimization to spiking neural networks. IEEE Signal Process. Mag. **36**(6), 51–63 (2019)
30. Opez-Vázquez, G., et al.: Evolutionary spiking neural networks for solving supervised classification problems. Comput. Intell. Neurosci. **2019**, 13 (2019). https://doi.org/10.1155/2019/4182639
31. Orchard, G., Jayawant, A., Cohen, G.K., Thakor, N.: Converting static image datasets to spiking neuromorphic datasets using saccades. Front. Neurosci. **9**, 437 (2015). https://doi.org/10.3389/fnins.2015.00437
32. Perot, E., De Tournemire, P., Nitti, D., Masci, J., Sironi, A.: Learning to detect objects with a 1 megapixel event camera. In: Larochelle, H., Ranzato, M., Hadsell, R., Balcan, M.F., Lin, H. (eds.) Advances in Neural Information Processing Systems, vol. 33, pp. 16639–16652. Curran Associates, Inc. (2020). https://proceedings.neurips.cc/paper/2020/file/c213877427b46fa96cff6c39e837ccee-Paper.pdf
33. Posch, C., Matolin, D., Wohlgenannt, R.: A qVGA 143 dB dynamic range frame-free PWM image sensor with lossless pixel-level video compression and time-domain CDS. IEEE J. Solid-State Circuits **46**(1), 259–275 (2011). https://doi.org/10.1109/JSSC.2010.2085952
34. Qiao, N., et al.: A reconfigurable on-line learning spiking neuromorphic processor comprising 256 neurons and 128k synapses. Front. Neurosci. **9**, 141 (2015). https://doi.org/10.3389/fnins.2015.00141
35. Rathi, N., Roy, K.: DIET-SNN: Direct input encoding with leakage and threshold optimization in deep spiking neural networks. arXiv 2008.03658 (2020)
36. Rebecq, H., Ranftl, R., Koltun, V., Scaramuzza, D.: Events-to-video: bringing modern computer vision to event cameras. In: Proceedings of the IEEE/CVF Conference on Computer Vision and Pattern Recognition (CVPR) (June 2019)
37. Rieke, F.: Spikes: Exploring the Neural Code. MIT Press, Bradford book, Cambridge (1999)
38. Rueckauer, B., Lungu, I.A., Hu, Y., Pfeiffer, M., Liu, S.C.: Conversion of continuous-valued deep networks to efficient event-driven networks for image classification. Front. Neurosci. **11**, 682 (2017). https://doi.org/10.3389/fnins.2017.00682
39. Schemmel, J., Brüderle, D., Grübl, A., Hock, M., Meier, K., Millner, S.: A wafer-scale neuromorphic hardware system for large-scale neural modeling. In: Proceedings of 2010 IEEE International Symposium on Circuits and Systems, pp. 1947–1950 (2010)
40. Sengupta, A., Ye, Y., Wang, R., Liu, C., Roy, K.: Going deeper in spiking neural networks: VGG and residual architectures. Front. Neurosci. **13**, 95 (2019). https://doi.org/10.3389/fnins.2019.00095
41. Serrano-Gotarredona, T., Linares-Barranco, B.: A 128 $\times$ 128 1.5% contrast sensitivity 0.9% FPN 3 μs latency 4 mW asynchronous frame-free dynamic vision sensor using transimpedance preamplifiers. IEEE J. Solid-State Circuits **48**(3), 827–838 (2013). https://doi.org/10.1109/JSSC.2012.2230553
42. Shrestha, S.B., Orchard, G.: SLAYER: Spike layer error reassignment in time. In: Bengio, S., Wallach, H., Larochelle, H., Grauman, K., Cesa-Bianchi, N., Garnett, R. (eds.) Advances in Neural Information Processing Systems, vol. 31, pp. 1412–1421. Curran Associates, Inc. (2018). http://papers.nips.cc/paper/7415-slayer-spike-layer-error-reassignment-in-time.pdf

43. Simonyan, K., Zisserman, A.: Very deep convolutional networks for large-scale image recognition. arXiv 1409.1556 (2015)
44. Sironi, A., Brambilla, M., Bourdis, N., Lagorce, X., Benosman, R.: HATS: histograms of averaged time surfaces for robust event-based object classification. In: Proceedings of the IEEE Conference on Computer Vision and Pattern Recognition (CVPR) (June 2018)
45. Stimberg, M., Brette, R., Goodman, D.F.: Brian 2, an intuitive and efficient neural simulator. eLife **8**, e47314 (2019). https://doi.org/10.7554/eLife.47314
46. Vidal, A.R., Rebecq, H., Horstschaefer, T., Scaramuzza, D.: Ultimate SLAM? Combining events, images, and IMU for robust visual SLAM in HDR and high-speed scenarios. IEEE Robot. Autom. Lett. **3**(2), 994–1001 (2018). https://doi.org/10.1109/LRA.2018.2793357
47. Wu, Y., Deng, L., Li, G., Zhu, J., Xie, Y., Shi, L.: Direct training of spiking neural networks: faster, larger, better. In: Proceedings of the AAAI Conference on Artificial Intelligence (2019)
48. Zhu, D., et al.: Neuromorphic visual odometry system for intelligent vehicle application with bio-inspired vision sensor. In: 2019 IEEE International Conference on Robotics and Biomimetics (ROBIO), pp. 2225–2232. IEEE (2019)

A Comparative Study of PnP and Learning Approaches to Super-Resolution in a Real-World Setting

Samim Zahoor Taray[1], Sunil Prasad Jaiswal[1(✉)], Shivam Sharma[1,2], Noshaba Cheema[2,3], Klaus Illgner-Fehns[1], Philipp Slusallek[2,3], and Ivo Ihrke[1,4]

[1] K|Lens GmbH, Saarbrücken, Germany
sunil.jaiswal@k-lens.de
[2] Saarland Informatics Campus, Saarbrücken, Germany
[3] German Research Center for Artificial Intelligence (DFKI), Kaiserslautern, Germany
[4] Center for Sensor Systems (ZESS), University of Siegen, Siegen, Germany

Abstract. Single-Image Super-Resolution has seen dramatic improvements due to the application of deep learning and commonly achieved results show impressive performance. Nevertheless, the applicability to real-world images is limited and expectations are often disappointed when comparing to the performance achieved on synthetic data. For improving on this aspect, we investigate and compare two extensions of orthogonal popular techniques, namely plug-and-play optimization with learned priors, and a single end-to-end deep neural network trained on a larger variation of realistic synthesized training data, and compare their performance with special emphasis on model violations. We observe that the end-to-end network achieves a higher robustness and flexibility than the optimization based technique. The key to this is a wider variability and higher realism in the training data than is commonly employed in training these networks.

Keywords: Deconvolution · Generalizability · Degradations · Super-resolution

1 Introduction

Single-image super-resolution (SISR) techniques, in particular, learning based or learning supported techniques, have evolved to a point where impressive image improvements can be achieved on a wide variety of scenes [8,9,22,32,41]. However, their applicability in real-world scenarios is reduced by an often inadequate modeling of the image formation in real imaging systems [25]. In particular, spatially varying blur is often ignored and the noise characteristics of real sensors, modified by the non-linear operations of image signal processors, are commonly inadequately treated as Gaussian processes. This limitation can be traced back to synthesized data being used for training. The general approach is to use images from existing data sets and to synthesize low resolution images by using bicubic downsampling.

Supplementary Information The online version contains supplementary material available at https://doi.org/10.1007/978-3-030-92659-5_20.

C. Bauckhage et al. (Eds.): DAGM GCPR 2021, LNCS 13024, pp. 313–327, 2021.
https://doi.org/10.1007/978-3-030-92659-5_20

This, in conjunction with the required generalizability to different camera and lens combinations, poses a challenge for the real-world application of SISR techniques. Our goal is to improve on this situation. For this purpose, we follow two orthogonal approaches: 1) we generalize a method based on the Plug-and-Play (PnP) framework [35] to handle spatially varying blur kernels. 2) we explore the potential of Deep Neural Networks (DNNs) [8] to model mappings of input images that are subject to a variety of possible degradations, including different blur kernels of the expected spatial variation due to the lens system, as well as different noise levels and compositions, i.e. different ratios of Gaussian and Poisson noise as well as correlations introduced by image processing. We compare the performance of these two approaches and assess their robustness against model violations.

Common knowledge holds that PnP techniques are more resilient against the outlined problems, whereas DNNs quickly deteriorate in performance when run on data outside the characteristics of the training set. The latter is often assumed to include training data for a single image formation model and a modest variability in the noise settings. Our investigations show that the training data set can include a sufficient variety of image formation settings such that the performance of the trained DNN approaches an effectively blind deconvolution setting.

In summary, our contributions are as follows:

- we generalize PnP optimization techniques for SISR to handle spatially varying blur kernels,
- we propose a data synthesis approach for generating realistic input data for DNN training that incorporates the major sources of variation in real-world data,
- we train a state-of-the-art DNN [12,36] on a suitably synthesized training data set that includes expected real-world image formation variations (varying blur and realistic noise distributions),
- experimentally compare the resulting techniques on different scenarios with special emphasis on a violation of the image formation model.

We find that the end-to-end learning technique yields superior results and robustness as compared to the extended plug-and-play technique, as well as compared against the state-of-the-art.

2 Related Work

The goal of SISR is to generate a high-resolution (HR) image from a single low-resolution (LR) image. Most methods are intended for photographic content and aim to hallucinate details and textures that fit nicely with the input LR image while producing a realistic looking higher resolution image.

Example-based methods can be divided into two categories namely Internal and External Example-based methods.

Internal Example-based Methods [13,17] rely on the assumption that a natural image has repetitive content and exploit this recurrence by replacing each patch within the LR image with a higher resolution version. Shocher et al. [33] propose Zero Shot Super-Resolution wherein they train a small CNN using patches taken only from the test image.

External Example-based methods rely on an external database of paired LR/HR images. Freeman et al. [11] first generate a dictionary of LR and HR patches that is then used similarly to the previously discussed techniques in applications. Chang et al. [6] extended this method using k nearest dictionary neighbors whereas Zeyde et al. [38] learn sparse representations of LR and HR patches. Dictionary-based methods were superseded by Convolutional Neural Network (CNN) based methods which have been very successful and continue to dominate standard SISR benchmarks.

CNN-based methods were introduced by Dong et al. [8] who proposed the shallow convolutional network SRCNN. Kim et al. [20] show significant performance gains by using deeper networks. Shi et al. [32] propose the Efficient Sub-Pixel Convolution Layer (ESPCN) to include an up-scaling step into the network that had previously been implemented as a pre-process. Typical training losses are the L_1 and L_2 losses. These loss functions tend to create unattractive structures in the super-resolved (SR) images [19,22]. Thus, several recent works [19,22,36] have aimed to devise loss functions which correlate better with human perception. Ledig et al. [22] propose SRGAN which uses a combination of three loss functions to achieve photo-realistic super-resolution. Wang et al. [36] introduce improvements by comprehensively studying the factors affecting the performance of SRGAN, creating the ESRGAN technique. It should be noted that SRGAN and ESRGAN tend to produce images with textures even in flat image regions, creating perceptually unpleasing artifacts. To overcome this problem, Fritsche et al. [12] propose frequency separation, i.e. the GAN cannot synthesize in low-frequency image regions. We base our end-to-end technique, Sect. 5 on a combination of ESRGAN trained with a perceptual loss and frequency separation.

Recently, some algorithms have been proposed to overcome the challenges of real-world super-resolution mentioned in the introduction [18,21,24,27,31,42]. Zhou and Süsstrunk [42] construct a synthetic data set by simulating degradation on LR and HR pairs and train a CNN to perform SR, whereas Lugmayr et al. [24] propose unsupervised learning for data set generation while employing a supervised network for SR. Some methods [21,27] exploit blind kernel estimation to generalize better to real images while maintaining a non-blind general approach, whereas Ji et al. [18] propose to use two differently trained networks with a final output fusion stage to solve the realistic image super-resolution problem.

Plug-and-Play (PnP) Approaches are, in contrast, based on an optimization formulation of the SISR problem [7,15,26,34,35,43]. Typically, convex optimization has been employed for this purpose in image restoration [7,35,43] and, due to its flexibility, the idea has been extended to different applications [15,26,34], including super-resolution [4,5]. Most of these methods are limited to Gaussian noise assumptions. Zhang et al. [39] proposed to include a learned CNN prior in the iterative solution using the MAP framework. We build on their technique in Sect. 4.

3 Overview

The important components of an imaging system are its optics, its sensor and its processing unit, respective the algorithm running on this unit. These components work together to give images the characteristics that we associate with natural real-world images, i.e. non-uniform lens blur and non-Gaussian noise afflicts real-world images. Optimization-based algorithms must properly model this image formation and we extend the PnP technique DPSR [39] to handle spatially varying blur in Sect. 4. Learning techniques rely on large amounts of training data that are most suitably generated synthetically. We describe the generation of suitable data with the outlined real-world characteristics and give details of the training process of the state-of-the-art network ESRGAN-FS [12,36] in Sect. 5. In Sect. 6, we compare the performance of the two approaches, first in synthetic experiments and later on actual real-world images.

4 Plug and Play Framework

Let the observed LR image be denoted by $\mathbf{y} \in R^{MN}$. It is related to the desired HR image $\mathbf{x} \in R^{s^2MN}$ via blurring, sampling and the addition of noise. Here M and N are the height and width of the LR image and $s \geq 1$ is the scaling factor. The effect of spatially varying blur is modeled by a linear operator H that operates on $\mathbf{x}$, while the sampling of the blurred image is modeled by the linear operator D. In this section, the sampled image is degraded by additive noise denoted by vector $\mathbf{n} \in R^{MN}$. In our implementations, the operators are realized as function evaluations rather than being discretized as sparse matrices.

According to Zhang et al. [39], it is advantageous, mathematically, to switch the procedures of blurring and sampling, because it enables a simple algorithmic decomposition that allows neural networks to be used as priors for super resolution.

$$\mathbf{y} = HD\mathbf{x} + \mathbf{n}. \tag{1}$$

The image formation model of Eq. 1, while physically incorrect, is an approximation that enables network-encoded priors to be easily incorporated into proximal optimization algorithms [3].

The resulting optimization problem is

$$\hat{\mathbf{x}} = \arg\min_{\mathbf{x}} \frac{1}{2\sigma^2}||\mathbf{y} - HD\mathbf{x}||^2 + \lambda\Phi(\mathbf{x}), \tag{2}$$

where the first term is the data-fidelity term and $\Phi(\mathbf{x})$ is the regularization term encoding the prior information, that is network-encoded in our setting. Solving the optimization problem via proximal algorithms consists of a variable splitting, that, in the case of the ADMM method [28] applied to Eq. 2 leads to the following iterative scheme:

$$\mathbf{x}^{k+1} = \arg\min_{\mathbf{x}} \lambda\Phi(\mathbf{x}) + \frac{\rho}{2}||D\mathbf{x} - \mathbf{z}^k + \mathbf{u}^k||_2^2, \tag{3}$$

$$\mathbf{z}^{k+1} = \arg\min_{\mathbf{z}} \frac{1}{2\sigma^2}||\mathbf{y} - H\mathbf{z}||_2^2 + \frac{\rho}{2}||D\mathbf{x}^{k+1} - \mathbf{z} + \mathbf{u}^k||_2^2, \tag{4}$$

$$\mathbf{u}^{k+1} = \mathbf{u}^k + \left(D\mathbf{x}^{k+1} - \mathbf{z}^{k+1}\right). \tag{5}$$

Here, $D\mathbf{x} = \mathbf{z}$ is the variable splitting, $\mathbf{u}$ is the dual variable, σ is the noise level under a Gaussian white noise assumption, λ is the regularization parameter and ρ is the penalty parameter of the associated augmented Lagrangian L_ρ [3]. For a detailed derivation, please see the supplemental material. Please observe that the linear operators H and D now appear in different sub-expressions and not, as before, in combination. This enables the plug-and-play nature of the algorithm.

The first update step, Eq. 3 is a super-resolution scheme with a network-based prior encoded in $\Phi(\mathbf{x})$ as can be seen by ignoring the dual variable $\mathbf{u}$ that tends to zero as the splitting equality $D\mathbf{x} = \mathbf{z}$ is approached by the iteration. We follow the procedure of Zhang et al. [39] to obtain $\mathbf{x}^{k+1}$ via a pre-trained CNN. We use the same architecture as [39], i.e. SRResnet+, and train it for variable levels of additive white Gaussian noise in the range $[0, 50]$ for joint denoising and super-resolution. We denote the network by $\mathcal{SR}_D$ to note that the network is trained for a fixed downsampling operator D. To adapt a single network for various noise levels, a noise level map of the same size as the input image is created and concatenated with the input image. The image and the noise level map concatenated together become the input to the network which outputs a super-resolved and denoised image $\mathbf{x}^{k+1}$, i.e. Eq. 3 is replaced by

$$\mathbf{x}^{k+1} = \mathcal{SR}_D\left(\mathbf{z}^k, \rho\right). \tag{6}$$

The second update step, Eq. 4 can be interpreted as a deblurring step at the size of the LR image, which results in an intermediate deblurred LR image $\mathbf{z}$. Previous work considers the blur to be spatially invariant [39]. However, we consider the more general problem where the blur is spatially varying. We discuss this minimization problem in detail in Sect. 4.1 because it contains our extension of the algorithm [39] to spatially varying blur kernels.

The third update step given in Eq. 5 is simply the update of the dual variable required in ADMM.

We set $\lambda = 0.3$ and vary ρ to affect regularization for all our experiments. In practice, we increase ρ on an exponential scale from $\rho = 1/2500$ to $\rho = 2/\sigma$, where σ is the noise level of the image (assumed to be in the range $[0, 255]$).

4.1 Spatially Varying De-blurring

In this section we describe our approach to tackle the minimization problem of Eq. 4 which can be re-written in the following way:

$$\mathbf{z}^{k+1} = \arg\min_{\mathbf{z}} \underbrace{\|\mathbf{y} - H\mathbf{z}\|_2^2 + \sigma^2\rho\|\tilde{\mathbf{x}}^{k+1} - \mathbf{z}\|_2^2}_{G(\mathbf{z})}. \tag{7}$$

For ease of notation, we have used $\tilde{\mathbf{x}}^{k+1} = D\mathbf{x}^{k+1} + \mathbf{u}^k$. Recall that $\mathbf{y}$ is the input LR image that we wish to super-resolve and σ is the noise level of each pixel in $\mathbf{y}$. As mentioned above, H now represents the spatially varying blur operator. The objective $G(\mathbf{z})$ is smooth and convex. To minimize the objective, we rely on the Gradient Descent with Momentum [30] method with constant step size and momentum parameters given by

$$\mathbf{w}_{t+1} = \mathbf{z}_t + \beta(\mathbf{z}_t - \mathbf{z}_{t-1}) \tag{8}$$

$$\mathbf{z}_{t+1} = \mathbf{w}_{t+1} - \alpha\nabla_{\mathbf{z}}G(\mathbf{w}_{t+1}), \tag{9}$$

Algorithm 1. $\mathcal{AG}$ solver

Input Operators H and H^T. Constant quantities $\mathbf{y}$, $\tilde{\mathbf{x}}^{k+1}$, ρ and σ. Parameters α and β and total iterations $N_{\mathcal{AG}}$

1: Initialize $\mathbf{z}_{-1} = \mathbf{y}$ and $\mathbf{z}_0 = \mathbf{0}$
2: **while** $t < N_{\mathcal{AG}}$ **do**
3: $\mathbf{w}_{t+1} = \mathbf{z}_t + \beta(\mathbf{z}_t - \mathbf{z}_{t-1})$
4: $\mathbf{z}_{t+1} = \mathbf{w}_{t+1} - 2\alpha\left(H^T H \mathbf{w}_{t+1} + \rho\sigma^2 \mathbf{w}_{t+1} - H^T y - \rho\sigma^2 \tilde{\mathbf{x}}^{k+1}\right)$

Output $\mathbf{z}_{N_{\mathcal{AG}}}$

where $\nabla_{\mathbf{z}}$ denotes the gradient operator with respect to $\mathbf{z}$ and $\mathbf{w}$ is an intermediate vector introduced to shorten the notation. We set $\alpha = 0.2$ and $\beta = 0.1$ for all our experiments. The details of the solver are given in Algorithm 1.

Note that the solver only relies on matrix vector multiplications of the form $H\mathbf{z}$ and $H^T\mathbf{z}$. We implement these efficiently using the *Filter Flow Framework* of Hirsch et al. [16]. Explicit formulas for evaluating H and H^T are given in the supplement. For convenience, let us denote the solver by $\mathcal{AG}$ which allows us to write Eq. 4 as:

$$\mathbf{z}^{k+1} = \mathcal{AG}(H, H^T, \tilde{\mathbf{x}}^{k+1}, \mathbf{y}, \sigma, \rho). \tag{10}$$

The solver $\mathcal{AG}$ takes as inputs the operators H and H^T along with the vectors $\mathbf{y}$, $\tilde{\mathbf{x}}^k$ and the constants ρ and σ, and outputs the deblurred image $\mathbf{z}^{k+1}$. Finally, the overall scheme to solve the original SISR problem of Eq. 2 is presented in Algorithm 2.

Algorithm 2. PnP Optimization with DNN prior for SISR

Input LR image $\mathbf{y}$, noise level σ, operators H and H^T total iterations T

1: Initialize $\mathbf{z}^0 = \mathbf{0}$ and $\mathbf{u}^0 = \mathbf{0}$
2: **while** $k < N_{\mathcal{SR}}$ **do**
3: $\mathbf{x}^{k+1} = \mathcal{SR}_D\left(\mathbf{z}^k, \rho\right)$ [Joint SR and denoising]
4: $\mathbf{z}^{k+1} = \mathcal{AG}(H, H^T, \mathbf{x}^{k+1}, \mathbf{y}, \sigma, \rho)$[Deblurring Step]
5: $\mathbf{u}^{k+1} = \mathbf{u}^k + \left(D\mathbf{x}^{k+1} - \mathbf{z}^{k+1}\right)$ [Dual var. update]

Output $\mathbf{x}^{N_{\mathcal{SR}}}$

5 End-to-End Learning with Input Variants

In the above section, we generalize PnP optimization techniques and present a non-blind learning-supported optimization algorithm to handle spatially varying blur kernels. Our goal in the current section is to train the state-of-the-art SISR end-to-end network ESRGAN [36] to include variations in expected input data both in terms of realistic noise models and levels as well as in expected blur variations. We use the Residual-in-Residual Dense Block Network (RRDBNet) architecture of [36] with 23 RRDBs, and train the network using either L_1 loss or the perceptually motivated L_{per} loss [22]. We refer to the respective training loss in discussing our experimental results in Sect. 6. We also employ the frequency separation technique of Fritsche et al. [12] to suppress texture generation in smooth image areas. The images are cropped to a size of 128×128 during training and data augmentation consisting of random horizontal and vertical flips is also applied for better generalization. We implement the network in PyTorch and use the built in Adam optimizer with $\beta_1 = 0.9$ and $\beta_2 = 0.99$ and an initial learning rate of 2×10^{-4} for optimization. We set the batch size to 24 and train the network for 500 epochs. Training the network on one Nvidia Quadro 6000 RTX takes around 10 h.

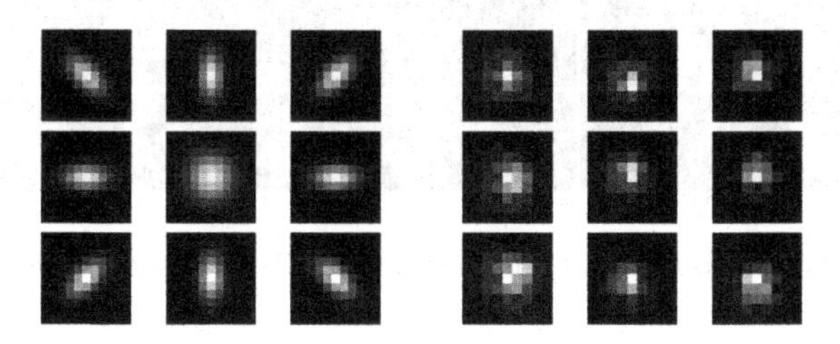

Fig. 1. Simulated spatially varying blur kernels used for the synthetic experiments (left). Measured spatially varying blur kernels of the target optical system (right). The spatial layout indicates the image regions affected by the corresponding kernels. The kernels are smoothly blended across the simulated images.

5.1 Training Data Synthesis

In order to generate a large amount of training data, we rely on LR image synthesis. To achieve realism, we aim to match the blurring and noise characteristics found in real world images. We use a data set containing high quality images [1,23] as a basis. Since the images in these data sets can contain residual noise, we clean these images prior to the synthesis operation by Gaussian filtering and downsampling, resulting in the clean HR image. Let this image be denoted by $\mathbf{x}$. The HR image is blurred to obtain a blurred image denoted by $\mathbf{x}_{\text{blurred}}$ by convolving $\mathbf{x}$ with a filter $\mathbf{h}$. This gives the clean but blurred image $\mathbf{x}_{\text{blurred}}$ which is then downsampled by the desired scale factor s giving the downsampled image denoted by $\mathbf{y}_s$ where s denotes, as before, the factor by which we ultimately wish to super-resolve the images. The filter $\mathbf{h}$ is obtained by randomly selecting one filter from a filter bank. The filter bank consists of filters obtained as explained below.

- We synthesize a set of filters consisting of 9×9, 7×7 and 5×5 Gaussian filters of different major axes and orientations. A subset of these filters are shown in Fig. 1 (left).
- We also measure PSFs by imaging an illuminated pinhole of $30\,\mu\text{m}$ diameter in a darkened photographic studio and add these to the filter bank. These are shown in Fig. 1 (right).

- In order to increase the size and diversity of the filter bank, we extract blur kernels from real-world images taken with the target optical system[1] using KernelGAN [21] and add them to the filter bank. We extracted around 3000 kernels using this approach and a few of them are shown in Fig. 2.

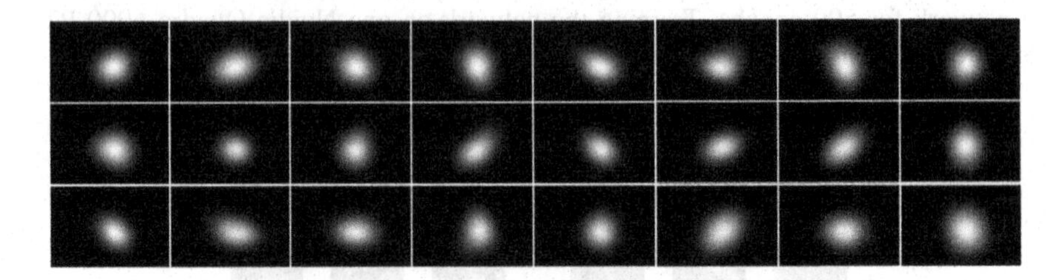

Fig. 2. Blur kernels extracted using KernelGAN [21] on 24 real-world images.

The next step consists of injecting realistic noise into $\mathbf{y}_s$ to finally arrive at the desired low resolution image. To achieve this we first convert $\mathbf{y}_s$ from sRGB to RAW space, yielding $\mathbf{r}_s$. This is in order to apply noise in the linear response regime of the sensor rather than in the non-linearly distorted processed image domain. Next, we generate noise according to the Poisson-Gaussian noise model of Foi et al. [10] by sampling from a heteroskedastic Gaussian distribution [14], i.e.

$$\mathbf{n}(\mathbf{r}_s) \sim \mathcal{N}(0, \sigma^2(\mathbf{r}_s)), \tag{11}$$

where $\sigma^2(\mathbf{r}_s)$ denotes the variance of the Gaussian distribution which is a function of the RAW image and is given by $\sigma^2(\mathbf{r}_s) = a\mathbf{r}_s + b$. The parameter a determines the amount of the Poisson and b the amount of the Gaussian noise component. The values of a and b are determined by the amount of noise present in the target images that we aim at super-resolving. As shown by Guo et al. [14], deep convolutional neural networks can effectively denoise images containing noise with different noise levels if the network is trained for these different noise levels. Therefore, we generate a variety of images containing different levels of Poisson and Gaussian noise by sampling a and b from a sensible range of values [14]. The last step consists of converting the noisy RAW image back into sRGB space which finally results in the noisy low resolution sRGB image $\mathbf{y}$ corresponding to the high resolution image $\mathbf{x}$. Figures 4 and 5 show examples of ground truth (GT) HR images and the corresponding synthesized LR images (LR input) with realistic blur and noise characteristics with this method. For the conversion steps, i.e. to convert the downsampled sRGB image to RAW space and then back to the sRGB space, we use CycleISP [37]. We note that our method would work with other methods for sRGB to RAW conversion.

[1] This way the SR scheme is tailored towards a particular type of optical system. In order to gain generality w.r.t. manufacturing differences for the optical system, the kernels have been extracted from pictures taken with different lenses of the same model.

6 Experimental Results

In the following, we evaluate and compare the two methods of Sect. 4 and Sect. 5 in terms of their performance with respect to real-world imperfections, i.e. realistic noise components and spatially varying blur. We first perform a set of quantitative synthetic experiments on a subset of 10 images taken from the DIV2K [1] validation set in Sect. 6.1 and finally show the qualitative performance on real-world data where no ground truth is available in Sect. 6.2.

6.1 Synthetic Experiments

Effect of Noise: To study and understand the effect of noise on the quality of SR we perform an experiment in which we synthesize test data only with different noise settings of a and b in Eq. 11 including the ISP-simulation (realistic noise, but no image formation model) and study the performance of the different algorithms. The results are shown in Fig. 3. Our test candidates are two CNN variants according to Sect. 5, trained with L_1 (red) and with L_{per} (green) loss and the PnP method of Sect. 4 (blue). We train the networks with training data up to the noise parameters $a = 0.03, b = 0.025$. The maximum training noise is marked by a dashed black vertical bar in Fig. 3. The figure shows three different performance metrics averaged over 10 test images. PSNR (dotted) and SSIM (dashed) curves refer to the right axis labels of the plots, whereas the perceptual LPIPS (solid) metric [40] uses the left axis labels. PSNR is plotted as relative PSNR which is a fraction of the maximum of 32.0 throughout the paper. Figure 4 shows visual examples on one of the test images. The red vertical dashed line in Fig. 3 indicates the noise settings used in Fig. 4.

As expected, with rising noise levels, the performance decreases for all methods and measures, both for Gaussian and Poissonian noise components. The L_1 trained models dominate the PSNR and SSIM scores in both cases while PnP is outperformed for PSNR. For SSIM, the L_{per} trained CNN and PnP compete in performance: for lower noise levels, the L_{per} trained model has advantages, whereas for larger levels, PnP performs better. In terms of Poisson noise, this trend is large, whereas for the Gaussian part (parameter b), the two methods behave very similarly for larger amounts of noise. In terms of the perceptual LPIPS score, a lower graph is better. Here, the L_{per} trained CNN has the best performance for low to medium noise levels in both cases, however, a cross-over is observed with the L_1 trained model. For the Poisson component (parameter a), the cross-over occurs, as expected, close to the maximum trained noise level. For the Gaussian component it occurs much earlier due to unknown reasons. The PnP method is not competitive in this metric which can be traced to its generation of flat structures in the images, see also Fig. 4.

Effect of Spatially Varying Blur: In this experiment, we test the importance of modeling the spatially varying blur in an imaging system. For this, we synthesize test images with known spatially varying blur kernels according to the methodology described in Sect. 4.1. An example of spatially varying blur kernels is provided in Fig. 1. For a fair comparison, the PnP method should run on Gaussian noise, whereas the CNNs should run on realistic noise. We solve this problem by fixing a realistic noise level

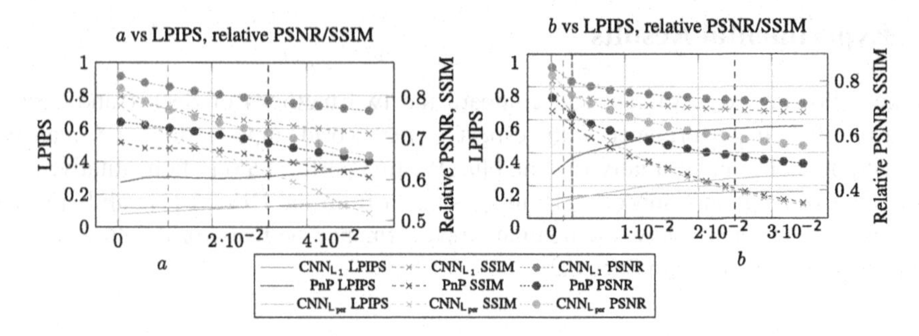

Fig. 3. Model performance against noise parameter variations of a Poisson-Gaussian model with ISP simulation. For PSNR and SSIM, larger values are better, for LPIPS, lower values indicate better performance. Curves are only to be compared for like scores (same decorator). The details of the plots are explained in the main text. (Color figure online)

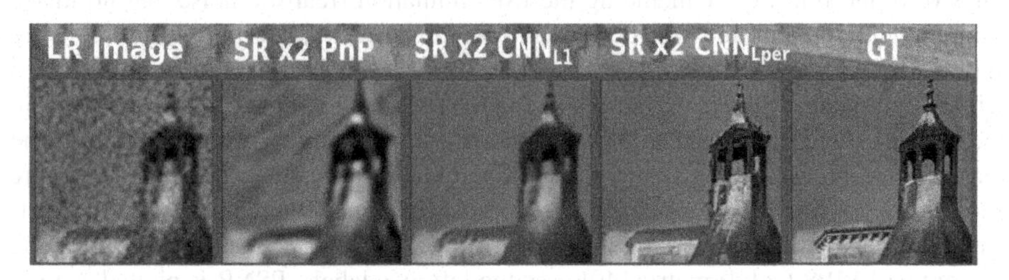

Fig. 4. Visual examples of the noise experiment on one of the test images degraded with the varying blur kernels shown in Fig. 1 (left) and realistic noise with noise levels given by $a = 0.011, b = 0.0001$.

Table 1. Numerical results for spatially varying blur. For the experimental setting, see the main text.

Method	Varying blur, Gaussian noise		
	PSNR↑	SSIM↑	LPIPS↓
PnP	23.20739	0.67589	0.3049
CNN L_1 (invariant)	26.3152	0.7699	0.281
CNN L_1	**26.5405**	**0.781**	0.266
CNN L_{per}	24.4716	0.7068	**0.0955**

of ($a = 0.01, b = 0.001$), synthesizing LR images as input for the CNN technique and estimating the standard deviation. We then synthesize another set of LR images as input for the PnP technique that uses this estimated Gaussian noise level. The numerical performance is shown in Table 1.

We again report averages over 10 test images synthesized with the fixed set of kernels in Fig. 1 (left). In addition to the previous contestants, we add an L_1 trained CNN that was only learned on a spatially invariant kernel (center kernel in Fig. 1 left) to demonstrate the importance of proper modeling. The PnP method is used with known

Fig. 5. Qualitative comparison of super-resolved images produced by the PnP approach with the L_1-trained CNN (spatial variation). The results from the CNN have sharper details and more realistic textures.

blur kernels (non-blind setting). Visual results for this experiment are shown in Fig. 5 for the L_1 trained CNN and the PnP method. PnP favors smooth regions with sharp edges, similar but not as noticable as in a standard Total Variation regularized setting. The variation-trained CNN recovers meaningful structures, but avoids hallucination of detail in smooth images regions thanks to frequency separation.

Model Violations. In this numerical experiment, we analyze the effect of mismatches in model assumptions. The PnP method of Sect. 4 has the strongest assumptions since it is a non-blind technique and is based on a Gaussian error assumption as well as resting on a non-physical image formation model, Eq. 1, whereas the network family of Sect. 5 is adjusted by the training set and has blind estimation capabilities due to training on a variety of kernels and noise settings.

We synthesize test data for two image formation models: space-variant blurring followed by downsampling (the physically correct model), and the non-physical inverse order of operations, Eq. 1. In addition, we compare performance in the assumed Gaussian noise setting vs. the realistic noise setting of Sect. 5.

6.2 Real-World Images

We perform a comparison of our methods (CNN L_{per}, blind = "Ours ESRGAN-FS", PnP with manually tuned kernels and noise settings = "Ours PnP") with recent state-of-the-art methods for real-world SR on a variety of real-world images taken using different cameras, lenses and under different settings. The methods we use for comparison are ESRGAN [36] which is trained on the default DF2K data set, i.e. with clean LR images. We also use the combination of CycleISP [37] + ESRGAN. CycleISP is the

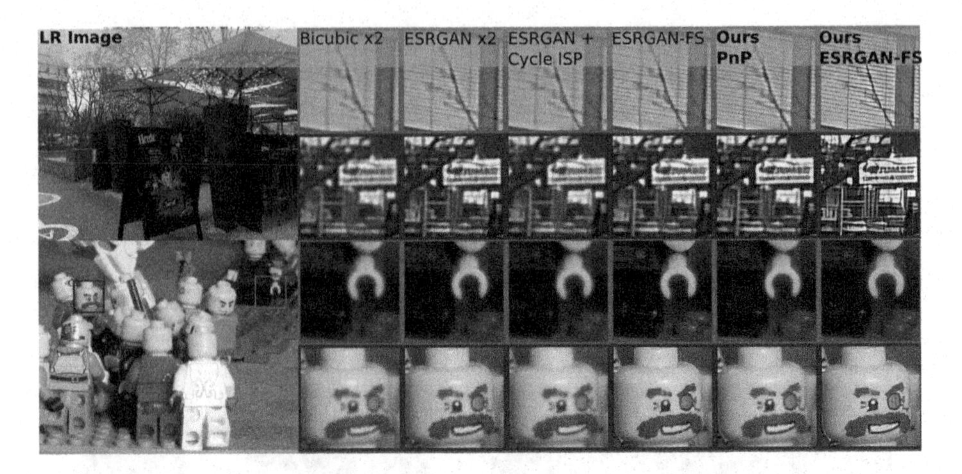

Fig. 6. Comparison with state of the art for a variation of scenes. Zoom into the digital version for details. The test images show a variation of artificial/man made and natural structures as well as smooth and highly textured imaged regions. We also include structures close to the sampling limit.

current state-of-the-art method for real-world denoising [29] and we use it to clean up the LR images first. Then, we use ESRGAN to perform SR. We also compare against the method of Fritsche et al. [12] (ESRGAN-FS), which is the current state of the art in real-world SR [25]. Since ground truth HR images are not available for these real-world images, we have to rely on a qualitative comparison. The results of our experiment are shown in Fig. 6.

ESRGAN can produce sharp images, however, it also enhances the noise and cannot effectively deal with blur in real world images. The CycleISP + ESRGAN method, on the other hand, is able to remove some noise but the resulting images look smooth and lack detailed textures. The results from the method of Fritsche et al. are sharp and contain a good amount of detail. However, it also enhances noise in parts of the images significantly. For "Ours PnP", we manually tuned the blur kernels and noise settings for $\approx$1 h per image and show the best result. As in the synthetic test, the contrast is good, but fine detail is smoothed out. "Ours ESRGAN-FS" produces the best results overall. The noise is effectively suppressed in all parts of the super-resolved image. The generated images are sharp and contain a good amount of texture and details. The images from this method are perceptually the most pleasing out of the methods compared.

7 Discussion and Conclusions

Our experiments show that a single end-to-end CNN for blind SISR trained on a suitably varied data set can match the performance of a non-blind learning-supported optimization algorithm that has full knowledge of both the PSFs and the noise level. In terms of PSNR and SSIM, the performance is slightly worse, but in the perceptually more relevant LPIPS metric [40], the blind end-to-end network out-performs the optimiza-

Table 2. Comparison of results on test images first blurred and then downsampled by a scale factor of 2. CNN L_1 achieves better results for realistic noise, whereas the PnP approach achieves better results for additive white Gaussian noise in terms of PSNR and SSIM. However, CNN L_1 achieves better results in terms of LPIPS score.

Method	Realistic noise			Gaussian noise		
	PSNR↑	SSIM↑	LPIPS↓	PSNR↑	SSIM↑	LPIPS↓
	Blurring followed by downsampling					
CNN L_1	**26.6646**	**0.7841**	**0.1212**	23.6885	0.6673	**0.1681**
PnP	23.0467	0.6845	0.3115	**25.9489**	**0.7617**	0.1820
	Downsampling followed by blurring					
CNN L_1	**25.1422**	**0.7342**	**0.1503**	24.4181	0.7029	**0.1438**
PnP	23.3231	0.6706	0.3080	**25.8148**	**0.7408**	0.1743

tion algorithm. This observation holds even when the image formation model of the optimization algorithm is matched exactly in the synthetic experiment.

The optimization scheme of Sect. 4 can, for our purposes, be interpreted as an upper bound on the performance that a blind learning-supported optimization scheme would be able to achieve. The conclusion from our experiments is then that the end-to-end network is more flexible and results in perceptually more meaningful super-resolved images. This experimental observation suggests that end-to-end networks are competitive for *blind* denoising, deblurring, and super-resolution tasks (Table 2).

SISR CNNs are mappings from an input manifold of blurred and noisy patches to an output manifold of clean super-resolved images. Training the network to generalize to a variation of possible input data involves the approximation of a many-to-one mapping: differently blurred versions of the same real-world patch (different noise levels being strictly analogous) are supposed to map to the same clean super-resolved image patch. This implies that several points that are well separated on the input manifold map to points that are very close to each other on the output manifold.

This is the defining property of an ill-posed problem, when considering the SISR network in reverse. In order to analyze this behavior more concretely, future work could use invertible neural networks [2] to visualize the posterior distribution of blurred patches that map to a clean patch. Given posterior-sampled blurred and corresponding clean patches, the associated blur kernels could be visualized and thus analyzed, possibly illuminating the blind super-resolution capabilities of networks as proposed here.

In summary, we have experimentally shown the ability of CNNs to generalize to blind settings in the context of SISR where they can meet or surpass the performance of dedicated non-blind schemes, depending on the employed quality measure. We thus believe that training on data variations can find application in other domains involving more complex image formation models such as microscopy, light field imaging, computed tomography and more. Key to success is the faithful modeling of the forward image formation process, including both the optical and the digital parts of the process. The required large amounts of training data to cover the input and output manifolds adequately can then be synthesized from commonly available data sources.

Acknowledgement. This work was partially funded by the German Ministry for Education and Research (BMBF) under the grant PLIMASC.

References

1. Agustsson, E., Timofte, R.: NTIRE 2017 challenge on single image super-resolution: dataset and study, July 2017
2. Ardizzone, L., et al.: Analyzing inverse problems with invertible neural networks. arXiv preprint arXiv:1808.04730 (2018)
3. Boyd, S., Boyd, S.P., Vandenberghe, L.: Convex Optimization. Cambridge University Press, Cambridge (2004)
4. Brifman, A., Romano, Y., Elad, M.: Turning a denoiser into a super-resolver using plug and play priors, pp. 1404–1408 (2016)
5. Chan, S.H., Wang, X., Elgendy, O.A.: Plug and-play ADMM for image restoration: fixed-point convergence and applications. IEEE Trans. Comput. Imaging **3**, 84–98 (2017)
6. Chang, H., Yeung, D.Y., Xiong, Y.: Super-resolution through neighbor embedding, vol. 1. IEEE (2004)
7. Danielyan, A., Katkovnik, V., Egiazarian, K.: Image deblurring by augmented Lagrangian with BM3D frame prior. In: Workshop on Information Theoretic Methods in Science and Engineering, pp. 16–18 (2010)
8. Dong, C., Loy, C.C., He, K., Tang, X.: Learning a deep convolutional network for image super-resolution. In: Fleet, D., Pajdla, T., Schiele, B., Tuytelaars, T. (eds.) ECCV 2014. LNCS, vol. 8692, pp. 184–199. Springer, Cham (2014). https://doi.org/10.1007/978-3-319-10593-2_13
9. Dong, C., Loy, C.C., Tang, X.: Accelerating the super-resolution convolutional neural network. In: Leibe, B., Matas, J., Sebe, N., Welling, M. (eds.) ECCV 2016. LNCS, vol. 9906, pp. 391–407. Springer, Cham (2016). https://doi.org/10.1007/978-3-319-46475-6_25
10. Foi, A., Trimeche, M., Katkovnik, V., Egiazarian, K.: Practical Poissonian-Gaussian noise modeling and fitting for single-image raw-data. IEEE Trans. Image Process. **17**(10), 1737–1754 (2008)
11. Freeman, W.T., Jones, T.R., Pasztor, E.C.: Example-based super-resolution. IEEE CG&A **22**(2), 56–65 (2002)
12. Fritsche, M., Gu, S., Timofte, R.: Frequency separation for real-world super-resolution. In: Proceedings of ICCV Workshops, pp. 3599–3608. IEEE (2019)
13. Glasner, D., Bagon, S., Irani, M.: Super-resolution from a single image, pp. 349–356. IEEE (2009)
14. Guo, S., Yan, Z., Zhang, K., Zuo, W., Zhang, L.: Toward convolutional blind denoising of real photographs, pp. 1712–1722 (2019)
15. Heide, F., et al.: FlexISP: a flexible camera image processing framework. ACM Trans. Graph. (ToG) **33**, 1–13 (2014)
16. Hirsch, M., Sra, S., Schölkopf, B., Harmeling, S.: Efficient filter flow for space-variant multiframe blind deconvolution, pp. 607–614. IEEE (2010)
17. Huang, J.B., Singh, A., Ahuja, N.: Single image super-resolution from transformed self-exemplars, pp. 5197–5206 (2015)
18. Ji, X., Cao, Y., Tai, Y., Wang, C., J. Li, F.H.: Real-world super-resolution via kernel estimation and noise injection. In: CVPRW (2020)
19. Johnson, J., Alahi, A., Fei-Fei, L.: Perceptual losses for real-time style transfer and super-resolution. In: Leibe, B., Matas, J., Sebe, N., Welling, M. (eds.) ECCV 2016. LNCS, vol. 9906, pp. 694–711. Springer, Cham (2016). https://doi.org/10.1007/978-3-319-46475-6_43

20. Kim, J., Kwon Lee, J., Mu Lee, K.: Accurate image super-resolution using very deep convolutional networks, pp. 1646–1654 (2016)
21. Kligler, S., Shocher, A., Irani, M.: Blind super-resolution kernel estimation using an internal-GAN, pp. 284–293 (2019)
22. Ledig, C., et al.: Photo-realistic single image super-resolution using a generative adversarial network, pp. 4681–4690 (2017)
23. Lim, B., Son, S., Kim, H., Nah, S., Lee, K.M.: Enhanced deep residual networks for single image super-resolution, July 2017
24. Lugmayr, A., Danelljan, M., Timofte, R.: Unsupervised learning for real-world super-resolution (2019)
25. Lugmayr, A., et al.: Aim 2019 challenge on real-world image super-resolution: methods and results, pp. 3575–3583. IEEE (2019)
26. Meinhardt, T., Moller, M., Hazirbas, C., Cremers, D.: Learning proximal operators: using denoising networks for regularizing inverse imaging problems, pp. 1781–1790 (2017)
27. Michaeli, T., Irani, M.: Nonparametric blind super-resolution, pp. 945–952 (2013)
28. Parikh, N., Boyd, S.: Proximal algorithms. Found. Trends Optim. **1**(3), 127–239 (2014)
29. Plotz, T., Roth, S.: Benchmarking denoising algorithms with real photographs. In: Proceedings of the IEEE Conference on Computer Vision and Pattern Recognition, pp. 1586–1595 (2017)
30. Qian, N.: On the momentum term in gradient descent learning algorithms. Neural Netw. **12**(1), 145–151 (1999)
31. Ren, H., Kheradmand, A., El-Khamy, M., Wang, S., Bai, D., Lee, J.: Real-world super-resolution using generative adversarial networks. In: CVPRW, pp. 1760–1768 (2020)
32. Shi, W., et al.: Real-time single image and video super-resolution using an efficient sub-pixel convolutional neural network, pp. 1874–1883 (2016)
33. Shocher, A., Cohen, N., Irani, M.: Zero-shot super-resolution using deep internal learning, pp. 3118–3126 (2018)
34. Tirer, T., Giryes, R.: Image restoration by iterative denoising and backward projections. IEEE Trans. Image Process. **28**, 1220–1234 (2019)
35. Venkatakrishnan, S.V., Bouman, C.A., Wohlberg, B.: Plug-and-play priors for model-based reconstruction, pp. 945–948. IEEE (2013)
36. Wang, X., et al.: ESRGAN: enhanced super-resolution generative adversarial networks. In: Leal-Taixé, L., Roth, S. (eds.) ECCV 2018. LNCS, vol. 11133, pp. 63–79. Springer, Cham (2019). https://doi.org/10.1007/978-3-030-11021-5_5
37. Zamir, S.W., et al.: CycleISP: real image restoration via improved data synthesis, pp. 2696–2705 (2020)
38. Zeyde, R., Elad, M., Protter, M.: On single image scale-up using sparse-representations. In: Boissonnat, J.-D., et al. (eds.) Curves and Surfaces 2010. LNCS, vol. 6920, pp. 711–730. Springer, Heidelberg (2012). https://doi.org/10.1007/978-3-642-27413-8_47
39. Zhang, K., Zuo, W., Zhang, L.: Deep plug-and-play super-resolution for arbitrary blur kernels, pp. 1671–1681 (2019)
40. Zhang, R., Isola, P., Efros, A.A., Shechtman, E., Wang, O.: The unreasonable effectiveness of deep features as a perceptual metric, pp. 586–595 (2018)
41. Zhang, Y., Li, K., Li, K., Wang, L., Zhong, B., Fu, Y.: Image super-resolution using very deep residual channel attention networks. In: Ferrari, V., Hebert, M., Sminchisescu, C., Weiss, Y. (eds.) ECCV 2018. LNCS, vol. 11211, pp. 294–310. Springer, Cham (2018). https://doi.org/10.1007/978-3-030-01234-2_18
42. Zhou, R., Süsstrunk., S.: Kernel modeling superresolution on real low-resolution images, pp. 2433–2443 (2019)
43. Zoran, D., Weiss, Y.: From learning models of natural image patches to whole image restoration, pp. 497–486 (2011)

Merging-ISP: Multi-exposure High Dynamic Range Image Signal Processing

Prashant Chaudhari[1](✉), Franziska Schirrmacher[2], Andreas Maier[3], Christian Riess[2], and Thomas Köhler[1]

[1] e.solutions GmbH, Erlangen, Germany
{prashant.chaudhari,thomas.koehler}@esolutions.de
[2] IT Security Infrastructures Lab, Friedrich-Alexander-Universität (FAU) Erlangen-Nürnberg, Erlangen, Germany
{franziska.schirrmacher,christian.riess}@fau.de
[3] Pattern Recognition Lab, Friedrich-Alexander-Universität (FAU) Erlangen-Nürnberg, Erlangen, Germany
andreas.maier@fau.de

Abstract. High dynamic range (HDR) imaging combines multiple images with different exposure times into a single high-quality image. The image signal processing pipeline (ISP) is a core component in digital cameras to perform these operations. It includes demosaicing of raw color filter array (CFA) data at different exposure times, alignment of the exposures, conversion to HDR domain, and exposure merging into an HDR image. Traditionally, such pipelines cascade algorithms that address these individual subtasks. However, cascaded designs suffer from error propagation, since simply combining multiple steps is not necessarily optimal for the entire imaging task.

This paper proposes a multi-exposure HDR image signal processing pipeline (Merging-ISP) to jointly solve all these subtasks. Our pipeline is modeled by a deep neural network architecture. As such, it is end-to-end trainable, circumvents the use of hand-crafted and potentially complex algorithms, and mitigates error propagation. Merging-ISP enables direct reconstructions of HDR images of dynamic scenes from multiple raw CFA images with different exposures. We compare Merging-ISP against several state-of-the-art cascaded pipelines. The proposed method provides HDR reconstructions of high perceptual quality and it quantitatively outperforms competing ISPs by more than 1 dB in terms of PSNR.

1 Introduction

Computational photography aims at computing displayable images of high perceptual quality from raw sensor data. In digital cameras, these operations are performed by the image signal processing pipeline (ISP). The ISP plays a particularly important role in commodity devices like smartphones, to overcome physical limitations of the sensors and the optical system. ISPs cascade processing steps that perform different computer

---

Supplementary Information The online version contains supplementary material available at https://doi.org/10.1007/978-3-030-92659-5_21.

C. Bauckhage et al. (Eds.): DAGM GCPR 2021, LNCS 13024, pp. 328–342, 2021.
https://doi.org/10.1007/978-3-030-92659-5_21

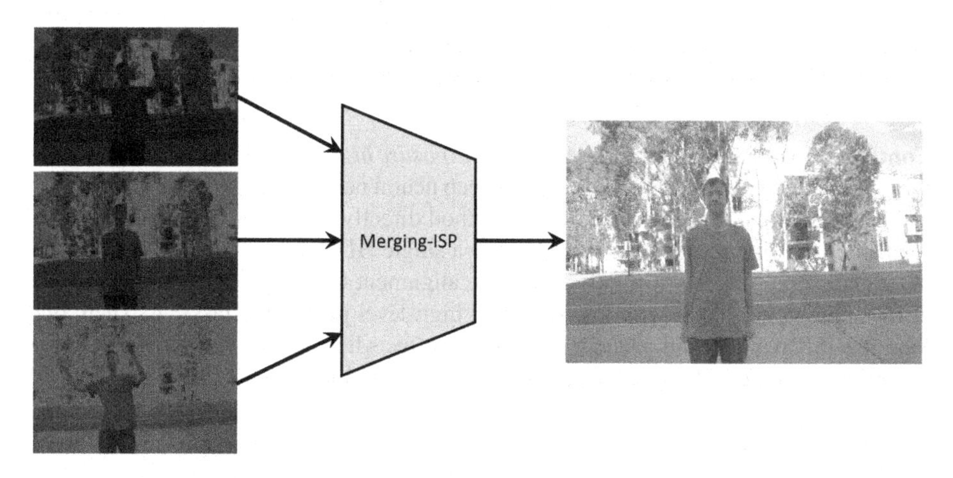

Fig. 1. We propose a camera image signal processing pipeline (ISP) using deep neural networks to directly merge multi-exposure Bayer color filter array data of low dynamic range (LDR) into a high dynamic range (HDR) image.

vision tasks. Initial steps address the low-level *reconstruction* of images in RGB space from raw data, comprising demosaicing, denoising, and deblurring. Specifically, demosaicing [21] creates full RGB images from sensor readouts behind a color filter array (CFA) that provides per pixel only a single spectral component. Later steps in the ISP comprise higher level *enhancement* operations like high dynamic range (HDR) imaging [12,30,39]. One effective approach adopted in this paper enhances dynamic ranges by merging multiple low dynamic range (LDR) images, where each LDR image is captured sequentially with different exposure time [35]. These exposures have additionally to be aligned in an intermediary step when the scene is dynamic.

For each of these individual processing tasks, there exist several algorithms with different quality criteria and computational requirements. This enables camera manufacturers to assemble an ISP that is tailored to the capabilities of the device. Nevertheless, the resulting ISP is not necessarily optimal. In particular, error propagation between these isolated processing steps oftentimes leads to suboptimal results [9]. For example, demosaicing artifacts can be amplified by image sharpening, or misalignments of different exposures can reduce the overall image quality.

Solving all tasks *jointly* is a principal approach to overcome error propagation. This can be achieved by formulating the ISP output as solution of an inverse problem [9]. The individual steps in the pipeline are modeled by a joint operator and signal processing is accomplished via non-linear optimization. While this yields globally optimal solutions, e.g. in a least-squares sense, analytical modeling for real cameras is overly complex. Also uncertainties due to simplifications cause similar effects as error propagation.

End-to-end learning of the ISP [22,28,29] avoids the need for analytical modeling. However, current methods in that category consider single exposures only. Different to merging multi-exposure data, they can only hallucinate HDR content. For instance, single-exposure methods often fail in highly over-saturated image regions [39]. Aligning exposures captured sequentially from dynamic scenes is, however, challenging due to inevitable occlusions and varying image brightness. This is particularly the case when using raw CFA data preventing the use of standard image alignment.

Contributions. This paper proposes a *multi-exposure high dynamic range image signal processing pipeline (Merging-ISP)* using a deep neural network architecture and end-to-end learning of its processing steps. Our method directly maps raw CFA data captured sequentially with multiple exposure times to a single HDR image, see Fig. 1. Contrary to related works [29,39], it jointly learns the alignment of multi-exposure data in case of dynamic scenes along with low-level and high-level processing tasks. As such, our Merging-ISP avoids common error propagation effects like demosaicing artifacts, color distortions, or image alignment errors amplified by HDR merging.

2 Related Work

Overall, we group ISPs into two different modules: 1) low-level vision aiming at image reconstruction on a pixel-level of CFA data to form full RGB images, and 2) high-level vision focusing on the recovery of HDR content from LDR observations. Our method also follows this modular design but couples all stages by end-to-end learning.

2.1 Low-Level Vision

Low-level vision comprises image reconstruction tasks like demosaicing of CFA data, defect pixel interpolation, denoising, deblurring, or super-resolution. Classical pipelines employ isotropic filters (e. g., linear interpolation of missing CFA pixels or linear smoothing). Edge-adaptive techniques [26,36] can avoid blurring or zippering artifacts of such non-adaptive filtering. Another branch of research approaches image reconstruction from the perspective of regularized inverse problems with suitable image priors [15,27]. These methods are, however, based on hand-crafted models. Also simply cascading them leads to accumulated errors.

Later, deep learning advanced the state-of-the-art in denoising [41], demosaicing [32], deblurring [33], or super-resolution [18]. Deep neural networks enable image reconstructions under non-linear models. Using generative adversarial networks (GANs) [17,38] or loss functions based on deep features [37] also allow to optimize such methods with regard to perceptual image quality. It also forms the base for multi-task learning of pipelines like demosaicing coupled with denoising [2,16] or super-resolution [43]. In contrast to cascading these steps, this can avoid error propagation. We extend this design principle by incorporating high-level vision, namely HDR reconstruction.

2.2 High-Level Vision

High-level vision focuses on global operations like contrast enhancement. In this context, we are interested in capturing HDR data. This can be done using special imaging technologies, e. g. beam splitter [34] or coded images [31], but such techniques are not readily available for consumer cameras due to cost or size constraints. There are also various inverse tone mapping methods estimating HDR data from single LDR acquisitions [4,5,19,20]. However, this can only hallucinate the desired HDR content.

The dynamic range can also be increased by aligning and merging frame bursts. Merging bursts captured with constant exposure times [8,42] enhances the signal-to-noise ratio and allows to increase dynamic ranges to a certain extent, while making alignment for dynamic scenes robust. In contrast, bursts with bracketed exposure images as utilized in this paper can expand dynamic ranges by much larger factors but complicates alignment. Therefore, multi-exposure alignment has been proposed [6,11,35], which is especially challenging in case of large object motions or occlusions.

Recently, joint alignment and exposure merging have been studied to improve robustness. Patch based approaches [10,30] fill missing under/over-exposed pixels in a reference exposure image using other exposures. Recent advances in deep learning enable merging with suppression of ghosting artifacts. In [12], Kalantari and Ramamoorthi have proposed a CNN for robust merging. Wu *et al.* [39] have proposed to tackle multi-exposure HDR imaging as image translation. Their U-Net considers alignment as part of its learning process. However, existing patch-based or learning-based methods necessitate full RGB inputs and cannot handle raw CFA data directly. With Merging-ISP, we aim at direct HDR reconstruction from multiple raw images.

2.3 End-to-End Coupling of Low-Level and High-Level Vision

Several attempts have been made to couple low-level and high-level vision in an end-to-end manner. FlexISP [9] is a popular model-based approach handling different sensors with their pixel layouts, noise models, processing tasks, and priors on natural images in a unified optimization framework. However, FlexISP and related approaches require analytical modeling of a given system as inverse problem, which can become complex.

Data-driven approaches learn an ISP from example data and circumvent this effort. The DeepISP as proposed by Schwartz *et al.* [29] enables direct mappings from low-light raw data to color corrected and contrast enhanced images. Ratnasingam [28] has combined defect pixel interpolation, demosaicing, denoising, white balancing, exposure correction, and contrast enhancement using a single CNN. In contrast to monolithic networks, CameraNet introduced by Liang *et al.* [22] comprises separate modules for these low-level reconstruction and high-level enhancement tasks. These frameworks are closely related to our proposed method but did not consider the reconstruction of HDR content. Our Merging-ISP uses multi-exposure data captured in burst mode for true HDR reconstruction rather than hallucinating such content from single images.

3 Proposed Merging-ISP Framework

In this section, we introduce our Merging-ISP. The input to our pipeline is a stack of M raw images $\mathbf{Y}_i \in \mathbb{R}^{N_x \times N_y \times 3}$, $i = 1, \dots, M$ captured with a known CFA (e. g., Bayer pattern) and ascending exposure times. The CFA masks pixels in RGB space according to its predefined pattern. Also, each raw image is defined in LDR domain with limited bit depth. We further consider the challenging situation of *dynamic scenes*, where these raw images are misaligned due to camera and/or object motion during their acquisition.

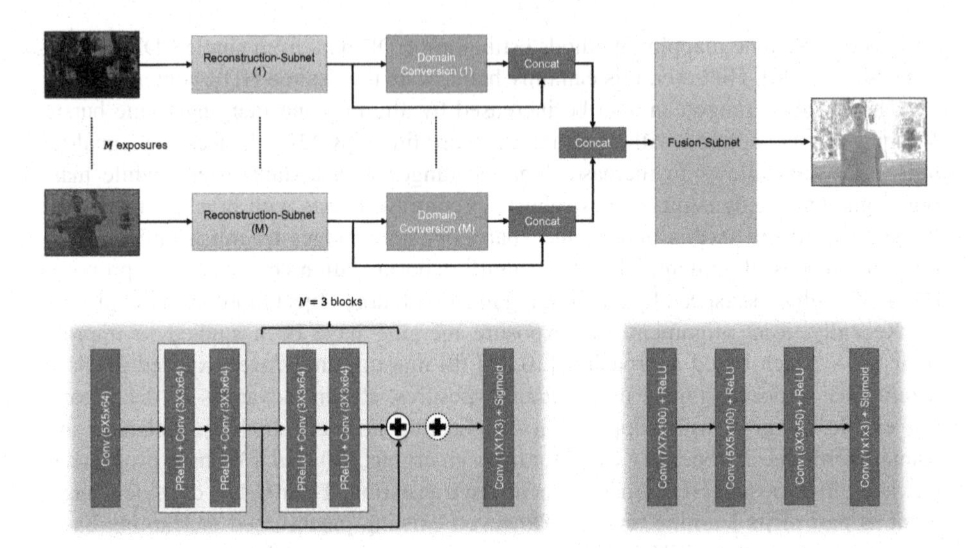

Fig. 2. Our overall Merging-ISP architecture. *Reconstruction-Subnets* (orange block) map raw CFA inputs into an intermediate feature space. *Domain conversion* transfers intermediate LDR features to HDR domain. *Fusion-Subnet* uses a series of convolutional layers to align and merge the channel-wise concatenated features from the previous stage into one HDR image. (Color figure online)

Given the raw inputs in LDR domain, we aim at learning the mapping:

$$\mathbf{X} = f(\mathbf{Y}_1, \ldots, \mathbf{Y}_M), \tag{1}$$

where $\mathbf{X} \in \mathbb{R}^{N_x \times N_y \times 3}$ is the target HDR image. To regularize the learning process of this general model, it is constrained such that the output HDR image is geometrically aligned to one of the inputs acting as a *reference exposure*. For instance, we can choose the medium exposure to be the reference. Other exposures are aligned with this reference, which is done either separately or as integral part of the pipeline. In classical approaches, Eq. (1) is further decomposed as $f(\mathbf{Y}) = (f_1 \odot f_2 \odot \ldots \odot f_K)(\mathbf{Y})$. This represents a camera ISP comprising K successive processing stages $f_j(\mathbf{Y})$ that can be developed independently. In contrast, we learn the entire pipeline including the alignment of different exposures for dynamic scenes in an end-to-end manner.

3.1 Network Architecture

Merging-ISP employs a modularized deep neural network to model the LDR-to-HDR mapping in Eq. (1). Figure 2 illustrates its architecture comprising three modules: parallel *Reconstruction-Subnets* to restore intermediate features from input CFA data for each exposure, *domain conversion* to transfer these features from LDR to HDR domain, and the *Fusion-Subnet* estimating the final HDR output image from the feature space.

Reconstruction-Subnet. This module implements the low-level vision tasks in our pipeline. Each raw CFA image is fed into one Reconstruction-Subnet instance, which consists of three fundamental stages as shown in Fig. 2. This subnet follows the general concept of a feed-forward denoising convolutional neural network (DnCNN) [40] and is a variation of the residual architecture in [16]. The first stage is a single convolutional layer with 64 filters of size 5×5. The basic building block of the second stage are two layers with parametric rectified linear unit (PReLU) activation followed by 64 convolutional filters of size 3×3. We always use one of these blocks, followed by N further blocks with skip connections. In our approach, we use $N = 3$. The final stage comprises a single convolution layer with 3 filters of size 1×1 and sigmoid activation. The role of this final stage is to reduce the depth of the feature volume of the second stage from 64 channels to 3-channels $\mathbf{Z}_i$ for each exposure. Note that [16] additionally links the input to that last layer, which we omit because we are fusing feature maps instead of restoring images. We use reflective padding for all convolutions such that $\mathbf{Z}_i$ has the same spatial dimensions as the raw CFA input.

Feature volumes provided by the Reconstruction-Subnet encodes demosaiced versions of raw inputs in an intermediate feature space. Our Merging-ISP employs instances of this network to process the different exposures independently without parameter sharing. As a consequence of this design choice, the pipeline learns the reconstruction of full RGB data with consideration of exposure-specific data characteristics.

Domain Conversion. As Reconstruction-Subnet features are defined in LDR while exposure merging needs to consider HDR we integrate domain conversion as intermediate stage. Following the notion of *precision learning* [24], we propose to formulate domain conversion using known operators. Since such existing operators are well understood, this can greatly reduce the overall number of trainable parameters and maximum error bounds of model training and thus the learning burden. Using the LDR feature volume $\mathbf{Z}_i$ obtained from the ith exposure $\mathbf{Y}_i$ and the conversion rule proposed in [12], the corresponding feature volume $\mathbf{Z}_i^H$ in the HDR domain is computed element-wise:

$$Z_{i,jkl}^H = \frac{{Z_{i,jkl}}^\gamma}{t_i}, \tag{2}$$

where t_i is the exposure time and $\gamma = 2.2$. This domain conversion does not involve additional trainable parameters, which greatly simplifies the training of our Merging-ISP. It is applied to each of the intermediate feature volumes provided by the Reconstruction-Subnets.

Fusion-Subnet. This is the high-level stage of our pipeline reconstructing HDR images. We construct its input by channel-wise concatenating LDR and HDR features as $\mathbf{U}_i = \text{concat}(\mathbf{Z}_i, \mathbf{Z}_i^H)$ for each exposure. Given M exposures and 3-channel images, we obtain the $N_x \times N_y \times 6M$ volume $\mathbf{U} = \text{concat}(\mathbf{U}_1, \dots, \mathbf{U}_M)$. This combined feature space facilitates the detection and removal of outliers like oversaturations to obtain artifact-free HDR content. Given the feature volume $\mathbf{U}$, the purpose of the Fusion-Subnet is a joint exposure alignment *and* HDR merging, which is translated into two subtasks: 1) Its input feature volume $\mathbf{U}$ needs to be aligned towards one reference exposure coordinate frame to compensate for motion in dynamic scenes. 2) The aligned features need to be fused into the output HDR image $\mathbf{X}$. Both subtasks are solved intrinsically by the Fusion-Subnet.

We adopt the CNN proposed in [12] for aligning and fusing multiple exposures in the Fusion-Subnet. Overall, the Fusion-Subnet comprises four convolutional layers with decreasing receptive fields of 7×7 for 100 filters in the first layer to 1×1 for 3 filters in the last layer. The input layer and hidden layers use rectified linear unit (ReLU) activation, while the output layer uses sigmoid activation to obtain linear-domain HDR data. Like in the Reconstruction-Subnet, each convolution uses reflective padding to preserve the spatial dimensions. One distinctive property compared to the baseline CNN in [12] is that we do not employ an additional pre-alignment of input exposures in the image space, e.g., via optical flow. Instead we allow the Fusion-Subnet to learn this alignment in the feature space formed by the volume $\mathbf{U}$, which can greatly reduce error propagation in the entire ISP (see Sect. 4.4).

3.2 Loss Function

While our proposed ISP provides predictions in a linear domain, HDR images are usually displayed after tonemapping. Hence, to train our model, we compute the loss on tonemapped HDR predictions. For tone mapping, we use the μ-law [12] defined as:

$$T(\mathbf{X}) = \frac{\log(1 + \mu\mathbf{X})}{\log(1 + \mu)}. \tag{3}$$

Here, μ is a hyperparameter controlling the level of compression and $\mathbf{X}$ is an estimated HDR image in linear domain as inferred by Fusion-Subnet. We set $\mu = 5 \cdot 10^3$.

This differentiable tonemapping allows to formulate loss functions either on the basis of pixel-wise or perceptual measures [37]. In this paper, the overall loss function is defined pixel-wise using the L_2 norm:

$$\mathcal{L}(\tilde{\mathbf{X}}, \mathbf{X}) = ||T(\tilde{\mathbf{X}}) - T(\mathbf{X})||_2^2, \tag{4}$$

where $\tilde{\mathbf{X}}$ denotes the ground truth and $T(\tilde{\mathbf{X}})$ is its tonemapped version.

4 Experiments and Results

In this section, we describe the training of the proposed Merging-ISP network and compare it against different ISPs for HDR imaging. In addition, we present an ablation study on the design of its deep neural network architecture.

4.1 Datasets

We include multiple databases to train and test our ISP ranging from synthetic to real data.

First, we use the HDR database collected by Kalantari *et al.* [12] comprising ground truth HDR images with corresponding captured LDR images of 89 dynamic scenes. Each scene contains three bracketed exposure images and the medium exposure is used as a reference. Misalignments between these exposures are related to real non-rigid motion like head or hand movements of human subjects. To obtain raw CFA data as input for our ISP, we mosaic the LDR images with an RGGB mask [13]. Overall, 74 scenes are used for training or validation and the remaining 15 scenes are used for testing.

To study the capability of our method to generalize to other sensors unseen during training, we additionally use examples of the multi-exposure dataset of Sen *et al.* [30]. The exposure values -1.3, 0, and +1.3 from each scene are used an input and the medium exposure is chosen as reference. The CFA for each exposure value employs an RGGB mask. This real-world dataset does not provide ground truths and is used for perceptual evaluation of HDR image quality in the wild.

4.2 Training and Implementation Details

Out of the 74 training scenes, we use 4 scenes to validate our model w.r.t. the loss function in Eq. (4). Random flipping (left-right and up-down) and rotation by 90° of the images is performed to augment the training set from 70 to 350 scenes. Since training on full images has a high memory footprint, we extract 210,000 non-overlapping patches of size 50×50 pixels using a stride of 50. The network weights are initialized using Xavier method [7] and training is done using Adam optimization [14] with $\beta_1 = 0.9$ and $\beta_2 = 0.999$. We perform training over 70 epochs with a constant learning rate of 0.0001 and batches of size 32. During each epoch, all batches are randomly shuffled.

We implemented Merging-ISP in Tensorflow [1] with a NVIDIA GeForce 1080 Ti GPU. The overall training takes roughly 18 h based on the architecture proposed in Sect. 3.1. The prediction of one HDR image from three input CFA images with a resolution of 1500×1000 pixels takes 1.1 s. The performances of alternative network architectures considering modifications of the used residual blocks and convolutional layers are reported in our supplementary material.

4.3 Comparisons with State-of-the-Art

We compare our proposed Merging-ISP with several ISP variants that comprise different state-of-the-art demosaicing, denoising, image alignment, HDR merging, and tone mapping techniques. In terms of low-level vision, we evaluate directional filtering based demosaicing [26] and deep joint demosaicing and denoising [16]. For the high-level stages, we use single-exposure HDR [4], patch-based HDR [30], and learning-based HDR with a U-net [39]. We use the publicly available source codes for all methods.

We conduct our benchmark by calculating PSNR and SSIM of the ISP outputs against ground truth HDR data. In addition to pixel-based measures, we use HDR-VDP-2 [25] that expresses mean opinion scores based on the probability that a human observer notices differences between ground truths and predictions. HDR-VDP-2 is evaluated assuming 24 in. displays and 0.5 m viewing distance. For fair qualitative comparisons, we show all outputs using Durand tonemapping with $\gamma = 2.2$ [3].

Table 1. Quantitative comparison of Merging-ISP to cascades of state-of-the-art demosaicing [16,26], single-exposure HDR reconstruction [4] and multi-exposure HDR [30,39] on the Kalantari test set. We report the mean PSNR, SSIM, and HDR-VDP-2 of output HDR images after tonemapping considering all combinations of cascaded demosaicing and HDR imaging.

Demosaicing	HDR merging	PSNR	SSIM	HDR-VDP-2
Menon *et al.* [26]	Sen *et al.* [30]	28.37	0.9633	62.27
	Eilertsen *et al.* [4]	17.34	0.7673	53.36
	Wu *et al.* [39]	28.28	0.9661	62.15
Kokkinos *et al.* [16]	Sen *et al.* [30]	40.07	0.9898	63.49
	Eilertsen *et al.* [4]	16.16	0.7689	53.81
	Wu *et al.* [39]	41.62	0.9942	64.99
Proposed Merging-ISP		**43.17**	**0.9951**	**65.29**

Comparison Against Cascaded ISPs. To the best of our knowledge, there are no related methods for direct reconstruction of HDR content from multiple CFA images. Thus, we compare against the design principle of cascading state-of-the-art algorithms for subtasks and combine different demosaicing/denoising methods with HDR merging.

In Table 1, we report mean quality measures of our method and six cascaded ISPs in the tonemapped domain on 15 scenes of the Kalantari test set. We found that Merging-ISP consistently outperforms the cascaded ISPs by a large margin. We compare on a test scene from the Kalantari dataset in Fig. 3. Here, Merging-ISP provides HDR content of high perceptual quality, while the cascades suffer from error propagation. For example, even with the integration of state-of-the-art demosaicing/denoising like [16], the cascaded designs lead to noise breakthroughs.

We also compare Merging-ISP to the most competitive cascaded ISP variant (demosaicing [16] + HDR [39]) on a challenging real raw scene from Sen *et al.* [30] in Fig. 4, where image regions are underexposed across all inputs and there is notable foreground motion between the exposures. Here, the cascaded design propagates demosaicing artifacts to exposure merging and thus to the final HDR output in static background (orange and red image patch). In the moving foreground (blue and black image patch), the cascaded design suffers from motion artifacts due to inaccurate alignments. Both types of error propagations are mitigated by Merging-ISP.

Comparison Against Single-Exposure HDR Imaging. In Fig. 5, we compare our multi-exposure approach against a recent deep learning method for single-exposure HDR reconstruction [4]. For fair comparisons, we demosaic the reference exposure raw image using the method in [16] and feed the preprocessed image into the HDR reconstruction developed in [4]. Overall, a single-exposure approach does not require alignments of multiple exposures in dynamic scenes. However, it fails to recover reliable color information, e. g. in high saturated regions, as depicted in Fig. 5b. Merging-ISP in Fig. 5c exploits multiple exposure inputs and avoids such color distortions.

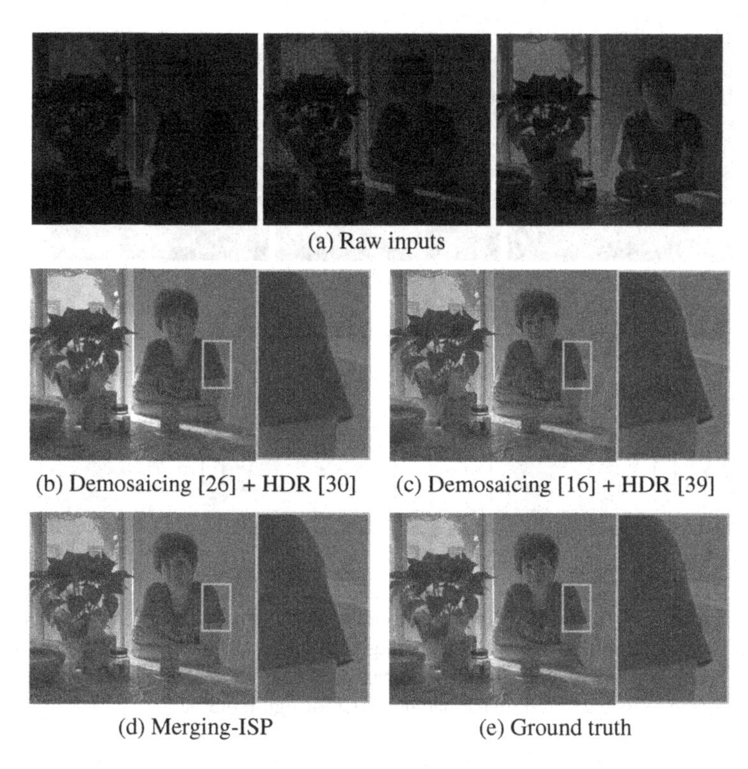

(a) Raw inputs

(b) Demosaicing [26] + HDR [30] (c) Demosaicing [16] + HDR [39]

(d) Merging-ISP (e) Ground truth

Fig. 3. Comparison of our Merging-ISP against different baseline ISPs formed by cascading state-of-the-art demosaicing [16,26] and multi-exposure HDR methods [30,39]. The cascaded methods shown in (b) and (c) suffer from error propagations like demosaicing artifacts causing residual noise in the output. In contrast, Merging-ISP shown in (d) avoids noise amplifications.

4.4 Ablation Study

We investigate the influence of different design choices of our Merging-ISP in an ablation study. To this end, we develop and compare several variants of this pipeline.

Learning Subtasks Separately vs. End-to-End Learning. In our method, all ISP subtasks are learned in an end-to-end fashion. To study the impact of this property, we evaluate variants of Merging-ISP, where subtasks are solved separately and cascaded to from an ISP. To this end, we trained the Reconstruction-Subnet for demosaicing and the Fusion-Subnet to merge demosaiced exposures. Different to our approach that intrinsically covers exposure alignment, the optical flow algorithm of Liu [23] is used in two variants for alignment: optical flow can be either computed on CFA input data (referred to as *pre-align cascaded* Merging-ISP) or on the Reconstruction-Subnet output (referred to as *post-align cascaded* Merging-ISP). Both approaches can be considered as extensions of the HDR merging proposed in [12] equipped with optical flow alignment and demosaicing implemented by our Reconstruction-Subnet to handle raw CFA inputs.

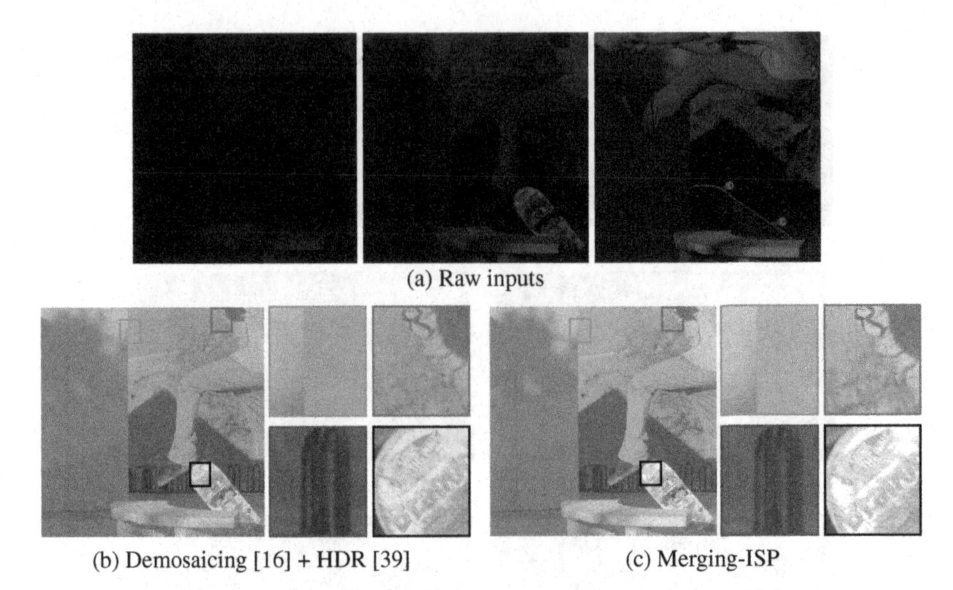

(a) Raw inputs

(b) Demosaicing [16] + HDR [39] (c) Merging-ISP

Fig. 4. Comparison of the best competing cascading ISP (demosaicing [16] + learning-based HDR [39]) against Merging-ISP on a challenging scene from the dataset by Sen *et al.* [30]. The used Bayer pattern appears as artifact in the cascaded ISP output, which is mitigated by the Merging-ISP (see orange and red image patches). (Color figure online)

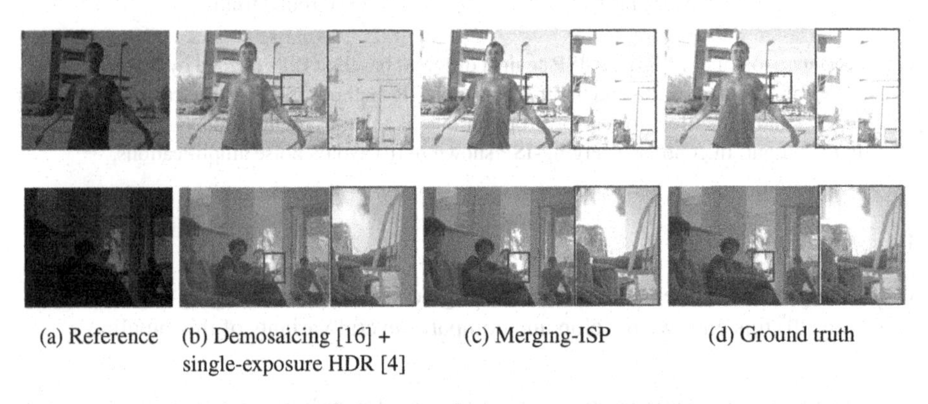

(a) Reference (b) Demosaicing [16] + single-exposure HDR [4] (c) Merging-ISP (d) Ground truth

Fig. 5. Comparison of our multi-exposure based Merging-ISP against state-of-the-art single-exposure HDR reconstruction [4] (cascaded with demosaicing [16]). In contrast to single-exposure methods, the proposed method mitigates color distortions. (Color figure online)

Figure 6a and 6b depict both cascaded ISP variants in the presence of non-rigid motion between exposures. Both cascaded architectures suffer from error propagations in the form of color distortions. In the *pre-align cascaded* approach, alignment is affected by missing pixel values resulting in ghosting artifacts. The *post-align cascaded* approach is degraded by demosaicing errors affecting optical flow alignment and exposure merging. The end-to-end learned Merging-ISP shows higher robustness and

Table 2. Ablation study of our Merging-ISP on the Kalantari data. We compare the proposed end-to-end learning of all subtasks against cascading our Reconstruction- and Fusion-Subnets with optical flow alignment [23] before and after the reconstruction (pre- and post-align cascaded Merging-ISP) as well as end-to-end learning of both networks with optical flow pre-alignment (pre-align end-to-end Merging-ISP).

	PSNR	SSIM	HDR-VDP-2
Pre-align cascaded Merging-ISP	41.16	0.9923	64.71
Post-align cascaded Merging-ISP	42.03	0.9937	65.09
Pre-align end-to-end Merging-ISP	42.56	0.9945	64.84
Proposed Merging-ISP	**43.17**	**0.9951**	**65.29**

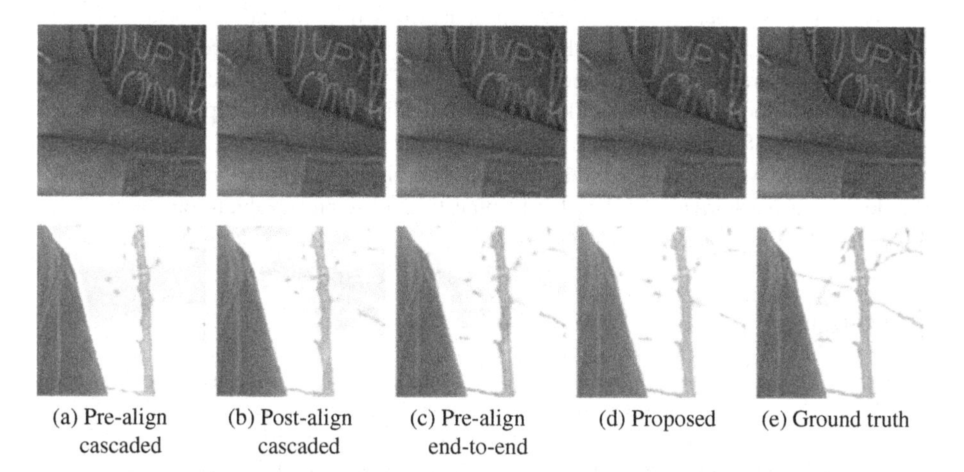

Fig. 6. Evaluation of different pipeline variants in an ablation study. Cascading separately trained Reconstruction- and Fusion-Subnets with optical flow based alignment (pre-align and post-align cascaded Merging-ISP) leads to accumulated ghosting artifacts related to misalignments. End-to-end learning both networks but replacing the alignment provided by Fusion-Subnet with optical flow (pre-align end-to-end Merging-ISP) leads to similar artifacts.

reconstructs HDR data with less artifacts as depicted in Fig. 6d. Table 2 confirms these observations quantitatively on the Kalantari test set. Overall, end-to-end learning leads to reconstructions with higher fidelity to the ground truth.

Pre-aligning Exposures vs. No Alignment. The Fusion-Subnet in the proposed method is used to jointly align and merge a feature volume associated with multiple exposures. To further analyze the impact of this property, we evaluate an additional variant of Merging-ISP that employs optical flow alignment on its raw CFA inputs while using the same end-to-end learning for the Reconstruction-Subnet and the Fusion-Subnet (referred to as *pre-align end-to-end* Merging-ISP). Figure 6c depicts this pre-alignment approach. It is interesting to note that pre-alignment can cause errors that are difficult to compensate in the subsequent ISP stages. In contrast, handling exposure alignment in conjunction with low-level and high-level vision by the proposed method in Fig. 6d mitigates error accumulations. The benchmark in Table 2 confirms that Merging-ISP without hand-crafted pre-alignment also leads to higher quantitative image quality.

5 Conclusion

We proposed an effective deep neural network architecture named *Merging-ISP* for multi-exposure HDR imaging of dynamic scenes. The proposed method outperforms the conventional approach of cascading different methods for image demosaicing, LDR to HDR conversion, exposure alignment, and merging both qualitatively and quantitatively. Our joint method avoids error propagations that typically appear in a cascaded camera ISP. The proposed Merging-ISP is also robust compared to state-of-the-art single-exposure HDR reconstruction in terms of obtaining HDR content. It avoids hand-crafted prepossessing steps like optical flow alignment of multiple exposures for merging in case of dynamic scenes.

In our future work, we aim at fine-tuning our architecture for additional ISP subtasks like denoising or sharpening. We also to plan to deploy and evaluate our method for HDR vision applications.

References

1. Abadi, M., et al.: TensorFlow: large-scale machine learning on heterogeneous systems (2015). http://tensorflow.org/. Software available from tensorflow.org
2. Dong, W., Yuan, M., Li, X., Shi, G.: Joint demosaicing and denoising with perceptual optimization on a generative adversarial network. arXiv preprint arXiv:1802.04723 (2018)
3. Durand, F., Dorsey, J.: Fast bilateral filtering for the display of high-dynamic-range images. ACM Trans. Graph. (TOG) **21**(3), 257–266 (2002)
4. Eilertsen, G., Kronander, J., Denes, G., Mantiuk, R.K., Unger, J.: HDR image reconstruction from a single exposure using deep CNNs. ACM Trans. Graph. (TOG) **36**(6), 178 (2017)
5. Endo, Y., Kanamori, Y., Mitani, J.: Deep reverse tone mapping. ACM Trans. Graph. **36**(6), 1–10 (2017)
6. Gallo, O., Troccoli, A., Hu, J., Pulli, K., Kautz, J.: Locally non-rigid registration for mobile HDR photography. In: IEEE Conference on Computer Vision and Pattern Recognition (CVPR) Workshops, pp. 49–56 (2015)
7. Glorot, X., Bengio, Y.: Understanding the difficulty of training deep feedforward neural networks. In: Teh, Y.W., Titterington, M. (eds.) Proceedings of the Thirteenth International Conference on Artificial Intelligence and Statistics. Proceedings of Machine Learning Research, Chia Laguna Resort, Sardinia, Italy, 13–15 May 2010, vol. 9, pp. 249–256. PMLR. http://proceedings.mlr.press/v9/glorot10a.html
8. Hasinoff, S.W., et al.: Burst photography for high dynamic range and low-light imaging on mobile cameras. ACM Trans. Graph. (TOG) **35**(6), 192 (2016)
9. Heide, F., et al.: FlexISP: a flexible camera image processing framework. ACM Trans. Graph. **33**(6), 1–13 (2014). Proceedings SIGGRAPH Asia 2014
10. Hu, J., Gallo, O., Pulli, K., Sun, X.: HDR deghosting: how to deal with saturation? In: 2013 IEEE Conference on Computer Vision and Pattern Recognition, pp. 1163–1170, June 2013. https://doi.org/10.1109/CVPR.2013.154

11. Jacobs, K., Loscos, C., Ward, G.: Automatic high-dynamic range image generation for dynamic scenes. IEEE Comput. Graph. Appl. **28**(2), 84–93 (2008). https://doi.org/10.1109/MCG.2008.23
12. Kalantari, N.K., Ramamoorthi, R.: Deep high dynamic range imaging of dynamic scenes. ACM Trans. Graph. **36**(4), 1–12 (2017). Proceedings of SIGGRAPH 2017
13. Khashabi, D., Nowozin, S., Jancsary, J., Fitzgibbon, A.W.: Joint demosaicing and denoising via learned nonparametric random fields. IEEE Trans. Image Process. **23**(12), 4968–4981 (2014). https://doi.org/10.1109/TIP.2014.2359774
14. Kingma, D.P., Ba, J.: Adam: a method for stochastic optimization. CoRR abs/1412.6980 (2014). http://dblp.uni-trier.de/db/journals/corr/corr1412.html#KingmaB14
15. Köhler, T., Huang, X., Schebesch, F., Aichert, A., Maier, A., Hornegger, J.: Robust multi-frame super-resolution employing iteratively re-weighted minimization. IEEE Trans. Comput. Imaging **2**(1), 42–58 (2016)
16. Kokkinos, F., Lefkimmiatis, S.: Iterative residual network for deep joint image demosaicking and denoising. CoRR abs/1807.06403 (2018). http://arxiv.org/abs/1807.06403
17. Kupyn, O., Budzan, V., Mykhailych, M., Mishkin, D., Matas, J.: DeblurGAN: blind motion deblurring using conditional adversarial networks. In: Proceedings of the IEEE Conference on Computer Vision and Pattern Recognition, pp. 8183–8192 (2018)
18. Lai, W.S., Huang, J.B., Ahuja, N., Yang, M.H.: Fast and accurate image super-resolution with deep Laplacian pyramid networks. IEEE Trans. Pattern Anal. Mach. Intell. **41**(11), 2599–2613 (2018)
19. Lee, S., An, G.H., Kang, S.J.: Deep chain HDRI: reconstructing a high dynamic range image from a single low dynamic range image. IEEE Access **6**, 49913–49924 (2018)
20. Lee, S., An, G.H., Kang, S.-J.: Deep recursive HDRI: inverse tone mapping using generative adversarial networks. In: Ferrari, V., Hebert, M., Sminchisescu, C., Weiss, Y. (eds.) ECCV 2018. LNCS, vol. 11206, pp. 613–628. Springer, Cham (2018). https://doi.org/10.1007/978-3-030-01216-8_37
21. Li, X., Gunturk, B., Zhang, L.: Image demosaicing: a systematic survey. In: Visual Communications and Image Processing 2008, vol. 6822, p. 68221J. International Society for Optics and Photonics (2008)
22. Liang, Z., Cai, J., Cao, Z., Zhang, L.: CameraNet: a two-stage framework for effective camera isp learning. arXiv preprint arXiv:1908.01481 (2019)
23. Liu, C.: Beyond pixels: exploring new representations and applications for motion analysis. Ph.D. thesis, Massachusetts Institute of Technology, January 2009
24. Maier, A.K., et al.: Learning with known operators reduces maximum error bounds. Nat. Mach. Intell. **1**(8), 373–380 (2019)
25. Mantiuk, R., Kim, K.J., Rempel, A.G., Heidrich, W.: HDR-VDP-2: a calibrated visual metric for visibility and quality predictions in all luminance conditions. ACM Trans. Graph. (TOG) **30**(4), 40 (2011)
26. Menon, D., Andriani, S., Calvagno, G.: Demosaicing with directional filtering and a posteriori decision. IEEE Trans. Image Process. **16**(1), 132–141 (2007). https://doi.org/10.1109/TIP.2006.884928
27. Pan, J., Hu, Z., Su, Z., Yang, M.H.: l_0-regularized intensity and gradient prior for deblurring text images and beyond. IEEE Trans. Pattern Anal. Mach. Intell. **39**(2), 342–355 (2016)
28. Ratnasingam, S.: Deep camera: a fully convolutional neural network for image signal processing. arXiv preprint arXiv:1908.09191 (2019)
29. Schwartz, E., Giryes, R., Bronstein, A.M.: DeepISP: toward learning an end-to-end image processing pipeline. IEEE Trans. Image Process. **28**(2), 912–923 (2018)
30. Sen, P., Kalantari, N.K., Yaesoubi, M., Darabi, S., Goldman, D.B., Shechtman, E.: Robust patch-based HDR reconstruction of dynamic scenes. ACM Trans. Graph. (TOG) **31**(6), 203:1-203:11 (2012). Proceedings of SIGGRAPH Asia 2012

31. Serrano, A., Heide, F., Gutierrez, D., Wetzstein, G., Masia, B.: Convolutional sparse coding for high dynamic range imaging. Comput. Graph. Forum **35**(2), 153–163 (2016)
32. Tan, D.S., Chen, W.Y., Hua, K.L.: Deepdemosaicking: adaptive image demosaicking via multiple deep fully convolutional networks. IEEE Trans. Image Process. **27**(5), 2408–2419 (2018)
33. Tao, X., Gao, H., Shen, X., Wang, J., Jia, J.: Scale-recurrent network for deep image deblurring. In: Proceedings of the IEEE Conference on Computer Vision and Pattern Recognition, pp. 8174–8182 (2018)
34. Tocci, M.D., Kiser, C., Tocci, N., Sen, P.: A versatile HDR video production system. ACM Trans. Graph. (TOG) **30**(4), 41 (2011)
35. Tomaszewska, A., Mantiuk, R.: Image registration for multi-exposure high dynamic range image acquisition. In: International Conference in Central Europe on Computer Graphics, Visualization and Computer Vision (WSCG), pp. 49–56 (2007)
36. Tsai, C.Y., Song, K.T.: A new edge-adaptive demosaicing algorithm for color filter arrays. Image Vis. Comput. **25**(9), 1495–1508 (2007)
37. Gondal, M.W., Schölkopf, B., Hirsch, M.: The unreasonable effectiveness of texture transfer for single image super-resolution. In: Leal-Taixé, L., Roth, S. (eds.) ECCV 2018. LNCS, vol. 11133, pp. 80–97. Springer, Cham (2019). https://doi.org/10.1007/978-3-030-11021-5_6
38. Wang, X., et al.: ESRGAN: enhanced super-resolution generative adversarial networks. In: Leal-Taixé, L., Roth, S. (eds.) ECCV 2018. LNCS, vol. 11133, pp. 63–79. Springer, Cham (2019). https://doi.org/10.1007/978-3-030-11021-5_5
39. Wu, S., Xu, J., Tai, Y.-W., Tang, C.-K.: Deep high dynamic range imaging with large foreground motions. In: Ferrari, V., Hebert, M., Sminchisescu, C., Weiss, Y. (eds.) ECCV 2018. LNCS, vol. 11206, pp. 120–135. Springer, Cham (2018). https://doi.org/10.1007/978-3-030-01216-8_8
40. Zhang, K., Zuo, W., Chen, Y., Meng, D., Zhang, L.: Beyond a Gaussian denoiser: residual learning of deep CNN for image denoising. CoRR abs/1608.03981 (2016). http://arxiv.org/abs/1608.03981
41. Zhang, K., Zuo, W., Chen, Y., Meng, D., Zhang, L.: Beyond a Gaussian denoiser: residual learning of deep CNN for image denoising. IEEE Trans. Image Process. **26**(7), 3142–3155 (2017)
42. Zhang, L., Deshpande, A., Chen, X.: Denoising vs. deblurring: HDR imaging techniques using moving cameras. In: IEEE Conference on Computer Vision and Pattern Recognition (CVPR), pp. 522–529. IEEE (2010)
43. Zhou, R., Achanta, R., Süsstrunk, S.: Deep residual network for joint demosaicing and super-resolution. CoRR abs/1802.06573 (2018). http://arxiv.org/abs/1802.06573

Spatiotemporal Outdoor Lighting Aggregation on Image Sequences

Haebom Lee[1,2(✉)], Robert Herzog[1], Jan Rexilius[3], and Carsten Rother[2]

[1] Computer Vision Research Lab, Robert Bosch GmbH, Hildesheim, Germany
[2] Computer Vision and Learning Lab, Heidelberg University, Heidelberg, Germany
[3] Campus Minden, Bielefeld University of Applied Sciences, Minden, Germany

Abstract. In this work, we focus on outdoor lighting estimation by aggregating individual noisy estimates from images, exploiting the rich image information from wide-angle cameras and/or temporal image sequences. Photographs inherently encode information about the scene's lighting in the form of shading and shadows. Whereas computer graphic (CG) methods target accurately reproducing the image formation process knowing the exact lighting in the scene, the inverse rendering is an ill-posed problem attempting to estimate the geometry, material, and lighting behind a recorded 2D picture. Recent work based on deep neural networks has shown promising results for single image lighting estimation despite its difficulty. However, the main challenge remains on the stability of measurements. We tackle this problem by combining lighting estimates from many image views sampled in the angular and temporal domain of an image sequence. Thereby, we make efficient use of the camera calibration and camera ego-motion estimation to globally register the individual estimates and apply outlier removal and filtering algorithms. Our method not only improves the stability for rendering applications like virtual object augmentation but also shows higher accuracy for single image based lighting estimation compared to the state-of-the-art.

Keywords: Lighting estimation · Spatio-temporal filtering · Virtual object augmentation

1 Introduction

Deep learning has shown its potential in estimating hidden information like depth from monocular images [4] by only exploiting learned priors. Accordingly, it has also been applied for the task of lighting estimation. The shading in a photograph captures the incident lighting (irradiance) on a surface point. It depends not only on the local surface geometry and material but also on the global (possibly occluded) lighting in a mostly unknown 3D scene. Different configurations of

Supplementary Information The online version contains supplementary material available at https://doi.org/10.1007/978-3-030-92659-5_22.

C. Bauckhage et al. (Eds.): DAGM GCPR 2021, LNCS 13024, pp. 343–357, 2021.
https://doi.org/10.1007/978-3-030-92659-5_22

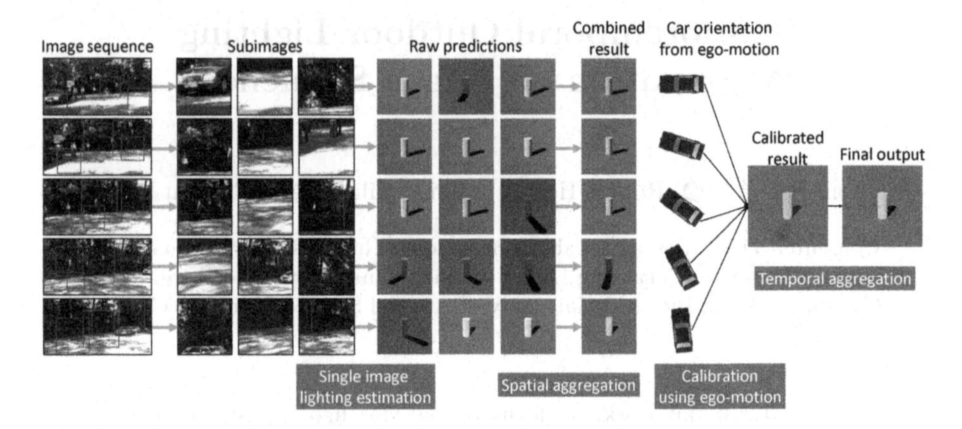

Fig. 1. Spatio-temporal outdoor lighting aggregation on an image sequence: individual estimates from each generated subimage are combined in the spatial aggregation step. Spatial aggregation results for each image in the sequence are then calibrated using camera ego-motion data and further refined in the temporal aggregation step to generate the final lighting estimate for the sequence.

material, geometry, and lighting parameters may lead to the same pixel color, which creates an ill-posed optimization problem without additional constraints. Hence, blindly estimating the lighting conditions is notoriously difficult, and we restrict ourselves to outdoor scenes considering only environment lighting where the incident lighting is defined to be spatially invariant.

Estimating environment lighting can be regarded as the first step towards holistic scene understanding and enables several applications [1,11,22,28]. It is essential for augmented reality (seamlessly rendering virtual objects into real background images) because photo-realistically inserting virtual objects in real images requires knowing not just the 3D geometry and camera calibration but also the lighting. The human eye quickly perceives wrong lighting and shadows as unrealistic, and it has also been shown [27] that shadows are essential for depth-from-mono estimation using convolutional neural networks.

There have been numerous studies on estimating the lighting from image data. Those methods mostly focus on estimating sky map textures [5] or locating the sun position from a single RGB image [6,8,34], calculating sun trajectories from time-lapse videos [1,19], or utilizing material information to conjecture the positions of multiple light sources [29].

In this paper, we propose a method to robustly estimate the global environment's sun direction by exploiting temporal and spatial coherency in outdoor lighting. The image cues for resolving the lighting in a scene appear sparsely (e.g., shadows, highlights, etc.) or very subtle and noisy (e.g., color gradients, temperature, etc.) and not all images provide the same quality of information for revealing the lighting parameters. For example, consider an image view completely covered in shadow. Hence, the predictions for the lighting on individual images of a sequence are affected by a large amount of noise and many outliers. To alleviate this issue we propose to sample many sub-views of an image sequence essentially sampling in the angular and temporal domain. This approach has two advantages. First, we effectively filter noise and detect outliers, and second, our neural network-based lighting estimator becomes invariant to the imaging parameters like size, aspect ratio, and camera focal length and can explore details in the high-resolution image content. To this end, the contributions of this paper are:

1. A single image based sunlight estimation using a deep artificial neural network that is on par or better than the current state-of-the-art,
2. A two-stage post-processing approach for spatial and temporal filtering with outlier detection that fully exploits the information from calibrated image sequences to overcome noisy, outlier-sensitive estimation methods.

2 Related Work

Outdoor lighting condition estimation has been studied in numerous ways because of its importance in computer graphics and computer vision applications [12,20]. Related techniques can be categorized into two parts, one that analyzes a single image [5,9,15,21] and the other that utilizes a sequence of images [1,16,19,22]. For example, the outdoor illumination estimation method presented in [23] belongs to the latter as the authors estimated the sun trajectory and its varying intensity from a sequence of images. Under the assumption that a static 3D model of the scene is available, they designed a rendering equation-based [10] optimization problem to determine the continuous change of the lighting parameters. On the other hand, Hold-Geoffroy et al. [6] proposed a method that estimates outdoor illumination from a single low dynamic range image using a convolutional neural network [14]. The network was able to classify the sun location on 160 evenly distributed positions on the hemisphere and estimated other parameters such as sky turbidity, exposure, and camera parameters.

Analyzing outdoor lighting conditions is further developed in [34] where they incorporated a more delicate illumination model [16]. The predicted parameters were numerically compared with the ground truth values and examined rather qualitatively by utilizing the render loss. Jin et al. [8] and Zhang et al. [35] also proposed single image based lighting estimation methods. While their predecessors [6,34] generated a probability distribution of the sun position on the discretized hemisphere, the sun position parameters were directly regressed from their networks. Recently, Zhu et al. [36] combined lighting estimation with intrinsic image decomposition. Although they achieved a noticeable outcome in the sun position estimation on a synthetic dataset, we were unable to compare it with ours due to the difference in the datasets.

The aforementioned lighting estimation techniques based on a single image often suffer from insufficient cues to determine the lighting condition, for example, when the given image is in a complete shadow. Therefore, several attempts were made to increase the accuracy and robustness by taking the temporal domain into account [1,16,22]. The method introduced in [19] extracts a set of features from each image frame and utilizes it to estimate the relative changes of the lighting parameters in an image sequence. Their method is capable of handling a moving camera and generating temporally coherent augmentations. However, the estimation process utilized only two consecutive frames and assumed that the sun position is given in the form of GPS coordinates and timestamps [25].

Lighting condition estimation is also crucial for augmented reality where virtual objects become more realistic when rendered into the background image using the correct lighting. Lu et al. [20], for instance, estimated a directional light vector from shadow regions and the corresponding objects in the scene to achieve realistic occlusion with augmented objects. The performance of the estimation depends solely on the shadow region segmentation and finding related items. Therefore, the method may struggle if a shadow casting object is not visible in the image. Madsen and Lal [22] utilize a stereo camera to extend [23] further. Using the sun position calculated from GPS coordinates and timestamps, they estimated the variances of the sky and the sun over an image sequence. The estimation is then combined with a shadow detection algorithm to generate plausible augmented scenes with proper shading and shadows.

Recently, there have been several attempts utilizing auxiliary information to estimate the lighting condition [11,33]. Such information may result in better performance but only with a trade-off in generality. Kán and Kaufmann [11] proposed a single RGB-D image-based lighting estimation method for augmented reality applications. They utilized synthetically generated scenes for training a deep neural network, which outputs the dominant light source's angular coordinates in the scene. Outlier removal and temporal smoothing processes were applied to make the method temporally consistent. However, their technique was demonstrated only on fixed viewpoint scenes. Our method, on the other hand, improves its estimation by aggregating observations from different viewpoints. We illustrate the consistency gained from our novel design by augmenting virtual objects in consecutive frames.

3 Proposed Method

We take advantage of different aspects of previous work and refine them into our integrated model. As illustrated in Fig. 1, our model is composed of four subprocesses. We first randomly generate several small subimages from an input image and upsample them to a fixed size. Since modern cameras are capable of capturing fine details of a scene, we found that lighting condition estimation can be done on a small part of an image. These spatial samples obtained from one image all share the same lighting condition and therefore yield more robustness compared to a single image view. Then, we train our lighting estimation network on each sample to obtain the global lighting for a given input image.

After the network estimates the lighting conditions for the spatial samples, we perform a spatial aggregation step to get a stable prediction for each image. Note that the estimate for each frame is based on its own camera coordinate system. Our third step is to unify the individual predictions into one global coordinate system using the camera ego-motion. Lastly, the calibrated estimates are combined in the temporal aggregation step. The assumption behind our approach is that distant sun-environment lighting is invariant to the location the picture was taken and that the variation in lighting direction is negligible for short videos. Through the following sections, we introduce the details of each submodule.

3.1 Lighting Estimation

There have been several sun and sky models to parameterize outdoor lighting conditions [7,16]. Although those methods are potentially useful to estimate complex lighting models consistently, in this work we focus only on the most critical lighting parameter: the sun direction. The rationale behind this is that ground-truth training data can easily be generated for video sequences with GPS and timestamp information (e.g., KITTI dataset [3]). Therefore, the lighting estimation network's output is a 3D unit vector $\vec{v}_{pred}$ pointing to the sun's location in the camera coordinate system.

Unlike our predecessors [6,34], we design our network as a direct regression model to overcome the need for a sensitive discretization of the hemisphere. The recent work of Jin et al. [8] presented a regression network estimating the sun direction in spherical coordinates (altitude and azimuth). Our method, however, estimates the lighting direction using Cartesian coordinates and does not suffer from singularities in the spherical parametrization and the ambiguity that comes from the cyclic nature of the spherical coordinates.

Since we train our network in a supervised manner, the loss function is defined to compare the estimated sun direction with the ground truth $\vec{v}_{gt}$:

$$L_{cosine} = 1 - \vec{v}_{gt} \cdot \vec{v}_{pred}/||\vec{v}_{pred}||, \tag{1}$$

with the two adjacent unit vectors having their inner product close to 1. To avoid the uncertainty that comes from the vectors pointing the same direction with different lengths, we apply another constraint to the loss function:

$$L_{norm} = (1 - ||\vec{v}_{pred}||)^2. \tag{2}$$

The last term of the loss function ensures that the estimated sun direction resides in the upper hemisphere because we assume the sun is the primary light source in the given scene:

$$L_{hemi} = max(0, -z_{pred}), \tag{3}$$

where z_{pred} is the third component of $\vec{v}_{pred}$, indicating the altitude of the sun. The final loss function is simply the sum of all terms as they share a similar range of values:

$$L_{light} = L_{cosine} + L_{norm} + L_{hemi}. \tag{4}$$

3.2 Spatial Aggregation

Using our lighting estimator, we gather several lighting condition estimates from different regions of the image. Some of those estimates may contain larger errors due to insufficient information in the given region to predict the lighting condition. We refer to such estimates as outliers. Our method's virtue is to exclude anomalies that commonly occur in single image-based lighting estimation techniques and deduce the best matching model that can explain the inliers.

Among various outlier removal algorithms, we employ the isolation forest (iForest) algorithm [18]. The technique is specifically optimized to isolate anomalies instead of building a model of inliers and eliminate samples not complying with it. In essence, the iForest algorithm recursively and randomly splits the feature space into binary decision trees (hence forming a forest). Since the outliers are outside of a potential inlier cluster, a sample is classified as an outlier if the sample's average path length is shorter than a threshold (*contamination ratio* [24]). We determine this value empirically and use it throughout all results.

On the remaining inliers, we apply the *mean shift algorithm* [2] to conjecture the most feasible lighting parameters. Unlike naive averaging over all inliers, this process further refines the lighting estimate by iteratively climbing to the maximum density in the distribution. Another experimentally discoverable parameter *bandwidth* determines the size of the Gaussian kernel to measure the samples' local density gradient. In the proposed method, we set the bandwidth as the median of all samples' pairwise distances. By moving the data points iteratively towards the closest peak in the density distribution, the algorithm locates the highest density within a cluster, our spatial aggregation result. We compare various aggregation methods in the ablation study in Sect. 4.4.

3.3 Calibration

Since our primary goal is to assess the sun direction for an input video, we perform a calibration step to align the estimates because the sun direction determined from each image in a sequence is in its own local camera coordinate system. The camera ego-motion data is necessary to transform the estimated sun direction vectors into the world coordinate system. We assume the noise and drift in the ego-motion estimation is small relative to the lighting estimation. Hence, we employ a state-of-the-art structure-from-motion (SfM) technique such as [26] to estimate the ego-motion from an image sequence. Then there exists a camera rotation matrix R_f for each frame f and the resulting calibrated vector $\hat{\vec{v}}_{pred}$ is computed as $R_f^{-1} \cdot \vec{v}_{pred}$.

3.4 Temporal Aggregation

Having the temporal estimates aligned in the same global coordinate system, we consider them as independent observations of the same lighting condition in the temporal domain. Although the lighting estimates from our regression network are not necessarily independent for consecutive video frames, natural image sequences, as shown empirically in our experiments, reveal a large degree

Table 1. Number of data and subimages for training and test

Dataset		SUN360	KITTI
Training	Data	16 891	3630
	Subimg	135 128	116 160
Test	Data	1688	281
	Subimg	108 032	17 984

of independent noise in the regression results, which is however polluted with a non-neglectable amount of outliers. Consequently, we apply a similar aggregation strategy as in the spatial domain also for the temporal domain. Therefore, the final output of our pipeline, the lighting condition for the given image sequence, is the mean shift algorithm's result on the inliers from all frames of the entire image sequence.

4 Experiments

4.1 Datasets

One of the common datasets considered in the outdoor lighting estimation methods is the SUN360 dataset [31]. Several previous methods utilized it in its original panorama form or as subimages by generating synthetic perspective images [6]. We follow the latter approach since we train our network using square images. We first divide 20267 panorama images into the training, validation, and test sets with a 10:1:1 ratio. For the training and the validation sets, 8 subimages from each panorama are taken by evenly dividing the azimuth range. To increase the diversity, 64 subimages with random azimuth values are generated from each panorama in the test set. Note that we introduce small random offsets on the camera elevation with respect to the horizon in $[-10°, 10°]$ and randomly select a camera field of view within a range $[50°, 80°]$. The generated images are resized to 256×256. In this way, we produced 135128, 13504, and 108032 subimages from 16891, 1688, and 1688 panoramas for the training, validation, and test sets, respectively. The ground truth labeling was given by the authors of [34].

The well-known KITTI dataset [3] has also attracted our attention. Since the dataset is composed of several rectified driving image sequences and provides the information required for calculating the ground truth sun directions [25], we utilize it for both training and test. Specifically, since the raw data was recorded at 10 Hz, we collect every 10th image to avoid severe repetition and split off five randomly chosen driving scenes for validation and test set. The resulting training set is composed of 3630 images. If we train our network using only one crop for each KITTI image, the network is likely to be biased to the SUN360 dataset due to the heavy imbalance in the amount of data. To match the number of samples, we crop 32 subimages from one image by varying the cropping location and the crop size. Each image in the test set is again cropped into 64 subimages and the cropped images are also resized to 256×256. In total, we train our network on about 250000 images. The exact numbers of samples are presented in Table 1 and Fig. 2 illustrates examples of the two datasets.

4.2 Implementation Details

Our lighting estimation model is a regression network with convolution layers. It accepts an RGB image of size 256×256 and outputs the sun direction estimate. We borrow the core structure from ResNeXt [32] and carefully determine the number of blocks, groups, and filters as well as the sizes of filters under extensive experiments. As illustrated in Fig. 3, the model is roughly composed of 8 bottleneck blocks, each of which is followed by a convolutional block attention module [30]. In this way, our network is capable of focusing on important spatial and channel features while acquiring resilience from vanishing or exploding gradients by using the shortcut connections. A global average pooling layer is adopted to connect the convolution network and the output layer and serves as a tool to mitigate possible overfitting [17]. The dense layer at the end then refines the encoded values into the sun direction estimate.

We train our model and test its performance on the SUN360 and the KITTI datasets (see Table 1). In detail, we empirically trained our lighting estimation network for 18 epochs using early stopping. The training was initiated with the Adam optimizer [13] using a learning rate of 1×10^{-4} and the batch size was 64. It took 12 h on a single Nvidia RTX 2080 Ti GPU. Prediction on a single image takes 42 ms. Our single image lighting estimation and spatial aggregation modules are examined upon 108 032 unobserved SUN360 crops generated from 1688 panoramas. The whole pipeline including the calibration and temporal aggregation modules is analyzed on five unseen KITTI sequences composed of 281 images.

4.3 Results

We evaluate the angular errors of the spatially aggregated sun direction estimates on the SUN360 test set. At first, single image lighting estimation results

Fig. 2. Examples of the two datasets [3,31]. From the original image (*top*), we generate random subimages (*bottom*).

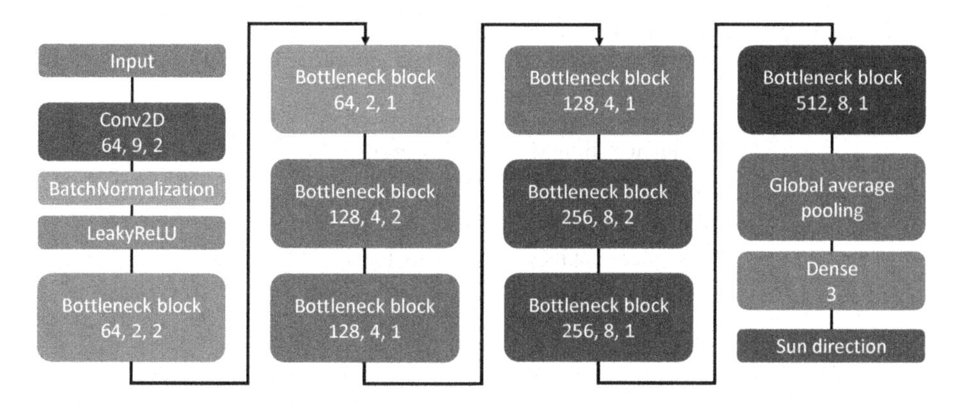

Fig. 3. The proposed lighting estimation network. The numbers on the *Conv2D* layer indicate the number of filters, the filter size, and the stride, whereas the numbers on each *Bottleneck block* depict the number of 3×3 filters, the cardinality, and the stride. A *Bottleneck block* is implemented following the structure proposed in [32] except for a convolutional block attention module [30] attached at the end of each block.

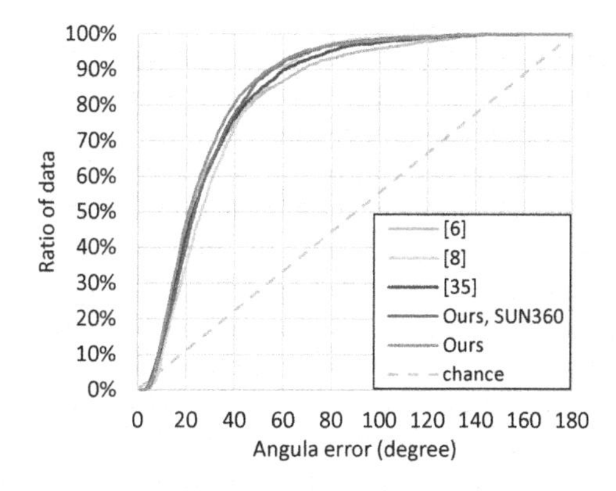

Fig. 4. The cumulative angular error for spatially aggregated sun direction estimates on the SUN360 test set. *Ours, SUN360* indicates our results when the network was only trained with the SUN360 dataset.

are gathered using [6,8,35], and our method. Then we compensate the camera angles and apply our spatial aggregation step on the subimages to acquire the spatially combined estimate for each panorama. The explicit spatial aggregation step involves two additional hyperparameters: the contamination ratio and the mean-shift kernel width. We found those parameters to be insensitive to different data sets and kept the same values in all our experiments. The *contamination ratio* is set to 0.5 because we assume the estimations with angular errors larger than an octant (22.5°) as outliers, which is roughly 50% of the data for our method when observing Fig. 7. As a result, we apply the mean shift algorithm on 50% potential inliers among the total observations.

Table 2. Angular errors of each aggregation step (from left to right: single image (baseline), spatial aggregation, spatio-temporal aggregation). Sequences correspond to Fig. 5.

Sequence	Single	Spatial	Spatiotemporal
(a)	13.43	6.76	3.54
(b)	26.06	7.81	6.87
(c)	34.68	24.83	13.17
(d)	23.03	10.04	3.27

Figure 4 illustrates the cumulative angular errors of the four methods. Since the previous methods were trained with only the SUN360 training set, due to the characteristics of their networks (requiring ground truth exposure and turbidity information which are lacked in the KITTI dataset), we also report our method's performance when it was trained only on SUN360 (see *Ours, SUN360* in Fig. 4). Our method performs better than the previous techniques even with the same training set. The detailed quantitative comparison is presented in Fig. 7. Note that all methods are trained and tested with subimages instead of full images.

For the KITTI dataset, we can further extend the lighting estimation to the temporal domain. Although the dataset provides the ground truth ego-motion, we calculated it using [26] to generalize our approach. The mean angular error of the estimated camera rotation using the default parameters was 1.01° over the five test sequences. We plotted the sun direction estimates of each step in our pipeline for four (out of five) test sequences in Fig. 5. Note that in the plots all predictions are registered to a common coordinate frame using the estimated camera ego-motion. Individual estimates of the subimages are shown with gray dots. Our spatial aggregation process refines the noisy observations using outlier removal and mean shift (black dots). Those estimates for each frame in a sequence are finally combined in the temporal aggregation step (denoted with the green dot). The ground truth direction is indicated by the red dot. Using the spatio-temporal filtering, the mean angular error over the five test sequences recorded 7.68°, which is a reduction of 69.94% (25.56° for single image based estimation). A quantitative evaluation of the performance gain for each aggregation step is presented in Table 2.

Our model's stability is better understood with a virtual object augmentation application as shown in Fig. 6. Note that other lighting parameters, such as the sun's intensity are manually determined. When the lighting conditions are estimated from only a single image on each frame, the virtual objects' shadows are fluctuating compared to the ground truth results. The artifact is less visible on our spatial aggregation results and entirely removed after applying the spatio-temporally aggregated lighting condition.

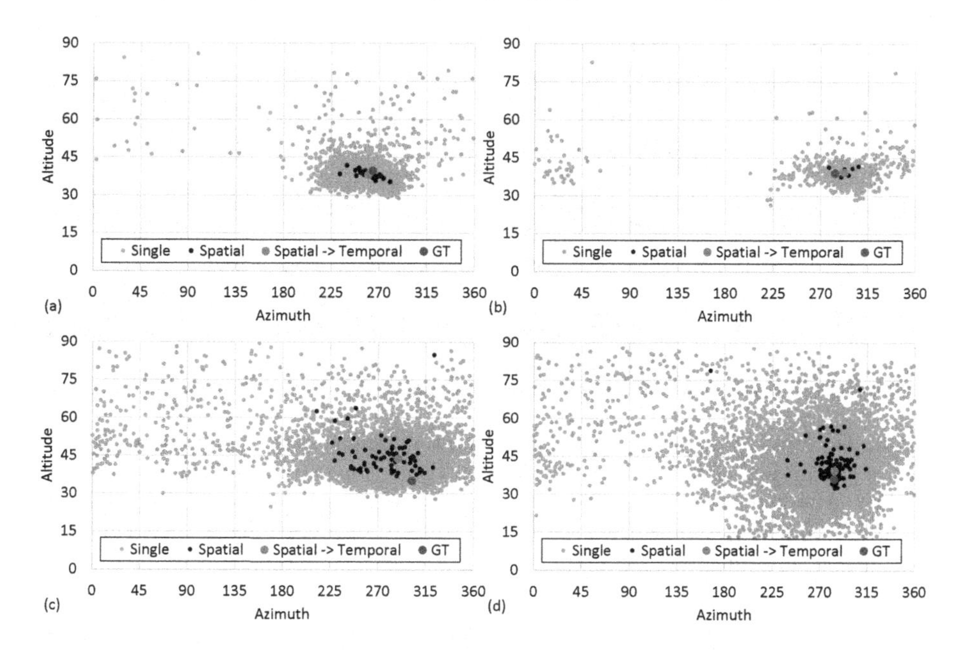

Fig. 5. Scatter plots representing sun direction estimates of individual subimages and the results of two aggregation steps. Each graph corresponds to an image sequence in the KITTI test set. Despite numerous outliers in the raw observations (the gray dots), our two-step aggregation determines the video's lighting condition with small margins to the ground truth sun direction (the black dots for spatial aggregation and the green dot for spatio-temporal aggregation). Angular errors for our spatio-temporal filtering results are (a) 3.54 (b) 6.87 (c) 13.17 and (d) 3.27°. (Color figure online)

4.4 Ablation Study

The performance gain of the spatial aggregation process is thoroughly analyzed by breaking down the individual filtering steps on the SUN360 test set. Figure 7 shows the cumulative angular error for the raw observations and compares the four lighting estimation methods with four different aggregation strategies:

- *Single*: unprocessed individual observations,
- *Mean all*: mean of all estimates from each panorama,
- *Mean inliers*: mean of inlier estimates,
- *Meanshift*: mean shift result of inlier estimates.

As illustrated in Fig. 7, the average angular error of each method is decreased by at most 10° after applying the proposed spatial aggregation. This result demonstrates our method's generality, showing that it can increase the accuracy of any lighting estimation method. We observe a slight increase in the average error for the *Mean all* metric due to the outlier observations. A similar analysis

Fig. 6. Demonstration of a virtual augmentation application. Fluctuations in the shadow of the augmented object decrease as the estimates are refined through our pipeline. After applying the spatio-temporal filtering, the results are fully stabilized and almost indistinguishable from the ground truth. Please also refer to the augmented video in the supplementary material.

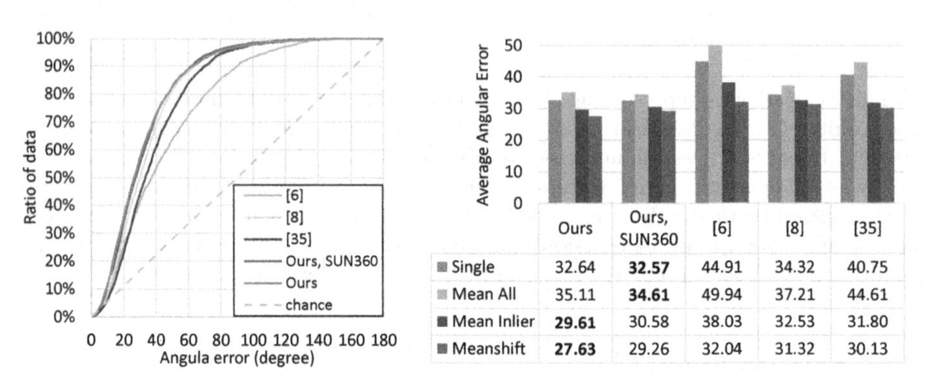

	Ours	Ours, SUN360	[6]	[8]	[35]
Single	32.64	**32.57**	44.91	34.32	40.75
Mean All	35.11	**34.61**	49.94	37.21	44.61
Mean Inlier	**29.61**	30.58	38.03	32.53	31.80
Meanshift	**27.63**	29.26	32.04	31.32	30.13

Fig. 7. (*left*) The cumulative angular error for the *single* estimates on the SUN360 test set. (*right*) Comparing average angular error for three methods with different spatial aggregation strategies. Our method achieved the best result when the mean shift is applied to the inliers. We outperform previous methods even without the KITTI dataset.

is done for the KITTI dataset with only our method. The cumulative angular error graphs for the four steps are presented in Fig. 8.

Our pipeline was also tested as an end-to-end model, but it failed to show comparable performance. We provide the details of this experiment as well as additional studies with different combinations of loss functions in the supplementary material.

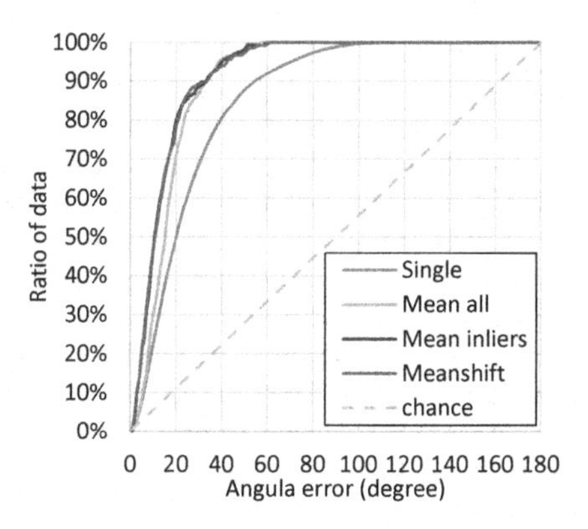

Fig. 8. The cumulative angular error on the KITTI test set with different spatial aggregation strategies. The best result is recorded when the mean shift result of the inlier estimates is utilized.

5 Conclusion

In this paper, we proposed a single image lighting estimation method and showed how its performance can be improved using spatial and temporal aggregation. Our method achieved state-of-the-art performance on outdoor lighting estimation for a given image sequence. We utilized 360° panoramas and wide view images in our work, but our spatial aggregation can be also applied to any image containing enough details. To this end, our spatio-temporal aggregation can be extended to different methods of gathering globally shared scene information.

Although we demonstrated noticeable outcomes in augmented reality applications, intriguing future research topics are remaining. We plan to extend our model to examine other factors such as cloudiness or exposure as it helps to accomplish diverse targets, including photorealistic virtual object augmentation over an image sequence. With such augmented datasets, we could enhance the performance of other deep learning techniques.

References

1. Balcı, H., Güdükbay, U.: Sun position estimation and tracking for virtual object placement in time-lapse videos. SIViP **11**(5), 817–824 (2017)
2. Comaniciu, D., Meer, P.: Mean shift: a robust approach toward feature space analysis. IEEE Trans. Pattern Anal. Mach. Intell. **24**(5), 603–619 (2002)
3. Geiger, A., Lenz, P., Urtasun, R.: Are we ready for autonomous driving? The KITTI vision benchmark suite. In: 2012 IEEE Conference on Computer Vision and Pattern Recognition, pp. 3354–3361. IEEE (2012)
4. Godard, C., Mac Aodha, O., Firman, M., Brostow, G.J.: Digging into self-supervised monocular depth estimation. In: Proceedings of the IEEE International Conference on Computer Vision, pp. 3828–3838 (2019)

5. Hold-Geoffroy, Y., Athawale, A., Lalonde, J.F.: Deep sky modeling for single image outdoor lighting estimation. In: Proceedings of the IEEE Conference on Computer Vision and Pattern Recognition, pp. 6927–6935 (2019)
6. Hold-Geoffroy, Y., Sunkavalli, K., Hadap, S., Gambaretto, E., Lalonde, J.F.: Deep outdoor illumination estimation. In: Proceedings of the IEEE Conference on Computer Vision and Pattern Recognition, pp. 7312–7321 (2017)
7. Hosek, L., Wilkie, A.: An analytic model for full spectral sky-dome radiance. ACM Trans. Graph. (TOG) **31**(4), 1–9 (2012)
8. Jin, X., et al.: Sun-sky model estimation from outdoor images. J. Ambient Intell. Hum. Comput. 1–12 (2020). https://doi.org/10.1007/s12652-020-02367-3
9. Jin, X., et al.: Sun orientation estimation from a single image using short-cuts in DCNN. Optics Laser Technol. **110**, 191–195 (2019)
10. Kajiya, J.T.: The rendering equation. In: Proceedings of the 13th Annual Conference on Computer Graphics and Interactive Techniques, pp. 143–150 (1986)
11. Kán, P., Kaufmann, H.: Deeplight: light source estimation for augmented reality using deep learning. Vis. Comput. **35**(6–8), 873–883 (2019)
12. Karsch, K., Hedau, V., Forsyth, D., Hoiem, D.: Rendering synthetic objects into legacy photographs. ACM Trans. Graph. (TOG) **30**(6), 1–12 (2011)
13. Kingma, D.P., Ba, J.: Adam: a method for stochastic optimization. arXiv preprint arXiv:1412.6980 (2014)
14. Krizhevsky, A., Sutskever, I., Hinton, G.E.: ImageNet classification with deep convolutional neural networks. In: Advances in Neural Information Processing Systems, pp. 1097–1105 (2012)
15. Lalonde, J.F., Efros, A.A., Narasimhan, S.G.: Estimating the natural illumination conditions from a single outdoor image. Int. J. Comput. Vis. **98**(2), 123–145 (2012)
16. Lalonde, J.F., Matthews, I.: Lighting estimation in outdoor image collections. In: 2014 2nd International Conference on 3D Vision, vol. 1, pp. 131–138. IEEE (2014)
17. Lin, M., Chen, Q., Yan, S.: Network in network. arXiv preprint arXiv:1312.4400 (2013)
18. Liu, F.T., Ting, K.M., Zhou, Z.H.: Isolation forest. In: 2008 Eighth IEEE International Conference on Data Mining, pp. 413–422. IEEE (2008)
19. Liu, Y., Granier, X.: Online tracking of outdoor lighting variations for augmented reality with moving cameras. IEEE Trans. Visual Comput. Graph. **18**(4), 573–580 (2012)
20. Lu, B.V., Kakuta, T., Kawakami, R., Oishi, T., Ikeuchi, K.: Foreground and shadow occlusion handling for outdoor augmented reality. In: 2010 IEEE International Symposium on Mixed and Augmented Reality, pp. 109–118. IEEE (2010)
21. Ma, W.C., Wang, S., Brubaker, M.A., Fidler, S., Urtasun, R.: Find your way by observing the sun and other semantic cues. In: 2017 IEEE International Conference on Robotics and Automation (ICRA), pp. 6292–6299. IEEE (2017)
22. Madsen, C.B., Lal, B.B.: Outdoor illumination estimation in image sequences for augmented reality. GRAPP **11**, 129–139 (2011)
23. Madsen, C.B., Störring, M., Jensen, T., Andersen, M.S., Christensen, M.F.: Real-time illumination estimation from image sequences. In: Proceedings: 14th Danish Conference on Pattern Recognition and Image Analysis, Copenhagen, Denmark, pp. 1–9 (2005)
24. Pedregosa, F., et al.: Scikit-learn: machine learning in Python. J. Mach. Learn. Res. **12**, 2825–2830 (2011)
25. Reda, I., Andreas, A.: Solar position algorithm for solar radiation applications. Sol. Energy **76**(5), 577–589 (2004)

26. Schonberger, J.L., Frahm, J.M.: Structure-from-motion revisited. In: Proceedings of the IEEE Conference on Computer Vision and Pattern Recognition, pp. 4104–4113 (2016)
27. Van Dijk, T., de Croon, G.C.H.E.: How do neural networks see depth in single images? In: Proceedings of the IEEE International Conference on Computer Vision, pp. 2183–2191 (2019)
28. Wei, H., Liu, Y., Xing, G., Zhang, Y., Huang, W.: Simulating shadow interactions for outdoor augmented reality with RGBD data. IEEE Access **7**, 75292–75304 (2019)
29. Whelan, T., Salas-Moreno, R.F., Glocker, B., Davison, A.J., Leutenegger, S.: Elasticfusion: real-time dense slam and light source estimation. Int. J. Robot. Res. **35**(14), 1697–1716 (2016)
30. Woo, S., Park, J., Lee, J.-Y., Kweon, I.S.: CBAM: convolutional block attention module. In: Ferrari, V., Hebert, M., Sminchisescu, C., Weiss, Y. (eds.) ECCV 2018. LNCS, vol. 11211, pp. 3–19. Springer, Cham (2018). https://doi.org/10.1007/978-3-030-01234-2_1
31. Xiao, J., Ehinger, K.A., Oliva, A., Torralba, A.: Recognizing scene viewpoint using panoramic place representation. In: 2012 IEEE Conference on Computer Vision and Pattern Recognition, pp. 2695–2702. IEEE (2012)
32. Xie, S., Girshick, R., Dollár, P., Tu, Z., He, K.: Aggregated residual transformations for deep neural networks. arXiv preprint arXiv:1611.05431 (2016)
33. Xiong, Y., Chen, H., Wang, J., Zhu, Z., Zhou, Z.: DSNet: deep shadow network for illumination estimation. In: 2021 IEEE Virtual Reality and 3D User Interfaces (VR), pp. 179–187. IEEE (2021)
34. Zhang, J., Sunkavalli, K., Hold-Geoffroy, Y., Hadap, S., Eisenman, J., Lalonde, J.F.: All-weather deep outdoor lighting estimation. In: Proceedings of the IEEE Conference on Computer Vision and Pattern Recognition, pp. 10158–10166 (2019)
35. Zhang, K., Li, X., Jin, X., Liu, B., Li, X., Sun, H.: Outdoor illumination estimation via all convolutional neural networks. Comput. Electr. Eng. **90**, 106987 (2021)
36. Zhu, Y., Zhang, Y., Li, S., Shi, B.: Spatially-varying outdoor lighting estimation from intrinsics. In: Proceedings of the IEEE/CVF Conference on Computer Vision and Pattern Recognition, pp. 12834–12842 (2021)

Generative Models and Multimodal Data

AttrLostGAN: Attribute Controlled Image Synthesis from Reconfigurable Layout and Style

Stanislav Frolov[1,2(✉)], Avneesh Sharma[1], Jörn Hees[2], Tushar Karayil[2], Federico Raue[2], and Andreas Dengel[1,2]

[1] Technical University of Kaiserslautern, Kaiserslautern, Germany
asharma@rhrk.uni-kl.de

[2] German Research Center for Artificial Intelligence (DFKI), Kaiserslautern, Germany
{stanislav.frolov,jorn.hees,tushar.karayil, federico.raue,andreas.dengel}@dfki.de

Abstract. Conditional image synthesis from layout has recently attracted much interest. Previous approaches condition the generator on object locations as well as class labels but lack fine-grained control over the diverse appearance aspects of individual objects. Gaining control over the image generation process is fundamental to build practical applications with a user-friendly interface. In this paper, we propose a method for attribute controlled image synthesis from layout which allows to specify the appearance of individual objects without affecting the rest of the image. We extend a state-of-the-art approach for layout-to-image generation to additionally condition individual objects on attributes. We create and experiment on a synthetic, as well as the challenging Visual Genome dataset. Our qualitative and quantitative results show that our method can successfully control the fine-grained details of individual objects when modelling complex scenes with multiple objects. Source code, dataset and pre-trained models are publicly available (https://github.com/stanifrolov/AttrLostGAN).

Keywords: Generative Adversarial Networks · Image synthesis

1 Introduction

The advent of Generative Adversarial Networks (GANs) [8] had a huge influence on the progress of image synthesis research and applications. Starting from low-resolution, gray-scale face images, current methods can generate high-resolution face images which are very difficult to distinguish from real photographs [20]. While unconditional image synthesis is interesting, most practical applications

Supplementary Information The online version contains supplementary material available at https://doi.org/10.1007/978-3-030-92659-5_23.

C. Bauckhage et al. (Eds.): DAGM GCPR 2021, LNCS 13024, pp. 361–375, 2021.
https://doi.org/10.1007/978-3-030-92659-5_23

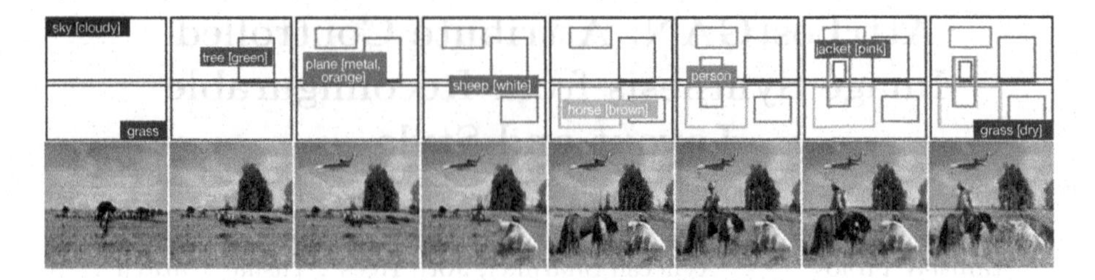

Fig. 1. Generated images using a reconfigurable layout and attributes to control the appearance of individual objects. *From left to right:* add tree [green], add plane [metal, orange], add sheep [white], add horse [brown], add person, add jacket [pink], grass → grass [dry]. (Color figure online)

require an interface which allows users to specify what the model should generate. In recent years, conditional generative approaches have used class labels [3,27], images [17,50], text [34,43,46], speech [4,42], layout [37,48], segmentation masks [6,41], or combinations of them [15], to gain control over the image generation process. However, most of these approaches are "one-shot" image generators which do not allow to reconfigure certain aspects of the generated image.

While there has been much progress on iterative image manipulation, researchers have so far not investigated how to gain better control over the image generation process of complex scenes with multiple interacting objects. To allow the user to create a scene that reflects what he/she has in mind, the system needs to be capable of iteratively and interactively updating the image. A recent approach by Sun and Wu [37] takes a major step towards this goal by enabling reconfigurable spatial layout and object styles. In their method, each object has an associated latent style code (sampled from a normal distribution) to create new images. However, this implies that users do not have true control over the specific appearance of objects. This lack of control also translates into the inability to specify a style (i.e., to change the color of a shirt from red to blue one would need to sample new latent codes and manually inspect whether the generated style conforms to the requirement). Being able to not just generate, but control individual aspects of the generated image without affecting other areas is vital to enable users to generate what they have in mind. To overcome this gap and give users control over style attributes of individual objects, we propose to extend their method to additionally incorporate attribute information. To that end, we propose Attr-ISLA as an extension of the Instance-Sensitive and Layout-Aware feature Normalization (ISLA-Norm) [37] and use an adversarial hinge loss on object-attribute features to encourage the generator to produce objects reflecting the input attributes. At inference time, a user can not only reconfigure the location and class of individual objects, but also specify a set of attributes. See Fig. 1 for an example of reconfigurable layout-to-image generation guided by attributes using our method. Since we continue to use latent codes for each object and the overall image, we can generate diverse images of objects with specific attributes. This approach not only drastically improves the flexibility but also allows the user to easily articulate the contents of his mind into the image generation process. Our contributions can be summarized as following:

- we propose a new method called AttrLostGAN, which allows attribute controlled image generation from reconfigurable layout;
- we extend ISLA to Attr-ISLA thereby gaining additional control over attributes;
- we create and experiment on a synthetic dataset to empirically demonstrate the effectiveness of our approach;
- we evaluate our model on the challenging Visual Genome dataset both qualitatively and quantitatively and achieve state-of-the-art performance.

2 Related Work

Class-Conditional Image Generation. Generating images given a class label is arguably the most direct way to gain control over what image to generate. Initial approaches concatenate the noise vector with the encoded label to condition the generator [27,30]. Recent approaches [3,33] have improved the image quality, resolution and diversity of generative models drastically. However, there are two major drawbacks that limit their practical application: they are based on single-object datasets, and do not allow reconfiguration of individual aspects of the image to be generated.

Layout-to-Image. The direct layout-to-image task was first studied in Layout2Im [48] using a VAE [21] based approach that could produce diverse 64×64 pixel images by decomposing the representation of each object into a specified label and an unspecified (sampled) appearance vector. LostGAN [37,38] allows better control over individual objects using a reconfigurable layout while keeping existing objects in the generated image unchanged. This is achieved by providing individual latent style codes for each object, wherein one code for the whole image allows to generate diverse images from the same layout when the object codes are fixed. We use LostGAN as our backbone to successfully address a fundamental problem: the inability to specify the appearance of individual objects using attributes.

Scene-Graph-to-Image. Scene graphs represent a scene of multiple objects using a graph structure where objects are encoded as nodes and edges represent the relationships between individual objects. Due to their convenient and flexible structure, scene graphs have recently been employed in multiple image generation approaches [1,11,18,39,45]. Typically, a graph convolution network (GCN) [10] is used to predict a scene layout containing segmentation masks and bounding boxes for each object which is then used to generate an image. However, scene graphs can be cumbersome to edit and do not allow to specify object locations directly on the image canvas.

Text-to-Image. Textual descriptions provide an intuitive way for conditional image synthesis [7]. Current methods [14,34,43,46,51] first produce a text

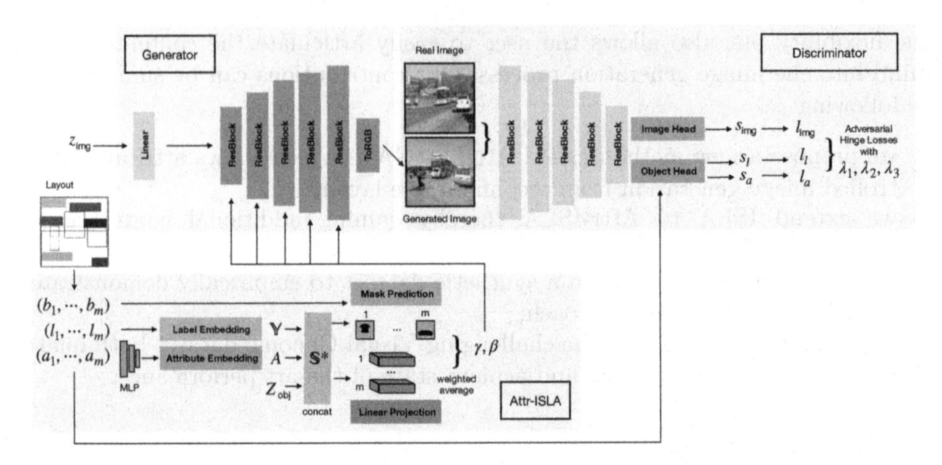

Fig. 2. Illustration of our proposed method for attribute controlled image synthesis from reconfigurable layout and style. Given a layout of object positions, class labels and attributes, we compute affine transformation parameters γ and β to condition the generator. Using separate latent codes Z_{obj} and z_{img} for individual objects and image, respectively, enables our model to produce diverse images. The discriminator minimizes three adversarial hinge losses on image features, object-label features, and object-attribute features.

embedding which is then input to a multi-stage image generator. In [13,14], additional layout information is used by adding an object pathway to learn the features of individual objects and control object locations. Decomposing the task into predicting a semantic layout from text, and then generate images conditioned on both text and semantic layout has been explored in [15,25]. Other works focus on disentangling content from style [23,49], and text-guided image manipulation [24,29]. However, natural language can be ambiguous and textual descriptions are difficult to obtain.

Usage of Attributes. Early methods used attributes to generate outdoor scene [5,19], human face and bird images [44]. In contrast, our method can generate complex scene images containing multiple objects from a reconfigurable layout. Most similar to our work are the methods proposed by Ke Ma et al. [26] as an extension of [26] using an auxiliary attribute classifier and explicit reconstruction loss for horizontally shifted objects, and [31] which requires semantic instance masks. To the best of our knowledge, [26] is currently the only other direct layout-to-image method using attributes. Our method improves upon [26] in terms of visual quality, control and image resolution using a straightforward, yet effective approach built on [37,38].

3 Approach

LostGAN [37,38] achieves remarkable results and control in the layout-to-image task, but it lacks the ability to specify the attributes of an object. While an object class label defines the high-level category (e.g., "car", "person", "dog", "building)", attributes refer to structural properties and appearance variations such as colors (e.g., "blue", "yellow"), sentiment (e.g., "happy", "angry"), and forms (e.g., "round", "sliced") which can be assigned to a variety of object classes [22]. Although one could randomly sample many different object latent codes to generate diverse outputs, it does not allow to provide specific descriptions of the appearance to enable users to generate "what they have in mind". To address this fundamental problem, we build upon [37] and additionally condition individual objects on a set of attributes. To that end, we create an attribute embedding, similar to the label embedding used in [37], and propose Attr-ISLA to compute affine transformation parameters which depend on object positions, class labels and attributes. Furthermore, we utilize a separate attribute embedding to compute an additional adversarial hinge loss on object-attribute features. See Fig. 2 for an illustration of our method.

3.1 Problem Formulation

Given an image layout $L = \{(l_i, b_i, a_i)_{i=1}^m\}$ of m objects, where each object is defined by a class label l_i, a bounding box b_i, and attributes a_i, the goal of our method is to generate an image I with accurate positioned and recognizable objects which also correctly reflect their corresponding input attributes. We use LostGAN [37,38] as our backbone, in which the overall style of the image is controlled by the latent z_{img}, and individual object styles are controlled by the set of latents $Z_{\text{obj}} = \{z_i\}_{i=1}^m$. Latent codes are sampled from the standard normal distribution $\mathcal{N}(0, 1)$. Note, the instance object style codes Z_{obj} are important even though attributes are provided to capture the challenging one-to-many mapping and enable the generation of diverse images (e.g., there are many possible images of a person wearing a blue shirt). In summary, we want to find a generator function G parameterized by Θ_G which captures the underlying conditional data distribution $p = (I|L, z_{\text{img}}, Z_{\text{obj}})$ such that we can use it to generate new, realistic samples. Similar to [37], the task we are addressing in this work can hence be expressed more formally as in Eq. 1

$$I = G(L, z_{\text{img}}, Z_{\text{obj}}; \Theta_G), \tag{1}$$

where all components of the layout L (i.e., class labels l_i, object positions b_i and attributes a_i) are reconfigurable to allow fine-grained control of diverse images using the randomly sampled latents z_{img} and Z_{obj}. In other words, our goals are 1) to control the appearance of individual objects using attributes, but still be able to 2) reconfigure the layout and styles to generate diverse objects corresponding to the desired specification.

3.2 Attribute ISLA (Attr-ISLA)

Inspired by [3,20,28], the authors of [37] extended the Adaptive Instance Normalization (AdaIN) [20] to object Instance-Sensitive and Layout-Aware feature Normalization (ISLA-Norm) to enable fine-grained and multi-object style control.

In order to gain control over the appearance of individual objects, we propose to additionally condition on object attributes using a simple, yet effective enhancement to the ISLA-Norm [37]. On a high-level, the channel-wise batch mean μ and variance σ are computed as in BatchNorm [16] while the affine transformation parameters γ and β are instance-sensitive (class labels and attributes) and layout-aware (object positions) per sample. Similar to [37], this is achieved in a multi-step process:

1) Label Embedding: Given one-hot encoded label vectors for m objects with d_l denoting the number of class labels, and d_e the embedding dimension, the one-hot label matrix Y of size $m \times d_l$ is transformed into the $m \times d_e$ label-to-vector matrix representation of labels $\mathbb{Y} = Y \cdot W$ using a learnable $d_l \times d_e$ size embedding matrix W.
2) Attribute Embedding: Given binary encoded attribute vectors for m objects, an intermediate MLP is used to map the attributes into an $m \times d_e$ size attribute-to-vector matrix representation A.
3) Joint Label, Attribute & Style Projection: The sampled object style noise matrix Z_{obj} of size $m \times d_{\text{noise}}$ is concatenated with the label-to-vector matrix $\mathbb{Y}$ and attribute-to-vector matrix A to obtain the $m \times (2 \cdot d_e + d_{\text{noise}})$ size embedding matrix $\mathbb{S}^* = (\mathbb{Y}, A, Z_{\text{obj}})$. The embedding matrix $\mathbb{S}^*$, which now depends on the class labels, attributes and latent style codes, is used to compute object attribute-guided instance-sensitive channel-wise γ and β via linear projection using a learnable $(2 \cdot d_e + d_{\text{noise}}) \times 2C$ projection matrix, with C denoting the number of channels.
4) Mask Prediction: A non-binary $s \times s$ mask is predicted for each object by a sub-network consisting of up-sample convolutions and a sigmoid transformation. Next, the masks are resized to the corresponding bounding box sizes.
5) ISLA γ, β Computation: The γ and β parameters are unsqueezed to their corresponding bounding boxes, weighted by the predicted masks, and finally added together with averaged sum used for overlapping regions.

Because the affine transformation parameters depend on individual objects in a sample (class labels, bounding boxes, attributes and styles), our AttrLostGAN achieves better and fine-grained control over the image generation process. This allows the user to create an image iteratively and interactively by updating the layout, specifying attributes, and sampling latent codes. We refer the reader to [37] for more details on ISLA and [38] for an extended ISLA-Norm which integrates the learned masks at different stages in the generator.

3.3 Architecture and Objective

We use LostGAN [37,38] as our backbone without changing the general architecture of the ResNet [9] based generator and discriminator. The discriminator $D(\cdot; \Theta_D)$ consists of three components: a shared ResNet backbone to extract features, an image head classifier, and an object head classifier. Following the design of a separate label embedding to compute the object-label loss, we create a separate attribute embedding to compute the object-attribute loss to encourage the generator G to produce objects with specified attributes. Similar to [37,38], the objective can be formulated as follows. Given an image I, the discriminator predicts scores for the image (s_{img}), and average scores for the object-label (s_l) and object-attribute (s_a) features, respectively:

$$(s_{\text{img}}, s_l, s_a) = D(I, L; \Theta_D) \tag{2}$$

We use the adversarial hinge losses

$$\mathcal{L}_t(I, L) = \begin{cases} \max(0, 1 - s_t); & \text{if } I \text{ is real} \\ \max(0, 1 + s_t); & \text{if } I \text{ is fake} \end{cases} \tag{3}$$

where $t \in \{\text{img}, l, a\}$. The objective can hence be written as

$$\mathcal{L}(I, L) = \lambda_1 \mathcal{L}_{\text{img}}(I, L) + \lambda_2 \mathcal{L}_l(I, L) + \lambda_3 \mathcal{L}_a(I, L), \tag{4}$$

where $\lambda_1, \lambda_2, \lambda_3$ are trade-off parameters between image, object-label, and object-attribute quality. The losses for the discriminator and generator can be written as

$$\begin{aligned} \mathcal{L}_D &= \mathbb{E}\left[\mathcal{L}\left(I^{\text{real}}, L\right) + \mathcal{L}\left(I^{\text{fake}}, L\right)\right] \\ \mathcal{L}_G &= -\mathbb{E}\left[\mathcal{L}\left(I^{\text{fake}}, L\right)\right] \end{aligned} \tag{5}$$

We set $\lambda_1 = 0.1$, $\lambda_2 = 1.0$, and $\lambda_3 = 1.0$ to obtain our main results in Table 1, and train our models for 200 epochs using a batch size of 128 on three NVIDIA V100 GPUs. Both λ_1, and λ_2 are as in [37]. We use the Adam optimizer, with $\beta_1 = 0$, $\beta_2 = 0.999$, and learning rates 10^{-4} for both generator and discriminator.

4 Experiments

Since we aim to gain fine-grained control of individual objects using attributes, we first create and experiment with a synthetic dataset to demonstrate the effectiveness of our approach before moving to the challenging Visual Genome [22] dataset.

4.1 MNIST Dialog

Dataset. We use the MNIST Dialog [36] dataset and create an annotated layout-to-image dataset with attributes. In MNIST Dialog each image is a 28×28 pixel MNIST image with three additional attributes, i.e., digit color (red, blue, green, purple, or brown), background color (cyan, yellow, white, silver, or salmon), and style (thick or thin). Starting from an empty 128×128 image canvas, we randomly select, resize and place 3–8 images on it thereby creating an annotated layout-to-image with attributes dataset, where each "object" in the image is an image from MNIST Dialog. While randomly placing the images on the canvas, we ensure that each image is sufficiently visible by allowing max. 40% overlap between any two images.

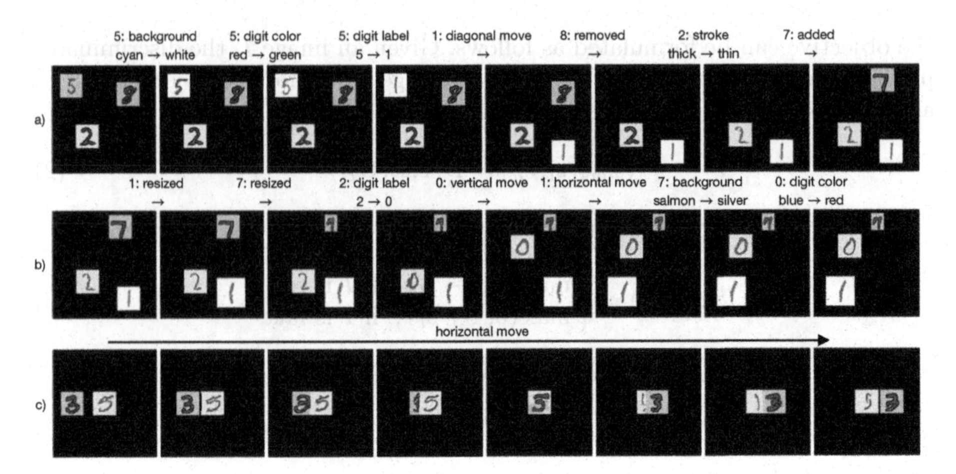

Fig. 3. Iterative reconfiguration example on the MNIST Dialog based dataset (all images are generated by our model). In a) and b), we reconfigure various aspects of the (not shown) input layout to demonstrate controlled image generation. Our approach allows fine-grained control over individual objects with no or minimal changes to other parts of the image. During reconfiguration, we can sometimes observe small style changes, indicating partial entanglement of latent codes and specified attributes. In c), we horizontally shift one object showing that nearby and overlapping objects influence each other which might be a desirable feature to model interactions in more complex settings.

Results. In Fig. 3, we depict generated images using a corresponding layout. Our model learned to generate sharp objects at the correct positions with corresponding labels and attributes, and we can successfully control individual object attributes without affecting other objects. When reconfiguring one object we can sometimes observe slight changes in the style of how a digit is drawn, indicating that the variation provided by the object latent codes is not fully disentangled from the attribute specification. However, we also observe that other objects remain unchanged, hence providing fine-grained control over the individual appearance. We further show how two nearby or even overlapping objects can influence each other which might be necessary to model interacting objects in more complex settings. We hypothesize this is due to the weighted average pool in ISLA which computes an average style for that position.

4.2 Visual Genome

Dataset. Finally, we apply our proposed method to the challenging Visual Genome [22] dataset. Following the setting in [18,37], we pre-process and split the dataset by removing small and infrequent objects, resulting in 62,565 training, 5,062 validation, and 5,096 testing images, with 3 to 30 objects from 178 class labels per image. We filter all available attributes to include only such that appear at least 2,000 times, and allow up to 30 attributes per image.

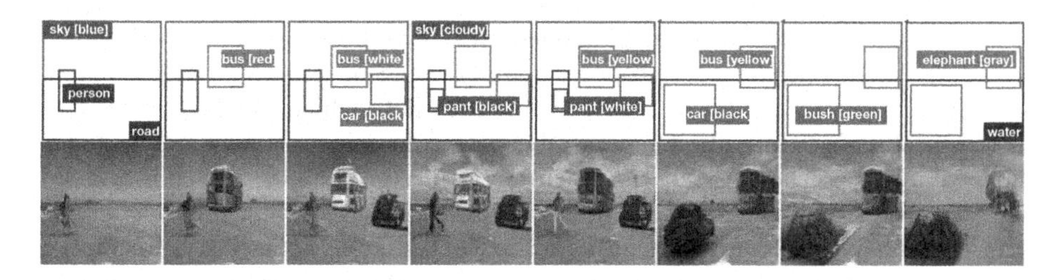

Fig. 4. Our model can control the appearance of generated objects via attributes. *From left to right:* add bus [red]; bus [red → white], add car [black]; sky [blue → cloudy], add pant [black]; bus [white → yellow], pant [black → white]; remove person, remove pant [white], reposition bus [yellow], reposition and resize car [black]; car [black] → bush [green]; road → water, bus [yellow] → elephant [gray]. (Color figure online)

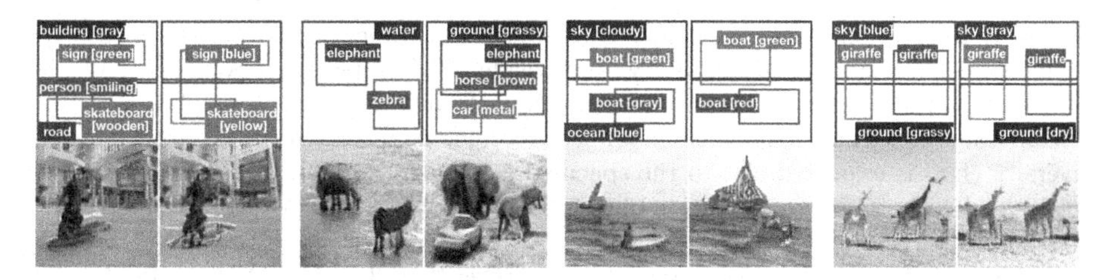

Fig. 5. More reconfiguration examples. *First pair:* sign [green → blue], skateboard [wooden → yellow]. *Second pair:* water → ground [grassy], zebra → horse [brown], repositioned horse [brown], resized elephant, added car [metal]. *Third pair:* boat [gray → red], resized boat [green]. *Fourth pair:* sky [blue → gray], ground [grassy → dry], resized both giraffes. (Color figure online)

Fig. 6. Generating images from a linear interpolation between attribute embeddings produces smooth transitions. From left to right, we interpolate between the following attribute specification: sky [white → blue], umbrella [purple → orange], surfboard [blue → red]. (Color figure online)

Metrics. Evaluating generative models is challenging because there are many aspects that would resemble a good model such as visual realism, diversity, and sharp objects [2,40]. Additionally, a good layout-to-image model should generate objects of the specified class labels and attributes at their corresponding locations. Hence, we choose multiple metrics to evaluate our model and compare with baselines.

To evaluate the image quality and diversity, we use the IS [35] and FID [12]. To assess the visual quality of individual objects we choose the SceneFID

[39] which corresponds to the FID applied on cropped objects as defined by the bounding boxes. Similarly, we propose to apply the IS on generated object crops, denoted as SceneIS. As in [37], we use the CAS [32], which measures how well an object classifier trained on generated can perform on real image crops. Note, this is different to the classification accuracy as used in [26,48], which is trained on real and tested on generated data, and hence might overlook the diversity of generated images [38]. Additionally, and in the same spirit as CAS, we report the micro F1 (Attr-F1) by training multi-label classification networks to evaluate the attribute quality by training on generated and test on real object crops. As in [26,37,38,48], we adopt the LPIPS metric as the Diversity Score (DS) [47] to compute the perceptual similarity between two sets of images generated from the same layout in the testing set.

Table 1. Results on Visual Genome. Our models (in italics) achieve the best scores on most metrics, and AttrLostGANv2 is considerably better than AttrLostGANv1. When trained on higher resolution, our model performs better in terms of object and attribute quality, while achieving similar scores on image quality metrics. Note, a lower diversity (DS) is expected due to the specified attributes. Models marked with $\diamond$, $\dagger$, $\star$ are trained with an image resolution of 64×64, 128×128, and 256×256, respectively.

Method	IS ↑	SceneIS ↑	FID ↓	SceneFID ↓	DS (↑)	CAS ↑	Attr-F1 ↑
Real images†	23.50	13.43	11.93	2.46	-	46.22	15.77
Layout2Im [48]$^\diamond$	8.10	-	40.07	-	0.17	-	-
LostGANv1 [37]†	10.30	9.07	35.20	11.06	0.47	31.04	-
LostGANv2 [38]†	10.25	9.15	34.77	15.25	0.42	30.97	-
Ke Ma et al. [26]†	9.57	8.17	43.26	16.16	0.30	**33.09**	12.62
AttrLostGANv1†	10.68	9.24	32.93	8.71	0.40	32.11	13.64
AttrLostGANv2†	**10.81**	**9.46**	**31.57**	**7.78**	0.28	32.90	**14.61**
Real images*	31.41	19.58	12.41	2.78	-	50.94	17.80
LostGANv2 [38]*	**14.88**	11.87	**35.03**	18.87	0.53	**35.80**	-
AttrLostGANv2*	14.25	**11.96**	35.73	**14.76**	0.45	35.36	**14.49**

Qualitative Results. Figure 1 and Fig. 4 depict examples of attribute controlled image generation from reconfigurable layout. Our model provides a novel way to iteratively reconfigure the properties of individual objects to generate images of complex scenes without affecting other parts of the image. Figure 5 shows more examples in which we reconfigure individual objects by changing attributes, class labels, object position and size. In Fig. 6, we linearly interpolate between two sets of attributes for the same layout. Our model learns a smooth transition between attributes. In Fig. 7 we compare generated images between our AttrLostGANv1 and [26] using layouts from the testing set. As can be seen, generating realistic images of complex scenes with multiple objects is still very difficult. Although the images generated by our model look more realistic, individual objects and details such as human faces are hard to recognize. In terms of attribute control, our images better depict the input specifications in general.

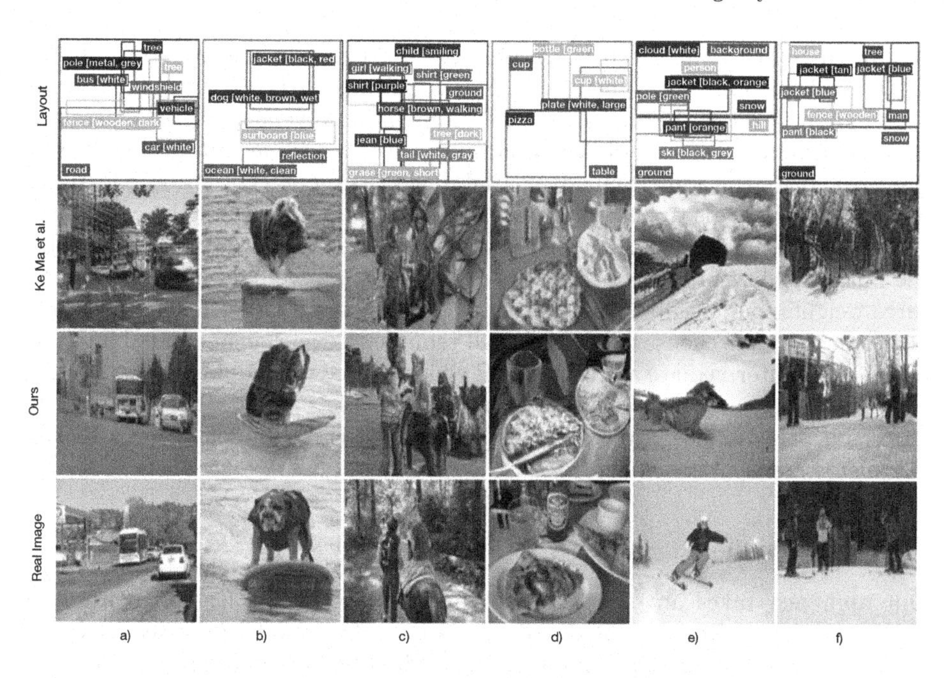

Fig. 7. Visual comparison between images generated by our AttrLostGANv1 and Ke Ma et al. [26] using the layouts shown in the first row. Our images are consistently better at reflecting the input attributes and individual objects have more details and better texture. For example, a) bus [white], b) jacket [black, red], c) shirt [purple], d) plate [white, large], e) pant [orange], f) jacket [blue]. (Color figure online)

Quantitative Results. Table 1 shows quantitative results. We train two variants of our approach: AttrLostGANv1 which is based on [37], and AttrLostGANv2 which is based on [38]. We compare against the recent and only other direct layout-to-image baseline proposed by Ke Ma et al. [26], which is an extension of Layout2Im [48] that can be conditioned on optional attributes. Since no pre-trained model was available at the official codebase of [26], we used the open-sourced code to train a model. For fair comparison, we evaluate all models trained by us. Our models achieve the best scores across most metrics, and AttrLostGANv2 is considerably better than AttrLostGANv1. [26] reaches a competing performance on attribute control, but inferior image and object quality. For example, our method increases the SceneIS from 8.17 to 9.46, and lowers the FID from 43.26 to 31.57. Furthermore, our method is better at generating the appearance specified by the attributes as indicated by the improvement of Attr-F1 from 12.62 to 14.61. In terms of CAS, our model performs slightly worse than [26], which might be due to the explicit attribute classifier used in [26] during training. Building upon [38] our method can also generate higher resolution images (256×256 compared to 128×128). By specifying attributes, a decreased DS is expected but we include it for completeness.

Ablation Study. We also perform ablations of our main changes to the LostGAN [37] backbone, see Appendix C. Starting from LostGAN we add attribute information to the generator and already gain an improvement over the baseline in terms of image quality, object discriminability, as well as attribute information. We ablate the additional adversarial hinge loss on object-attribute features λ_3. A higher λ_3 leads to better Attr-F1, but decreased image and object quality. Interestingly, a high $\lambda_3 = 2.0$ achieves the best SceneFID on object crops, while the image quality in terms of FID is worst. Although we only have to balance three weights, our results show that there exists a trade-off between image and object quality. We choose $\lambda_3 = 1.0$ for all remaining experiments and ablate the depth of the intermediate attribute MLP which is used to compute the attribute-to-vector matrix representation for all objects. While a shallow MLP leads to a decreased performance, a medium deep MLP with three hidden layers achieves the best overall performance.

4.3 Discussion

Our approach takes an effective step towards reconfigurable, and controlled image generation from layout of complex scenes. Our model provides unprecedented control over the appearance of individual objects without affecting the overall image. Although the quantitative as well as visual results are promising current approaches require attribute annotations which are time-consuming to obtain. While the attribute control is strong when fixing the object locations, as demonstrated in our results, the object styles can change when target objects are nearby or overlap. We hypothesize that this might be due to the average pool in ISLA when combining label and attribute features of individual objects and might hence lead to entangled representations. At the same time, such influence might be desirable to model object interactions in complex settings. Despite clearly improving upon previous methods both quantitatively and qualitatively, current models are still far from generating high-resolution, realistic images of complex scenes with multiple interacting objects which limits their practical application.

5 Conclusion

In this paper, we proposed AttrLostGAN, an approach for attribute controlled image generation from reconfigurable layout and style. Our method successfully addresses a fundamental problem by allowing users to intuitively change the appearance of individual object details without changing the overall image or affecting other objects. We created and experimented on a synthetic dataset based on MNIST Dialog to analyze and demonstrate the effectiveness of our approach. Further, we evaluated our method against the recent, and only other baseline on the challenging Visual Genome dataset both qualitatively and quantitatively. We find that our approach not only outperforms the existing method in most common measures while generating higher resolution images, but also that it provides users with intuitive control to update the generated image to their needs. In terms of future work, our first steps are directed towards enhancing the image quality and resolution. We would also like to investigate unsupervised methods to address the need of attribute annotations and whether we can turn attribute labels into textual descriptions.

Acknowledgments. This work was supported by the BMBF projects ExplAINN (Grant 01IS19074), XAINES (Grant 01IW20005), the NVIDIA AI Lab (NVAIL) and the TU Kaiserslautern PhD program.

References

1. Ashual, O., Wolf, L.: Specifying object attributes and relations in interactive scene generation. In: Proceedings of the IEEE International Conference on Computer Vision, pp. 4561–4569 (2019)
2. Borji, A.: Pros and cons of GAN evaluation measures. Comput. Vis. Image Underst. **179**, 41–65 (2018)
3. Brock, A., Donahue, J., Simonyan, K.: Large scale GAN training for high fidelity natural image synthesis. In: International Conference on Learning Representations (2018)
4. Choi, H.S., Park, C.D., Lee, K.: From inference to generation: end-to-end fully self-supervised generation of human face from speech. In: International Conference on Learning Representations (2020)
5. Choi, Y., Choi, M., Kim, M., Ha, J.W., Kim, S., Choo, J.: StarGAN: unified generative adversarial networks for multi-domain image-to-image translation. In: Proceedings of the IEEE Computer Vision and Pattern Recognition, pp. 8789–8797 (2018)
6. Dong, H., Yu, S., Wu, C., Guo, Y.: Semantic image synthesis via adversarial learning. In: Proceedings of the IEEE International Conference on Computer Vision (2017)
7. Frolov, S., Hinz, T., Raue, F., Hees, J., Dengel, A.: Adversarial text-to-image synthesis: a review. arXiv:2101.09983 (2021)
8. Goodfellow, I.J., et al.: Generative adversarial nets. In: Advances in Neural Information Processing Systems, pp. 2672–2680 (2014)
9. He, K., Zhang, X., Ren, S., Sun, J.: Deep residual learning for image recognition. In: Proceedings of the IEEE Computer Vision and Pattern Recognition, pp. 770–778 (2016)
10. Henaff, M., Bruna, J., LeCun, Y.: Deep convolutional networks on graph-structured data. arXiv:1506.05163 (2015)
11. Herzig, R., Bar, A., Xu, H., Chechik, G., Darrell, T., Globerson, A.: Learning canonical representations for scene graph to image generation. In: Vedaldi, A., Bischof, H., Brox, T., Frahm, J.-M. (eds.) ECCV 2020. LNCS, vol. 12371, pp. 210–227. Springer, Cham (2020). https://doi.org/10.1007/978-3-030-58574-7_13
12. Heusel, M., Ramsauer, H., Unterthiner, T., Nessler, B., Hochreiter, S.: GANs trained by a two time-scale update rule converge to a local nash equilibrium. In: Advances in Neural Information Processing Systems, pp. 6626–6637 (2017)
13. Hinz, T., Heinrich, S., Wermter, S.: Generating multiple objects at spatially distinct locations. In: International Conference on Learning Representations (2019)
14. Hinz, T., Heinrich, S., Wermter, S.: Semantic object accuracy for generative text-to-image synthesis. IEEE Trans. Pattern Anal. Mach. Intell. **14**, 1–14 (2020)

15. Hong, S., Yang, D., Choi, J., Lee, H.: Inferring semantic layout for hierarchical text-to-image synthesis. In: Proceedings of the IEEE Computer Vision and Pattern Recognition, pp. 7986–7994 (2018)
16. Ioffe, S., Szegedy, C.: Batch normalization: accelerating deep network training by reducing internal covariate shift. In: International Conference on Machine Learning, pp. 448–456 (2015)
17. Isola, P., Zhu, J.Y., Zhou, T., Efros, A.A.: Image-to-image translation with conditional adversarial networks. In: Proceedings of the IEEE Computer Vision and Pattern Recognition, pp. 1125–1134 (2016)
18. Johnson, J., Gupta, A., Fei-Fei, L.: Image generation from scene graphs. In: Proceedings of the IEEE Computer Vision and Pattern Recognition, pp. 1219–1228 (2018)
19. Karacan, L., Akata, Z., Erdem, A., Erdem, E.: Learning to generate images of outdoor scenes from attributes and semantic layouts. arXiv:1612.00215 (2016)
20. Karras, T., Laine, S., Aila, T.: A style-based generator architecture for generative adversarial networks. In: Proceedings of the IEEE Computer Vision and Pattern Recognition, pp. 4401–4410 (2018)
21. Kingma, D.P., Welling, M.: Auto-encoding variational Bayes. CoRR arXiv:1312.6114 (2013)
22. Krishna, R., et al.: Visual genome: connecting language and vision using crowdsourced dense image annotations. Int. J. Comput. Vision **123**(1), 32–73 (2017)
23. Li, B., Qi, X., Lukasiewicz, T., Torr, P.H.S.: Controllable text-to-image generation. In: Advances in Neural Information Processing Systems (2019)
24. Li, B., Qi, X., Lukasiewicz, T., Torr, P.H.: ManiGAN: text-guided image manipulation. In: Proceedings of the IEEE Computer Vision and Pattern Recognition, pp. 7880–7889 (2020)
25. Li, W., et al.: Object-driven text-to-image synthesis via adversarial training. In: Proceedings of the IEEE Computer Vision and Pattern Recognition, pp. 12166–12174 (2019)
26. Ma, K., Zhao, B., Sigal, L.: Attribute-guided image generation from layout. In: British Machine Vision Virtual Conference (2020). arXiv:2008.11932
27. Mirza, M., Osindero, S.: Conditional generative adversarial nets. arXiv:1411.1784 (2014)
28. Miyato, T., Koyama, M.: cGANs with projection discriminator. arXiv:1802.05637 (2018)
29. Nam, S., Kim, Y., Kim, S.J.: Text-adaptive generative adversarial networks: manipulating images with natural language. In: Advances in Neural Information Processing Systems, pp. 42–51 (2018)
30. Odena, A., Olah, C., Shlens, J.: Conditional image synthesis with auxiliary classifier GANs. In: International Conference on Machine Learning, pp. 2642–2651 (2016)
31. Pavllo, D., Lucchi, A., Hofmann, T.: Controlling style and semantics in weakly-supervised image generation. In: Vedaldi, A., Bischof, H., Brox, T., Frahm, J.-M. (eds.) ECCV 2020. LNCS, vol. 12351, pp. 482–499. Springer, Cham (2020). https://doi.org/10.1007/978-3-030-58539-6_29
32. Ravuri, S., Vinyals, O.: Classification accuracy score for conditional generative models. In: Advances in Neural Information Processing Systems, pp. 12268–12279 (2019)
33. Razavi, A., van den Oord, A., Vinyals, O.: Generating diverse high-fidelity images with VQ-VAE-2. In: Advances in Neural Information Processing Systems, pp. 14866–14876 (2019)

34. Reed, S.E., Akata, Z., Yan, X., Logeswaran, L., Schiele, B., Lee, H.: Generative adversarial text to image synthesis. In: International Conference on Machine Learning, pp. 1060–1069 (2016)
35. Salimans, T., Goodfellow, I., Zaremba, W., Cheung, V., Radford, A., Chen, X.: Improved techniques for training GANs. In: Advances in Neural Information Processing Systems, pp. 2234–2242 (2016)
36. Seo, P.H., Lehrmann, A., Han, B., Sigal, L.: Visual reference resolution using attention memory for visual dialog. In: Advances in Neural Information Processing Systems, pp. 3719–3729 (2017)
37. Sun, W., Wu, T.: Image synthesis from reconfigurable layout and style. In: Proceedings of the IEEE International Conference on Computer Vision (2019)
38. Sun, W., Wu, T.: Learning layout and style reconfigurable GANs for controllable image synthesis. arXiv:2003.11571 (2020)
39. Sylvain, T., Zhang, P., Bengio, Y., Hjelm, R.D., Sharma, S.: Object-centric image generation from layouts. arXiv:2003.07449 (2020)
40. Theis, L., van den Oord, A., Bethge, M.: A note on the evaluation of generative models. CoRR arXiv:1511.01844 (2015)
41. Wang, T.C., Liu, M.Y., Zhu, J.Y., Tao, A., Kautz, J., Catanzaro, B.: High-resolution image synthesis and semantic manipulation with conditional GANs. In: Proceedings of the IEEE Computer Vision and Pattern Recognition, pp. 8798–8807 (2017)
42. Wang, X., Qiao, T., Zhu, J., Hanjalic, A., Scharenborg, O.: S2IGAN: speech-to-image generation via adversarial learning. In: INTERSPEECH (2020)
43. Xu, T., et al.: AttnGAN: fine-grained text to image generation with attentional generative adversarial networks. In: Proceedings of the IEEE Computer Vision and Pattern Recognition, pp. 1316–1324 (2017)
44. Yan, X., Yang, J., Sohn, K., Lee, H.: Attribute2Image: conditional image generation from visual attributes. In: Leibe, B., Matas, J., Sebe, N., Welling, M. (eds.) ECCV 2016. LNCS, vol. 9908, pp. 776–791. Springer, Cham (2016). https://doi.org/10.1007/978-3-319-46493-0_47
45. Yikang, L., Ma, T., Bai, Y., Duan, N., Wei, S., Wang, X.: PasteGAN: a semi-parametric method to generate image from scene graph. In: Advances in Neural Information Processing Systems, pp. 3948–3958 (2019)
46. Zhang, H., et al.: StackGAN++: realistic image synthesis with stacked generative adversarial networks. IEEE Trans. Pattern Anal. Mach. Intell. **41**, 1947–1962 (2017)
47. Zhang, R., Isola, P., Efros, A.A., Shechtman, E., Wang, O.: The unreasonable effectiveness of deep features as a perceptual metric. In: Proceedings of the IEEE Computer Vision and Pattern Recognition (2018)
48. Zhao, B., Meng, L., Yin, W., Sigal, L.: Image generation from layout. In: Proceedings of the IEEE Computer Vision and Pattern Recognition (2019)
49. Zhou, X., Huang, S., Li, B., Li, Y., Li, J., Zhang, Z.: Text guided person image synthesis. In: Proceedings of the IEEE Computer Vision and Pattern Recognition, pp. 3663–3672 (2019)
50. Zhu, J.Y., Park, T., Isola, P., Efros, A.A.: Unpaired image-to-image translation using cycle-consistent adversarial networks. In: Proceedings of the IEEE International Conference on Computer Vision, pp. 2223–2232 (2017)
51. Zhu, M., Pan, P., Chen, W., Yang, Y.: DM-GAN: dynamic memory generative adversarial networks for text-to-image synthesis. In: Proceedings of the IEEE Computer Vision and Pattern Recognition, pp. 5802–5810 (2019)

Learning Conditional Invariance Through Cycle Consistency

Maxim Samarin[1(✉)], Vitali Nesterov[1], Mario Wieser[1], Aleksander Wieczorek[1], Sonali Parbhoo[2], and Volker Roth[1]

[1] Department of Mathematics and Computer Science, University of Basel, Spiegelgasse 1, 4051 Basel, Switzerland
{maxim.samarin,vitali.nesterov,mario.wieser,aleksander.wieczorek, volker.roth}@unibas.ch

[2] Harvard John A. Paulson School of Engineering and Applied Sciences, Harvard University, 150 Western Avenue, Boston, MA 02134, USA
sparbhoo@seas.harvard.edu

Abstract. Identifying meaningful and independent factors of variation in a dataset is a challenging learning task frequently addressed by means of deep latent variable models. This task can be viewed as learning symmetry transformations preserving the value of a chosen property along latent dimensions. However, existing approaches exhibit severe drawbacks in enforcing the invariance property in the latent space. We address these shortcomings with a novel approach to cycle consistency. Our method involves two separate latent subspaces for the target property and the remaining input information, respectively. In order to enforce invariance as well as sparsity in the latent space, we incorporate semantic knowledge by using cycle consistency constraints relying on property side information. The proposed method is based on the deep information bottleneck and, in contrast to other approaches, allows using continuous target properties and provides inherent model selection capabilities. We demonstrate on synthetic and molecular data that our approach identifies more meaningful factors which lead to sparser and more interpretable models with improved invariance properties.

Keywords: Sparsity · Cycle consistency · Invariance · Deep variational information bottleneck · Variational autoencoder · Model selection

1 Motivation

Understanding the nature of a generative process for observed data typically involves uncovering explanatory factors of variation responsible for the obser-

M. Samarin and V. Nesterov—Both authors contributed equally.

Supplementary Information The online version contains supplementary material available at https://doi.org/10.1007/978-3-030-92659-5_24.

C. Bauckhage et al. (Eds.): DAGM GCPR 2021, LNCS 13024, pp. 376–391, 2021.
https://doi.org/10.1007/978-3-030-92659-5_24

vations. But the relationship between these factors and our observation usually remains unclear. A common assumption is that the relevant factors can be expressed by a low-dimensional latent representation Z [25]. Therefore, popular machine learning methods involve learning of appropriate latent representations to *disentangle* factors of variation. Learning disentangled representations is often considered in an unsupervised setting which does not rely on the prior knowledge about the data such as labels [7,8,13,17,24]. However, it was shown that inductive bias on the dataset and learning approach is necessary to obtain disentanglement [25]. Inductive biases allow us to express assumptions about the generative process and to prioritise different solutions not only in terms of disentanglement [5,13,21,35,44], but also in terms of constrained latent space structures [15,16], preservation of causal relationships [40], or interpretability [45].

We consider a supervised setting where semantic knowledge about the input data allows structuring the latent representation in disjoint subspaces Z_0 and Z_1 of the latent space Z by enforcing conditional invariance. In such supervised settings, disentanglement can be viewed as an extraction of level sets or symmetries inherent to our data X which leave a specified property Y invariant. An important application in that direction is the generation of diverse molecular structures with similar chemical properties [44]. The goal is to disentangle factors of variation relevant for the property. Typically, level sets L_y are defined implicitly through $L_y(f) = \{(x_1, ..., x_d) | f(x_1, ..., x_d) = y\}$ for a property y which implicitly describes the level curve or surface w.r.t. inputs $(x_1, ..., x_d) \in \mathbb{R}^d$. The topic of this paper is to identify a sparse parameterisation of level sets which encodes conditional invariances and thus selects a correct model. Several techniques have been developed to steer model selection by sparsifying the number of features, e.g. [38,39], or compressing features into a low-dimensional feature space, e.g. [4,33,42]. These methods improve generalisation by focusing on only a subset of relevant features and using these to explain a phenomenon. Existing methods for including such prior knowledge in the model usually do not include dimensionality reduction techniques and perform a hand-tuned selection [15,21,44].

In this paper, we introduce a novel approach to cycle consistency, relying on property side information Y as our semantic knowledge, to provide conditional invariance in the latent space. With this we mean that conditioning on part of the latent space, i.e. Z_0, allows property-invariant sampling in the latent space Z_1. By ensuring that our method consistently performs on generated samples when fed back to the network, we achieve more disentangled and sparser representations. Our work builds on [42], where a general sparsity constraint on latent representations is provided, and on [21,44], where conditional invariance is obtained through adversarial training. We show that our approach addresses some drawbacks in previous approaches and allows us to identify more meaningful factors for learning better models and achieve improved invariance performance. Our contributions may thus be summarised as follows:

- We propose a novel approach for supervised disentanglement where conditional invariance is enforced by a novel cycle consistency on property side

information. This facilitates the guided exploration of the latent space and improves sampling with a fixed property.
- Our model inherently favours sparse solutions, leading to more interpretable latent dimensions and facilitates built-in model selection.
- We demonstrate that our method improves on the state-of-the-art performance for conditional invariance as compared to existing approaches on both synthetic and molecular benchmark datasets.

2 Related Work

2.1 Deep Generative Latent Variable Models and Disentanglement

Because of its flexibility, the variational autoencoder (VAE) [20,34] is a popular deep generative latent variable model in many areas such as fairness [27], causality [26], semi-supervised learning [19], and design and discovery of novel molecular structures [11,22,28]. The VAE is closely related to the Information Bottleneck (IB) principle [4,39]. Various approaches exploit this relation, e.g. the deep variational information bottleneck (DVIB) [2,4]. Further extensions were proposed in the context of causality [9,29,30] or archetypal analysis [15,16].

The β-VAE [13] extends the standard VAE approach and allows unsupervised disentanglement. In unsupervised settings, there exists a great variety of approaches based on VAEs and generative adversarial networks (GANs) to achieve disentanglement such as FactorVAE [17], β-TCVAE [7] or InfoGAN [8,24]. Partitioning the latent space into subspaces is inspired by the multi-level VAE [5], where the latent space is decomposed into a local feature space that is only relevant for a subgroup and a global feature space. In supervised settings, several approaches such as [10,21,23,44] achieve disentanglement by applying adversarial information elimination to select a model with partitioned feature and property space. In such a setting, different to unsupervised disentanglement, our goal is supervised disentanglement with respect to a particular target property.

Another important line of research employs the idea of cycle consistency for learning disentangled representations. Presumably the most closely related work to this study is conducted by [14,43,44]. Here, the authors employ a cycle-consistent loss on the latent representations to learn symmetries and disentangled representations in weakly supervised settings, respectively. Moreover, in [44], the authors use adversarial training and mutual information estimation to learn symmetry transformations instead of explicitly modelling them. In contrast, our work replaces adversarial training by using cycle consistency.

2.2 Model Selection via Sparsity

Several works perform model selection by introducing sparsity constraints which penalise the model complexity. A common sparsity constraint is the Least Absolute Shrinkage and Selection Operator (LASSO) [38]. Extensions of the LASSO propose a log-penalty to obtain even sparser solutions in the compressed IB setting [33] and generalise it further to deep generative models [42]. Furthermore, the LASSO has been extended to the group LASSO, where combinations of covariates are set to zero, the sparse group LASSO [36], and the Bayesian group LASSO [32]. Perhaps most closely related to our work is the oi-VAE [3], which incorporates a group LASSO prior in deep latent variable models. These methods employ a general sparsity constraint to achieve a sparse representation. Our model extends these ideas and imposes a semantic sparsity constraint in the form of cycle consistency that performs regularisation based on prior knowledge.

3 Preliminaries

3.1 Deep Variational Information Bottleneck

We focus on the DVIB [4] which is a method for information compression based on the IB principle [39]. The objective is to compress a random variable X into a random variable Z while being able to predict a third random variable Y. The DVIB is closely related to the VAE [20,34]. The optimal compression is achieved by solving the parametric problem

$$\min_{\phi,\theta} I_\phi(Z;X) - \lambda I_{\phi,\theta}(Z;Y), \tag{1}$$

where I is the mutual information between two random variables. Hence, the DVIB objective balances maximisation of $I_{\phi,\theta}(Z;Y)$, i.e. Z being informative about Y, and minimisation of $I_\phi(Z;X)$, i.e. compression of X into Z. We assume a parametric form of the conditionals $p_\phi(Z|X)$ and $p_\theta(Y|Z)$ with ϕ and θ representing the parameters of the encoder and decoder network, respectively. Parameter λ controls the degree of compression and is closely related to β in the β-VAE [13]. The relationship to the VAE becomes more apparent with the definition of the mutual information terms:

$$I_\phi(Z;X) = \mathbb{E}_{p(X)} D_{KL}(p_\phi(Z|X) \| p(Z)), \tag{2}$$

$$I_{\phi,\theta}(Z;Y) \geq \mathbb{E}_{p(X,Y)} \mathbb{E}_{p_\phi(Z|X)} \log p_\theta(Y|Z) + h(Y), \tag{3}$$

with D_{KL} being the Kullback-Leibler divergence, and $h(Y)$ the entropy. Note that we write Eq. (3) as an inequality which uses the insight of [41] that the RHS is in fact a lower bound to $I_\theta(Z;Y)$; see [41] for details.

3.2 Cycle Consistency

We use the notion of cycle consistency similar to [14,46]. The CycleGAN [46] performs unsupervised image-to-image translation, where a data point is mapped to its initial position after being transferred to a different space. For instance, suppose that domain X consists of summer landscapes, while domain Y consists of winter landscapes (see Appendix Fig. A1). A function $f(x)$ may be used to

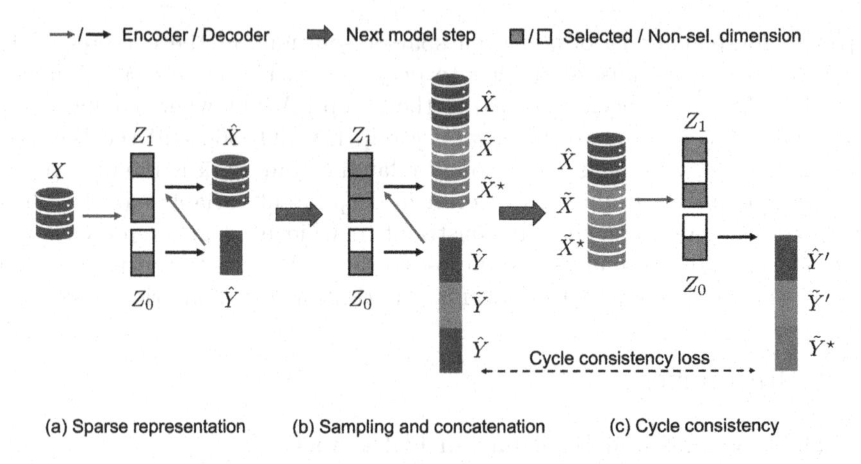

Fig. 1. Model illustration. (a) Firstly, we learn a sparse representation Z from our input data X which we separate into a property space Z_0 and an invariant space Z_1. Given this representation, we try to predict the property $\hat{Y}$ and reconstruct our input $\hat{X}$. Grey arrows indicate that $\hat{Y} = \text{dec}_Y(Z_0)$ instead of Z_0 is used for decoding $\hat{X}$ (see Sect. 4.3). (b) Secondly, we sample new data in two ways: (i) uniformly in Z to get new data points $\tilde{X}$ and $\tilde{Y}$ (orange data), (ii) uniformly in Z_1 with fixed Z_0 to get $\tilde{X}^\star$ at fixed $\hat{Y}$ (cyan data). We concatenate the respective decoder outputs. (c) Lastly, we feed the concatenated input batch X^c into our model and calculate the cycle consistency loss between the properties. (Color figure online)

transform a summer landscape x to a corresponding winter landscape y. Similarly, function $g(y)$ maps y back to the domain X. The goal of cycle consistency is to learn a mapping to $\hat{x}$, which is close to the initial x. In most cases, there is a discrepancy between x and $\hat{x}$ referred to as the cycle consistency loss. In order to obtain an almost invertible mapping, the loss $\|(g(f(x)) - x\|_1$ is minimised.

4 Model

Our model is based on the DVIB to learn a compact latent representation. The input X and the output Y may be complex objects and take continuous values, such as molecules with their respective molecular properties. Unlike the standard DVIB, we do not only want to predict Y from an input X, but also want to generate new $\tilde{X}$ by sampling from our latent representation. As a consequence, we add an additional second decoder that reconstructs X from Z (similar to [11] for decoder Y in the VAE setting), leading to the adjusted parametric objective

$$\min_{\phi,\theta,\tau} I_\phi(Z;X) - \lambda\big(I_{\phi,\theta}(Z;Y) + I_{\phi,\tau}(Z;X)\big), \quad (4)$$

where ϕ are the encoder parameters, and θ and τ describe network parameters for decoding Y and X, respectively.

4.1 Learning a Compact Representation

Formulating our model as a DVIB allows leveraging properties of the mutual information with respect to learning compact latent representations. To see this, first assume that X and Y are jointly Gaussian-distributed which leads to the *Gaussian Information Bottleneck* [6] where the solution Z can be found analytically and proved to be Gaussian. In particular, for $X \sim \mathcal{N}(0, \Sigma_X)$, the optimal Z is a noisy projection of X: $Z = AX + \xi$, where $\xi \sim \mathcal{N}(0, I)$. The mutual information between X and Z is then equal to

$$I(X; Z) = \tfrac{1}{2} \log |A\Sigma_X A^\top + I|. \tag{5}$$

If we now assume A to be diagonal, the model becomes sparse [33]. This is because a full-rank projection AX' of X' does not change the mutual information since $I(X; X') = I(X; AX')$. A reduction in mutual information can only be achieved by a rank-deficient matrix A. In general, the conditionals $Z|X$ and $Y|Z$ in Eq. (1) may be parameterised by neural networks with X and Z as input. The diagonality constraint on A does not cause any loss of generality of the DVIB solution as long as the neural network encoder f_ϕ makes it possible to diagonalise $Af_\phi(X)f_\phi(X)^\top A^\top$ (see [42] for more details). In the following, we consider A to be diagonal and define the sparse representation as the dimensions of the latent space Z selected by the non-zero entries of A. Recalling Eq. (5), this allows us to approximate the mutual information for the encoder in Eq. (2) in a sparse manner

$$I_\phi(X; Z) = \tfrac{1}{2} \log |\mathrm{diag}(f_\phi(X)f_\phi(X)^\top) + \mathbf{1}|, \tag{6}$$

where $\mathbf{1}$ is the all-one vector and the diagonal elements of A are subsumed in the encoder f_ϕ.

4.2 Conditional Invariance and Informed Sparsity

A general sparsity constraint is not sufficient to ensure that latent dimensions indeed represent independent factors. In a supervised setting, our target Y conveys semantic knowledge about the input X, e.g. a chemical property of a molecule. To incorporate semantic knowledge into our model, we require a mechanism that partitions the representation such that it encodes the semantic meaning not only sparsely but preferably independently of other information concerning the input.

To this end, the central element of our approach is cycle consistency with respect to target property Y, which is illustrated in steps (b) and (c) in Fig. 1. The idea is, that reconstructed $\hat{X}$ or newly sampled $\tilde{X}$ with associated prediction $\hat{Y}$ and $\tilde{Y}$ are expected to provide matching predictions $\hat{Y}'$ and $\tilde{Y}'$ when $\hat{X}$ and $\tilde{X}$ are used as an input to the network. This means, if we perform another cycle through the network with sampled or reconstructed inputs, the property prediction should stay consistent. The partitioning of the latent space Z in the property subspace Z_0 and the *invariant* subspace Z_1 is crucial. The property

Y is predicted from Z_0, while the input is reconstructed from the full latent space Z. Ensuring cycle consistency with respect to the property allows putting property-relevant information into the property subspace Z_0. Furthermore, the latent space is regularised by drawing samples which adhere to cycle consistency and provide additional sparsity. If information about Y is encoded in Z_1, this will lead to a higher cycle consistency loss. In this way, cycle consistency enforces invariance in subspace Z_1. By fixing coordinates in Z_0, and thus fixing a property, sampling in Z_1 results in newly generated $\tilde{X}$ with the same property $\tilde{Y}$. More formally, fixing Z_0 renders random variables X and Y conditionally independent, i.e. $X \perp\!\!\!\perp Y|Z_0$ (see Appendix Fig. A2). We ensure conditional invariance with a particular sampling: We fix the Z_0 coordinates and sample in Z_1 to obtain generated $\tilde{X}^{\star}$ all with a fixed property $\hat{Y}$. Using these inputs allows to obtain a new prediction $\tilde{Y}^{\star}$ which should be close to the fixed target property $\hat{Y}$. We choose the L_2 norm for convenience and define the full cycle consistency loss by

$$\mathcal{L}_{\text{cycle}} = \|\hat{Y} - \hat{Y}'\|_2 + \|\tilde{Y} - \tilde{Y}'\|_2 + \|\hat{Y} - \tilde{Y}^{\star}\|_2. \tag{7}$$

4.3 Proposed Framework

The resulting model in Eq. (8) combines sparse DVIBs with partitioned latent space and a novel approach to cycle consistency, which drives conditional invariance and informed sparsity in the latent structure. This allows latent dimensions in Z_0 relevant for prediction of Y to disentangle of latent dimensions in Z_1 which encode remaining input information of X.

$$\begin{aligned}\mathcal{L} &= I_\phi(X;Z) - \lambda\Big(I_{\phi,\tau}(Z_0,Z_1;X) + I_{\phi,\theta}(Z_0;Y) \\ &\qquad -\beta\big(\|\hat{Y} - \hat{Y}'\|_2 + \|\tilde{Y} - \tilde{Y}'\|_2 + \|\hat{Y} - \tilde{Y}^{\star}\|_2\big)\Big)\end{aligned} \tag{8}$$

The proposed model performs model selection as it inherently favours sparser latent representations. This in turn facilitates easier interpretation of latent factors because of the built-in conditional independence between property space Z_0 and invariant space Z_1. These adjustments address some of the issues of the STIB [44] relying on adversarial training, mutual information estimation (which can be difficult in high-dimensions [37]) and bijective mapping which can make the training challenging. In contrast to the work of [14], we impose a novel cycle consistency loss on the predicted outputs Y instead of the latent representation Z. A reason to consider rather Y than Z is that varying latent dimensionality leads to severe problems in the optimisation process as it requires an adaptive rescaling of the different loss weights. To overcome this drawback, we close the full cycle and define the loss on the outputs. Appendix Sec. A.3 and Algorithm A.1 provide more information on the implementation.[1] As an implementation detail, we choose to concatenate the decoded Z_0 code with Z_1 in order to decode $\hat{X}$, i.e. $\hat{X} = \text{dec}_X(Z_1, \hat{Y} = \text{dec}_Y(Z_0))$. This is an additional measure to ensure that Z_0 contains information relevant for property prediction Y and prevent superfluous remaining information about the input X in property space Z_0.

[1] Implementation [1,18]: https://github.com/bmda-unibas/CondInvarianceCC.

5 Experimental Evaluation

We evaluate the effectiveness of our proposed method w.r.t. (i) selection of a sparse representation with meaningful factors of variation (i.e. model selection) and (ii) enforcing conditional independence in the latent space between these factors. To this end, we conduct experiments on a synthetic dataset with knowledge about appropriate parameterisations to highlight the differences to existing models. Additionally, we evaluate our model on a real-world application with a focus on conditional invariance and generation of novel samples. To assess the performance of our model, we compare our approach to two state-of-the-art baselines: (i) the β-VAE [13] which is a typical baseline model in disentanglement studies and (ii) the symmetry-transformation information bottleneck (STIB) [44] which ensures conditional invariance through adversarial training and is the direct competitor to our model. We adapt the β-VAE by adding a decoder for property Y (similar to [11]) which takes only subspace Z_0 as input. The latent space of the adapted β-VAE is split into two subspaces as in the STIB and our model, but has no explicit mechanisms to enforce invariance. This setup can be viewed as an ablation study in which the β-VAE is the basis model of our approach without cycle consistency and sparsity constraints. The STIB provides an alternative approach for the same goal but with a different mechanism.

The objective of the supervised disentanglement approach is to ensure disentanglement of a fixed property with respect to variations in the invariant space Z_1. This is a slightly different setting than in standard unsupervised disentanglement and therefore standard disentanglement metrics might be less insightful. Instead, in order to test the property invariance, we first encode the inputs of the test set and fix the coordinates in the property subspace Z_0 which provides prediction $\hat{Y}$. Then we sample uniformly at random in Z_1 (plus/minus one standard deviation), decode the generated $\tilde{X}$ and perform a cycle through the network to obtain $\tilde{Y}$. This provides the predicted property for the generated $\tilde{X}$. If conditional invariance between X and Y at a fixed Z_0 is warranted, the mean absolute error (MAE) between $\hat{Y}$ and $\tilde{Y}$ should be close to zero. Thus, all models are trained to attain similar MAEs for reconstructing X and, in particular, predicting Y, to ensure a fair comparison.

5.1 Synthetic Dataset

In the first experiments, we focus on learning level sets of ellipses and ellipsoids mapped into five dimensions. We consider these experiments as they allow a clear interpretation and visualisation of fixing a property, i.e. choosing the ellipse curve or ellipsoid surface, and known low-dimensional parameterisations are readily available. To this end, we sample uniformly at random data points X_{original} from $\mathcal{U}([-1,1]^{d_{\text{X}}})$ and calculate as the corresponding one-dimensional properties Y_{original} the ellipse curves ($d_X = 2$) and ellipsoid surfaces ($d_X = 3$) rotated by 45° in the X_1X_2-plane. In addition, we add Gaussian noise to the property Y_{original}. In a real-world scenario, we typically do not have access to the underlying generating process providing X_{original} and property Y_{original} but a transformed view on these quantities. To reflect this, we map the input X_{original} into a five dimensional space ($d'_X = 5$), i.e. $X_{\text{original}} \in [-1,1]^{N \times d_X} \rightarrow X \in \mathbb{R}^{N \times 5}$, and property Y_{original} into three dimensional space ($d'_Y = 3$), i.e. $Y_{\text{original}} \in \mathbb{R}_+^{N \times 1} \rightarrow Y \in \mathbb{R}^{N \times 3}$, with N data points and dimensions $d_X = \{2, 3\}$. See Appendix Sec. A.4 for more details and Fig. A3 for an illustration of the dataset.

Level sets are usually defined implicitly (see Appendix Eq. (A.9)). Common parameterisations consider polar coordinates $(x, y) = (r \cos \varphi, r \sin \varphi)$ for the ellipse and spherical coordinates $(x, y, z) = (r \cos \varphi \sin \theta, r \sin \varphi \sin \theta, r \cos \theta)$ for the ellipsoid, with radius $r \in [0, \infty)$, (azimuth) angle $\varphi \in [0, 2\pi)$ in the X_1X_2-plane, and polar angle $\theta \in [0, \pi]$ measured from the X_3 axis. The goal of our experiment is to identify a low-dimensional parameterisation which captures the underlying radial and angular components, i.e. identify latent dimensions which correspond to parameters (r, φ) and (r, φ, θ).

Details on the architecture and training can be found in Appendix Sec. A.4. We use fully-connected layers for our encoder and decoder networks. Note that in our model, the noise level is fixed at $\sigma_{\text{noise}} = 1$ w.l.o.g. (see Sect. 4.1). We choose an 8-dim. latent space, with 3 dimensions reserved for property subspace Z_0 and 5 dimensions for invariant subspace Z_1. We consider a generous latent space with $d_{Z_1} = d'_X = 5$ and $d_{Z_0} = d'_Y = 3$ to evaluate the sparsity and model selection.

Results: All models attain similar MAEs for X reconstruction and Y prediction but differ in the property invariance as summarised in Table 1. Our model learns more invariant representations with several factors difference w.r.t. the property invariance in both experiments. In Fig. 2(a), signal vs. noise for the different models is presented. The standard deviation σ_{signal} is calculated as the sample standard deviation of the learned means in the respective latent dimension. The sampling noise σ_{noise} is optimised as a free parameter during training. We consider a latent dimension to be informative or selected if the signal exceeds the noise. The sparest solution is obtained in our model with one latent dimension selected in the property subspace Z_0 and one in the invariant subspace Z_1. In Fig. 2(b), we examine the obtained solution more closely in the original data space by mapping back from $d'_X = 5$ to $d_X = 2$ dimensions. We consider ten equidistant values in the selected Z_0 dim. 1 and sample points in the selected Z_1 dim. 8. The different colours represent fixed values in Z_0, with latent traversal in Z_1 dim. 8 reconstructing the full ellipse. This means, the selected latent dim. 8 contains all relevant information at a given coordinate in Z_0, while dim. 4 to 7 do not contain any relevant information. We can relate the selected dim. 1 in Z_0 to the radius r and dim. 8 in Z_1 to the angle φ. For the ellipsoid ($d_X = 3$) we obtain qualitatively the same results as for the ellipse. Again, only our model selects the correct number of latent factors with one in Z_0 and two in Z_1 (see Fig. 2(c)). The latent traversal results in Fig. 2(d) are more intricate to interpret. For latent dim. 6, we obtain a representation which can be interpreted as encoding the polar angle θ. Traversal in latent dim. 8 yields closed curves in three dimensions which can be viewed as on orthogonal representation to

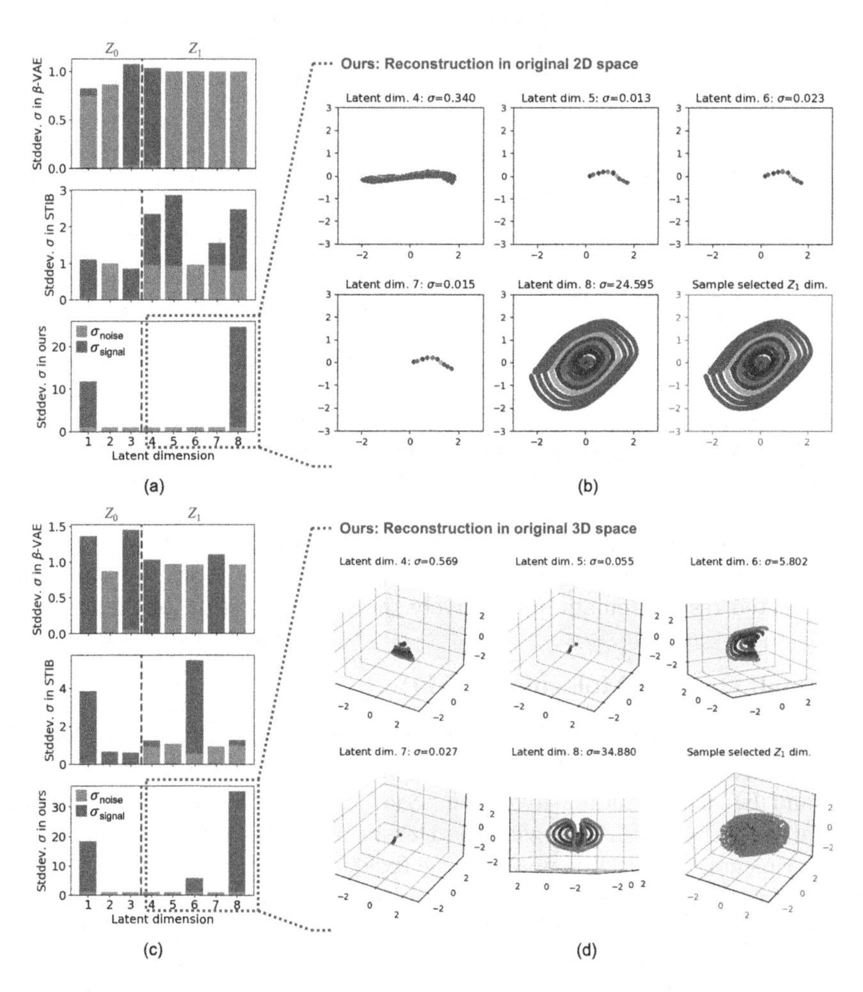

Fig. 2. Results for ellipse and ellipsoid in original input space ($d_X = \{2, 3\}$). (a, c) Illustration of standard deviation in the different latent dimensions, where property subspace Z_0 spans dimensions 1–3 and invariant subspace Z_1 spans dimensions 4–8. Grey bars indicate the sampling noise σ_{noise} and orange bars the sample standard deviation σ_{signal} in the respective dimension. We consider a latent dimension to be selected if the signal exceeds the noise, i.e. orange bars are visible. Only our model selects the expected numbers of parameters. (b, d) Illustration of latent traversal in our model in latent dimensions 4 to 8 in our model in the original input space for fixed values in the property space dimension 1 (different colours). (b) The selected dimension 8 represents the angular component φ and reconstructs the full ellipse curves. (d) The selected dimension 6 represents the polar angle θ, while dimension 8 can be related to the azimuth angle φ. (b, d) The last plot (red borders) samples in all selected dimensions, which reconstructs the full ellipse and ellipsoid, respectively. We intentionally did not sample the ellipsoid surfaces completely to allow seeing surfaces underneath. (Color figure online)

Table 1. Mean absolute errors (MAE) for reconstruction of input X, prediction of property Y, and property invariance. Ellipse/Ellipsoid: MAEs on 5-dim. input X and 3-dim. property Y are depicted. Molecules: MAEs on input X and property Y as the band gap energy in kcal mol^{-1}.

Model	Ellipse			Ellipsoid			Molecules		
	X	Y	Invar.	X	Y	Invar.	X	Y	Invar.
β-VAE	0.03	0.25	0.058	0.02	0.25	0.153	0.01	4.01	5.66
STIB	0.03	0.25	0.027	0.03	0.25	0.083	0.01	4.08	3.05
Ours	0.04	0.25	**0.006**	0.05	0.25	**0.006**	0.01	4.06	**1.34**

dim. 6 and be interpreted as an encoding of the azimuth angle φ. In both Fig. 2(b) and 2(d), the last plot shows sampling in the selected Z_1 dimensions for fixed Z_0 (i.e. property Y) and reconstructs the full ellipse and ellipsoid. Although β-VAE and STIB perform equally well on reconstructing and predicting on the test set, these models do not consistently lead to sparse and easily interpretable representations which allow direct traversal on the level sets as shown for our model. The presented results remain qualitatively the same for reruns of the models.

5.2 Small Organic Molecules (QM9)

As a more challenging example, we consider the QM9 dataset [31] which includes 133,885 organic molecules. The molecules consist of up to nine heavy atoms (C, O, N, and F), not including hydrogen. Each molecule includes corresponding chemical properties computed with the Density Functional Theory methods. In our experiments, we select a subset with a fixed stoichiometry ($C_7O_2H_{10}$) which consists of 6,093 molecules. We choose the band gap energy as the property.

Details on the architecture and training can be found in Appendix Sec. A.5. We use fully-connected layers for our encoder and decoder. For the input X we use the bag-of-bonds [12] descriptor as a translation, rotation, and permutation invariant representation of molecules, which involves 190 dimensions. The latent space size is 17, where Z_0 is 1-dimensional and Z_1 is 16-dimensional. To evaluate the invariance, we first adjust the regularisation loss weights for a fair comparison of the models. The weights for the irrelevance loss in the STIB and the invariance loss terms in our model were increased until a drop in reconstruction and prediction performances compared to the β-VAE results was noticeable.

Results: Table 1 summarises the results. On a test set of 300 molecules, all models achieve similar MAE of 0.01 for the reconstruction of X. For prediction of the band gap energies Y a MAE of approx. 4 kcal mol^{-1} is achieved. The invariance is computed on the basis of 25 test molecules and 400 samples generated for each reference molecule. Similarly to the synthetic experiments, the STIB model performs almost twice as well as the β-VAE, while our model yields a distinctly better invariance of 1.34 kcal mol^{-1} among both models. With this

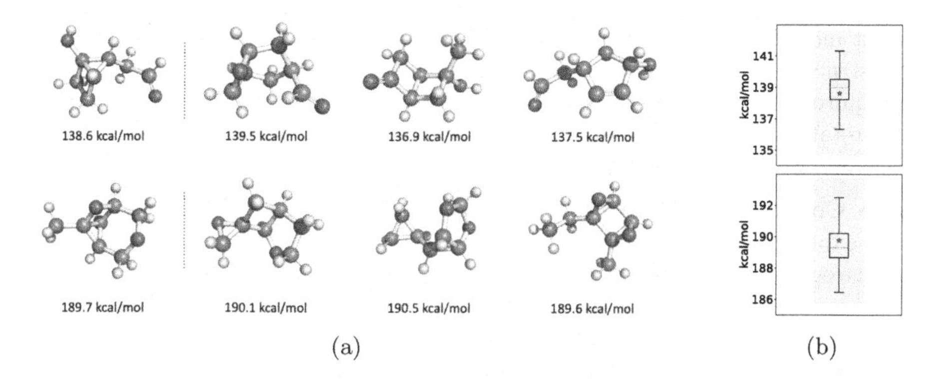

Fig. 3. Illustration of the generative capability of our model for two reference molecules (rows). (a) The first molecule is the reference molecule with a fixed reference band gap energy. We display three samples and their predicted band gap energies out of 2,000 samples. (b) Boxplots for distribution of predicted property. The star symbol marks the fixed reference band gap energy. The shaded background depicts the prediction error range of the model. (Color figure online)

result, we can generate novel molecules which are very close to a fixed property. This capability is illustrated in Fig. 3. For two reference molecules in the test set, we generate 2,000 new molecules by sampling uniformly at random with one standard deviation in the invariant subspace Z_1 and keeping the reference property value, i.e. fixed Z_0 coordinates. We show three such examples in Fig. 3a and select the nearest neighbours in the test set for visualisation of the molecular structure. For all samples, the boxplots in Fig. 3b illustrate the distribution in the predicted property values. The spread of predicted property values is generally smaller than the model prediction error of 4.06 kcal mol^{-1} and the predicted property of a majority of samples is close to the target property value.

6 Discussion

Sparsity constraints and cycle consistency lead to sparse and interpretable models facilitating model selection. The results in Fig. 2(a, c) demonstrate that our method identifies the sparsest solution in comparison to the standard disentanglement baseline β-VAE and the direct competitor STIB, which do not address sparsity explicitly. Furthermore, the experiments on ellipses and ellipsoids show that only our model also identifies a correct parameterisation. It correctly learns the radius r in the property subspace Z_0 as it encodes the level set, i.e. the ellipse curve or ellipsoid surface given by property Y. The angular components φ and θ are correctly – and in particular independently – learned in the invariant subspace Z_1 (see Fig. 2(b, d)). This is a direct consequence of the cycle consistency on the property Y. It allows for semantically structuring the latent space on the basis of the semantic knowledge on property Y. Finally, these results highlight that our method is able to inherently select the correct model. Although the β-VAE and STIB are capable of attaining similar reconstruction and prediction errors, a reconstruction of level sets in these models requires a more complicated combination of latent dimensions and hinders interpretation. Therefore, only our model makes an interpretation of the learned latent representation feasible.

Cycle Consistency Enforces Conditional Invariance. Table 1 shows that for all experiments, our model exhibits the best property invariance at otherwise similar reconstruction and prediction errors. The β-VAE has no mechanisms to ensure invariance and thus performs worst. But although the STIB relies on adversarial training to minimise mutual information (MI) between Z_1 and Y, the alternating training and MI estimation can pose practical obstacles, especially in cases with high-dimensional latent spaces. Our cycle-consistency-based approach has the same benefits and is more feasible. In particular, our approach can operate on arbitrarily large latent spaces in both Z_0 and Z_1, because of the inherent sparsity of the solution. Typically, an upper limit for the size of property subspace Z_0 and invariant subspace Z_1 can be defined by the dimensionality of the property Y and input X (see Fig. 2). Noteworthy – although our model is trained and tested on data in the interval $[-1, 1]^{d_X}$, $d_X = \{2, 3\}$ – the results generalise well beyond this interval, as long as a part of the level curve or surface was encountered during training (see Fig. 2(b)). This can be directly attributed to the regularisation of the latent space through additional sampling and cycle consistency of generated samples. These mechanisms impose conditional invariance which, in turn, facilitates generalisation and exploration of new samples by sharing the same level set or symmetry-conserved property.

Conditional Invariance Improves Targeted Molecule Discovery. Conditional invariance is of great importance for the generative potential of our model. In Fig. 3 we exemplary explored the molecular structures for two reference molecules. By sampling in the invariant space Z_1, we discover molecular structures with property values which are very close to the fixed targets, i.e. the mean absolute deviation is below the model prediction error. Our experiment demonstrates the ability to generate molecules with self-consistent properties which rely on the improved conditional invariance provided by our model. This facilitates the discovery of novel molecules with desired chemical properties.

In conclusion, we demonstrated on synthetic and real-world use cases that our method allows selecting a correct model and improve interpretability as well as exploration of the latent representation. In our synthetic study, we focused on simple cases of connected and convex level sets. To generalise these findings, more general level sets are interesting to be investigated in order to relate to more real-world scenarios. In addition, our approach could be applied to medical applications where a selection of interpretable models is of particular relevance.

Acknowledgements. This research was supported by the Swiss National Science Foundation through projects No. 167333 within the National Research Programme 75 "Big Data" (M.S.), No. P2BSP2 184359 (S.P.) and the NCCR MARVEL (V.N., M.W., A.W.). Furthermore, the authors would like to thank the anonymous reviewers for their valuable comments and suggestions.

References

1. Abadi, M., et al.: TensorFlow: large-scale machine learning on heterogeneous systems (2015). https://www.tensorflow.org/. Software available from tensorflow.org
2. Achille, A., Soatto, S.: Information dropout: learning optimal representations through noisy computation. IEEE Trans. Pattern Anal. Mach. Intell. **40**(12), 2897–2905 (2018)
3. Ainsworth, S.K., Foti, N.J., Lee, A.K.C., Fox, E.B.: oi-VAE: output interpretable VAEs for nonlinear group factor analysis. In: Proceedings of the 35th International Conference on Machine Learning (2018)
4. Alemi, A.A., Fischer, I., Dillon, J.V., Murphy, K.: Deep variational information bottleneck. In: 5th International Conference on Learning Representations, ICLR 2017, Toulon, France, 24–26 April 2017, Conference Track Proceedings. OpenReview.net (2017). https://openreview.net/forum?id=HyxQzBceg
5. Bouchacourt, D., Tomioka, R., Nowozin, S.: Multi-level variational autoencoder: learning disentangled representations from grouped observations. In: AAAI Conference on Artificial Intelligence (2018)
6. Chechik, G., Globerson, A., Tishby, N., Weiss, Y.: Information bottleneck for Gaussian variables. J. Mach. Learn. Res. **6**, 165–188 (2005)
7. Chen, R.T., Li, X., Grosse, R., Duvenaud, D.: Isolating sources of disentanglement in variational autoencoders. arXiv preprint arXiv:1802.04942 (2018)
8. Chen, X., Duan, Y., Houthooft, R., Schulman, J., Sutskever, I., Abbeel, P.: InfoGAN: interpretable representation learning by information maximizing generative adversarial nets. arXiv preprint arXiv:1606.03657 (2016)
9. Chicharro, D., Besserve, M., Panzeri, S.: Causal learning with sufficient statistics: an information bottleneck approach. arXiv preprint arXiv:2010.05375 (2020)
10. Creswell, A., Mohamied, Y., Sengupta, B., Bharath, A.A.: Adversarial information factorization (2018)
11. Gómez-Bombarelli, R., et al.: Automatic chemical design using a data-driven continuous representation of molecules. ACS Cent. Sci. **4**(2), 268–276 (2018)
12. Hansen, K., et al.: Machine learning predictions of molecular properties: accurate many-body potentials and nonlocality in chemical space. J. Phys. Chem. Lett. **6**(12), 2326–2331 (2015)
13. Higgins, I., et al.: β-VAE: learning basic visual concepts with a constrained variational framework. In: International Conference on Learning Representations (2017)
14. Jha, A.H., Anand, S., Singh, M., Veeravasarapu, V.S.R.: Disentangling factors of variation with cycle-consistent variational auto-encoders. In: Ferrari, V., Hebert, M., Sminchisescu, C., Weiss, Y. (eds.) ECCV 2018. LNCS, vol. 11207, pp. 829–845. Springer, Cham (2018). https://doi.org/10.1007/978-3-030-01219-9_49
15. Keller, S.M., Samarin, M., Torres, F.A., Wieser, M., Roth, V.: Learning extremal representations with deep archetypal analysis. Int. J. Comput. Vision **129**(4), 805–820 (2021)
16. Keller, S.M., Samarin, M., Wieser, M., Roth, V.: Deep archetypal analysis. In: Fink, G.A., Frintrop, S., Jiang, X. (eds.) DAGM GCPR 2019. LNCS, vol. 11824, pp. 171–185. Springer, Cham (2019). https://doi.org/10.1007/978-3-030-33676-9_12
17. Kim, H., Mnih, A.: Disentangling by factorising. In: International Conference on Machine Learning, pp. 2649–2658. PMLR (2018)
18. Kingma, D.P., Ba, J.: Adam: a method for stochastic optimization. In: International Conference on Learning Representations (2015)

19. Kingma, D.P., Mohamed, S., Rezende, D.J., Welling, M.: Semi-supervised learning with deep generative models. In: Advances in Neural Information Processing Systems, pp. 3581–3589 (2014)
20. Kingma, D.P., Welling, M.: Auto-encoding variational Bayes. In: Bengio, Y., LeCun, Y. (eds.) 2nd International Conference on Learning Representations, ICLR 2014, Banff, AB, Canada, 14–16 April 2014, Conference Track Proceedings (2014). http://arxiv.org/abs/1312.6114
21. Klys, J., Snell, J., Zemel, R.: Learning latent subspaces in variational autoencoders. In: Advances in Neural Information Processing Systems (2018)
22. Kusner, M.J., Paige, B., Hernández-Lobato, J.M.: Grammar variational autoencoder. In: International Conference on Machine Learning (2017)
23. Lample, G., Zeghidour, N., Usunier, N., Bordes, A., Denoyer, L., Ranzato, M.: Fader networks: manipulating images by sliding attributes. In: Advances in Neural Information Processing Systems (2017)
24. Lin, Z., Thekumparampil, K., Fanti, G., Oh, S.: InfoGAN-CR and modelcentrality: self-supervised model training and selection for disentangling GANs. In: International Conference on Machine Learning, pp. 6127–6139. PMLR (2020)
25. Locatello, F., et al.: Challenging common assumptions in the unsupervised learning of disentangled representations. In: International Conference on Machine Learning, pp. 4114–4124. PMLR (2019)
26. Louizos, C., Shalit, U., Mooij, J.M., Sontag, D., Zemel, R., Welling, M.: Causal effect inference with deep latent-variable models. In: Guyon, I., et al. (eds.) Advances in Neural Information Processing Systems 30, pp. 6446–6456. Curran Associates, Inc. (2017)
27. Louizos, C., Swersky, K., Li, Y., Welling, M., Zemel, R.S.: The variational fair autoencoder. In: Bengio, Y., LeCun, Y. (eds.) 4th International Conference on Learning Representations, ICLR 2016, San Juan, Puerto Rico, 2–4 May 2016, Conference Track Proceedings (2016). http://arxiv.org/abs/1511.00830
28. Nesterov, V., Wieser, M., Roth, V.: 3DMolNet: a generative network for molecular structures (2020)
29. Parbhoo, S., Wieser, M., Roth, V., Doshi-Velez, F.: Transfer learning from well-curated to less-resourced populations with HIV. In: Proceedings of the 5th Machine Learning for Healthcare Conference (2020)
30. Parbhoo, S., Wieser, M., Wieczorek, A., Roth, V.: Information bottleneck for estimating treatment effects with systematically missing covariates. Entropy **22**(4), 389 (2020)
31. Ramakrishnan, R., Dral, P.O., Rupp, M., Von Lilienfeld, O.A.: Quantum chemistry structures and properties of 134 kilo molecules. Sci. Data **1**(1), 1–7 (2014)
32. Raman, S., Fuchs, T.J., Wild, P.J., Dahl, E., Roth, V.: The Bayesian group-lasso for analyzing contingency tables. In: Proceedings of the 26th Annual International Conference on Machine Learning (2009)
33. Rey, M., Roth, V., Fuchs, T.: Sparse meta-gaussian information bottleneck. In: International Conference on Machine Learning, pp. 910–918. PMLR (2014)
34. Rezende, D.J., Mohamed, S., Wierstra, D.: Stochastic backpropagation and approximate inference in deep generative models. In: International Conference on Machine Learning (2014)
35. Robert, T., Thome, N., Cord, M.: DualDis: dual-branch disentangling with adversarial learning (2019)
36. Simon, N., Friedman, J., Hastie, T., Tibshirani, R.: A sparse-group lasso. J. Comput. Graph. Stat. **22**(2), 231–245 (2013)

37. Song, J., Ermon, S.: Understanding the limitations of variational mutual information estimators. arXiv preprint arXiv:1910.06222 (2019)
38. Tibshirani, R.: Regression shrinkage and selection via the lasso. J. R. Stat. Soc. (Ser. B) **58**(1), 267–288 (1996)
39. Tishby, N., Pereira, F.C., Bialek, W.: The information bottleneck method. In: Allerton Conference on Communication, Control and Computing (1999)
40. Wieczorek, A., Roth, V.: Causal compression. arXiv preprint arXiv:1611.00261 (2016)
41. Wieczorek, A., Roth, V.: On the difference between the information bottleneck and the deep information bottleneck. Entropy **22**(2), 131 (2020)
42. Wieczorek, A., Wieser, M., Murezzan, D., Roth, V.: Learning sparse latent representations with the deep copula information bottleneck. In: 6th International Conference on Learning Representations, ICLR 2018, Vancouver, BC, Canada, 30 April–3 May 2018, Conference Track Proceedings. OpenReview.net (2018). https://openreview.net/forum?id=Hk0wHx-RW
43. Wieser, M.: Learning invariant representations for deep latent variable models. Ph.D. thesis, University of Basel (2020)
44. Wieser, M., Parbhoo, S., Wieczorek, A., Roth, V.: Inverse learning of symmetries. In: Advances in Neural Information Processing Systems (2020)
45. Wu, M., Hughes, M.C., Parbhoo, S., Zazzi, M., Roth, V., Doshi-Velez, F.: Beyond sparsity: tree regularization of deep models for interpretability (2017)
46. Zhu, J., Park, T., Isola, P., Efros, A.A.: Unpaired image-to-image translation using cycle-consistent adversarial networks. In: International Conference on Computer Vision (2017)

CAGAN: Text-To-Image Generation with Combined Attention Generative Adversarial Networks

Henning Schulze(✉), Dogucan Yaman, and Alexander Waibel

Institute for Anthropomatics and Robotics, Karlsruhe Institute of Technology, Karlsruhe, Germany
henning.schulze1@web.de, {dogucan.yaman,alexander.waibel}@kit.edu

Abstract. Generating images according to natural language descriptions is a challenging task. Prior research has mainly focused to enhance the quality of generation by investigating the use of spatial attention and/or textual attention thereby neglecting the relationship between channels. In this work, we propose the Combined Attention Generative Adversarial Network (CAGAN) to generate photo-realistic images according to textual descriptions. The proposed CAGAN utilises two attention models: word attention to draw different sub-regions conditioned on related words; and squeeze-and-excitation attention to capture non-linear interaction among channels. With spectral normalisation to stabilise training, our proposed CAGAN achieves state-of-the-art FID and comparative IS scores on the CUB dataset and on the more challenging COCO dataset. Furthermore, we demonstrate that judging a model by a single evaluation metric can be misleading by developing an additional model adding local self-attention which scores a higher IS than our other model, but generates unrealistic images through feature repetition.

Keywords: Text-to-image synthesis · Generative adversarial network (GAN) · Attention

1 Introduction

Generating images according to natural language descriptions spans a wide range of difficulty, from generating synthetic images to simple and highly complex real-world images. It has tremendous applications such as photo-editing, computer-aided design, and may be used to reduce the complexity of or even replace rendering engines [28]. Furthermore, good generative models involve learning new representations. These are useful for a variety of tasks, for example classification, clustering, or supporting transfer among tasks.

Although generating images highly related to the meanings embedded in a natural language description is a challenging task due to the gap between text and image modalities, there has been exciting recent progress in the field using numerous techniques and different inputs [3–5,12,18–21,29,38,39,45,46,

C. Bauckhage et al. (Eds.): DAGM GCPR 2021, LNCS 13024, pp. 392–404, 2021.
https://doi.org/10.1007/978-3-030-92659-5_25

this bird has wings that are grey and has a yellow belly

Fig. 1. Example results of the proposed CAGAN (SE). The generated images are of 64×64, 128×128, and 256×256 resolutions respectively, bilinearly upsampled for visualization.

49] yielding impressive results on limited domains. A majority of approaches are based on Generative Adversarial Networks (GANs) [8]. Zhang et al. introduced Stacked GANs [47] which consist of two GANs generating images in a low-to-high resolution fashion. The second generator receives the image encoding of the first generator and the text embedding as input to correct defects and generate higher resolution images. Further research has mainly focused to enhance the quality of generation by investigating the use of spatial attention and/or textual attention thereby neglecting the relationship between channels.

In this work, we propose Combined Attention Generative Adversarial Network (CAGAN) that combines multiple attention models, thereby paying attention to word, channel, and spatial relationships. First, the network uses a deep bi-directional LSTM encoder [45] to obtain word and sentence features. Then, the images are generated in a coarse to fine fashion (see Fig. 1) by feeding the encoded text features into a three stage GAN. Thereby, we utilise local-self attention [27] mainly during the first stage of generation; word attention at the beginning of the second and the third generator; and squeeze-and-excitation attention [13] throughout the second and the third generator. We use the publicly available CUB [41] and COCO [22] datasets to conduct the experimental analysis. Our experiments show that our network generates images of similar quality as previous work while either advancing or competing with the state of the art on the Inception Score (IS) [35] and the Fréchet Inception Distance (FID) [11].

The main contributions of this paper are threefold:

(1) We incorporate multiple attention models, thereby reacting to subtle differences in the textual input with fine-grained word attention; modelling long-range dependencies with local self-attention; and capturing non-linear interaction among channels with squeeze-and-excitation (SE) attention. SE attention can learn to learn to use global information to selectively emphasise informative features and suppress less useful ones.

(2) We stabilise the training with spectral normalisation [24], which restricts the function space from which the discriminators are selected by bounding the Lipschitz norm and setting the spectral norm to a designated value.
(3) We demonstrate that improvements on single evaluation metrics have to be viewed carefully by showing that evaluation metrics may react oppositely.

The rest of the paper is organized as follows: In Sect. 2, we give a brief overview of the literature. In Sect. 3, we explain the presented approach in detail. In Sect. 4, we mention the employed datasets and experimental results. Then, we discuss the outcomes and we conclude the paper in Sect. 5.

2 Related Work

While there has been substantial work for years in the field of image-to-text translation, such as image caption generation [1,7,44], only recently the inverse problem came into focus: text-to-image generation. Generative image models require a deep understanding of spatial, visual, and semantic world knowledge. A majority of recent approaches are based on GANs [8].

Reed et al. [32] use a GAN with a direct text-to-image approach and have shown to generate images highly related to the text's meaning. Reed et al. [31] further developed this approach by conditioning the GAN additionally on object locations. Zhang et al. built on Reed et al.'s direct approach developing StackGAN [47] generating 256×256 photo-realistic images from detailed text descriptions. Although StackGAN yields remarkable results on specific domains, such as birds or flowers, it struggles when many objects and relationships are involved. Zhang et al. [48] improved StackGAN by arranging multiple generators and discriminators in a tree-like structure, allowing for more stable training behaviour by jointly approximating multiple distributions. Xu et al. [45] introduced a novel loss function and fine-grained word attention into the model.

Recently, a number of works built on Xu et al.'s [45] approach: Cheng et al. [5] employed spectral normalisation [24] and added global self-attention to the first generator; Qiao et al. [30] introduced a semantic text regeneration and alignment module thereby learning text-to-image generation by redescription; Li et al. [18] added channel-wise attention to Xu et al.'s spatial word attention to generate shape-invariant images when changing text descriptions; Cai et al. [3] enhanced local details and global structures by attending to related features from relevant words and different visual regions; Tan et al. [38] introduced image-level semantic consistency and utilised adaptive attention weights to differentiate keywords from unimportant words; Yin et al. [46] focused on disentangling the semantic-related concepts and introduced a contrasive loss to strengthen the image-text correlation; Zhu et al. [49] refined Xu et al.'s fine-grained word attention by dynamically selecting important words based on the content of an initial image; and Cheng et al. [4] enriched the given description with prior knowledge and then generated an image from the enriched multi-caption.

Instead of using multiple stages or multiple GANs, Li et al. [20] used one generator and three independent discriminators to generate multi-scale images conditioned on text in an adversarial manner. Tao et al. [39] discarded the stacked architecture approach, proposing a GAN to directly synthesize images without extra networks. Johnson et al. [14] introduced a GAN that receives a scene graph consisting of objects and their relationships as input and generates complex images with many recognizable objects. However, the images are not photo-realistic. Qiao et al. [29] introduced LeicaGAN which adopts text-visual co-embeddings to convey the visual information needed for image generation.

Other approaches are based on autoencoders [6,36,42], autoregressive models [9,26,33], or other techniques such as generative image modelling using an RNN with spatial LSTM neurons [40]; multiple layers of convolution and deconvolution operators trained with Stochastic Gradient Variational Bayes [17]; a probabilistic programming language for scene understanding with fast general-purpose inference machinery [16]; and generative ConvNets [43].

We propose to expand the focus of attention to channel, word, and spatial relationships instead of a subset of these thereby enhancing the quality of generation.

3 The Framework of Combined Attention Generative Adversarial Networks

3.1 Combined Attention Generative Adversarial Networks

The proposed CAGAN utilises three attention models: word attention to draw different sub-regions conditioned on related words, local self-attention to model long-range dependencies, and squeeze-and-excitation attention to capture non-linear interaction among channels.

The attentional generative model consists of three generators, which receive image feature vectors as input and generate images of small-to-large scales. First, a deep bidirectional LSTM encoder encodes the input sentence into a global sentence vector s and a word matrix. Conditioning augmentation F^{CA} [47] converts the sentence vector into the conditioning vector. A first network receives the conditioning vector and noise, sampled from a standard normal distribution, as input and computes the first image feature vector. Each generator is a simple 3x3 convolutional layer that receives the image feature vector as input to compute an image. The remaining image feature vectors are computed by networks receiving the previous image feature vector and the result of the ith attentional model F_i^{attn} (see Fig. 2), which uses the word matrix computed by the text encoder.

To compute word attention, the word vectors are converted into a common semantic space. For each subregion of the image a word-context vector is computed, dynamically representing word vectors that are relevant to the subregion of the image, i.e., indicating the weight the word attention model attends to the lth word when generating a subregion. The final objective function of the attentional generative network is defined as:

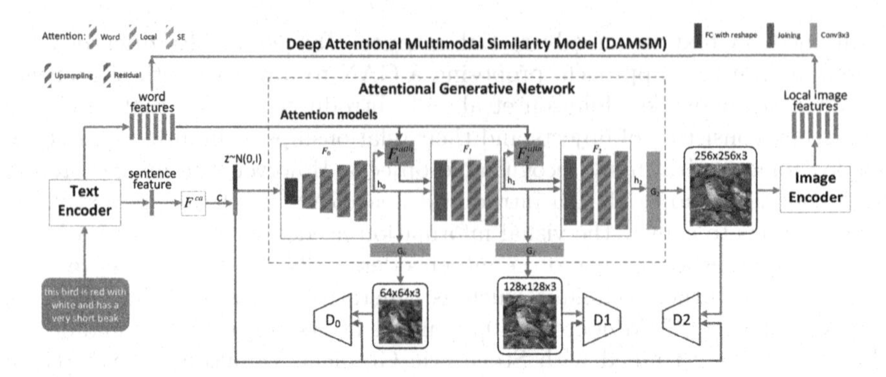

Fig. 2. The architecture of the proposed CAGAN with word, SE, and local attention. When omitting local attention, local attention is removed from the F_n^{attn} networks. In the upsampling blocks it is replaced by SE attention.

$$L = L_G + \lambda L_{\text{DAMSM}}, \ \ where \ L_G = \sum_{i=0}^{m-1} L_{G_i}. \tag{1}$$

Here, λ is a hyperparameter to balance the two terms. The first term is the GAN loss that jointly approximates conditional and unconditional distributions [48]. At the ith stage, the generator G_i has a corresponding discriminator D_i. The adversarial loss for G_i is defined as:

$$L_{G_i} = \underbrace{-\frac{1}{2}\mathbb{E}_{\hat{y_i}\sim P_{G_i}}\left[log(D_i(\hat{y_i}))\right]}_{\text{unconditional loss}} \underbrace{-\frac{1}{2}\mathbb{E}_{\hat{y_i}\sim P_{G_i}}\left[log(D_i(\hat{y_i}, s))\right]}_{\text{conditional loss}}, \tag{2}$$

where $\hat{y_i}$ are the generated images. The unconditional loss determines whether the image is real or fake while the conditional loss determines whether the image and the sentence match or not. Alternately to the training of G_i, each discriminator D_i is trained to classify the input into the class of real or fake by minimizing the cross-entropy loss.

The second term of Eq. 1, L_{DAMSM}, is a fine-grained word-level image-text matching loss computed by the DAMSM [45]. The DAMSM learns two neural networks that map subregions of the image and words of the sentence to a common semantic space, thus measuring the image-text similarity at the word level to compute a fine-grained loss for image generation. The image encoder prior to the DAMSM is built upon a pretrained Inception-v3 model [37] with added perceptron layers to extract visual feature vectors for each subregion of the image and a global image vector.

3.2 Attention Models

Local Self-attention. Similar to a convolution, local self-attention [27] extracts a local region of pixels $ab \in \mathcal{N}_k(i,j)$ for each pixel x_{ij} and a given spatial extent k. An output pixel y_{ij} computes as follows:

$$y_{ij} = \sum_{a,b \in \mathcal{N}_k(i,j)} \text{softmax}_{ab}(q_{ij}^T k_{ab}) v_{ab}. \tag{3}$$

$q_{ij} = W_Q x_{ij}$ denotes the queries, $k_{ab} = W_K x_{ab}$ the keys, and $v_{ab} = W_V x_{ab}$ the values, each obtained via linear transformations W of the pixel ij and their neighbourhood pixels. The advantage over a simple convolution is that each pixel value is aggregated with a convex convolution of value vectors with mixing weights (softmax_{ab}) parametrised by content interactions.

Squeeze-and-Excitation (SE) Attention. Instead of focusing on the spatial component of CNNs, SE attention [13] aims to improve the channel component by explicitly modelling interdependencies among channels via channel-wise weighting. Thus, they can be interpreted as a light-weight self-attention function on channels.

First, a transformation, which is typically a convolution, outputs the feature map U. Because convolutions use local receptive fields, each entry of U is unaware of contextual information outside its region. A corresponding SE-block addresses this issue by performing a feature recalibration.

A squeeze operation aggregates the feature maps of U across the spatial dimension ($H \times W$) yielding a channel descriptor. The proposed squeeze operation is mean-pooling across the entire spatial dimension of each channel. The resulting channel descriptor serves as an embedding of the global distribution of channel-wise features.

A following excitation operation F_{ex} aims to capture channel-wise dependencies, specifically non-linear interaction among channels and non-mutually exclusive relationships. The latter allows multiple channels to be emphasized. The excitation operation is a simple self-gating operation with a sigmoid activation function:

$$F_{ex}(z, W) = \sigma(g(z, W)) = \sigma(W_2 \delta(W_1 z)), \tag{4}$$

where δ refers to the ReLU activation function, $W_1 \in \mathbb{R}^{\frac{C}{r} \times C}$, and $W_2 \in \mathbb{R}^{C \times \frac{C}{r}}$. To limit model complexity and increase generalisation, a bottleneck is formed around the gating mechanism: a Fully Connected (FC) layer reduces the dimensionality by a factor of r. A second FC layer restores the dimensionality after the gating operation. The authors recommend an r of 16 for a good balance between accuracy and complexity (~10% parameter increase on ResNet-50 [10]). Ideally, r should be tuned for the intended architecture.

The excitation operation F_{ex} computes per-channel modulation weights. These are applied to the feature maps U performing an adaptive recalibration.

4 Experiments

Dataset. We employed CUB [41] and COCO [22] datasets for the experiments. The CUB dataset [41] consists of 8855 train and 2933 test images. To perform evaluation, one image per caption in the test set is computed since each image has ten captions. The COCO dataset [22] with the 2014 split consists of 82783 train and 40504 test images. We randomly sample 30000 captions from the test set for the evaluation.

Evaluation Metric. In this work, we utilized the Inception Score (IS) [35] and The Fréchet Inception Distance (FID) [11] to evaluate the performance of proposed method. The IS [35] is a quantitative metric to evaluate generated images. It measures two properties: highly classifiable and diverse with respect to class labels. Although the IS is the most widely used metric in text-to-image generation, it has several issues [2,25,34] regarding the computation of the score itself and the usage of the score. According to the authors of [2] it: "fails to provide useful guidance when comparing models".

The FID [11] views features as a continuous multivariate Gaussian and computes a distance in the feature space between the real data and the generated data. A lower FID implies a closer distance between the generated image distribution and the real image distribution. The FID is consistent with human judgment and more consistent to noise than the IS [11] although it has a slight bias [23]. Please note that there is some inconsistency in how the FID is calculated in prior work, originating from different pre-processing techniques that significantly impact the score. We use the official implementation[1] of the FID. To ensure a consistent calculation of all of our evaluation metrics, we replace the generic Inception v3 network with the pre-trained Inception v3 network we used for computing the IS of the corresponding dataset. We re-calculate the FID scores of papers with an official model to provide a fair comparison.

Implementation Detail. We employ spectral normalisation [24], a weight normalisation technique to stabilise the training of the discriminator, during training. To compute the semantic embedding for text descriptions, we employ a pre-trained bi-direction LSTM encoder by Xu et al. [45] with a dimension of 256 for the word embedding. The sentence length was 18 for the CUB dataset and 12 for the COCO dataset.

All networks are trained using the Adam optimiser [15] with a batch size of 20, a learning rate of 0.0002, and $\beta_1 = 0.5$ and $\beta_2 = 0.999$. We train for 600 epochs on the CUB and for 200 epochs on the COCO dataset. For the model utilising squeeze-and-excitation attention we use $r = 1$, and $\lambda = 0.1$ and $\lambda = 50.0$, respectively for the CUB and the COCO dataset. For the model utilising local self-attention as well we use $r = 4$, and $\lambda = 5.0$ and $\lambda = 50.0$.

4.1 Results

Quantitative Results. As Table 1 and Fig. 3 show, our model utilising squeeze-and-excitation attention outperforms the baseline AttnGAN [45] in both metrics

[1] https://github.com/bioinf-jku/TTUR.

Table 1. Fréchet Inception Distance (FID) and Inception Score (IS) of state-of-the-art models and our two CAGAN models on the CUB and COCO dataset with a 256×256 image resolution. The unmarked scores are those reported in the original papers, note that the reported FID scores may be inconsistent (see Sect. 4 Evaluation Metric). Scores marked with † were re-calculated by us using the pre-trained model provided by the respective authors. $\uparrow$ ($\downarrow$) means the higher (lower), the better.

Model	CUB dataset		COCO dataset	
	IS↑	FID↓	IS↑	FID↓
Real data	25.52 ± .09	0.00	37.97 ± .88	0.00
AttnGAN [45]	4.36 ± .04	47.76†	25.89 ± .47	31.05†
PPAN [20]	4.38 ± .05	-	-	-
HAGAN [5]	4.43 ± .03	-	-	-
MirrorGAN [30]	4.56 ± .05	-	26.47 ± .41	-
ControlGAN [18]	4.58 ± .09	49.18†	24.06 ± .60	-
DualAttn-GAN [3]	4.59 ± .07	-	-	-
LeicaGAN [29]	4.62 ± .06	-	-	-
SEGAN [38]	4.67 ± .04	-	27.86 ± .31	32.28
SD-GAN [46]	4.67 ± .09	-	35.69 ± .50	-
DM-GAN [49]	4.75 ± .07	43.20†	30.49 ± .57	22.84†
DF-GAN [39]	5.10	–	–	21.42
RiFeGAN [4]	**5.23 ± .09**	–	31.70	-
Obj-GAN [19]	-	-	30.29 ± .33	25.64
OP-GAN [12]	-	-	27.88 ± .12	23.29†
CPGAN [21]	-	-	**52.73 ± .61**	49.92†
CAGAN_SE (ours)	4.78 ± .06	**42.98**	32.60 ± .75	**19.88**
CAGAN_L+SE (ours)	4.96 ± .05	61.06	33.89 ± .69	27.40

on both datasets. The IS is improved by 9.6% ± 2.4% and 25.9% ± 5.3% and the FID by 10.0% and 36.0% on the CUB and the COCO dataset, respectively. Our approach also achieves the best FID on both datasets though not all listed models could be fairly compared (see Sect. 4 Evaluation Metric).

Our second model, utilising squeeze-and-excitation attention and local self-attention, shows better IS scores than our other model. However, it generates completely unrealistic images through feature repetitions (see Fig. 4) and has a major negative impact on the FID throughout training (see Fig. 3). This behaviour is similar to [21] on the COCO dataset and demonstrates that a single score can be misleading and thus the importance of reporting both scores.

In summary, according to the experimental results, our proposed CAGAN achieved state-of-the-art results on both the CUB dataset and COCO dataset based on the FID metric and comparative results on the IS metric. All these results indicate how our CAGAN model is effective for the text-to-image generation task.

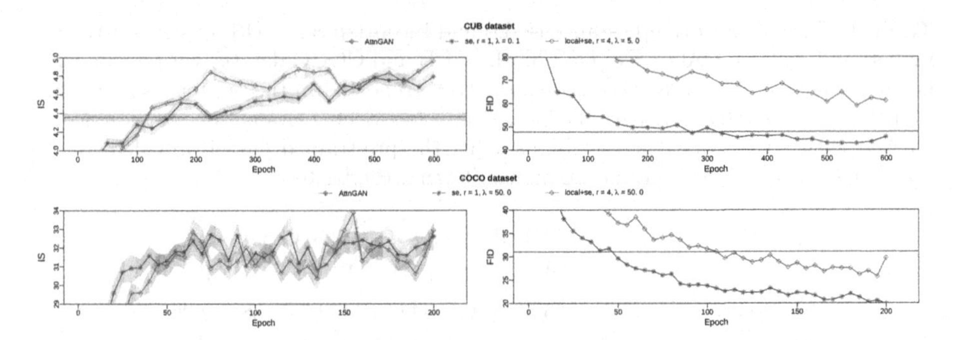

Fig. 3. IS and FID of the AttnGAN [45], our model utilising squeeze-and-excitation attention, and our model utilising squeeze-and-excitation attention and local self-attention on the CUB and the COCO dataset. The IS of the AttnGAN is the reported score and the FID was re-evaluated using the official model. The IS of the AttnGAN on the COCO dataset is with $25.89 \pm .47$ significantly lower than our models. We omitted the score to highlight the distinctions between our two models.

Qualitative Results: Figure 4 shows images generated by our models and by several other models [12,45,49] on the CUB dataset and on the more challenging COCO dataset. On the CUB dataset, our model utilising SE attention generates images of vivid details (see 1st, 4th, 5th, and 6th row), demonstrating a strong text-image correlation (see 3th, 4th, and 5th row), avoiding feature repetitions (see double beak, DM-GAN $6th$ row), and managing the difficult scene (see 7th row) best. Cut-off artefacts occur in all presented models.

Our model incorporating local self-attention fails to produce realistic looking image, despite scoring higher ISs than the AttnGAN and our model utilising SE attention. Instead, it draws repetitive features manifesting in the form of multiple birds, drawn out birds, multiple heads, or strange patterns. The drawn features mostly match the textual descriptions. This provides a possible explanation why the model has a high IS despite scoring poorly on the FID: the IS cares mainly about the images being highly classifiable and diverse. Thereby, it presumes that highly classifiable images are of high quality. Our network demonstrates that high classify-ability and diversity and therefore a high IS can be achieved through completely unrealistic, repetitive features of the correct bird class. This is further evidence that improvements solely based on the IS have to be viewed sceptically.

On the more challenging COCO dataset, our model utilising SE attention demonstrates semantic understanding by drawing features that resemble the object, for example, the brown-white pattern of a giraffe (1st row), umbrellas (4th row), and traffic lights (5th row). Furthermore, our model draws distinct shapes for the bathroom (2nd row), broccoli (3rd row), and is the only one that properly approximates a tower building with a clock (7th row). Generally speaking, the results on the COCO dataset are not as realistic and robust as on the CUB dataset. We attribute this to the more complex scenes coupled with more abstract descriptions that focus rather on the category of objects

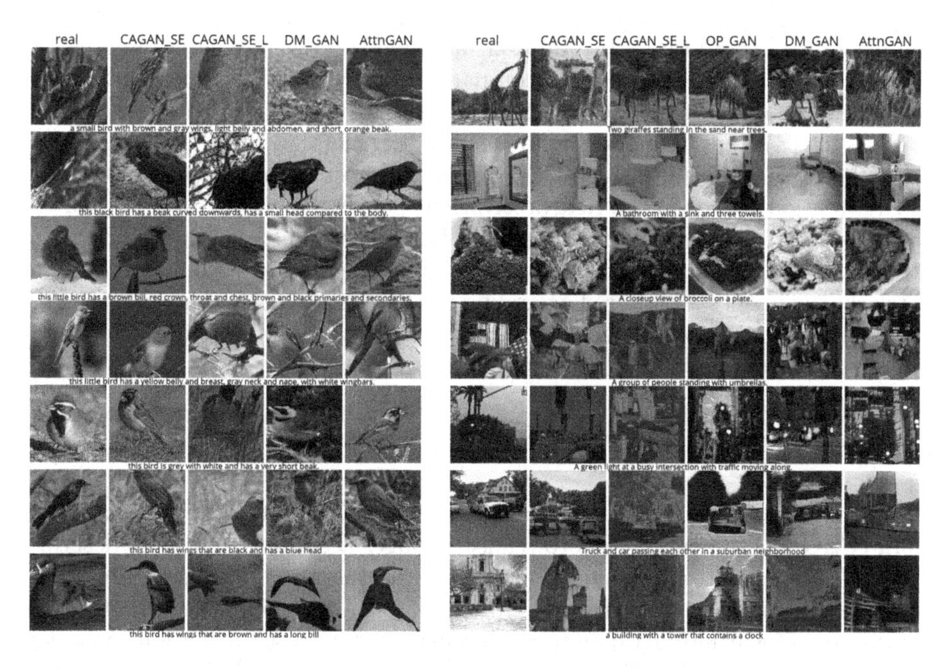

Fig. 4. Comparison of images generated by our models (CAGAN_SE and CAGAN_SE_L) with images generated by other current models [12,45,49] on the CUB dataset (left) and on the more challenging COCO dataset (right).

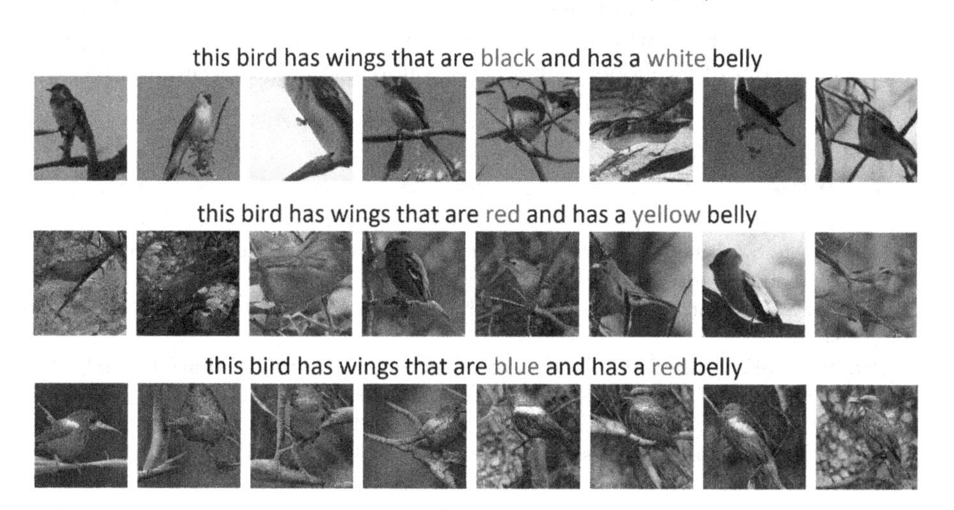

Fig. 5. Example results of our SE attention model with $r = 1, \lambda = 0.1$ trained on the CUB dataset while changing some most attended, in the sense of word attention, words in the text descriptions.

than detailed descriptions. In addition, although there are a large number of categories, each category only has comparatively few examples thereby further increasing the difficulty for text-to-image-generation.

For our SE attention model we further test its generalisation ability by testing how sensitive the outputs are to changes in the most attended, in the sense of word attention, words in the text descriptions (see Fig. 5). The test is similar to the one performed on the AttnGAN [45]. The results illustrate that adding SE attention and spectral normalisation do not harm the generalisation ability of the network: the images are altered according to the changes in the input sentences, showing that the network retains its ability to react to subtle semantic differences in the text descriptions.

5 Conclusion

In this paper, we propose the Combined Attention Generative Adversarial Network (CAGAN) to generate photo-realistic images according to textual descriptions. We utilise attention models such as, word attention to draw different sub-regions conditioned on related words; squeeze-and-excitation attention to capture non-linear interaction among channels; and local self-attention to model long-range dependencies. With spectral normalisation to stabilise training, our proposed CAGAN achieves state-of-the-art FID and comparative IS scores on the CUB dataset and on the more challenging COCO dataset. Furthermore, we demonstrate that judging a model by a single evaluation metric can be misleading by developing an additional model adding local self-attention which scores a higher IS than our other model, but generates unrealistic images through feature repetition.

References

1. Bai, S., An, S.: A survey on automatic image caption generation. Neurocomputing **311**, 291–304 (2018)
2. Barratt, S.T., Sharma, R.: A note on the inception score. CoRR abs/1801.01973 (2018)
3. Cai, Y., et al.: Dualattn-GAN: text to image synthesis with dual attentional generative adversarial network. IEEE Access **7**, 183706–183716 (2019)
4. Cheng, J., Wu, F., Tian, Y., Wang, L., Tao, D.: RiFeGAN: rich feature generation for text-to-image synthesis from prior knowledge. In: CVPR, pp. 10908–10917 (2020)
5. Cheng, Q., Gu, X.: Hybrid attention driven text-to-image synthesis via generative adversarial networks. In: Tetko, I.V., Kůrková, V., Karpov, P., Theis, F. (eds.) ICANN 2019. LNCS, vol. 11731, pp. 483–495. Springer, Cham (2019). https://doi.org/10.1007/978-3-030-30493-5_47
6. Dorta, G., Vicente, S., Agapito, L., Campbell, N.D.F., Prince, S., Simpson, I.: Laplacian pyramid of conditional variational autoencoders. In: CVMP, pp. 7:1–7:9 (2017)

7. Fang, H., et al.: From captions to visual concepts and back. In: CVPR, pp. 1473–1482 (2015)
8. Goodfellow, I.J., et al.: Generative adversarial nets. In: NIPS, pp. 2672–2680 (2014)
9. Gupta, T., Schwenk, D., Farhadi, A., Hoiem, D., Kembhavi, A.: Imagine this! Scripts to compositions to videos. In: Ferrari, V., Hebert, M., Sminchisescu, C., Weiss, Y. (eds.) ECCV 2018. LNCS, vol. 11212, pp. 610–626. Springer, Cham (2018). https://doi.org/10.1007/978-3-030-01237-3_37
10. He, K., Zhang, X., Ren, S., Sun, J.: Deep residual learning for image recognition. In: CVPR, pp. 770–778 (2016)
11. Heusel, M., Ramsauer, H., Unterthiner, T., Nessler, B., Hochreiter, S.: GANs trained by a two time-scale update rule converge to a local nash equilibrium. In: NIPS, pp. 6626–6637 (2017)
12. Hinz, T., Heinrich, S., Wermter, S.: Semantic object accuracy for generative text-to-image synthesis. CoRR abs/1910.13321 (2019)
13. Hu, J., Shen, L., Sun, G.: Squeeze-and-excitation networks. In: CVPR, pp. 7132–7141 (2018)
14. Johnson, J., Gupta, A., Fei-Fei, L.: Image generation from scene graphs. In: CVPR, pp. 1219–1228 (2018)
15. Kingma, D.P., Ba, J.: Adam: a method for stochastic optimization. In: ICLR (Poster) (2015)
16. Kulkarni, T.D., Kohli, P., Tenenbaum, J.B., Mansinghka, V.K.: Picture: a probabilistic programming language for scene perception. In: CVPR, pp. 4390–4399 (2015)
17. Kulkarni, T.D., Whitney, W.F., Kohli, P., Tenenbaum, J.B.: Deep convolutional inverse graphics network. In: NIPS, pp. 2539–2547 (2015)
18. Li, B., Qi, X., Lukasiewicz, T., Torr, P.H.S.: Controllable text-to-image generation. In: NIPS, pp. 2063–2073 (2019)
19. Li, W., et al.: Object-driven text-to-image synthesis via adversarial training. In: CVPR, pp. 12174–12182 (2019)
20. Li, Z., Wu, M., Zheng, J., Yu, H.: Perceptual adversarial networks with a feature pyramid for image translation. IEEE CG&A **39**(4), 68–77 (2019)
21. Liang, J., Pei, W., Lu, F.: CPGAN: full-spectrum content-parsing generative adversarial networks for text-to-image synthesis. CoRR abs/1912.08562 (2019)
22. Lin, T.-Y., et al.: Microsoft COCO: common objects in context. In: Fleet, D., Pajdla, T., Schiele, B., Tuytelaars, T. (eds.) ECCV 2014. LNCS, vol. 8693, pp. 740–755. Springer, Cham (2014). https://doi.org/10.1007/978-3-319-10602-1_48
23. Lucic, M., Kurach, K., Michalski, M., Gelly, S., Bousquet, O.: Are GANs created equal? A large-scale study. In: NIPS, pp. 698–707 (2018)
24. Miyato, T., Kataoka, T., Koyama, M., Yoshida, Y.: Spectral normalization for generative adversarial networks. In: ICLR, Conference Track Proceedings. OpenReview.net (2018)
25. Odena, A., Olah, C., Shlens, J.: Conditional image synthesis with auxiliary classifier GANs. In: Proceedings of Machine Learning Research, ICML, vol. 70, pp. 2642–2651. PMLR (2017)
26. van den Oord, A., Kalchbrenner, N., Espeholt, L., Kavukcuoglu, K., Vinyals, O., Graves, A.: Conditional image generation with PixelCNN decoders. In: NIPS, pp. 4790–4798 (2016)
27. Parmar, N., Ramachandran, P., Vaswani, A., Bello, I., Levskaya, A., Shlens, J.: Stand-alone self-attention in vision models. In: NIPS, pp. 68–80 (2019)
28. Pharr, M., Jakob, W., Humphreys, G.: Physically Based Rendering: From Theory to Implementation. Morgan Kaufmann, Burlington (2016)

29. Qiao, T., Zhang, J., Xu, D., Tao, D.: Learn, imagine and create: text-to-image generation from prior knowledge. In: NIPS, pp. 885–895 (2019)
30. Qiao, T., Zhang, J., Xu, D., Tao, D.: MirrorGAN: learning text-to-image generation by redescription. In: CVPR, pp. 1505–1514 (2019)
31. Reed, S.E., Akata, Z., Mohan, S., Tenka, S., Schiele, B., Lee, H.: Learning what and where to draw. In: NIPS, pp. 217–225 (2016)
32. Reed, S.E., Akata, Z., Yan, X., Logeswaran, L., Schiele, B., Lee, H.: Generative adversarial text to image synthesis. In: JMLR Workshop and Conference Proceedings, ICML, vol. 48, pp. 1060–1069 (2016)
33. Reed, S.E., et al.: Parallel multiscale autoregressive density estimation. In: Proceedings of Machine Learning Research, ICML, vol. 70, pp. 2912–2921 (2017)
34. Rosca, M., Lakshminarayanan, B., Warde-Farley, D., Mohamed, S.: Variational approaches for auto-encoding generative adversarial networks. CoRR abs/1706.04987 (2017)
35. Salimans, T., Goodfellow, I.J., Zaremba, W., Cheung, V., Radford, A., Chen, X.: Improved techniques for training GANs. In: NIPS, pp. 2226–2234 (2016)
36. Snell, J., Ridgeway, K., Liao, R., Roads, B.D., Mozer, M.C., Zemel, R.S.: Learning to generate images with perceptual similarity metrics. In: ICIP, pp. 4277–4281 (2017)
37. Szegedy, C., Vanhoucke, V., Ioffe, S., Shlens, J., Wojna, Z.: Rethinking the inception architecture for computer vision. In: CVPR, pp. 2818–2826 (2016)
38. Tan, H., Liu, X., Li, X., Zhang, Y., Yin, B.: Semantics-enhanced adversarial nets for text-to-image synthesis. In: ICCV, pp. 10500–10509 (2019)
39. Tao, M., Tang, H., Wu, S., Sebe, N., Wu, F., Jing, X.: DF-GAN: deep fusion generative adversarial networks for text-to-image synthesis. CoRR abs/2008.05865 (2020)
40. Theis, L., Bethge, M.: Generative image modeling using spatial LSTMs. In: NIPS, pp. 1927–1935 (2015)
41. Wah, C., Branson, S., Welinder, P., Perona, P., Belongie, S.: The Caltech-UCSD birds-200-2011 dataset. Technical report CNS-TR-2011-001 (2011)
42. Wu, J., Tenenbaum, J.B., Kohli, P.: Neural scene de-rendering. In: CVPR (2017)
43. Xie, J., Lu, Y., Zhu, S., Wu, Y.N.: A theory of generative convnet. In: JMLR Workshop and Conference Proceedings, ICML, vol. 48, pp. 2635–2644 (2016)
44. Xu, K., et al.: Show, attend and tell: neural image caption generation with visual attention. In: JMLR Workshop and Conference Proceedings, ICML, vol. 37, pp. 2048–2057 (2015)
45. Xu, T., et al.: AttnGAN: fine-grained text to image generation with attentional generative adversarial networks. In: CVPR, pp. 1316–1324 (2018)
46. Yin, G., Liu, B., Sheng, L., Yu, N., Wang, X., Shao, J.: Semantics disentangling for text-to-image generation. In: CVPR, pp. 2327–2336 (2019)
47. Zhang, H., Xu, T., Li, H.: StackGAN: text to photo-realistic image synthesis with stacked generative adversarial networks. In: ICCV, pp. 5908–5916 (2017)
48. Zhang, H., et al.: StackGAN++: realistic image synthesis with stacked generative adversarial networks. CoRR abs/1710.10916 (2017)
49. Zhu, M., Pan, P., Chen, W., Yang, Y.: DM-GAN: dynamic memory generative adversarial networks for text-to-image synthesis. In: CVPR, pp. 5802–5810 (2019)

TxT: Crossmodal End-to-End Learning with Transformers

Jan-Martin O. Steitz[1(✉)], Jonas Pfeiffer[1], Iryna Gurevych[1,2], and Stefan Roth[1,2]

[1] Department of Computer Science, TU Darmstadt, Darmstadt, Germany
jan-martin.steitz@visinf.tu-darmstadt.de
[2] hessian.AI, Darmstadt, Germany

Abstract. Reasoning over multiple modalities, *e.g.* in Visual Question Answering (VQA), requires an alignment of semantic concepts across domains. Despite the widespread success of end-to-end learning, today's multimodal pipelines by and large leverage pre-extracted, fixed features from object detectors, typically Faster R-CNN, as representations of the visual world. The obvious downside is that the visual representation is not specifically tuned to the multimodal task at hand. At the same time, while transformer-based object detectors have gained popularity, they have not been employed in today's multimodal pipelines. We address both shortcomings with *TxT*, a transformer-based crossmodal pipeline that enables fine-tuning both language and visual components on the downstream task in a fully end-to-end manner. We overcome existing limitations of transformer-based detectors for multimodal reasoning regarding the integration of global context and their scalability. Our transformer-based multimodal model achieves considerable gains from end-to-end learning for multimodal question answering.

Keywords: Vision and language · Transformers · End-to-end learning

1 Introduction

Vision and language tasks, such as Visual Question Answering (VQA) [2,11,44], are inherently multimodal and require aligning visual perception and textual semantics. Many recent language models are based on BERT [6], which is based on transformers [45] and pre-trained solely on text. The most common strategy to turn such a language model into a multimodal one is to modify it to reason additionally over image embeddings, which are typically obtained through some bottom-up attention mechanism [1] and pre-extracted. The consequence is that the visual features cannot be adjusted to the necessities of the multimodal reasoning task during training. Jiang *et al.* [17] recently showed that fine-tuning the visual representation together with the language model can benefit the accuracy in a VQA task. Pre-trained image features are

Supplementary Information The online version contains supplementary material available at https://doi.org/10.1007/978-3-030-92659-5_26.

C. Bauckhage et al. (Eds.): DAGM GCPR 2021, LNCS 13024, pp. 405–420, 2021.
https://doi.org/10.1007/978-3-030-92659-5_26

thus not optimal for the multimodal setting, even when pre-training with semantically diverse datasets, e.g. Visual Genome [20]. In this work, we follow [17] to ask how the visual representation can be trained end-to-end within a multimodal reasoning pipeline. In contrast, however, we focus on recent transformer-based architectures [5,6] with their specific requirements.

Multimodal pipelines typically [44] rely on visual features from the popular Faster R-CNN object detector [37]. As Faster R-CNN makes heavy use of subsampling to balance positive and negative examples during training and needs non-maximum suppression to select good predictions, it is not easily amenable to end-to-end training as desired here. Therefore, [17] remodels Faster R-CNN into a CNN that produces dense feature maps, which are combined with an LSTM-based model [48] and trained end-to-end. It is difficult though to combine such dense features with the latest transformer-based language models, *e.g.* [5,22,29,43], as the large number of image features causes scalability issues in the underlying pairwise attention mechanism. We thus follow a different route, and instead propose to employ an alternate family of object detectors, specifically the recent Detection Transformers (DETR) [4] and variants, which treat the detection task as a set-prediction problem. DETR produces only a small set of object detections that do not need any treatment with non-maximum suppression as its transformer decoder allows for the interaction between the detections. We are thus able to employ the full object detector and only need to reason over a comparatively small set of image features in the BERT model. However, a number of technical hurdles have to be overcome.

Specifically, we make the following contributions: *(i)* We introduce alternative region features for image embeddings produced by DETR [4] and Deformable-DETR [49], two modern, transformer-based approaches to object detection. Out of the box and despite their competitive accuracy for pure object detection, these transformer-based detectors deteriorate the VQA accuracy compared to the common Faster R-CNN bottom-up attention mechanism [1]. *(ii)* We show that this accuracy loss stems from the global context of the detected objects not being sufficiently represented in the bottom-up features. We mitigate this effect through additional global context features, which bring the VQA accuracy much closer to the baseline. While transformer-based detectors, such as DETR, are powerful, they are also very computationally heavy to train; faster to train alternatives, *i.e.* Deformable-DETR [49], are less efficient at test time, which is also undesirable. *(iii)* We address this using a more scalable variant of Deformable-DETR that still leverages multi-scale information by querying the multi-scale deformable attention module with only one selected feature map instead of a full multi-scale self-attention. This allows retaining the training efficiency of [49], yet is comparably fast at test time as DETR. *(iv)* Our analysis shows that Deformable-DETR, while being a stronger object detector than DETR, also modifies the empirical distribution of the number of detected objects per image. We find that this negatively impacts the VQA accuracy and trace this effect to the use of the focal loss [24] within Deformable-DETR. *(v)* Our final transformer-based detection pipeline enables competitive VQA accuracy when combined with transformer-based language models despite having a shorter CNN backbone compared to [37]. More importantly, the full model including the visual features can be trained end-to-end, leading to accuracy gains of 2.3% over pre-extracted DETR features and 1.1% over Faster R-CNN features on the VQAv2 dataset [11].

2 Related Work

With the advancement of transfer learning, the current modus operandi for multimodal learning is to combine pre-trained models of the respective modality by learning a shared multimodal space. We thus review approaches for single modalities before turning to crossmodal reasoning.

2.1 Learning to Analyze Single Modalities

Language Understanding. Recently, transfer learning has dominated the domain of natural language processing, where pre-trained models achieve state-of-the-art results on the majority of language tasks. These models are predominantly trained with self-supervised objectives on large unlabeled text corpora, and are subsequently fine-tuned on a downstream task [14,31]. Most recent language models leverage the omnipresent transformer architectures [45] and are trained predominantly with Masked-Language-Modelling (MLM) objectives as *encoders* [6,27], with next-word prediction objectives as *generative/decoder* models [21,35], or as *sequence-to-sequence* models [3,33,34].

Visual Scene Analysis. Convolutional neural networks (*e.g.*, ResNet [12]) are widely used to encode raw image data. While originating in image classification tasks on datasets like ImageNet [38], they are much more broadly deployed with transfer learning as backbone for other tasks like object detection [10] or semantic segmentation [41].

The dominant methods for object detection are Faster R-CNN [37] and variants like Feature Pyramid Networks (FPN) [23] due to their high accuracy [15]. Faster R-CNN has a two stage architecture with a shared backbone: First, a region proposal network suggests regions by a class-agnostic distinction between foreground and background. After non-maximum suppression, the highest scoring regions are fed to the region-of-interest (RoI) pooling layer. RoI pooling extracts fixed-sized patches from the feature map of the shared backbone, spanning the respective regions. These are sent to the second stage for classification and bounding box regression. A final non-maximum suppression step is necessary to filter out overlapping predictions. Single-stage object detectors, *e.g.* SSD [26] and YOLO [36], offer an alternative, but still need sampling of positive and negative examples during training, hand-crafted parts for *e.g.* anchor generation, and a final non-maximum suppression step. More recently, Detection Transformers (DETR) [4] were proposed with a transformer-based architecture, which is conceptually much simpler and can be trained without any hand-crafted components. Deformable-DETR [49] introduced a multi-scale deformable attention module to address DETR's long training time and its low accuracy for small objects.

2.2 Learning Multimodal Tasks

Overview. Most recent work on crossmodal learning relies on combining pre-trained single-modality models to learn a shared multimodal space. To do so, both image and text representations are passed into a transformer model [5,22,29,43], where a multi-head attention mechanism reasons over the representations of both modalities. The transformer model is initialized with the weights of a pre-trained language model. While word-embeddings represent the text input, raw images are passed through pre-trained vision models that generate encoded representations, which are passed into the transformer. While ResNet encodings of the entire image can be leveraged [19], it has been shown that utilizing object detection models (*i.e.* Faster R-CNN [37]), which provide encoded representations of multiple regions of interest [1], benefits the downstream task [29,43, *inter alia*]. Here, the image features are passed through an affine-transformation layer, which learns to align the visual features with the pre-trained transformer. Similarly, the pixel offsets[1] are used to generate positional embeddings. By combining these two representations, each object region is passed into the transformer separately.[2]

Datasets. To learn a shared multimodal representation space, image captioning datasets such as COCO [25], Flicker30k [32], Conceptual Captions (CC) [40], and SBU [30]) are commonly utilized. The self-supervised objectives are largely the same among all approaches: next to MLM on the text part, masked feature regression, masked object detection, masked attribute detection, and cross-modality matching.

Transformer Approaches. Recent multimodal models initialize the transformer parameters with BERT [6] weights and leverage the Faster R-CNN object detection model: LXMERT [43] and ViLBERT [29] propose a dual-stream architecture, which provides designated language and vision transformer weights. A joint multi-head attention component attends over both modalities at every layer. UNITER [5] and Oscar [22] propose a single-stream architecture, which shares all transformer weights among both modalities. Oscar additionally provides detected objects as input to the transformer, and argues that this allows for better multimodal grounding. VILLA [8] proposes to augment and perturb the embedding space for improved pre-training.

End-to-End Training. Above approaches combine pre-trained vision and language models, however, they do not back-propagate into the vision component. Thus, no capacity is given to the vision model to reason over the raw image data in terms of the downstream task; the assumption is that the pre-encoded representations are sufficient for the downstream crossmodal task. Kamath *et al.* [18] avoid this by incorporating multimodality into a transformer-based object detector. It is end-to-end trainable but needs computationally heavy pre-training. Jiang *et al.* [17] address this by proposing to extract the Faster R-CNN weights from [1] into a CNN. They are able to leverage multimodal end-to-end training, but use a model based on long short-term memory (LSTM) [13]. The sequential processing of LSTM models hinders parallelization and thus limits the sequence length. Pixel-BERT [16] proposes to embed the image with a CNN and reasons over all resulting features with a multimodal transformer. As such their model is end-to-end trainable, but very heavy on computational resources because of the pairwise attention over these dense features. For VL-BERT [42], a Faster R-CNN is used

[1] Depending on the model, this includes relative position, width, and height of the original image.

[2] Different approaches for combining pixel offsets with object features have been proposed, such as adding the representation (*e.g.*, used by LXMERT [43]) or transforming them jointly (*e.g.*, used by Oscar [22]).

to pre-extract object bounding boxes. The region proposal network of Faster R-CNN is then excluded in the further training process and the language model is jointly trained with the object detector in a Fast R-CNN [9] setting. This separates region proposals from object classification and localization for test time, hence multi-stage inference is necessary and no full end-to-end training is possible.

To mitigate this, we propose TxT, a transformer-based architecture that combines transformer-based object detectors with transformer-based language models and can be trained end-to-end for the crossmodal reasoning task at hand.

3 Transformers for Images and Text

Transformers [45] not only quickly developed into a standard architecture for language modeling [6], but their benefits have recently been demonstrated also in visual scene analysis [4,7,49]. We aim to bring these two streams of research together, which aside from conceptual advantages will allow for end-to-end learning of multimodal reasoning.

3.1 Transformer-Based Object Detection

DETR. The Detection Transformer (DETR) [4] is a recent transformer-based object detector, which treats object detection as a set-prediction problem and, therefore, obviates the need for non-maximum suppression. It outperforms the very widely used Faster R-CNN [37] for object detection on standard datasets like COCO [25]. DETR first extracts a feature map from the input image using a standard CNN backbone. The features are then projected to a lower dimensionality through 1×1 convolutions. A subsequent transformer encoder uses multi-head self-attention layers and pointwise feed-forward networks to encode the features. These are then passed to the decoder, where a fixed set of learned object queries is used to embed object information by cross-attention with the encoder output. Finally, feed-forward networks predict classes and bounding boxes. The predictions are connected by bi-partite matching with the ground truth and a set-based loss is used for training. As such, DETR is completely end-to-end trainable without hand-crafted parts.

Deformable-DETR. Deformable-DETR [49] addresses two important drawbacks of the original DETR architecture: The long required training time (500 epochs are needed for full training) and its relatively low detection accuracy for small objects. Deformable-DETR replaces the standard multi-head attention layers with multi-scale deformable attention modules in its transformer layers. Instead of a pairwise interaction between all queries and keys, they only attend to a small set of sampling points around each reference point, interpolated at learned position offsets. To incorporate multi-scale information, the attention modules work on a set of feature maps of different scales and each query feature attends across all scales. Additionally, in Deformable-DETR the focal loss [24] is used for matching between predictions and ground truth as well as scoring the classification. Deformable-DETR is not only able to reduce the total training time to only one 10th of that of DETR, but even outperforms it for object detection on standard datasets [25].

Table 1. *Comparison of object detector architectures* on the COCO validation set [25] in terms of average precision (AP, higher is better) using the COCO detection evaluation metrics. The results for DETR, DETR-DC5, and Faster R-CNN are quoted from [4] and the multi-scale Deformable-DETR evaluation is taken from [49]. The (+) in Faster R-CNN indicates that the models were trained with an additional GIoU loss. The GFLOPs are measured over the first 100 images of the COCO validation set. Deviating from [4,49], which estimate the number of floating point operations and omit part of the model, we log all CUDA calls with NVIDIA Nsight Compute, therefore recording higher – but more reliable – numbers.

Model	GFLOPs	#params	Epochs	AP	AP_{50}	AP_{75}	AP_S	AP_M	AP_L
Faster R-CNN-FPN+	253	41M	109	42.0	62.1	45.5	26.6	45.4	53.4
Faster R-CNN-R101-FPN+	360	60M	109	44.0	63.9	47.8	27.2	48.1	56.0
DETR	161	41M	500	42.0	62.4	44.2	20.5	45.8	61.1
DETR-DC5	490	41M	500	43.3	63.1	45.9	22.5	47.3	61.1
Deformable-DETR *query stride 16*	177	40M	50	42.9	62.9	45.9	24.2	46.1	60.4
Deformable-DETR *multi-scale*	337	41M	50	45.4	64.7	49.0	26.8	48.3	61.7

3.2 Global Context for DETR Models

Before we can bring together transformer architectures for object detection and language modeling in an end-to-end fashion, we first need to assess whether transformer-based detectors yield a competitive baseline. To that end, we use pre-extracted features from pre-trained detectors in visual question answering.

Surprisingly, we find that replacing Faster R-CNN features by those from DETR leads to a noticeable 2.5% accuracy loss in terms of VQAv2 [11], even if DETR is a stronger detector on standard object detection benchmarks [25]. Looking at this more closely, we find that in contrast to Faster R-CNN, DETR and Deformable-DETR encode the semantic information in a single feature vector for each query in the transformer-based decoder. We posit that their transformer encoder-decoder structure leads to learning more object-specific information in its feature vectors and global scene context not being captured. As this global scene context is arguably important for VQA and other crossmodal tasks, we compensate for the loss in global context by *augmenting each output feature vector with context information from global average pooling*. We analyze different positions for extracting this contextual information and find a *substantial gain of 1.3% accuracy* for VQAv2 from the global context. Full details are described in Sect. 4.4. While not completely closing the accuracy gap to Faster R-CNN features, we note that these are more high-dimensional and not amenable to end-to-end crossmodal training. Also note that adding the same contextual information to Faster R-CNN features, in contrast, does not aid VQA accuracy. Consequently, we conclude that *global contextual information about the scene is important for using transformer-based detectors for VQA*.

3.3 Scalable Deformable-DETR

Another significant impediment in the deployment of transformer-based architectures for multimodal tasks are the very significant training times of DETR [4]. This has been

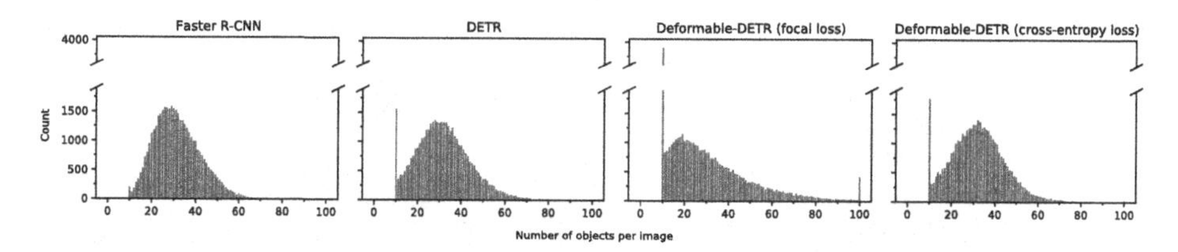

Fig. 1. *Number of extracted objects per image for COCO validation.* Only predictions with a confidence exceeding a threshold are extracted, but a minimum of 10 and a maximum of 100 objects are selected per image. Thresholds are chosen such that the total number of extracted regions is the same for all methods. Deformable-DETR with focal loss produces a skewed distribution while using a cross-entropy loss resembles the detection distribution of Faster R-CNN and DETR.

very recently addressed with Deformable-DETR [49], which is much faster to train, but is slower at test time, requiring more than double the computations compared to DETR. We address this through a *more scalable Deformable-DETR* method, which can still use the multi-scale information but is computationally more lightweight. Deformable-DETR extracts the features maps of the last three ResNet blocks with strides of 8, 16, and 32. It also learns an additional convolution with stride 2 to produce a fourth feature map with stride 64. Those four feature maps are then fed to the multi-scale deformable attention layers of the encoder. In the deformable self-attention, each feature then attends to a fixed number of interpolated features of each of the feature maps, incorporating multi-scale information in the process.

Instead of performing a full self-attention, where query and keys are the same set, we use the feature maps of all scales as keys but propose to query only with the feature map of one chosen scale. This results in *a single feature map that still incorporates multi-scale information*, because the query features attend to the keys across all scales. All further attention layers then work on this single feature map, reducing the amount of computational operations needed. By choosing feature maps of different strides as the query, we can scale the computational complexity and accuracy corresponding to our task. See Appendix A in the supplemental material for detailed results.

For a good trade-off between computational cost and AP to be used for our cross-modal architecture, we query with feature maps of stride 16. In Table 1 we compare our model to Faster R-CNN with ResNet-50 and ResNet-101 backbones [12], DETR, DETR-DC5 (with dilated convolutions in the last ResNet block), and multi-scale Deformable-DETR. Compared to the standard DETR model, our proposed scalable Deformable-DETR maintains a comparable number of model parameters and reaches a *0.9% higher AP* with only 10% more computational expense while still being trained in only 50 instead of 500 epochs. It is also faster at test time than Faster R-CNN, while being more accurate when compared with the same backbone. Our approach thus provides a favorable trade-off as a detector basis.

3.4 Detection Losses for Crossmodal Tasks

Deformable-DETR not only aims to address the lacking training efficiency of DETR, but also its limitations for hard-to-detect objects, *e.g.* small objects or objects with rare

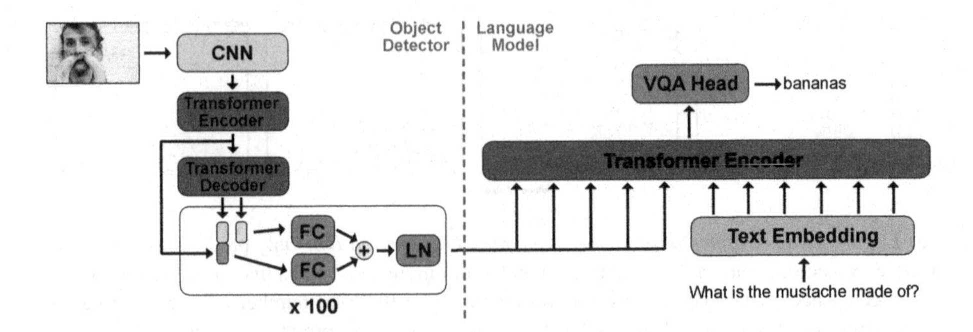

Fig. 2. *Overview of our proposed TxT model:* A transformer-based object detector [4,49] produces feature representations and bounding box coordinates for its predictions *(left)*. Following [5], each feature and position pair is combined by fully connected (FC) layers. The aggregated representations are passed through a layer norm (LN) and utilized as object embeddings; concatenated with the sequence of text embeddings, the multimodal instance is passed into the transformer-based language model *(right)*. A task specific head (here for VQA) predicts the answer. The model is end-to-end trainable, including both the vision and language components.

classes. To that end, a focal loss [24] is employed. To assess if this impacts its use in downstream tasks, such as VQA, we analyze both the detection output as well as its use as pre-extracted feature for VQA. To compare the suitability of the features extracted by different object detectors, we only extract the features of objects with a confidence above a certain threshold. A minimum of 10 objects and a maximum of 100 objects per image are extracted following Anderson *et al.* [1]. The thresholds of DETR and Deformable-DETR are chosen such that all models produce the same total number of objects on the COCO validation set. When we plot the distribution of the number of objects per image across the validation dataset, shown in Fig. 1, we see that DETR, which is also trained with a cross-entropy loss, produces a distribution with a peak at a similar position as Faster R-CNN, although a little broader. Deformable-DETR trained with a focal loss to improve on difficult objects, on the other hand, has a distribution shifted to a lower number of regions and notably more spread out and skewed. Since small or rarely occuring objects (currently) do not play a significant role in VQA,[3] this property may not necessarily benefit the crossmodal task. Indeed, as shown in Table 2, the *focal loss degrades the VQAv2 performance*. Since we are less interested in pure detection accuracy, but rather in crossmodal performance, we therefore train our Deformable-DETR with a *cross-entropy loss for the class labels*. As can be seen in Fig. 1 *(right)*, this makes the detection statistics more closely resemble those of Faster R-CNN and DETR, and leads to a *clear increase of 0.6% on VQAv2 test-dev* (*cf*. Table 2).

[3] Verified by measuring object sizes via gradient backtracing in UNITER [5] using the Grad-CAM [39] approach and selecting image regions with normalized activations greater than 0.9.

3.5 The TxT Model

With the proposed *TxT model*, we now combine transformer-based object detection (T) and transformer-based language models (T) in a crossmodal approach (x). It enables us to bridge the information gap between language and vision models and perform crossmodal end-to-end training while avoiding computationally heavy dense features [16]. We base the language model on BERT-base, pre-trained with MLM on text corpora only [6]. For the object detector, we employ either DETR or Deformable-DETR models as described in Sects. 3.1 to 3.4. The structure of TxT is shown in Fig. 2. In the object detector a CNN backbone (ResNet-50 [12]) computes feature representations from the input image. They are passed to the transformer encoder, where they interact in its multi-head attention layers. In the decoder, the encoded feature map gets queried for objects and for each object the feature representation and the bounding box coordinates are extracted [4] (see Sect. 3.3 for details). The global average-pooled encoder output is concatenated to the feature representation of each object to provide global context (*cf*. Sect. 3.2). We use the complete set of predictions generated by the object detector for TxT. Following [5], features and bounding box coordinates are projected by fully connected layers to the dimension of the BERT model and combined. After passing a normalization layer, they are used as object embeddings in the language model.

The text input is run through a tokenizer and both embeddings, visual and textual, are concatenated to a sequence so that the self-attention of the transformer encoder is able to reason over both modalities. The encoder output is passed through a task-specific head, in our case a VQA head, consisting of a small multi-layer perceptron (MLP) to produce the answer predictions, again following [5]. The TxT model is trained fully end-to-end with a binary cross-entropy loss across the border between the vision and the language part. Thus, we are able to fine-tune the object detector specifically for the crossmodal task.

4 Experiments

To evaluate our TxT model, we perform experiments on a visual questioning answering task. Specifically, we use the VQAv2 training set [11] and add additional question and answer pairs from Visual Genome [20] for augmentation, following [1,5], but omit using the validation split for training. We follow standard practice and only classify for the 3129 most frequent answers [1,5,47]. All results are evaluated with the standard VQA accuracy metrics [2] on the test-dev set from the 2021 challenge for VQAv2 [11].

4.1 Technical Details

Object Detector. The transformer-based object detectors in our TxT model employ a ResNet-50 [12] backbone. The extracted features are projected to a hidden dimension of 256 of the transformer layers and then passed to an encoder consisting of 6 transformer layers. The transformer layers consist of multi-head attention layers with 8 heads and feed-forward networks with a dimension of 2048 for DETR and 1024 for Deformable-DETR, as described in [45]. Learned object queries generate 100 object features with an alternation of self-attention and cross-attention in the 6 transformer layers of the decoder. A linear projection is used to classify the object and a 3-layer perceptron is used to generate bounding box coordinates [4]. Deformable-DETR uses 300 queries in its default configuration. For end-to-end learning in TxT, we use a variant with 100 object queries. Deformable-DETR uses multi-scale deformable attention in its transformer layers, which employs 4 keys interpolated around each reference point of the query. We use Deformable-DETR with iterative bounding box refinement [49]. Both DETR and Deformable-DETR use global context as discussed in Sect. 3.2.

Pre-training. Following Anderson *et al.* [1], we pre-train all object detectors on the Visual Genome dataset [20] with a maximum input image size of 600×1000 and add an additional head for the prediction of object attributes. We train DETR for 300 epochs on 8 NVIDIA V100 GPUs with 4 images per GPU. We use a learning rate of 1e−4 for DETR and a learning rate of 1e−5 for the ResNet backbone. After 200 epochs, we drop the learning rate by a factor of 10. For all other settings, we follow [4]. The training protocol for Deformable-DETR follows [49].

Language Model. The language model of TxT is based on BERT-base [6] with 12 transformer layers and 768 hidden dimensions. BERT is pre-trained on the combination of the BooksCorpus [50] and the entire English Wikipedia corpus, consisting of 3.3 billion words in total. A WordPiece tokenizer [46] is trained on the entire corpus, amounting to a vocabulary size of 30k, with each token corresponding to a position in the designated word-embedding matrix. *Positional* embeddings are trained to preserve the sentence order, and *segment* embeddings are trained to indicate if a token belongs to sentence A or B. The embedding representation, which is passed into the transformer, thus amounts to the sum over *word-*, *positional-*, and *segment*-embedding of each respective token. BERT is pre-trained with an MLM,[4] as well as a next-sentence prediction objective.[5]

VQA Training. The TxT model is trained for 10k iterations on 4 GPUs and a learning rate of 8e−5. We employ a binary cross-entropy loss and the AdamW [28] optimizer. We use a linear warmup of 600 iterations and a linear decay of the learning rate afterwards. In case of pre-extracted, fixed object features, we train the TxT model similar to [5] with 5120 input tokens and for each iteration we accumulate the gradients over 5 steps. For end-to-end training, we pre-train the TxT model with fixed features for 4k iterations for DETR and 6k iterations for Deformable-DETR with the above settings. Then, we train the model fully end-to-end for 10k iterations with 8 images and questions per batch and accumulate the gradients for 32 steps for each iteration to maintain the effective batch size of the fixed-feature setting. When we employ TxT with DETR as the object detector, we initialize the learning rate for the DETR and CNN backbone components with 3e−4 and 3e−5, respectively. Deformable-DETR is trained with a learning

[4] Input tokens are randomly masked, with the objective of the model being to predict the missing token given the surrounding context.

[5] Two sentences A and B are passed into the model, where 50% of the time B follows A in the corpus, and 50% of the time B is randomly sampled. The objective is to predict whether or not B is a negative sample.

Table 2. *Comparison of VQA accuracy on pre-extracted features from different object detectors* evaluated on VQAv2 test-dev [11] (higher is better). Deformable-DETR refers to our variant, where we query the attention module with a feature map of stride 16 (see Sect. 3.3 for details). The best results in each column are bold, the 2nd best underlined.

Method	number	yes/no	other	overall
Faster R-CNN	**51.18**	**85.04**	**59.00**	**68.84**
DETR	<u>49.49</u>	83.88	<u>57.81</u>	67.60
Deformable-DETR (focal loss)	47.59	83.87	57.18	67.09
Deformable-DETR (cross-entropy loss)	48.66	<u>84.46</u>	57.62	<u>67.66</u>

rate of 6e−4 and 6e−5 for the backbone. When TxT is used with DETR and thresholding the predictions, we apply a factor of 10 to the learning rate of the class-prediction layer of DETR. For end-to-end training of TxT, the learning rates and training schedule were determined with a coarse grid search over the hyperparameters.

4.2 Representational Power of Visual Features

In a first step, we investigate if DETR and Deformable-DETR are able to produce object feature descriptors whose representational power for VQA tasks is comparable to the widely used Faster R-CNN; we assess this through the VQA accuracy. We pre-extract object features and bounding box coordinates with all object detectors. Following standard practice (*e.g.* [1,5,42]), all objects with a class confidence above a certain threshold are extracted, but a minimum of 10 and a maximum of 100 objects are used per image. We calibrate the confidence threshold of DETR and Deformable-DETR on COCO validation so that the total number of extracted object features is 1.3 million, comparable to Faster R-CNN. The TxT model is then trained with the pre-extracted features. When used with Faster R-CNN features, this is equivalent to the UNITER-base model [5] with only MLM pre-training, *i.e.* initalized with BERT weights. We note that the numbers differ slightly from those in [5], because we omit the validation set when training for test-dev submission and use the most recent VQA challenge 2021.

The results in Table 2 show that DETR and Deformable-DETR are able to produce object features with a representational power that *competitively supports VQA tasks*, especially considering their low dimensionality of 256 (compared to 2048 dimensions of Faster R-CNN features). Also, these results are achieved with only a ResNet-50 backbone (compared to ResNet-101 for Faster R-CNN). While DETR leads to a $\sim$1.2% loss in VQA accuracy compared to Faster R-CNN, it is more efficient at test time (see Table 1) and, more importantly, allows for end-to-end training (*cf*. Sect. 4.3).

Moreover, we find that Deformable-DETR (with query stride 16) as the object detector for VQA performs less well than DETR, about 0.5% worse in VQA accuracy. As analyzed in Sect. 3.4, this is traceable to the focal loss. We find that a cross-entropy loss is a more suitable pre-training objective for Deformable-DETR as object detector for the VQA task, leading to $\sim 0.6\%$ higher accuracy on VQA while maintaining the same total number of predictions. With these modifications, *Deformable-DETR performs comparatively with DETR, yet is still much faster to train.*

Table 3. *Results of the multimodal end-to-end trainable TxT model* on VQAv2 test-dev [11] (higher is better). *e2e* denotes end-to-end training. In the upper part of the table, we show the results of counterparts with pre-extracted features and Faster R-CNN features. The best results in each column are bold, the 2nd best underlined.

Method	e2e	number	yes/no	other	overall
Faster R-CNN		51.18	85.04	59.00	68.84
DETR		49.49	83.88	57.81	67.60
DETR (all predictions)		**52.33**	85.57	<u>59.84</u>	<u>69.59</u>
Deformable-DETR (100 queries)		47.19	84.05	56.92	66.99
Deformable-DETR (100 q., all predictions)		48.54	83.79	57.82	67.47
TxT-DETR (thresholded)	✓	50.59	85.66	59.23	69.14
TxT-DETR	✓	<u>51.73</u>	**86.33**	**60.04**	**69.93**
TxT-Deformable-DETR	✓	49.70	<u>85.76</u>	59.19	69.06

4.3 Multimodal End-to-End Training

As a preliminary experiment, referred to as TxT-DETR (thresholded), we only pass objects to the multimodal transformer layers where the class confidence exceeds a threshold of 0.5. To accomplish this, we generate a mask from thresholding the class confidence. We then multiply the object features with the mask, thus allowing the gradient to backpropagate to the class prediction layer of DETR. Only features and related bounding box coordinates according to the mask are then passed from the DETR part of TxT to its language model. At the beginning of training, DETR selects around 30 objects per image, equivalent to the setting of pre-extracted features. During training, the TxT model learns to select more objects, saturating at around 80 per image. This suggests that *more objects are beneficial for solving the VQA task*, which confirms an empirical observation of Jiang *et al.* [17], who showed that the accuracy starts to saturate between 100 and 200 regions. As Table 3 shows, including *all* predicted objects in the multimodal reasoning (DETR *all predictions*) without end-to-end training indeed leads to a ∼0.5% higher accuracy than TxT-DETR (thresholded), which leverages the end-to-end training.

For multimodal end-to-end training of our TxT model, we therefore employ all 100 object predictions, eliminating the need to threshold and enabling a gradient for object predictions that would be discarded otherwise. In order to alleviate the computational and memory cost of multimodal self-attention, we pre-train Deformable-DETR with only 100 object queries (instead of the standard 300), which reduces the VQA accuracy by ∼0.7% when evaluating in the pre-extracted features setup. To make the effect of the higher object number distinguishable from the end-to-end benefit, we also show results for pre-extracted features training with all object predictions in Table 3. With DETR as object detector, the gain from using all predictions is ∼2% and for Deformable-DETR it is less pronounced with a benefit of ∼0.5%.

Finally, we assess our full end-to-end trainable model. We find that *multimodal end-to-end training substantially improves the accuracy of TxT*: The TxT model with

Table 4. *Global context features* for DETR [4] and Faster R-CNN [37] from different locations in the network. The results on a VQA task are evaluated on VQAv2 [11] test-dev (in %, higher is better). We also report the total dimensionality d of the extracted features.

Model	d	number	yes/no	other	overall
DETR (no context)	256	47.65	82.67	56.61	66.33
DETR (backbone context)	2304	48.80	83.68	57.38	67.24
DETR (projected backbone context)	512	48.69	83.43	57.36	67.12
DETR (encoder context)	512	49.49	83.88	57.81	67.60
Faster R-CNN (no context)	2048	51.18	85.04	59.00	68.84
Faster R-CNN (backbone context)	3072	49.86	84.29	58.37	68.08

Deformable-DETR improves by $\sim$2.1% from 66.99% for pre-extracted Deformable-DETR features to 69.06% accuracy on VQAv2 test-dev. The TxT model in combination with DETR achieves an overall gain in accuracy of $\sim$2.3% from 67.60% accuracy for pre-extracted DETR features to 69.93% accuracy for multimodal end-to-end training, thus improving by $\sim$1.1% over the Faster R-CNN features. While our results are competitive with current models on an equal footing, they could be improved further with pre-training tasks like image-text matching and word-region alignment [5] at the expense of significant computational overhead.

4.4 Ablation Study

As discussed in Sect. 3.2, adding global context is helping DETR [4] to produce visual features that are better suited for VQA. We now investigate the suitable position from which to source this global information. We consider three positions at which to apply a global average pooling of the feature map in the DETR network: *(i)* pooling the feature map produced by the CNN backbone, *(ii)* after the backbone feature map is projected to the lower dimensions of the transformer layers, and *(iii)* pooling the encoder output. As we can see in Table 4, *all variants with added global context lead to a gain in accuracy*. We obtain the best results when using the *encoder output*, leading to a gain of $\sim$1.3% in terms of VQAv2 [11] accuracy over using no global context. Pooling the feature map produced by the CNN backbone gives a slightly lower gain of $\sim$0.9% and requires 2304-dimensional features compared to only 512 dimensions for the encoder output. Projecting the backbone feature map first to the transformer dimension also gives 512-dimensional features, but yields only a $\sim$0.8% gain.

To ensure a fair comparison, we also verify if Faster R-CNN [37] features similarly benefit from global context. To that end, we concatenate the Faster R-CNN features with global average-pooled features of its shared backbone. In contrast to DETR, this does not lead to better VQA results, rather to a slight degradation in accuracy of $\sim$0.7%. We conclude that global context is important for VQA, yet is sufficiently represented in the common Faster R-CNN features. Transformer-based detectors, in contrast, benefit significantly from the added context.

5 Conclusion

In this paper we proposed TxT, an end-to-end trainable transformer-based architecture for crossmodal reasoning. Starting from the observation that transformer-based architectures have not only been successfully used for language modeling, but more recently also for object detection, we investigated the use of Detection Transformers and variants as source of visual features for visual question answering. We found that global context information as well as an adaptation of the loss function are needed to yield rich visual evidence for crossmodal tasks. We also proposed a speed-up mechanism for multi-scale attention. Our final end-to-end trainable architecture yields a clear improvement in terms of VQA accuracy over standard pre-trained visual features and paves the way for a tighter integration of visual and textual reasoning with a unified network architecture.

Acknowledgement. This work has been funded by the LOEWE initiative (Hesse, Germany) within the emergenCITY center.

References

1. Anderson, P., et al.: Bottom-up and top-down attention for image captioning and visual question answering. In: CVPR, pp. 6077–6086 (2018)
2. Antol, S., et al.: VQA: visual question answering. In: ICCV, pp. 2425–2433 (2015)
3. Brown, T.B., et al.: Language models are few-shot learners. arXiv:2005.14165 [cs.CL] (2020)
4. Carion, N., Massa, F., Synnaeve, G., Usunier, N., Kirillov, A., Zagoruyko, S.: End-to-end object detection with transformers. In: Vedaldi, A., Bischof, H., Brox, T., Frahm, J.-M. (eds.) ECCV 2020. LNCS, vol. 12346, pp. 213–229. Springer, Cham (2020). https://doi.org/10.1007/978-3-030-58452-8_13
5. Chen, Y.-C., et al.: UNITER: UNiversal image-TExt representation learning. In: Vedaldi, A., Bischof, H., Brox, T., Frahm, J.-M. (eds.) ECCV 2020. LNCS, vol. 12375, pp. 104–120. Springer, Cham (2020). https://doi.org/10.1007/978-3-030-58577-8_7
6. Devlin, J., Chang, M., Lee, K., Toutanova, K.: BERT: pre-training of deep bidirectional transformers for language understanding. In: NAACL-HLT, pp. 4171–4186 (2019)
7. Dosovitskiy, A., et al.: An image is worth 16×16 words: transformers for image recognition at scale. In: ICLR (2021)
8. Gan, Z., Chen, Y., Li, L., Zhu, C., Cheng, Y., Liu, J.: Large-scale adversarial training for vision-and-language representation learning. In: NeurIPS, pp. 6616–6628 (2020)
9. Girshick, R.B.: Fast R-CNN. In: ICCV. pp. 1440–1448 (2015)
10. Girshick, R.B., Donahue, J., Darrell, T., Malik, J.: Region-based convolutional networks for accurate object detection and segmentation. IEEE Trans. Pattern Anal. Mach. Intell. **38**(1), 142–158 (2016)
11. Goyal, Y., Khot, T., Agrawal, A., Summers-Stay, D., Batra, D., Parikh, D.: Making the V in VQA matter: elevating the role of image understanding in visual question answering. Int. J. Comput. Vis. **127**(4), 398–414 (2019)
12. He, K., Zhang, X., Ren, S., Sun, J.: Deep residual learning for image recognition. In: CVPR, pp. 770–778 (2016)
13. Hochreiter, S., Schmidhuber, J.: Long short-term memory. Neural Comput. **9**(8), 1735–1780 (1997)

14. Howard, J., Ruder, S.: Universal language model fine-tuning for text classification. In: ACL, pp. 328–339 (2018)
15. Huang, J., et al.: Speed/accuracy trade-offs for modern convolutional object detectors. In: CVPR, pp. 3296–3297 (2017)
16. Huang, Z., Zeng, Z., Liu, B., Fu, D., Fu, J.: Pixel-BERT: aligning image pixels with text by deep multi-modal transformers. arXiv:2004.00849 [cv.CV] (2020)
17. Jiang, H., Misra, I., Rohrbach, M., Learned-Miller, E.G., Chen, X.: In defense of grid features for visual question answering. In: CVPR, pp. 10264–10273 (2020)
18. Kamath, A., Singh, M., LeCun, Y., Misra, I., Synnaeve, G., Carion, N.: MDETR - modulated detection for end-to-end multi-modal understanding. arXiv:2104.12763 [cs.CV] (2021)
19. Kiela, D., Bhooshan, S., Firooz, H., Testuggine, D.: Supervised multimodal bitransformers for classifying images and text. In: Visually Grounded Interaction and Language (ViGIL), NeurIPS 2019 Workshop (2019)
20. Krishna, R., et al.: Visual Genome: Connecting language and vision using crowdsourced dense image annotations. Int. J. Comput. Vis. **123**(1), 32–73 (2017)
21. Lewis, M., et al.: BART: denoising sequence-to-sequence pre-training for natural language generation, translation, and comprehension. In: ACL, pp. 7871–7880 (2020)
22. Li, X., et al.: Oscar: object-semantics aligned pre-training for vision-language tasks. In: Vedaldi, A., Bischof, H., Brox, T., Frahm, J.-M. (eds.) ECCV 2020. LNCS, vol. 12375, pp. 121–137. Springer, Cham (2020). https://doi.org/10.1007/978-3-030-58577-8_8
23. Lin, T., Dollár, P., Girshick, R.B., He, K., Hariharan, B., Belongie, S.J.: Feature pyramid networks for object detection. In: CVPR, pp. 936–944 (2017)
24. Lin, T., Goyal, P., Girshick, R.B., He, K., Dollár, P.: Focal loss for dense object detection. IEEE Trans. Pattern Anal. Mach. Intell. **42**(2), 318–327 (2020)
25. Lin, T.-Y., et al.: Microsoft COCO: common objects in context. In: Fleet, D., Pajdla, T., Schiele, B., Tuytelaars, T. (eds.) ECCV 2014. LNCS, vol. 8693, pp. 740–755. Springer, Cham (2014). https://doi.org/10.1007/978-3-319-10602-1_48
26. Liu, W., et al.: SSD: single shot multibox detector. In: Leibe, B., Matas, J., Sebe, N., Welling, M. (eds.) ECCV 2016. LNCS, vol. 9905, pp. 21–37. Springer, Cham (2016). https://doi.org/10.1007/978-3-319-46448-0_2
27. Liu, Y., et al.: RoBERTa: a robustly optimized BERT pretraining approach. arXiv:1907.11692 [cs.CL] (2019)
28. Loshchilov, I., Hutter, F.: Decoupled weight decay regularization. In: ICLR (2019)
29. Lu, J., Batra, D., Parikh, D., Lee, S.: ViLBERT: pretraining task-agnostic visiolinguistic representations for vision-and-language tasks. In: NeurIPS, pp. 13–23 (2019)
30. Ordonez, V., Kulkarni, G., Berg, T.L.: Im2Text: describing images using 1 million captioned photographs. In: NIPS, pp. 1143–1151 (2011)
31. Peters, M., Ammar, W., Bhagavatula, C., Power, R.: Semi-supervised sequence tagging with bidirectional language models. In: ACL, pp. 1756–1765, July 2017
32. Plummer, B.A., Wang, L., Cervantes, C.M., Caicedo, J.C., Hockenmaier, J., Lazebnik, S.: Flickr30k entities: collecting region-to-phrase correspondences for richer image-to-sentence models. In: ICCV, pp. 2641–2649 (2015)
33. Radford, A., Narasimhan, K., Salimans, T., Sutskever, I.: Improving language understanding by generative pre-training. Technical report, OpenAI (2018)
34. Radford, A., Wu, J., R., C., Luan, D., Amodei, D., Sutskever, I.: Language models are unsupervised multitask learners. Technical report, OpenAI (2019)
35. Raffel, C., et al.: Exploring the limits of transfer learning with a unified text-to-text transformer. J. Mach. Learn. Res. **21**(140), 1–67 (2020)
36. Redmon, J., Divvala, S.K., Girshick, R.B., Farhadi, A.: You only look once: unified, real-time object detection. In: CVPR, pp. 779–788 (2016)

37. Ren, S., He, K., Girshick, R.B., Sun, J.: Faster R-CNN: towards real-time object detection with region proposal networks. IEEE Trans. Pattern Anal. Mach. Intell. **39**(6), 1137–1149 (2017)
38. Russakovsky, O., et al.: ImageNet large scale visual recognition challenge. Int. J. Comput. Vis. **115**(3), 211–252 (2015)
39. Selvaraju, R.R., Cogswell, M., Das, A., Vedantam, R., Parikh, D., Batra, D.: Grad-CAM: visual explanations from deep networks via gradient-based localization. Int. J. Comput. Vis. **128**(2), 336–359 (2020)
40. Sharma, P., Ding, N., Goodman, S., Soricut, R.: Conceptual captions: a cleaned, hypernymed, image alt-text dataset for automatic image captioning. In: ACL, pp. 2556–2565, July 2018
41. Shelhamer, E., Long, J., Darrell, T.: Fully convolutional networks for semantic segmentation. IEEE Trans. Pattern Anal. Mach. Intell. **39**(4), 640–651 (2017)
42. Su, W., Zhu, X., Cao, Y., Li, B., Lu, L., Wei, F., Dai, J.: VL-BERT: pre-training of generic visual-linguistic representations. In: ICLR (2020)
43. Tan, H., Bansal, M.: LXMERT: learning cross-modality encoder representations from transformers. In: EMNLP-IJCNLP, pp. 5099–5110 (2019)
44. Teney, D., Anderson, P., He, X., van den Hengel, A.: Tips and tricks for visual question answering: learnings from the 2017 challenge. In: CVPR, pp. 4223–4232 (2018)
45. Vaswani, A., et al.: Attention is all you need. In: NIPS, pp. 5998–6008 (2017)
46. Wu, Y., et al.: Google's neural machine translation system: bridging the gap between human and machine translation. arXiv:1609.08144 cs.[CL] (2016)
47. Yu, Z., Yu, J., Cui, Y., Tao, D., Tian, Q.: Deep modular co-attention networks for visual question answering. In: CVPR, pp. 6281–6290 (2019)
48. Yu, Z., Yu, J., Xiang, C., Fan, J., Tao, D.: Beyond bilinear: generalized multimodal factorized high-order pooling for visual question answering. IEEE Trans. Neural Netw. Learn. Syst. **29**(12), 5947–5959 (2018)
49. Zhu, X., Su, W., Lu, L., Li, B., Wang, X., Dai, J.: Deformable DETR: deformable transformers for end-to-end object detection. In: ICLR (2021)
50. Zhu, Y., et al.: Aligning books and movies: towards story-like visual explanations by watching movies and reading books. In: ICCV, pp. 19–27 (2015)

Diverse Image Captioning with Grounded Style

Franz Klein[1], Shweta Mahajan[1(✉)], and Stefan Roth[1,2]

[1] Department of Computer Science, TU Darmstadt, Darmstadt, Germany
shweta.mahajan@visinf.tu-darmstadt.de
[2] hessian.AI, Darmstadt, Germany

Abstract. Stylized image captioning as presented in prior work aims to generate captions that reflect characteristics beyond a factual description of the scene composition, such as sentiments. Such prior work relies on *given* sentiment identifiers, which are used to express a certain global style in the caption, *e.g.* positive or negative, however without taking into account the stylistic content of the visual scene. To address this shortcoming, we first analyze the limitations of current stylized captioning datasets and propose COCO attribute-based augmentations to obtain varied stylized captions from COCO annotations. Furthermore, we encode the stylized information in the latent space of a Variational Autoencoder; specifically, we leverage extracted image attributes to explicitly structure its sequential latent space according to different localized style characteristics. Our experiments on the Senticap and COCO datasets show the ability of our approach to generate accurate captions with diversity in styles that are grounded in the image.

Keywords: Diverse image captioning · Stylized captioning · VAEs

1 Introduction

Recent advances in deep learning and the availability of multi-modal datasets at the intersection of vision and language [26,47] have led to the successful development of image captioning models [3,13,29,31,39]. Most of the available datasets for image captioning, *e.g.* COCO [26], consist of several ground-truth captions per image from different human annotators, each of which factually describes the scene composition. In general, captioning frameworks leveraging such datasets deterministically generate a single caption per image [5,11,21,23, 24,28,32,45]. However, it is generally not possible to express the entire content of an image in a single, human-sounding sentence. Diverse image captioning aims to address this limitation with frameworks that are able to generate several *different* captions for a single image [4,30,43]. Nevertheless, these approaches

Supplementary Information The online version contains supplementary material available at https://doi.org/10.1007/978-3-030-92659-5_27.

C. Bauckhage et al. (Eds.): DAGM GCPR 2021, LNCS 13024, pp. 421–436, 2021.
https://doi.org/10.1007/978-3-030-92659-5_27

largely ignore image and text properties that go beyond reflecting the scene composition; in fact, most of the employed training datasets hardly consider such properties.

Stylized image captioning summarizes these properties under the term *style*, which includes variations in linguistic style through variations in language, choice of words and sentence structure, expressing different emotions about the visual scene, or by paying more attention to one or more localized concepts, *e.g.* attributes associated with objects in the image [35]. To fully understand and reproduce the information in an image, it is inevitable to consider these kinds of characteristics. Existing image captioning approaches [19,33,35] implement style as a global sentiment and strictly distinguish the sentiments into 'positive', 'negative', and sometimes 'neutral' categories. This simplification ignores characteristics of styles that are crucial for the comprehensive understanding and reproduction of visual scenes. Moreover, they are designed to produce one caption based on a given sentiment identifier related to one sentiment category, ignoring the actual stylistic content of the corresponding image [22,33,35].

In this work, we attempt *(1)* to obtain a more diverse representation of style, and *(2)* ground this style in attributes from localized image regions. We propose a Variational Autoencoder (VAE) based framework, Style-SeqCVAE, to generate stylized captions with styles expressed in the corresponding image. To this end, we address the lack of image-based style information in existing captioning datasets [23,33] by extending the ground-truth captions of the COCO dataset [23], which focus on the scene composition, with localized attribute information from different image regions of the visual scene [38]. This style information in the form of diverse attributes is encoded in the latent space of the Style-SeqCVAE. We perform extensive experiments to show that our approach can indeed generate captions with image-specific stylized information with high semantic accuracy *and* diversity in stylistic expressions.

2 Related Work

Image Captioning. A large proportion of image captioning models rely on Long Short-Term Memories (LSTMs) [20] as basis for language modeling in an encoder-decoder or compositional architecture [16,23,27,32,42,44]. These methods are designed to generate a single accurate caption and, therefore, cannot model the variations in stylized content for an image. Deep generative model-based architectures [4,13,30,31,43] aim to generate multiple captions, modeling feature variations in a low-dimensional space. Wang et al. [43] formulate constrained latent spaces based on object classes in a Conditional Variational Autoencoder (CVAE) framework. Aneja et al. [4] use a sequential latent space to model captions with Gaussian priors and Mahajan et al. [30] learn domain-specific variations of images and text in the latent space. All these approaches, however, do not allow to directly control the intended style to be reflected in each of the generated captions. This is crucial to generate captions with stylistic variation grounded in the images. In contrast, here we aim to generate diverse captions with the many localized styles representative of the image.

- A nice man is stretching his arms with a frisbee out in the beautiful woods.
- A good man standing in a nice area of woods stretching his arms.
- A man standing in a forest with rotten wood is holding up a frisbee.
- A dead man stretching his arms holding a frisbee in the woods.

(a) Example from the Senticap dataset.

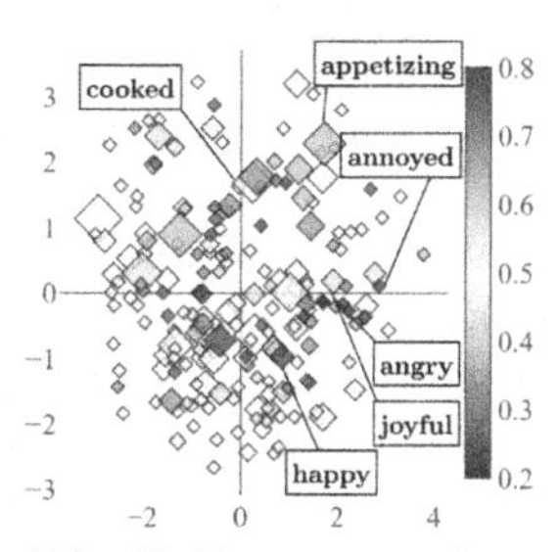

(b) GloVe vectors of attributes from COCO Attributes [38].

Fig. 1. *(a)* Senticap example ground-truth captions with globally positive or negative sentiment. *(b)* GloVe vectors of captions with adjectives from COCO Attributes. The color indicates the attribute SentiWordNet score [6] and the scale indicates the frequency of occurrence in the entire dataset. The original SentiWordNet scores are rescaled between 0 for the most negative and 1 for the most positive sentiment.

Stylized Image Captioning. Recent work has considered the task of stylized caption generation [9,33,40,46]. Mathews et al. [33] utilize 2000 stylized training captions in addition to those available from the COCO dataset to generate captions with either positive or negative sentiments. Shin et al. [40] incorporate an additional CNN, solely trained on weakly supervised sentiment annotations of a large image corpus scraped from different platforms. These approaches, however, rely on *given* style indicators during inference to generate new captions, to emphasize sentiment words in stylized captions, which may not reflect the true sentiment of the image. A series of works attempts to overcome the lack of available stylized captioning training data by separating the style components and the remaining information from the textual description such that they can be trained on both factual caption-image pairs and a distinct stylized text corpus. Gan et al. [17] share components in the LSTM over sentences of a particular style. Similarly, Chen et al. [9] use self-attention to adaptively consider semantics or style in each time step. You et al. [46] add a sentiment cell to the LSTM and Nezami et al. [35] apply the well established semantic attention idea to stylized captioning. Various approaches utilize adversarial formulations to generate stylized captions [19,22,36]. However, they also rely on given globally positive or negative sentiment identifiers to generate varied captions. Since an image can contain a variety of localized styles, in this work, we instead focus on generating diverse captions where the style information is *locally grounded in the image.*

3 Image Captioning Datasets with Stylized Captions

We begin by discussing the limitations of existing image captioning datasets with certain style information in the captions, specifically the Senticap dataset

[33]. Further, we introduce an attribute (adjective) insertion method to extend the COCO dataset with captions containing different localized styles.

Senticap. Besides its limited size (1647 images and 4419 captions in the training set), the Senticap dataset [33] does not provide a comprehensive impression of the image content in combination with different sentiments anchored in the image (Fig. 1a). The main issue in the ground-truth captions is that the positive or negative sentiments may not be related to the sentiment actually expressed in the image. Some adjectives may even distort the image semantics. For instance, the man shown in Fig. 1a is anything but dead, though that is exactly what the last ground-truth caption describes. Another issue is the limited variety: There are 842 positive adjective-noun pairs (ANPs) composed of 98 different adjectives combined with one out of 270 different nouns. For the negative set only 468 ANPs exist, based on 117 adjectives and 173 objects. These adjectives, in turn, appear with very different frequencies in the caption set, *cf.* Fig. 3a.

COCO Attributes. The attributes of COCO Attributes [38] have the big advantage over Senticap that they actually reflect image information. Furthermore, the average number of 9 attribute annotations per object may reflect different possible perceptions of object characteristics. On the downside, COCO contains fairly neutral, rather positive than negatively connoted images: for instance, many image scenes involve animals, children, or food. This is reflected in the sentiment intensity of the associated attributes, visualized in Fig. 1b. Most of them tend to be neutral or positive; the negative ones are underrepresented.

3.1 Extending Ground-Truth Captions with Diverse Style Attributes

To address the lack of stylized ground-truth captions, we combine COCO captions [26], focusing on the scene composition, with style-expressive adjectives in COCO Attributes [38]. We remove 98 attribute categories that are less relevant for stylized captioning (*e.g.*, "cooked") and define sets of synonyms for the remaining attributes to increase diversity. Some of the neutral adjective attributes are preserved since recognizing and dealing with object attributes having a neutral sentiment is also necessary to fully solve captioning with image-grounded styles. However, most of the 185 adjectives are either positive or negative. Likewise for the various COCO object categories, we initially define sets of nouns that appear interchangeably in the corresponding captions to name this object category. Subsequently, we iterate over all COCO images with available COCO Attributes annotations. Given a COCO image, associated object/attribute labels, and the existing ground-truth captions, we locate nouns in the captions that also occur in the sets of object categories to insert a sampled adjective of the corresponding attribute annotations in front of it. A part-of-speech tagger [8] helps to insert the adjective at the right position. This does not protect against an adjective being associated with the wrong noun if it occurs multiple times in the same caption. However, our observations suggest this

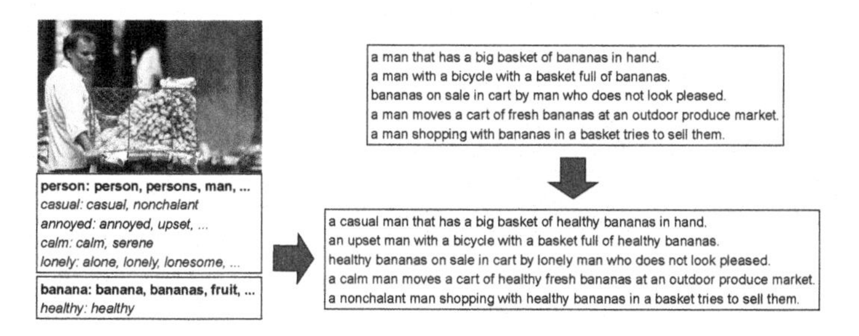

Fig. 2. Example of COCO caption augmentation by insertion of random samples from the COCO Attributes synonym sets in front of the nouns of COCO object categories.

to be rare. The example of the augmentation in Fig. 2 illustrates that inserting various adjectives could reflect a certain level of ambiguity in perception. Based on the COCO 2017 train split [23], this gives us a set of 266K unique, adjective-augmented and image-grounded training captions, which potentially reflect the image style.

3.2 Extending the Senticap Dataset

Since most prior work is evaluated on the Senticap dataset [22,33–35], whose characteristics strongly differ from COCO Attributes, using COCO Attributes-augmented captions for training and the Senticap test split for evaluation would result in scores that provide only little information about the actual model capabilities. We thus generate a second set of augmented COCO captions, based on the Senticap ANPs. These ANPs indicate which nouns and adjectives jointly appear in Senticap captions, independent of the sentiment actually expressed by the underlying image. Hence, to insert these sentiment adjectives into captions such that they represent the image content, we locate nouns appearing in the ANPs to sample and insert one of the adjectives corresponding to the detected nouns, utilizing the method from above. We thus obtain two different sets of training captions: COCO Attributes-augmented captions paired with COCO Attributes and Senticap-augmented captions, which are composed of 520K captions with positive adjectives and 485K captions with negative adjectives.

4 The Style-SeqCVAE Approach

To obtain image descriptions with styles grounded in images, we first extract the attributes associated with objects in the visual scene. Equipped with these style features, we formalize our Style-SeqCVAE with a structured latent space to encode the localized image-grounded style information.

4.1 Image Semantics and Style Information

Dense image features and corresponding object detections are extracted using a Faster R-CNN [18]. In order to obtain the different object attributes, we add a

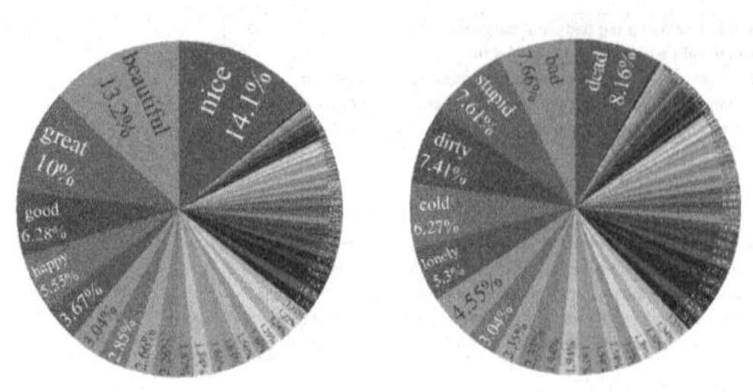

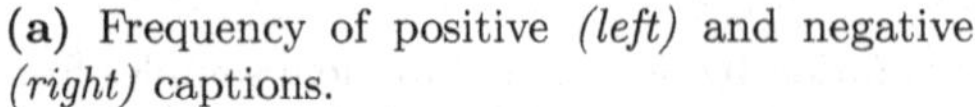

(a) Frequency of positive *(left)* and negative *(right)* captions.

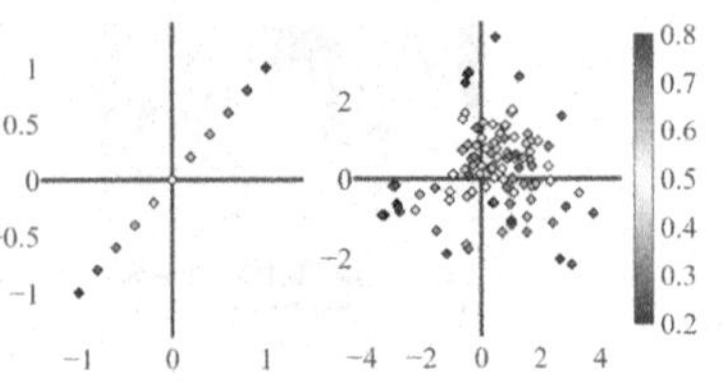

(b) SentiWordNet scores of detected object attributes *(left)*. GloVe dimensions that encode the most sentiment information *(right)*.

Fig. 3. *(a)* Adjective frequency in Senticap captions. *(b)* Two different latent space structuring approaches.

dense layer with sigmoid activation for multi-label classification. The loss term of the classification layer L_{CN}, which traditionally consists of a loss term for background/foreground class scores (L_{cls}), and a term for regression targets per anchor box (L_{reg}), is extended with another loss L_{att}:

$$\begin{aligned} L_{\text{CN}}(\{p_i\},\{b_i\},\{a_i\}) = \frac{1}{N_{\text{cls}}}\sum_i L_{\text{cls}}\left(p_i, p_i^*\right) + \lambda_1 \frac{1}{N_{\text{reg}}}\sum_i p_i^* L_{\text{reg}}\left(b_i, b_i^*\right) \\ + \lambda_2 \frac{1}{N_{\text{att}}}\sum_i \beta_i L_{\text{att}}\left(a_i, a_i^*\right), \end{aligned} \tag{1}$$

where a_i^* are ground-truth attribute annotations and a_i denotes the predicted probabilities for each particular attribute category being present at anchor i. L_{cls} is the binary cross-entropy loss between the predicted probability p_i of anchor i being an object and the ground-truth label p_i^*. The regression loss L_{reg} is the smooth L_1 loss [18] between the predicted bounding box coordinates b_i and the ground-truth coordinates b_i^*. L_{att} is the class-balanced softmax cross-entropy loss [12]. N_{cls}, N_{reg}, and N_{att} are the normalization terms. λ_1 and λ_2 are the regularization parameters. Here, $\beta_i = 1$ if there is an attribute associated with anchor i and $\beta_i = 0$ otherwise.

4.2 The Style-Sequential Conditional Variational Autoencoder

The goal of our Style-SeqCVAE framework is to generate diverse captions that reflect different perceivable style expressions in an image. We illustrate the proposed model in Fig. 4a. Consider an image I and caption sequence $x = (x_1, \ldots, x_T)$, the visual features $\{v_1, \ldots, v_K\}$ for K regions of the image are extracted from a Faster R-CNN (*cf.* Eq. 1) and the mean-pooled image features $\bar{v} = \frac{1}{K}\sum_k v_k$ are input to the attention LSTM [3]. In this work, we propose to further encode region-level style information in $c(I)_t$ (as discussed below), and update it at each time step using the attention weights (α_t). This is based on the observation that the image styles can vary greatly across different

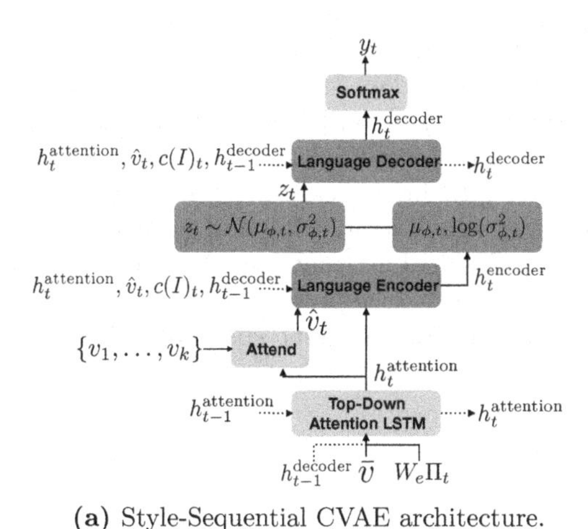

(a) Style-Sequential CVAE architecture.

- A smiling man in a nice hat is holding a frisbee.
- a happy young man holding a frisbee in his gentle hand.
- an annoying man is holding a frisbee in his hand.
- a weird person holding a frisbee in a field.

(b) Generated example captions with positive and negative styles.

Fig. 4. *(a)* Style-Sequential CVAE for stylized image captioning: overview of one time step. *(b)* Captions generated with Style-SeqCVAE on Senticap.

regions. To take this into account, we model a sequential VAE with explicit latent space structuring with an LSTM-based language encoder and language decoder (Fig. 4a highlights this in yellow). $h_t^{\text{attention}}$, h_t^{encoder}, and h_t^{decoder} denote the hidden vectors of the respective LSTMs at a time step t. Encoding the latent vectors z_t at each time step based on the reweighted image regions and corresponding component vectors $c(I)_t$ enables the model to structure the latent space at the image region level instead of only globally. Similar to Anderson et al. [3], the relevance of the input features at a specific time step t depends on the generated word $W_e\Pi_t$, where W_e is a word embedding matrix and Π_t is the one-hot encoding of the input word. Given the attended image feature $\hat{v}_t$, the latent vectors z are encoded in the hidden states $h_{t-1}^{\text{attention}}$ and h_{t-1}^{decoder}. Moreover, to allow for image-specific localized style information, we enforce an attribute-based structured latent space.

The log-evidence lower bound at time step t is given by

$$\begin{aligned} \log p\left(x_t|I, x_{<t}, z_{\leq t}, c(I)_t\right) \geq \mathbb{E}_{q_\phi}\left[\log p_\theta\left(x_t|I, x_{<t}, z_{\leq t}, c(I)_t\right)\right] \\ - D_{KL}[q_\phi(z_t|I, x_{<t}, z_{<t}, c(I)_t) \,\|\, p_\theta(z_t|c(I)_t)]. \end{aligned} \tag{2}$$

Here, p_θ is the prior distribution parameterized by θ and q_ϕ denotes the variational posterior distribution with parameters ϕ.

Attribute-Specific Latent Space Structuring. The choice of the prior contributes significantly to how the latent space is structured. The additive Gaussian prior has proven to be beneficial for both diversity and controllability of the caption generation process [4,43]. Unlike [43], where the latent space is constrained based on the objects, we instead leverage attributes to encode the styles in the image. Specifically, available attributes are explicitly assigned to one of the image features $\{v_1, ..., v_k\}$ and can thus be divided into subsets $A_k = \{a_{k,1}, ..., a_{k,j}\}$ containing J_k different attributes. Furthermore, we adopt the attention mechanism of [2], which provides a set of weights $\alpha_t = \{\alpha_{t,1}, ..., \alpha_{t,k}\}$ to readjust the impact of each image feature v_k at every time step t. Accordingly, a weight $\alpha_{t,k}$ is also mapped to the subset A_k, which belongs to the image feature v_k. This property is exploited for the attribute-specific latent space structuring and makes it possible to weight the contribution of each individual attribute set A_k. We assume that the approximate style of an image region v_t is represented by the total of all associated attributes A_k. The additive Gaussian latent space can then be reformulated to calculate a μ_t at each time step t as

$$\mu_t = \sum_{k=1}^{K} \frac{\alpha_{t,k}}{J_k} \sum_{i=1}^{J_k} \mu_{i,k}. \tag{3}$$

The prior mean μ_t is thus composed of the attention-weighted average of each image region-specific linear combination of $\mu_{i,k}$. The variance σ_t^2 does not need to be calculated explicitly as σ_i^2 is equal for all attribute categories i. However, randomly initialized, attribute category-specific Gaussian components μ_i do not reflect any semantic or contextual similarities between different attributes. Therefore, in this work two alternative attribute category-specific μ_i initialization approaches are pursued: In the first case, we uniformly initialize each μ_i to the SentiWordNet score [6] corresponding to attribute category i. In the following, this latent space is referred to as the *SentiWordNet* latent space. In the second case, both sentimental and semantic characteristics of different attributes are taken into account by initializing each μ_i with the n dimensions of GloVe vectors corresponding to the attribute category i that most strongly encode its word sentiments. These dimensions are identified by an application of PCA on the 20 COCO Attributes labels with the strongest SentiWordNet scores. If the number of extracted dimensions is less than the desired dimensionality z, its elements are simply repeated to upscale it. We refer to this setup as *SentiGloVe* latent space. Both latent space structuring approaches are exemplified in Fig. 3b. With μ_i already encoding attribute-specific information, it serves as explicit input $c(I)_t$ to the encoder and decoder, thus $c(I)_t = \mu_t$. While training, the encoder produces Gaussian parameters $\mu_{\phi,t}$ and $\log \sigma_{\phi,t}^2$, which are used to encode a latent z_t and calculate the KL-Divergence

$$\begin{aligned} D_{KL}[q_\phi(z_t|I, x_{<t}, z_{<t}, c(I)_t) \,||\, p(z_t|c(I)_t)] &= \log\left(\frac{\sigma_t}{\sigma_{\phi,t}}\right) + \frac{1}{2\sigma_t^2}\mathbb{E}_{q_\phi}\left[||z_t - \mu_t||^2\right] - \frac{1}{2} \\ &= \log\left(\frac{\sigma_t}{\sigma_{\phi,t}}\right) + \frac{\sigma_{\phi,t}^2 + ||\mu_{\phi,t} - \mu_t||^2}{2\sigma_t^2} - \frac{1}{2}, \end{aligned} \tag{4}$$

which is used to maximize the variational lower bound. Furthermore, z_t is provided to the decoder to produce the output word y_t of the current time step. During generation, the encoder is dropped and z_t values are sampled from the same attribute-specific additive Gaussian prior that is used while training. The attributes attached to the image features are actual (hard) detections from the image feature extractor instead of ground-truth annotations.

Sentiment-Specific Latent Space Structuring. Furthermore, we extend our approach to the Senticap dataset available for stylized image captioning and thus allow for a quantitative comparison of the proposed framework with existing work. The Senticap dataset provides positive as well as negative captions for images in the dataset. This implies that the captions in the Senticap dataset do not represent a variety of image-anchored styles and therefore, we cannot expect the latent space structure defined above to be particularly helpful. Thus, we take into account the COCO captions which are augmented with Senticap adjectives and are either labeled as "positive" or "negative" and define a simple latent space structure around the two clusters. Additionally, a third cluster is defined for captions with a neutral sentiment, *e.g.* original COCO captions. In contrast to the attention-weighted, attribute-based latent space structures defined above, the mean remains constant over all time steps and only depends on the provided sentiment identifier in this setup. And since each caption is distinctively assigned to one of these three clusters, the additive Gaussian structuring is not suitable. Instead, the prior mean is fully defined by one of the three predefined clusters with values $c(I)_t \in \{-0.5, 0.0, 0.5\}$ depending on the negative, neutral, or positive sentiment identifier. In this case, $\mu_t = c(I)_t$ and $p(z_t|c(I)_t) = \mathcal{N}\left(z_t \mid \mu_t, \sigma^2 I\right)$. Similar to the latent space structuring approaches described above, the prior variance σ_t^2 is initially set and identical for all three clusters. Having obtained these prior parameters and $c(I)_t$, the training procedure is identical to that of the attribute-specific latent space. During evaluation, we generate diverse captions for a given sentiment by selecting one of the three clusters (by choosing the prior mean associated with the intended sentiment).

Style-Based Constrained Beam Search. Since only a fraction of COCO images are annotated with attributes, we rely on COCO Attributes-augmented captions for a fraction of training images. When the model is now trained using a combination of the original and attribute-augmented COCO captions, the attribute words occur rarely in the decoding procedure, especially the ones that are extremely underrepresented in the training data. This issue is optionally addressed by presenting the detected object attributes as constraints during the decoding procedure using constrained beam search (CBS) [1].

5 Experiments

We next evaluate and analyze our approach for the generation of a diverse set of image captions for a particular image with image-grounded styles. We evaluate on the Senticap and COCO datasets. On Senticap, we use the Senticap-augmented captions for training. When evaluating on the standard COCO dataset, we utilize the COCO Attributes-augmented captions for training in order to encode image-specific style information in the latent space.

Evaluation Metrics. The accuracy of the captions is evaluated with standard metrics – Bleu (B) 1–4 [37], CIDEr (C) [10], ROUGE (R) [25], and METEOR

Table 1. Top-1 oracle scores on Senticap test split captions with positive and negative sentiments. n denotes the number of captions generated per image.

		Positive							Negative						
Method	n	B1	B2	B3	B4	R	C	M	B1	B2	B3	B4	R	C	M
Senticap [33]	1	49.1	29.1	17.5	10.8	36.5	54.4	16.8	50.0	31.2	20.3	13.1	37.9	61.8	16.8
StyleNet [17]	1	45.3	–	12.1	–	–	36.3	12.1	43.7	–	10.6	–	–	36.6	10.9
You et al. [46]	1	51.2	31.4	19.4	12.3	38.6	61.1	17.2	52.2	33.6	22.2	14.8	39.8	70.1	17.1
Chen et al. [9]	1	50.5	30.8	19.1	12.1	38.0	60.0	16.6	50.3	31.0	20.1	13.3	38.0	59.7	16.2
Senti-Attend [35]	1	57.6	34.2	20.5	12.7	45.1	68.6	18.9	58.6	35.4	22.3	14.7	45.7	71.9	19.0
MSCap [19]	1	46.9	–	16.2	–	–	55.3	16.8	45.5	–	15.4	–	–	51.6	16.2
AttendGAN [36]	1	56.9	33.6	20.3	12.5	44.3	61.6	18.8	56.2	34.1	21.3	13.6	44.6	64.1	17.9
Memcap [48]	1	51.1	–	17.0	–	–	52.8	16.6	49.2	–	18.1	–	–	59.4	15.7
Karayil et al. [22]	1	54.7	34.6	22.0	14.4	41.8	46.1	18.5	57.0	36.2	23.4	15.1	44.5	50.9	19.9
	10	65.6	43.9	29.5	20.2	48.8	63.1	22.1	**67.6**	46.3	31.9	21.9	50.4	68.8	23.5
Style-SeqCVAE	1	53.8	33.2	20.1	12.5	41.5	71.1	19.7	55.2	34.5	21.5	13.4	41.5	75.5	19.4
	10	**66.3**	**46.3**	**32.2**	**22.2**	**51.2**	**110.5**	**25.2**	66.1	**46.8**	**33.3**	**23.7**	**50.9**	**111.9**	**24.4**

(M) [7]. Similar to Mathews et al. [33], we consider the percentage of candidate captions containing at least one of the Senticap ANPs as part of the evaluation, referred to as SEN%. Additionally, we report the precision (SP) and recall (SR) of sentiment adjectives occurring in candidate and reference captions.

5.1 Evaluation on the Senticap Dataset

In this setting, the original COCO 2017 train split is combined with the COCO captions augmented with Senticap adjectives. The available captions express either a negative, neutral, or positive sentiment and, therefore, the latent space is explicitly structured around three different clusters encoding the sentiment expressed in the generated captions. Unless otherwise stated, constrained beam search is not used for caption generation. Since the Senticap dataset consists of positive and negative ground-truth captions for images, not related to the actual image sentiment, prior work [9,17,19,22,33,35,36,46] generates a positive as well as a negative caption for a given image based on the style indicator. Therefore, in order to compare our approach on the Senticap evaluation dataset, we generate positive and negative captions for a given image based on the sentiment-specific latent space as discussed above (Sect. 4.2). The quality of the generated positive and negative captions is reported in Table 1. When generating only one caption per image ($n = 1$), the achieved scores are comparable to the best-performing existing work. When generating $n = 10$ captions, the presented approach performs clearly better than the only related work [22] that generates diverse captions conditioned on the style indicator. This implies that unlike our Style-SeqCVAE approach, [22] does not encode as many variations in style content for a given image. The fact that we obtain high scores with our approach on metrics that take longer n-grams into account indicates that appropriate adjectives related to style are inserted into the captions in a

Table 2. Additional evaluation on the Senticap test set. SP denotes the sentiment precision and SR the sentiment recall.

n	Method	B1	B2	B3	B4	R	C	M	%SEN	SP	SR
1	Senticap, Mathews et al. [33]	48.8	29.8	18.7	11.8	37.2	56.6	16.8	87.5	**0.33**	0.14
	COCO + Senticap-augm.	54.5	33.9	20.8	13.0	41.5	73.3	19.6	93.6	0.30	0.15
	COCO + Senticap-augm. + CBS	54.7	34.0	21.1	13.2	41.7	74.1	19.7	100.0	0.30	0.15
10	COCO + Senticap-augm.	66.2	46.6	32.8	23.0	51.0	111.2	24.8	99.3	0.26	0.30
	COCO + Senticap-augm. + CBS	**66.3**	**47.2**	**33.6**	**23.9**	**51.5**	**114.5**	**25.3**	**100.0**	0.25	**0.32**

suitable place. This also suggests that our proposed caption augmentation approach preserves the syntactic correctness of the resulting training captions. Most striking is the significant performance increase in the CIDEr score, which particularly rewards the use of n-grams that only rarely appear in the reference captions. This implies that explicitly structuring the latent space around fixed style-based clusters encourages even underrepresented style adjectives to prevail during decoding, especially when multiple captions are generated for an image.

In Table 2, we show that the diversity in the generated captions is the result of the different ways of expressing sentiment (and not solely based on diversity from the factual descriptions). In this setting, we include the Senticap training set without decoding constraints (COCO + Senticap-augm) and with decoding constraints (COCO + Senticap-augm + CBS). The proportion of reference sentiment adjectives appearing in the candidate captions doubles from $\sim$0.15 to $\sim$0.3 when producing 10 captions per image, which shows clearly that the diversity in the captions also has an effect on sentiment expression. Captions generated by [33] exclusively consider the most dominant adjectives in the training/test captions without much diversity. Thus, its slightly higher SP score is expected, given the massive adjective imbalance of Senticap. The Senticap-augmented model (COCO + Senticap-augm) inserts at least one ANP with matching sentiment in almost every caption (>93.6%). Furthermore, we do not observe a significant improvement in any metric when attribute-based CBS constraints are applied. This shows that the latent space of our Style-SeqCVAE can effectively model the style information in the latent space. Additionally, the high CIDEr score supports that the captions generated by our approach are accurate. Qualitative examples in Fig. 4b show the captions with varied styles generated with Style-SeqCVAE. For the given image, the diverse captions reflect the positive as well as negative sentiments showing that the latent space of our approach effectively captures style information of the data distribution.

5.2 Evaluation on the COCO Dataset

We now show that our approach along with the extended COCO Attributes-augmented captions can generate diverse captions with styles anchored in the image. Since the COCO test split contains only descriptions of the scene composition, it is difficult to quantitatively evaluate for stylized caption generation on the COCO dataset. Therefore, we first present a qualitative analysis of the

['angry', 'alone / lonely', 'furry', 'hairy', 'bulky', 'fluffy', 'fuzzy','playful', 'wild', 'natural', 'soft', 'strong', 'dangerous', 'heavy']
weak constraining
• a furry large brown bear sitting in the water.
• a close shot of a furry bear in the water with a fish.
• a furry young bear is swimming in the water.
individual constraining
• a playful adult bear in a body of water.
• a dangerous bear that is sitting in the water.
• a furry bear swimming alone in water.

Fig. 5. Examples of captions generated by either constraining the decoding procedure on the whole set of detected image attributes (weak constraining) vs. one specific attribute as constraint per caption (individual constraining).

• black and white photo of a clean man talking on a cell phone.
• a black and white photo of a lonely man talking on a cell.
• a busy man with a cell phone in his hand.
• a man with a soft face talking on a cell phone.

• a serious man and his tame dog sit on a useful small boat in the water.
• a clean man and his tame dog on a simple boat in the ocean.
• a nonchalant man on a useful boat with his tame dog.
• a serious man on a useful boat with a tame dog on the front.

Fig. 6. Generated example captions with SentiGloVe-structured latent space and individual decoding constraints for images with multiple objects.

Style-SeqCVAE approach. For this, we consider captions generated directly from the latent space (without CBS) as well as with CBS constraints to account for the rarity of the attributes in the dataset. Unlike [1], which presents class labels as constraints, in this work, we enforce style information from automatically extracted image attributes as constraint to the caption generator. For an effective application of CBS constraints in conjunction with Style-SeqCVAE, two different decoding strategies are considered: In weak constraining, the constraint is to use at least one attribute from the set of detected attributes during decoding for a given image. The detected attributes are obtained directly from the image extractor (Faster R-CNN) trained using Eq. 1. In individual constraining, the attribute to be inserted is explicitly selected from the set of detected attributes for an image and provided as a constraint. Since the model is forced to take this attribute into account, it enables to also consider attributes that are under-represented in the training data and which might otherwise be ignored. When generating multiple captions for a given image, a constraint is randomly selected from the set of detected attributes for each of the captions. This encourages diversity in the style of the generated captions for each image.

With the weak constraining mechanism, we observe that attributes that occur in high frequency in the training data are repeated in the generated captions. As shown in Fig. 5, the detected attribute “furry” associated with the object “bear” occurs across all the generated captions for the given image. In case of individual constraining, where an attribute is randomly provided as constraint from the set of detected attributes, this effect is not present. For example, in Fig. 5, we observe diversity in style with attributes like “playful” and “dangerous” describing the object “bear”. In Fig. 6, we show an extension of the individual constraining procedure to occurrences of multiple objects in an image. Here, the model is forced to insert an attribute for at least two detected objects. For example, in

Table 3. Evaluation of semantic accuracy.

Method	std	B1	B2	B3	B4	R	C	M
Div-BS [41]	–	83.7	68.7	53.8	38.3	65.3	140.5	35.7
AG-CVAE [43]	–	83.4	69.8	57.3	47.1	63.8	125.9	30.9
POS [14]	–	**87.4**	**73.7**	**59.3**	44.9	**67.8**	**146.8**	**36.5**
Seq-CVAE [4]	–	87.0	72.7	59.1	44.5	67.1	144.8	35.6
Style-SeqCVAE	1	84.2	69.5	56.0	44.7	63.4	130.4	31.2
	2	86.6	72.0	58.8	**47.6**	65.9	137.2	32.8

Table 4. Caption diversity.

Method	std	Div-1	Div-2
Div-BS [41]	–	0.20	0.26
AG-CVAE [43]	–	0.24	0.34
POS [14]	–	0.24	0.35
Seq-CVAE [4]	–	0.25	**0.54**
Style-SeqCVAE	1	0.24	0.31
	2	**0.29**	0.43

Fig. 6 the diverse attributes are successfully inserted for the objects "man" and "face" or for "man", "dog", and "boat" in the captions of the respective images.

The quantitative evaluation of the proposed framework on the COCO dataset is limited due to the lack of ground-truth captions for direct comparison purposes (see supplemental). To obtain a better assessment in spite of that, we compare our method with various established approaches for diverse image captioning. Here, we use the SentiGloVe latent space to generate diverse captions. Following the standard evaluation protocol of [4,30,43], in Table 3 we show the top-1 oracle performance using 20 samples on various metrics for caption evaluation. We observe generally competitive performance, especially in the case where a standard deviation of 2 is used when sampling from the latent space. This is despite the fact that, unlike previous work on diverse image captioning, our model focuses on stylistic diversity, which is not represented in the ground-truth captions of the test set. The high accuracy scores, moreover, demonstrate the ability of the approach to successfully model the attribute-based style information to generate semantically coherent captions (*cf.* Fig. 5).

Furthermore, we quantitatively compare our approach against [4,14,41,43] for diversity. In Table 4, we use Div-1 and Div-2 for evaluation, where Div-n is the ratio of distinct n-grams per caption to the total number of words generated per set of diverse captions. Following prior work, the scores are based on the top-5 captions with highest CIDEr scores of the COCO validation split [23] and consensus re-ranking [15] is applied before selecting the top-5 captions. Owing to the unconstrained latent space, Seq-CVAE [4] generates captions with high diversity. The scores on the diversity metrics indicate that in comparison to other approaches that also impose constraints in the latent space, *e.g.* AG-CVAE [43], our attribute-based latent space exhibits a higher diversity in style. This can be attributed to the structured sequential latent space, where the distribution of attributes is better captured conditioned on the image. This highlights the advantages of our style-specific latent space for diverse caption generation.

6 Conclusion

We present Style-SeqCVAE, a variational autoencoder framework to encode localized style information representative of the visual scene. The structured latent space exploits the object attributes from the associated images to model the characteristic styles. In particular, we leverage the attribute information in different image regions to express different styles present in the image. The key to the success of the proposed latent space is the combination of attribute-based style information and region-based image features via an attention mechanism for coherent caption generation. Our experiments demonstrate that our approach generates diverse and accurate captions with varied styles expressed in the image.

Acknowledgement. This project has received funding from the European Research Council (ERC) under the European Union's Horizon 2020 research and innovation programme (grant agreement No. 866008).

References

1. Anderson, P., Fernando, B., Johnson, M., Gould, S.: Guided open vocabulary image captioning with constrained beam search. In: EMNLP, pp. 936–945 (2017)
2. Anderson, P., Gould, S., Johnson, M.: Partially-supervised image captioning. In: NeurIPS, pp. 1875–1886 (2018)
3. Anderson, P., et al.: Bottom-up and top-down attention for image captioning and visual question answering. In: CVPR, pp. 6077–6086 (2018)
4. Aneja, J., Agrawal, H., Batra, D., Schwing, A.: Sequential latent spaces for modeling the intention during diverse image captioning. In: ICCV, pp. 4261–4270 (2019)
5. Aneja, J., Deshpande, A., Schwing, A.G.: Convolutional image captioning. In: CVPR, pp. 5561–5570 (2018)
6. Baccianella, S., Esuli, A., Sebastiani, F.: SentiWordNet 3.0: an enhanced lexical resource for sentiment analysis and opinion mining. In: LREC, pp. 2200–2204 (2010)
7. Banerjee, S., Lavie, A.: METEOR: an automatic metric for MT evaluation with improved correlation with human judgments. In: ACL Workshop on Intrinsic and Extrinsic Evaluation Measures for Machine Translation and/or Summarization, pp. 65–72 (2005)
8. Bird, S., Klein, E., Loper, E.: Natural Language Processing with Python: Analyzing Text with the Natural Language Toolkit. O'Reilly Media, Inc. (2009)
9. Chen, C.K., Pan, Z., Liu, M.Y., Sun, M.: Unsupervised stylish image description generation via domain layer norm. In: AAAI, pp. 8151–8158 (2019)
10. Chen, X., et al.: Microsoft COCO captions: data collection and evaluation server. arXiv:1504.00325 (2015)
11. Chen, X., Zitnick, C.L.: Mind's eye: a recurrent visual representation for image caption generation. In: CVPR, pp. 2422–2431 (2015)
12. Cui, Y., Jia, M., Lin, T.Y., Song, Y., Belongie, S.: Class-balanced loss based on effective number of samples. In: CVPR, pp. 9268–9277 (2019)
13. Dai, B., Fidler, S., Urtasun, R., Lin, D.: Towards diverse and natural image descriptions via a conditional GAN. In: ICCV, pp. 2970–2979 (2017)

14. Deshpande, A., Aneja, J., Wang, L., Schwing, A.G., Forsyth, D.: Fast, diverse and accurate image captioning guided by part-of-speech. In: CVPR, pp. 10695–10704 (2019)
15. Devlin, J., Gupta, S., Girshick, R., Mitchell, M., Zitnick, C.L.: Exploring nearest neighbor approaches for image captioning. arXiv:1505.04467 (2015)
16. Donahue, J., et al.: Long-term recurrent convolutional networks for visual recognition and description. TPAMI **39**(4), 677–691 (2017)
17. Gan, C., Gan, Z., He, X., Gao, J., Deng, L.: StyleNet: generating attractive visual captions with styles. In: CVPR, pp. 3137–3146 (2017)
18. Girshick, R.: Fast R-CNN. In: ICCV, pp. 1440–1448 (2015)
19. Guo, L., Liu, J., Yao, P., Li, J., Lu, H.: MSCap: multi-style image captioning with unpaired stylized text. In: CVPR, pp. 4204–4213 (2019)
20. Hochreiter, S., Schmidhuber, J.: Long short-term memory. Neural Comput. **9**(8), 1735–1780 (1997)
21. Johnson, J., Karpathy, A., Fei-Fei, L.: DenseCap: fully convolutional localization networks for dense captioning. In: CVPR, pp. 4565–4574 (2016)
22. Karayil, T., Irfan, A., Raue, F., Hees, J., Dengel, A.: Conditional GANs for image captioning with sentiments. In: Tetko, I.V., Kůrková, V., Karpov, P., Theis, F. (eds.) ICANN 2019. LNCS, vol. 11730, pp. 300–312. Springer, Cham (2019). https://doi.org/10.1007/978-3-030-30490-4_25
23. Karpathy, A., Fei-Fei, L.: Deep visual-semantic alignments for generating image descriptions. TPAMI **39**(4), 664–676 (2017)
24. Kulkarni, G., et al.: BabyTalk: understanding and generating simple image descriptions. TPAMI **35**(12), 2891–2903 (2013)
25. Lin, C.Y.: ROUGE: a package for automatic evaluation of summaries. In: ACL Text Summarization Branches Out, pp. 74–81 (2004)
26. Lin, T.-Y., et al.: Microsoft COCO: common objects in context. In: Fleet, D., Pajdla, T., Schiele, B., Tuytelaars, T. (eds.) ECCV 2014. LNCS, vol. 8693, pp. 740–755. Springer, Cham (2014). https://doi.org/10.1007/978-3-319-10602-1_48
27. Lu, J., Xiong, C., Parikh, D., Socher, R.: Knowing when to look: Adaptive attention via a visual sentinel for image captioning. In: CVPR, pp. 3242–3250. IEEE Computer Society (2017)
28. Lu, J., Yang, J., Batra, D., Parikh, D.: Neural baby talk. In: CVPR, pp. 7219–7228 (2018)
29. Mahajan, S., Botschen, T., Gurevych, I., Roth, S.: Joint Wasserstein autoencoders for aligning multimodal embeddings. In: ICCVW, pp. 4561–4570 (2019)
30. Mahajan, S., Gurevych, I., Roth, S.: Latent normalizing flows for many-to-many cross-domain mappings. In: ICLR (2020)
31. Mahajan, S., Roth, S.: Diverse image captioning with context-object split latent spaces. In: NeurIPS, pp. 3613–3624 (2020)
32. Mao, J., Xu, W., Yang, Y., Wang, J., Yuille, A.L.: Deep captioning with multimodal recurrent neural networks (m-RNN). In: ICLR (2015)
33. Mathews, A., Xie, L., He, X.: SentiCap: generating image descriptions with sentiments. In: AAAI, pp. 3574–3580 (2016)
34. Mathews, A.P., Xie, L., He, X.: SemStyle: learning to generate stylised image captions using unaligned text. In: CVPR, pp. 8591–8600 (2018)
35. Nezami, O.M., Dras, M., Wan, S., Paris, C.: Senti-attend: image captioning using sentiment and attention. arXiv:1811.09789 (2018)

36. Mohamad Nezami, O., Dras, M., Wan, S., Paris, C., Hamey, L.: Towards generating stylized image captions via adversarial training. In: Nayak, A.C., Sharma, A. (eds.) PRICAI 2019. LNCS (LNAI), vol. 11670, pp. 270–284. Springer, Cham (2019). https://doi.org/10.1007/978-3-030-29908-8_22
37. Papineni, K., Roukos, S., Ward, T., Zhu, W.J.: BLEU: a method for automatic evaluation of machine translation. In: ACL, pp. 311–318 (2002)
38. Patterson, G., Hays, J.: COCO attributes: attributes for people, animals, and objects. In: Leibe, B., Matas, J., Sebe, N., Welling, M. (eds.) ECCV 2016. LNCS, vol. 9910, pp. 85–100. Springer, Cham (2016). https://doi.org/10.1007/978-3-319-46466-4_6
39. Rennie, S.J., Marcheret, E., Mroueh, Y., Ross, J., Goel, V.: Self-critical sequence training for image captioning. In: CVPR, pp. 7008–7024 (2017)
40. Shin, A., Ushiku, Y., Harada, T.: Image captioning with sentiment terms via weakly-supervised sentiment dataset. In: BMVC (2016)
41. Vijayakumar, A.K., et al.: Diverse beam search: decoding diverse solutions from neural sequence models. arXiv:1610.02424 (2016)
42. Vinyals, O., Toshev, A., Bengio, S., Erhan, D.: Show and tell: a neural image caption generator. In: CVPR, pp. 3156–3164 (2015)
43. Wang, L., Schwing, A., Lazebnik, S.: Diverse and accurate image description using a variational auto-encoder with an additive Gaussian encoding space. In: NIPS, pp. 5756–5766 (2017)
44. Xu, K., et al.: Show, attend and tell: neural image caption generation with visual attention. In: ICML, pp. 2048–2057 (2015)
45. Yao, T., Pan, Y., Li, Y., Mei, T.: Hierarchy parsing for image captioning. In: ICCV, pp. 2621–2629 (2019)
46. You, Q., Jin, H., Luo, J.: Image captioning at will: a versatile scheme for effectively injecting sentiments into image descriptions. arXiv:1801.10121 (2018)
47. Young, P., Lai, A., Hodosh, M., Hockenmaier, J.: From image descriptions to visual denotations: new similarity metrics for semantic inference over event descriptions. TACL **2**, 67–78 (2014)
48. Zhao, W., Wu, X., Zhang, X.: MemCap: memorizing style knowledge for image captioning. In: AAAI, pp. 12984–12992 (2020)

Labeling and Self-Supervised Learning

Leveraging Group Annotations in Object Detection Using Graph-Based Pseudo-labeling

Daniel Pototzky[1,2(✉)], Matthias Kirschner[1], and Lars Schmidt-Thieme[2]

[1] Robert Bosch GmbH, Gerlingen, Germany
{daniel.pototzky,matthias.kirschner}@de.bosch.com
[2] Information Systems and Machine Learning Lab, University of Hildesheim, Hildesheim, Germany
schmidt-thieme@ismll.de

Abstract. We address the problem of dealing with group annotations in object detection where a multitude of items is included in a single bounding box. The standard training protocols in use for most datasets either ignore anchors overlapping with group annotations in the loss function or discard group annotations completely. In this paper, we argue that group annotations contain unexplored potential in many commonly used datasets. We propose to leverage group annotations by generating pseudo-labels on top of them. We develop a novel graph-based pseudo-label generation method that interprets pseudo-label creation as a graph clustering problem that is suited to deal with overlapping objects. In experiments on CityPersons, MS COCO, and a subset of OpenImages we show that our approach outperforms the usual training strategies on the respective datasets when dealing with group annotations as well as other pseudo-label generation methods. We find that the greater the share of group annotations, the larger the increase in performance.

Keywords: Semi-supervised learning · Weakly-supervised learning · Object detection · Pseudo-labeling

1 Introduction

Group annotations are present in many object detection datasets, however, properly dealing with them has barely been investigated. Usually, they contain multiple semantically similar and highly overlapping objects of a single class. Group annotations can be seen as a way of reducing annotation costs. Instead of labeling every object individually which would be time-consuming and expensive, multiple items are included in a single bounding box.

The share of group annotations strongly differs between datasets, ranging from 12.6% in CityPersons (class people) [20], to 6.0% in OpenImages (isgroup = True) [6] and 1.2% in MS COCO (iscrowd = True) [8]. Also within each dataset, the share per class differs substantially. For example, more than 68% of the

C. Bauckhage et al. (Eds.): DAGM GCPR 2021, LNCS 13024, pp. 439–452, 2021.
https://doi.org/10.1007/978-3-030-92659-5_28

bounding boxes of class grape in OpenImages are group annotations compared to 0% for the class vehicle registration plate. Each group annotation contains two or more (OpenImages: five or more) objects. That means that even if the overall share of group annotations is small, they contain a substantial amount of objects.

For commonly used datasets, standard procedures have been established of how to deal with group annotations: In CityPersons and OpenImages regions overlapping group annotations are ignored in the loss function (or at least the corresponding classes) [3,9,11]; In MS COCO, crowd annotations are usually discarded [13]. We argue that these approaches are suboptimal as they do not fully exploit the information contained in group annotations. Each group bounding box specifies that two or more objects of a class are located in a certain region of an image. We propose to leverage this information by generating pseudo-labels on top of group annotations. We develop a novel pseudo-label-based method that can deal with common features of group annotations such as overlapping objects (see Fig. 1). Thereby, we interpret pseudo-label creation as a graph clustering problem and show that our setup is superior to standard approaches of dealing with group annotations as well as to other pseudo-label generation methods used in object detection such as data distillation [14]. Overall, the main contributions of our work can be summarized as follows:

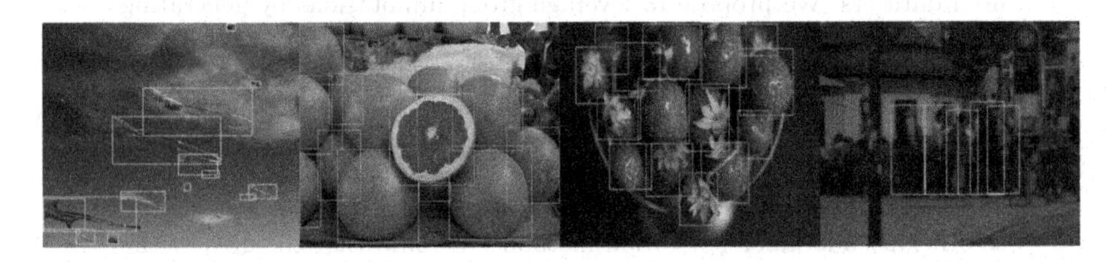

Fig. 1. Examples for pseudo-labels generated on top of group annotations in MS COCO, OpenImages, and CityPersons for classes kite, orange, strawberry, and person.

- We show that standard strategies for dealing with group annotations on commonly used datasets such as CityPersons, MS COCO, and OpenImages are suboptimal and that generating pseudo-labels on top of group annotations consistently improves performance.
- We propose a novel pseudo-label generation method that is suited to deal with overlapping objects. We interpret pseudo-label creation as a graph clustering problem and experimentally demonstrate that our method performs better than commonly used pseudo-label generation methods.
- We demonstrate that the greater the share of group annotations, the larger the increase in performance by pseudo-label-based methods.

2 Related Work

In this section, we contrast our setup of leveraging group annotations with related problems where not every object in the dataset has a bounding box label. These include semi-supervised object detection, weakly-supervised object detection and dealing with missing annotations.

2.1 Semi-supervised Object Detection

In semi-supervised object detection, some images are fully labeled whereas others are unlabeled. One proven strategy is to generate pseudo-labels on the unlabeled images and to use the combined dataset for training [10,14,16,17,22]. Radosavovic et al. [14] introduced a data distillation method where pseudo-labels were created by bounding box voting using predictions generated from a baseline model on multiple augmentations of every image. Unbiased teacher [10] employed a teacher-student framework in which the teacher is an estimated moving average of the students' weights. Zopf et al. [22] trained a detector on MS COCO, created pseudo-labels on ImageNet, and retrained using both datasets jointly while injecting noise in the process. The authors showed an increase in performance regardless of what share of the MS COCO dataset was used.

Pseudo-labels can also be used for enforcing consistency in the model output, e.g. if pseudo-labels are created on a non-augmented image whereas a strongly augmented version of the same image is used in training [16]. Consistency in the output can also be enforced directly, e.g. if a term is included in the loss function that ensures that the network output stays the same even if the input image is augmented (e.g. flipped) [5].

2.2 Weakly-Supervised Object Detection

In weakly-supervised object detection only per-image labels are available. Therefore, no information is provided about the number and location of objects. To address this difficult problem, many methods employ a two-step procedure [21]. First, pseudo-labels are generated using Multiple Instance Learning. Second, the resulting pseudo-annotations are used in a supervised training. Due to the difficulty of the problem, the resulting detectors do not match the performance of supervised training with bounding box labels [4,15,18,19,21].

2.3 Dealing with Missing Annotations

Some datasets contain a substantial amount of missing bounding box annotations. OpenImages, for example, has an image-level recall of only 43% [6]. An established strategy for dealing with missing annotations is to ignore anchors in the loss which are likely to contain an unlabeled object. Niitani et al. [11] first trained a detector on the original data. Then, the predictions of the model were used to determine locations in images that are likely to contain an unlabeled object. Those were subsequently ignored in a second training. This setup won first place in the OpenImages challenge 2018. Using pseudo-labels as positive labels for training on a sparsely labeled version of MS COCO diminished performance.

2.4 Handling Group Annotations

Handling group annotations for training object detectors has not received specific coverage in the literature yet. We argue that it is a special case of semi-supervised learning, weakly-supervised learning as well as missing annotations, namely one where we have at least two non-labeled instances of a given class in specified regions of an image. A core difference between our use case and the others is that objects in group boxes usually strongly overlap. We argue that existing pseudo-label generation methods are not suited for dealing with group annotations because they do not properly handle overlapping objects. Instead, our graph-based method is specifically designed for dealing with crowd scenes. As we show in the experimental section, a representative semi-supervised method, data distillation, performs consistently lower than our approach.

3 Method

3.1 Overview

In this section, we interpret pseudo-label creation as a graph clustering problem. We suggest applying a modified version of a seed-based clustering algorithm to generate pseudo-labels. This concept is embedded in the setup visualized in Fig. 2. In the first iteration, a baseline model is trained on the given dataset using the respective standard strategy for dealing with group annotations. Using

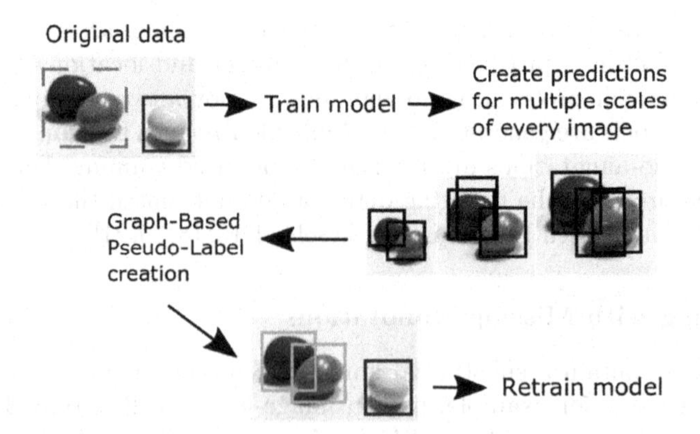

Fig. 2. Overview pseudo-label generation. A model is trained, predictions are made, pseudo-labels are generated, combined with the ground truth, and used for training a new model. The dashed box indicates a group annotation. The step of graph-based pseudo-label creation, our core innovation, is visualized in more detail in Fig. 3.

this model, predictions are generated for multiple scaled and flipped versions of every image in the training set. The multitude of predictions above a confidence threshold γ is interpreted as a graph $G_c(V_c, E_c)$ with V_c being bounding box predictions of class c and weighted edges defined as $E_c(u,v) = IOU(u,v)$ (see Fig. 3). Using the graph clustering algorithm outlined below, pseudo-labels are generated on top of group annotations and combined with the known ground truth. Then, the model is retrained with the resulting dataset. Thereby, the strategy of dealing with group annotations from the first training is kept, e.g. ignore overlapping anchors in the loss, but the regions of a group annotation where pseudo-labels were created are used in training as positive labels.

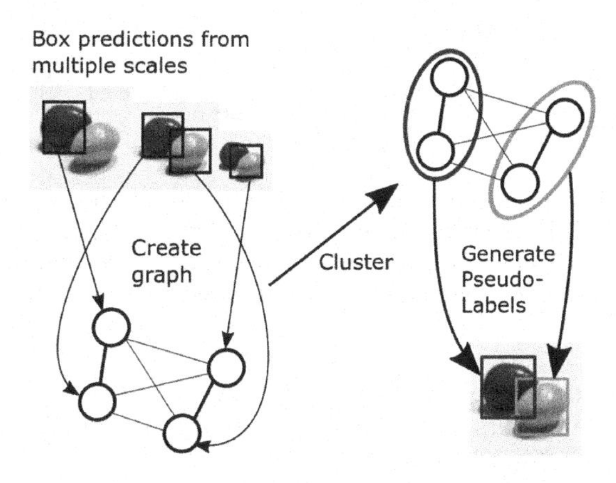

Fig. 3. Graph-based pseudo-label creation. Bounding box predictions are generated, combined and a graph is created. Clusters are extracted and used for pseudo-label generation. The width of the edges in the graph indicates the IOU between predictions. A clustering example is visualized in Fig. 4.

3.2 Graph-Based Pseudo-labels

We interpret pseudo-label creation as a graph clustering problem. To solve the problem efficiently, the chosen clustering algorithm needs to fulfill the following properties: create a varying number of clusters of different sizes, properly deal with outliers and use weighted edges. Among a large number of clustering algorithms, several fulfill the outlined conditions. We chose to modify a seed-based algorithm, DPClus [1,12], as the concept of choosing a seed node in every iteration resembles the NMS step used in bounding box voting [2]. However, by iteratively growing clusters more fine-grained decisions are made which predicted box to use for which pseudo-label, resulting in superior performance.

In our setup node weights are equivalent to the sum of edge weights connected to this node. At the start of the algorithm (see Algorithm 1), the highest weighted node V_{max} is chosen as seed (see Fig. 4). Starting from there, new nodes are

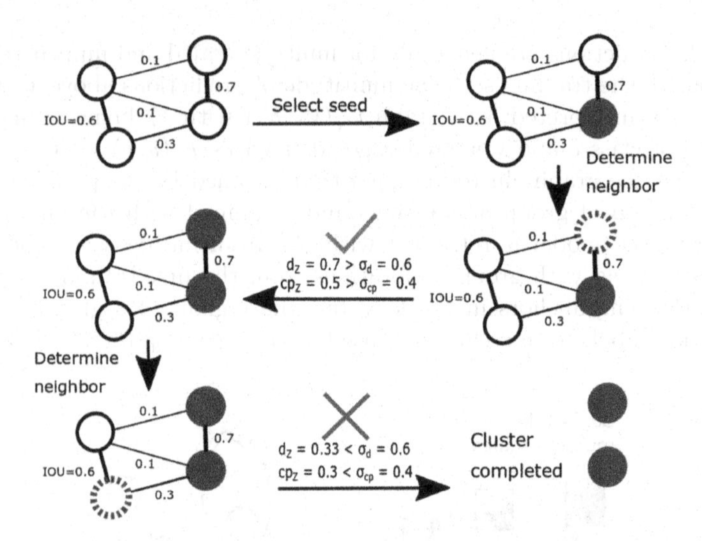

Fig. 4. Example for cluster creation. After a seed node is chosen a first neighbor is determined and added to the cluster because none of the stopping criteria is met. A second neighbor is not added because the stopping criteria are fulfilled and the cluster is completed. The width of the edges indicates the IOU between predictions.

iteratively added to the cluster. The order of new nodes to consider is determined by the sum of edge weights of neighbor nodes with cluster nodes. New nodes are added to the cluster Z until one of the following two conditions is violated: Either the density of the cluster d_Z computed as Eq. (1) drops below σ_d or the cluster property cp_Z (see Eq. (2)) gets lower than σ_{cp}. The cluster density is defined as two times the ratio between the actual sum of edge weights (The sum of edge weights W_Z in current cluster Z, the sum of edge weights W_{Zk} between selected neighbor node V_k and the nodes in the cluster V_Z) and the possible maximum of edge weights (n being the actual number of nodes in the cluster). The cluster density d_Z is a measure that indicates whether or not to add a node from the perspective of the cluster, ensuring that its nodes have strong edge connections with each other. The cluster property denotes the ratio between the sum of edge weights between the selected neighbor node and nodes in the cluster W_{Zk} and the cluster density d_Z multiplied by the number of nodes in the cluster n plus one for the neighbor to add. The cluster property considers how strong a neighbor is connected to every node in the cluster. Once a cluster is completed, its nodes are removed from the graph and a new seed node is determined.

$$d_Z = \frac{2 * (W_Z + W_{Zk})}{n * (n+1)} < \sigma_d \tag{1}$$

$$cp_Z = \frac{W_{Zk}}{d_Z * (n+1)} < \sigma_{cp} \tag{2}$$

Algorithm 1: Generate Clusters for Pseudo-Labels

Initialization: $G_c(V_c, E_c)$ with bounding box predictions V_c and weighted edges $E_c(u,v) = IOU(u,v)$, σ_d, σ_{cp}

```
1  Clusters = ∅
2  while G_c(V_c, E_c) ≠ ∅ do
3      Set V_max as seed
4      Single cluster Z = {V_max}
5      while True do
6          Select V_k which is the node with largest W_Zk
7          d_Z = 2*(W_Z + W_Zk) / (n*(n+1))
8          cp_Z = W_Zk / (d_Z*(n+1))
9          if d_Z > σ_d and cp_Z > σ_cp then
10             Add V_k to Z
11         else
12             if Z ≠ {V_max} then
13                 Add Z to Clusters
14             end
15             Remove nodes in Z from G_c(V_c, E_c)
16             break
17         end
18     end
19 end
```

3.3 Relation to Prior Work

Data distillation [14] uses a similar setup as the one shown in Fig. 2. The main difference, however, is that pseudo-labels are generated by bounding box voting [2], which starts with a step of non-maximum suppression. In the case of a graph interpretation, this means that the highest scoring node V_{max} is selected and every node which is connected by an edge with a weight of more than λ_{nms} is deleted from the graph. Afterwards, the second-highest scoring node is chosen and the process continued until no unvisited node in the graph remains. Finally, the NMS output is selected as 'seed nodes' and bounding box predictions which overlap greater than a threshold are used for refining the boxes by computing the confidence score weighted mean. This can be interpreted as a simple way to create clusters and to use them for pseudo-label creation.

We argue that, for multiple augmented images where many confident and highly overlapping box predictions are available, NMS leads to a loss of information. Especially in the case of overlapping objects, NMS introduces a crude trade-off between precision and recall. Furthermore, using every prediction with an IOU above some threshold for refining bounding boxes is suboptimal if objects are near each other.

4 Experiments

4.1 Datasets

Our approach was tested on CityPersons, MS COCO, and a subset of OpenImages.

CityPersons is a pedestrian detection dataset. Performance on CityPersons is evaluated using the log-average miss rate on subset reasonable [20] which we denote as MR. The share of group annotations is 12.6%.

For MS COCO we utilized the train2017 split which contains 115k labeled images and 80 different classes of objects. We report final results on test-dev. 1.2% of the labels in the training set are of type crowd.

Due to the size of OpenImages, we did not train on the full dataset. Instead, we conducted experiments on the 15 child classes of parent class fruit which contain around 6500 images. We chose this subset because the classes are semantically similar, yet the share of group annotations ranges from 0.9% (orange) to 25.4% (peach) and 68.8% (grape) with an average of 16%. Final results are reported on the test split.

Overall, we provide experiments for a diverse set of problems, namely single-class (CityPersons), multi-class with few group annotations (MS COCO), and multi-class with many group annotations (OpenImages).

4.2 Implementation Details

In all experiments, RetinaNet with a ResNet50 backbone was used. Most of the hyperparameters were chosen according to Lin et al. [7].

For generating pseudo-labels, the baseline model was used to create predictions for multiple augmented versions of the images. On MS COCO, we applied both scaling and flipping like described in [14]. On CityPersons and OpenImages we adapted the scaling to the size of the respective images (CityPersons 816–1310; OpenImages 400–1200) while keeping the aspect ratio fixed. In case of MS COCO and OpenImages, for example, this resulted in 18 different augmentations per image. The threshold for cluster density σ_d was set to 0.6 and for cluster property σ_{cp} to 0.4 based on experiments on MS COCO. To show the robustness of our method, we used the same values for CityPersons and OpenImages.

As a baseline for pseudo-label generation we used data distillation [14]. To ensure a fair comparison, we applied the setup described in Fig. 2 and only replaced graph-based pseudo-label creation with bounding box voting [2].

4.3 CityPersons

Our default RetinaNet setup with the settings mentioned above reaches a miss rate of 12.6 (see Table 1). Using data distillation, performance can be improved to 12.0 while our graph-based method even reaches a miss rate of 11.7. This means that pseudo-label methods improve upon standard training. Furthermore, our graph-based approach outperforms data distillation.

Table 1. Results using RetinaNet on CityPersons. MR denotes log-average miss rate on subset reasonable.

Method	MR
RetinaNet	12.6
Data distillation	12.0
Graph-based	**11.7**

4.4 MS COCO

Table 2 compares pseudo-label-based approaches with the baseline. Among the datasets that are included in this study, MS COCO contains by far the smallest share of crowd annotations. Therefore, it is not surprising that our approach only leads to a modest improvement. Nevertheless, when analyzing the results in more detail (see Table 3), we found that the larger the share of crowd annotations, the greater the improvement. If only 2–3% of annotations were of type crowd, mAP improved by 0.5%. Furthermore, the classes which achieve the greatest increase in mAP are also those which entail the greatest share of pseudo-labels (see Table 4). Table 5 shows statistics of the generated pseudo-labels. The number of pseudo-labels generated by our method exceeds the number of crowd regions by a factor of 5, indicating that on average five pseudo-labels are created per group region. The overall share of pseudo-labels among all bounding box annotations is almost 6%. Thereby, the numbers differ substantially between classes. The mean share of pseudo-labels among all bounding box annotations is 3.0%, the median 0.8%, and the maximum 15.4% (class banana).

Table 2. Results using RetinaNet on COCO test-dev.

Method	COCO test-dev mAP
RetinaNet [7]	35.7
RetinaNet	35.7
Data distillation	**35.8**
Graph-based	**35.8**

4.5 OpenImages

Table 6 shows results on the 15 fruit classes in OpenImages. As can be seen, data distillation improves upon the standard training strategy. However, our graph-based setup reaches an even higher mAP.

Table 3. Change in mAP on COCO test-dev depending on share of group annotations for our graph-based method compared to standard training. 'Share of crowd annot.' indicates the percentage of crowd annotations among all annotations. 'Change in mAP': Average increase of mAP for categories with given share of crowd annotations. 'Nb of classes': Number of classes in COCO in given range of crowd annotations

Share of crowd annot	Change in mAP	Nb of classes
0.0–0.9%	+0.0	61
1.0–1.9%	+0.1	12
2.0–3.0%	+0.5	7

Table 4. Top performing classes on MS COCO and their share of pseudo-labels among bounding boxes of the respective class

Class name	Increase in mAP	Share of Pseudo-labels
Donut	+1.1	12.68%
Banana	+0.9	15.44%
Sheep	+0.9	14.34%
Person	+0.7	10.96%
Mean	+0.1	3.0%

Table 5. Pseudo-label statistics on MS COCO for our graph-based method

Metric	Value
Total number of bboxes	850000
Total number of crowd annotations	10000
Total number of pseudo-labels	52500
Share crowd annotations among all bboxes	1.2%
Share pseudo-labels among all bboxes	5.8%
Max share of pseudo-labels per class	15.4%
Mean share of pseudo-labels per class	3.0%
Median share of pseudo-labels per class	0.8%

The change in performance depending on the share of group annotations is further investigated in table 7. It can be seen that for classes with a small share of group annotations (0–10%) a modest increase (+0.8) is achieved, whereas for those with a large proportion (>10%) mAP significantly rises (+3.2).

4.6 Qualitative Evaluation of Pseudo-labels

In this section, we conduct a qualitative analysis of pseudo-labels generated by data distillation and graph-based clustering. We visualize examples of pseudo-

Table 6. Results using RetinaNet on OpenImages test.

Method	Fruits Test mAP
Standard training	51.0
Data Distillation	52.2
Graph-Based	**53.3**

Table 7. Change in mAP for OpenImages Fruits depeding on share of group annotations. Comparison between graph-based pseudo-labeling and standard training

Share of group annot	Change in mAP	Nb of classes
0–10%	+0.8	8
>10%	+3.2	7

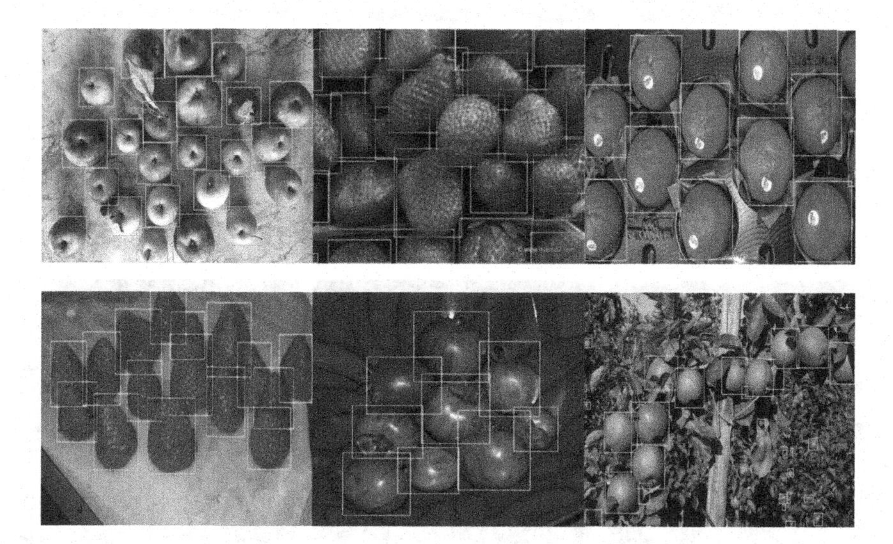

Fig. 5. Examples for pseudo-labels (green boxes) generated by our graph-based method on top on group annotations in OpenImages (Color figure online)

labels (green boxes) generated by our graph-based method (see Fig. 5). For most objects located on a group annotation, correct pseudo-labels are created.

In the case of data distillation, bounding box voting is applied which results in a trade-off between precision and recall introduced by the usage of NMS. A higher NMS threshold λ_{nms} increases recall (see Fig. 6) at the expense of false positives like double detections (see Fig. 7). Whereas NMS solves only one of the two example cases regardless of the chosen λ_{nms}, our graph-based clustering method creates correct pseudo-labels for both.

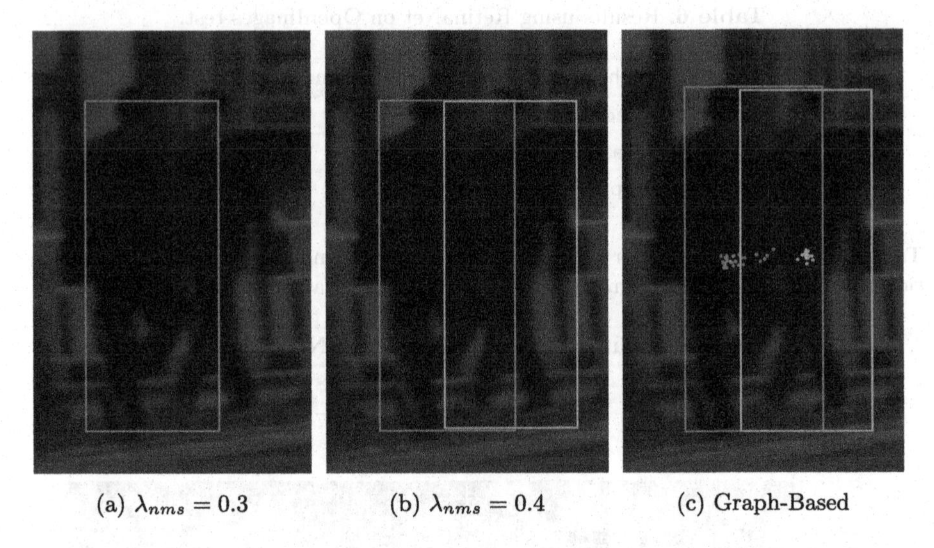

(a) $\lambda_{nms} = 0.3$ (b) $\lambda_{nms} = 0.4$ (c) Graph-Based

Fig. 6. Example for a missed detection (a) due to a low λ_{nms} used in bounding box voting. Our method successfully generates two correct pseudo-labels (c). The dots indicate the center of the predicted bounding boxes and the respective color shows the cluster assignment (Color figure online)

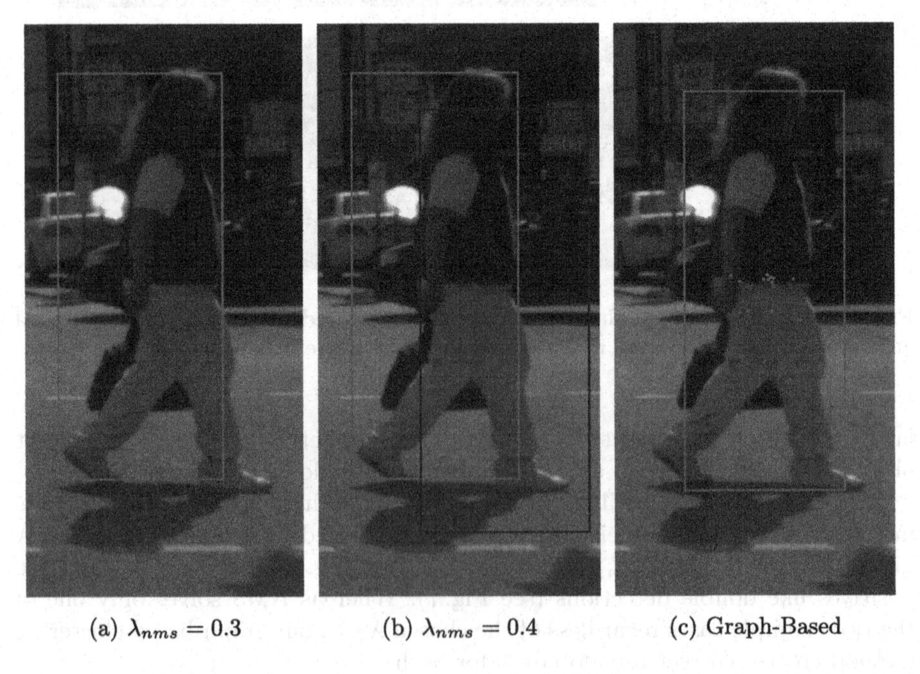

(a) $\lambda_{nms} = 0.3$ (b) $\lambda_{nms} = 0.4$ (c) Graph-Based

Fig. 7. Double detection in case the λ_{nms} is too high (b). Again, the clustering method (c) generates a correct pseudo-label

5 Conclusion

In this work, we address the problem of leveraging group annotations for training object detectors. We challenge the standard training protocol of several commonly used datasets, namely CityPersons, MS Coco, and OpenImages, and show that pseudo-label-based approaches consistently improve performance. Because previous pseudo-label methods do not account for strongly overlapping objects, we develop a new approach for dealing with those cases. Thereby, we interpret pseudo-label creation as a graph clustering problem and show that it results in superior performance compared to data distillation, a competitive pseudo-label generation method. Furthermore, we find that the greater the share of group annotations, the larger the performance improvement.

References

1. Altaf-Ul-Amin, M., Shinbo, Y., Mihara, K., Kurokawa, K., Kanaya, S.: Development and implementation of an algorithm for detection of protein complexes in large interaction networks. BMC Bioinform. **7**, 1–13 (2007)
2. Gidaris, S., Komodakis, N.: Object detection via a multi-region and semantic segmentation-aware CNN model. In: 2015 IEEE International Conference on Computer Vision (ICCV), pp. 1134–1142 (2015)
3. Hasan, I., Liao, S., Li, J., Akram, S.U., Shao, L.: Generalizable pedestrian detection: The elephant in the room. In: Proceedings of the IEEE/CVF Conference on Computer Vision and Pattern Recognition (CVPR), pp. 11328–11337 (2021)
4. Huang, Z., Zou, Y., Kumar, B.V.K.V., Huang, D.: Comprehensive attention self-distillation for weakly-supervised object detection. In: Advances in Neural Information Processing Systems, vol. 33, pp. 16797–16807 (2020)
5. Jeong, J., Lee, S., Kim, J., Kwak, N.: Consistency-based semi-supervised learning for object detection. In: Advances in Neural Information Processing Systems, vol. 32 (2019)
6. Kuznetsova, A., et al.: The open images dataset V4. Int. J. Comput. Vis. **128**, 956–1981 (2020)
7. Lin, T.Y., Goyal, P., Girshick, R., He, K., Dollar, P.: Focal loss for dense object detection. In: Proceedings of the IEEE International Conference on Computer Vision (ICCV) (2017)
8. Lin, T.Y., et al.: Microsoft coco: common objects in context (2015). arXiv:1405.0312
9. Liu, W., Liao, S., Ren, W., Hu, W., Yu, Y.: High-level semantic feature detection: a new perspective for pedestrian detection. In: Proceedings of the IEEE/CVF Conference on Computer Vision and Pattern Recognition (CVPR) (2019)
10. Liu, Y.C., et al.: Unbiased teacher for semi-supervised object detection. In: International Conference on Learning Representations ICLR (2021)
11. Niitani, Y., Akiba, T., Kerola, T., Ogawa, T., Sano, S., Suzuki, S.: Sampling techniques for large-scale object detection from sparsely annotated objects. In: Proceedings of the IEEE/CVF Conference on Computer Vision and Pattern Recognition (CVPR) (2019)
12. Price, T., Pen, F., Cho, Y.R.: Survey: enhancing protein complex prediction in PPI networks with GO similarity weighting. Interdisc. Sci. Computat. Life Sci. **5**, 196–210 (2013)

13. Qiao, S., Chen, L.C., Yuille, A.: DetectoRS: detecting objects with recursive feature pyramid and switchable atrous convolution. In: Proceedings of the IEEE/CVF Conference on Computer Vision and Pattern Recognition (CVPR), pp. 10213–10224 (2021)
14. Radosavovic, I., Dollár, P., Girshick, R., Gkioxari, G., He, K.: Data distillation: towards omni-supervised learning. In: Proceedings of the IEEE Conference on Computer Vision and Pattern Recognition (CVPR) (2018)
15. Ren, Z., et al.: Instance-aware, context-focused, and memory-efficient weakly supervised object detection. In: Proceedings of the IEEE/CVF Conference on Computer Vision and Pattern Recognition (CVPR) (2020)
16. Sohn, K., Zhang, Z., Li, C.L., Zhang, H., Lee, C.Y., Pfister, T.: A simple semi-supervised learning framework for object detection (2020). arXiv:2005.04757
17. Tang, P., Ramaiah, C., Wang, Y., Xu, R., Xiong, C.: Proposal learning for semi-supervised object detection. In: Proceedings of the IEEE/CVF Winter Conference on Applications of Computer Vision (WACV), pp. 2291–2301 (2021)
18. Wan, F., Liu, C., Ke, W., Ji, X., Jiao, J., Ye, Q.: C-MIL: continuation multiple instance learning for weakly supervised object detection. In: Proceedings of the IEEE/CVF Conference on Computer Vision and Pattern Recognition (CVPR) (2019)
19. Zeng, Z., Liu, B., Fu, J., Chao, H., Zhang, L.: WSOD2: learning bottom-up and top-down objectness distillation for weakly-supervised object detection. In: Proceedings of the IEEE/CVF International Conference on Computer Vision (ICCV) (2019)
20. Zhang, S., Benenson, R., Schiele, B.: CityPersons: a diverse dataset for pedestrian detection (2017). arXiv:1702.05693
21. Zhang, Y., Bai, Y., Ding, M., Li, Y., Ghanem, B.: W2F: a weakly-supervised to fully-supervised framework for object detection. In: Proceedings of the IEEE Conference on Computer Vision and Pattern Recognition (CVPR) (2018)
22. Zoph, B., et al.: Rethinking pre-training and self-training (2020). arXiv:2006.06882

Quantifying Uncertainty of Image Labelings Using Assignment Flows

Daniel Gonzalez-Alvarado[1,2](✉), Alexander Zeilmann[1], and Christoph Schnörr[1,2]

[1] Image and Pattern Analysis Group, Heidelberg University, Heidelberg, Germany
daniel.gonzalez@iwr.uni-heidelberg.de
[2] Heidelberg Collaboratory for Image Processing, Heidelberg University, Heidelberg, Germany

Abstract. This paper introduces a novel approach to uncertainty quantification of image labelings determined by assignment flows. Local uncertainties caused by ambiguous data and noise are estimated by fitting Dirichlet distributions and pushed forward to the tangent space. The resulting first- and second-order moments are then propagated using a linear ODE parametrization of assignment flows. The corresponding moment evolution equations can be solved in closed form and numerically evaluated using iterative Krylov subspace techniques and low-rank approximation. This results in a faithful representation and quantification of uncertainty in the output space of image labelings, which is important in all applications where confidence in pixelwise decisions matters.

Keywords: Image labeling · Uncertainty quantification · Assignment flows

1 Introduction

1.1 Overview, Motivation

Quantifying the uncertainty of image segmentations and labelings is an important topic in the field of image analysis and computer vision. For more than three decades, probabilistic graphical models [35] and Bayesian inference was the method of choice. Due to the intractability of exact inference, however, quantifying uncertainty of inference, e.g. by evaluating marginal probabilities of the posterior distribution, requires either computational expensive MCMC-based sampling methods or variational approximations [9] whose performance is difficult to assess. Another line of research relying on continuous variational approaches to image segmentation, like the relaxed Mumford-Shah functional [3], employs polynomial chaos expansions for uncertainty propagation [22], a framework for approximate stochastic computing that is widely used in scientific computing [36]. The recent work [7] focuses on binary image segmentation from the viewpoint of semi-supervised learning using an approach that combines

C. Bauckhage et al. (Eds.): DAGM GCPR 2021, LNCS 13024, pp. 453–466, 2021.
https://doi.org/10.1007/978-3-030-92659-5_29

ideas from phase-field models [3], classical Gaussian processes for classification [23] and spectral graph theory [12].

In summary, before the era of deep networks, methods for quantifying the uncertainty of image segmentations mainly stressed the *Bayesian viewpoint* and employed established methods for *approximate* inference. Accordingly, there are two aspects that put into question the trustworthiness of corresponding uncertainty estimates: On the one hand, uncertainty quantification using methods where exact inference is intractable defines an intractable problem as well, and quantifying the approximation error is a hard task. On the other hand, Bayesian inference itself has been increasingly criticized recently as being too sensitive to misspecification of models and priors [13,20].

For several years, the state of the art in image segmentation is based on deep networks [19]. It is well-known too, however, that current deep network architectures come along with deficiencies: sensitivity against tiny perturbations (e.g. adversarial attacks) with unpredictable consequences, and with insufficient theoretical understanding of how predictions are generated. It is not surprising, therefore, that uncertainty quantification based on deep networks is an unsolved problem that has been tackled by hundreds of different architectures and heuristics during the last years – see [1] and more than 700 references therein.

1.2 Contribution, Organization

This paper is based on the *assignment flow* approach to image labeling [4,30] that provides a framework for studying the mathematics of deep networks in 'small controlled' steps, yet outperforms already in its present form traditional variational and graphical models (cf., e.g. [31,32]). Specifically, we consider the *linearized assignment flow* [38] which enjoys a parametrization through a *linear ODE* (ordinary differential equation) on the tangent space and reasonably approximates the (full) assignment flow [37,38].

We employ a basic method for estimating *locally* initial first- and second-order statistical moments from given input data. Next, we exploit the specific structure of the linearized assignment flow for propagating these initial uncertainties to the *nonlocal* labelings that result from integrating numerically the flow.

Our approach should not be confused with 'normalizing flows' [18,21] that estimate a generative model of the data distribution through optimal transport of a reference distribution to a given empirical distribution (data). Rather, we *quantify* the *uncertainty of label assignments* performed by a context-sensitive nonlocal process (linearized assignment flow), caused by uncertainties of noisy and ambiguous initial data.

Section 2 collects basic material required for presenting our contribution in Sect. 3. Details of our implementation are specified in Sect. 4. Proof-of-concept experiments validate and illustrate our approach in Sect. 5. We conclude in Sect. 6.

2 Preliminaries

2.1 Categorial Distributions, Dirichlet Distribution

We denote in this paper by c the number of classes (labels, categories). *Categorial distributions* are points $p \in \Delta_{c-1} = \{q \in \mathbb{R}^c_+ : \langle \mathbb{1}_c, q\rangle = 1\}$ in the probability simplex. The subset of strictly positive vectors

$$\mathcal{S} = \{p \in \Delta_{c-1} : p_1 > 0, \dots, p_c > 0\} \tag{1}$$

together with the Fisher-Rao metric $g_p(u,v) = \langle u, \mathrm{Diag}(p)^{-1} v\rangle,\ u, v \in T_0$ [2] becomes a Riemannian manifold $(\mathcal{S}, g)$ with trivial tangent bundle $T\mathcal{S} = \mathcal{S} \times T_0$ and tangent space $T_0 = \{v \in \mathbb{R}^c : \langle \mathbb{1}_c, v\rangle = 0\}$. The orthogonal projection onto T_0 is denoted by

$$\Pi_0 : \mathbb{R}^c \to T_0, \qquad v \mapsto v - \langle \mathbb{1}_{\mathcal{S}}, v\rangle \mathbb{1}_c, \qquad \mathbb{1}_{\mathcal{S}} = \frac{1}{c}\mathbb{1}_c. \tag{2}$$

The *Dirichlet distribution* [8,17]

$$\mathcal{D}_\alpha(p) = \frac{\Gamma(\alpha_0)}{\Gamma(\alpha_1)\cdots\Gamma(\alpha_c)} p_1^{\alpha_1 - 1} \cdots p_c^{\alpha_c - 1} \mathbb{I}_{\mathcal{S}}(p), \quad \alpha \in \mathbb{R}^c_{>0}, \quad \alpha_0 = \langle \mathbb{1}_c, \alpha\rangle \tag{3}$$

with $\mathbb{I}_{\mathcal{S}}(p) = 1$ if $p \in \mathcal{S}$ and 0 otherwise, belongs to the exponential family of distributions [5,35]. It is strictly unimodal as long as the concentration parameters satisfy $\alpha_1, \dots, \alpha_c \geq 1$. We denote by $\mathbb{E}_\alpha[\cdot], \mathrm{Cov}_\alpha[\cdot]$ the expectation and covariance operator with respect to $\mathcal{D}_\alpha$ and record the relations for a random vector $p \sim \mathcal{D}_\alpha$

$$\mathbb{E}_\alpha[\log p] = \psi(\alpha) - \psi(\alpha_0)\mathbb{1}_c, \tag{4a}$$

$$\mathrm{Cov}_\alpha[\log p, \log p] = \mathrm{Diag}(\psi'(\alpha)) - \psi'(\alpha_0)\mathbb{1}_c\mathbb{1}_c^\top, \tag{4b}$$

where $\psi(\alpha) := (\psi(\alpha_1), \dots, \psi(\alpha_c))^\top$ and ψ, ψ' are the digamma and trigamma functions, respectively [16, pp. 8–9].

We will use Dirichlet distributions in Sect. 3.1 for estimating local uncertainties of input data, to be propagated to uncertainties of non-local labelings by linearized assignment flows in Sects. 3.2 and 3.3.

2.2 Linearized Assignment Flows

Linearized assignment flows (LAFs) were introduced by [38] as approximations of (full) assignment flows [4,30] for metric data labeling on graphs $G = (I, E)$. The linearization concerns the parametrization of assignment flows on the tangent space T_0 through a *linear* ODE with respect to a linearization point $W_0 \in \mathcal{W} := \mathcal{S} \times \cdots \times \mathcal{S}$ ($n = |I|$ factors). Each point $W \in \mathcal{W} \subset \mathbb{R}^{n \times c}$ of the assignment manifold $\mathcal{W}$ comprises as row vector a categorial distribution $W_i \in \mathcal{S}$, $i \in I$, with $\mathcal{S}$ defined by (1). Adopting the barycenter $W_0 = \mathbb{1}_{\mathcal{W}}$ of the assignment manifold defined by $(\mathbb{1}_{\mathcal{W}})_i = \mathbb{1}_{\mathcal{S}}$ (cf. (2)) as simplest choice of the point of linearization, the LAF takes the form

$$\dot{V} = V_D + \Omega V, \qquad V_D = \Pi_0 L_D, \quad V(0) = 0, \tag{5}$$

where $V(t) \in \mathcal{T}_0 = T_0 \times \cdots \times T_0 \subset \mathbb{R}^{n \times c}$ ($|I|$ factors), i.e. each row vector $V_i(t)$ evolves on T_0, $\Omega \in \mathbb{R}^{n \times n}_{+}$ is a parameter matrix that defines the regularization properties of the LAF, and

$$L_D \in \mathcal{W}, \qquad L_{D;i} = \exp_{\mathbb{1}_{\mathcal{S}}}(-D_i/\rho) := \frac{e^{-D_i/\rho}}{\langle \mathbb{1}_c, e^{-D_i/\rho} \rangle}, \quad \rho > 0, \quad i \in I \tag{6}$$

encodes the input data as point on the assignment manifold (here, the exponential function applies componentwise). The input data are given as distance vector field $D \in \mathbb{R}^{n \times c}$, where each vector $D_i \in \mathbb{R}^c$ encodes the distances D_{ij}, $j \in [c]$ of the datum (feature) observed at vertex $i \in I$ to c pre-specified labels (class prototypes). Both integration (inference) of (5) and supervised learning of the parameter matrix Ω by minimizing a LAF-constrained loss function can be computed efficiently [37,38]. Once $V(T)$ has been computed for a sufficiently large time $t = T$, almost hard label assignments to the data are given by the assignment vectors

$$W(T), \qquad W_i(T) = \exp_{\mathbb{1}_{\mathcal{S}}}(V_i(T)) = \frac{e^{V_i(T)}}{\langle \mathbb{1}_c, e^{V_i(T)} \rangle}, \quad i \in I. \tag{7}$$

Our contribution in this paper to be developed subsequently is to augment labelings (7) with uncertainty estimates, by propagating initial local uncertainties that are estimated from the data using the LAF.

3 Modeling and Propagating Uncertainties

3.1 Data-Based Local Uncertainty Estimation

Local initial labeling information is given by the input data (6) and the weight patches

$$\Omega_i = \{\omega_{ik} \colon k \in \mathcal{N}_i\}, \quad \mathcal{N}_i = \{i\} \cup \{k \in I : i \sim k\}, \quad i \in I, \tag{8}$$

that form sparse row vectors of the weight matrix Ω of the LAF equation (5). We denote the weighted geometric averaging of the categorial distributions (6) by

$$S_i := \Omega_i * L_D := \frac{\prod_{k \in \mathcal{N}_i} (L_{D;k})^{w_{ik}}}{\langle \mathbb{1}_c, \prod_{k \in \mathcal{N}_i} (L_{D;k})^{w_{ik}} \rangle}, \quad i \in I, \tag{9}$$

where exponentiation is done componentwise. This formula can be derived by performing a single step towards the Riemannian center of mass of the vectors (6), but with the e-connection from information geometry in place of the Riemannian (Levi-Civita) connection [30, Sect. 2.2.2].

Local data variations can thus be represented at every pixel $i \in I$ by the set

$$\mathcal{F}_i := \{S_j : j \in \mathcal{N}_i\}, \quad \text{where} \quad S_j = \Omega_j * L_D. \tag{10}$$

We use Dirichlet distributions as a natural statistical model to represent the initial uncertainties of these data, i.e. $\mathcal{F}_i \sim \mathcal{D}_{\alpha^i}$ at every pixel $i \in I$. We estimate the parameters α^i by maximizing the corresponding log-likelihood. To this end, we exploit the structure of the Dirichlet distribution as member of the exponential family and rewrite the densities (3) in the form

$$\mathcal{D}_{\alpha^i} : S \mapsto e^{\langle T(S), F(\alpha^i)\rangle - \zeta(\alpha^i)}, \quad i \in I, \tag{11}$$

where

$$T(S) = \log(S), \quad F(\alpha^i) = \alpha^i - \mathbb{1}_c, \quad \zeta(\alpha^i) = \sum_{k=1}^{c} \log \Gamma(\alpha^i_k) - \log \Gamma(\alpha^i_0). \tag{12}$$

Assuming that $S_1, \ldots, S_{|\mathcal{N}_i|}$ are i.i.d. samples at every pixel $i \in I$, for simplicity, it follows that $S_{\mathcal{F}_i} := \sum_{j \in \mathcal{N}_i} T(S_j)$ is a sufficient statistic for the Dirichlet class $\mathcal{D}_{\alpha^i}$ [10, Theorem 1.11]. Hence the parameters α^i which best fit the data $\mathcal{F}_i$ can be obtained by maximizing the log-likelihood functions

$$\mathcal{L}_i : \mathbb{R}^c_{>0} \to \mathbb{R}, \quad \alpha \mapsto \mathcal{L}_i(\alpha) = \langle S_{\mathcal{F}_i}, F(\alpha)\rangle - |\mathcal{N}_i|\zeta(\alpha), \quad i \in I. \tag{13}$$

This is a concave optimization problem [25], [10, Lemma 5.3] with derivatives of the objective function given by

$$\nabla \mathcal{L}_i(\alpha) = |\mathcal{N}_i|\Big(\psi(\alpha_0)\mathbb{1}_c - \psi(\alpha) + \frac{1}{|\mathcal{N}_i|}\sum_{j \in \mathcal{N}_i} \log(S_j)\Big), \tag{14a}$$

$$\mathcal{H}(\mathcal{L}_i)(\alpha) = |\mathcal{N}_i|\left(-\mathrm{Diag}(\psi'(\alpha)) + \psi'(\alpha_0)\mathbb{1}_c\mathbb{1}_c^\top\right). \tag{14b}$$

We solve this problem numerically as specified in Sect. 4 and point here merely out that Hessian-based algorithms are convenient to apply due to the structure of (14a), since $\mathcal{H}(\mathcal{L}_i)^{-1}(\alpha)$ can be computed in linear time [15].

3.2 Local Uncertainty Representation

We transfer the local data uncertainties obtained in the preceding section in terms of the Dirichlet distributions $\mathcal{D}_{\alpha^i}$, $i \in I$ to the tangent space T_0, in order to compute first- and second-order moments. These moments will be propagated to labelings using the LAF in Sect. 3.3.

The following proposition uses the inverse of the diffeomorphism $\exp_{\mathbb{1}_\mathcal{S}}$ defined by (6),

$$\phi\colon \mathcal{S} \to T_0, \qquad p \mapsto \phi(p) := \exp^{-1}_{\mathbb{1}_\mathcal{S}}(p) = \Pi_0 \log p, \tag{15}$$

where the log applies componentwise. See [29, Lemma 3.1] for more details.

Proposition 1. *Let $\mathcal{D}_\alpha$ be a given Dirichlet distribution, and let $\nu = \phi_\sharp \mathcal{D}_\alpha$ denote the pushforward measure of $\mathcal{D}_\alpha$ by ϕ and $\mathbb{E}_\nu, \mathrm{Cov}_\nu$ the corresponding expectation and covariance operators. Then, for a random vector $v \sim \nu$,*

$$\mathbb{E}_\nu[v] = \Pi_0 \psi(\alpha), \tag{16a}$$

$$Cov_\nu[v, v] = \Pi_0 Diag(\psi'(\alpha)) \Pi_0. \tag{16b}$$

Proof. By definition of the pushforward operation, for any integrable function $f: T_0 \to \mathbb{R}$ and $v = \phi(p)$, one has $\mathbb{E}_\nu[f(v)] = \mathbb{E}_\alpha[f \circ \phi(p)]$ and consequently

$$\mathbb{E}_\nu[v_j] = \mathbb{E}_\alpha[\phi(p)_j] \overset{(15),(2)}{=} \mathbb{E}_\alpha[\log p_j - \langle \mathbb{1}_S, \log p \rangle] \overset{4a}{=} \psi(\alpha_j) - \langle \mathbb{1}_S, \psi(\alpha) \rangle, \tag{17}$$

which implies (16a). Regarding (16b), we compute

$$\mathrm{Cov}_\nu[v, v] = \mathbb{E}_\nu[(v - \mathbb{E}_\nu[v])(v - \mathbb{E}_\nu[v])^\top] = \Pi_0 \mathrm{Cov}_\alpha[\log p, \log p] \Pi_0 \tag{18}$$

which implies (16b) by (4b) and by the equation $\Pi_0 \mathbb{1}_c = 0$. □

Applying this proposition with $V_i(0)$ and α^i in place of v and α, respectively, we get the first- and second-order Dirichlet moments lifted to T_0,

$$m_{0,i} := \Pi_0 \psi(\alpha^i), \tag{19a}$$

$$\Sigma_{0,i} := \Pi_0 \mathrm{Diag}(\psi'(\alpha^i)) \Pi_0. \tag{19b}$$

3.3 Uncertainty Propagation

In this section, we model the initial data of the LAF (5) as Gaussian Markov random field (GMRF) and study the Gaussian random process induced by the LAF, as a model for propagating the initial data uncertainties to the resulting labeling.

Given the lifted Dirichlet moments (19), we regard each initial vector $V_i(0)$, $i \in I$ of the LAF (5) as normally distributed vector

$$V_i(0) \sim \mathcal{N}(m_{0,i}; \Sigma_{0,i}). \tag{20}$$

The formulae (19) show that these normal distributions are supported on the tangent space T_0.

Next, we stack the row vectors $V_i(0)$, $i \in I$ of $V(0)$ and denote the resulting vector by

$$v(0) := \mathrm{vec}_r(V(0)) := \big(V_1(0)^\top, V_2(0)^\top, \dots, V_n(0)^\top\big)^\top, \quad n = |I|. \tag{21}$$

Thus, $v(0)$ is governed by the GMRF

$$v(0) \sim \mathcal{N}(m_0, \Sigma_0), \qquad m_0 = \mathrm{vec}_r(m_{0,i})_{i \in I}, \quad \Sigma_0 = \mathrm{BlockDiag}(\Sigma_{0,i})_{i \in I}. \tag{22}$$

In fact, since the covariance matrices $\Sigma_{0,i}$ are singular with rank $c-1$, this is an *intrinsic* GMRF of first order – cf. [26, Chap. 3]. Conforming to (21), we transfer the LAF equation to its vectorized version

$$\dot{v} = v_D + \Omega_c v, \quad v(0) \sim \mathcal{N}(m_0, \Sigma_0), \tag{23}$$

where $\dot{v} = \mathrm{vec}_r(\dot{V})$, $v = \mathrm{vec}_r(V)$, $v_D = \mathrm{vec}_r(V_D)$ and $\Omega_c = \Omega \otimes I_c$ using the stacking convention $\mathrm{vec}_r(ABC) = (A \otimes C^\top)\mathrm{vec}_r(B)$ using the Kronecker product $\otimes$ [34] for arbitrary matrices A, B, C with compatible dimensions.

The evolution of the initial probability distribution that governs the evolving state $v(t)$ of (23) is generally described by the Fokker-Planck equation [24,28]. For the specific simple case considered here, i.e. a linear deterministic ODE with random initial conditions, this is a Gaussian random process with moments determined by the differential equations

$$\dot{m}(t) = \mathbb{E}[\dot{v}(t)], \quad m(0) = m_0 \tag{24a}$$

$$\dot{\Sigma}(t) = \mathbb{E}[\dot{v}(t)(v(t) - m(t))^\top] + \mathbb{E}[(v(t) - m(t))\dot{v}(t)^\top], \quad \Sigma(0) = \Sigma_0, \tag{24b}$$

with initial moments given by (23). Equations (24) propagate the uncertainty on the tangent space until the time of point $t = T$ and to the labeling in terms of the distribution governing the state $v(T) = \mathrm{vec}_r(V(T)) \sim \mathcal{N}(m(T), \Sigma(T))$ and Eq. (7). Regarding the quantitative evaluation of these uncertainties, we focus on the *marginal* distributions of the subvectors $V_i(t)$, $i \in I$.

Proposition 2. *Let $v(t)$ be the random vector solving (23). Then the first- and second-order moments of the marginal distributions of $V_i(t) = (vec_r^{-1}(v(t)))_i$ are given by*

$$(m_i(t))_{i \in I} = m(t) = \mathrm{expm}(t(\Omega_c))m_0 + t\varphi(t\Omega_c)v_D \tag{25a}$$

$$\Sigma_i(t) = \sum_{j \in I} \left((\mathrm{expm}(t\Omega))_{ij} \right)^2 \Sigma_{0,j}, \quad i \in I, \tag{25b}$$

where expm *denotes the matrix exponential and φ the matrix-valued function given by the entire function $\varphi : z \mapsto \frac{e^z - 1}{z}$.*

Proof. The solution (25a) results from the linear equation

$$\dot{m}(t) = \mathbb{E}[\dot{v}(t)] = \mathbb{E}[\Omega_c v(t) + v_D] = \Omega_c m(t) + v_D \tag{26}$$

and applying Duhamel's formula [33, Eq. (3.48)], to obtain

$$m(t) = \mathrm{expm}(t\Omega_c)m_0 + \mathrm{expm}(t\Omega_c) \int_0^t \mathrm{expm}(-s\Omega_c)v_D \, ds \tag{27a}$$

$$= \mathrm{expm}(t\Omega_c)m_0 + t\,\mathrm{expm}(t\Omega_c) \sum_{k=0}^{\infty} \left[\frac{(-t\Omega_c)^k}{(k+1)!} \right] v_D \tag{27b}$$

$$= \mathrm{expm}(t(\Omega_c))m_0 + t\varphi(t\Omega_c)v_D. \tag{27c}$$

In order to derive (25b), we first compute separately the two terms on the r.h.s. of (24b) which yields

$$\dot{\Sigma}(t) = \Omega_c \Sigma(t) + \Sigma(t) \Omega_c^\top . \tag{28}$$

This differential Lyapunov equation has the solution [6, Theorem 1]

$$\Sigma(t) = \mathrm{expm}(t\Omega_c) \Sigma_0 \left(\mathrm{expm}(t\Omega_c)\right)^\top . \tag{29}$$

In order to compute the marginal covariance matrices, we use properties of the Kronecker product [34] and compute

$$\mathrm{expm}((t\Omega_c)) = \sum_{k=0}^{\infty} \frac{(t\Omega)^k \otimes (I_c)^k}{k!} = \sum_{k=0}^{\infty} \left(\frac{(t\Omega)^k}{k!}\right) \otimes I_c = \mathrm{expm}(t\Omega) \otimes I_c, \tag{30}$$

which together with $\mathrm{expm}(t(\Omega \otimes I_c))^\top = \mathrm{expm}(t(\Omega^\top \otimes I_c))$ transforms (29) into

$$\Sigma(t) = \left(\mathrm{expm}(t\Omega) \otimes I_c\right) \Sigma_0 \left(\mathrm{expm}(t\Omega^\top) \otimes I_c\right) . \tag{31}$$

Since the matrix $\Sigma(0) = \Sigma_0$ is block diagonal, we have

$$\left(\mathrm{expm}(t\Omega) \otimes I_c\right) \Sigma_0 = \begin{pmatrix} \mathrm{expm}(t\Omega)_{1,1} \Sigma_{0,1} & \cdots & \mathrm{expm}(t\Omega)_{1,n} \Sigma_{0,n} \\ \vdots & \ddots & \vdots \\ \mathrm{expm}(t\Omega)_{n,1} \Sigma_{0,1} & \cdots & \mathrm{expm}(t\Omega)_{n,n} \Sigma_{0,n}, \end{pmatrix}, \quad n = |I| \tag{32}$$

and thus can extract from (31) the covariance matrices

$$\Sigma_i(t) = \sum_{j \in I} \left(\mathrm{expm}(t\Omega)\right)_{ij} \Sigma_{0,j} \left(\mathrm{expm}(t\Omega^\top)\right)_{ji} = \sum_{j \in I} \left(\left(\mathrm{expm}(t\Omega)\right)_{ij}\right)^2 \Sigma_{0,j} \tag{33}$$

□

of the marginal distributions. The direct numerical evaluation of the equations (25) is infeasible for typical problem sizes, but can be conveniently done using a low-rank approximation; see Sect. 4.2.

4 Algorithms

4.1 Estimating Local Uncertainties and Moments

We use gradient ascent to maximize numerically the log-likelihoods $\mathcal{L}_i$ in (13) for every pixel $i \in I$. The corresponding global maxima correspond to parameter values α^i which best represent local data variations S_j in (10) in terms of Dirichlet distributions $\mathcal{D}_{\alpha^i}$ given by (3) and (11), respectively. While numerical gradient ascent is safe, Hessian-based optimization method (cf. (14a)) may be more time-efficient but require numerical damping to ensure convergence [15,25].

4.2 Moment Evolution

The explicit computation of the moments $m_i(t)$ and $\Sigma_i(t)$ of the marginal uncertainty distributions in (25) require the evaluation of the matrix exponential followed by a matrix-vector multiplication. Such operations are typically intractable numerically due to the large size of the involved matrices. Recall that the size of $\Omega \in \mathbb{R}^{|I|\times|I|}$ is quadratic in the number of pixels. We overcome this numerical problem by using low-rank Krylov subspace approximation [14,27] that has shown to perform very well in connection with the linearized assignment flow [37,38].

Evolution of Means. We first describe the numerical evaluation of the mean $m_0 := m(0)$ in (25a) at any pixel $i \in I$. Using the Arnoldi iteration, we compute an orthonormal basis for each of the Krylov subspaces

$$\mathcal{K}_d(\Omega, m_0) := \operatorname{span}\{m_0, \Omega m_0, \dots, \Omega^{d-1} m_0\}, \tag{34}$$

$$\mathcal{K}_d(\Omega, v_D) := \operatorname{span}\{v_D, \Omega v_D, \dots, \Omega^{d-1} v_D\}, \tag{35}$$

of dimension $d \ll |I|$. This iterative method yields a basis $V_d = (v_1, \dots, v_d)$ together with the upper Hessenberg matrix

$$H_d = V_d^\top \Omega V_d, \tag{36}$$

without having to compute explicitly the right-hand side of (36). We use this Hessenberg matrix, which represents the projection of Ω onto the Krylov subspace, to approximate the action of the matrix exponential [27]

$$\operatorname{expm}(t\Omega) m_0 \approx \|m_0\| V_d \operatorname{expm}(tH_d) e_1, \tag{37}$$

where e_1 denotes the first unit vector. In agreement with [37,38], our experience is that the computational costs are reduced remarkably while still very accurate approximations are obtained even when working with low-dimensional Krylov subspaces ($d \approx 10$).

In order to approximate the second term on the right-hand side of (25a), we consider the matrices H_d, V_d corresponding to the Krylov space (35) and employ the approximation

$$t\varphi(t\Omega) v_D \approx t\|v_D\| V_d\, \varphi(tH_d) e_1, \tag{38}$$

where the action of the matrix-valued φ-function on the vector e_1 can be derived using an evaluation of the matrix exponential of the form [27, Proposition 2.1]

$$\operatorname{expm}\left(t \begin{pmatrix} H_d & e_1 \\ 0 & 0 \end{pmatrix}\right) = \begin{pmatrix} \operatorname{expm}(tH_d) & t\varphi(tH_d) e_1 \\ 0 & 1 \end{pmatrix}. \tag{39}$$

Evolution of Covariance Matrices. For the evolution of each covariance matrix (25b), we compute the vectors

$$\sigma_i = \mathrm{expm}(t\Omega^\top)e_i \in \mathbb{R}^{|I|}, \quad i \in I \tag{40}$$

using the Krylov approximation (37) as described above and evaluate the covariance matrices of the marginal distributions by

$$\Sigma_i(t) = \sum_{j\in I} \sigma_{i,j}^2 \Sigma_{0,j}, \quad i \in I. \tag{41}$$

Here, the matrices $\Sigma_{0,j}$ correspond to the initial local uncertainties estimated at pixel $j \in I$, as described in connection with (19b).

5 Experiments and Findings

We illustrate our approach (Sect. 3) in a number of proof-of-concept experiments. We refer to Sect. 4 for implementation details and to the figure captions for a description of the experiments and our findings.

Unless otherwise specified, we used the following options in the experiments:

- The linear assignment flow was integrated up to time $t = T := 15$. This suffices to obtain almost integral assignments by (7).
- All Krylov dimensions (cf. (34), (35)) were set to $d = 10$.
- We used $|\mathcal{N}_i| = 3 \times 3$ neighborhoods for specifying the regularizing parameters (weights) by (8).
- The weight matrix Ω with sparse row vectors due to (8) were either taken as uniform weights or computed using a nonlocal means algorithm [11].

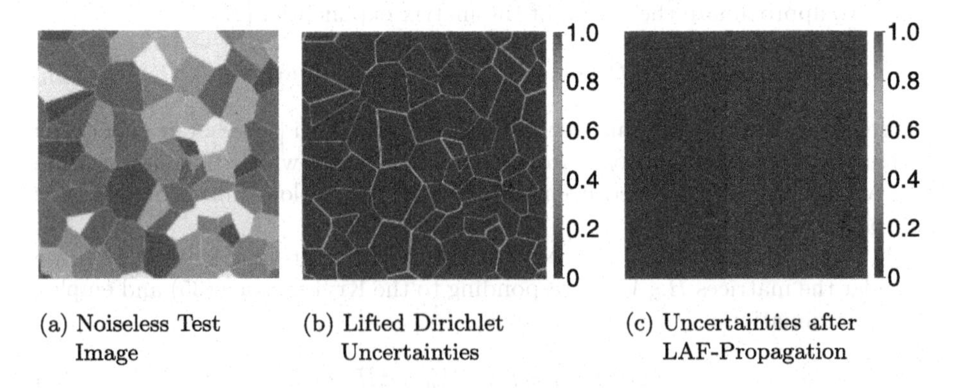

(a) Noiseless Test Image (b) Lifted Dirichlet Uncertainties (c) Uncertainties after LAF-Propagation

Fig. 1. Uncertainties for labeling a noiseless image. (a) Noiseless test image, where each of the eight colors (■, ■, ■, ■, ■, ■, ■, ■) represents a label. **(b)** Initial uncertainties at $t = 0$ based on the lifted Dirichlet moments (19). Label assignments in the interior of the cells have uncertainty zero, in agreement with the fact that local rounding already produces the correct labeling. The initial decisions at the boundaries between cells are uncertain, however. **(c)** Uncertainties of final label assignments through the LAF. Regularization performed by the LAF largely removed the initial uncertainties shown by (b), up to pixels in (a) with color values that do *not* correspond to *any* label due to rasterization effects (visible after zooming in on the screen). (Color figure online)

Computation of Uncertainties. The *uncertainty* of assigning a label to the data observed at pixel $i \in I$ is the probability that the random vector $V_i \sim \mathcal{N}(m_i(T), \Sigma_i(T))$ yields a *different* label assignment by (7), where $m_i(T), \Sigma_i(T)$

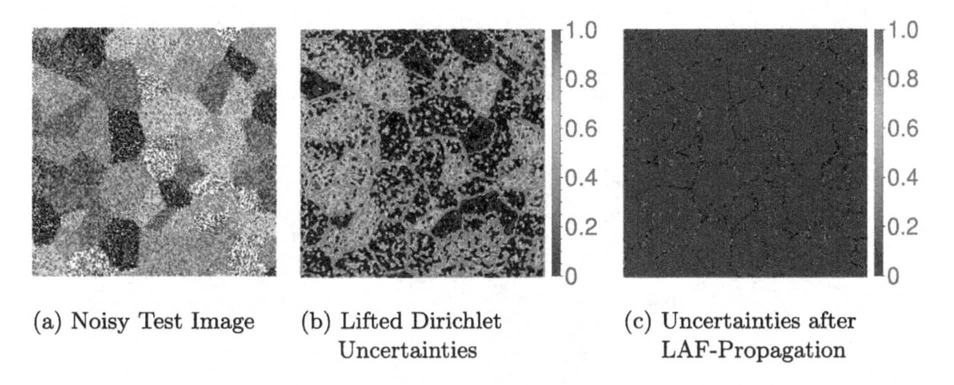

(a) Noisy Test Image (b) Lifted Dirichlet Uncertainties (c) Uncertainties after LAF-Propagation

Fig. 2. Uncertainties for labeling noisy data. (a) Noisy version of the image from Fig. 1a, **(b)** Initial uncertainties at $t = 0$ based on the lifted Dirichlet moments (19). Smaller uncertainties in yellow regions reflects the fact that the yellow label has the largest distance to all other labels. **(c)** Uncertainties of final label assignments through the LAF. The LAF effectively removes noise and feels confident, except for pixels at signal transitions. (Color figure online)

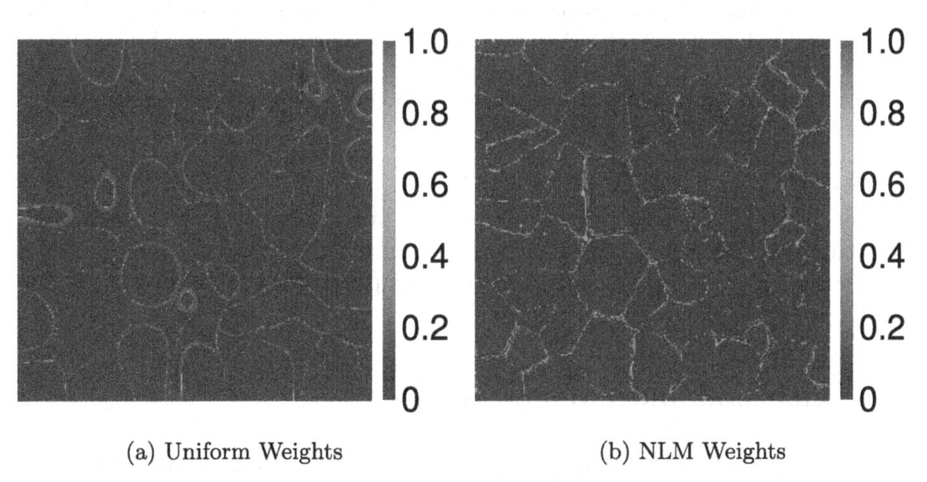

(a) Uniform Weights (b) NLM Weights

Fig. 3. Uncertainties for labeling with different weight types. Both panels display uncertainties of final label assignments through the LAF, obtained using weights computed in two ways, for the noisy data depicted by Fig. 2a. **(a)** Uniform weights size using large patches of size 11×11 pixels, i.e., each weight has the value $\frac{1}{121}$. Such 'uninformed' weights (regularization parameters) increase the uncertainty of both label assignments and localization, in particular at small-scale image structures relative to the patch size that determines the strength of regularization of the LAF. **(b)** Using nonlocal means weights in patches of size 11×11 avoids the effects displayed by (a), despite using the same patch size.

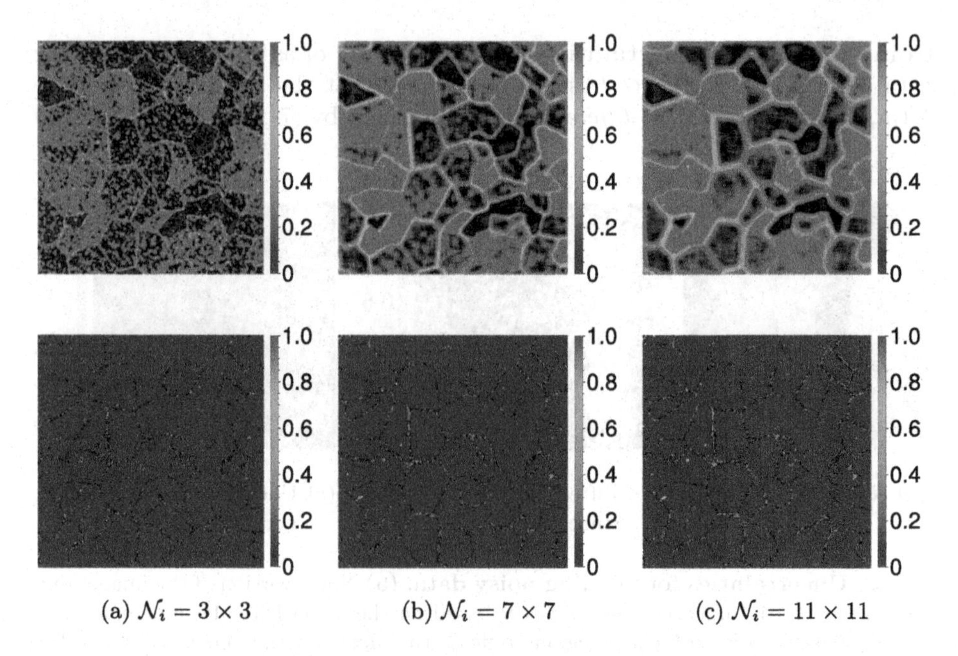

Fig. 4. Uncertainties of labelings using different neighborhood sizes. This figure illustrates the influence of the neighborhood size $|\mathcal{N}_i|$, $i \in I$ on the uncertainties of *initial* label assignments **(top row)** and on the uncertainties of *final* label assignments through the LAF **(bottom row)**, for the noisy data depicted by Fig. 2a: **(a)** $|\mathcal{N}_i| = 3 \times 3$ (left column), **(b)** $|\mathcal{N}_i| = 7 \times 7$ (center column), and **(c)** $|\mathcal{N}_i| = 11 \times 11$ (right column). Larger neighborhoods increase the initial uncertainties, whereas final uncertainties are fairly insensitive against variation of the neighborhood size due to regularization performed by the LAF.

are given by (25) at time $t = T$. In practice, we compute these uncertainties as normalized frequencies of label confusion using a sufficiently large number of samples $V_i \sim \mathcal{N}(m_i(T), \Sigma_i(T))$. These computations can be carried out in parallel for each pixel $i \in I$.

6 Conclusion and Further Work

We introduced a novel approach to uncertainty quantification of image labelings. The approach is based on simplifying assumptions, like the i.i.d. assumption underlying maximum-likelihood parameter estimation in Sect. 4.1, regarding the estimation and representation of initial uncertainties of *local* label assignments using the given data. These initial uncertainties are *rigorously* propagated using the mathematical representation of assignment flows. Numerical results validate and illustrate the approach.

Our further work will mainly focus (i) on studying how *learning* regularization parameters of the LAF [37] affects uncertainty quantification, and (ii) on an extension to the *full* assignment flow approach by approximating the flow through a composition of the linearized assignment flow (LAF) determined at few points of time.

Acknowledgement. This work is supported by the Deutsche Forschungsgemeinschaft (DFG, German Research Foundation) under Germany's Excellence Strategy EXC 2181/1 - 390900948 (the Heidelberg STRUCTURES Excellence Cluster).

References

1. Abdar, M., et al.: A review of uncertainty quantification in deep learning: techniques, applications and challenges. Inf. Fusion **76**, 243–297 (2021)
2. Amari, S.I., Nagaoka, H.: Methods of Information Geometry. American Mathematical Society and Oxford University Press, Oxford (2000)
3. Ambrosio, L., Tortorelli, V.M.: Approximation of functional depending on jumps by elliptic functional via Γ-convergence. Comm. Pure Appl. Math. **43**(8), 999–1036 (1990)
4. Åström, F., Petra, S., Schmitzer, B., Schnörr, C.: Image labeling by assignment. J. Math. Imaging Vis. **58**(2), 211–238 (2017). https://doi.org/10.1007/s10851-016-0702-4
5. Barndorff-Nielsen, O.E.: Information and Exponential Families in Statistical Theory. Wiley, Chichester (1978)
6. Behr, M., Benner, P., Heiland, J.: Solution formulas for differential Sylvester and Lyapunov equations. Calcolo **56**(4), 1–33 (2019)
7. Bertozzi, A., Luo, X., Stuart, A., Zygalakis, K.: Uncertainty quantification in graph-based classification of high dimensional data. SIAM/ASA J. Uncertain. Quantif. **6**(2), 568–595 (2018)
8. Bishop, C.: Pattern Recognition and Machine Learning. Springer, Heidelberg (2006)
9. Blei, D.M., Kucukelbir, A., McAuliffe, J.D.: Variational inference: a review for statisticians. J. Am. Stat. Assoc. **112**(518), 859–877 (2017)
10. Brown, L.D.: Fundamentals of Statistical Exponential Families. Institute of Mathematical Statistics, Hayward (1986)
11. Buades, A., Coll, B., Morel, J.M.: Image denoising methods. A new nonlocal principle. SIAM Rev. **52**(1), 113–147 (2010)
12. Chung, F.: Spectral Graph Theory. American Mathematical Society (1997)
13. Grünwald, P., van Ommen, T.: Inconsistency of Bayesian inference for misspecified linear models, and a proposal for repairing it. Bayesian Anal. **12**(4), 1069–1103 (2017)
14. Hochbruck, M., Lubich, C.: On Krylov Subspace approximations to the matrix exponential operator. SIAM J. Numer. Anal. **34**(5), 1911–1925 (1997)
15. Huang, J.: Maximum likelihood estimation of Dirichlet distribution parameters. CMU Technique report (2005)
16. Johnson, N.L., Kemp, A.W., Kotz, S.: Univariate Discrete Distributions, 3rd edn. Wiley-Interscience, Hoboken (2005)
17. Kotz, S., Balakrishnan, N., Johnson, N.L.: Continuous Multivariate Distributions: Models and Applications, vol. 1, 2nd edn. Wiley, Hoboken (2000)
18. Marzouk, Y., Moselhy, T., Parno, M., Spantini, A.: An introduction to sampling via measure transport. In: Handbook of Uncertainty Quantification, pp. 1–41. Springer, Heidelberg (2017)

19. Minaee, S., Boykov, Y.Y., Porikli, F., Plaza, A.J., Kehtarnavaz, N., Terzopoulos, D.: Image segmentation using deep learning: a survey. IEEE Trans. Pattern Anal. Mach. Intell. (2021). https://ieeexplore.ieee.org/document/9356353
20. Owhadi, H., Scovel, C.: Brittleness of Bayesian inference under finite information in a continuous world. Electr. J. Stat. **9**, 1–79 (2015)
21. Papamakarios, G., Nalisnick, E., Rezende, D.J., Mohamed, S., Lakshminarayanan, B.: Normalizing flows for probabilistic modeling and inference. J. Mach. Learn. Res. **22**(57), 1–64 (2021)
22. Pätz, T., Kirby, R.M., Preusser, T.: Ambrosio-Tortorelli segmentation of stochastic images: model extensions, theoretical investigations and numerical methods. Int. J. Comput. Vis. **103**, 190–212 (2013)
23. Rasmussen, C.E., Williams, C.: Gaussian Processes for Machine Learning. MIT Press, Cambridge (2006)
24. Risken, H.: The Fokker-Planck Equation: Methods of Solution and Applications, 2nd edn. Springer, Heidelberg (1989)
25. Ronning, G.: Maximum likelihood estimation of Dirichlet distributions. J. Stat. Comput. Simul. **32**(4), 215–221 (1989)
26. Rue, H., Held, L.: Gaussian Markov Random Fields: Theory and Applications. CRC Press (2005)
27. Saad, Y.: Analysis of some Krylov subspace approximations to the matrix exponential operator. SIAM J. Numer. Anal. **29**(1), 209–228 (1992)
28. Särkkä, S., Solin, A.: Applied Stochastic Differential Equations, vol. 10. Cambridge University Press, Cambridge (2019)
29. Savarino, F., Schnörr, C.: Continuous-domain assignment flows. Eur. J. Appl. Math. **32**(3), 570–597 (2021)
30. Schnörr, C.: Assignment flows. In: Grohs, P., Holler, M., Weinmann, A. (eds.) Variational Methods for Nonlinear Geometric Data and Applications, pp. 235–260. Springer, Heidelberg (2020)
31. Sitenko, D., Boll, B., Schnörr, C.: Assignment flow for order-constrained OCT segmentation. In: GCPR (2020)
32. Sitenko, D., Boll, B., Schnörr, C.: Assignment flow for order-constrained OCT segmentation. Int. J. Comput. Vis. **129**, 3088–3118 (2021). https://link.springer.com/content/pdf/10.1007/s11263-021-01520-5.pdf
33. Teschl, G.: Ordinary Differential Equations and Dynamical Systems, Grad. Studies Mathematics, vol. 140. American Mathematical Society (2012)
34. Van Loan, C.F.: The ubiquitous Kronecker product. J. Comput. Appl. Math. **123**, 85–100 (2000)
35. Wainwright, M., Jordan, M.: Graphical models, exponential families, and variational inference. Found. Trends Mach. Learn. **1**(1–2), 1–305 (2008)
36. Xiu, D., Karniadakis, G.E.: The Wiener-Askey polynomial chaos for stochastic differential equations. SIAM J. Sci. Comput. **24**(2), 619–644 (2002)
37. Zeilmann, A., Petra, S., Schnörr, C.: Learning linear assignment flows for image labeling via exponential integration. In: Elmoataz, A., Fadili, J., Quéau, Y., Rabin, J., Simon, L. (eds.) SSVM 2021. LNCS, vol. 12679, pp. 385–397. Springer, Cham (2021). https://doi.org/10.1007/978-3-030-75549-2_31
38. Zeilmann, A., Savarino, F., Petra, S., Schnörr, C.: Geometric numerical integration of the assignment flow. Inverse Probl. **36**(3), 034004 (2020). (33pp)

Implicit and Explicit Attention for Zero-Shot Learning

Faisal Alamri(✉) and Anjan Dutta

University of Exeter, Streatham Campus, Exeter EX4 4RN, UK
{F.Alamri2,A.Dutta}@exeter.ac.uk

Abstract. Most of the existing Zero-Shot Learning (ZSL) methods focus on learning a compatibility function between the image representation and class attributes. Few others concentrate on learning image representation combining local and global features. However, the existing approaches still fail to address the bias issue towards the seen classes. In this paper, we propose implicit and explicit attention mechanisms to address the existing bias problem in ZSL models. We formulate the implicit attention mechanism with a self-supervised image angle rotation task, which focuses on specific image features aiding to solve the task. The explicit attention mechanism is composed with the consideration of a multi-headed self-attention mechanism via Vision Transformer model, which learns to map image features to semantic space during the training stage. We conduct comprehensive experiments on three popular benchmarks: AWA2, CUB and SUN. The performance of our proposed attention mechanisms has proved its effectiveness, and has achieved the state-of-the-art harmonic mean on all the three datasets.

Keywords: Zero-shot learning · Attention mechanism · Self-supervised learning · Vision transformer

1 Introduction

Most of the existing Zero-Shot Learning (ZSL) methods [38,44] depend on pretrained visual features and necessarily focus on learning a compatibility function between the visual features and semantic attributes. Recently, attention-based approaches have got a lot of popularity, as they allow to obtain an image representation by directly recognising object parts in an image that correspond to a given set of attributes [50,53]. Therefore, models capturing global and local visual information have been quite successful [50,51]. Although visual attention models quite accurately focus on object parts, it has been observed that often recognised parts in image and attributes are biased towards training (or *seen*) classes due to the learned correlations [51]. This is mainly because the model fails to decorrelate the visual attributes in images.

Therefore, to alleviate these difficulties, in this paper, we consider two alternative attention mechanisms for reducing the effect of bias towards training

C. Bauckhage et al. (Eds.): DAGM GCPR 2021, LNCS 13024, pp. 467–483, 2021.
https://doi.org/10.1007/978-3-030-92659-5_30

classes in ZSL models. The first mechanism is via the self-supervised pretext task, which implicitly attends to specific parts of an image to solve the pretext task, such as recognition of image rotation angle [27]. For solving the pretext task, the model essentially focuses on learning image features that lead to solving the pretext task. Specifically, in this work, we consider rotating the input image concurrently by four different angles $(0°, 90°, 180°, 270°)$ and then predicting the rotation class. Since pretext tasks do not involve attributes or class-specific information, the model does not learn the correlation between visual features and attributes. Our second mechanism employs the Vision Transformer (ViT) [13] for mapping the visual features to semantic space. ViT having a rich multi-headed self-attention mechanism explicitly attends to those image parts related to class attributes. In a different setting, we combine the implicit with the explicit attention mechanism to learn and attend to the necessary object parts in a decorrelated or independent way. We attest that incorporating the rotation angle recognition in a self-supervised approach with the use of ViT does not only improve the ZSL performance significantly, but also and more importantly, contributes to reducing the bias towards seen classes, which is still an open challenge in the Generalised Zero-Shot Learning (GZSL) task [43]. Explicit use of attention mechanism is also examined, where the model is shown to enhance the visual feature localisation and attends to both global and discriminative local features guided by the semantic information given during training. As illustrated in Fig. 1, images fed into the model are taken from two different sources: 1) labelled images, which are the training images taken from the *seen* classes, shown in green colour, and 2) other images, which could be taken from any source, shown in blue. The model is donated as $(\mathcal{F}(.))$, in this paper, we implement $\mathcal{F}(.)$ either by ViT or by ResNet-101 [22] backbones. The first set of images is used to train the model to predict class attributes leading to the class labels via nearest search. However, the second set of images is used for rotation angle recognition, guiding the model to learn visual representations via implicit attention mechanism.

To summarise, in this paper, we make the following contributions: (1) We propose the utilisation of alternative attention mechanisms for reducing the bias towards the seen classes in zero-shot learning. By involving self-supervised pretext task, our model implicitly attends decorrelated image parts aiding to solve the pretext task, which learns image features independent of the training classes. (2) We perform extensive experiments on three challenging benchmark datasets, i.e. AWA2, CUB and SUN, in the generalised zero-shot learning setting and demonstrate the effectiveness of our proposed alternative attention mechanisms. We also achieve consistent improvement over the state-of-the-art methods. (3) The proposed method is evaluated with two backbone models: ResNet-101 and ViT, and shows significant improvement in the model performances, and reduces the issue of bias towards seen classes. We also show the effectiveness of our model qualitatively by plotting the attention maps.

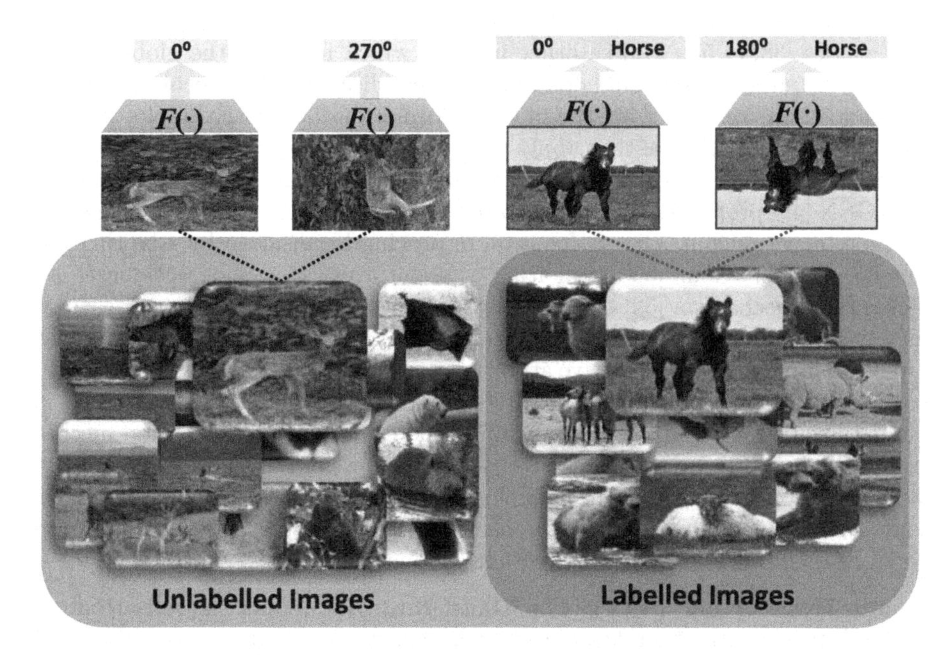

Fig. 1. Our method maps the visual features to the semantic space provided with two different input images (unlabelled and labelled data). Green represents the labelled images provided to train the model to capture visual features and predict object classes. Blue represents the unlabelled images that are rotated and attached to the former set of images to recognise rotated image angles in a self-supervised task. The model learns the visual representations of the rotated images implicitly via the use of attention. (Best viewed in colour)

2 Related Work

In this section we briefly review the related arts on zero-shot learning, Vision Transformer and self-supervised learning.

Zero-Shot Learning (ZSL): Zero-Shot Learning (ZSL) uses semantic side information such as attributes and word embeddings [4,14,16,32,36,47] to predict classes that have never been presented during training. Early ZSL models train different attribute classifiers assuming independence of attributes and then estimate the posterior of the test classes by combining attribute prediction probabilities [28]. Others do not follow the independence assumption and learn a linear [2,3,17] or non-linear [45] compatibility function from visual features to semantic space. There are some other works that learn an inverse mapping from semantic to visual feature space [39,55]. Learning a joint mapping function for each space into a common space (i.e. a shared latent embedding) is also investigated in [20,23,45]. Different from the above approaches, generative models synthesise samples of unseen classes based on information learned from seen classes and their semantic information, to tackle the issue of bias towards the seen classes [38,44,58]. Unlike other models, which focus on the global visual features, attention-based methods aim to learn discriminative local visual features and then combine with the global information [53,59]. Examples include S^2GA [53] and AREN [50] that apply an attention-based network to incorporate discriminative regions to provide rich visual expression automatically. In addition, GEN [49] proposes a graph reasoning method to learn relationships among multiple image regions. Others focus on improving localisation by adapting the human gaze behaviour [30], exploiting a global average pooling scheme as an aggregation mechanism [52] or by jointly learning both global and local features [59]. Inspired by the success of the recent attention-based ZSL models, in this paper, we propose two alternative attention mechanisms to capture robust image features suitable to ZSL task. Our first attention mechanism is implicit and is based on self-supervised pretext task [27], whereas the second attention mechanism is explicit and is based on ViT [13]. To the best of our knowledge, both of these attention models are still unexplored in the context of ZSL. Here we also point out that the inferential comprehension of visual representations upon the use of SSL and ViT is a future direction to consider for ZSL task.

Vision Transformer (ViT): The Transformer [41] adopts the self-attention mechanism to weigh the relevance of each element in the input data. Inspired by its success, it has been implemented to solve many computer vision tasks [5,13,25] and many enhancements and modifications of Vision Transformer (ViT) have been introduced. For example, CaiT [40] introduces deeper transformer networks, Swin Transformer [31] proposes a hierarchical Transformer capturing visual representation by computing self-attention via shifted windows, and TNT [21] applies the Transformer to compute the visual representations using both patch-level and pixel-level information. In addition, CrossViT [9] proposes a dual-branch Transformer with different sized image patches. Recently, TransGAN [24] proposes a completely free of convolutions generative adversarial network solely based on pure transformer-based architectures. Readers are referred to [25], for further reading about ViT based approaches. The applicability of ViT-based models is growing, but it has remained relatively unexplored to the zero-shot image recognition tasks where attention based models have already attracted a lot of attention. Therefore employing robust attention based models, such as ViT is absolutely timely and justified for improving the ZSL performance.

Self-Supervised Learning (SSL): Self-Supervised Learning (SSL) is widely used for unsupervised representation learning to obtain robust representations of samples from raw data without expensive labels or annotations. Although the recent SSL methods use contrastive objectives [10,19], early works used to focus on defining pretext tasks, which typically involves defining a surrogate task on a domain with ample weak supervision labels, such as predicting the rotation of images [27], relative positions of patches in an image [11,33], image colours [29, 56] etc. Encoders trained to solve such pretext tasks are expected to learn general features that might be useful for other downstream tasks requiring expensive annotations (e.g. image classification). Furthermore, SSL has been widely used in various applications, such as few-shot learning [18], domain generalisation [7] etc. In contrast, in this paper, we utilise the self-supervised pretext task of image rotation prediction for obtaining implicit image attention to solve ZSL.

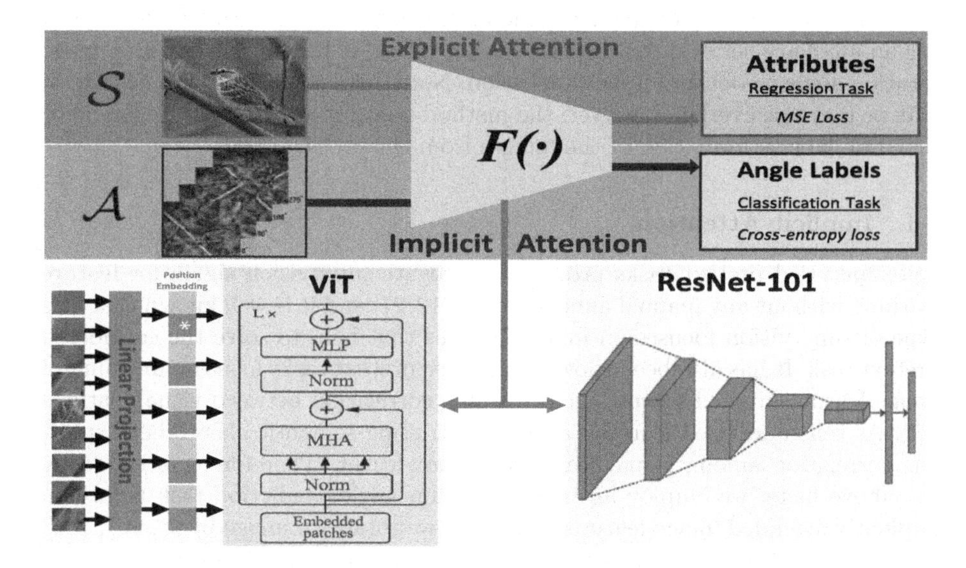

Fig. 2. IEAM-ZSL architecture. IEAM-ZSL consists of two pipelines represented in Green and Blue colours, respectively. The former takes images from the ZSL datasets with their class-level information input to the Transformer Encoder for attributes predictions. Outputs are compared with semantic information of the corresponding images using MSE loss as a regression task. The latter, shown in Blue colour, is fed with images after generating four rotations for each (i.e. 0°, 90°, 180°, and 270°), to predict the rotation angle. At inference, solely the ZSL test datasets, with no data augmentation, are inputted to the model to predict the class-level attributes. A search for the nearest class label is then conducted.

3 Implicit and Explicit Attention for Zero-Shot Learning

In this work, we propose an Implicit and Explicit Attention mechanism-based Model for solving image recognition in Zero-Shot Learning (IEAM-ZSL). We utilise self-supervised pretext tasks, such as image rotation angle recognition, for obtaining image attention in an implicit way. Here the main rational is for predicting the correct image rotation angle, the model needs to focus on image features with discriminative textures, colours etc., which implicitly attend to specific regions in an image. For having explicit image attention, we utilise the multi-headed self-attention mechanism involved in Vision Transform model.

From ZSL perspective, we follow the inductive approach for training our model, i.e. during training, the model only has access to the training set (*seen* classes), consisting of only the labelled images and continuous attributes of the seen classes ($\mathcal{S} = \{\mathbf{x}, \mathbf{y} | \mathbf{x} \in \mathcal{X}, \mathbf{y} \in \mathcal{Y}^s\}$). An RGB image in image space $\mathcal{X}$ is denoted as $\mathbf{x}$, where $\mathbf{y} \in \mathcal{Y}$ is the class-level semantic vector annotated with M different attributes. As depicted in Fig. 2, a 224×224 image $\mathbf{x} \in \mathbb{R}^{H \times W \times C}$ with resolution $H \times W$ and C channels is fed into the model. Addition to $\mathcal{S}$, we also use an auxiliary set of unlabelled images $\mathcal{A} = \{\mathbf{x} \in \mathcal{X}\}$ for predicting the image rotation angle to obtain implicit attention. Note, here the images from $\mathcal{S}$ and $\mathcal{A}$ may or may not overlap, however, the method does not utilise the categorical or semantic label information of the images from the set $\mathcal{A}$.

3.1 Implicit Attention

Self-supervised pretext tasks provide a surrogate supervision signal for feature learning without any manual annotations [1,12,27] and it is well known that this type of supervision focuses on image features that help to solve the considered pretext task. It has also been shown that these pretext tasks focus on meaningful image features and effectively avoid learning correlation between visual features [27]. As self-supervised learning avoids considering semantic class labels, spurious correlation among visual features are not learnt. Therefore, motivated by the above facts, we employ an image rotation angle prediction task to obtain implicitly attended image features. For that, we rotate an image by 0°, 90°, 180° and 270°, and train the model to correctly classify the rotated images. Let $g(\cdot|a)$ be an operator that rotates an image $\mathbf{x}$ by an angle $90° \times a$, where $a \in \{0, 1, 2, 3\}$. Now let $\hat{\mathbf{y}}_a$ be the predicted probability for the rotated image $\mathbf{x}_a$ with label a, then the loss for training the underlying model is computed as follows:

$$\mathcal{L}_{\text{CE}} = -\sum_{a=1}^{4} \log(\hat{\mathbf{y}}_a) \tag{1}$$

In our case, the task of predicting image rotation angle trains the model to focus on specific image regions having rich visual features (for example, textures or colours). This procedure implicitly learns to attend image features.

3.2 Explicit Attention

For obtaining explicit attention, we employ Vision Transformer model [13], where each image $\mathbf{x} \in \mathbb{R}^{H \times W \times C}$ with resolution $H \times W$ and C channels is fed into the model after resizing it to 224×224. Afterwards, the image is split into a sequence of N patches denoted as $\mathbf{x}_p \in \mathbb{R}^{N \times (P^2.C)}$, where $N = \frac{H.W}{P^2}$. Patch embeddings (small red boxes in Fig. 2) are encoded by applying a trainable 2D convolution layer with kernel size = (16, 16) and stride = (16, 16)). An extra learnable classification token ($\mathbf{z}_0^0 = \mathbf{x}_{\text{class}}$) is appended at the beginning of the sequence to encode the global image representation, which is donated as (∗). Position embeddings (orange boxes) are then attached to the patch embeddings to obtain the relative positional information. Patch embeddings ($\mathbf{z}$) are then projected through a linear projection $\mathbf{E}$ to D dimension (i.e. $D = 1024$) as in Eq. 2. Embeddings are then passed to the Transformer encoder, which consists of Multi-Head Attention (MHA) (Eq. 3) and MLP blocks (Eq. 4). A layer normalisation (Norm) is applied before every block, and residual connections after every block. The image representation ($\hat{\mathbf{y}}$) is then produced as in Eq. 5.

$$\mathbf{z}_0 = [\mathbf{x}_{\text{class}}; \mathbf{x}_p^1\mathbf{E}; \mathbf{x}_p^2\mathbf{E}; \ldots; \mathbf{x}_p^N\mathbf{E}] + \mathbf{E}_{\text{pos}}, \quad \mathbf{E} \in \mathbb{R}^{(P^2.C)\times D}, \mathbf{E}_{\text{pos}} \in \mathbb{R}^{(N+1)\times D} \tag{2}$$

$$\mathbf{z}'_\ell = \text{MHA}(\text{Norm}(\mathbf{z}_{\ell-1})) + \mathbf{z}_{\ell-1}, \quad \ell = 1 \ldots L \; (L = 24) \tag{3}$$

$$\mathbf{z}_\ell = \text{MLP}(\text{Norm}(\mathbf{z}'_\ell)) + \mathbf{z}'_\ell, \quad \ell = 1 \ldots L \tag{4}$$

$$\hat{\mathbf{y}} = \text{Norm}(\mathbf{z}_L^0) \tag{5}$$

Below we provide details of our multi-head attention mechanism within the ViT model.

Multi-Head Attention (MHA): Patch embeddings are fed into the transformer encoder, where the multi-head attention takes place. Self-attention is performed for every patch in the sequence of the patch embeddings independently; thus, attention works simultaneously for all the patches, leading to multi-headed self-attention. This is computed by creating three vectors, namely Query (Q), Key (K) and Value (V). They are created by multiplying the patch embeddings by three trainable weight matrices (i.e. W^Q, W^K and W^V) applied to compute the self-attention. A dot-product operation is performed on the Q and K vectors, calculating a scoring matrix that measures how much a patch embedding has to attend to every other patch in the input sequence. The score matrix is then scaled down and converted into probabilities using a softmax. Probabilities are then multiplied by the V vectors, as in Eq. 6, where d_k is the dimension of the vector K. Multi-headed self-attention mechanism produces a number of self-attention matrices which are concatenated and fed into a linear layer and passed sequentially to 1) regression head and 2) classification head.

$$\text{Attention}(Q, K, V) = \text{softmax}(\frac{QK^T}{\sqrt{d_k}})V \tag{6}$$

The multi-headed self-attention mechanism involved in the Vision Transformer guides our model to learn both the global and local visual features. It is worth noting that the standard ViT has only one classification head implemented by an MLP, which is changed in our model to two heads to meet the two different underlying objectives. The first head is a regression head applied to predict M different class attributes, whereas the second head is added for rotation angle classification. For the former task, the objective function employed is the Mean Squared Error (MSE) loss as in Eq. 7, where $\mathbf{y}_i$ is the target attributes, and $\hat{\mathbf{y}}_i$ is the predicted ones. For the latter task, cross-entropy (Eq. 1) objective is applied.

$$\mathcal{L}_{\text{MSE}} = \frac{1}{M}\sum_{i=1}^{M}(\mathbf{y}_i - \hat{\mathbf{y}}_i)^2 \tag{7}$$

The total loss used for training our model is defined in Eq. 8, where $\lambda_1 = 1$ and $\lambda_2 = 1$.

$$\mathcal{L}_{\text{TOT}} = \lambda_1 \mathcal{L}_{\text{CE}} + \lambda_2 \mathcal{L}_{\text{MSE}} \tag{8}$$

During the inference phase, original test images from the seen and unseen classes are inputted. Class labels are then determined using the cosine similarity between the predicted attributes and every target class embeddings predicted by our model.

4 Experiments

Datasets: We have conducted our experiments on three popular ZSL datasets: AWA2, CUB, and SUN, whose details are presented in Table 1. The main aim of this experimentation is to validate our proposed method IEAM-ZSL, demonstrating its effectiveness and comparing it with the existing state-of-the-art methods. Among these datasets, AWA2 [47] consists of 37, 322 images of 50 categories (40 seen + 10 unseen). Each category contains 85 binary as well as continuous class attributes. CUB [42] contains 11, 788 images forming 200 different types of birds, among them 150 classes are considered as seen, and the other 50 as unseen, which is split by [2]. Together with images CUB dataset also contains 312 attributes describing birds. Finally, SUN [35] has the largest number of classes among others. It consists of 717 types of scene images, which divided into 645 seen and 72 unseen classes. The SUN dataset contains 14, 340 images with 102 annotated attributes.

Table 1. Dataset statistics: the number of classes (seen + unseen classes shown within parenthesis), the number of attributes and the number of images per dataset.

Datasets	AWA2 [47]	CUB [42]	SUN [35]
Number of classes (seen + unseen)	50 (40 + 10)	200 (150 + 50)	717 (645 + 72)
Number of attributes	85	312	102
Number of images	37, 322	11, 788	14, 340

Implementation Details: In our experiment, we have used two different backbones: (1) ResNet-101 and (2) Vision Transformer (ViT), both of which are pretrained on ImageNet and then finetuned for the ZSL tasks on the datasets mentioned above. We resize the image to 224×224 before inputting it into the model. For ViT, the primary baseline model employed uses an input patch size 16×16, with 1024 hidden dimension, and having 24 layers and 16 heads on each layer, and 24 series encoder. We use the Adam optimiser for training our model with a fixed learning rate of 0.0001 and a batch size of 64. In the setting where we use self-supervised pretext task, we construct the batch with 32 *seen* training images from set $\mathcal{S}$ and 32 rotated images (i.e. eight images, where each image is rotated to 0°, 90°, 180° and 270°) from set $\mathcal{A}$. We have implemented our model with PyTorch[1] deep learning framework and trained the model on a GeForce RTX 3090 GPU on a workstation with Xeon processor and 32 GB of memory.

Evaluation: The proposed model is evaluated on the three above mentioned datasets. We have followed the inductive approach for training our model, i.e. our model has no access to neither visual nor side-information of unseen classes during training. During the evaluation, we have followed the GZSL protocol. Following [46], we compute the top-1 accuracy for both seen and unseen classes. In addition, the harmonic mean of the top-1 accuracies on the seen and unseen classes is used as the main evaluation criterion. Inspired by the recent works [8,50,52], we have used the Calibrated Stacking [8] for evaluating our model under GZSL setting. The calibration factor γ is dataset-dependent and decided based on a validation set. For AWA2 and CUB, the calibration factor γ is set to 0.9 and for SUN, it is set to 0.4.

Table 2. Generalised zero-shot classification performance on AWA2, CUB and SUN. Reported models are ordered in terms of their publishing dates. Results are reported in %.

Models	AWA2			CUB			SUN		
	S	U	H	S	U	H	S	U	H
DAP [28]	84.7	0.0	0.0	67.9	1.7	3.3	25.1	4.2	7.2
IAP [28]	87.6	0.9	1.8	72.8	0.2	0.4	37.8	1.0	1.8
DeViSE [17]	74.7	17.1	27.8	53.0	23.8	32.8	30.5	14.7	19.8
ConSE [34]	90.6	0.5	1.0	72.2	1.6	3.1	39.9	6.8	11.6
ESZSL [37]	77.8	5.9	11.0	63.8	12.6	21.0	27.9	11.0	15.8
SJE [3]	73.9	8.0	14.4	59.2	23.5	33.6	30.5	14.7	19.8
SSE [57]	82.5	8.1	14.8	46.9	8.5	14.4	36.4	2.1	4.0
LATEM [45]	77.3	11.5	20.0	57.3	15.2	24.0	28.8	14.7	19.5
ALE [2]	81.8	14.0	23.9	62.8	23.7	34.4	33.1	21.8	26.3
*GAZSL [58]	86.5	19.2	31.4	60.6	23.9	34.3	34.5	21.7	26.7
SAE [26]	82.2	1.1	2.2	54.0	7.8	13.6	18.0	8.8	11.8
*f-CLSWGAN [44]	64.4	57.9	59.6	57.7	43.7	49.7	36.6	42.6	39.4
AREN [50]	79.1	54.7	64.7	63.2	69.0	66.0	40.3	32.3	35.9
*f-VAEGAN-D2 [48]	76.1	57.1	65.2	75.6	63.2	68.9	50.1	37.8	43.1
SGMA [59]	87.1	37.6	52.5	71.3	36.7	48.5	–	–	–
IIR [6]	83.2	48.5	61.3	52.3	55.8	53.0	30.4	47.9	36.8
*E-PGN [54]	83.5	52.6	64.6	61.1	52.0	56.2	–	–	–
SELAR [52]	78.7	32.9	46.4	76.3	43.0	55.0	37.2	23.8	29.0
ResNet-101 [22]	66.7	40.1	50.1	59.5	52.3	55.7	35.5	28.8	31.8
ResNet-101 with Implicit Attention	**74.1**	**45.9**	**56.8**	**62.7**	**54.5**	**58.3**	**36.3**	**31.9**	**33.9**
Our model (ViT)	**90.0**	**51.9**	**65.8**	**75.2**	**67.3**	**71.0**	**55.3**	**44.5**	**49.3**
Our model (ViT) with Implicit Attention	**89.9**	**53.7**	**67.2**	**73.8**	**68.6**	**71.1**	**54.7**	**48.2**	**51.3**

S, U, H denote Seen classes ($\mathcal{Y}^s$), Unseen classes ($\mathcal{Y}^u$), and the Harmonic mean, respectively. For each scenario, the best is in red and the second-best is in blue. * indicates generative representation learning methods.

[1] Our code is available at: https://github.com/FaisalAlamri0/IEAM-ZSL.

4.1 Quantitative Results

Table 2 illustrates a quantitative comparison between the state-of-the-art methods and the proposed method using two different backbones: (1) ResNet-101 [22] and (2) ViT [13]. The baseline models performance without the employment of the SSL approach is also reported. The performance of each model is shown in % in terms of Seen (S) and Unseen (U) classes and their harmonic mean (H). As reported, the classical ZSL models [2,17,28,34,34,45] show good performance in terms of seen classes. However, they perform poorly on unseen classes and encounter the bias issue, resulting in a very low harmonic mean. Among the classical approaches, [2] performs the best on all the three datasets, as it overcomes the shortcomings of the previous models and considers the dependency between attributes. Among generative approaches, f-VAEGAN-D2 [48] performs the best. Although f-CLSWGAN [44] achieves the highest score on AWA2 unseen classes, it shows lower harmonic means on all the datasets than [48]. As noticed, the first top scores for the AWA2 unseen classes accuracy are obtained by generative models [44,48], which we assume is because they include both seen and synthesised unseen features during the training phase. Moreover, attention-based models, such as [50,59] are the closest to our proposed model, perform better than the other models due to the inclusion of global and local representations. [50] outperforms all reported models on the unseen classes of the CUB dataset, but still has low harmonic means on all the datasets. SGMA [59] performs poorly on both AWA2 and CUB, and it clearly suffers from the bias issue, where its performance on unseen classes is considered deficient compared to other models. Recent models such as SELAR [52] uses global maximum pooling as an aggregation method and achieves the best scores on CUB seen classes, but achieves low harmonic means. In addition, its performance is seen to be considerably impacted by the bias issue.

ResNet-101: For a fair evaluation of the robustness and effectiveness of our proposed alternative attention-based approach, we consider the ResNet-101 [22] as one of our backbones, which is also used in prior related arts [2,17,26,52, 54]. We have used the ResNet-101 backbone as a baseline model, where we only consider the global representation. Moreover, we also use this backbone with implicit attention, i.e. during training, we simultaneously impose a self-supervised image rotation angle prediction task for training the model. Note, for producing the results in Table 2, we only use the images from the seen classes as set $\mathcal{A}$, which is used for rotation angle prediction task. As presented in Table 2, our model with ResNet-101 backbone has performed inferiorly compared to our implicit and explicit variant, which will be discussed in the next paragraph. However, even with the ResNet-101 backbone, the contribution of our implicit attention mechanism should be noted, which provides a substantial boost to the model performance. For example, on AWA2, a considerable increment is observed on both seen and unseen classes, leading to a significant increase in the harmonic mean (i.e. 50.1% to 56.8%). The performance of the majority of the related arts seems to suffer from bias towards the seen classes. We argue that our method tends to mitigate this issue on all the three datasets. Our method enables the model to learn the visual representations of unseen classes implicitly; hence, the performance is increased, and the bias issue is alleviated. Similarly, on the SUN dataset, although this dataset consists of 717 classes, the proposed implicit attention mechanism illustrates the capability of providing ResNet-101 with an increase in the accuracy in terms of both seen and unseen classes, leading to an increase of ≈2 points in the harmonic mean, i.e. from 31.8% to 33.9%.

Vision Transformer (ViT): We have used Vision Transformer (ViT) as another backbone to enable explicit attention in our model. Similar to the ResNet-101 backbone, we use the implicit attention mechanism with ViT backbone as well. During training, we simultaneously impose self-supervised image rotation angle prediction task for training the model. Here also we only use the images from the seen classes for image rotation angle task. As shown in Table 2, consideration of explicit attention performs very well on all the three datasets and it outperforms all the previously reported results with a significant margin. Such results are expected due to the involvement of self-attention employed in ViT. It captures both the global and local features explicitly guided by the class attributes given during training. Furthermore, attention focuses to each element of the input patch embeddings after the image is split, which effectively weigh the relevance of different patches, resulting in more compact representations. Although explicit attention mechanism is seen to provide better visual understanding, the effectiveness of the implicit attention process in terms of recognising the image rotation angle is also quite important. It does not only improve the performance further but also reduces the bias issue considerably, which can be seen in the performance of the unseen classes. In addition, it allows the model via an implicit use of self-attention to encapsulate the visual features and regions that are semantically relevant to the class attributes. Our model achieves the highest harmonic mean among all the reported models on all the three datasets. In terms of AWA2, our approach scores the third highest accuracy on both seen and unseen classes, but the highest harmonic mean. Note that on AWA2 dataset, our model still suffers from bias towards seen classes. We speculate that is due to the lack of the co-occurrence of some vital and identical attributes between seen and unseen classes. For example, attributes *nocturnal* in bat, *longneck* in giraffe or *flippers* in seal score the highest attributes in the class-attribute vectors, but rarely appear among other classes. However, on CUB dataset, this issue seems to be mitigated, as our model scores the highest harmonic mean (i.e. $H = 71.1\%$), where the performance on unseen classes is increased compared to our model with explicit attention. Finally, our model with implicit and explicit attention achieves the highest scores on classes on the SUN dataset, resulting in the best achieved harmonic mean. In summary, our proposed implicit and explicit attention mechanism proves to be very effective across all the three considered datasets. Explicit attention using the ViT backbone with multi-head self-attention is quite important for the good performance of the ZSL model. Implicit attention in terms of self-supervised pretext task is another important mechanism to look at, as it boosts the performance on the unseen classes and provides better generalisation.

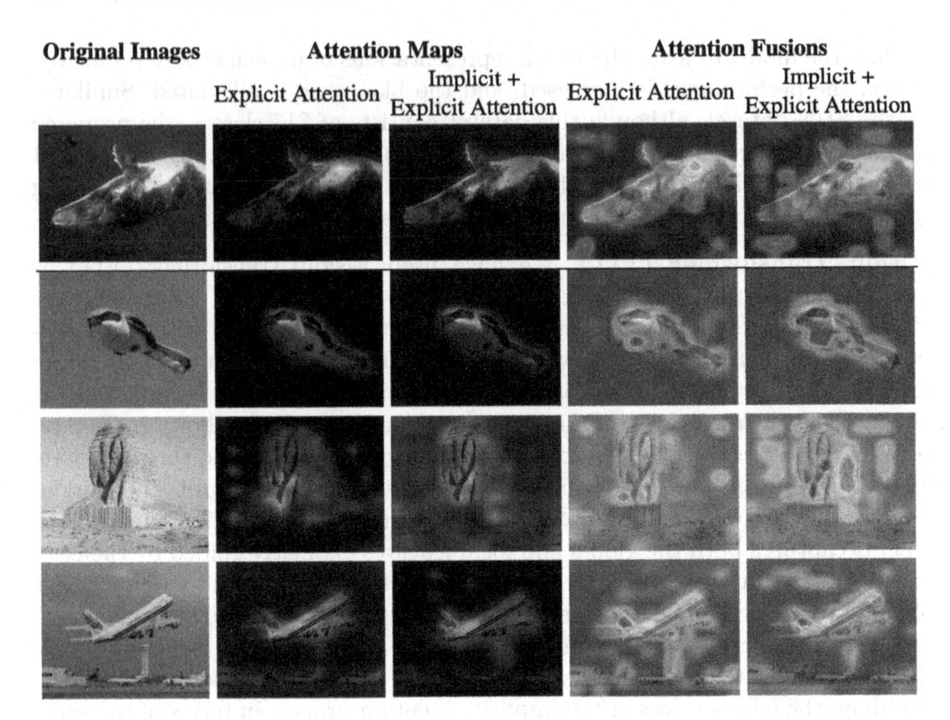

Fig. 3. Examples of implicit and explicit attention. First column: original images, Second and third: attention maps without and with SSL, respectively, Four and fifth: attention fusions without and with SSL, respectively. Our model benefits from using the attention mechanism and can implicitly learn object-level attributes and their discriminative features.

Attention Maps: Figure 3 presents some qualitative results, i.e. attention maps and fusions obtained by our proposed implicit and explicit attention-based model. For generating these qualitative results, we have used our model with explicit attention mechanism, i.e. we have used the ViT backbone. Attention maps and fusions are presented on four randomly chosen images from the considered datasets. Explicit attention with ViT backbone seems to be quite important for the ZSL tasks as it can perfectly focus on the object appearing in the image, which justifies the better performance obtained by our model with ViT backbone. Inclusion of implicit attention mechanism in terms of self-supervised rotated image angle prediction further enhances the attention maps and particularly focuses on specific image parts important for that object class. For example, as shown in the first row of Fig. 3, our model with implicit and explicit attention mechanism focuses on both global and local features of the *Whale* (i.e. water, big, swims, hairless, bulbous, flippers, etc.). Similarly, on the CUB dataset, the model pays attention to objects' global features, and more importantly, the discriminative local features (i.e. *loggerhead shrike* has a white belly, breast and throat, and a black crown forehead and bill). For natural images taken from the

SUN dataset, our model with implicit attention is seen to focus on the *ziggurat* paying more attention to its global features. Furthermore, as in the *airline* image illustrated in the last row, our model considers both global and discriminative features, leading to precise attention map that focuses accurately on the object.

Table 3. Ablation performance of our model with ResNet-101 and ViT backbone on AWA2, CUB and SUN datasets. Here we use the training images from the seen classes as $\mathcal{S}$ and varies $\mathcal{A}$ as noted in the first column of the following table. S, U and PASCAL respectively denote the training images from the seen classes, test images from the unseen classes, and PASCAL VOC2012 training set images.

Source of rotated Images ($\mathcal{A}$)	Backbone (implicit attention)	AWA2			CUB			SUN		
		S	U	H	S	U	H	S	U	H
S & U	ResNet-101	79.9	44.2	56.4	60.1	56.0	58.0	35.0	33.1	33.7
	ViT	87.3	56.8	68.8	74.2	68.9	71.1	54.7	50.0	52.2
PASCAL	ResNet-101	72.0	44.3	54.8	62.5	53.1	57.4	35.6	30.3	33.1
	ViT	88.1	51.8	65.2	73.4	68.0	70.6	55.2	46.3	50.6
PASCAL & U	ResNet-101	75.1	46.5	57.4	62.9	54.4	58.4	33.7	32.7	33.2
	ViT	89.8	53.2	66.8	73.02	69.7	71.3	53.9	51.0	52.4
PASCAL & S	ResNet-101	73.1	44.5	55.4	62.5	53.2	57.5	36.6	30.1	33.1
	ViT	91.2	51.6	65.9	73.7	68.8	71.1	54.19	46.9	50.9

4.2 Ablation Study

Our ablation study evaluates the effectiveness of our proposed implicit and explicit attention-based model for the ZSL tasks. Here we mainly analyse the outcome of our proposed approach if we change the set $\mathcal{A}$ which we use for sampling images for self-supervised image angle prediction task during training. In Sect. 4.1, we have only used the seen images for this purpose; however, we have also noted important observation if we change the set $\mathcal{A}$. Note, here we can use any collection of images as $\mathcal{A}$, since it does not need any annotation regarding its semantic class, because in this case, the only supervision used is the class corresponds to image angle rotation which can be generated online during training. In Table 3, we present results on all three considered datasets with the above mentioned evaluation metric, where we only vary $\mathcal{A}$ as noted in the first column of Table 3. Note, in all these settings $\mathcal{S}$ remains fixed, and it is set to the set of images from the seen classes. In all the settings, we observe that explicit attention in terms of ViT backbone performs significantly better than the classical CNN backbone, such as ResNet-101. We also observe that the inclusion of unlabelled images from unseen classes (can be considered as transductive ZSL [2]) significantly boosts the performance on all the datasets (see rows 1 and 3 in Table 3). Moreover, we also observe that including datasets that contain diverse images, such as PASCAL [15] improve the performance on unseen classes and increase generalisation.

5 Conclusion

This paper has proposed implicit and explicit attention mechanisms for solving the zero-shot learning task. For implicit attention, our proposed model has imposed self-supervised rotated image angle prediction task, and for the purpose of explicit attention, the model employs the multi-head self-attention mechanism via the Vision Transformer model to map visual features to the semantic space. We have considered three publicly available datasets: AWA2, CUB and SUN, to show the effectiveness of our proposed model. Throughout our extensive experiments, explicit attention via the multi-head self-attention mechanism of ViT is revealed to be very important for the ZSL task. Additionally, the implicit attention mechanism is also proved to be effective for learning image representation for zero-shot image recognition, as it boosts the performance on unseen classes and provides better generalisation. Our proposed model based on implicit and explicit attention mechanism has provided very encouraging results for the ZSL task and particularly has achieved state-of-the-art performance in terms of harmonic mean on all the three considered benchmarks, which shows the importance of attention-based models for ZSL task.

Acknowledgement. This work was supported by the Defence Science and Technology Laboratory (Dstl) and the Alan Turing Institute (ATI). The TITAN Xp and TITAN V used for this research were donated by the NVIDIA Corporation.

References

1. Agrawal, P., Carreira, J., Malik, J.: Learning to see by moving. In: ICCV (2015)
2. Akata, Z., Perronnin, F., Harchaoui, Z., Schmid, C.: Label-embedding for image classification. IEEE TPAMI **38**, 1425–1438 (2016)
3. Akata, Z., Reed, S.E., Walter, D., Lee, H., Schiele, B.: Evaluation of output embeddings for fine-grained image classification. In: CVPR (2015)
4. Alamri, F., Dutta, A.: Multi-head self-attention via vision transformer for zero-shot learning. In: IMVIP (2021)
5. Alamri, F., Kalkan, S., Pugeault, N.: Transformer-encoder detector module: Using context to improve robustness to adversarial attacks on object detection. In: ICPR (2021)
6. Cacheux, Y.L., Borgne, H.L., Crucianu, M.: Modeling inter and intra-class relations in the triplet loss for zero-shot learning. In: ICCV (2019)
7. Carlucci, F.M., D'Innocente, A., Bucci, S., Caputo, B., Tommasi, T.: Domain generalization by solving jigsaw puzzles. In: CVPR (2019)
8. Chao, W.-L., Changpinyo, S., Gong, B., Sha, F.: An empirical study and analysis of generalized zero-shot learning for object recognition in the wild. In: Leibe, B., Matas, J., Sebe, N., Welling, M. (eds.) ECCV 2016. LNCS, vol. 9906, pp. 52–68. Springer, Cham (2016). https://doi.org/10.1007/978-3-319-46475-6_4
9. Chen, C., Fan, Q., Panda, R.: CrossViT: cross-attention multi-scale vision transformer for image classification. In: ICCV (2021)
10. Chen, T., Kornblith, S., Norouzi, M., Hinton, G.: A simple framework for contrastive learning of visual representations. In: ICML (2020)

11. Doersch, C., Gupta, A., Efros, A.A.: Unsupervised visual representation learning by context prediction. In: ICCV (2015)
12. Dosovitskiy, A., Springenberg, J.T., Riedmiller, M.A., Brox, T.: Discriminative unsupervised feature learning with convolutional neural networks. In: NIPS (2014)
13. Dosovitskiy, A., et al.: An image is worth 16x16 words: transformers for image recognition at scale. In: ICLR (2021)
14. Dutta, A., Akata, Z.: Semantically tied paired cycle consistency for any-shot sketch-based image retrieval. Int. J. Comput. Vis. **128**, 2684–2703 (2020). https://doi.org/10.1007/s11263-020-01350-x
15. Everingham, M., Van Gool, L., Williams, C.K.I., Winn, J., Zisserman, A.: The PASCAL visual object classes challenge 2012 (VOC2012) results (2012). http://www.pascal-network.org/challenges/VOC/voc2012/workshop/index.html
16. Federici, M., Dutta, A., Forré, P., Kushman, N., Akata, Z.: Learning robust representations via multi-view information bottleneck. In: ICLR (2020)
17. Frome, A., Corrado, G.S., Shlens, J., Bengio, S., Dean, J., Ranzato, M.A., Mikolov, T.: DeViSE: a deep visual-semantic embedding model. In: NIPS (2013)
18. Gidaris, S., Bursuc, A., Komodakis, N., Perez, P., Cord, M.: Boosting few-shot visual learning with self-supervision. In: ICCV (2019)
19. Grill, J.B., et al.: Bootstrap your own latent: A new approach to self-supervised Learning. In: NeurIPS (2020)
20. Gune, O., Banerjee, B., Chaudhuri, S.: Structure aligning discriminative latent embedding for zero-shot learning. In: BMVC (2018)
21. Han, K., Xiao, A., Wu, E., Guo, J., Xu, C., Wang, Y.: Transformer in transformer. arXiv (2021)
22. He, K., Zhang, X., Ren, S., Sun, J.: Deep residual learning for image recognition. In: CVPR (2016)
23. Jiang, H., Wang, R., Shan, S., Yang, Y., Chen, X.: Learning discriminative latent attributes for zero-shot classification. In: ICCV (2017)
24. Jiang, Y., Chang, S., Wang, Z.: TransGAN: two pure transformers can make one strong GAN, and that can scale up. In: CVPR (2021)
25. Khan, S., Naseer, M., Hayat, M., Zamir, S.W., Khan, F., Shah, M.: Transformers in vision: a survey. arXiv (2021)
26. Kodirov, E., Xiang, T., Gong, S.: Semantic autoencoder for zero-shot learning. In: CVPR (2017)
27. Komodakis, N., Gidaris, S.: Unsupervised representation learning by predicting image rotations. In: ICLR (2018)
28. Lampert, C.H., Nickisch, H., Harmeling, S.: Learning to detect unseen object classes by between-class attribute transfer. In: CVPR (2009)
29. Larsson, G., Maire, M., Shakhnarovich, G.: Learning representations for automatic colorization. In: Leibe, B., Matas, J., Sebe, N., Welling, M. (eds.) ECCV 2016. LNCS, vol. 9908, pp. 577–593. Springer, Cham (2016). https://doi.org/10.1007/978-3-319-46493-0_35
30. Liu, Y., et al.: Goal-oriented gaze estimation for zero-shot learning. In: CVPR (2021)
31. Liu, Z., et al.: Swin transformer: hierarchical vision transformer using shifted windows. arXiv (2021)
32. Mikolov, T., Sutskever, I., Chen, K., Corrado, G., Dean, J.: Distributed representations of words and phrases and their compositionality. In: NIPS (2013)
33. Noroozi, M., Favaro, P.: Unsupervised learning of visual representations by solving jigsaw puzzles. In: Leibe, B., Matas, J., Sebe, N., Welling, M. (eds.) ECCV 2016.

LNCS, vol. 9910, pp. 69–84. Springer, Cham (2016). https://doi.org/10.1007/978-3-319-46466-4_5
34. Norouzi, M., et al.: Zero-shot learning by convex combination of semantic embeddings. In: ICLR (2014)
35. Patterson, G., Hays, J.: Sun attribute database: discovering, annotating, and recognizing scene attributes. In: CVPR (2012)
36. Pennington, J., Socher, R., Manning, C.D.: GloVe: global vectors for word representation. In: EMNLP (2014)
37. Romera-Paredes, B., Torr, P.: An embarrassingly simple approach to zero-shot learning. In: ICML (2015)
38. Schönfeld, E., Ebrahimi, S., Sinha, S., Darrell, T., Akata, Z.: Generalized zero- and few-shot learning via aligned variational autoencoders. In: CVPR (2019)
39. Shigeto, Y., Suzuki, I., Hara, K., Shimbo, M., Matsumoto, Y.: Ridge regression, hubness, and zero-shot learning. In: Appice, A., Rodrigues, P.P., Santos Costa, V., Soares, C., Gama, J., Jorge, A. (eds.) ECML PKDD 2015. LNCS (LNAI), vol. 9284, pp. 135–151. Springer, Cham (2015). https://doi.org/10.1007/978-3-319-23528-8_9
40. Touvron, H., Cord, M., Sablayrolles, A., Synnaeve, G., Jégou, H.: Going deeper with image transformers. arXiv (2021)
41. Vaswani, A., et al.: Attention is all you need. In: NIPS (2017)
42. Wah, C., Branson, S., Welinder, P., Perona, P., Belongie, S.: The Caltech-UCSD Birds-200-2011 dataset. Technical report, California Institute of Technology (2011)
43. Wang, W., Zheng, V., Yu, H., Miao, C.: A survey of zero-shot learning. ACM-TIST **10**, 1–37 (2019)
44. Xian, Y., Lorenz, T., Schiele, B., Akata, Z.: Feature generating networks for zero-shot learning. In: CVPR (2018)
45. Xian, Y., Akata, Z., Sharma, G., Nguyen, Q., Hein, M., Schiele, B.: Latent embeddings for zero-shot classification. In: CVPR (2016)
46. Xian, Y., Lampert, C.H., Schiele, B., Akata, Z.: Zero-shot learning-a comprehensive evaluation of the good, the bad and the ugly. IEEE TPAMI **41**, 2251–2265 (2019)
47. Xian, Y., Schiele, B., Akata, Z.: Zero-shot learning - the good, the bad and the ugly. In: CVPR (2017)
48. Xian, Y., Sharma, S., Schiele, B., Akata, Z.: F-VAEGAN-D2: a feature generating framework for any-shot learning. In: CVPR (2019)
49. Xie, G.-S., et al.: Region graph embedding network for zero-shot learning. In: Vedaldi, A., Bischof, H., Brox, T., Frahm, J.-M. (eds.) ECCV 2020. LNCS, vol. 12349, pp. 562–580. Springer, Cham (2020). https://doi.org/10.1007/978-3-030-58548-8_33
50. Xie, G.S., et al.: Attentive region embedding network for zero-shot learning. In: CVPR (2019)
51. Xu, W., Xian, Y., Wang, J., Schiele, B., Akata, Z.: Attribute prototype network for zero-shot learning. In: NIPS (2020)
52. Yang, S., Wang, K., Herranz, L., van de Weijer, J.: On implicit attribute localization for generalized zero-shot learning. IEEE SPL **28**, 872–876 (2021)
53. Yu, Y., Ji, Z., Fu, Y., Guo, J., Pang, Y., Zhang, Z.M.: Stacked semantics-guided attention model for fine-grained zero-shot learning. In: NeurIPS (2018)
54. Yu, Y., Ji, Z., Han, J., Zhang, Z.: Episode-based prototype generating network for zero-shot learning. In: CVPR (2020)
55. Zhang, L., Xiang, T., Gong, S.: Learning a deep embedding model for zero-shot learning. In: CVPR (2017)

56. Zhang, R., Isola, P., Efros, A.A.: Colorful image colorization. In: Leibe, B., Matas, J., Sebe, N., Welling, M. (eds.) ECCV 2016. LNCS, vol. 9907, pp. 649–666. Springer, Cham (2016). https://doi.org/10.1007/978-3-319-46487-9_40
57. Zhang, Z., Saligrama, V.: Zero-shot learning via semantic similarity embedding. In: ICCV (2015)
58. Zhu, Y., Elhoseiny, M., Liu, B., Peng, X., Elgammal, A.: Imagine it for me: generative adversarial approach for zero-shot learning from noisy texts. In: CVPR (2018)
59. Zhu, Y., Xie, J., Tang, Z., Peng, X., Elgammal, A.: Semantic-guided multi-attention localization for zero-shot learning. In: NeurIPS (2019)

Self-supervised Learning for Object Detection in Autonomous Driving

Daniel Pototzky[1,2(✉)], Azhar Sultan[1], Matthias Kirschner[1], and Lars Schmidt-Thieme[2]

[1] Robert Bosch GmbH, Gerlingen, Germany
{daniel.pototzky,azhar.sultan,matthias.kirschner}@de.bosch.com
[2] Information Systems and Machine Learning Lab, University of Hildesheim, Hildesheim, Germany
schmidt-thieme@ismll.de

Abstract. Recently, self-supervised pretraining methods have achieved impressive results, matching ImageNet weights on a variety of downstream tasks including object detection. Despite their success, these methods have some limitations. Most of them are optimized for image classification and compute only a global feature vector describing an entire image. On top of that, they rely on large batch sizes, a huge amount of unlabeled data and vast computing resources to work well.

To address these issues, we propose SSLAD, a **S**elf-**S**upervised **L**earning approach especially suited for object detection in the context of **A**utonomous **D**riving. SSLAD computes local image descriptors that are invariant to augmentations and scale. In experiments, we show that our method outperforms state-of-the-art self-supervised pretraining methods, on various object detection datasets from the automotive domain. By leveraging just 20,000 unlabeled images taken from a video camera on a car, SSLAD almost matches ImageNet weights trained on 1,200,000 labeled images. If as few as 20 unlabeled images are available, SSLAD generates weights that are far better than a random initialization, whereas competing self-supervised methods just do not work given so little data. Furthermore, SSLAD is very robust with respect to batch size. Even with a batch size of one, SSLAD generates weights that are clearly superior to a random initialization while greatly outperforming other self-supervised methods. This property of SSLAD even allows for single-GPU training with only a minor decrease in performance.

Keywords: Self-supervised learning · Object detection · Autonomous driving · Pedestrian detection

1 Introduction

Supervised learning has shown impressive performance for a broad range of tasks, yet it has one core downside. It requires large amounts of labeled data. For each new task, a new labeled dataset is needed which is usually costly to obtain. One

C. Bauckhage et al. (Eds.): DAGM GCPR 2021, LNCS 13024, pp. 484–497, 2021.
https://doi.org/10.1007/978-3-030-92659-5_31

approach for mitigating these labeling costs is self-supervised learning, which tries to learn useful representations given unlabeled images. Once a network is pretrained, it can be finetuned on a downstream task using a small, labeled dataset.

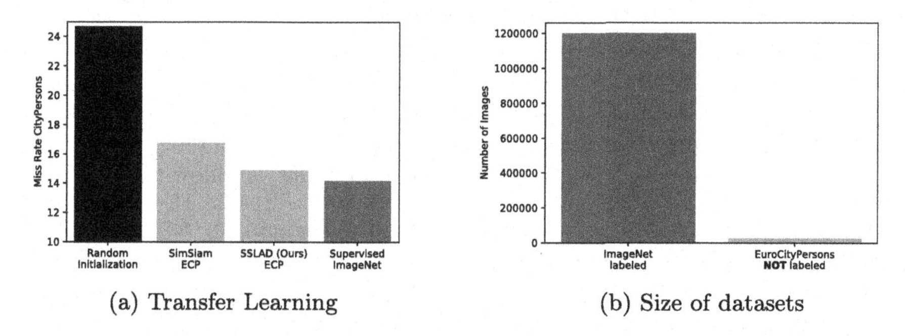

(a) Transfer Learning (b) Size of datasets

Fig. 1. (a) Comparison of pretrained weights by finetuning on CityPersons. The blue color indicates that the pretraining was done on full ImageNet with labels. The green color shows that the models were pretrained on EuroCityPersons (ECP) without labels. (b) Illustration of dataset size for ImageNet with labels and EuroCityPersons without labels. (Color figure online)

A particularly successful type of self-supervised learning algorithms in computer vision are methods that compare different views of the same image [4,6,8,14,15]. Among them are, for example, contrastive methods that maximize feature similarity between views of the same image while maximizing dissimilarity between views of different images. By focusing on representations for entire images, high-quality global features can be learned. However, local information is lost.

To address these issues, we develop SSLAD, a self-supervised learning method for object detection. Our method computes features locally, which is especially important for object detection downstream tasks. Furthermore, SSLAD matches corresponding regions in views, thereby ensuring that maximizing their feature similarity is desirable.

Pretrained weights for object detection are typically evaluated on Pascal VOC [11] and MS COCO [18]. Similar to ImageNet, these datasets contain a large number of object-centric images, whereas many other object detection datasets do not have these properties. We argue that it is a limitation of previous research on self-supervised methods that both pretraining and finetuning are done on object-centric datasets. Conversely, we investigate pretraining and finetuning on datasets that contain a variety of non-centric objects from different classes. In the automotive domain, for example, several object detection datasets are readily available, which have these properties. Due to these characteristics, we select several automotive datasets for pretraining and finetuning. In experiments on CityPersons [26], EuroCityPersons [2] and DHD traffic [22], we show that SSLAD outperforms state-of-the-art self-supervised methods. Given just 20,000 unlabeled images, SSLAD almost matches ImageNet weights trained on 1,200,000 labeled images (see Fig. 1). If supervised ImageNet pretraining is only done on a subset of ImageNet, the resulting weights are inferior to those generated by SSLAD. Moreover, SSLAD is very robust with respect to dataset size and generates weights that are far better than a random initialization, even if only 20 unlabeled images are available for pretraining. In addition, SSLAD works even with a batch size of 1. Conversely, other self-supervised methods just collapse given so little data or such a small batch size. These properties make SSLAD especially powerful if only limited resources are available.

Overall, our main contributions can be summarized as follows:

- We propose SSLAD, a self-supervised learning method for object detection, which computes local features and works well if multiple objects are present per image. SSLAD outperforms state-of-the-art self-supervised methods in experiments on CityPersons, EuroCityPersons, and DHD Traffic.
- Although self-supervised methods conventionally use massive amounts of compute and data, we demonstrate that applying SSLAD to just 20,000 images of street scenes almost matches ImageNet weights trained on 1,200,000 labeled images. Even if only 20 unlabeled images are available for pretraining, SSLAD generates weights that are far better than a random initialization.
- We show that SSLAD works even with a batch size of one, whereas other self-supervised methods require very large batch sizes.

2 Related Work

Self-supervised learning aims to learn generic representations without using any human-labeled data. Most traditional self-supervised methods for vision rely on handcrafted pretext tasks, such as predicting the relative patch location [10], rotation [13], solving jigsaw puzzles [20], colorizing grey-scale images [16], and clustering [3]. More recently, methods that compare different views of images [4,6–8,12,14,15,25] have gained momentum. Among them are contrastive methods which are based on the idea that two views of an image should have similar features and two views of different images should have dissimilar features. Convolutional weights pretrained by these methods have been shown to match ImageNet weights on a variety of downstream tasks.

SimCLR [6] emphasizes the benefit of strong data augmentation, large batch sizes and a long training schedule. In addition, adding a nonlinear transformation between the network output and the loss function improves downstream performance. MOCO [15] employs a momentum encoder which makes it possible to achieve competitive performance even with a smaller batch size. SWAV [4] does not use output features directly, but instead clusters them and enforces consistency in the cluster assignment. BYOL [14] illustrates that using negative examples explicitly is not essential. An online network is trained to generate feature vectors that match the output of a target network. The target network itself is a moving average of the online network's weights. Making online and target networks asymmetric is beneficial. SimSiam [8] shows that using a moving average as a target network like in BYOL is not necessary and just employing a siamese architecture with a stop-gradient operation works very well. The suggested changes greatly simplify the overall setup. They allow for training with smaller batch sizes and typically result in faster convergence than other contrastive methods while not requiring a memory bank or negative examples.

One core downside of SimSiam and most contrastive methods is that they were optimized for image classification tasks. By computing only a global feature vector, local information gets lost. Conversely, some methods take local information into account [5,21]. VADeR [21] learns representations on a pixel-level that are invariant to augmentations. Matching pixel representations taken from two views of the same image are trained to be close in embedding space while pixel representations from other images should be far apart. Overall, VADeR generates features per-pixel whereas conventional methods like SimSiam compute per-image representations. In contrast, SSLAD aims at generating features per patch. This can be seen as a compromise between local per-pixel and global per-image features.

In parallel work, DenseCL [24] was developed which bears some similarity to SSLAD in that it matches local regions. Different from DenseCL, SLLAD does not require negative examples or a memory bank.

3 Method

3.1 Overview

SSLAD is a self-supervised learning method developed for object detection downstream tasks. While conventional self-supervised methods compute feature vectors describing entire images, SSLAD computes fine-grained local features.

3.2 Architecture

SSLAD takes two random views of an image as input, $x_1 \in \mathbb{R}^{H \times W \times C}$ and $x_2 \in \mathbb{R}^{H \times W \times C}$. Both are processed in parallel by a backbone network f such as ResNet but without using a pooling layer at the end. For each input, the output $w \in \mathbb{R}^{(H/k) \times (W/k) \times D}$ is a feature map that is downscaled by factor k compared to the input image size. w is further processed by a projector network g, which has the same structure as the projector in SimSiam, but where fully-connected layers are replaced by 1×1 convolutions. The resulting output z is of shape $(H/k) \times (W/k) \times D$. Finally, the output of one branch is further processed through a predictor head h. h mimics the predictor in SimSiam, in which fully-connected layers are replaced by 1×1 convolutions. The output p is a feature map of shape $(H/k) \times (W/k) \times D$.

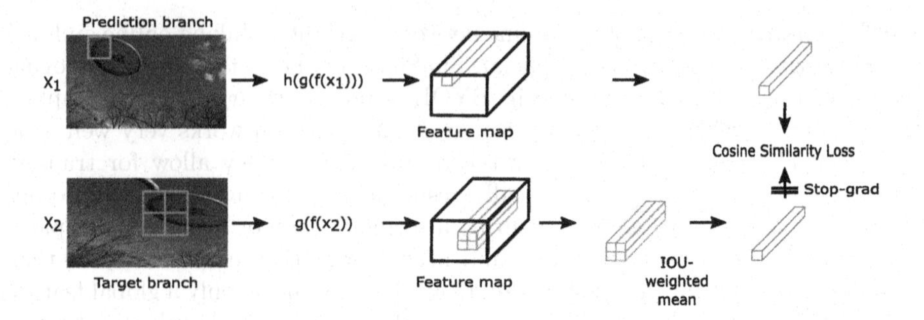

Fig. 2. Underlying concept of SSLAD. The visualization shows the exemplary calculation of the loss for **one** feature vector from the prediction branch with the corresponding target, which is an IOU-weighted mean of overlapping vectors. In SSLAD, this calculation is done for **every** feature vector in the prediction branch with the corresponding target. On the target branch, the gradient is not backpropagated as indicated by the crossed-out arrow.

3.3 Target Computation

x_1 and x_2 are fed to both branches each. $z_1 \triangleq g(f(x_1))$ and $z_2 \triangleq g(f(x_2))$ are used as targets for $p_2 \triangleq h(g(f(x_2)))$ and $p_1 \triangleq h(g(f(x_1)))$ respectively. Since these are feature maps, they do not directly match given that x_1 and x_2 are random views. However, we do know in what way x_1 and x_2 were augmented and cropped. That means we know what regions in the original image the two views x_1 and x_2 correspond to. Because of that, we can compute what regions of the feature map in p_1 (prediction) correspond to the feature map in z_2 (target). Since there is no perfect matching between the feature maps, we compute the target as an IOU-weighted mean of overlapping regions. Overall, we train the network to output similar features for overlapping regions regardless of input scale and augmentations. Figure 2 visualizes an example and formula 1 shows the loss function that is optimized.

$$L(p, z) = \sum_{i}^{R} -\frac{p_i}{\|p_i\|_2} \cdot \frac{\sum_{m}^{R} IOU(z_m, p_i) z_m}{\|\sum_{n}^{R} IOU(z_n, p_i) z_n\|_2} \cdot J(p_i) \tag{1}$$

$R = (H/k) \cdot (W/k)$ is the number of output vectors of the prediction and target branch. The function called IOU computes the intersection over union in input pixel space for the given output vectors. $J(p_i)$ is a binary indicator that determines whether an output vector from the prediction branch has a sufficiently high overlap with output vectors from the target branch. If the overlap is below threshold T, no proper target can be computed and the corresponding image region is not included in the loss (see formula 2).

$$J(p_i) = \begin{cases} 1, & \text{if } \sum_{m}^{R} IOU(z_m, p_i) \geq T \\ 0, & \text{otherwise} \end{cases} \tag{2}$$

Formula 3 shows the overall symmetric loss function. On the target branch, the gradient is not backpropagated to avoid collapsing solutions. Therefore, z_1 and z_2 are constants.

$$TotalLoss = \frac{1}{2} \cdot L(p_1, stopgrad(z_2)) + \frac{1}{2} \cdot L(p_2, stopgrad(z_1)) \tag{3}$$

4 Experiments

4.1 Datasets

To evaluate our approach we use several datasets from the automotive domain. Most images in each dataset contain a variety of classes and are typically not object-centric. Some classes occur much more often than others. Therefore, the chosen datasets differ substantially from ImageNet, which contains 1,000 balanced classes and object-centric images.

CityPersons is a pedestrian detection dataset derived from **Cityscapes** [9], which has 3,000 labeled training images. In addition, Cityscapes contains 20,000 coarsely labeled images that we include without labels. For SSLAD we use the combination of the two or, if specified, a subset of them. During supervised fine-tuning, we only use the labeled training set. We evaluate the performance using the log-average miss rate on the reasonable subset (size ≥ 50 pixels, occlusion ratio < 0.35) which we denote as MR.

EuroCityPersons is a large and diverse automotive dataset that contains images from all seasons, multiple countries and different weather conditions. As commonly done, we use the same evaluation protocol as for CityPersons and evaluate on the reasonable subset.

DHD Pedestrian Traffic is a recently introduced dataset from the automotive domain. The dataset contains a diverse set of scenes and images have a high resolution. Like for CityPersons, we evaluate the performance using the log-average miss rate on the reasonable subset.

4.2 Pretraining Methods

We contrast SSLAD with different methods described below. To ensure a fair comparison, settings for pretraining weights are identical unless specified otherwise.

ImageNet Supervised: For pretraining on full ImageNet, we use the official Pytorch [23] checkpoint. When pretraining on a subset of the classes, we follow the official schedule used in Pytorch, but increase the number of epochs. If 10%

of the data is included, we train for 200 epochs. If only 1% is used, we train for 500 epochs.

Random Initialization: We follow the default for network initialization in Pytorch. The same initialization strategy is also used in SimSiam [8] and our reimplementation of it. Furthermore, all other experiments including supervised pretraining and SSLAD are also initialized using this strategy.

SimSiam: We choose SimSiam [8] as our main self-supervised baseline. According to the literature, SimSiam works especially well on object detection downstream tasks [8,25]. In addition, SimSiam does not depend on very large batch sizes, a memory network or negative examples like other state-of-the-art self-supervised methods do. Specifically, a batch size of 256 is optimal for SimSiam [8], whereas competing methods typically require a batch size of 4096 [4,14] with batch-norm statistics synchronized across GPUs.

4.3 Implementation Details

Pretraining: In both SimSiam and SSLAD we use the optimal hyperparameters as described in Chen et al. [8] including optimizer, learning rate, batch size, and preprocessing. Different from that, we increase the number of epochs in pretraining, because the datasets we use are smaller than ImageNet and we notice improvements until 1,000 epochs of training. In some of the ablations (e.g., see Fig. 4) only a subset of the data is used. In that case we extend the training protocol, e.g., if only 20 images are available, we compute 40,000 update steps. In preprocessing, we use different augmentations. These are shown in Pytorch notation [23]. RandomResizedCrop with scale in [0.2, 1.0], ColorJitter with {brightness, contrast, saturation, hue} strength of {0.4, 0.4, 0.4, 0.1} with an applying probability of 0.8, and RandomGrayscale with an applying probability of 0.2.

Finetuning: For finetuning, we use a RetinaNet with a ResNet-50 backbone. The weights of the backbone are initialized with pretrained weights. We use the default hyperparameters for this architecture as described in Lin et al. [17].

4.4 Results for Finetuning

We pretrain a ResNet-50 either with supervision or self-supervised on ImageNet, EuroCityPersons, and CityPersons/Cityscapes. The resulting weights are then used as initialization for finetuning a RetinaNet with a Resnet-50 backbone on CityPersons, EuroCityPersons, and DHD.

As Table 1 shows, pretraining on full ImageNet with labels is clearly superior to a random initialization. Our implementation with ImageNet weights compares favorably to the results reported in the literature for the same backbone and architecture. When using only a subset of ImageNet for supervised pretraining, finetuning results become much worse. Thereby, reducing the number of ImageNet classes is much more harmful than decreasing the number of images while

Table 1. Miss rate for finetuning a RetinaNet with a ResNet-50 backbone on CityPersons, EuroCityPersons, and DHD. The backbone is initialized with pretrained weights. These are trained with labels or self-supervised (SimSiam and SSLAD) on ImageNet, EuroCityPersons, and CityPersons/Cityscapes. The number of images in the pretraining dataset and whether or not labels were available is explicitly mentioned. 'cl.' is short for classes, 'img.' is an abbreviation for images.

Method	Pretraining dataset	Labels	#Images	MR City	MR ECP	MR DHD
Random Init.	–	–	–	24.66	21.37	35.19
Supervised [19]	ImageNet 1,000 cl.	Yes	1,300,000	15.6	–	–
Supervised [22]	ImageNet 1,000 cl.	Yes	1,300,000	15.99	–	23.89
Supervised	ImageNet 1,000 cl.	Yes	1,300,000	**14.16**	**10.91**	**23.36**
Supervised	ImageNet 100 cl.	Yes	130,000	18.06	14.36	29.68
Supervised	ImageNet 10 cl.	Yes	13,000	28.01	26.02	37.83
Supervised	ImageNet 10% img.	Yes	130,000	17.01	12.97	25.49
Supervised	ImageNet 1% img.	Yes	13,000	27.64	18.67	31.68
SimSiam	EuroCityPersons	No	23,000	16.74	13.72	27.29
SSLAD	EuroCityPersons	No	23,000	**14.87**	**12.59**	**26.71**
SimSiam	CityPersons/Citysc.	No	23,000	15.18	14.49	27.55
SSLAD	CityPersons/Citysc.	No	23,000	**14.50**	**14.03**	**26.45**

keeping all classes. The reason for that might be that more fine-grained features are necessary for distinguishing 1,000 classes. Conversely, 10 classes can be discriminated given relatively crude features which may not generalize well to other downstream tasks.

SimSiam and SSLAD are trained without human labels on the 23,000 images from EuroCityPersons and CityPersons/Cityscapes. Weights trained by SimSiam are superior to those generated with supervision on labeled subsets from ImageNet, including ImageNet 100 classes. SSLAD consistently outperforms SimSiam, almost matching weights trained on full ImageNet with labels.

4.5 Ablations on CityPersons

In this section, we analyze factors that impact performance on downstream tasks. First, we investigate the effect that batch size in pretraining has on downstream performance. Most self-supervised methods including SimCLR, BYOL, and SWAV rely on very large batches to work well, i.e. they are typically trained with a batch size of 4096 with batch-norm statistics synchronized across GPUs. Conversely, SimSiam works best with a batch size of 256 [8]. As can be seen in Fig. 3, SSLAD is far more robust than SimSiam with respect to the batch size. Even training with a batch size of one only leads to a modest decrease in performance. This finding is of great practical importance, as it shows that SSLAD can even be trained on a single GPU if necessary, making it usable given limited amounts of computing.

We believe that SSLAD's robustness regarding batch size is related to the training target. Even with a batch size of one, up to 64 feature vectors are included in the loss, compared to only one in SimSiam.

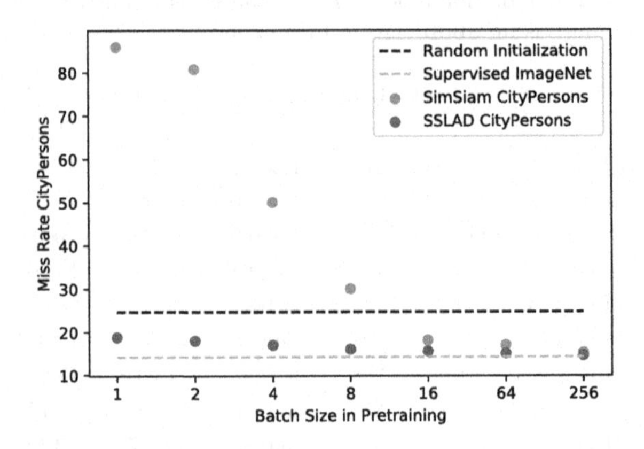

Fig. 3. Effect of batch size in pretraining on finetuning performance. SSLAD is far more robust with respect to the batch size than SimSiam.

In addition, we analyze the effect that dataset size in pretraining has on downstream performance. We pretrain SSLAD and SimSiam on various subsets of CityPersons/Cityscapes. The resulting weights are used as initialization for finetuning on CityPersons with labels. Figure 4 shows that SSLAD is far more robust regarding the size of the pretraining dataset than SimSiam. Even if just 20 unlabeled images are available, SSLAD generates weights that are clearly better than a random initialization. Conversely, the performance of SimSiam drops substantially if the dataset size is small. It is worth noting that also some of the traditional self-supervised pretraining methods have been found to work given limited data [1].

Moreover, we analyze the effect of using the same images for pretraining and finetuning (see Table 2). That means, we first pretrain SSLAD on the labeled subset of CityPersons, but without using the labels. In a second step, we use the same 2975 images with labels for finetuning. It is noteworthy, that including this pretraining step is very beneficial, even if the same images with labels are used during finetuning. Miss rate is significantly lower than without this pretraining step (17.23 vs. 24.66). What is even more beneficial is to pretrain SSLAD on different images, e.g. unlabeled Cityscapes images not used during finetuning. Doing this further reduces the miss rate (16.26 vs. 17.23).

4.6 On the Necessity of the Predictor Head

For several self-supervised learning methods [8,14], using an asymmetric architecture is necessary. For example, SimSiam does not work at all without a predictor head [8]. We observe the same issue for SSLAD. We identify that one

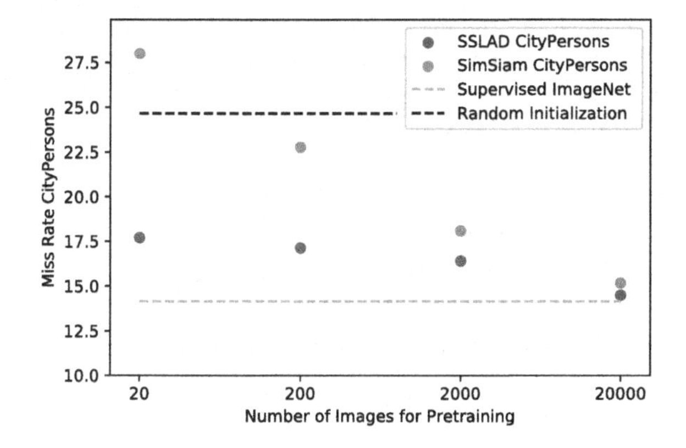

Fig. 4. Effect of dataset size for self-supervised pretraining. We contrasted SSLAD and SimSiam when using different amounts of data.

Table 2. Overlap between data for pretraining and finetuning. 'Same dataset' indicates that we use the same images in self-supervised pretraining and supervised finetuning. 'Different dataset' means that there is no overlap in data between pretraining and finetuning. If we use the same images for pretraining with SSLAD as for finetuning, performance is lower compared to using different images.

Data for pretraining	Images in pretraining	MR CityPersons
Random initialization	–	24.66
SSLAD; Same dataset	2975	17.23
SSLAD; Different dataset	2975	16.26
Supervised ImageNet	1 200 000	14.16

reason for that is the way the network is initialized. At the beginning of training, z_1 and z_2 that are generated from different views of the same image are very similar (see Table 3). Cosine similarity is close to 1. Therefore, training the network to make the two similar to each other is not a useful training task, as it is already fulfilled at the beginning of training. Conversely, the output of the predictor head p is very different from the projector output z. Cosine similarity is almost 0. Making these two more similar to each other is at least potentially a useful pretraining task.

To better understand what the predictor head does, consider the following: The target z_2 is created based on a random view of an image by the backbone network and the projector head. On the prediction branch, z_1 is generated using the same backbone and projector but another random view. The task of the predictor then comes down to how to modify z_1 to match z_2. This task depends on what view was sampled for z_2. However, this information is not provided to the prediction branch. This means that over the course of many epochs, the predictor is trained to modify z_1 in a way that it is likely to match z_2. Seen over multiple epochs, the predictor is trained to modify z_1 to match the average of z_2 for the given image.

Table 3. Cosine similarity of the network output of SSLAD and SimSiam for different augmentations of the same image. At the beginning of training, z_1 and z_2 are very similar. Training the network to make z_1 and z_2 similar to each other is therefore not a useful training task, as it is already fulfilled at the beginning of training.

Method	$Sim(z_1, z_2)$	$Sim(p_1, z_2)$
SSLAD (initialized)	0.99	0.01
SSLAD (converged)	0.97	0.95
SimSiam (initialized)	0.99	0.02
SimSiam (converged)	0.97	0.96

4.7 Qualitative Analysis of Features

Finally, we qualitatively analyze the properties of features generated by supervised pretraining, SSLAD, and random weights. For this, we input images from CityPersons into a pretrained ResNet-50 and extract the final feature map of the last residual block. This feature map $m \in \mathbb{R}^{H \times W \times C}$ contains $H \times W$ local feature vectors having C dimensions each. The distance between the feature vectors is measured by the Manhattan distance. For example, the distance from the top left corner to the bottom right corner of the feature map is $(H - 1) + (W - 1)$.

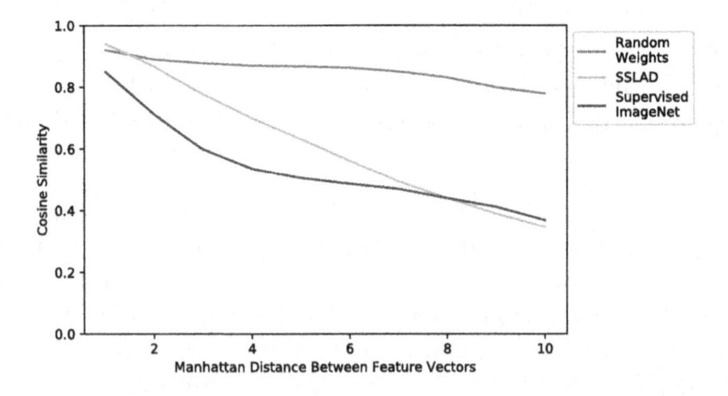

Fig. 5. Similarity of feature vectors depending on their distance in feature space. A feature vector can be seen as a local descriptor of one region in an image. Intuitively, the further two regions of an image are apart, the lower their feature similarity should become. A steep decline of cosine similarity given greater distance might be an indication that the model created discriminatory local features.

A feature vector can be seen as a local descriptor of one region in an image. One might assume, that the greater the distance between two patches, the more different their respective feature vectors should become. After all, the greater the spatial distance, the more likely it is that the two patches contain different objects. Based on this assumption, a steep decline of cosine similarity given greater distance might be an indication that the model created discriminatory local features.

Figure 5 shows the cosine similarity of feature vectors depending on their respective Manhattan distance. Given random weights, feature vectors are very similar even if they are far away from each other. This is an indication that random weights do not produce discriminative local features. Conversely, both SSLAD and weights trained on ImageNet with supervision generate discriminative features. Given a low spatial distance, cosine similarity is very high. Increasing distance results in strongly decreasing cosine similarity. The decline is larger and more consistent for SSLAD, indicating more discriminative local features.

5 Conclusion

We introduced SSLAD, a self-supervised pretraining method especially suited for object detection downstream tasks. SSLAD computes local image descriptors that are invariant to augmentations and scale. On a variety of datasets from the automotive domain, we showed that SSLAD outperforms SimSiam, a state-of-the-art self-supervised method. By leveraging only 20,000 images taken from a video camera on a car, SSLAD almost matches ImageNet weights which are pretrained on 1,200,000 labeled images. Even if just 20 images are available for pretraining, SSLAD generates weights that are far better than a random initialization. Other self-supervised methods such as SimSiam just collapse given so little data. Furthermore, SSLAD works even with a batch size of 1, allowing for low-resource training. This is quite different from other self-supervised pretraining methods, which are known to rely on very large batches. Overall, SSLAD makes it possible to pretrain weights even if little unlabeled data and computational resources are available. These properties make SSLAD interesting for a wider audience of people who can access only limited amounts of computing.

References

1. Asano, Y.M., Rupprecht, C., Vedaldi, A.: A critical analysis of self-supervision, or what we can learn from a single image. In: International Conference on Learning Representations ICLR (2020)
2. Braun, M., Krebs, S., Flohr, F., Gavrila, D.M.: Eurocity persons: a novel benchmark for person detection in traffic scenes. IEEE Trans. Pattern Anal. Mach. Intell. **41**, 1844–1861 (2019)
3. Caron, M., Bojanowski, P., Joulin, A., Douze, M.: Deep clustering for unsupervised learning of visual features. In: Proceedings of the European Conference on Computer Vision (ECCV) (2018)
4. Caron, M., Misra, I., Mairal, J., Goyal, P., Bojanowski, P., Joulin, A.: Unsupervised learning of visual features by contrasting cluster assignments. In: Advances in Neural Information Processing Systems, vol. 33, pp. 9912–9924 (2020)

5. Chaitanya, K., Erdil, E., Karani, N., Konukoglu, E.: Contrastive learning of global and local features for medical image segmentation with limited annotations. In: Larochelle, H., Ranzato, M., Hadsell, R., Balcan, M.F., Lin, H. (eds.) Advances in Neural Information Processing Systems, vol. 33, pp. 12546–12558. Curran Associates, Inc. (2020)
6. Chen, T., Kornblith, S., Norouzi, M., Hinton, G.: A simple framework for contrastive learning of visual representations. In: Proceedings of the 37th International Conference on Machine Learning. Proceedings of Machine Learning Research, vol. 119, pp. 1597–1607 (2020)
7. Chen, X., Fan, H., Girshick, R., He, K.: Improved baselines with momentum contrastive learning (2020). arXiv:2003.04297
8. Chen, X., He, K.: Exploring simple siamese representation learning. In: Proceedings of the IEEE/CVF Conference on Computer Vision and Pattern Recognition (CVPR), pp. 15750–15758 (2021)
9. Cordts, M., et al.: The cityscapes dataset for semantic urban scene understanding (2016). arXiv:1604.01685
10. Doersch, C., Gupta, A., Efros, A.A.: Unsupervised visual representation learning by context prediction. In: Proceedings of the IEEE International Conference on Computer Vision (ICCV) (2015)
11. Everingham, M., Gool, L.V., Williams, C., Winn, J., Zisserman, A.: The pascal visual object classes (VOC) challenge. Int. J. Comput. Vis. (IJCV) **88**, 303–338 (2010)
12. Gidaris, S., Bursuc, A., Puy, G., Komodakis, N., Cord, M., Perez, P.: OBoW: online bag-of-visual-words generation for self-supervised learning. In: Proceedings of the IEEE/CVF Conference on Computer Vision and Pattern Recognition (CVPR), pp. 6830–6840 (2021)
13. Gidaris, S., Singh, P., Komodakis, N.: Unsupervised representation learning by predicting image rotations. In: International Conference on Learning Representations ICLR (2018)
14. Grill, J.B., et al.: Bootstrap your own latent: a new approach to self-supervised learning (2020). arXiv:2006.07733
15. He, K., Fan, H., Wu, Y., Xie, S., Girshick, R.: Momentum contrast for unsupervised visual representation learning. In: Proceedings of the IEEE/CVF Conference on Computer Vision and Pattern Recognition (CVPR) (2020)
16. Larsson, G., Maire, M., Shakhnarovich, G.: Colorization as a proxy task for visual understanding. In: Proceedings of the IEEE Conference on Computer Vision and Pattern Recognition (CVPR) (2017)
17. Lin, T.Y., Goyal, P., Girshick, R., He, K., Dollar, P.: Focal loss for dense object detection. In: Proceedings of the IEEE International Conference on Computer Vision (ICCV) (2017)
18. Lin, T.Y., et al.: Microsoft coco: common objects in context (2015). arXiv:1405.0312
19. Liu, W., Liao, S., Ren, W., Hu, W., Yu, Y.: High-level semantic feature detection: a new perspective for pedestrian detection. In: Proceedings of the IEEE/CVF Conference on Computer Vision and Pattern Recognition (CVPR) (2019)
20. Noroozi, M., Favaro, P.: Unsupervised learning of visual representations by solving jigsaw puzzles. In: Leibe, B., Matas, J., Sebe, N., Welling, M. (eds.) ECCV 2016. LNCS, vol. 9910, pp. 69–84. Springer, Cham (2016). https://doi.org/10.1007/978-3-319-46466-4_5

21. Pinheiro, P.O., Almahairi, A., Benmalek, R., Golemo, F., Courville, A.C.: Unsupervised learning of dense visual representations. In: Advances in Neural Information Processing Systems, vol. 33, pp. 4489–4500 (2020)
22. Pang, Y., Cao, J., Li, Y., Xie, J., Sun, H., Gong, J.: TJU-DHD: a diverse high-resolution dataset for object detection. IEEE Trans. Image Process. **30**, 207–219 (2021)
23. Paszke, A., et al.: PyTorch: an imperative style, high-performance deep learning library. In: Neural Information Processing Systems NeurIPS (2019)
24. Wang, X., Zhang, R., Shen, C., Kong, T., Li, L.: Dense contrastive learning for self-supervised visual pre-training. In: Proceedings of the IEEE/CVF Conference on Computer Vision and Pattern Recognition (CVPR), pp. 3024–3033 (2021)
25. Zbontar, J., Jing, L., Misra, I., LeCun, Y., Deny, S.: Barlow twins: self-supervised learning via redundancy reduction. In: International Conference on Machine Learning (ICML) (2021)
26. Zhang, S., Benenson, R., Schiele, B.: CityPersons: a diverse dataset for pedestrian detection (2017). arXiv:1702.05693

Assignment Flows and Nonlocal PDEs on Graphs

Dmitrij Sitenko([✉]), Bastian Boll, and Christoph Schnörr

Image and Pattern Analysis Group (IPA) and Heidelberg Collaboratory for Image Processing (HCI), Heidelberg University, Heidelberg, Germany
dmitrij.sitenko@iwr.uni-heidelberg.de

Abstract. This paper employs nonlocal operators and the corresponding calculus in order to show that assignment flows for image labeling can be represented by a nonlocal PDEs on the underlying graph. In addition, for the homogeneous Dirichlet condition, a tangent space parametrization and geometric integration can be used to solve the PDE numerically. The PDE reveals a nonlocal balance law that governs the spatially distributed dynamic mass assignment to labels. Numerical experiments illustrate the theoretical results.

Keywords: Image labeling · Assignment flows · Nonlocal PDEs

1 Introduction

Overview, Motivation. As in most research areas of computer vision, state of the art approaches to image segmentation are based on *deep networks*. A recent survey [18] reviews a vast number of different network *architectures* and their empirical performance on various benchmark data sets. Among the challenges discussed in [18, Section 6], the authors write: "... a concrete study of the underlying behavior/dynamics of these models is lacking. A *better understanding* of the *theoretical aspects* of these models can enable the development of better models curated toward various segmentation scenarios."

Among the various approaches towards a better understanding of the *mathematics* of deep networks, the connection between general continuous-times ODEs and deep networks, in terms of so-called *neural ODEs*, has been picked out as a central them [8,17]. A particular class of neural ODEs derived from first principles of information geometry, so-called *assignment flows*, were introduced recently [3,20]. The connection to deep networks becomes evident by applying the simplest *geometric* numerical integration scheme (cf. [22]) to the system of ODEs (23), which results in the *discrete-time* dynamical system (see Sect. 3 for definitions of the mappings involved)

$$W^{(t+1)}(x) = \mathrm{Exp}_{W^{(t)}(x)} \circ R_{W^{(t)}(x)}\big(h_{(t)} S(W^{(t)})(x)\big), \quad x \in \mathcal{V}. \tag{1}$$

Here, $t \in \mathbb{N}$ is the discrete time index (iteration counter), x is any vertex of an underlying graph, and $W^{(t)}(x)$ is the assignment vector that converges for

C. Bauckhage et al. (Eds.): DAGM GCPR 2021, LNCS 13024, pp. 498–512, 2021.
https://doi.org/10.1007/978-3-030-92659-5_32

$t \to \infty$ to some unit vector e_j and thus assigns label j to the data observed at x. The key observation to be made here is that the right-hand side of (1) comprises two major ingredients of any deep network: (i) a *context-sensitive interaction* of the variables over the underlying graph in terms of the mapping $S(\cdot)$, and (ii) a *pointwise nonlinearity* in terms of the exponential map Exp_W corresponding to the e-connection of information geometry [1]. Thus, implementing each iteration of (1) as layer of a networks yields a *deep network*.

The **aim of this paper** is to contribute to the mathematics of deep networks by *representing assignment flows* through a *nonlocal PDE* on the underlying graph. This *establishes a link* between traditional *local* PDE-based image *processing* [21] to modern advanced *nonlocal* schemes of image *analysis*. Regarding the latter nonlocal approaches, we mention the seminal work of Gilboa and Osher [16], a PDE [6] resulting as zero-scale limit of the nonlocal means neighborhood filter [7], nonlocal Laplacians on graphs for image denoising, enhancement and for point cloud processing [14,15], and variational phase-field models [5] motivated by both total-variation based image denoising and the classical variational relaxation of the Mumford-Shah approach to image segmentation [2]. However, regarding image *segmentation*, while some of the afore-mentioned approaches apply to binary fore-/background separation, none of them was *specifically* designed for image segmentation and labeling, with an arbitrary number of labels.

Contribution. Our *theoretical* paper introduces a *nonlocal PDE for image labeling* related to assignment flows on arbitrary graphs. Starting point is the so-called S-parametrization of assignment flows introduced by [19]. The mathematical basis is provided by the nonlocal calculus developed in [12,13], see also [11], whose operators include as *special cases* the mappings employed in the above-mentioned works (graph Laplacians etc.). We show, in particular, that the geometric integration schemes of [22] can be used to *solve* the novel nonlocal PDE, which additionally *generalizes* the *local* PDE derived in [19] corresponding to the zero-scale limit of the assignment flow. In addition, the novel PDE reveals in terms of *nonlocal balance laws* the flow of information between label assignments across the graph.

Organization. Section 2 presents required concepts of nonlocal calculus. Section 3 summarizes the assignment flow approach. Our contribution is presented and discussed in Sect. 4, and illustrated in Sect. 5. We conclude in Sect. 6.

2 Nonlocal Calculus

In this section, following [13], we collect some basic notions and nonlocal operators which will be used throughout this paper. See [11] for a more comprehensive exposition.

Let $(\mathcal{V}, \mathcal{E}, \Omega)$ be a weighted connected graph consisting of $|\mathcal{V}| = n$ nodes with no self-loops, where $\mathcal{E} \subset \mathcal{V} \times \mathcal{V}$ denotes the edge set. In what follows we focus on *undirected* graphs with connectivity given by the neighborhoods $\mathcal{N}(x) = \{y \in$

$\mathcal{V}\colon \Omega(x,y) > 0\}$ of each node $x \in \mathcal{V}$ through nonnegative symmetric weights $\Omega(x,y)$ satisfying

$$0 \leq \Omega(x,y) \leq 1, \qquad \Omega(x,y) = \Omega(y,x). \tag{2}$$

The weighting function $\Omega(x,y)$ serves as the similarity measure between two vertices x and y on the graph. We define the function spaces

$$\mathcal{F}_{\mathcal{V}} := \{f\colon \mathcal{V} \to \mathbb{R}\}, \qquad \mathcal{F}_{\mathcal{V}\times\mathcal{V}} := \{F\colon \mathcal{V}\times\mathcal{V} \to \mathbb{R}\}, \tag{3a}$$

$$\mathcal{F}_{\mathcal{V},E} := \{F\colon \mathcal{V} \to E\}, \qquad \mathcal{F}_{\mathcal{V}\times\mathcal{V},E} := \{F\colon \mathcal{V}\times\mathcal{V} \to E\}, \tag{3b}$$

where E denotes any subset of an Euclidean space. The spaces $\mathcal{F}_{\mathcal{V}}$ and $\mathcal{F}_{\mathcal{V}\times\mathcal{V}}$ respectively are equipped with the inner products

$$\langle f,g\rangle_{\mathcal{V}} = \sum_{x,y\in\mathcal{V}} f(x)g(y), \qquad \langle F,G\rangle_{\mathcal{V}\times\mathcal{V}} = \sum_{x,y\in\mathcal{V}\times\mathcal{V}} F(x,y)G(x,y). \tag{4}$$

Given an *antisymmetric* mapping $\alpha \in \mathcal{F}_{\overline{\mathcal{V}}\times\overline{\mathcal{V}}}$ with $\alpha(x,y) = -\alpha(y,x)$ and $\overline{\mathcal{V}}$ defined by (5) and (6), and assuming the vertices $x \in \mathcal{V}$ correspond to points in the Euclidean space $\mathbb{R}^d$, the *nonlocal interaction domain with respect to* α is defined as

$$\mathcal{V}_{\mathcal{I}}^{\alpha} := \{x \in \mathbb{Z}^d \setminus \mathcal{V} : \alpha(x,y) > 0 \text{ for some } y \in \mathcal{V}\}. \tag{5}$$

The set (5) serves as discrete counterpart of the nonlocal boundary on a Euclidean underlying domain with positive measure, as opposed to the traditional local formulation of a boundary that has measure zero. See Figure 1 for a schematic illustration of possible nonlocal boundary configuration. Introducing the abbreviation

$$\overline{\mathcal{V}} = \mathcal{V}\dot{\cup}\mathcal{V}_{\mathcal{I}}^{\alpha} \qquad \text{(disjoint union)}, \tag{6}$$

we state the following identity

$$\sum_{x,y\in\overline{\mathcal{V}}} F(x,y)\alpha(x,y) - F(y,x)\alpha(y,x) = 0, \qquad \forall F \in \mathcal{F}_{\overline{\mathcal{V}}\times\overline{\mathcal{V}}}. \tag{7}$$

Based on (5), the *nonlocal divergence operator* $\mathcal{D}^{\alpha} : \mathcal{F}_{\overline{\mathcal{V}}\times\overline{\mathcal{V}}} \to \mathcal{F}_{\overline{\mathcal{V}}}$ and the *nonlocal interaction operator* $\mathcal{N}^{\alpha}\colon \mathcal{F}_{\overline{\mathcal{V}}\times\overline{\mathcal{V}}} \to \mathcal{F}_{\mathcal{V}_{\mathcal{I}}^{\alpha}}$ are defined by

$$\begin{aligned} \mathcal{D}^{\alpha}(F)(x) &:= \sum_{y\in\overline{\mathcal{V}}} \big(F(x,y)\alpha(x,y) - F(y,x)\alpha(y,x)\big), \qquad x \in \overline{\mathcal{V}}, \\ \mathcal{N}^{\alpha}(G)(x) &:= -\sum_{y\in\overline{\mathcal{V}}} \big(G(x,y)\alpha(x,y) - G(y,x)\alpha(y,x)\big), \quad x \in \mathcal{V}_{\mathcal{I}}^{\alpha}. \end{aligned} \tag{8}$$

Due to the identity (7), these operators (8) satisfy the *nonlocal Gauss theorem*

$$\sum_{x\in\mathcal{V}} \mathcal{D}^{\alpha}(F)(x) = \sum_{y\in\mathcal{V}_{\mathcal{I}}^{\alpha}} \mathcal{N}^{\alpha}(F)(y). \tag{9}$$

The operator $\mathcal{D}^\alpha$ maps two-point functions $F(x,y)$ to $\mathcal{D}^\alpha(F) \in \mathcal{F}_{\overline{\mathcal{V}}}$, whereas $\mathcal{N}^\alpha$ is defined on the domain $\mathcal{V}_\mathcal{I}^\alpha$ given by (5) where nonlocal boundary conditions are imposed. In view of (4), the adjoint mapping $(\mathcal{D}^\alpha)^*$ is determined by the relation

$$\langle f, \mathcal{D}^\alpha(F)\rangle_{\overline{\mathcal{V}}} = \langle (\mathcal{D}^\alpha)^*(f), F\rangle_{\overline{\mathcal{V}}\times\overline{\mathcal{V}}}, \quad \forall f \in \mathcal{F}_{\overline{\mathcal{V}}}, \quad F \in \mathcal{F}_{\overline{\mathcal{V}}\times\overline{\mathcal{V}}}, \tag{10}$$

which yields the operator $(\mathcal{D}^\alpha)^* \colon \mathcal{F}_{\overline{\mathcal{V}}} \to \mathcal{F}_{\overline{\mathcal{V}}\times\overline{\mathcal{V}}}$ acting on $\mathcal{F}_{\overline{\mathcal{V}}}$ by

$$(\mathcal{D}^\alpha)^*(f)(x,y) := -(f(y) - f(x))\alpha(x,y), \qquad x, y \in \overline{\mathcal{V}}. \tag{11}$$

The *nonlocal gradient operator* is defined as

$$\mathcal{G}^\alpha(x,y) := -(\mathcal{D}^\alpha)^*(x,y). \tag{12}$$

For *vector-valued* mappings, the operators (8) and (11) naturally extend to $\mathcal{F}_{\overline{\mathcal{V}}\times\overline{\mathcal{V}},E}$ and $\mathcal{F}_{\overline{\mathcal{V}},E}$, respectively, by acting *componentwise.*

Using definitions (11), (12), the nonlocal Gauss theorem (9) implies for $u \in \mathcal{F}_{\overline{\mathcal{V}}}$ *Greens nonlocal first identity*

$$\sum_{x\in\mathcal{V}} u(x)\mathcal{D}^\alpha(F)(x) - \sum_{x\in\overline{\mathcal{V}}}\sum_{y\in\overline{\mathcal{V}}} \mathcal{G}^\alpha(u)(x,y)F(x,y) = \sum_{x\in\mathcal{V}_\mathcal{I}^\alpha} u(x)\mathcal{N}^\alpha(F)(x). \tag{13}$$

Given a function $f \in \mathcal{F}_{\overline{\mathcal{V}}}$ and a mapping $\Theta \in \mathcal{F}_{\overline{\mathcal{V}}\times\overline{\mathcal{V}}}$ with $\Theta(x,y) = \Theta(y,x)$, we define the linear *nonlocal diffusion operator* as the composition of (8) and $\mathcal{G}^\alpha$,

$$\mathcal{D}^\alpha\big(\Theta\mathcal{G}^\alpha(f)\big)(x) = 2\sum_{y\in\overline{\mathcal{V}}} (\mathcal{G}^\alpha)(f)(x,y)\big(\Theta(x,y)\alpha(x,y)\big). \tag{14}$$

For the particular case with no interactions, i.e. $\alpha(x,y) = 0$ if $x \in \mathcal{V}$ and $y \in \mathcal{V}_\mathcal{I}^\alpha$, expression (14) reduces to

$$\mathcal{L}_\omega f(x) = \sum_{y\in\mathcal{N}(x)} \omega(x,y)\big(f(y) - f(x)\big), \qquad \omega(x,y) = 2\alpha(x,y)^2, \tag{15}$$

which coincides with the *combinatorial Laplacian* [9,10] after reversing the sign.

3 Assignment Flows

We summarize the assignment flow approach introduced by [3] and refer to the recent survey [20] for more background and a review of related work.

Assignment Manifold. Let $(\mathcal{F}, d_\mathcal{F})$ be a metric space and $\mathcal{F}_n = \{f(x) \in \mathcal{F} \colon x \in \mathcal{V}\}$ be given data on a graph $(\mathcal{V}, \mathcal{E}, \Omega)$ with $|\mathcal{V}| = n$ nodes and associated neighborhoods $\mathcal{N}(x)$ as specified in Sect. 2. We encode assignment of nodes $x \in \mathcal{V}$ to a set $\mathcal{F}_* = \{f_j^* \in \mathcal{F}, j \in \mathcal{J}\}$, $|\mathcal{J}| = c$ of predefined prototypes by *assignment vectors*

$$W(x) = (W_1(x), \dots, W_c(x))^\top \in \mathcal{S}, \tag{16}$$

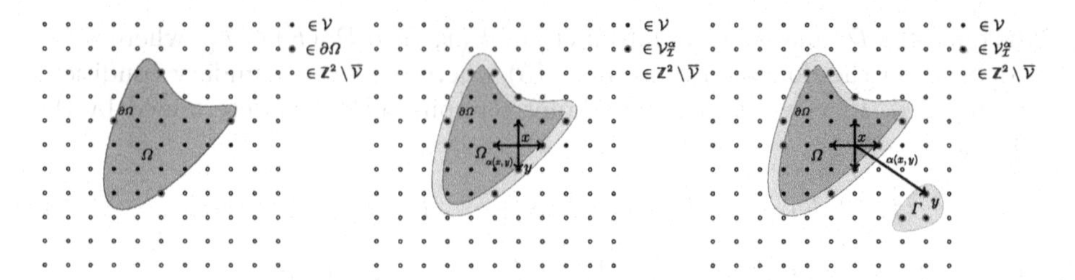

Fig. 1. *Schematic visualization of a nonlocal boundary: From left to right:* A bounded open domain $\Omega \subset \mathbb{R}^2$ with continuous *local* boundary $\partial\Omega$ overlaid by the grid $\mathbb{Z}^2$. A bounded open domain Ω with nonlocal boundary depicted by light gray color with nodes • and • representing the vertices of the graph $\mathcal{V}$ and the interaction domain $\mathcal{V}_\mathcal{I}^\alpha$, respectively. Possible decomposition of $\mathcal{V}_\mathcal{I}^\alpha$ satisfying (6). In contrast to the center configuration, this configuration enables nonlocal interactions between nodes • $\in \Gamma$.

where $\mathcal{S} = \operatorname{rint} \Delta_c$ denotes the relative interior of the probability simplex $\Delta_c \subset \mathbb{R}^c_+$ that we turn into a Riemannian manifold $(\mathcal{S}, g)$ with the Fisher-Rao metric g from information geometry [1,4] at each $p \in \mathcal{S}$

$$g_p(u,v) = \sum_{j\in\mathcal{J}} \frac{u^j v^j}{p^j}, \quad u,v \in T_0 = \{v \in \mathbb{R}^c \colon \langle \mathbb{1}_c, v\rangle = 0\}. \tag{17}$$

The *assignment manifold* $(\mathcal{W}, g)$ is defined as the product space $\mathcal{W} = \mathcal{S}\times\cdots\times\mathcal{S}$ of $n = |\mathcal{V}|$ such manifolds. Points on the assignment manifold have the form

$$W = (\dots, W(x)^\top, \dots)^\top \in \mathcal{W} \subset \mathbb{R}^{n\times c}, \qquad x \in \mathcal{V}. \tag{18}$$

The assignment manifold has the trivial tangent bundle $T\mathcal{W} = \mathcal{W} \times \mathcal{T}_0$ with tangent space

$$\mathcal{T}_0 = T_0 \times \cdots \times T_0. \tag{19}$$

The orthogonal projection onto T_0 is given by $\Pi_0 \colon \mathbb{R}^c \to T_0$, $u \mapsto u - \langle \mathbb{1}_\mathcal{S}, u\rangle \mathbb{1}_c$, $\mathbb{1}_\mathcal{S} := \frac{1}{c}\mathbb{1}_c$. We denote also by Π_0 the orthogonal projection onto $\mathcal{T}_0$, for notational simplicity.

Assignment Flows. Based on the given data and prototypes, we define the distance vector field on $\mathcal{V}$ by $D_\mathcal{F}(x) = \big(d_\mathcal{F}(f(x), f_1^*), \dots, d_\mathcal{F}(f(x), f_c^*)\big)^\top$, $x \in \mathcal{V}$. These data are lifted to $\mathcal{W}$ to obtain the *likelihood vectors*

$$L(x) \colon \mathcal{S} \to \mathcal{S}, \quad L(W)(x) = \frac{W(x)e^{-\frac{1}{\rho}D_\mathcal{F}(x)}}{\langle W(x), e^{-\frac{1}{\rho}D_\mathcal{F}(x)}\rangle}, \quad x \in \mathcal{V}, \quad \rho > 0, \tag{20}$$

where the exponential function applies componentwise. This map is based on the affine e-connection of information geometry, and the scaling parameter $\rho > 0$ normalizes the a priori unknown scale of the components of $D_\mathcal{F}(x)$. Likelihood vectors are spatially regularized by the *similarity map* and the *similarity vectors*, respectively, given for each $x \in \mathcal{V}$ by

$$S(x)\colon \mathcal{W} \to \mathcal{S}, \quad S(W)(x) = \mathrm{Exp}_{W(x)}\Big(\sum_{y \in \mathcal{N}(x)} \Omega(x,y) \mathrm{Exp}^{-1}_{W(x)}\big(L(W)(y)\big)\Big), \tag{21}$$

where $\mathrm{Exp}\colon \mathcal{S} \times T_0 \to \mathcal{S}$, $\mathrm{Exp}_p(v) = \frac{pe^{v/p}}{\langle p, e^{v/p}\rangle}$ is the exponential map corresponding to the e-connection. Hereby, the weights $\Omega(x,y)$ determine the regularization of the dynamic label assignment process (23) and satisfy (2) with the additional constraint

$$\sum_{y \in \mathcal{N}(x)} \Omega(x,y) = 1, \quad \forall x \in \mathcal{V}. \tag{22}$$

The *assignment flow* is induced on the assignment manifold $\mathcal{W}$ by solutions $W(t,x) = W(x)(t)$ of the system of nonlinear ODEs

$$\dot{W}(x) = R_{W(x)} S(W)(x), \qquad W(0,x) = W(x)(0) \in \mathbb{1}_{\mathcal{S}}, \quad x \in \mathcal{V}, \tag{23}$$

where the map $R_p = \mathrm{Diag}(p) - pp^\top$, $p \in \mathcal{S}$ turns the right-hand side into the tangent vector field $\mathcal{V} \ni x \mapsto \mathrm{Diag}(W(x)) - \langle W(x), S(W)(x)\rangle W(x) \in T_0$.

S-Flow Parametrization. In the following, it will be convenient to adopt from [19, Prop. 3.6] the *S-parametrization* of the assignment flow (23)

$$\dot{S} = R_S(\Omega S), \qquad S(0) = \exp_{\mathbb{1}_{\mathcal{W}}}(-\Omega D), \tag{24a}$$

$$\dot{W} = R_W(S), \qquad W(0) = \mathbb{1}_{\mathcal{W}}, \quad \mathbb{1}_{\mathcal{W}}(x) = \mathbb{1}_{\mathcal{S}}, \quad x \in \mathcal{V}, \tag{24b}$$

where both S and W are points on $\mathcal{W}$ and hence have the format (18) and

$$R_S(\Omega S)(x) = R_{S(x)}\big((\Omega S)(x)\big), \quad (\Omega S)(x) = \sum_{y \in \mathcal{N}(x)} \Omega(x,y) S(y), \tag{25}$$

$$\exp_{\mathbb{1}_{\mathcal{W}}}(-\Omega D)(x) = \big(\dots, \mathrm{Exp}_{\mathbb{1}_{\mathcal{S}}} \circ R_{\mathbb{1}_{\mathcal{S}}}(-(\Omega D)(x)), \dots\big)^\top \in \mathcal{W}, \tag{26}$$

with the mappings Exp_p, R_p, $p \in \mathcal{S}$ defined after (21) and (23), respectively.

Parametrization (24) has the advantage that $W(t)$ depends on $S(t)$, but not vice versa. As a consequence, it suffices to focus on (24a) since its solution $S(t)$ determines the solution to (24b) by [23, Prop. 2.1.3] $W(t) = \exp_{\mathbb{1}_{\mathcal{W}}}\big(\int_0^t \Pi_0 S(\tau) \mathrm{d}\tau\big)$. In addition, (24a) was shown in [19] to be the Riemannian gradient flow with respect to the potential $J\colon \mathcal{W} \to \mathbb{R}$ given by

$$J(S) = -\frac{1}{2}\langle S, \Omega S\rangle = \frac{1}{4}\sum_{x \in \mathcal{V}} \sum_{y \in \mathcal{N}(x)} \Omega(x,y) ||S(x) - S(y)||^2 - \frac{1}{2}||S||^2. \tag{27}$$

Convergence and stability results for this gradient flow were established by [23].

4 Relating Assignment Flows and Nonlocal PDEs

Using the nonlocal concepts from Sect. 2, we show that the assignment flow introduced in Sect. 3 corresponds to a particular nonlocal diffusion process. This provides an equivalent formulation of the Riemannian flow (24a) in terms of a composition of operators of the form (14).

4.1 $\mathcal{S}$-Flow: Non-local PDE Formulation

We begin by first specifying a general class of parameter matrices Ω satisfying (2) and (22), in terms of an anti-symmetric mapping $\alpha \in \mathcal{F}_{\overline{\mathcal{V}}\times\overline{\mathcal{V}}}$.

Lemma 1. *Let $\alpha, \Theta \in \mathcal{F}_{\overline{\mathcal{V}}\times\overline{\mathcal{V}}}$ be nonnegative anti-symmetric and symmetric mappings, respectively, i.e. $\alpha(y,x) = -\alpha(x,y)$ and $\Theta(x,y) = \Theta(y,x)$, and assume α, Θ satisfy*

$$\lambda(x) = \sum_{y\in\mathcal{N}(x)} \Theta(x,y)\alpha^2(x,y) + \Theta(x,x) \leq 1, \qquad x \in \mathcal{V}, \tag{28a}$$

$$\text{and} \quad \alpha(x,y) = 0, \qquad x,y \in \mathcal{V}^{\alpha}_{\mathcal{I}}. \tag{28b}$$

Then, for arbitrary neighborhoods $\mathcal{N}(x)$, the parameter matrix Ω given by

$$\Omega(x,y) = \begin{cases} \Theta(x,y)\alpha^2(x,y), & \text{if } x \neq y, \\ \Theta(x,x), & \text{if } x = y. \end{cases} \qquad x,y \in \mathcal{V}, \tag{29}$$

satisfies property (2) and (22) and achieves equality in (28). Additionally, for each function $f \in \mathcal{F}_{\overline{\mathcal{V}}}$ with $f|_{\mathcal{V}^{\alpha}_{\mathcal{I}}} = 0$, the identity

$$\sum_{y\in\mathcal{V}} \Omega(x,y)f(y) = \frac{1}{2}\mathcal{D}^{\alpha}(\Theta\mathcal{G}^{\alpha}(f))(x) + \lambda(x)f(x), \qquad x \in \mathcal{V}, \tag{30}$$

holds with $\mathcal{D}^{\alpha}, \mathcal{G}^{\alpha}$ given by (8), (12) and $\lambda(x)$ by (28a).

Proof. For any $x \in \mathcal{V}$, we directly compute using assumption (28) and definition (29),

$$\sum_{y\in\mathcal{V}} \Omega(x,y)f(y) = \sum_{y\in\overline{\mathcal{V}}} \Theta(x,y)\alpha(x,y)^2(f(y)) + \Theta(x,x)f(x) \tag{31a}$$

$$= -\sum_{y\in\overline{\mathcal{V}}} \Theta(x,y)\alpha(x,y)^2\Big(-(f(y)-f(x))\Big) + \lambda(x)f(x) \tag{31b}$$

$$\overset{(11)}{=} -\sum_{y\in\overline{\mathcal{V}}} \Theta(x,y)\big((\mathcal{D}^{\alpha})^*(f)(x,y)\big)\alpha(x,y) + \lambda(x)f(x) \tag{31c}$$

$$= \sum_{y\in\overline{\mathcal{V}}} \frac{1}{2}\Big(\Theta(x,y)\big(-2(\mathcal{D}^{\alpha})^*(f)(x,y)\alpha(x,y)\big)\Big) + \lambda(x)f(x). \tag{31d}$$

Using (12) and (14) yields Eq. (30). □

Next, we generalize the common *local* boundary conditions for PDEs to *volume constraints* for *nonlocal* PDEs on discrete domains. Following [13], given an antisymmetric mapping α as in (5) and Lemma 1, the natural domains $\mathcal{V}^{\alpha}_{\mathcal{I}_N}, \mathcal{V}^{\alpha}_{\mathcal{I}_D}$ for imposing nonlocal *Neumann-* and *Dirichlet* volume constraints are given by a disjoint decomposition $\mathcal{V}^{\alpha}_{\mathcal{I}} = \mathcal{V}^{\alpha}_{\mathcal{I}_N} \dot{\cup} \mathcal{V}^{\alpha}_{\mathcal{I}_D}$ of the interaction domain (5). The following proposition reveals how the flow (24a) with Ω satisfying the assumptions of Lemma 1 can be reformulated as a nonlocal partial difference equation with *Dirichlet* boundary condition imposed on the entire interaction domain, i.e. $\mathcal{V}^{\alpha}_{\mathcal{I}} = \mathcal{V}^{\alpha}_{\mathcal{I}_D}$. Recall the definition of $\mathcal{S}$ in connection with Eq. (16).

Proposition 1 ***(S-Flow as nonlocal PDE).*** *Let $\alpha, \Theta \in \mathcal{F}_{\overline{\mathcal{V}} \times \overline{\mathcal{V}}}$ be as in (28). Then the flow* (24a) *with Ω given through* (29) *admits the representation*

$$\partial_t S(x,t) = R_{S(x,t)}\Big(\frac{1}{2}\mathcal{D}^{\alpha}\big(\Theta\mathcal{G}^{\alpha}(S)\big) + \lambda S\Big)(x,t), \qquad \text{on } \mathcal{V} \times \mathbb{R}_+, \tag{32a}$$

$$S(x,t) = 0, \qquad \text{on } \mathcal{V}^{\alpha}_{\mathcal{I}} \times \mathbb{R}_+, \tag{32b}$$

$$S(x,0) = \overline{S}(x)(0), \qquad \text{on } \mathcal{V} \dot{\cup} \mathcal{V}^{\alpha}_{\mathcal{I}} \times \mathbb{R}_+, \tag{32c}$$

where $\lambda = \lambda(x)$ is given by (28) and $\overline{S}(x)(0)$ denotes the zero extension of the $\mathcal{S}$-valued vector field $S \in \mathcal{F}_{\mathcal{V},\mathcal{S}}$ to $\mathcal{F}_{\overline{\mathcal{V}},\mathcal{S}}$.

Proof. The system (32) follows directly from applying Lemma 1 and

$$\sum_{y \in \mathcal{N}(x) \setminus \mathcal{V}^{\alpha}_{\mathcal{I}}} S(y)\Theta(x,y)\alpha^2(x,y) + \big(\lambda(x) - \sum_{y \in \mathcal{N}(x)} \Theta(x,y)\alpha(x,y)^2\big)S(x) = \Omega S(x). \tag{33}$$

□

Proposition 1 clarifies the connection of the potential flow (24a) with Ω satisfying (22) and the nonlocal diffusion process (32). Specifically, for a nonnegative, symmetric and row-stochastic matrix Ω as in Sect. 3, let the mappings $\widetilde{\alpha}, \widetilde{\Theta} \in \mathcal{F}_{\mathcal{V} \times \mathcal{V}}$ be defined on $\mathcal{V} \times \mathcal{V}$ by

$$\widetilde{\Theta}(x,y) = \begin{cases} \Omega(x,y) & \text{if } y \in \mathcal{N}(x), \\ 0 & \text{else} \end{cases}, \quad \widetilde{\alpha}^2(x,y) = 1, \quad x, y \in \mathcal{V}. \tag{34}$$

Further, denote by $\Theta, \alpha \in \mathcal{F}_{\overline{\mathcal{V}} \times \overline{\mathcal{V}}}$ the extensions of $\widetilde{\alpha}, \widetilde{\Theta}$ to $\overline{\mathcal{V}} \times \overline{\mathcal{V}}$ by 0, that is

$$\Theta(x,y) = \big(\delta_{\mathcal{V}\times\mathcal{V}}(\widetilde{\Theta})\big)(x,y), \quad \alpha(x,y) := \big(\delta_{\mathcal{V}\times\mathcal{V}}(\widetilde{\alpha})\big)(x,y) \quad x, y \in \overline{\mathcal{V}}, \tag{35}$$

where $\delta_{\mathcal{V}\times\mathcal{V}} \colon \mathbb{Z}^d \times \mathbb{Z}^d \to \{0,1\}$ is the indicator function of the set $\mathcal{V} \times \mathcal{V} \subset \mathbb{Z}^d \times \mathbb{Z}^d$. Then, for α, Θ as above, the potential flow (24a) is equivalently represented by the system (32). In particular, Proposition 1 shows that the assignment flow introduced in Sect. 3 is a special case of the system (32) with an empty interaction domain (5).

Now we focus on the connection of (32) and the *continuous local* formulation of (24a) on an open, simply connected bounded domain $\mathcal{M} \subset \mathbb{R}^2$, as introduced by [19], that characterizes solutions $S^* = \lim_{t\to\infty} S(t) \in \overline{\mathcal{W}}$ by the PDE

$$R_{S^*(x)}\big(-\Delta S^*(x) - S^*(x)\big) = 0, \quad x \in \mathcal{M}. \tag{36}$$

On the discrete Cartesian mesh $\mathcal{M}^h$ with boundary $\partial\mathcal{M}^h$ specified by a small spatial scale parameter $h > 0$, identify each *interior* grid point $(hk, hl) \in \mathcal{M}^h$ (grid graph) with a node $x = (k, l) \in \mathcal{V}$ of the graph. As in [19], assume (36) is complemented by zero local boundary conditions imposed on S^* on $\partial\mathcal{M}^h$ and adopt the sign convention $L^h = -\Delta^h$ for the basic discretization of the *continuous negative Laplacian* on $\mathcal{M}^h$ by the five-point stencil which leads to strictly positive entries $L^h(x, x) > 0$ on the diagonal. Further, introduce the interaction domain (5) as the local discrete boundary, i.e. $\mathcal{V}^{\alpha}_{\mathcal{I}} = \partial\mathcal{M}^h$ and neighborhoods $\widetilde{\mathcal{N}}(x) = \mathcal{N}(x) \setminus \{x\}$. Finally, let the parameter matrix Ω be defined by (29) in terms of the mappings $\alpha, \Theta \in \mathcal{F}_{\overline{\mathcal{V}}\times\overline{\mathcal{V}}}$ given by

$$\alpha^2(x,y) = \begin{cases} 1, & y \in \widetilde{\mathcal{N}}(x), \\ 0, & \text{else}, \end{cases}, \quad \Theta(x,y) = \begin{cases} -L^h(x,y), & y \in \widetilde{\mathcal{N}}(x), \\ 1 - L^h(x,x), & x = y, \\ 0, & \text{else}. \end{cases} \tag{37}$$

Then, for each $x \in \mathcal{V}$, the action of Ω on S reads

$$\sum_{y\in\widetilde{\mathcal{N}}(x)} -L^h(x,y)S(y) + \big(1 - L^h(x,x)\big)S(x) = -\big(-\Delta^h(S) - S\big)(x), \tag{38}$$

which is the discretization of (36) by L^h multiplied by the minus sign. In particular due to relation $R_S(-W) = -R_S(W)$ for $W \in \mathcal{W}$, the solution to the *nonlocal* formulation (32) satisfies Eq. (36). We conclude that the novel approach (32) includes the *local* PDE (36) as special case and hence provides a *natural nonlocal extension.*

4.2 Tangent-Space Parametrization of the S-Flow PDE

Because $S(x, t)$ solving (32) evolves on the non-Euclidean space $\mathcal{S}$, applying some standard discretization to (32) directly will not work. We therefore work out a parametrization of (32) on the *flat* tangent space (19) by means of the equation

$$S(t) = \exp_{S_0}(V(t)) \in \mathcal{W}, \quad V(t) \in \mathcal{T}_0, \qquad S_0 = S(0) \in \mathcal{W}. \tag{39}$$

Applying $\frac{d}{dt}$ to both sides and using the expression of the differential of the mapping $\exp_{S_0}$ due to [19, Lemma 3.1], we get $\dot{S}(t) = R_{\exp_{S_0}(V(t))}\dot{V}(t) \overset{(39)}{=} R_{S(t)}\dot{V}(t)$. Comparing this equation and (24a) shows that $V(t)$ solving the nonlinear ODE

$$\dot{V}(t) = \Omega \exp_{S_0}(V(t)), \qquad V(0) = 0 \tag{40}$$

determines by (39) $S(t)$ solving (24a). Hence, it suffices to focus on (40) which evolves on the flat space $\mathcal{T}_0$. Repeating the derivation above that resulted in the PDE representation (32) of the S-flow (24a), yields the nonlinear PDE representation of (40)

$$\begin{aligned}
\partial_t V(x,t) &= \Big(\frac{1}{2}\mathcal{D}^{\alpha}\big(\Theta\mathcal{G}^{\alpha}(\exp_{S_0}(V))\big) + \exp_{S_0}(V)\Big)(x,t) && \text{on } \mathcal{V}\times\mathbb{R}_+, && \text{(41a)}\\
V(x,t) &= 0 && \text{on } \mathcal{V}^{\alpha}_{\mathcal{I}}\times\mathbb{R}_+, && \text{(41b)}\\
V(x,0) &= \overline{V}(x)(0) && \text{on } \mathcal{V}\dot{\cup}\mathcal{V}^{\alpha}_{\mathcal{I}}\times\mathbb{R}_+. && \text{(41c)}
\end{aligned}$$

From the numerical point of view, this new formulation (41) has the following expedient properties. Firstly, using a parameter matrix as specified by (29) and (35) enables to define the entire system (41) on $\mathcal{V}$. Secondly, since $V(x,t)$ evolves on the flat space T_0, numerical techniques of geometric integration as studied by [22] can here be applied as well. We omit details due to lack of space.

4.3 Non-local Balance Law

A key property of PDE-based models are balance laws implied by the model; see [12, Section 7] for a discussion of various scenarios. The following proposition reveals a *nonlocal* balance law of the assignment flow based on the novel PDE-based parametrization (41), that we express for this purpose in the form

$$\partial_t V(x,t) + \mathcal{D}^{\alpha}(F(V))(x,t) = b(x,t) \quad x\in\mathcal{V}, \tag{42a}$$
$$F(V(t))(x,y) = -\tfrac{1}{2}\Big(\Theta\mathcal{G}^{\alpha}\big(\exp_{S_0}(V(t))\big)\Big)(x,y), \quad b(x,t) = \lambda(x)S(x,t), \tag{42b}$$

where $S(x,t) = \exp_{S_0}(V(x,t))$ by (39).

Proposition 2 (nonlocal balance law of S-flows). *Under the assumptions of Lemma 1, let $V(t)$ solve (41). Then, for each component $S_j(t) = \{S_j(x,t)\colon x\in\mathcal{V}\}$, $j\in[c]$, of $S(t) = \exp_{S_0}(V(t))$, the following identity holds*

$$\begin{aligned}
\frac{1}{2}\frac{d}{dt}\langle S_j,\mathbb{1}\rangle_{\mathcal{V}} + \frac{1}{2}\langle\mathcal{G}^{\alpha}(S_j),\Theta\mathcal{G}^{\alpha}(S_j)\rangle_{\overline{\mathcal{V}}\times\overline{\mathcal{V}}} + \langle S_j,\phi_S-\lambda S_j\rangle_{\mathcal{V}}&\\
+\langle S_j,\mathcal{N}^{\alpha}(\Theta\mathcal{G}^{\alpha}(S_j))\rangle_{\mathcal{V}_{\mathcal{I}^{\alpha}}} = 0&,
\end{aligned}\tag{43}$$

where $\phi_S(\cdot)\in\mathcal{F}_{\mathcal{V}}$ is defined in terms of $S(t)\in\mathcal{W}$ by

$$\phi_S:\mathcal{V}\to\mathbb{R},\qquad x\mapsto\big\langle S(x),\Pi_0\Omega\exp_{S_0}\big(S(x)\big)\big\rangle. \tag{44}$$

Proof. For brevity, we omit the argument t and simply write $S = S(t), V = V(t)$. In the following, $\odot$ denotes the componentwise multiplication of vectors, e.g. $(S\odot V)_j(x) = S_j(x)V_j(x)$ for $j\in[c]$, and $S^2(x) = (S\odot S)(x)$.

Multiplying both sides of (42a) with $S(x) = \exp_{S_0}(V(x))$ and summing over $x\in\mathcal{V}$ yields

$$\sum_{x\in\mathcal{V}}\big(S\odot\dot{V}\big)_j(x) - \sum_{x\in\mathcal{V}}\frac{1}{2}\Big(S\odot\mathcal{D}^{\alpha}\big(\Theta\mathcal{G}^{\alpha}(S)\big)\Big)_j(x) = \sum_{x\in\mathcal{V}}\big(\lambda S^2\big)_j(x). \tag{45}$$

Applying Greens nonlocal first identity (13) with $u(x) = S_j(x)$ to the second term on the left-hand side yields with (6)

$$\sum_{x\in\mathcal{V}} \big(S \odot \dot{V}\big)_j(x) + \frac{1}{2} \sum_{x\in\mathcal{V}\cup\mathcal{V}_I^\alpha} \sum_{y\in\mathcal{V}\cup\mathcal{V}_I^\alpha} \big(\mathcal{G}^\alpha(S) \odot (\Theta\mathcal{G}^\alpha(S))\big)_j(x,y) \tag{46a}$$

$$+ \sum_{y\in\mathcal{V}_I^\alpha} S_j(y)\mathcal{N}^\alpha\big(\Theta\mathcal{G}^\alpha(S_j)\big)(y) = \sum_{x\in\mathcal{V}} \big(\lambda S^2\big)_j(x). \tag{46b}$$

Now, using the parametrization (39) of S, with the right-hand side defined analogous to (26) and componentwise application of the exponential function to the row vectors of V, we compute at each $x \in \mathcal{V}$: $\dot{S} = S \odot \dot{V} - \langle S, \dot{V}\rangle S$. Substitution into (46) gives for each $S_j = \{S_j(x) \colon x \in \mathcal{V}\}$, $j \in [c]$

$$\frac{1}{2}\frac{d}{dt}\Big(\sum_{x\in\mathcal{V}} S_j(x)\Big) + \frac{1}{2}\langle\mathcal{G}^\alpha(S_j), \Theta\mathcal{G}^\alpha(S_j)\rangle_{\overline{\mathcal{V}}\times\overline{\mathcal{V}}} + \Big(\sum_{x\in\mathcal{V}}\langle S(x), \dot{V}(x)\rangle\Big) S_j(x) \tag{47a}$$

$$+ \sum_{y\in\mathcal{V}_{\mathcal{I}^\alpha}} S_j \mathcal{N}^\alpha\big(\Theta\mathcal{G}^\alpha(S_j)\big)(y) = \sum_{x\in\mathcal{V}} \big(\lambda S_j^2\big)(x) \tag{47b}$$

which after rearranging the terms yields (43). □

We briefly inspect the *nonlocal balance law* (43) that comprises *four terms*. Since $\sum_{j\in[c]} S_j(x) = 1$ at each vertex $x \in \mathcal{V}$, the *first term* of (43) measures the *rate* of 'mass' assigned to label j over the entire image. This rate is governed by two interacting processes corresponding to the *three remaining terms*:

(i) *spatial* propagation of assignment mass through the nonlocal diffusion process including nonlocal boundary conditions: *second and fourth term*;
(ii) *exchange* of assignment mass with the remaining labels $\{l \in [c] \colon l \neq j\}$: *third term* comprising the function ϕ_S (44).

We are not aware of any other approach, including Markov random fields and deep networks, that makes explicit the flow of information during inference in such an *explicit* manner.

5 Numerical Experiments

In this section, we report numerical results in order to demonstrate two aspects of the mathematical results presented above:

(1) Using geometric integration for numerically solving nonlocal PDEs with appropriate boundary conditions;
(2) zero vs. nonzero interaction domain and the affect of corresponding nonlocal boundary conditions on image labeling.

Numerically Solving Nonlocal PDEs By Geometric Integration. According to Sect. 4.2, imposing the homogeneous Dirichlet condition via the interaction domain (5) makes the right-hand side of (41a) equivalent to (40). Applying a simple explicit time discretization with stepsize h to (41a) results in the iterative update formula

$$V(x,t+h) \approx V(x,t) + h\Pi_0 \exp_{S_0(x)}(\Omega V(x,t)), \qquad h > 0. \tag{48}$$

By virtue of the parametrization (39), one recovers with any nonnegative symmetric mapping Ω as in Lemma 1 the *explicit geometric Euler* scheme on $\mathcal{W}$

$$S(t+h) \approx \exp_{S_0}\big(V(t) + h\dot{V}(t)\big) = \exp_{S(t)}\big(h\Omega S(t)\big), \tag{49}$$

where the last equality is due to property $\exp_S(V_1 + V_2) = \exp_{\exp_S(V_1)}(V_2)$ of the lifting map $\exp_S$ and due to the equation $\dot{V} = \Omega S$ implied by (39) and (40). Higher order geometric integration methods [22] generalizing (49) can be applied in a similar way. This provides new perspective on solving a certain class of nonlocal PDEs numerically, conforming to the underlying geometry.

Influence of the Nonlocal Interaction Domain. We considered two different scenarios and compared corresponding image labelings obtained by solving (32a) using the scheme (49), uniform weight parameters but different boundary conditions: (i) *zero-extension* to the interaction domain according to (35) which makes (32) equivalent to (24a) according to Proposition 1; (ii) *uniform extension* in terms of uniform mappings $\Theta, \alpha \in \mathcal{F}_{\overline{\mathcal{V}}\times\overline{\mathcal{V}}}$ given for each $x \in \mathcal{V}$ by fixed neighborhood sizes $|\mathcal{N}(x)| = 7 \times 7$ and $\alpha^2(x,y) = \frac{1}{7^2}$ if $y \in \mathcal{N}(x)$ and 0 otherwise; $\Theta(x,y) = \frac{1}{7^2}$ if $x = y$ and 1 otherwise. We iterated (49) with step size $h = 1$ until assignment states (24b) of low average entropy 10^{-3} were reached. To ensure a fair comparison and to assess solely the effects of the boundary conditions through nonlocal regularization, we initialized (32) in the same way as (24a) and adopted an uniform encoding of the 31 labels as described by [3, Figure 6].

Figure 2 depicts the results. Closely inspecting panels (c) (*zero extension*) and (d) (*uniform extension*) shows that using a nonempty interaction domain may improve the labeling near the boundary (cf. close-up views), and almost equal performance in the interior domain.

Figure 3 shows the decreasing average entropy values for each iteration (left panel) and the number of iterations required to converge (right panel), for different neighborhood sizes. We observe, in particular, that integral label assignments corresponding to zero entropy are achieved in both scenarios, at comparable computational costs.

A more general system (32) with *nonuniform* interaction is defined through mappings $\alpha, \Theta \in \mathcal{F}_{\overline{\mathcal{V}}\times\overline{\mathcal{V}}}$ measuring similarity of pixel patches analogous to [7]

$$\Theta(x,y) = e^{-G_{\sigma_p} * ||S(x+\cdot) - S(y+\cdot)||^2}, \quad \alpha^2(x,y) = e^{\frac{-|x-y|^2}{2\sigma_s^2}} \quad \sigma_p, \sigma_s > 0, \tag{50}$$

where G_{σ_p} denotes the Gaussian kernel. Decomposition (34) yields a symmetric parameter matrix Ω which not necessarily satisfies property (22). Iterating (49) with step size $h = 0.1$ and $\sigma_s = 1, \sigma_p = 5$ in (50) Fig. 4 visualizes labeling results for different patch sizes. In particular defining Θ by (50) implies a non-zero interaction domain $\mathcal{V}_{\mathcal{I}}^{\alpha}$ as depicted by the right image in Fig. 4.

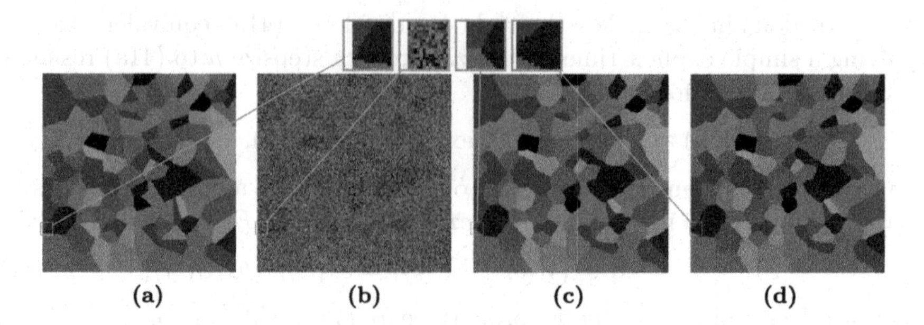

Fig. 2. Labeling through nonlocal geometric flows. **(a)** Ground truth with 31 labels. **(b)** Noisy input data used to evaluate (24a) and (32). **(c)** Labeling returned by (24a) corresponding to a zero extension to the interaction domain. **(d)** Labeling returned by (41) with a uniform extension to the interaction domain in terms of Θ, α specified above. The close-up view show differences close to the boundary, whereas the results in the interior domain are almost equal.

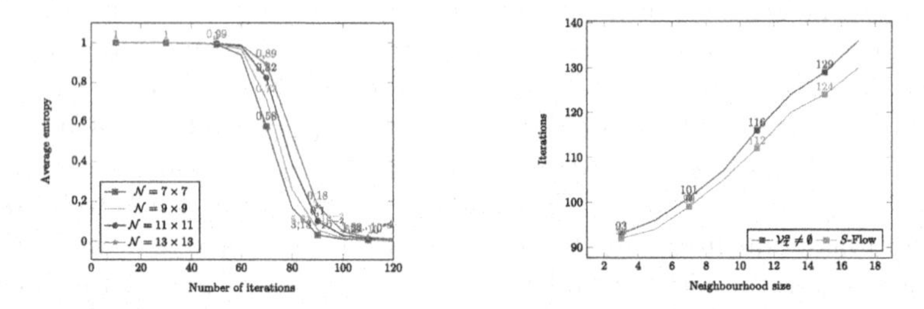

Fig. 3. Left: Convergence rates of the scheme (49) solving (32) with nonempty interaction domain specified by Θ, α above. The convergence behavior is rather insensitive with respect to the neighborhood size. **Right:** Number of iterations until convergence for (32) (●) and (24a) (●). This result shows that different nonlocal boundary conditions have only a minor influence on the convergence of the flow to labelings.

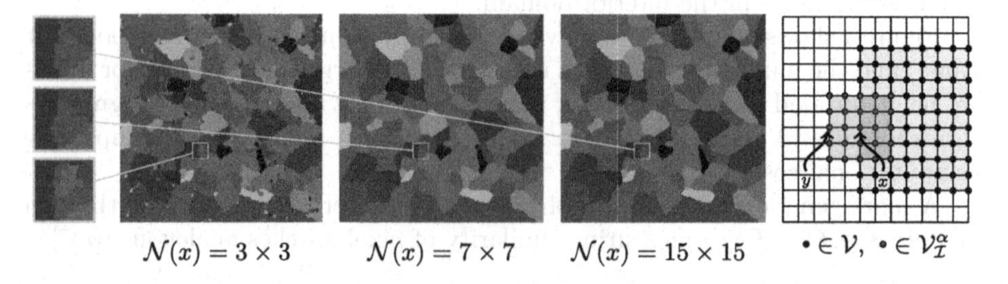

Fig. 4. *From left to right:* Labeling results using (32) for nonuniform interaction domains of size $\mathcal{N}(x) = 3 \times 3$, 7×7 and 15×15, with close up views indicating the regularization properties of the nonlocal PDE (32) with zero Dirichlet conditions. Schematic illustration of the nonlocal interaction domain $y \in \mathcal{V}_{\mathcal{I}}^{\alpha}$ (red area) induced by nodes (blue area) according to (50) with a Gaussian window of size 5×5 centered at $x \in \mathcal{V}$. (Color figure online)

6 Conclusion

We introduced a novel nonlocal PDE motivated by the assignment flow approach. Our results extend established PDE approaches from denoising and image enhancement to image labeling and segmentation, and to a class of *nonlocal* PDEs with nonlocal boundary conditions.

Our future work will study the nonlocal balance law in connection with parameter learning and image structure recognition.

Acknowledgement. This work is supported by the Deutsche Forschungsgemeinschaft (DFG, German Research Foundation) under Germany's Excellence Strategy EXC 2181/1 - 390900948 (the Heidelberg STRUCTURES Excellence Cluster).

References

1. Amari, S.I., Nagaoka, H.: Methods of Information Geometry. American Mathematical Society and Oxford University Press, Oxford (2000)
2. Ambrosio, L., Tortorelli, V.M.: Approximation of functional depending on jumps by elliptic functional via Γ-convergence. Commun. Pure Appl. Math. **43**(8), 999–1036 (1990)
3. Åström, F., Petra, S., Schmitzer, B., Schnörr, C.: Image labeling by assignment. J. Math. Imaging Vis. **58**(2), 211–238 (2017)
4. Ay, N., Jost, J., Lê, H.V., Schwachhöfer, L.: Information Geometry. Springer, Heidelberg (2017)
5. Bertozzi, A., Flenner, A.: Diffuse interface models on graphs for classification of high dimensional data. SIAM Rev. **58**(2), 293–328 (2016)
6. Buades, A., Coll, B., Morel, J.M.: Neighborhood filter and PDEs. Numer. Math. **105**, 1–34 (2006)
7. Buades, A., Coll, B., Morel, J.M.: Image denoising methods. A new nonlocal principle. SIAM Rev. **52**(1), 113–147 (2010)
8. Chen, R.T.Q., Rubanova, Y., Bettencourt, J., Duvenaud, D.: Neural ordinary differential equations. In: Proceedings of the NeurIPS (2018)
9. Chung, F.: Spectral Graph Theory. American Mathematical Society, Ann Arbor (1997)
10. Chung, F., Langlands, R.P.: A combinatorial Laplacian with vertex weights. J. Comb. Theory Seri. A **5**(2), 316–327 (1996)
11. Du, Q.: Nonlocal Modeling, Analysis, and Computation. SIAM, Philadelphia (2019)
12. Du, Q., Gunzburger, M., Lehoucq, R.B., Zhou, K.: A nonlocal vector calculus, nonlocal volume-constrained problems, and nonlocal balance laws. Math. Models Meth. Appl. Sci. **23**(3), 493–540 (2013)
13. Du, Q., Gunzburger, M., Lehoucq, R.B., Zhou, K.: Analysis and approximation of nonlocal diffusion problems with volume constraints. SIAM Rev. **54**(4), 667–696 (2012)
14. Elmoataz, A., Lezoray, O., Bougleux, S.: Nonlocal discrete regularization on weighted graphs: a framework for image and manifold processing. IEEE Trans. Image Proc. **17**(7), 1047–1059 (2008)

15. Elmoataz, A., Toutain, M., Tenbrinck, D.: On the p-Laplacian and ∞-Laplacian on graphs with applications in image and data processing. SIAM J. Imag. Sci. **8**(4), 2412–2451 (2015)
16. Gilboa, G., Osher, S.: Nonlocal operators with applications to image processing. Multiscale Model. Simul. **7**(3), 1005–1028 (2009)
17. Haber, E., Ruthotto, L.: Stable architectures for deep neural networks. Inverse Prob. **34**(1), 014004 (2017)
18. Minaee, S., Boykov, Y.Y., Porikli, F., Plaza, A.J., Kehtarnavaz, N., Terzopoulos, D.: Image segmentation using deep learning: a survey. IEEE Trans. Pattern Anal. Mach. Intell. (2021). https://ieeexplore.ieee.org/document/9356353
19. Savarino, F., Schnörr, C.: Continuous-domain assignment flows. Eur. J. Appl. Math. **32**(3), 570–597 (2021)
20. Schnörr, C.: Assignment flows. In: Grohs, P., Holler, M., Weinmann, A. (eds.) Handbook of Variational Methods for Nonlinear Geometric Data, pp. 235–260. Springer, Cham (2020). https://doi.org/10.1007/978-3-030-31351-7_8
21. Weickert, J.: Anisotropic Diffusion in Image Processing. B.G. Teubner, Stuttgart (1998)
22. Zeilmann, A., Savarino, F., Petra, S., Schnörr, C.: Geometric numerical integration of the assignment flow. Inverse Probl. **36**(3), 034004 (2020). (33pp)
23. Zern, A., Zeilmann, A., Schnörr, C.: Assignment Flows for Data Labeling on Graphs: Convergence and Stability. CoRR abs/2002.11571 (2020)

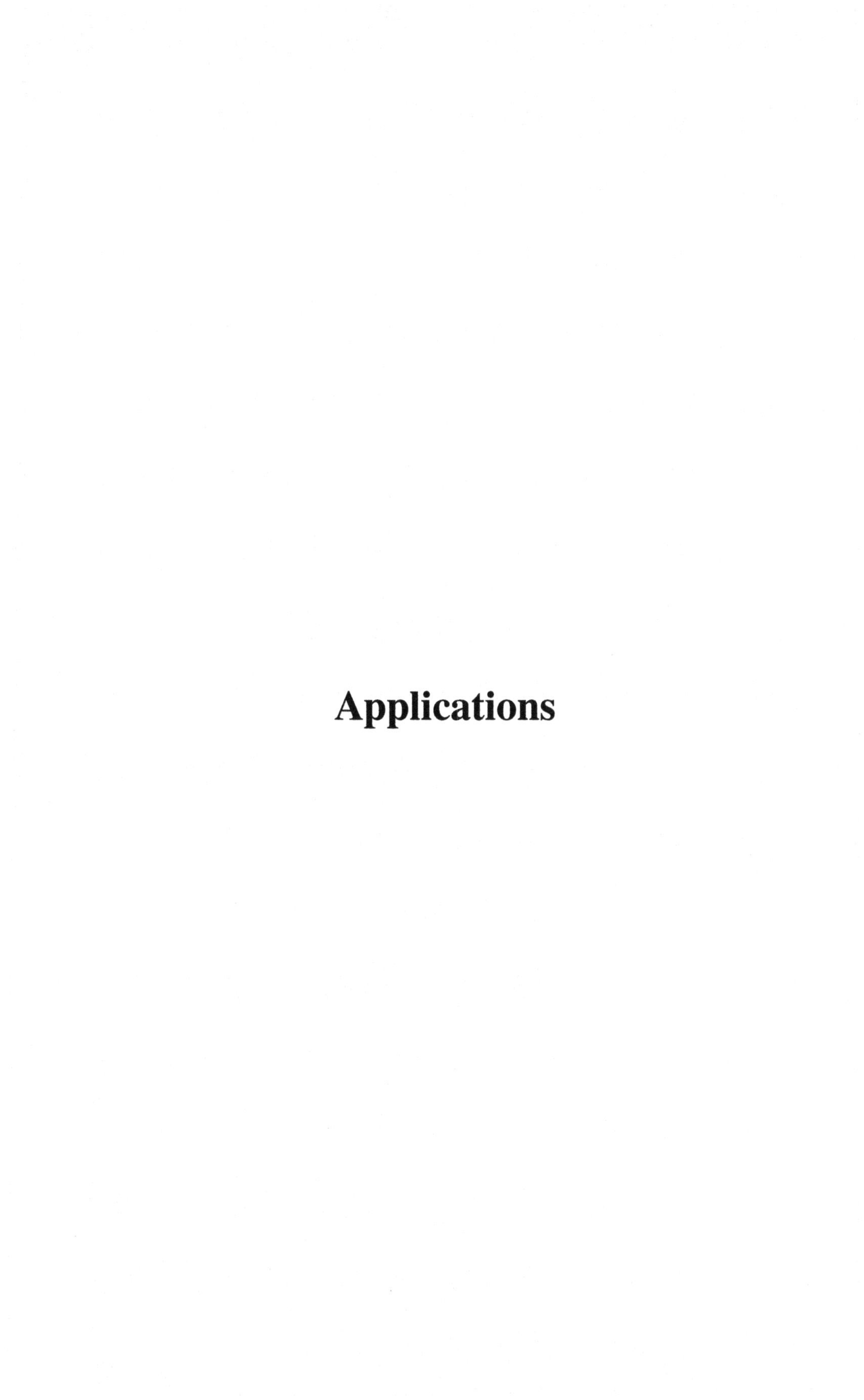

Applications

Viewpoint-Tolerant Semantic Segmentation for Aerial Logistics

Shiming Wang[1,2](✉), Fabiola Maffra[1], Ruben Mascaro[1], Lucas Teixeira[1], and Margarita Chli[1]

[1] Vision for Robotics Lab, ETH Zurich, Zurich, Switzerland
shimwang@student.ethz.ch
[2] RWTH Aachen University, Aachen, Germany

Abstract. Semantic segmentation is fundamental for enabling scene understanding in several robotics applications, such as aerial delivery and autonomous driving. While scenarios in autonomous driving mainly comprise roads and small viewpoint changes, imagery acquired from aerial platforms is usually characterized by extreme variations in viewpoint. In this paper, we focus on aerial delivery use cases, in which a drone visits the same places repeatedly from distinct viewpoints. Although such applications are already under investigation (e.g. transport of blood between hospitals), current approaches depend heavily on ground personnel assistance to ensure safe delivery. Aiming at enabling safer and more autonomous operation, in this work, we propose a novel deep-learning-based semantic segmentation approach capable of running on small aerial vehicles, as well as a practical dataset-capturing method and a network-training strategy that enables greater viewpoint tolerance in such scenarios. Our experiments show that the proposed method greatly outperforms a state-of-the-art network for embedded computers while maintaining similar inference speed and memory consumption. In addition, it achieves slightly better accuracy compared to a much larger and slower state-of-the-art network, which is unsuitable for small aerial vehicles, as considered in this work.

Keywords: Semantic segmentation · Viewpoint tolerance · Multi-task learning · Aerial logistics · Aerial delivery

1 Introduction

Unmanned Aerial Vehicles (UAVs) have recently emerged as a potential solution for freight and logistics in a variety of scenarios. Although applications, such as blood transport between hospitals have already become a reality[1], most often, heavy expert intervention is still required to guarantee safe delivery. A key step towards reaching the ultimate goal of safe fully-autonomous operation is to provide such robots with scene understanding capabilities, enabling effective

[1] flyzipline.com.

C. Bauckhage et al. (Eds.): DAGM GCPR 2021, LNCS 13024, pp. 515–529, 2021.
https://doi.org/10.1007/978-3-030-92659-5_33

onboard decision-making on-the-fly. This is of special interest in regions of operation, which can be explored in advance in order to capture and label multiple views of the scene promising to increase the robustness of semantic segmentation algorithms.

The state of the art in semantic segmentation is nowadays based on machine learning, which typically requires large training datasets that cover most of the possible input space. However, UAVs can observe places from extremely different viewpoints and even training a semantic segmentation network attached to a specific place can be quite challenging, as collecting datasets exhibiting enough variability in viewpoints is a difficult task.

In this paper, we propose an efficient approach for semantic segmentation of images of a region of interest achieving unprecedented viewpoint tolerance. Moreover, we demonstrate that it is possible to successfully train a semantic segmentation network with a limited set of images of the region of interest, captured during a reconnaissance flight with a simple circular pattern. For this task, we use data augmentation and a knowledge distillation strategy together with a novel and lightweight network that can be deployed onboard a small UAV.

In brief, the main contributions of this paper are:

- a novel, Viewpoint-Tolerant Multi-task network (VTM-Net), that computes semantics as the main task, with local feature detection and description as secondary tasks supervised by a knowledge distillation approach and used only during training for improving viewpoint tolerance. In our network, we also adopt an improved decoder based on LR-ASPP [10];
- a novel data generation strategy for training semantic segmentation algorithms on UAVs;
- a thorough quantitative and qualitative evaluation of the performance of the proposed approach, showing increased accuracy when compared to a state-of-the-art network, while reaching real-time capabilities comparable to a lightweight network.

2 Related Work

Since fully convolutional neural networks were firstly utilized for semantic segmentation [13], they have become the state-of-the-art method for this problem. Most of the semantic segmentation networks, such as U-Net [20] and DeepLabV3-Plus [3], adopt the encoder-decoder architecture. The encoder usually relies on a common classification method, such as VGG [24], ResNet [9], and GoogLeNet [25], to extract features. To further recover the spatial and detail information in the decoder, skip-connection structures are widely used in semantic segmentation networks [3,20]. Moreover, networks, such as PSPNet [27] and DeepLab family networks [2,3], apply spatial pyramid pooling modules to enrich the receptive fields in the network and reach the top performance in most benchmarks. However, these networks are very large and slow, and therefore, not suitable for deployment on a small UAV of limited computational capabilities. Here, we use

DeepLabV3Plus as baseline because its fast-deployment code is available [5] and other state-of-the-art approaches in semantic segmentation present only marginal improvements when compared to it. For example, [28] adopts a video prediction-based data augmentation method and a novel boundary label relaxation technique to boost semantic segmentation performance based on DeepLabV3Plus and reaches top performance on Cityscapes [6].

There is also a body of works focusing on semantic segmentation of aerial imagery, with some of them aiming to address specifically, the issues with overhead or remote sensing images employing the commonly used deep CNNs. For example, Mou et al. [17] introduced the spatial and channel relation modules to enhance the learning of global relationships, while Marmanis et al. [15] proposed a method to improve the object boundaries recognition by combining boundary detection. Despite good progress, however, the viewpoint tolerance performance has rarely been discussed in segmentation literature. In this spirit, this publication aims to push the boundaries of view-point tolerant segmentation from aerial imagery.

Real-time semantic segmentation has been actively researched for a few years now. Fast-SCNN [19] is developed based on a two-branch structure and it is suitable for real-time and efficient computation on low-memory devices. The MobileNet family [10,11,21], on the other hand, is a series of the most widely used neural network backbones specially designed for mobile devices. They adopt depth-wise separable convolutions, inverted residual blocks and light attention modules to improve the computation efficiency. MobileNetV3 also proposed the lite-designed segmentation decoder, LR-ASPP, to balance the efficiency and accuracy in semantic segmentation. However, the boost in inference speed and memory efficiency that such approaches offer come at the expense of lower effectiveness in complex scenarios and less accurate boundary recovery. In this work, we use MobileNetV3 as the baseline for mobile networks as it is widely used in state-of-the-art real-time segmentation applications, such as Panoptic-DeepLab [4].

Furthermore, as the great majority of datasets for semantic segmentation [6,8,14,16] lack viewpoint variance, which is ubiquitous in aerial imagery, they are not suitable for training and evaluating our network. Given the lack of large scale datasets, in this work, we focus on a place-specific solution rather than on a generic model, that requires less data for training.

3 Proposed Method

Typically, local feature detectors and descriptors exhibit viewpoint tolerance, making the interest point detections (e.g. corners) remain stable when the viewpoint changes. In aerial navigation, the variability of viewpoints, from which the same place can be experienced is vast and most often, results in varying spatial structural information in the captured images. In an encoder-decoder based neural network architecture, spatial context information will be heavily compressed as the layers deepen in the encoder. Therefore, the scene elements

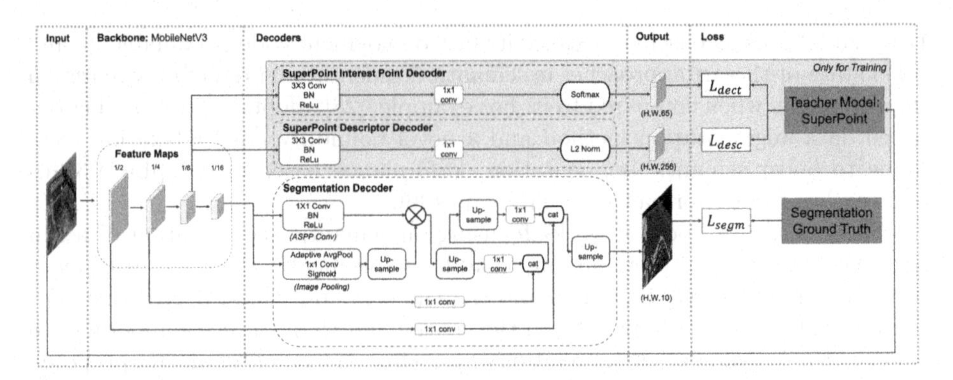

Fig. 1. VTM-Net has three outputs from a single image: a semantic segmentation, interest point detection and descriptors. The segmentation head is supervised by annotated ground truth, while the decoder heads from SuperPoint are trained under distillation from the teacher network, the pre-trained SuperPoint.

with varying spatial relationships are hard to get recognized in the decoder. With this in mind, here we propose to minimize this effect by using data augmentation and a novel network, the Viewpoint-Tolerant Multi-task Network, VTM-Net (see Fig. 1), which adds the decoding of local features and descriptors as training-only tasks. We used the decoders from SuperPoint [7] that currently comprise the best performing state-of-the-art method in keypoint repeatability and description matching. To this end, we adopt a teacher-student approach to force our network to learn how to mimic SuperPoint's local features, while also computing semantics in each image.

3.1 VTM-Net Architecture

The encoder of VTM-Net is MobilNetV3-Large [10] backbone, a specially designed architecture for mobile devices. The backbone adopts the lite attention Squeeze-and-Excite Module [12], to increase the representational power.

We utilize the decoding scheme architecture of SuperPoint [7] as the local feature decoders of VTM-Net. The non-upsampling decoder design of SuperPoint enables real-time inference and effectiveness in viewpoint stability. As the local features are on a lower semantic level than semantic segmentation, the local feature heads branch out from the 1/8-stride-out resolution feature maps to maintain more spatial context information.

The decoder of semantic segmentation is improved based on the Lite Reduced-ASPP (LR-ASPP) [10]. To pursue the extreme inference time and efficient memory consumption, the original LR-ASPP introduced the low-level features from 1/8 resolution and utilized element-wise addition to joint the high-level and low-level features. However, in this way, the decoder loses the majority of discriminative spatial information and performs weakly in semantic segmentation. Thus, inspired by MMSegmentation [5] and FastSeg [26], we adopt two

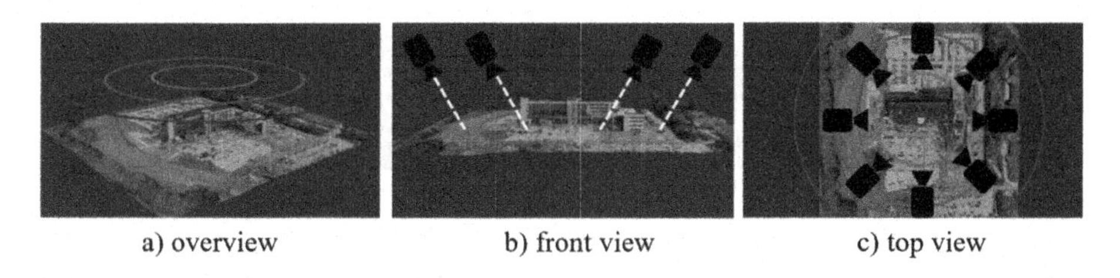

Fig. 2. Illustration of the circular sampling method. The height and pitch angle of the drone are fixed in the sampling process, and the yaw angles are adjusted to face the center of the circles.

main improvements on the LR-ASPP decoder. Firstly, instead of the skip connection from the 1/8 resolution features, two lower-level features are introduced from, respectively, the feature maps with the output stride of 2 and 4. In addition, the two low-level features and high-level features are all upsampled to the identical dimension and then fused with concatenation to retain the spatial features in the calculation instead of performing element-wise addition as in the original LR-ASPP decoder.

3.2 Data Sampling

For training, we capture the place-specific images in a manner that is as close to reality as possible, so that this can be reproduced in reality too. Hence, we use a circular sampling method (Fig. 2) of concentric circles for the drone's sample trajectories. Throughout the sampling procedure, the drone's height and pitch angles remain fixed, but the yaw angle gets adjusted to always face the circles' center. Thus, the dataset captured using this circular sampling method has a limited number of viewpoints, but it is very fast and easy to be captured. Semantic-label ground truth is created by labeling the 3D model mesh. This can be replicated also in real experiments by building a 3D model of the scene from the collected images (e.g. using COLMAP [23]).

For testing, as the drone's viewpoints are determined by its height and pitch angles, the whole test dataset is composed of a collection of sub-datasets sampled at various heights and pitch angles. Thus, each test sub-dataset only contains the single viewpoint information and the total test dataset covers plenty of viewpoint information. As the drone flies in a star-shaped trajectory, as shown in Fig. 3, for localization in the real-world scenario, we deploy this trajectory in the test datasets sampling.

To cover most viewpoints over the model, the total test dataset contains 24 sub-datasets by combining four different heights and six different pitch angles. For example, we sampled evaluation datasets over the hospital model at a height of 30 m, 60 m, 90 m, and 120 m at the pitch angles of 15°, 30°, 45°, 60°, 75°, and 90°. In Fig. 3, two of these sub-datasets are illustrated.

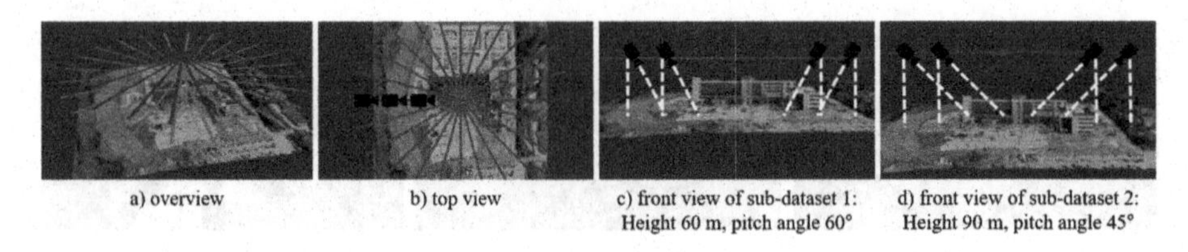

a) overview
b) top view
c) front view of sub-dataset 1: Height 60 m, pitch angle 60°
d) front view of sub-dataset 2: Height 90 m, pitch angle 45°

Fig. 3. Illustration of the star sampling method. Here, the sampling process of two sub-datasets is shown with two different viewpoints.

3.3 Training Process

Viewpoints Augmentation with Geometric Transformation: A viewpoint change can be generally represented by a combination of affine and homography transformations. Therefore, besides the usual data augmentation for semantic segmentation, such as random crop and random flip, here we apply two advanced data augmentation approaches, namely, random affine transformation and homography transformation (Fig. 4), in the training process as well. These data augmentation strategies do not only enrich the viewpoint information in the training dataset, but also introduce noise-like image distortion, which can enhance the network's robustness and reduce the overfitting during the training process.

Multi-task Training and Distillation: While the semantic labels are manually annotated in the 3D model, this not available for the local features. Inspired by HF-Net [22], we adopt Multi-task Distillation (Fig. 1) to train our multi-task network, so that the detector and descriptor can directly learn the representation from a well-trained teacher model. This technique forces the student model, our VTM-Net, to mimic the teacher output, enabling a much simpler training process than the original SuperPoint training. As SuperPoint outperforms other algorithms when viewpoint changes are present [7,22], we select it as the teacher model for the local features of VTM-Net.

The Cross-Entropy Loss is utilized as the loss function of the semantic segmentation task, and this can be expressed as

$$L_{segm} = CrossEntropy(\boldsymbol{y}_s, \boldsymbol{y}_{gt}), \tag{1}$$

where $\boldsymbol{y}_s$ denotes the output from the segmentation head of VTM-Net, and y_{gt} denotes the ground truth annotation of its corresponding training batches. The interest point detection and local feature descriptor tasks are supervised by the output of the teacher model, expressed, respectively as follows:

$$L_{dect} = CrossEntropy(\boldsymbol{p}_s, \boldsymbol{p}_t), \tag{2}$$

and

$$L_{desc} = \|\boldsymbol{d}_s - \boldsymbol{d}_t\|_2^2, \tag{3}$$

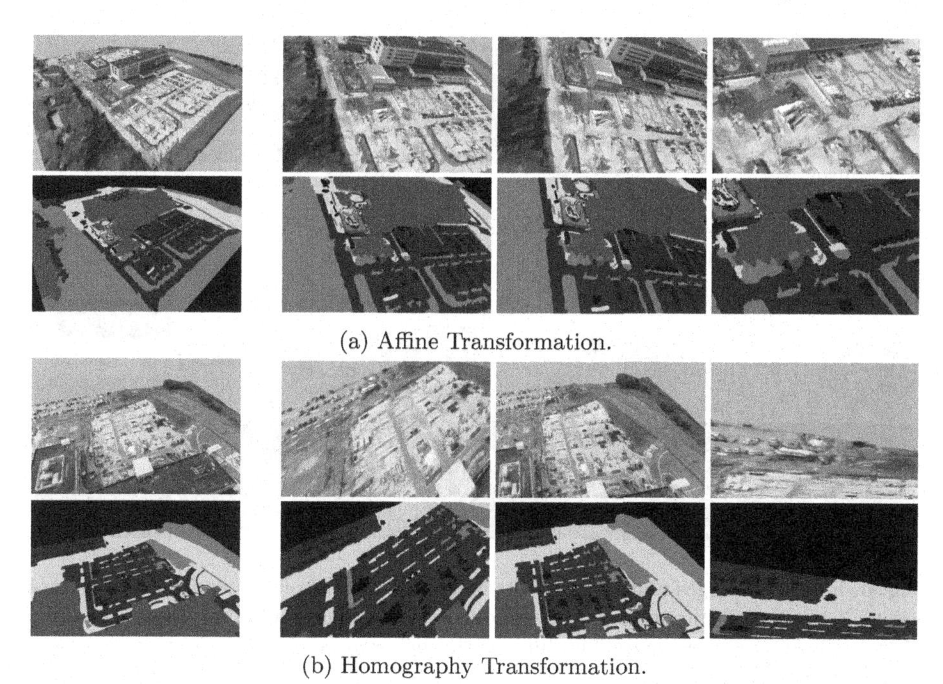

(a) Affine Transformation.

(b) Homography Transformation.

Fig. 4. The illustration of data augmentation techniques: affine transformation and homography transformation. The original images and annotations (first column) are applied with three times random affine or homography transformations. The transformed images (last three columns) can be regarded as the images captured from different viewpoints.

where $\boldsymbol{p}_s$ and $\boldsymbol{p}_t$ denote the interest point score outputs of the student model, VTM-Net, and the teacher model, SuperPoint, respectively, and $\boldsymbol{d}_s$ and $\boldsymbol{d}_t$ are the local descriptor head outputs of student and teacher model. We utilize fixed equal weights to join the three losses, thus the joint loss is defined as:

$$L_{total} = L_{segm} + L_{dect} + L_{desc}. \tag{4}$$

4 Experiments

In order to demonstrate the effectiveness of VTM-Net for semantic segmentation in viewpoint-changing scenarios, we present experiments on large-scale 3D models downloaded from the internet as meshes and we manually annotate them with semantic labels. The models used here are depicted in Fig. 5. The input images and ground-truth labels were generated by a custom-made OpenGL render. Figure 6 shows some examples of the rendered images.

Our training dataset contains very limited viewpoint information, while our test datasets contain a wide range of viewpoints with 24 distinct camera setups.

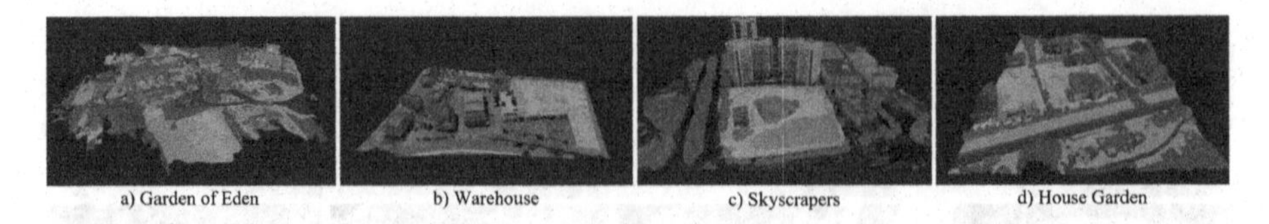

Fig. 5. Several examples of 3D models with the annotation texture. Different classes are annotated with different color.

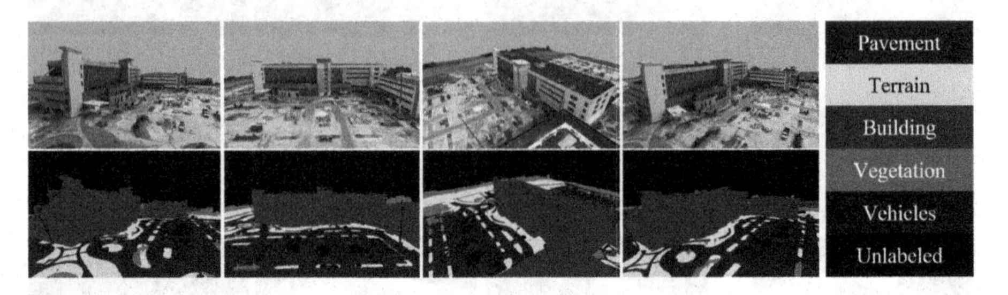

Fig. 6. Example images and their annotations over the Hospital scene and the defined palette for semantic annotations. The resolutions of images and annotations are 480×752. As illustrated, the main building of the hospital is pictured from a variety of viewpoints.

The network's performance on the test dataset can demonstrate whether it is robust to viewpoint changes.

Following the main paradigm of this work of a UAV repeatedly visiting the same place from very distinct viewpoints, for each evaluation experiment, the training dataset and test data are only sampled on a single model. To validate the universality of our method, we independently evaluate it on several 3D models.

We use the standard metric of the mean Intersection of Union (mIoU) to evaluate the semantic segmentation performance. In the test dataset, the mIoU value of each single sub-dataset will be individually calculated. The overall mean mIoU, defined as the arithmetic mean of the mIoU values in 24 sub-datasets, can be used to quantify the general performance of semantic segmentation. Meanwhile, the distribution of mIoU values from different viewpoints demonstrates the viewpoint-tolerance performance of semantic segmentation. A more concentrated distribution implies a better viewpoint-tolerance performance. The standard deviation (SD) of mIoU values from all the sub-datasets is used to assess the distribution.

4.1 Training Setup

All the networks are trained on the GPU NVIDIA Titan X Pascal with 12G memory. All training is conducted using PyTorch [18] with mini-batch size of 36. We adopt Adam as the optimizer with default parameters and the polynomial learning rate decay strategy [27] $lr = base\_lr * (1 - \frac{iter}{total_{iter}})^{power}$. The base learning rate is set to 0.001 and the power is set to 0.9. The networks are trained for 10000 epochs on all the training datasets regardless of the dataset size. In the training process, we use ImageNet pre-trained weights for all the networks. We use the image augmentation library Albumentations [1], to apply all the data augmentations.

4.2 Evaluation Results

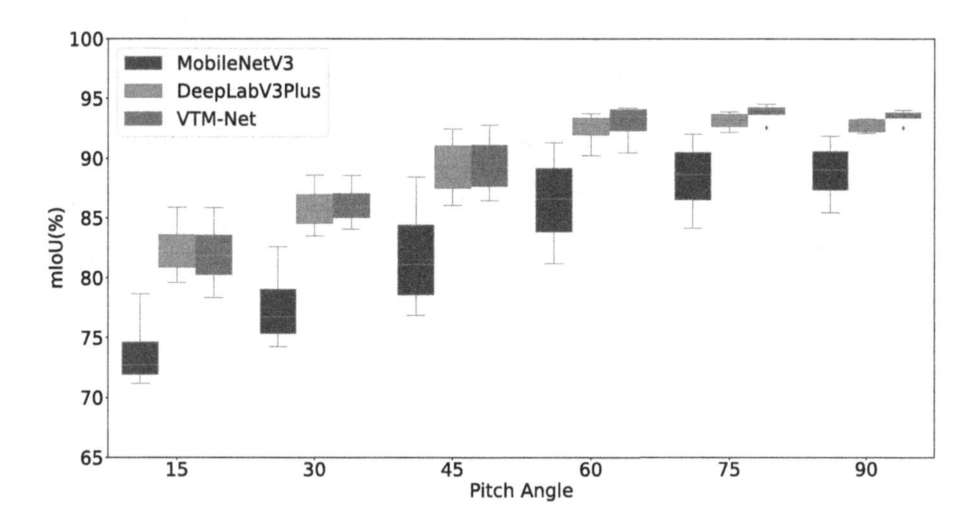

Fig. 7. Evaluation results over star sampled dataset on Hospital model. All the networks are trained with circular sampled dataset on the same model, which contains 288 images.

In order to compare the performance of the proposed VTM-Net with the state of the art, we use DeepLabV3Plus (MobileNetV2 as backbone) and MobileNetV3 as baseline networks with different training and test datasets as currently the DeepLabV3Plus(MobileNetV2) is one of the most powerful and widely used networks in edge applications, while MobileNetV3 achieves one of the best performances in real-time performance and memory efficiency.

At first, we start with the training dataset and the corresponding test dataset of Hospital model. The training dataset contains 288 images from the circular sampling method. The evaluation results are shown in the box plot in Fig. 7. The mIoU values of the datasets sampled at specific pitch angles but at different height are summarized in each single box. The performance differences among different pitch angles are explicitly compared, while the different performances resulted by various heights are implicitly represented by the width of each box. We can observe that VTM-Net exhibits the best performance compared to other networks. Moreover, the widths of boxes of VTM-Net are generally narrow, especially at pitch angles 75° and 90°, which means that the proposed VTM-Net

Table 1. Quantitative analysis of evaluation results on hospital model.

Method	Mean value of mIoU	Standard deviation of mIoU
MobileNetV3	82.82	6.67
DeepLabV3Plus	89.29	**4.52**
VTM-Net (ours)	**89.63**	4.81

reaches top performance in semantic segmentation accuracy and viewpoint tolerance. MobileNetV3, the most lightweight network, performs much worse than the others, while DeepLabV3Plus performs a little worse than VTM-Net in both aspects.

Table 1 records the arithmetic mean value and standard deviation of the mIoU values from the 24 sub-datasets to represent the general semantic segmentation and the viewpoint tolerance performance. These quantitative results confirm that MobileNetV3 exhibits the worst performance in both mean value and standard deviation, while the other networks achieve similar results in both aspects. Compared to MobileNetV3, VTM-Net achieves 8% improvement in mean mIoU value and almost 30% reduced standard deviation.

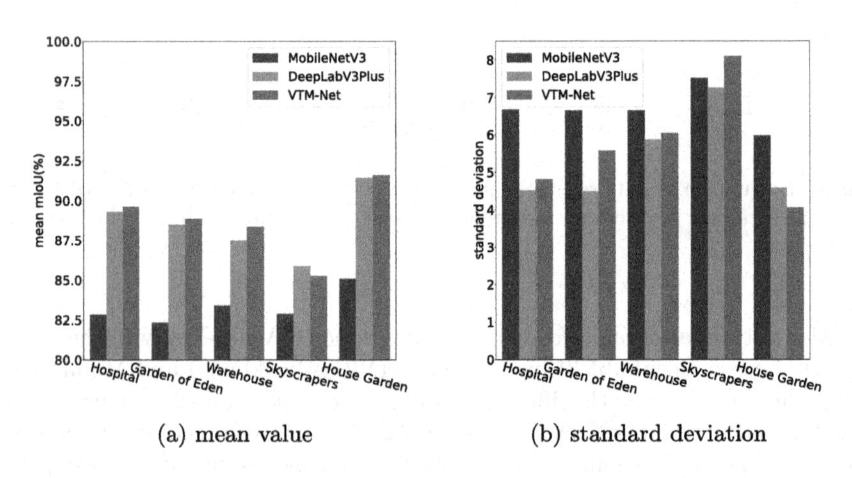

(a) mean value (b) standard deviation

Fig. 8. Evaluation results on multi models.

To test the performance of these networks in different scenes, we further evaluate them on four more models: Garden of Eden, Warehouse, Skyscrapers and House Garden. Figure 8 illustrates the resulting mean values and standard deviations of mIoU from the 24 test sub-datasets on these models. As before, VTM-Net and DeepLabV3Plus have similar semantic segmentation performance and greatly outperform the lightweight network, MobileNetV3. Meanwhile, VTM-Net achieves slightly improved performance compared to DeepLabV3Plus in most models. Regarding the standard deviation, VTM-Net reaches comparable

Table 2. Comparison of the number of parameters required, the inference time and the memory consumption of each method.

Methods	Params	Inference time	Memory
MobileNetV3	3.2 M	0.007 s (142 fps)	436.78 MB
DeepLabV3Plus	15.24 M	0.125 s (8 fps)	1,824.98 MB
VTM-Net (ours)	3.28 M	0.008 s (125 fps)	666.30 MB

performance to DeepLabV3Plus and a much smaller value than MobileNetV3 in almost all the models. From the results on multiple models, we can conclude that the proposed VTM-Net achieves equivalent viewpoint-tolerant semantic segmentation performance when compared to DeepLabV3Plus.

4.3 Runtime Evaluation

The requirements for inference and memory usage are crucial when deployed on a drone. As such, we examine various critical indications of the networks in order to demonstrate their utility in real-world applications. The inference evaluation is carried out on a GPU GTX 1070 Mobile with 8 GB of RAM. The performance on other GPUs can be easily extrapolated from the MobileNetV3 inference time.

Regarding the parameters and inference time, the proposed VTM-Net is as lightweight as MobileNetV3 and consumes almost 1/15 of the inference time of DeepLabV3Plus with 1/5 of parameters. While VTM-Net's memory consumption is approximately 1/3 that of DeepLabV3Plus, it is still significantly increased compared to MobileNetV3 (almost 50%).

To sum up, the proposed VTM-Net achieves a remarkable performance with a real-time and memory-efficient architecture. Figure 9 shows qualitative results of MobileNetV3 and VTM-Net on two models, demonstrating that VTM-Net provides better level of detail than MobileNetV3.

4.4 Ablation Study

For ablation study we propose a variant of VTM-Net consisting of the encoder and only the segmentation head of VTM-Net. As this is then a single-task network with improved decoder, we call it VTS-Net.

Improved Decoder. The impact of the improved decoder can be interpreted by the comparison between MobileNetV3 and VTS-Net. The detailed evaluation results of each viewpoint of hospital model are shown in Fig. 10a. As it can be seen, the enhanced decoder improves semantic segmentation performance substantially. Meanwhile, viewpoint tolerance is greatly increased as well, as indicated by the gentleness of lines and the spacing between lines. More specifically, the improvement at smaller pitch angles and at higher heights is greater than

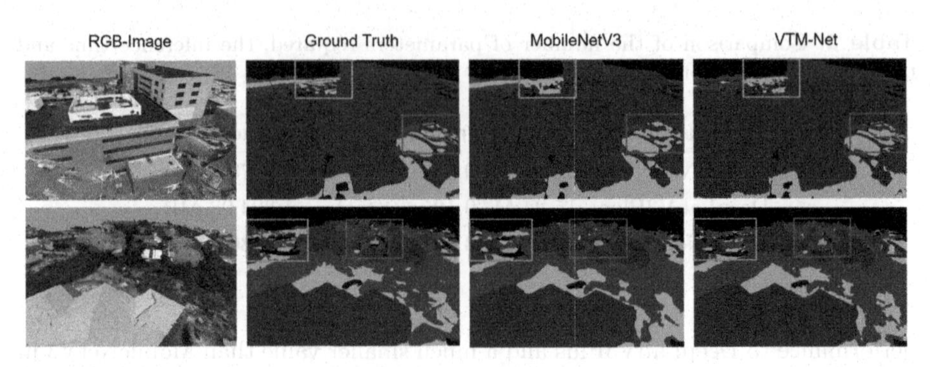

Fig. 9. Qualitative results on the Hospital (first row) and the Garden of Eden (second row) models. Compared to MobileNetV3, VTM-Net infers better in details and small elements with comparable inference time and memory consumption.

at bigger pitch angles and lower heights. This can be explained by considering that the image sampled at a small pitch angle comprises a large number of pixels belonging to the "unlabeled" class, resulting in small effective elements in the image and complex border information. In this situation, it is hard for MobileNetV3 to achieve a good image segmentation. With the introduction of lower-level features, the receptive fields are enriched and the segmentation performance is significantly improved. With increasing pitch angle, the effective elements become clearer and MobileNetV3 can already reach an equivalent performance and the improved decoder cannot help much more with it. The improvement at different heights follows the same principle. The boundaries of elements are hard to get recognized at bigger heights, such as at 75 m and 90 m, where the improved decoder can help a lot.

Multi-task Architecture. In order to assess the effectiveness of multi-task architecture, we compare the detailed evaluation results of VTM-Net and VTS-Net. The split results are presented in Fig. 10b. The results at a 15° pitch angle of VTM-Net are worse than VTS-Net, while the results improve as the pitch angle increases. The reason is that at pitch angle 15°, the background class occupies most pixels in the image and there are not many local features to be detected, thus the local feature heads cannot help much. More effective elements appear in the image when the pitch angle increases, and the local features can be detected by the interest point detector and descriptor, thus the improvement becomes more significant. Similarly, the benefit of multi-task architecture becomes more pronounced as the height increases.

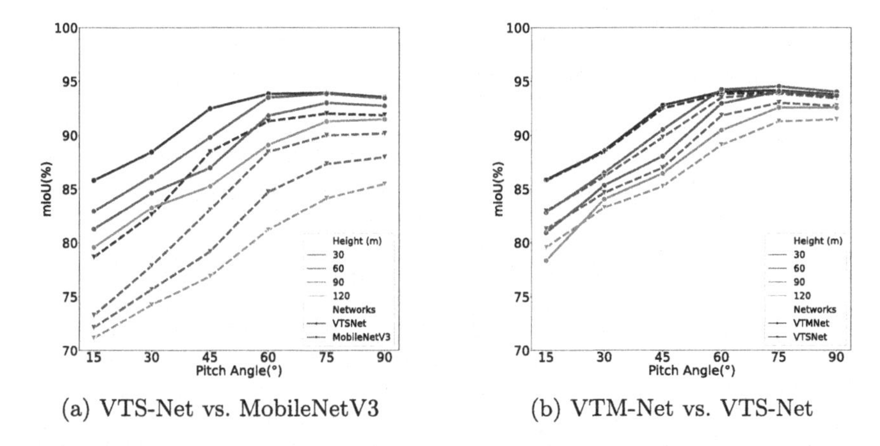

(a) VTS-Net vs. MobileNetV3 (b) VTM-Net vs. VTS-Net

Fig. 10. Ablation study for different methods based on the evaluation results of Hospital model.

5 Conclusions and Future Work

In this paper, we introduced the multi-task network, dubbed "VTM-Net", for viewpoint-tolerant semantic segmentation in the specific scenario of aerial delivery, where a UAV revisits the same place from very different viewpoints. We demonstrated that VTM-Net achieves state-of-the-art performance using a lightweight, fast and memory-efficient architecture, and thoroughly validated the impact of the newly introduced components. Specifically, the proposed decoder increases segmentation accuracy at small pitch angles, while the multi-task architecture with the integration of local features improves the performance at big pitch angles.

Future work will investigate the development and training of a model-indepen-dent network in order to achieve good segmentation performance on similar but not identical environments. In addition, to fill the knowledge gap between the synthetic and the real data domains, a knowledge adaptation method should be further researched. Lastly, a place recognition algorithm combining all the outputs of the network, i.e. interest point detection, local feature description and semantic segmentation, will be developed, resulting in a more advanced end-to-end scene understanding framework suitable for drone delivery applications.

Acknowledgments. This work was supported by the Swiss National Science Foundation (SNSF, NCCR Robotics, NCCR Digital Fabrication), the Amazon Research Awards and IDEA League Student Grant.

References

1. Buslaev, A., Iglovikov, V.I., Khvedchenya, E., Parinov, A., Druzhinin, M., Kalinin, A.A.: Albumentations: fast and flexible image augmentations. Information (2020)

2. Chen, L.C., Papandreou, G., Schroff, F., Adam, H.: Rethinking atrous convolution for semantic image segmentation. arXiv preprint arXiv:1706.05587 (2017)
3. Chen, L.C., Zhu, Y., Papandreou, G., Schroff, F., Adam, H.: Encoder-decoder with atrous separable convolution for semantic image segmentation. In: Proceedings of the European Conference on Computer Vision (ECCV) (2018)
4. Cheng, B., et al.: Panoptic-deeplab: a simple, strong, and fast baseline for bottom-up panoptic segmentation. In: Proceedings of the IEEE/CVF Conference on Computer Vision and Pattern Recognition (2020)
5. Contributors, M.: MMSegmentation: openmmlab semantic segmentation toolbox and benchmark (2020). https://github.com/open-mmlab/mmsegmentation
6. Cordts, M., et al.: The cityscapes dataset for semantic urban scene understanding. In: Proceedings of the IEEE/CVF International Conference on Computer Vision (CVPR) (2016)
7. DeTone, D., Malisiewicz, T., Rabinovich, A.: Superpoint: self-supervised interest point detection and description. In: Proceedings of the IEEE/CVF International Conference on Computer Vision (CVPR) (2018)
8. Geiger, A., Lenz, P., Urtasun, R.: Are we ready for autonomous driving? The kitti vision benchmark suite. In: Proceedings of the IEEE/CVF International Conference on Computer Vision (CVPR) (2012)
9. He, K., Zhang, X., Ren, S., Sun, J.: Deep residual learning for image recognition. In: Proceedings of the IEEE/CVF International Conference on Computer Vision (CVPR) (2016)
10. Howard, A., et al.: Searching for mobilenetv3. In: Proceedings of the IEEE/CVF International Conference on Computer Vision (CVPR) (2019)
11. Howard, A.G., et al.: Mobilenets: Efficient convolutional neural networks for mobile vision applications. arXiv preprint arXiv:1704.04861 (2017)
12. Hu, J., Shen, L., Sun, G.: Squeeze-and-excitation networks. In: Proceedings of the IEEE/CVF International Conference on Computer Vision (CVPR) (2018)
13. Long, J., Shelhamer, E., Darrell, T.: Fully convolutional networks for semantic segmentation. In: Proceedings of the IEEE/CVF International Conference on Computer Vision (CVPR) (2015)
14. Lyu, Y., Vosselman, G., Xia, G.S., Yilmaz, A., Yang, M.Y.: Uavid: a semantic segmentation dataset for uav imagery. ISPRS J. Photogramm. Remote Sens. (2020)
15. Marmanis, D., Schindler, K., Wegner, J.D., Galliani, S., Datcu, M., Stilla, U.: Classification with an edge: improving semantic image segmentation with boundary detection. ISPRS J. Photogramm. Remote Sens. (2018)
16. Mottaghi, R., et al.: The role of context for object detection and semantic segmentation in the wild. In: Proceedings of the IEEE/CVF International Conference on Computer Vision (CVPR) (2014)
17. Mou, L., Hua, Y., Zhu, X.X.: A relation-augmented fully convolutional network for semantic segmentation in aerial scenes. In: Proceedings of the IEEE/CVF Conference on Computer Vision and Pattern Recognition (2019)
18. Paszke, A., et al.: Pytorch: an imperative style, high-performance deep learning library. In: Wallach, H., Larochelle, H., Beygelzimer, A., d' Alché-Buc, F., Fox, E., Garnett, R. (eds.) Advances in Neural Information Processing Systems, vol. 32. Curran Associates Inc. (2019)
19. Poudel, R.P., Liwicki, S., Cipolla, R.: Fast-scnn: fast semantic segmentation network. In: Proceedings of the IEEE/CVF International Conference on Computer Vision (CVPR) (2019)

20. Ronneberger, O., Fischer, P., Brox, T.: U-net: convolutional networks for biomedical image segmentation. In: International Conference on Medical Image Computing and Computer-assisted Intervention (2015)
21. Sandler, M., Howard, A., Zhu, M., Zhmoginov, A., Chen, L.C.: Mobilenetv 2: inverted residuals and linear bottlenecks. In: Proceedings of the IEEE/CVF International Conference on Computer Vision (CVPR) (2018)
22. Sarlin, P.E., Cadena, C., Siegwart, R., Dymczyk, M.: From coarse to fine: robust hierarchical localization at large scale. In: Proceedings of the IEEE/CVF International Conference on Computer Vision (CVPR) (2019)
23. Schönberger, J.L., Frahm, J.M.: Structure-from-motion revisited. In: Proceedings of the IEEE/CVF International Conference on Computer Vision (CVPR) (2016)
24. Simonyan, K., Zisserman, A.: Very deep convolutional networks for large-scale image recognition. CoRR, abs/1409.1556 (2014)
25. Szegedy, C., et al.: Going deeper with convolutions. In: Proceedings of the IEEE/CVF International Conference on Computer Vision (CVPR) (2015)
26. Zhang, E.: Fast semantic segmentation (2020). https://github.com/ekzhang/fastseg
27. Zhao, H., Shi, J., Qi, X., Wang, X., Jia, J.: Pyramid scene parsing network. In: Proceedings of the IEEE/CVF International Conference on Computer Vision (CVPR) (2017)
28. Zhu, Y., et al.: Improving semantic segmentation via video propagation and label relaxation. In: Proceedings of the IEEE/CVF Conference on Computer Vision and Pattern Recognition (2019)

T6D-Direct: Transformers for Multi-object 6D Pose Direct Regression

Arash Amini, Arul Selvam Periyasamy(✉), and Sven Behnke

Autonomous Intelligent Systems, University of Bonn, Bonn, Germany
amini@uni-bonn.de, {periyasa,behnke}@ais.uni-bonn.de

Abstract. 6D pose estimation is the task of predicting the translation and orientation of objects in a given input image, which is a crucial prerequisite for many robotics and augmented reality applications. Lately, the Transformer Network architecture, equipped with multi-head self-attention mechanism, is emerging to achieve state-of-the-art results in many computer vision tasks. DETR, a Transformer-based model, formulated object detection as a set prediction problem and achieved impressive results without standard components like region of interest pooling, non-maximal suppression, and bounding box proposals. In this work, we propose T6D-Direct, a real-time single-stage direct method with a transformer-based architecture built on DETR to perform 6D multi-object pose direct estimation. We evaluate the performance of our method on the YCB-Video dataset. Our method achieves the fastest inference time, and the pose estimation accuracy is comparable to state-of-the-art methods.

Keywords: Pose estimation · Transformer · Self-attention

1 Introduction

6D object pose estimation in clutter is a necessary prerequisite for autonomous robot manipulation tasks and augmented reality. Given the complex nature of the task, methods for object pose estimation—both traditional and modern—are multi-staged [7,19,26,35]. The standard pipeline consists of an object detection and/or instance segmentation, followed by the region of interest cropping and processing the cropped patch to estimate the 6D pose of an object. Convolutional neural networks (CNNs) are the basic building blocks of the deep learning models for computer vision tasks. CNN's strength lies in the ability to learn local spatial features. Motivated by the success of deep learning methods for computer vision, in a strive for end-to-end differentiable pipelines, many of the traditional components like non-maximum suppression (NMS) and region of interest cropping (RoI) have been replaced by their differentiable counterparts [4,8,24]. Despite these advancements, the pose estimation accuracy still heavily depends on the initial object detection stage.

C. Bauckhage et al. (Eds.): DAGM GCPR 2021, LNCS 13024, pp. 530–544, 2021.
https://doi.org/10.1007/978-3-030-92659-5_34

Recently, Transformer, an architecture based on self-attention mechanism, is achieving state-of-the-art results in many natural language processing tasks. Transformers are efficient in modeling long-range dependencies in the data, which is also beneficial for many computer vision tasks. Some recent works achieved state-of-the-art results in computer vision tasks using the Transformer architecture to supplement CNNs or to completely replace CNNs [1,3,12,30,34,37].

Carion et al. introduced DETR [1], an object detection pipeline using Transformer in combination with a CNN backbone model and achieved impressive results. DETR is a simple architecture without any handcrafted procedures like NMS and anchor generation. It formulates object detection as a set prediction problem and uses bipartite matching and Hungarian loss to implement an end-to-end differentiable pipeline for object detection.

In this paper, we present T6D-Direct, an extension to the DETR architecture to perform multi-object 6D pose direct regression in real-time. T6D-Direct enables truly end-to-end pipeline for 6D object pose estimation where the accuracy of the pose estimation is not reliant on object detection and the subsequent cropping. In contrast to the standard methods of 6D object pose estimation that are multi-staged, our method is direct single-stage and estimates the pose of all the objects in a given image in one forward pass. In short, our contributions include:

1. An elegant real-time end-to-end differentiable architecture for multi-object 6D pose direct regression.
2. Evaluation of different design choices for implementing multi-object 6D pose direct regression as a set prediction problem.

2 Related Work

In this section, we review the state-of-the-art methods for 6D object pose estimation and DETR, the transformer architecture our proposed method is based on, in detail.

2.1 Pose Estimation

Like most other computer vision tasks, the state-of-the-art methods for 6D object pose estimation from RGB images are predominantly convolutional neural network (CNN)-based. The standard CNN architectures for object pose estimation are multi-staged. The first stage is object detection and/or instance segmentation. In the second stage, using the object bounding boxes, predicted an image patch containing the target object, is extracted and the 6D pose of the object is estimated. The common methods for object pose estimation can be broadly classified into three categories: direct, indirect, and refinement-based.

Direct methods regress for the translation and orientation components of the object pose directly from the RGB images [9,21,35]. Kehl et al. [11], Sundermeyer et al. [28] discretized the orientation component of the 6D pose and performed classification instead of regression.

Indirect approaches aim to recover the 6D pose from the 2D-3D correspondences using the PnP algorithm, where PnP is often used in combination with the RANSAC algorithm to increase the robustness against outliers in correspondence prediction [10,20,23,29]. Although indirect methods outperform direct methods in the recent benchmarks [7], indirect models are significantly larger, and the model size grows exponentially with the number of objects. One common solution to keep the model size small is to train one lighter model for each object. This approach, however, introduces significant overhead for many real-world applications. Direct models, on the other hand, are lighter, and their end-to-end differentiable nature is desirable in many applications [32]. Li et al. [33], Wang et al. [15] unified direct regression and dense estimation methods by introducing a learnable PnP module.

Refinement-based methods formulate 6D pose estimation as an iterative refinement problem where in each step given the observed image and the rendered image according to the current pose estimate, the model predicts a pose update that aligns the observed and the rendered image better. The process is repeated until the estimated pose update is negligibly small. Refinement-based methods are orthogonal to the direct and indirect methods and are often used in combination with these methods [13,14,18,22,27], i.e., direct or indirect methods produce initial pose estimate and the refinement-based methods are used to refine the initial pose estimate to predict the final accurate pose estimate.

2.2 DETR

Carion et al. [1] introduced DETR, an end-to-end differentiable object detection model using the Transformer architecture. They formulated object detection, the problem of estimating the bounding boxes and class label probabilities, as a set prediction problem. Given an RGB input image, the DETR model outputs a set of tuples with fixed cardinality. Each tuple consists of the bounding box and class label probability of an object. To allow an output set with a fixed cardinality, a larger cardinality is chosen, and a special class id $\emptyset$ is used for padding the rest of the tuples in addition to the actual object detections. The tuples in the predicted set and the ground truth target set are matched by bipartite matching using the Hungarian algorithm. The DETR model achieved competitive results on the COCO dataset [16] compared to standard CNN-based architectures.

3 Method

In this section, we describe our approach of formulating 6D object pose estimation as a set prediction problem and describe the extensions we made to the DETR model and the bipartite matching process to enable the prediction of a set of tuples of bounding boxes, class label probabilities, and 6D object poses. Figure 1 provides an overview of the proposed T6D-Direct model.

3.1 Pose Estimation as Set Prediction

Inspired by the DETR model, we formulate 6D object pose direct regression as a set prediction problem. We call our method T6D-Direct. In the following

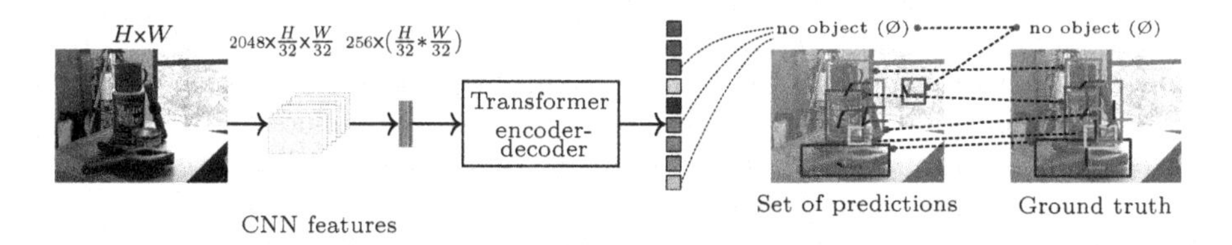

Fig. 1. T6D-Direct overview. Given an RGB image, we use a CNN backbone to extract lower-resolution image features and flatten them to create feature vectors suitable for a standard Transformer model. The Transformer model generates a set of predictions with a fixed cardinality N. To facilitate the prediction of a varying number of objects in an image, we choose N to be much larger than the expected number of objects in an image and pad the rest of the tuples in the set with Ø object predictions. We perform bipartite matching between the predicted and ground truth sets to find the matching pairs and train the pipeline to minimize the Hungarian loss between the matched pairs.

sections, we describe the individual components of the T6D-Direct model in detail.

Set Prediction. Given an RGB input image, our model generates a set of tuples. Each tuple consists of a bounding box, represented as center coordinates, height and width, class label probabilities, translation and orientation components of the 6D object pose. The height and width of the bounding boxes are proportional to the size of the image. For the orientation component, we opt for the 6D continuous representation as shown to yield the best performance in practice [36]. To facilitate the 6D pose prediction of a varying number of objects in an image, we fix the cardinality of the predicted set to N, which is a hyper-parameter, and we chose it to be larger than the expected maximum number of objects in the image. In this way, the network has enough options to embed each object freely. The T6D-Direct model is trained to predict the tuples corresponding to the objects in the image and predict Ø class to pad the rest of the tuples in the fixed size set.

Bipartite Matching. Given n ground truth objects $y_1, y_2, ..., y_n$, we pad Ø objects to create a ground truth set y of cardinality N. To match the predicted set $\hat{y}$, generated by our T6D-Direct model, with the ground truth set y, we perform bipartite matching. Formally, we search for the permutation of elements between the two sets $\sigma \in \mathfrak{S}_N$ that minimizes the matching cost:

$$\hat{\sigma} = \underset{\sigma \in \mathfrak{S}_N}{\arg\min} \sum_{i}^{N} \mathcal{L}_{match}(y_i, \hat{y}_{\sigma(i)}), \tag{1}$$

where $\mathcal{L}_{match}(y_i, \hat{y}_{\sigma(i)})$ is the pair-wise matching cost between the ground truth tuple y_i and the prediction at index $\sigma(i)$. DETR model included bounding boxes

b_i and class probabilities p_i in their cost function. In the case of T6D-Direct model, we have two options for defining $\mathcal{L}_{match}(y_i, \hat{y}_{\sigma(i)})$. One option is to use the same definition used by the DETR model, i.e., we include only bounding boxes and class probabilities and ignore pose predictions in the matching cost. We call this variant of matching cost as $\mathcal{L}_{match\_object}$.

$$\mathcal{L}_{match\_object}(y_i, \hat{y}_{\sigma(i)}) = -\mathbb{1}_{c_i \neq \varnothing} \hat{p}_{\sigma(i)}(c_i) + \mathbb{1}_{c_i \neq \varnothing} \mathcal{L}_{box}(b_i, \hat{b}_{\sigma(i)}). \tag{2}$$

The second option is to include the pose predictions in the matching cost as well. We call this variant $\mathcal{L}_{match\_pose}$.

$$\mathcal{L}_{match\_pose}(y_i, \hat{y}_{\sigma(i)}) = \mathcal{L}_{match\_object}(y_i, \hat{y}_{\sigma(i)}) + \mathcal{L}_{rot}(R_i, \hat{R}_{\sigma(i)}) + \mathcal{L}_{trans}(t_i, \hat{t}_{\sigma(i)}), \tag{3}$$

where $\mathcal{L}_{rot}$ is the angular distance between the ground truth and predicted rotations, and $\mathcal{L}_{trans}$ is the ℓ_2 loss between the ground truth and estimated translations. We experimented with both variants, and we opted for the former method.

Hungarian Loss. After establishing the matching pairs using the bipartite matching, the T6D-Direct model is trained to minimize the *Hungarian loss* between the predicted and ground truth target sets consisting of probability loss, bounding box loss, and pose loss:

$$\mathcal{L}_{Hungarian}(y, \hat{y}) = \sum_{i}^{N} [-\log \hat{p}_{\hat{\sigma}(i)}(c_i) + \mathbb{1}_{c_i \neq \varnothing} \mathcal{L}_{box}(b_i, \hat{b}_{\hat{\sigma}(i)}) + \lambda_{pose} \mathbb{1}_{c_i \neq \varnothing} \mathcal{L}_{pose}(R_i, t_i, \hat{R}_{\hat{\sigma}(i)}, \hat{t}_{\hat{\sigma}(i)})]. \tag{4}$$

Class Probability Loss. The first component in the *Hungarian loss* is the class probability loss. We use the standard negative log-likelihood loss as the class probabilities loss function. Additionally, the number of $\varnothing$ classes in a set is significantly larger than the other object classes. To counter this class imbalance, we weight the log probability loss for the $\varnothing$ class by a factor of 0.4.

Bounding Box Loss. The second component in the *Hungarian loss* is bounding box loss $\mathcal{L}_{box}(b_i, \hat{b}_{\sigma(i)})$. We use a weighted combination of generalized IOU [25] and ℓ_1 loss.

$$\mathcal{L}_{box}(b_i, \hat{b}_{\sigma(i)}) = \alpha \mathcal{L}_{iou}(b_i, \hat{b}_{\sigma(i)}) + \beta ||b_i - \hat{b}_{\sigma(i)}||, \tag{5}$$

$$\mathcal{L}_{iou}(b_i, \hat{b}_{\sigma(i)}) = 1 - \left(\frac{|b_i \cap \hat{b}_{\sigma(i)}|}{|b_i \cup \hat{b}_{\sigma(i)}|} - \frac{|B(b_i, \hat{b}_{\sigma(i)}) \setminus b_i \cup \hat{b}_{\sigma(i)}|}{|B(b_i, \hat{b}_{\sigma(i)})|} \right), \tag{6}$$

where α, β are hyperparameters and $B(b_i, \hat{b}_{\sigma(i)})$ is the largest box containing both the ground truth b_i and the prediction $\hat{b}_{\sigma(i)}$.

Pose Loss. The third component of the *Hungarian loss* is the pose loss. Inspired by Wang et al. [33], we use the disentangled loss to individually supervise the translation t and rotation R via employing symmetric aware loss [35] for the rotation, and ℓ_2 loss for the translation.

$$\mathcal{L}_{pose}(R_i, t_i, \hat{R}_{\sigma(i)}, \hat{t}_{\sigma(i)}) = \mathcal{L}_R(R_i, \hat{R}_{\sigma(i)}) + ||t_i - \hat{t}_{\sigma(i)}||, \tag{7}$$

$$\mathcal{L}_R = \begin{cases} \frac{1}{|\mathcal{M}|} \sum\limits_{\mathrm{x}_1 \in \mathcal{M}} \min\limits_{\mathrm{x}_2 \in \mathcal{M}} ||(R_i \mathrm{x}_1 - \hat{R}_{\sigma(i)} \mathrm{x}_2)|| & \text{if symmetric,} \\ \frac{1}{|\mathcal{M}|} \sum\limits_{\mathrm{x} \in \mathcal{M}} ||(R_i \mathrm{x} - \hat{R}_{\sigma(i)} \mathrm{x})|| & \text{otherwise,} \end{cases} \tag{8}$$

where $\mathcal{M}$ indicates the set of 3D model points. Here, we subsample 1500 points from provided meshes. R_i is the ground truth rotation and t_i is the ground truth translation. $\hat{R}_{\sigma(i)}$ and $\hat{t}_{\sigma(i)}$ are the predicted rotation and translation, respectively.

3.2 T6D-Direct Architecture

The proposed T6D-Direct architecture for 6D pose estimation is largely based on DETR architecture. We use the same backbone CNN (ResNet50), positional encoding, and the transformer encoder and decoder components of the DETR architecture. The only major modification is adding feed-forward prediction heads to predict the translation and rotation components of 6D object poses in addition to the bounding boxes and the class probabilities. We discuss the individual components of the T6D-Direct architecture in detail in the following sections.

CNN Feature Extraction and Positional Encoding. We use ResNet50 [5] model pretrained on ImageNet [2] with frozen batch normalization layers to extract features from the input RGB image. Given an image of height H and width W, the ResNet50 backbone model extracts a lower-resolution feature maps of dimension $2048 \times H/32 \times W/32$. We reduce the dimension of the feature maps to d using 1×1 convolution and vectorize the features into $d \times HW$ feature vectors. Transformer architecture is permutation-invariant and while processing the feature vectors, the spatial information is lost. To tackle this, similar to Transformer architectures for NLP problems, we use fixed positional encoding.

Transformer Encoder. The supplemented feature vector with the fixed sine positional encoding [31] is provided as input to each layer of the encoder. Each encoder layer consists of multi-headed self-attention with 256-dimensional *query, key, and value* vectors and a feed-forward network (FFN). The self-attention mechanism equipped with positional encoding enables learning the spatial relationship between pixels. Unlike CNNs which model the spatial relationship between pixels in a small fixed neighborhood defined by the kernel size, the

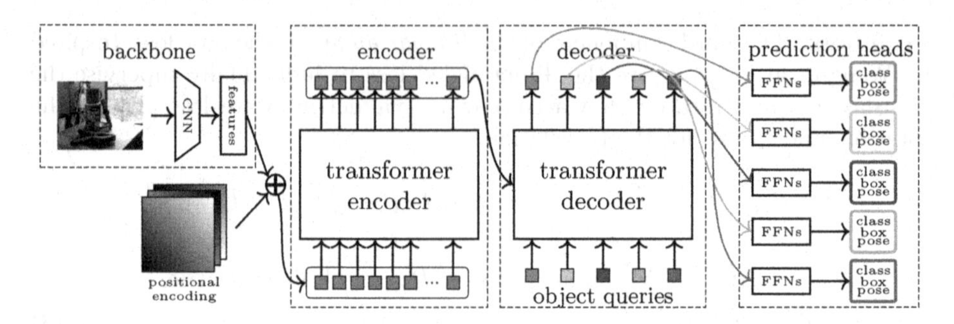

Fig. 2. T6D-Direct architecture in detail. Flattened positional encoded image features from a backbone model are made available to each layer of the transformer encoder. The output of the encoder is provided as input to the decoder along with positional encoding. But, unlike the encoder that takes fixed sine positional encoding, we provide learned positional encoding to the decoder. We call these learned positional encoding *object queries*. Each output of the decoder is processed independently in parallel by shared prediction heads to generate a set of N tuples each containing class probabilities, bounding boxes and translation and orientation components of the 6D object pose. Since the cardinality of the set is fixed, after predicting all the objects in the given image, we train the model to predict Ø object for the rest of the tuples.

self-attention mechanism enables learning spatial relationships between pixels over the entire image (Fig. 2).

Transformer Decoder. On the decoder part, from the encoder output embedding and N positional embedding inputs, we generate N decoder output embeddings using standard multi-head attention mechanism. N is the cardinality of the set we predict. Unlike the fixed sine positional encoding used in the encoder, we use learned positional encoding in the decoder. We call this encoding *object queries*. From the N decoder output embeddings, we use feed-forward prediction heads to generate set of N output tuples. Note that each tuple in the set is generated from a decoder output embedding independently—lending itself for efficient parallel processing.

Prediction Heads. For each decoder output (object query), we use four feed forward prediction heads to predict the class probability, bounding box modeled as the center and scale, translation and orientation components of 6D pose independently. Prediction heads are straightforward three-layer MLPs with 256 neurons in each hidden layer.

4 Experiments

4.1 Dataset

The YCB-Video (YCB-V) dataset [35] is a benchmark dataset for the 6D pose estimation task. The dataset consists of 92 video sequences of random subset of objects from a total of 21 objects arranged in random configurations. In total, the dataset consists of 133,936 images in 640×480 resolution with segmentation masks, depths, bounding boxes, and 6D object pose annotations. Twelve video sequences are held out for the test set with 20,738 images, and the rest images are used for training. Additionally, PoseCNN [35] provides 80K synthetic images for training. For the validation set, we adopt the BOP test set of YCB-V [7], a subset of 75 images from each of the 12 test scenes totaling 900 images. For the final evaluation, we follow the same approach as [35] and report the results on the subset of 2,949 key frames from 12 test scenes.

4.2 Metrics

For the model evaluation, the average distance (ADD) metric is employed from [6]. Given the predicted $\hat{R}$ and $\hat{t}$ and their corresponding ground-truths, ADD calculates the mean pairwise distance between transformed 3D model points ($\mathcal{M}$). If the ADD is below 0.1 m we consider the pose prediction to be correct.

$$\text{ADD} = \frac{1}{|\mathcal{M}|} \sum_{x \in \mathcal{M}} \|(Rx + t) - (\hat{R}x + \hat{t})\| \tag{9}$$

For symmetric objects, instead of using ADD metric, the average closest pairwise distance (ADD-S) metric is computed as follows:

$$\text{ADD-S} = \frac{1}{|\mathcal{M}|} \sum_{x_1 \in \mathcal{M}} \min_{x_2 \in \mathcal{M}} \|(Rx_1 + t) - (\hat{R}x_2 + \hat{t})\| \tag{10}$$

Following [35], we aggregate all results and measure the area under the accuracy-threshold curve (AUC) for distance thresholds of maximum 0.1 m.

4.3 Training

The DETR architecture suffers from the drawback of having a slow convergence [37]. To tackle this issue, we initialize the model using the provided pre-trained weights on the COCO dataset [16] and then train the complete T6D-Direct model on the YCB-V dataset. After initializing our model with the pre-trained weights, there are two possible strategies while training for the pose estimation task. In the first approach, we train the complete model for both object detection and pose estimation tasks simultaneously; therefore, the total loss function is the Hungarian loss brought in Eq. (4). In the second approach, we employ a multi-stage scheme to train only the pose prediction heads and freeze the rest of the network. Investigation on these methods are conducted in Sect. 5.

To further understand the behavior of the mentioned approaches, we visualize the decoder attention maps for the object queries corresponding to the predictions. In Fig. 3, the top row consists of the object predictions. The middle and bottom rows consist of the attention maps from the complete and partial trained models, respectively, corresponding to the object predictions in the top row. The partial trained model has higher activations along the object boundaries. These activations are the result of training the partial model only on the object detection task. When freezing the transformer model and training only the prediction heads, the prediction heads have to rely on the features already learned, whereas the complete trained model has denser activations compared to the partial trained model and the activations are spread over the whole object and not just the object boundaries. Thus, training the complete model helps learn features more suitable for pose estimation than the features learned for object detection.

Hyperparameters. α and β hyperparameters in computing $\mathcal{L}_{box}$ (Eq. (5)) are set to 2 and 5, respectively. The λ_{pose} hyperparameter in computing $\mathcal{L}_{Hungarian}$ (Eq. (4)) is set to 0.05, and the cardinality of the predicted set N is set to 20. The model takes the image of the size 640×480 as input and is trained using the AdamW optimizer [17] with an initial learning rate of 10^{-4} and for 78K iterations. The learning rate is decayed to 10^{-5} after 70K iterations, and the batch size is 32. Moreover, gradient clipping with a maximal gradient norm of 0.1 is applied. In addition to YCB-V dataset images, we use the synthetic dataset provided by PoseCNN for training our model.

4.4 Results

In Table 1, we present the per object quantitative results of T6D-Direct on the YCB-V dataset. We compare our results with PoseCNN [35], PVNet [20] and DeepIM [14]. In terms of the approach, T6D-Direct is comparable to PoseCNN; both are direct regression methods, whereas PVNet is an indirect method, and DeepIM is a refinement-based approach. In terms of both the AUC of ADDS and AUC of ADD(-S) metrics, T6D-Direct outperforms PoseCNN and outperforms the AUC of ADD(-S) results of PVNet. For a fair comparison, we follow the same object symmetry definition and evaluation procedure described by the YCB-Video dataset [35].

Some qualitative results are shown in Fig. 5. To demonstrate the ability of the Transformer architecture to model dependencies between pixels over the whole image instead of a just small local neighborhood, in Fig. 4, we visualize the self-attention maps for three pixels belonging to three objects in the image. All three pixels lie on the same horizontal line but attend to different parts of the image.

4.5 Inference Time Analysis

Since the prediction heads generate N predictions in parallel, the inference of our model is not dependent on the number of objects in an image. However, having a smaller cardinality of the prediction set requires estimating fewer object queries and facilitates faster inference time. Thus, we set N to 20. On an NVIDIA 3090 GPU and Intel 3.70 GHz CPU, our model runs at 58 fps which makes our model ideal for real-time applications.

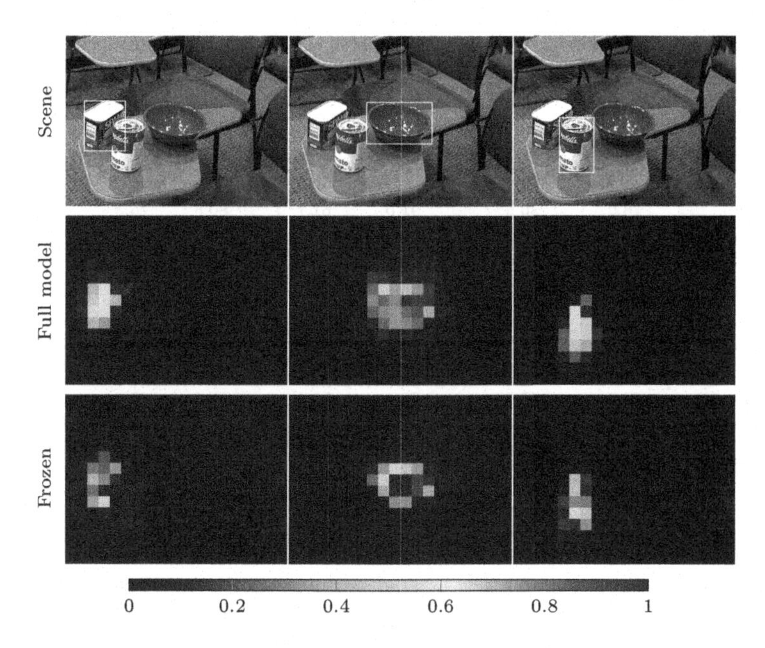

Fig. 3. Object predictions of a given image (first row) and decoder attention maps for the object queries (second and third rows). Training the complete model for both object detection and pose estimation tasks (second row). Training the model first on the object detection task, and then training the frozen model on the pose estimation task (third row). Attention maps are visualized using the jet color map (shown above for reference).

5 Ablation Study

In this section, we explore the effect of various training strategies, different loss functions, and egocentric *vs.* allocentric rotation representations on the T6D-Direct model performance for the YCB-V dataset.

Effectiveness of Loss Functions. In Table 3, we examine the performance of our model using the symmetry aware version of Point Matching loss with ℓ_2 norm [14,35] which, in contrast to the disentangled loss presented in Sect. 3.1, couples the rotation and translation components. This loss function results in the best AUC of ADD(-S) metric. Moreover, since the symmetry aware SLoss component of the Point Matching loss is computationally expensive, we experimented with training our model using only the PLoss component. Interestingly, the ADD(-S) result of the model trained using only the PLoss component (row 5) is only slightly worse than the model trained using the both components (row 1) (Table 2).

Table 1. Pose prediction results on the YCB-V Dataset. The symmetric objects are denoted by *.

Metric	AUC of ADD-S			AUC of ADD(-S)			
Object	PoseCNN	T6D-Direct	DeepIM	PoseCNN	PVNet	T6D-Direct	DeepIM
master_chef_can	84.0	91.9	93.1	50.9	81.6	61.5	71.2
cracker_box	76.9	86.6	91.0	51.7	80.5	76.3	83.6
sugar_box	84.3	90.3	96.2	68.6	84.9	81.8	94.1
tomato_soup_can	80.9	88.9	92.4	66.0	78.2	72.0	86.1
mustard_bottle	90.2	94.7	95.1	79.9	88.3	85.7	91.5
tuna_fish_can	87.9	92.2	96.1	70.4	62.2	59.0	87.7
pudding_box	79.0	85.1	90.7	62.9	85.2	72.7	82.7
gelatin_box	87.1	86.9	94.3	75.2	88.7	74.4	91.9
potted_meat_can	78.5	83.5	86.4	59.6	65.1	67.8	76.2
banana	85.9	93.8	72.3	91.3	51.8	87.4	81.2
pitcher_base	76.8	92.3	94.6	52.5	91.2	84.5	90.1
bleach_cleanser	71.9	83.0	90.3	50.5	74.8	65.0	81.2
bowl*	69.7	91.6	81.4	69.7	89.0	91.6	81.4
mug	78.0	89.8	91.3	57.7	81.5	72.1	81.4
power_drill	72.8	88.8	92.3	55.1	83.4	77.7	85.5
wood_block*	65.8	90.7	81.9	65.8	71.5	90.7	81.9
scissors	56.2	83.0	75.4	35.8	54.8	59.7	60.9
large_marker	71.4	74.9	86.2	58.0	35.8	63.9	75.6
large_clamp*	49.9	78.3	74.3	49.9	66.3	78.3	74.3
extra_large_clamp*	47.0	54.7	73.2	47.0	53.9	54.7	73.3
foam_brick*	87.8	89.9	81.9	87.8	80.6	89.9	81.9
MEAN	75.9	86.2	88.1	61.3	73.4	74.6	81.9

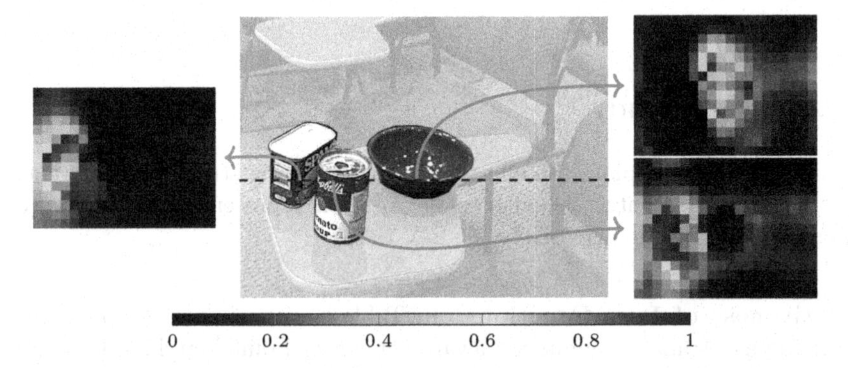

Fig. 4. Encoder self-attention. We visualize the self-attention maps for three pixels belonging to three objects in the image. All three pixels lie on the same horizontal line but attend to different parts of the image. Attention maps are visualized using the jet color map (shown above for reference).

Effectiveness of Training Strategies. As discussed in Sect. 4.3, there are two training schemes: single-stage and multi-stage. In the multi-stage scheme, we train the Transformer model for object detection and only train the FFNs for pose estimation, whereas in the single-stage scheme, we train the complete

Fig. 5. Qualitative examples from the YCB-V Dataset. Top row: PoseCNN [35]. Bottom row: our predictions.

Table 2. Comparison with state-of-the-art methods on YCB-V. In terms of the ADD (-S) 0.1d metric, we achieve the state-of-the-art result. † indicates that the method is refinement-based. Inference time is the average time taken for processing all objects in an image.

Method	ADD(-S)	AUC of ADD-S	AUC of ADD(-S)	Inference Time [s]
CosyPose† [13]	–	**89.8**	**84.5**	0.395
PoseCNN [35]	21.3	75.9	61.3	–
SegDriven [10]	39.0	–	–	–
Single-Stage [9]	**53.9**	–	–	–
GDR-Net [33]	49.1	89.1	80.2	0.065
T6D-Direct (Ours)	48.7	86.2	74.6	**0.017**

model in one stage. In our experiments, as shown in Table 3, multi-stage training (row 2) yielded inferior results, although both schemes were pretrained on the COCO dataset. This demonstrates that the Transformer model is learning the features specific to the 6D object pose estimation task on YCB-V, and COCO fine-tuning mainly helps in faster convergence during training and not in more accurate pose estimations. We thus believe that most large-scale image datasets can serve as pretraining data source. We also provide the results of including the pose component in the bipartite matching cost mentioned in Eq. (3). Including the pose component (row 3) does not provide any considerable advantage; thus, we include only the class probability and bounding box components in the bipartite matching cost in all further experiments. Further, egocentric rotation representation (row 1) performed slightly better than allocentric representation (row 4). We hypothesize that supplementing RGB images with positional encoding allows the Transformer model to learn spatial features efficiently. Therefore, the allocentric representation does not have any advantage over the egocentric representation.

Table 3. Ablation study on YCB-V. We provide results of our method with different loss functions and training schemes.

Row	Method	ADD(-S)	AUC of ADD(-S)
1	T6D-Direct with Point Matching loss	47.0	**75.6**
2	1 + multi-stage training	20.5	59.1
3	1 + pose matching cost component	42.8	71.7
4	1 + allocentric R_{6d}	42.9	74.4
5	T6D-Direct with PLoss	45.8	74.4
6	T6D-Direct	**48.7**	74.6

6 Conclusion

We introduced T6D-Direct, a transformer-based architecture for multi-object 6D pose estimation. Equipped with multi-head attention mechanism, our model obtains competitive results in the task of direct 6D pose estimation without any dense features. Unlike the standard multi-staged methods, our formulation of multi-object 6D pose estimation as a set prediction problem allows estimating the 6D pose of all the objects in a given image in one forward pass. Furthermore, our model is real-time capable. In the future, we plan to explore the possibilities of incorporating dense estimation features into our architecture and improve the performance further.

Acknowledgment. This research has been supported by the Competence Center for Machine Learning Rhine Ruhr (ML2R), which is funded by the Federal Ministry of Education and Research of Germany (grant no. 01—S18038A).

References

1. Carion, N., Massa, F., Synnaeve, G., Usunier, N., Kirillov, A., Zagoruyko, S.: End-to-end object detection with transformers. In: Vedaldi, A., Bischof, H., Brox, T., Frahm, J.-M. (eds.) ECCV 2020. LNCS, vol. 12346, pp. 213–229. Springer, Cham (2020). https://doi.org/10.1007/978-3-030-58452-8_13
2. Deng, J., Dong, W., Socher, R., Li, L.J., Li, K., Fei-Fei, L.: Imagenet: a large-scale hierarchical image database. In: CVPR, pp. 248–255 (2009)
3. Dosovitskiy, A., et al.: An image is worth 16 × 16 words: transformers for image recognition at scale. In: ICLR (2021)
4. Girshick, R.: Fast r-cnn. In: ICCV, pp. 1440–1448 (2015)
5. He, K., Zhang, X., Ren, S., Sun, J.: Deep residual learning for image recognition. In: CVPR, pp. 770–778 (2016)
6. Hinterstoisser, S., et al.: Model based training, detection and pose estimation of texture-less 3d objects in heavily cluttered scenes. In: Lee, K.M., Matsushita, Y., Rehg, J.M., Hu, Z. (eds.) ACCV 2012. LNCS, vol. 7724, pp. 548–562. Springer, Heidelberg (2013). https://doi.org/10.1007/978-3-642-37331-2_42

7. Hodaň, T., et al.: BOP challenge 2020 on 6d object localization. In: Bartoli, A., Fusiello, A. (eds.) ECCV 2020. LNCS, vol. 12536, pp. 577–594. Springer, Cham (2020). https://doi.org/10.1007/978-3-030-66096-3_39
8. Hosang, J., Benenson, R., Schiele, B.: Learning non-maximum suppression. In: CVPR, pp. 4507–4515 (2017)
9. Hu, Y., Fua, P., Wang, W., Salzmann, M.: Single-stage 6d object pose estimation. In: CVPR (2020)
10. Hu, Y., Hugonot, J., Fua, P., Salzmann, M.: Segmentation-driven 6D object pose estimation. In: CVPR, pp. 3385–3394 (2019)
11. Kehl, W., Manhardt, F., Tombari, F., Ilic, S., Navab, N.: SSD-6D: making RGB-based 3D detection and 6D pose estimation great again. In: CVPR, pp. 1521–1529 (2017)
12. Khan, S., Naseer, M., Hayat, M., Zamir, S.W., Khan, F.S., Shah, M.: Transformers in vision: A survey. arXiv:2101.01169 (2021)
13. Labbé, Y., Carpentier, J., Aubry, M., Sivic, J.: CosyPose: consistent multi-view multi-object 6d pose estimation. In: Vedaldi, A., Bischof, H., Brox, T., Frahm, J.-M. (eds.) ECCV 2020. LNCS, vol. 12362, pp. 574–591. Springer, Cham (2020). https://doi.org/10.1007/978-3-030-58520-4_34
14. Li, Y., Wang, G., Ji, X., Xiang, Yu., Fox, D.: DeepIM: deep iterative matching for 6d pose estimation. In: Ferrari, V., Hebert, M., Sminchisescu, C., Weiss, Y. (eds.) ECCV 2018. LNCS, vol. 11210, pp. 695–711. Springer, Cham (2018). https://doi.org/10.1007/978-3-030-01231-1_42
15. Li, Z., Wang, G., Ji, X.: Cdpn: coordinates-based disentangled pose network for real-time rgb-based 6-dof object pose estimation. In: ICCV (2019)
16. Lin, T.Y., et al.: Microsoft COCO: common objects in context. In: Fleet, D., Pajdla, T., Schiele, B., Tuytelaars, T. (eds.) ECCV 2014. LNCS, vol. 8693, pp. 740–755. Springer, Cham (2014). https://doi.org/10.1007/978-3-319-10602-1_48
17. Loshchilov, I., Hutter, F.: Decoupled weight decay regularization. In: ICLR (2017)
18. Manhardt, F., Kehl, W., Navab, N., Tombari, F.: Deep model-based 6d pose refinement in rgb. In: Ferrari, V., Hebert, M., Sminchisescu, C., Weiss, Y. (eds.) Computer Vision – ECCV 2018. LNCS, vol. 11218, pp. 833–849. Springer, Cham (2018). https://doi.org/10.1007/978-3-030-01264-9_49
19. Oberweger, M., Rad, M., Lepetit, V.: Making deep heatmaps robust to partial occlusions for 3d object pose estimation. In: Ferrari, V., Hebert, M., Sminchisescu, C., Weiss, Y. (eds.) ECCV 2018. LNCS, vol. 11219, pp. 125–141. Springer, Cham (2018). https://doi.org/10.1007/978-3-030-01267-0_8
20. Peng, S., Liu, Y., Huang, Q., Zhou, X., Bao, H.: PVNet: pixel-wise voting network for 6DOF pose estimation. In: CVPR, pp. 4561–4570 (2019)
21. Periyasamy, A.S., Schwarz, M., Behnke, S.: Robust 6D object pose estimation in cluttered scenes using semantic segmentation and pose regression networks. In: IROS (2018)
22. Periyasamy, A.S., Schwarz, M., Behnke, S.: Refining 6D object pose predictions using abstract render-and-compare. In: Humanoids, pp. 739–746 (2019)
23. Rad, M., Lepetit, V.: BB8: a scalable, accurate, robust to partial occlusion method for predicting the 3D poses of challenging objects without using depth. In: ICCV, pp. 3828–3836 (2017)
24. Ren, S., He, K., Girshick, R., Sun, J.: Faster r-cnn: towards real-time object detection with region proposal networks. In: NeurIPS, vol. 28 (2015)
25. Rezatofighi, H., Tsoi, N., Gwak, J., Sadeghian, A., Reid, I., Savarese, S.: Generalized intersection over union: a metric and a loss for bounding box regression. In: CVPR, pp. 658–666 (2019)

26. Schwarz, M., et al.: Fast object learning and dual-arm coordination for cluttered stowing, picking, and packing. In: ICRA, pp. 3347–3354 (2018)
27. Shao, J., Jiang, Y., Wang, G., Li, Z., Ji, X.: PFRL: pose-Free reinforcement learning for 6D pose estimation. In: CVPR (2020)
28. Sundermeyer, M., Marton, Z.-C., Durner, M., Brucker, M., Triebel, R.: Implicit 3d orientation learning for 6d object detection from rgb images. In: Ferrari, V., Hebert, M., Sminchisescu, C., Weiss, Y. (eds.) ECCV 2018. LNCS, vol. 11210, pp. 712–729. Springer, Cham (2018). https://doi.org/10.1007/978-3-030-01231-1_43
29. Tekin, B., Sinha, S.N., Fua, P.: Real-time seamless single shot 6D object pose prediction. In: CVPR (2018)
30. Touvron, H., Cord, M., Douze, M., Massa, F., Sablayrolles, A., Jégou, H.: Training data-efficient image transformers & distillation through attention. In: ICML, pp. 10347–10357 (2021)
31. Vaswani, A., et al.: Attention is all you need. In: NeurIPS, vol. 30 (2017)
32. Wang, G., Manhardt, F., Shao, J., Ji, X., Navab, N., Tombari, F.: Self6D: self-supervised monocular 6d object pose estimation. In: Vedaldi, A., Bischof, H., Brox, T., Frahm, J.-M. (eds.) ECCV 2020. LNCS, vol. 12346, pp. 108–125. Springer, Cham (2020). https://doi.org/10.1007/978-3-030-58452-8_7
33. Wang, G., Manhardt, F., Tombari, F., Ji, X.: GDR-Net: geometry-guided direct regression network for monocular 6D object pose estimation. In: CVPR (2021)
34. Wang, H., Zhu, Y., Green, B., Adam, H., Yuille, A., Chen, L.-C.: Axial-DeepLab: stand-alone axial-attention for panoptic segmentation. In: Vedaldi, A., Bischof, H., Brox, T., Frahm, J.-M. (eds.) ECCV 2020. LNCS, vol. 12349, pp. 108–126. Springer, Cham (2020). https://doi.org/10.1007/978-3-030-58548-8_7
35. Xiang, Y., Schmidt, T., Narayanan, V., Fox, D.: Posecnn: a convolutional neural network for 6d object pose estimation in cluttered scenes. In: RSS (2018)
36. Zhou, Y., Barnes, C., Lu, J., Yang, J., Li, H.: On the continuity of rotation representations in neural networks. In: CVPR, pp. 5745–5753 (2019)
37. Zhu, X., Su, W., Lu, L., Li, B., Wang, X., Dai, J.: Deformable detr: deformable transformers for end-to-end object detection. In: ICLR (2021)

TetraPackNet: Four-Corner-Based Object Detection in Logistics Use-Cases

Laura Dörr, Felix Brandt(✉), Alexander Naumann, and Martin Pouls

FZI Forschungszentrum Informatik, Haid-und-Neu Straße 10-14,
76131 Karlsruhe, Germany
{doerr,brandt}@fzi.de

Abstract. While common image object detection tasks focus on bounding boxes or segmentation masks as object representations, we consider the problem of finding objects based on four arbitrary vertices. We propose a novel model, named TetraPackNet, to tackle this problem. TetraPackNet is based on CornerNet and uses similar algorithms and ideas. It is designated for applications requiring high-accuracy detection of regularly shaped objects, which is the case in the logistics use-case of packaging structure recognition. We evaluate our model on our specific real-world dataset for this use-case. Baselined against a previous solution, consisting of an instance segmentation model and adequate post-processing, TetraPackNet achieves superior results (9% higher in accuracy) in the sub-task of four-corner based transport unit side detection.

1 Introduction

While common image recognition tasks like object detection, semantic segmentation or instance segmentation are frequently investigated in literature, some applications could greatly benefit from more specialized approaches. In this work, we investigate how such specialized algorithms and neural network designs can improve the performance of visual recognition systems. For this purpose, we consider the use-case of logistics packaging structure recognition.

Fig. 1. Illustration of the use-case of packaging structure recognition (taken from [2])

The use-case of logistics packaging structure recognition aims at inferring the number, type and arrangement of standardized load carriers in uniform logistics

C. Bauckhage et al. (Eds.): DAGM GCPR 2021, LNCS 13024, pp. 545–558, 2021.
https://doi.org/10.1007/978-3-030-92659-5_35

transport units from a single image of that unit. It is illustrated in Fig. 1. In an approach to design a robust solution to this task, we identified the recognition of two visible transport unit side faces, by finding the exact positions of their four corner points, as a reasonable sub-task [2]. Notably, our objects of interest, i.e. transport unit side faces, are of rectangular shape in real world. As the perspective projection is the main component of the imaging transformation, we can assume, that transport unit side faces can be accurately segmented by four image pixel coordinates in regular images of logistics transport units. Such assumptions are also valid in other logistics use-cases such as package detection or transport label detection. The same holds for non-logistics applications like license plate or document recognition and other cases where objects of regular geometric shapes need to be segmented accurately to perform further downstream processing, like image rectifications or perspective transforms.

To solve the challenge of detecting an object by finding a previously known number of feature points (e.g. four vertices), various approaches are thinkable. For instance, the application of standard instance segmentation methods and adequately designed post-processing algorithms, simplifying the obtained pixel masks, may be a viable solution. We aim to incorporate the geometrical a-priori knowledge into a deep-learning model by designing a convolutional neural network (CNN) detecting objects by four arbitrary feature points. To achieve that, we build upon existing work by Law et al. [7,8], enhancing the ideas of CornerNet. CornerNet finds the two bounding box corners to solve the task of classic object detection. We extend this idea to design a model detecting the four vertices of tetragonal shaped objects. Figure 2 illustrates the difference between common bounding box object representations and our four-point based representations. The example image is taken from our use-case specific dataset. The objects indicated are transport unit faces which need to be precisely localized.

Fig. 2. Sample annotations. Left: Bounding box. Right: Four vertices.

In this paper, we present a redesigned version of CornerNet, namely TetraPackNet, which segments objects by four object vertices instead of bounding boxes or pixel-masks. We evaluate the approach on data concerning the use-case of logistics packaging structure recognition. Notably, TetraPackNet cannot be baselined against its role model CornerNet, as it solves a fundamentally different task on differently annotated data. Baselined against an existing solution to the sub-task of transport unit side corner detection, we show that TetraPackNet achieves improved results. We observe that TetraPackNet is able to predict and localize tetragonal objects accurately.

The rest of the paper is organized as follows: We summarize related work in Sect. 2. The model itself is explained in Sect. 3. Section 4 concerns the example application of logistics packaging structure recognition and the corresponding dataset. We evaluate our approach in Sect. 5. Finally, Sect. 6 concludes our work with a summary and outlook.

2 Related Work

The primary use-case pushing our work is the one of logistics packaging structure detection, which is introduced in [2]. Apart from that work, we are not aware of any other publications considering the same use-case.

As the number and frequency of publications regarding image object detection is enormous, we refer to dedicated survey papers as introductory material. For instance, Wu et al. [14] or Liu et al. [10] give comprehensive overviews.

Our work builds on CornerNet [7], a recent work by Law et al., which aims to perform object detection without incorporating anchor boxes or other object position priors. Instead, corner positions of relevant objects' bounding boxes are predicted using convolutional feature maps as corner heat maps. Corners of identical objects are grouped based on predicted object embeddings. This approach, which outperformed all previous one-stage object detection methods on COCO [9], was further developed and improved by Duan et al. [4] and Zhou et al. [15]. The follow-up work by Law et al. [8], CornerNet-Lite, introduced faster and even more accurate variations of the original CornerNet method. These advancements of the original CornerNet are not in our scope, we build upon the original work [7].

Another approach relevant for this work, is the deep learning based cuboid detection by Dwibedi et al. [5]. In this work in the context of 3D-reconstruction, cuboid shaped, class agnostic objects are detected and their vertices are precisely localized. We refrain from comparing to this work for several reasons, one of which is the requirement for richer image annotations (cuboid based, eight vertices per object). Further, we do not aim to reconstruct 3D-models from our images but aim to classify and interpret intra-cuboid information.

3 Method

We present a novel method for four-corner based object detection based on CornerNet [7], a recent work of Law et al. Whereas in traditional object detection, object locations are referenced by bounding boxes (i.e. top left and bottom right corner position), we work with more detailed locations described by four independent corner points. The resulting shapes are not limited to rectangles but comprise a large variety of tetragons, i.e. four-cornered polygons. In this work, we focus on such tetragonal shapes where the four vertices can be identified as "top left", "top right", "bottom left", and "bottom right" corners respectively.

We use model, ground-truth and loss function designs very similar to those proposed in CornerNet [7]. These components, and especially our modifications for TetraPackNet, targeting tetragon-based object detection, are explained in the following sections.

3.1 Backbone Network

As suggested and applied by Law et al. [8], we use an hourglass network [12], namely Hourglass-54, consisting of 3 hourglass modules and 54 layers, as backbone network. Hourglass networks are fully convolutional neural networks. They are shaped like hourglasses in that regard, that input images are downsampled throughout the first set of convolutional and max pooling layers. Subsequently, they are upsampled to the original resolution in a similar manner. Skip layers are used to help conserve detailed image features, which may be lost by the network's convolutional downsampling. In TetraPackNet's network design, two instances of the hourglass network are stacked atop each other to improve result quality.

3.2 Corner Detection and Corner Modules

Following the backbone network's hourglass modules, so-called corner modules are applied to predict precise object corner positions. CornerNet utilizes two such corner modules to detect top-left and bottom-right corners of objects' bounding boxes. Our architecture includes four corner detection modules for the four object corner types top-left, top-right, bottom-left and bottom-right. It is important to note that we do not detect corners of bounding boxes, but vertices of tetragon-shaped objects.

Analogously to the original CornerNet approach, each corner module is fully convolutional and consists of specific corner pooling layers as well as a set of output feature maps of identical dimensions. These outputs are corner heat maps, offset maps and embedding. They each work in parallel on identical input information: the corner-pooled convolutional feature maps.

We shortly revisit CornerNet's specific pooling strategy. It is based on the idea that important object features can be found when starting at a bounding box top left corner and moving in horizontal right or vertical bottom direction. More precisely, by this search strategy and directions, object boundaries will be found by bounding box definition. In CornerNet, max pooling is performed in the corresponding two directions for both bounding box corner types. The pooling ouputs are added to one another and the results are used as input for corner prediction components. The authors show the benefits of this approach in several detailed evaluations. In our case, where precise object corners are predicted, instead of bounding box corners, one may argue that pooling strategies should be reconsidered. Still, for our first experiments, we retain this pooling approach.

3.3 Ground-truth

Required image annotations are object positions described by the object's four corner points, i.e. top left, top right, bottom left, bottom right corner. It is required that both right corners are further right as their counterparts and, equivalently, both top corners are further up as the corresponding bottom corners. For each ground-truth object one single positive location is added to each of the four ground-truth heatmaps. To allow for minor deviations of corner detections from these real corner locations, the ground-truth heatmaps' values are set to positive values in a small region around every corner location. As proposed by CornerNet, we use a Gaussian function centered at the true corner position to determine ground-truth heatmap values in the vicinity of that corner.

In Fig. 3 ground-truth and detected heatmaps, and embeddings are illustrated. The top row shows, cross-faded on the original input image, the ground-truth heatmaps for the four different corner types. There are two Gaussian circles in each corner type heatmap as there are two annotated ground-truth objects, i.e. two transport unit sides, in the image. The bottom row shows TetraPackNet's detected heatmaps (for object type transport unit side) and embeddings in a single visualization: Black regions indicate positions where the predicted heat is smaller than 0.1. Wherever the detected heat value exceeds this threshold, the color indicates the predicted embedding value. To map embedding values to colors, the range of all embeddings for this instance was normalized to the interval from 0 to 1. Afterward Open CV's Rainbow colormap was applied [1].

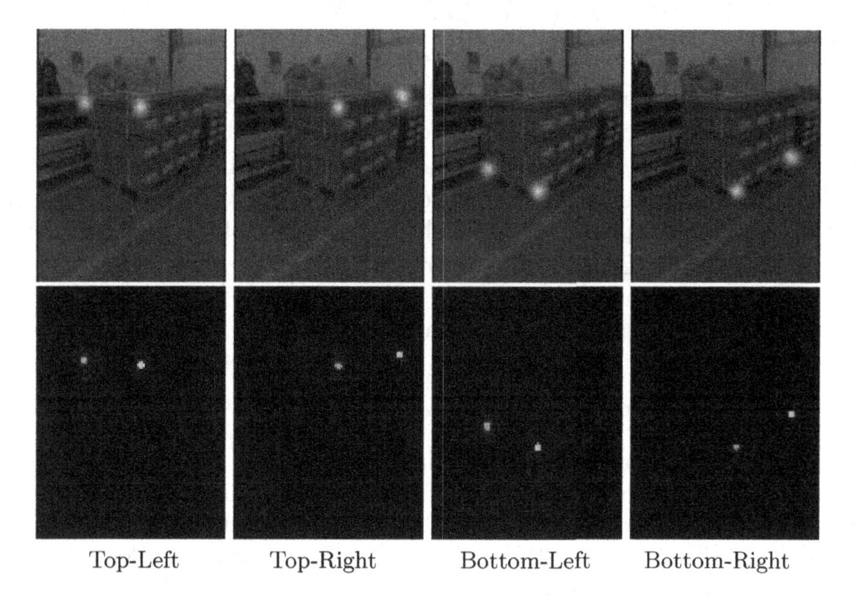

Fig. 3. Example heatmaps. Top row: Groundtruth. Bottom row: Detected heats and color-encoded embeddings.

3.4 Loss Function

The loss function used in training of our TetraPackNet model is structurally identical to that of CornerNet and consists of several components:

$$L = L_{\text{det}} + w_{\text{off}} \cdot L_{\text{off}} + (w_{\text{pull}} \cdot L_{\text{pull}} + w_{\text{push}} \cdot L_{\text{push}}) \tag{1}$$

The first component is the focal loss L_{det}, which aims to optimize heatmap corner detections by penalizing high heatmap values at points where there is no ground-truth corner location. Analogously, low heat values at ground-truth positive locations are penalized. Secondly, the offset loss component L_{off} is used to penalize corner offset predictions which do not result in accurate corner positions. Both focal loss and offset loss are adopted as proposed by CornerNet.

The only loss components we slightly modified to transition from CornerNet to TetraPackNet are the pull and push losses L_{pull} and L_{push}. These components are used to optimize the embedding values predicted at each potential corner location. Modifications were made to account for the increased number of feature points to be detected. Our version of pull and push loss are computed as follows

$$L_{\text{pull}} = \frac{1}{N} \sum_{k=1}^{N} \sum_{i \in \{tl,tr,bl,br\}} (e_{k_i}^{(i)} - e_k)^2 \tag{2}$$

$$L_{\text{push}} = \frac{1}{N(N-1)} \sum_{k=1}^{N} \sum_{\substack{j=1 \\ j \neq k}}^{N} \max\{0, 1 - |e_k - e_j|\} \tag{3}$$

where $\{1, ..., N\}$ enumerates the ground-truth objects and $i = \{tl, tr, bl, br\}$ indicate the four corner types top left, top right, bottom left, bottom right. The position of corner i of ground-truth object k is denoted by k_i. Further $e_{k_i}^{(i)} \in \mathbb{R}$ is the embedding value for corner i at position k_i, i.e. at the position of ground-truth object k's corresponding vertex. The average value of the embedding values of all four corners of a ground-truth object k is given by $e_k = \frac{1}{4} \sum_{i \in \{tl,tr,bl,br\}} e_{k_i}^{(i)}$. In our experiments, the loss component weights were set to $w_{\text{pull}} = w_{\text{push}} = 0.1$ and $w_{\text{off}} = 1.0$, as proposed by Law et al.

3.5 Assembling Corner Detections to Objects

Once corner positions and their embeddings are predicted, these predictions need to be aggregated to form tetragon object detections. Compared to the CornerNet setup, this task is slightly more complex as each object is composed of four vertices instead of only two. However, the original grouping implementation is based on Associative Embeddings [11], which is suitable for multiple data points in general, i.e. more than two. The same approach can be applied in our case.

To obtain an overall ranking for all detected and grouped objects, the four corner detection scores as well as the similarity of their embeddings, i.e. the corresponding pull loss values, are considered. This final score for a detection p of class c consisting of four corners $p_{tl}, p_{tr}, p_{bl}, p_{br}$ is computed as

$$\frac{1}{4} \sum_{i \in \{tl,tr,bl,br\}} h_{p_i}^{(i,c)} \cdot \max\{0, 1 - (e_{p_i}^{(i)} - e_p)^2\} \tag{4}$$

with e_p being the average embedding for a set of four corners as before. Further, $h_{p_i}^{(i,c)}$ denotes the predicted heat value for class c and corner type $i \in \{tl, tr, bl, br\}$ at position p_i.

Additionally, we only allow corners to be grouped which comply with the condition, that right corners are further right in the image as their left counterparts. Analogously bottom corners are required to be further down in the image as corresponding top corners.

4 Use-Case and Data

4.1 Logistics Unit Detection and Packaging Structure Analysis

Our work was developed in context of the logistics task of automated packaging structure recognition. The aim of logistics packaging structure recognition is to infer the number and arrangement of a well standardized logistics transport unit by analyzing a single RGB image of that unit. Figure 1 illustrates this use-case. To infer a transport unit's packaging structure from an image, the target unit, the unit's two visible faces and the faces of all contained load carriers are detected using learning-based detection models. Several restrictions and assumptions regarding materials, packaging order and imaging are incorporated to assure feasibility of the task: First of all, all materials like load carriers and base pallets are known. Further, transport units must be uniformly and regularly packed. Each image shows relevant transport units in their full extend and in an upright orientation and in such a way, that two faces of each transport unit are visible. The use-case and its setting, as well as limitations and assumptions are thoroughly explained in [2] and [3].

We designed a multi-step image processing pipeline to solve the task of logistics packaging structure recognition. The process' individual steps can be summarized as follows:

1. Transport unit detection
2. Transport unit side and package unit face segmentation
3. Transport unit side analysis
4. Information consolidation

In step 1), whole transport units are localized within the image and input images are cropped correspondingly. (see Fig. 4 (a)). As a result, the input for step 2) is an image crop showing exactly one transport unit to be analyzed. Subsequently, transport unit sides (and package units) are detected precisely. This is illustrated in Fig. 4 (b). Step 3) aims to analyze both transport unit sides within the image. This involves a rectification of the image's transport unit side region, in such a way to reconstruct a frontal, image boundary aligned view of each transport unit side. To perform such a rectification, the precise locations of the side's four corner points are required. Figure 4 (c) shows the rectified image patch of one transport unit side and illustrates package pattern analysis. Each transport unit side is analyzed independently and its packaging pattern is determined. In a last step, the information of both transport unit sides are consolidated. The pipeline's overall results are the precise number and arrangement of packages for each transport unit within an image.

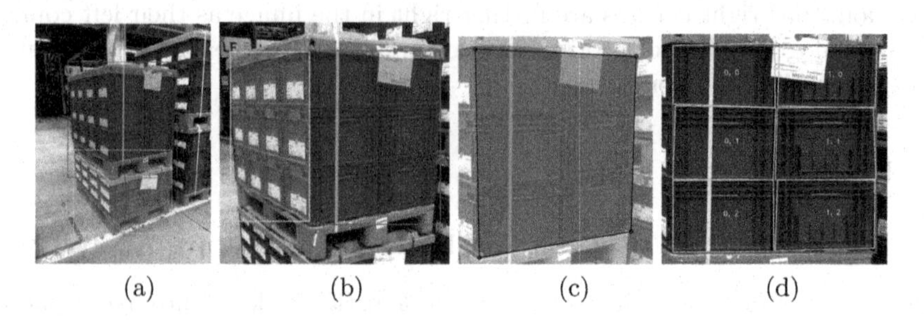

(a) (b) (c) (d)

Fig. 4. Method Visualization. (a) Transport unit identification. (b) Transport unit side face segmentation. (c) Side detection post-processing: Approximation of segmentation mask by four corner points. (d) Rectified transport unit side.

In the context of this work, we focus on step 2) of our overall packaging structure recognition pipeline. Importantly, the simple detection and localization of these components represented by bounding boxes is not sufficient. As we aim to rectify transport unit face image patches in a succeeding step, the precise locations of the four corners of both transport unit faces are required. In our previous solution to this task, which was introduced in [2], a standard Mask R-CNN model [6] is used to segment the transport unit faces in the image. Afterwards, in a post-processing step, the segmentation masks, described by arbitrary polygons, are simplified to consist of four points only. These points are refined by solving an optimization problem in such a way that the tetragon described by these four points has the highest possible overlap with the originally detected region. Figure 4 (c) illustrates this four-corner approximation of segmentation masks.

In this work, we aim to replace the previously described procedure by TetraPackNet. Instead of taking the detour via overly complex mask representations, we employ our specialized method TetraPackNet, which directly outputs the required four-corner based object representations.

4.2 Specific Dataset

For training and evaluation of our TetraPackNet model, a custom use-case specific dataset of 1267 images was used. The dataset was acquired in a German industrial plant of the automotive sector. Each image shows one or multiple stacked transport units in a logistics setting. The rich annotations for each image also include four-corner based transport unit side annotations. The dataset was split in three sub-sets: training, validation and test data. The test data set contains a handpicked selection of 163 images. For a set of images showing identical transport units it was ensured that either all or none of these images was added to the test data set. The remaining 1104 images were split into train and validation data randomly, using a train-validation ratio of 75–25.

5 Experiments

In this section, we examine the performance of TetraPackNet. We evaluate TetraPackNet on our use-case data and compare the results to baseline models and procedures, using standard and use-case specific metrics. In a standard metric evaluation, the four-point regions found by TetraPackNet are compared to segmentation mask regions found by a baseline instance segmentation model. This is done using the standard metric for the task of instance segmentation: mean Average Precision (mAP). To evaluate TetraPackNet more specifically for our use-case, we compare its performance for the task of transport unit side detection with our previously implemented solution. The previous solution, which was introduced in [2] and described in Sect. 4.1, consists, again, of an instance segmentation model and adequate post-processing steps. In this evaluation, four-point represented regions are compared using use-case specific metrics.

Setup. Both models, TetraPackNet and the instance segmentation model, were trained for the single-class problem of transport unit side detection, as described in Sect. 4.1. In both cases, the same dedicated training, validation and tests splits were used. Training and evaluation were performed on an Ubuntu 18.04 machine on a single GTX 1080 Ti GPU unit. Two different training scenarios are evaluated for both models: First, the models are trained to localize transport unit sides within the full images. In a second scenario, the cropped images are used as input instead: as implemented in our packaging structure recognition pipeline (see Sect. 4.1), all images are cropped in such ways that each crop shows exactly one whole transport unit. For each original image, one or multiple such crops can be generated, depending on the number of transport units visible within the image. This second scenario is comparatively easier as exactly two transport unit sides are present in each image and the variance of the scales of transport units within the image is minimal.

Training Details. In both trainings, we tried to find training configurations and hyperparameter asssignments experimentally. However, due to the high complexity of CNN training and its time consumption, an exhaustive search for ideal configurations could not be performed. To achieve fair preconditions for both training tasks, the following prerequisites were fixed. Both models were trained for the same amount of epochs: The training of the Mask R-CNN baseline model included 200.000 training steps using a batch size of 1, whereas the TetraPackNet training included 100.000 training steps with a batch size of 2. As backbone network, the Mask R-CNN model uses a standard Inception-ResNet-v2 [13] architecture. Input resolution for both models was limited to 512 pixels per dimension. Images are resized such that the larger dimension measures 512 pixels and aspect ratio is preserved. Subsequently, padding to quadratic shape is performed. In both trainings, we considered similar image augmentation methods: random flip, crop, color distortions and conversion to gray values. Of these options, only the ones yielding improved training results were retained.

5.1 Standard Metric Results

First of all, we compare TetraPackNet to a model performing classic image instance segmentation, a more complex task than is solved by our novel model TetraPackNet. Note that such a comparison only makes sense on data regarding four-point based object detection as TetraPackNet does not aim to solve the general task of instance segmentation. Still this comparison is meaningful as instance segmentation models are capable of detecting arbitrary shapes, including tetragonal ones. Additionally, we are not aware of other models performing four-point object detection which could be used as baseline methods.

As standard evaluation metric, the COCO dataset's [9] standards are used. We report average precision (AP) at intersection over union (IoU) threshold of 0.5 ($AP_{0.5}$), 0.75 ($AP_{0.75}$) and averaged for ten equidistant IoU thresholds from 0.5 to 0.95 (AP). Evaluations are performed on our dedicated 163-image test dataset. Table 1 shows the corresponding results.

Table 1. Evaluation results for the whole image scenario.

Model	AP	$AP_{0.5}$	$AP_{0.75}$	$AP_{0.9}$
Mask R-CNN	61.5	89.6	70.3	19.4
TetraPackNet	75.6	83.6	83.5	66.7

Table 2. Evaluation results for the cropped image scenario.

Model	AP	$AP_{0.5}$	$AP_{0.75}$	$AP_{0.9}$
Mask R-CNN	80.0	98.9	92.3	55.8
TetraPackNet	91.1	96.0	95.0	85.2

Considering only the values at the lowest IoU threshold examined (0.5), the baseline Mask R-CNN outperforms TetraPackNet by visible margins: Mask R-CNN's $AP_{0.5}$-value is 6 points higher than that of TetraPackNet (89.6 vs. 83.6).

However, as the IoU threshold for detections to be considered correct increases, TetraPackNet gains the advantage. When regarding performance values at IoU threshold 0.75 instead, TetraPackNet achieves a significantly higher precision of 83.5, compared to 70.3 for Mask R-CNN. This observation can be expanded for the average precision scores at higher IoU thresholds: TetraPackNet begins to gain advantage over our baseline method as detection accuracy requirements increase. This is illustrated and visible in the top-part of Fig. 5, which visualizes the same evaluation results shown in Table 1.

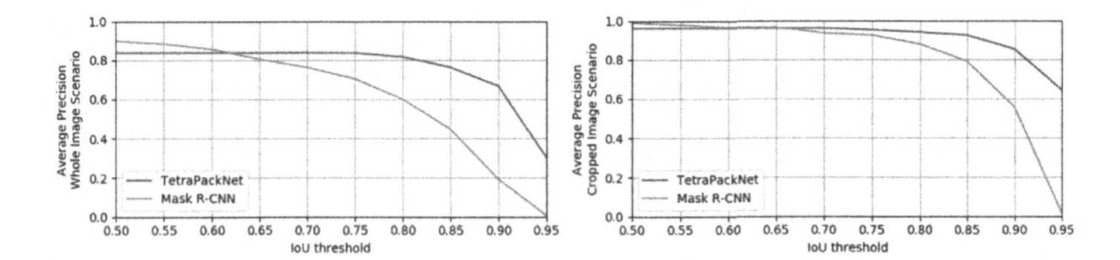

Fig. 5. Average precisions at different IoU thresholds for TetraPackNet and Mask R-CNN baseline model. Left: Whole image scenario. Right: Cropped image scenario.

Very similar observations can be made for the cropped image scenario: TetraPackNet clearly outperforms the reference model Mask R-CNN when high accuracy is required. The corresponding evaluation results are shown in Table 2, and in the bottom part of Fig. 5.

Overall, the results suggest that TetraPackNet does not detect quite as many ground-truth transport unit sides as our Mask R-CNN baseline model on a low accuracy basis. At the same time, the predictions made by TetraPackNet appear to be very precise as average precision steadily remains on a high level as IoU accuracy requirements are increased. For our use-case of packaging structure recognition, these are desirable conditions, as our processing pipeline requires very accurate four-point based transport unit side predictions.

5.2 Use-Case Specific Results

Within our use-case of packaging structure detection, we aim to localize each transport unit side by four corner points. In this section, we baseline our results against our previous approach to transport unit side detection, which consists of a segmentation model and a succeeding post-processing procedure. The former is a Mask R-CNN model with an Inception-ResNet-v2 backbone network, as described above. The latter is necessary to obtain the required four-point object representations from the segmentation model's output masks. This is performed by solving a suitable optimization model, which outputs four vertices giving the best approximation of a segmentation masks in terms of region overlap. Input to the task in our image processing pipeline, and for these evaluation, are cropped images showing exactly one full transport unit. In these evaluations, only the two highest-ranked detections of each method are considered, as there are exactly two transport unit sides in each image.

First of all, we report average precision values for the tetragonal detection of the baseline model (Mask R-CNN with post-processing), and compare them to TetraPackNet's average precision values (as already reported in the previous evaluation). The results can be found in the corresponding columns of Table 3. It shows that the mask-based average precision of the Mask R-CNN baseline model is slightly decreased due to the application of the necessary post-processing steps.

To further investigate the performance of TetraPackNet for our specific use-case, other metrics (than standard COCO Average Precision) are considered additionally. To this end, two different metrics are computed, both based on the standard value of intersection over union (IoU). For each ground-truth transport unit side, the IoUs with its assigned detection, represented by four vertices, is computed. Detections are assigned to ground-truth annotations from left to right, based on the a-priori knowledge, that one transport unit side is clearly positioned further left than the other one. Only for cases where only one instead of two transport unit sides are detected by the methods under consideration, the detected side is assigned to that ground-truth side with which it shares a greater IoU. If no detection was matched to a ground-truth object, the IoU value for this side is set to 0.

The first custom metric we compute is an overall accuracy value. We assume a ground-truth transport unit side to be detected correctly if it has an IoU of at least 0.8 with its assigned detection. The accuracy of transport unit side detection is equal to the percentage of annotated sides detected correctly.

Additionally, the average IoU for correctly assigned detections is evaluated. For the use-case at hand this is a reasonable and important assessment, as the succeeding packaging structure analysis steps rely on very accurate four-corner based transport unit side segmentations [2]. The two rightmost columns of Table 3 show the corresponding results.

Overall, the use-case specific evaluation requiring a high detection accuracy is dominated by our novel approach TetraPackNet. The latter achieves higher rates of high-accuracy transport unit side detections: TetraPackNet correctly detects 95.7% of transport unit sides, whereas the baseline method only achieves 86.6% in this metric. Additionally, if only considering sufficiently accurate detections of both models, the average IoU of the detections output by TetraPackNet was significantly higher (0.05 IoU points on average) than for those of our baseline method. We deduce the suitability of TetraPackNet for our application.

Table 3. Use-case specific evaluation results on our 163-image test dataset.

Model	AP	$AP_{0.5}$	$AP_{0.75}$	$AP_{0.9}$	Accuracy	Average IoU (Positives only)
Mask R-CNN & post-processing	78.9	99.0	92.1	50.7	86.6	0.908
TetraPackNet	91.1	96.0	95.0	85.2	95.7	0.958

6 Summary and Outlook

6.1 Results Summary

We presented TetraPackNet, a new detection model outputting objects based on a novel four-vertex representation. We showed that the use of such a specialized model, instead of an instance segmentation model employing more complex object representations, can lead to superior results in cases where corresponding object representations are reasonable and necessary. This is the case in our presented use-case of logistics packaging structure recognition, but may also apply to numerous other tasks, as for instance document or license plate detection.

We trained and evaluated TetraPackNet on our own use-case specific dataset. For the dedicated task, the observed results indicate a higher accuracy (superior by 9% points) compared to a previous approach involving a standard Mask R-CNN model and suitable post-processing.

6.2 Future Work

The applicability of TetraPackNet to our use-case of logistics packaging structure recognition will be evaluated further. First of all, we plan to apply the model to package unit detection. This task is additionally challenging as, in general, a large number of densely arranged package faces of very similar appearance need to be detected. On the other hand, a-priori knowledge about the regular package arrangement might allow for specific corner detection interpretations and even interpolations (e.g. in case of single missing corner detections). Additionally, we plan to extend TetraPackNet to specialized detection tasks including even more than four corner or feature points. In our case of logistics transport unit detections, a lot of a-priori knowledge about object structure, shape and posture is given. This can be exploited by integrating specific transport unit templates into the detection model: one can define multiple additional characteristic points (e.g. base pallet or packaging lid corners) which are to be detected by a highly specialized deep learning model.

Further, we will investigate different algorithmic and architectural choices. As mentioned before, corner pooling strategies adopted from CornerNet might not be ideal for TetraPackNet and its applications. Thus, for instance, experiments with different corner pooling functions will be performed.

At the same time, TetraPackNet is not necessarily limited to the use-case of packaging structure recognition or logistics in general. We plan to affirm our positive results by evaluating TetraPackNet on additional datasets of different use-cases.

References

1. Bradski, G., Kaehler, A.: Learning OpenCV: Computer Vision with the OpenCV Library. O'Reilly Media Inc., Sebastopol (2008)
2. Dörr, L., Brandt, F., Pouls, M., Naumann, A.: Fully-automated packaging structure recognition in logistics environments. arXiv preprint arXiv:2008.04620 (2020)
3. Dörr, L., Brandt, F., Pouls, M., Naumann, A.: An Image Processing Pipeline for Automated Packaging Structure Recognition. In: Forum Bildverarbeitung 2020, p. 239. KIT Scientific Publishing (2020)
4. Duan, K., Bai, S., Xie, L., Qi, H., Huang, Q., Tian, Q.: Centernet: keypoint triplets for object detection. In: Proceedings of the IEEE/CVF International Conference on Computer Vision. pp. 6569–6578 (2019)
5. Dwibedi, D., Malisiewicz, T., Badrinarayanan, V., Rabinovich, A.: Deep cuboid detection: Beyond 2d bounding boxes. arXiv preprint arXiv:1611.10010 (2016)
6. He, K., Gkioxari, G., Dollár, P., Girshick, R.: Mask r-cnn. In: Proceedings of the IEEE International Conference on Computer Vision, pp. 2961–2969 (2017)
7. Law, H., Deng, J.: Cornernet: detecting objects as paired keypoints. In: Proceedings of the European Conference on Computer Vision (ECCV) (September 2018)
8. Law, H., Teng, Y., Russakovsky, O., Deng, J.: Cornernet-lite: Efficient keypoint based object detection. arXiv preprint arXiv:1904.08900 (2019)
9. Lin, T.Y., et al.: Microsoft COCO: common objects in context. In: Fleet, D., Pajdla, T., Schiele, B., Tuytelaars, T. (eds.) ECCV 2014. LNCS, vol. 8693, pp. 740–755. Springer, Cham (2014). https://doi.org/10.1007/978-3-319-10602-1_48
10. Liu, L., et al.: Deep learning for generic object detection: a survey. Int. J. Comput. Vis. **128**(2), 261–318 (2020)
11. Newell, A., Huang, Z., Deng, J.: Associative embedding: End-to-end learning for joint detection and grouping. arXiv preprint arXiv:1611.05424 (2016)
12. Newell, A., Yang, K., Deng, J.: Stacked hourglass networks for human pose estimation. In: Leibe, B., Matas, J., Sebe, N., Welling, M. (eds.) ECCV 2016. LNCS, vol. 9912, pp. 483–499. Springer, Cham (2016). https://doi.org/10.1007/978-3-319-46484-8_29
13. Szegedy, C., Ioffe, S., Vanhoucke, V., Alemi, A.: Inception-v4, inception-resnet and the impact of residual connections on learning. In: Proceedings of the AAAI Conference on Artificial Intelligence, vol. 31 (2017)
14. Wu, X., Sahoo, D., Hoi, S.C.: Recent advances in deep learning for object detection. Neurocomputing **396**, 39–64 (2020)
15. Zhou, X., Wang, D., Krähenbühl, P.: Objects as points. arXiv preprint arXiv:1904.07850 (2019)

Detecting Slag Formations with Deep Convolutional Neural Networks

Christian von Koch[1], William Anzén[1], Max Fischer[1], and Raazesh Sainudiin[1,2,3(✉)]

[1] Combient Mix AB, Stockholm, Sweden
{christian.von.koch,william.anzen,max.fischer}@combient.com
[2] Combient Competence Centre for Data Engineering Sciences, Uppsala University, Uppsala, Sweden
[3] Department of Mathematics, Uppsala University, Uppsala, Sweden
raazesh.sainudiin@math.uu.se
https://combient.com/mix, https://math.uu.se/

Abstract. We investigate the ability to detect slag formations in images from inside a Grate-Kiln system furnace with two deep convolutional neural networks. The conditions inside the furnace cause occasional obstructions of the camera view. Our approach suggests dealing with this problem by introducing a convLSTM-layer in the deep convolutional neural network. The results show that it is possible to achieve sufficient performance to automate the decision of timely countermeasures in the industrial operational setting. Furthermore, the addition of the convLSTM-layer results in fewer outlying predictions and a lower running variance of the fraction of detected slag in the image time series.

Keywords: Image segmentation · Deep neural network · Iron ore pelletising plant · Furnace slag-detection

1 Introduction

The mining industry is an ancient industry, with the first recognition of underground mining dating back to approximately 40,000 BC [5]. Historically, mining consisted of heavy labour without the aid of modern technology, but today the industry portrays a different story. Large machines and production lines are responsible for carrying out the mining process more efficiently and at scale. This is evident in the process of transforming raw iron ore into iron ore pellets – small iron ore "balls" used as the core component in steel production.

Iron ore pellets are produced in large pelletising plants, where iron ore concentrate is mixed with bentonite, a clay mineral acting as a binder. The ore and bentonite mix is then formed into pellets in a rotating drum. At the final stage of this process, the wet iron ore pellets are pre-heated and dried in a large furnace called the Grate-Kiln system. The Grate-Kiln system [14] consists of three main units: 1. Drying and pre-heating (Grate), 2. Heating (Kiln) and 3. Cooling.

When the pellets flow between the Grate and the Kiln, a common challenge in pelletising plants is that an unwanted by-product known as slag slowly builds

C. Bauckhage et al. (Eds.): DAGM GCPR 2021, LNCS 13024, pp. 559–573, 2021.
https://doi.org/10.1007/978-3-030-92659-5_36

up over time. The slag gets attached to the base (slas) of the Kiln and restricts the flow of the pellets, worsening the quality of the end-product and, if it breaks loose, it can potentially damage the equipment. Thus, it is of vital importance to track the slag build-up over time, as timely countermeasures have to be taken to remove the slag before it grows too large.

One solution used to track the build-up is to attach a video camera positioned to record the slag inside the furnace. The video stream is then manually and visually analysed by employees at the plant. While this process enables plant operators to get an instant snapshot of the current state of the slag, the process is not only labour-intensive but also prone to inconsistencies due to individual human biases. Moreover, it does not provide a historical systems-level overview of how the amount of slag has changed over time.

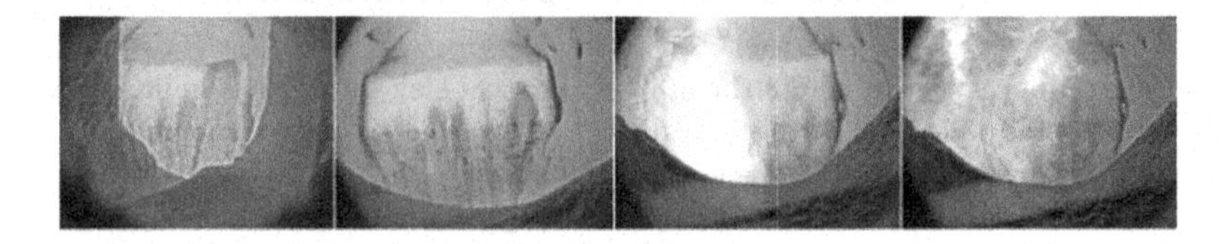

Fig. 1. Sample images with varying conditions from within the Kiln. The two images to the left are occlusion-free and the two images to the right contain occlusions.

Our aim is to research the possibility of measuring the amount of slag that has accumulated in the Kiln through image segmentation with deep convolutional neural networks (DCNNs). Furthermore, we want to assess the effects of utilising consecutive pairs of images in the time series of images when segmenting possibly occluded images (e.g. flames and smoke in the third and fourth images of Fig. 1). This will be done by introducing a convLSTM-layer in the DCNN. Our work contributes with the following novel findings:

1. We show that it is possible to measure the amount of slag that has accumulated in a Grate-Kiln system through image segmentation with DCNNs.
2. We show that it is possible to reduce variance in the predicted slag fraction by using consecutive pairs of images through a convLSTM-layer in the DCNN.
3. We demonstrate an approach for handling images containing occlusion in the video stream from the Grate-Kiln system. We believe that this approach can be applied to other applications where occasional obstructions of the camera view are considered problematic.
4. We have developed image segmentation models to predict slag formation through a scalable [24] delta lake architecture [2] and deployed in production, thus showing that our work is not only sufficient in theory, but also brings value in practise.

In Sect. 3, we describe the raw dataset and models used for conducting our experiments. In Sect. 4, we describe the first of our experiments which compares two DCNNs at the task of identifying slag formations in images. We show that it is possible to detect slag formations with both models, which to our knowledge has not been shown in the literature before. In Sect. 5, we analyse how utilising temporal information might be beneficial for lowering prediction variance and handling images with occlusions in them. Finally, we summarise our results and findings in Sect. 6.

2 Related Work

Image segmentation has been used successfully in a vast number of applications such as medical image analysis, video surveillance, scene understanding and others [17]. The most significant model and architecture contributions within the image segmentation field have been reviewed and evaluated by Minaee et al. [17], whose article covers more than 100 state-of-the-art algorithms for segmentation purposes, many of which are evaluated on image dataset benchmarks.

The use of image segmentation models in the field of mining is limited in the literature, but some research has been conducted on measuring the size of pellets and other mining-related particles through computer vision. Hamzeloo, et al. used a combination of Principal Component Analysis (PCA) and neural networks to estimate particle size distributions on industrial conveyor belts [8]. Support Vector Machines (SVMs) have been successfully used to estimate iron ore green pellet size distributions in a steel plant [10] and a DCNN (U-Net architecture) has been proposed for the same purpose [6].

To our knowledge, no research has been conducted on detecting slag formation through computer vision within furnaces used in the mining industry. A reason for this might be due to the Grate-Kiln system previously being regarded as a black-box process [26]. Gathering data through measuring devices that can withstand temperatures above 1000 °C has been difficult historically.

When evaluating videos or sequences of images, where potential insights from temporal information can be obtained, there are multiple approaches that can be used. For example, one can concatenate the current frame with the previous N frames to a common input and process these frames with an ordinary CNN. However, Karpathy et al. [12] showed that this only results in a slight improvement and that the use of a temporal model proved to be a more successful approach. Multiple authors have since then proposed the use of recurrent neural networks (RNNs) when performing image segmentation on videos or sequences of images. For example, Valipour et al. [22] proposed a modification of a Fully Convolutional Network (FCN) [15] with an added RNN unit between the encoder and decoder, achieving a higher performance on the Moving MNIST dataset [20] and other popular benchmarks. Yurdakul et al. [7] researched different combinations of convolutional and recurrent structures, finding that an additional convLSTM cell yielded the best result. Pfeuffer et al. [18] investigated where to place the convLSTM cell in the DCNN model architecture, finding that the largest increase in performance was achieved by placing it just before the final softmax layer. We use insights from such recent work on image segmentation, but with novel adaptations to our problem of slag detection above 1300 °C.

3 Raw Dataset and Models

In this section, the raw dataset and the deep convolutional neural network (DCNN) models for image segmentation used in our experiments are introduced.

3.1 Raw Dataset

All the data used in this report was supplied by LKAB (a Swedish mining company) and collected from one of their pelletising plants located in Kiruna, Sweden. The dataset at our disposal consisted of a stream or time series of 640 × 480 pixel RGB images. Each image was captured inside the Kiln with a 10-second time lag, and included metadata about when each image had been taken. The state of the furnace as well as the angle and position of the camera affect the characteristics of the images. For example, the furnace might be in its initial stages of heating causing a different lighting than when it has just been turned off. Furthermore, the camera sometimes had its position moved due to the vibrations within the furnace, causing the previous landscape to change. Lastly, the images captured sometimes contain a flame, used for heating the Kiln, with subsequent smoke. Both flame and smoke can occlude the view of the slag in the image. The images were ingested into a delta lake [2] to easily train our models.

3.2 Models

U-Net. The first model implemented in this project was the U-Net [19]. The U-Net is an early adaptation of the encoder-decoder architecture outperforming other state-of-the-art architectures of that time and field by a large margin. The U-Net produces high accuracy segmentation while only requiring a small training dataset. By utilising the feature maps of the encoder combined with the up-sampled feature maps of the decoder, localisation of features can be made in a higher resolution, thus increasing performance.

PSPNet. The second model implemented for our experiments was the Pyramid Scene Parsing Network (PSPNet) [25]. The PSPNet is a network that aims to utilise the global prior of a scene by downsampling the image into different sized sub-regions through average pooling. By doing this, both global and local information of each pixel can be exploited.

The network uses a backbone of a Residual Network (ResNet) [9] with a dilated strategy, i.e., replacing some of its convolutional operations with dilated convolutions and a modified stride setting, as described in [4,23]. ResNets are based on the idea of a residual learning framework that allows training of substantially deeper networks than previously explored [9]. Following the backbone, the architecture uses its proposed pyramid pooling module, a concatenating step, and finally a convolutional layer to create the final predicted feature maps.

4 Detecting Slag Accumulation with Deep Neural Networks

Here, we discuss the experiment with the aim of researching the potential of measuring slag accumulation using image segmentation with the U-Net and the PSPNet, two DCNN models.

4.1 Dataset

To train the DCNN models, a subset of the raw images was extracted. These images were manually and carefully chosen to ensure a high variance that is representative of the full distribution of images. Furthermore, the images chosen did not contain any occlusions to reduce interference and noise in the training process. Finally, the subset of images were manually labelled into four classes using the COCO-annotator tool [3]: *background*, *slag*, *camera edge* and *wall* (see Fig. 2). The classes were chosen based on discussions with domain experts at LKAB. *Background* and *slag* were needed to compute the percentage of slag stuck on the flat surface of the Kiln (i.e. the slas). *Camera edge* was needed to alert local plant maintenance when the camera hole needed cleaning. *Wall* simply represented the rest of the pixels. This resulted in a dataset consisting of 530 images, with 339 used for training, 85 for validation, and 106 for testing.

4.2 Implementation Details

U-Net. The U-Net implemented in our work was built as described in the original paper, with the exception of using less filters in its layers. The number of filters used was reduced by a factor of 16 in comparison to the original architecture, e.g., the first layer of the original architecture used 64 filters and our implementation used 4 filters in the same layer. The decision of using fewer filters in the implementation was made since no significant improvement in performance was observed by using the full-scale U-Net.

PSPNet. The PSPNet was built with a backbone of ResNet-50, a version of the ResNet model [9] with 50 convolutional layers. This was done by utilising the pre-built ResNet-50 model available in TensorFlow [21], with slight modifications to suit the dilated approach of the PSPNet, as described in [4,23]. The weights of the backbone ResNet-50 was initialised using pre-trained weights, trained on the ImageNet dataset.

Loss Functions and Class-Weighting Schemes. When analysing the class distribution of the dataset defined in Sect. 4.1, it became evident that some classes were under-represented, causing a *class imbalance problem*; see Fig. 2. Class imbalance within a dataset has been shown to decrease performance for some classifiers, including deep neural networks, causing the model to be more

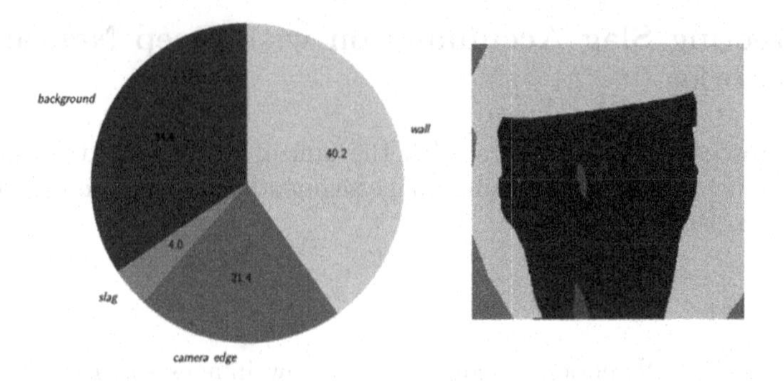

Fig. 2. Visualisation of the skewed class distribution present in the dataset used for training the models in experiment 1. The classes in the image are background (purple), slag (blue), camera edge (green) and wall (yellow). (*Left*): Class distribution of the data. (*Right*): Example ground truth image showing the skewed distribution of pixels. (Color figure online)

biased towards the more frequently seen classes [11]. To tackle this problem, we experimented with three loss functions; the *Tanimoto loss* [5], the *dice loss* [16], and the *cross-entropy loss*. We also experimented with two class-weighting schemes; the *inverse square volume weighting* (ISV) and the *inverse square root volume weighting* (ISRV), in comparison to no class-weighting scheme. The ISV weighting scheme was calculated by $w_c = 1/f_c^2$ and the ISRV weighting scheme was calculated by $w_c = 1/\sqrt{f_c}$, for each segmentation class c, where f_c denotes the frequency of pixels labelled as class c in the dataset used for training the model. The ISV weighting scheme together with the Tanimoto loss as well as the ISRV weighting scheme together with the cross-entropy loss function have been successfully used for class imbalance problems [1,5]. We found that the best performance in our trials was achieved by the dice loss with ISV and the cross-entropy loss with ISRV for the U-Net and the PSPNet models, respectively. For more details on ablation studies please see Sect. 5.2 of [13].

4.3 Evaluation Metrics

Pixel Accuracy. The most common evaluation metric used for classification problems is accuracy. For image segmentation tasks, pixel accuracy is frequently used. It is defined as the proportion of correctly classified pixels, as follows:

$$\text{accuracy} := \frac{\text{TP} + \text{TN}}{\text{TP} + \text{FP} + \text{TN} + \text{FN}}, \tag{1}$$

where TP, FP, TN and FN refer to the *true positives*, *false positives*, *true negatives* and *false negatives*, respectively.

Intersection over Union. One of the most commonly used metrics for image segmentation is the *intersection over union*, denoted by IoU [17]. This metric measures the overlap of pixel classifications, making it less sensitive to class imbalances within the image, in comparison to accuracy. The equation for IoU for each class c is defined below, where A denotes the predicted segmentation map over the classified pixels and B denotes the ground truth segmentation map.

$$\mathrm{IoU}_c := \frac{|A \cap B|}{|A \cup B|} \tag{2}$$

The IoU of the class labelled *slag* was used for validation since this was the class of interest in this problem.

Mean Intersection over Union. The *mean intersection over union* metric, denoted by mIoU, averages over the IoU metric for each of the classes and is defined as follows:

$$\mathrm{mIoU} := \frac{1}{|\mathbb{C}|} \sum_{c \in \mathbb{C}} \mathrm{IoU}_c \ , \tag{3}$$

where $|\mathbb{C}|$ is the number of classes in set $\mathbb{C}$ and IoU_c is the IoU for class c.

4.4 Results and Analysis

Table 1. Class-wise IoU_c scores of the slag detection experiment with mIoU and accuracy calculated over all classes. All metrics are computed on the test set.

Model	Class c	IoU_c(%)	mIoU(%)	accuracy(%)
PSPNet	*background*	94.06	87.72	97.32
	slag	65.33		
	camera edge	94.63		
	wall	96.85		
U-Net	*background*	92.54	85.37	96.38
	slag	62.67		
	camera edge	91.61		
	wall	94.67		

The results of this experiment are presented in Table 1. The results show that both models are able to segment the images in the test set, especially for the more frequent class labels: *background*, *camera edge* and *wall*. Both the U-Net and PSPNet are also able to segment slag at an IoU of 62.67% and 65.33%, respectively, thus showing that it is possible to utilise image segmentation when detecting slag. This was also visually verified by domain experts at LKAB, verifying that the achieved IoU of slag was sufficient in an industry setting.

The lower IoU for the class *slag* in comparison to the other classes we believe is due to the class imbalance problem, mentioned in Sect. 4.2. Another reason could be an effect of the manual labelling process, as it is sometimes difficult for the human eye to detect the slag islands due to their small size and sometimes indistinguishable edges. This is evident in the worst prediction yielded by the PSPNet visualised in Fig. 3, which also shows that even though the prediction of the model yielded a low IoU score in this case, the *amount* of slag predicted can be fairly accurate.

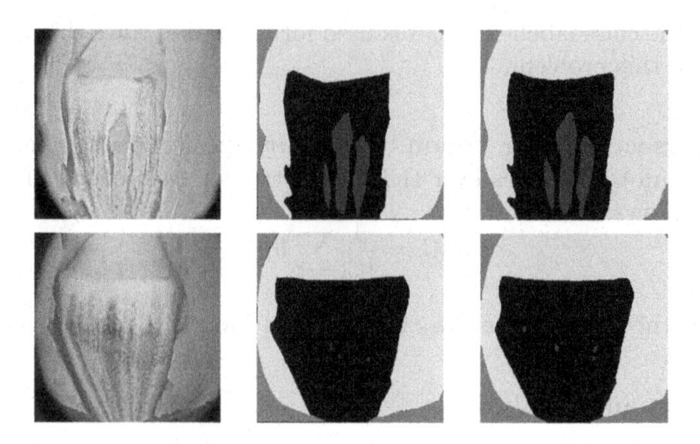

Fig. 3. Two examples of predictions yielded by the PSPNet with weighted loss. (*Top*): The best prediction, yielding 86.8% IoU of *slag*. (*Bottom*): The worst prediction, yielding 8.4% IoU of *slag*. (*Left*): The input images. (*Middle*): The ground truth images. (*Right*): The predictions made by the PSPNet.

Although both of the models showed potential in segmenting the images, the superiority of the PSPNet over the U-Net in this particular context is evident from looking at the results in Table 1, as it outperforms the U-Net in all of the metrics measured. Figure 3 displays some example predictions made by the PSPNet which we believe shows, together with the results presented in Table 1, that it is possible to measure the amount of slag that has accumulated in a Grate-Kiln system through image segmentation with a DCNN. We will in the next experiment look into whether the better performing PSPNet is consistent in its predictions by analysing its performance on unlabelled images over a longer time-frame.

5 Variance and Occlusion Experiment

Here, we present the experiment with the aim of researching the performance in a productionised setting. It includes analysing the prediction variance and the ability to deal with occluded images hindering the view of the camera.

5.1 Datasets

Training Dataset. The training dataset, used in Sect. 4, was extended by adding the subsequently captured images, captured with a 10 s time lag. This resulted in a time series dataset containing consecutive pairs of images. As the images captured in the Kiln sometimes include flames and subsequent smoke obstructing the view, the resulting dataset only contained 420 pairs of images, i.e., 110 out of the 530 subsequent images were discarded due to the presence of an occlusion. The newly added images were labelled in the same manner as was described in the first experiment; see Sect. 4.1. Out of the 420 pairs of images, 269 pairs were used for training, 67 used for validation, and 84 used for testing. After testing the performance of the models on unseen data, we retrained the models prior to evaluation on the following *evaluation datasets*, using 336 pairs for training and 84 for validation.

Evaluation Datasets. To evaluate the robustness of the models in a productionised setting, raw consecutive images taken in sequence from two separate days were extracted into two datasets, containing 7058 and 8632 images respectively. The two days were chosen because of their different characteristics. One of the days contained images that showed a slow and gradual slag build-up, whereas the other day contained images where it was known beforehand that slag had been removed throughout the day, resulting in a varying slag build-up. These images were not labelled.

5.2 Implementation Details

Occlusion Discriminator. In order to evaluate the DCNN models on the unlabelled evaluation datasets, the images input to the models needed to be occlusion-free. To replicate a production setting and to reduce the manual work of removing images containing occlusions, an *occlusion discriminator* model was implemented and trained. The model was a DCNN classifier of three convolutional blocks, each block containing two convolutional layers each of which was followed by a batch normalisation and ReLU activation functions. Each convolutional block was then followed by a max pooling and dropout layer. Finally, a dense layer was followed by a sigmoid function, yielding binary predictions on whether the image contains occlusions or not. After training the model, it achieved a precision and recall score of 99.7% and 83.7% respectively. This model was used to filter the evaluation datasets.

Two PSPNet Models. The best performing hyperparameters of the PSPNet from the previous experiment, described in Sect. 4, was used in this experiment. It was trained using the cross-entropy loss with the ISRV class-weighting scheme, and is referred to as the PSPNet. The PSPNet was evaluated in comparison to a modified version of the same PSPNet model, referred to as the PSPNet-LSTM. The PSPNet-LSTM was modified by replacing the final convolutional layer with a convLSTM-layer with two cells, with shared weights, to enable the model to process pairs of images. Furthermore, the layers prior to the convLSTM-layer were implemented to ensure that both of the input images were processed by the same weights.

The convLSTM-layer utilised information from a sequence of images. When predicting a segmentation mask of an image I_t of time step t, it used information from both I_t and the hidden state of the previous LSTM cell utilising information from $I_{t-\Delta}$, with Δ set to 10 s in our experiments. The output of the PSPNet-LSTM was a segmentation mask prediction of the image I_t. See a visualisation of the model in Fig. 4. In summary, these modifications enabled the PSPNet-LSTM to train on pairwise images.

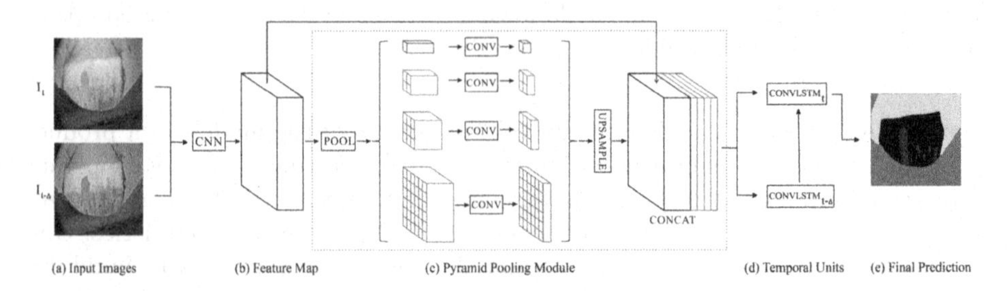

Fig. 4. Visualisation of the PSPNet-LSTM, our modified PSPNet with convLSTM cells. The images I_t and $I_{t-\Delta}$, seen in (a), are both processed by the same weights in the network. The feature maps produced from I_t are fed into the convLSTM unit at time step t, and the feature maps produced from $I_{t-\Delta}$ are fed into the convLSTM unit at time step $t - \Delta$. Finally, a segmentation map prediction of the image I_t is made.

To be able to compare the performance of the two models, the PSPNet model was retrained using the images (only I_t) defined in Sect. 5.1. The weights of PSPNet-LSTM were then initialised using the frozen pre-trained weights from the PSPNet. By using the same weights in both models and only training the additional convLSTM-layer in PSPNet-LSTM (using both I_t and $I_{t-\Delta}$), a comparison between the predictive ability of the two models on the images, with and without temporal information, could be made.

5.3 Evaluation Metrics

IoU, Fraction of Classified Slag and Running Variance. The IoU of the class *slag*, defined in (2), was used for validation and testing, when training the PSPNet. Since the purpose of this experiment was to evaluate the stability of the models in a productionised setting, raw images from two days were used for evaluation, defined in Sect. 5.1. These images were not labelled, thus the IoU could not be used. As a measurement of stability over time, the fraction of classified slag was used for evaluation:

$$\mathfrak{f}_t = \frac{n_s^{(t)}}{H \times W} \tag{4}$$

where $n_s^{(t)}$ is the number of pixels classified as slag and H and W represent the dimensions of the image at time t.

The running variance was calculated for each measurement of the fraction of classified slag at time t, i.e., $\mathfrak{f}_t$, over a fixed window of size k, prior to t. We set k to 60 amounting to 600 s or 10 min of the natural operational window for counter-measures..

$$\text{Variance}(\mathfrak{f}_{t-k+1} + \mathfrak{f}_{t-k+2} + ... + \mathfrak{f}_t) \tag{5}$$

5.4 Results and Analysis

When comparing the best performing PSPNet with the best performing PSPNet-LSTM, evaluated on the unseen test data held out during training (Sect. 5.1), the PSPNet outperformed the PSPNet-LSTM with a small margin in terms of the IoU for the class *slag*, yielding 62.39% and 61.76% respectively. The models were then evaluated on the two days of images to measure their robustness in a productionised setting by using the running variance of slag fraction—a domain-specific metric that can highlight discrepancies in performance by manual examination before operational countermeasures.

Figure 5 shows the predicted fraction of slag on the raw evaluation dataset (two full days of unlabelled images) and the corresponding running variance. The figures show that the predictions yielded by the PSPNet-LSTM has fewer outliers and is more consistent compared to the PSPNet predictions. Furthermore, the running variance is consistently lower for the PSPNet-LSTM during the period when the slag is slowly accumulating. A low and stable running variance is preferred since it should resemble the slow process of slag build-up.

We believe that the reason for the more stable predictions made by the PSPNet-LSTM is its ability to produce an adequate segmentation map surrounding one occluded image where it can partly disregard an occlusion obstructing the view. Even with an occlusion discriminator in production filtering out most occlusions, false negatives are likely to occur when used over longer periods of time. Figure 6 shows an image that was wrongly classified and passed through the trained occlusion discriminator filter (Sect. 5.2). Looking at the predictions made by the two models, combined with the fewer outlying predictions of the PSPNet-LSTM in Fig. 5, we arrive at the conclusion that the PSPNet-LSTM is more robust in its predictions and manages to predict slag regions behind the smoke, which obstructs the view, even though it has not been trained on images containing occlusions.

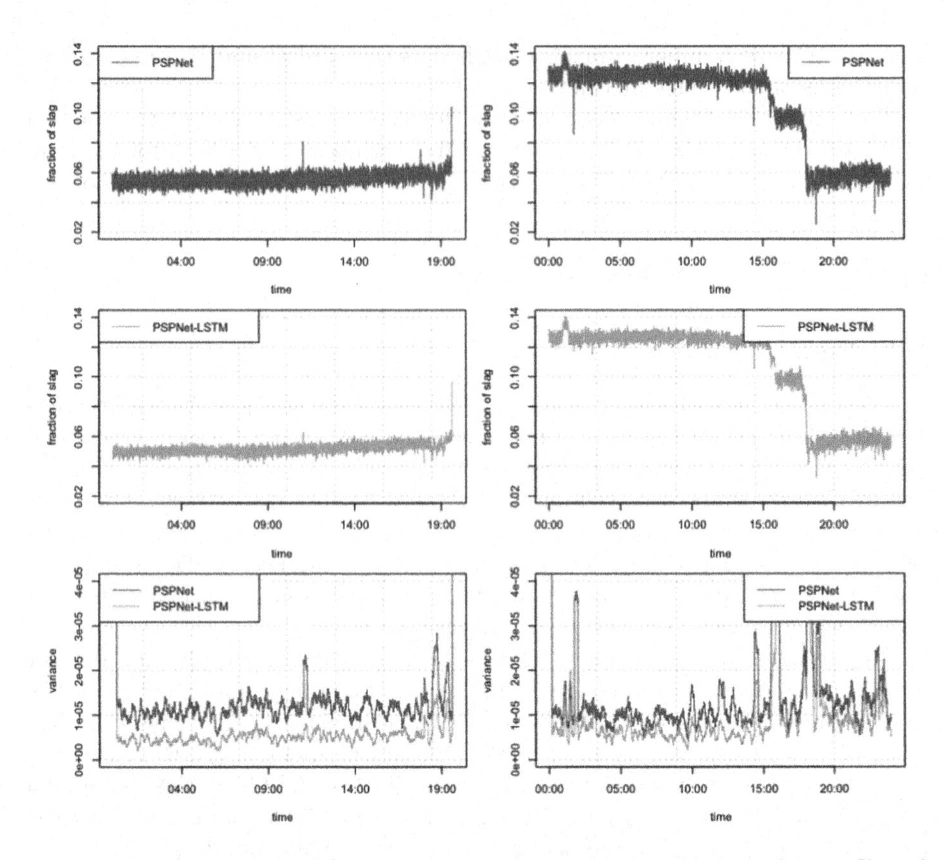

Fig. 5. Predicted fraction of slag and running variance on the evaluation dataset filtered by the occlusion discriminator.

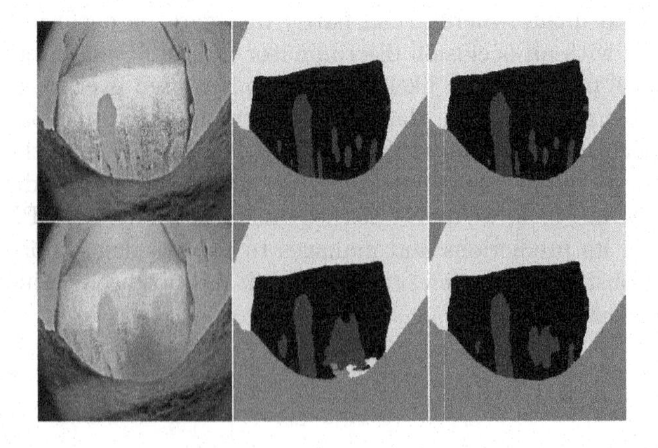

Fig. 6. (Top row): Image, $I_{t-\Delta}$, with no visible occlusion. (Bottom row): Consecutive image, I_t, with a visible occlusion that passed through the occlusion discriminator. (*Left*): The input image. (*Middle*): The prediction made by the PSPNet. (*Right*): The prediction made by the PSPNet-LSTM.

6 Conclusions

In this paper, we have researched the possibilities of measuring slag accumulation with the use of two DCNNs, the U-Net and the PSPNet. Furthermore, we have looked into how to handle images containing occlusions and proposed a novel approach to tackle this problem. We have shown that it is possible to measure slag accumulation in a Grate-Kiln system using a DCNN. The best performing model, the PSPNet, yielded an IoU of *slag* of 65.33%, an mIoU of 87.72% and an accuracy of 97.32%, all measured on test data. Finally, we have shown that a DCNN with an additional convLSTM-layer with two cells can increase the stability and lower the variance of the predictions by exploiting temporal information in consecutive images, thus making it more suitable for a production environment.

Based on interviews with domain experts at LKAB, having access to continuous measurement of slag formation will support their effort in putting in timely countermeasures to remove slag, as well as getting a better understanding of when enough slag has been removed. Additionally, having a historical and systemic view of slag formation will enable LKAB to correlate the slag build-up with other parts of the Grate-Kiln system – an important step towards understanding the root cause of the slag build-up. Such a view can be obtained by LKAB due to the raw images over multiple years being made readily available in a delta lake house to quickly build new AI models over GPU-enabled Apache Spark [24] clusters, and by integrating image data with other data sources.

Future work will focus on how to further enhance the ability of the model to deal with occluded images. We believe that this could be done by utilising longer sequences of images when making predictions. This will increase the probability of having multiple consecutive images without any occlusion, thus hopefully aiding the model in making accurate predictions. Moreover, our image delta lake architecture, over long time scales, allows to take a systems-level approach to the problem by incorporating other dependent variables.

Acknowledgements. This research was partially supported by the Wallenberg AI, Autonomous Systems and Software Program funded by Knut and Alice Wallenberg Foundation and Databricks University Alliance with AWS credits. We thank Gustav Häger and Michael Felsberg at Computer Vision Laboratory, Department of Electrical Engineering, Linköping University for their support. Many thanks to Håkan Tyni, Peter Alex, David Björnström and the rest at LKAB for answering all of our questions and supplying us with the raw data. We are grateful to Ammar Aldhahyani for the custom illustration in Fig. 4 with the kind permission of Zhao to modify the original image [25].

References

1. Aksoy, E.E., Baci, S., Cavdar, S.: Salsanet: fast road and vehicle segmentation in lidar point clouds for autonomous driving. In: 2020 IEEE Intelligent Vehicles Symposium (IV), pp. 926–932 (2020). https://doi.org/10.1109/IV47402.2020.9304694

2. Armbrust, M., et al.: Delta lake: high-performance acid table storage over cloud object stores. Proc. VLDB Endow. **13**(12), 3411–3424 (2020). https://doi.org/10.14778/3415478.3415560
3. Brooks, J.: COCO Annotator (2019). https://github.com/jsbroks/coco-annotator/
4. Chen, L.C., Papandreou, G., Kokkinos, I., Murphy, K., Yuille, A.L.: Semantic image segmentation with deep convolutional nets and fully connected crfs (2016)
5. Diakogiannis, F.I., Waldner, F., Caccetta, P., Wu, C.: Resunet-a: a deep learning framework for semantic segmentation of remotely sensed data. ISPRS J. Photogramm. Remote Sens. **162**, 94–114 (2020). https://doi.org/10.1016/j.isprsjprs.2020.01.013, https://www.sciencedirect.com/science/article/pii/S0924271620300149
6. Duan, J., Liu, X., Mau, C.: Detection and segmentation of iron ore green pellets in images using lightweight u-net deep learning network. Neural Comput. Appl. **32**, 5775 – 5790 (2020). https://doi.org/10.1007/s00521-019-04045-8, https://link.springer.com/article/10.1007/s00521-019-04045-8
7. Emre Yurdakul, E., Yemez, Y.: Semantic segmentation of rgbd videos with recurrent fully convolutional neural networks. In: Proceedings of the IEEE International Conference on Computer Vision (ICCV) Workshops (October 2017)
8. Hamzeloo, E., Massinaei, M., Mehrshad, N.: Estimation of particle size distribution on an industrial conveyor belt using image analysis and neural networks. Powder Technol. **261**, 185 – 190 (2014). https://doi.org/10.1016/j.powtec.2014.04.038, http://www.sciencedirect.com/science/article/pii/S0032591014003465
9. He, K., Zhang, X., Ren, S., Sun, J.: Deep residual learning for image recognition. In: Proceedings of the IEEE Conference on Computer Vision and Pattern Recognition (CVPR) (2016). 6
10. Heydari, M., Amirfattahi, R., Nazari, B., Rahimi, P.: An industrial image processing-based approach for estimation of iron ore green pellet size distribution. Powder Technol. **303**, 260 – 268 (2016). https://doi.org/10.1016/j.powtec.2016.09.020, http://www.sciencedirect.com/science/article/pii/S0032591016305885
11. Japkowicz, N., Stephen, S.: The class imbalance problem: a systematic study. Intell. Data Anal. **6**, 429–449 (2002). https://doi.org/10.3233/IDA-2002-6504
12. Karpathy, A., Toderici, G., Shetty, S., Leung, T., Sukthankar, R., Fei-Fei, L.: Large-scale video classification with convolutional neural networks. In: Proceedings of the IEEE Conference on Computer Vision and Pattern Recognition (CVPR) (June 2014)
13. von Koch, C., Anzén, W.: Detecting Slag Formation with Deep Learning Methods: An experimental study of different deep learning image segmentation models. Master's thesis, Linköping University (2021)
14. LKAB: Pelletizing (2020). https://www.lkab.com/en/about-lkab/from-mine-to-port/processing/pelletizing/
15. Long, J., Shelhamer, E., Darrell, T.: Fully convolutional networks for semantic segmentation. In: Proceedings of the IEEE Conference on Computer Vision and Pattern Recognition (CVPR) (June 2015)
16. Milletari, F., Navab, N., Ahmadi, S.: V-net: fully convolutional neural networks for volumetric medical image segmentation. In: 2016 Fourth International Conference on 3D Vision (3DV), pp. 565–571 (2016). https://doi.org/10.1109/3DV.2016.79
17. Minaee, S., Boykov, Y.Y., Porikli, F., Plaza, A.J., Kehtarnavaz, N., Terzopoulos, D.: Image segmentation using deep learning: a survey. IEEE Trans. Pattern Anal. Mach. Intell. 1 (2021). https://doi.org/10.1109/TPAMI.2021.3059968

18. Pfeuffer, A., Schulz, K., Dietmayer, K.: Semantic segmentation of video sequences with convolutional lstms. In: 2019 IEEE Intelligent Vehicles Symposium (IV), pp. 1441–1447 (2019). https://doi.org/10.1109/IVS.2019.8813852
19. Ronneberger, O., Fischer, P., Brox, T.: U-Net: convolutional networks for biomedical image segmentation. In: Navab, N., Hornegger, J., Wells, W.M., Frangi, A.F. (eds.) MICCAI 2015. LNCS, vol. 9351, pp. 234–241. Springer, Cham (2015). https://doi.org/10.1007/978-3-319-24574-4_28
20. Srivastava, N., Mansimov, E., Salakhudinov, R.: Unsupervised learning of video representations using lstms. In: International Conference on Machine Learning, pp. 843–852 (2015)
21. TensorFlow: tf.keras.applications.resnet50.ResNet50 (2021). https://www.tensorflow.org/api_docs/python/tf/keras/applications/resnet50/ResNet50
22. Valipour, S., Siam, M., Jagersand, M., Ray, N.: Recurrent fully convolutional networks for video segmentation. In: 2017 IEEE Winter Conference on Applications of Computer Vision (WACV), pp. 29–36 (2017). https://doi.org/10.1109/WACV.2017.11
23. Yu, F., Koltun, V.: Multi-scale context aggregation by dilated convolutions (2016)
24. Zaharia, M., et al.: Apache spark: a unified engine for big data processing. Commun. ACM **59**(11), 56–65 (2016). https://doi.org/10.1145/2934664
25. Zhao, H., Shi, J., Qi, X., Wang, X., Jia, J.: Pyramid scene parsing network. In: Proceedings of the IEEE Conference on Computer Vision and Pattern Recognition (CVPR) (July 2017)
26. Zhu, D., Zhou, X., Luo, Y., Pan, J., Zhen, C., Huang, G.: Monitoring the ring formation in rotary kiln for pellet firing. In: Battle, T.P., et al. (eds.) Drying, Roasting, and Calcining of Minerals, pp. 209–216. Springer, Cham (2015). https://doi.org/10.1007/978-3-319-48245-3_26

Virtual Temporal Samples for Recurrent Neural Networks: Applied to Semantic Segmentation in Agriculture

Alireza Ahmadi(✉), Michael Halstead, and Chris McCool

University of Bonn, Nussallee 5, 53115 Bonn, Germany
{alireza.ahmadi,michael.halstead,cmccool}@uni-bonn.de
http://agrobotics.uni-bonn.de

Abstract. This paper explores the potential for performing temporal semantic segmentation in the context of agricultural robotics without temporally labelled data. We achieve this by proposing to generate *virtual* temporal samples from labelled still images. By exploiting the relatively static scene and assuming that the robot (camera) moves we are able to generate virtually labelled temporal sequences with no extra annotation effort. Normally, to train a recurrent neural network (RNN), labelled samples from a video (temporal) sequence are required which is laborious and has stymied work in this direction. By generating virtual temporal samples, we demonstrate that it is possible to train a lightweight RNN to perform semantic segmentation on two challenging agricultural datasets. Our results show that by training a temporal semantic segmenter using virtual samples we can increase the performance by an absolute amount of 4.6 and 4.9 on sweet pepper and sugar beet datasets, respectively. This indicates that our *virtual* data augmentation technique is able to accurately classify agricultural images temporally without the use of complicated synthetic data generation techniques nor with the overhead of labelling large amounts of temporal sequences.

Keywords: Temporal data augmentation · Spatio-temporal segmentation · Agricultural robotics

1 Introduction

In recent years, agricultural robotics has received considerable attention from the computer vision, robotics, and machine learning communities due, in part, to its impact on the broader society. Agricultural applications, such as weeding and harvesting are demanding more automation due to the ever increasing need for both quality and quantity of crops. This increased attention has partly driven the development of agricultural platforms and expanded their capabilities, including quality detection [9], autonomous weeding [26] and automated crop harvesting [17].

C. Bauckhage et al. (Eds.): DAGM GCPR 2021, LNCS 13024, pp. 574–588, 2021.
https://doi.org/10.1007/978-3-030-92659-5_37

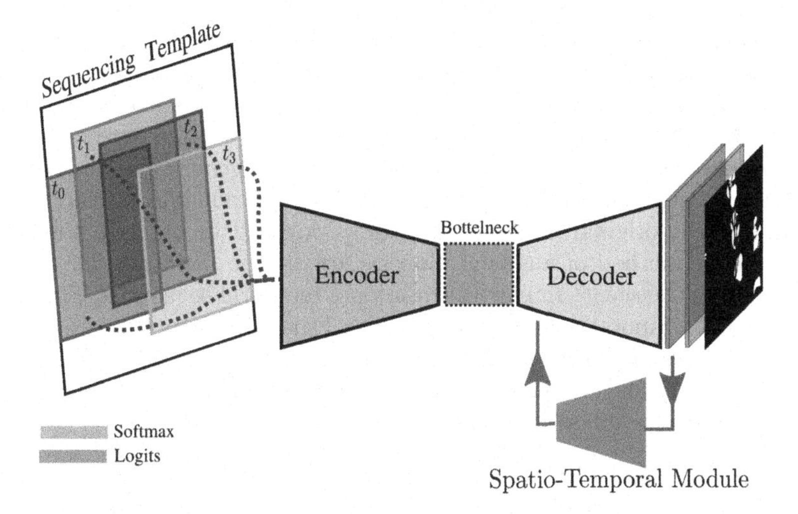

Fig. 1. Overall structure of our proposed spatio-temporal network. The input to this network is being drawn out of the sequencing template, considering the order of virtual sequences based on direction of motion and time. The spatio-temporal module feeds back feature maps to network and forces network to learn frames dependency.

An advantage of surveying an agricultural scene over a pedestrian scene is the structured and relatively static nature of crops, particularly with respect to the moving platform. While the scene is somewhat static and structured there are a number of challenges an automatic agent needs to overcome due to the complicated nature of the scene, including, illumination variation, and occlusion. Recent advances in agricultural robotics have exploited convolutional neural networks (CNNs) [15] to both alleviate some of these challenges and achieve high performance.

Modern machine learning techniques rely on CNNs to perceive useful visual information about a scene. Most of the successful deep learning techniques utilize a paradigm of multi-layer representation learning from which semantic segmentation [15,16] has evolved. These segmentation networks are able to classify on a pixel level [18] the appearance of a specific class, creating class based output maps. From a spatio-temporal perspective RNNs [32] are able to exploit previous information to improve performance in the current frame. Despite these advances, only feed-forward networks have been predominately used to generate the network parameters in each layer [5]. In Agriculture one of the few examples which uses spatio-temporal information is [19] where they exploit the regular planting intervals of crop rows. However, this approach is still ill suited to more generalised agricultural environments without such structure like fruit segmentation in horticulture. In this paper, we implement a lightweight RNN based on the UNet [23] architecture by employing "feedback" in the decoder layers. These "feedback" layers are used to perceive spatio-temporal information in an effort to improve segmentation accuracy.

We demonstrate that it is possible to train an RNN by generating virtual temporal sequences from annotated still images. This is important as the majority of agricultural datasets either do not contain temporal information or perform sparse labelling of the frames (consecutive image frames are rarely annotated). While creating annotations between the labelled frames can be achieved through weakly labelled techniques [24], these approaches are often noisy and introduce unwanted artifacts in the data. We are able to generate virtual temporal sequences by augmenting the annotated still images via successive crops and shifts inside the original annotated image as outlined in Fig. 1. By utilising a crop and shift method of augmentation we are able to maintain the structural information of the scene, similar to traditional temporal sequences. The validity of this virtual temporal sequence approach and the RNN structure are evaluated using two annotated agricultural datasets which represent vastly different scenes (field vs glasshouse), motions, crops, and camera orientations; see Sect. 5.1.

In this work the following contributions are made:

1. We propose a method for generating virtual temporal sequences from a single annotated image, creating spatio-temporal information for training an RNN model;
2. We explore different methods to perceive spatio-temporal information in an RNN-UNet structure and show that a convolution based module outperforms feature map downsizing through bi-linear interpolation;
3. Our proposed lightweight RNN architecture along with our novel data augmentation technique is able to improve semantic segmentation performance on different datasets regardless of the distribution of objects in the scene and considerably different nature of the scenes.

This paper is organised in the following manner: Sect. 2 reviews the prior work; Sect. 3 discusses our proposed virtual samples generation method; the proposed temporal approach is explained in Sect. 4; Sect. 5 describes our experimental setup and implementation details; the results are detailed in Sect. 6; and finally the conclusions Sect. 7.

2 Related Works

Recently, the problem of semantic segmentation has been addressed by a number of different approaches including [20,23,29]. Most of the state-of-the-art approaches used fully convolutional networks (FCNs) [18] as an end-to-end trainable, pixel-level classifier [25]. However, in an agricultural robotic setting a number of extra challenging factors need to be overcome including: non-static environments, highly complex scenes and significant variation in illumination.

In an effort to circumvent these factors researchers have enhanced standard semantic segmentation by embedding spatio-temporal information [10,30] into their architectures. In these cases, where spatio-temporal information was available, integrating this information improved performance [1,19]. Furthermore, Jarvers and Neumann [12] found that by incorporating sequential information, errors which occur in one frame could be recovered in subsequent frames.

In an agricultural context, Lottes *et al.* [19] improved their semantic segmentation network performance by incorporating spatial information about the sequence of images. By exploiting the crop arrangement information (geometric pattern of sowing plants in the field) they improved segmentation performances. While this created promising results in field settings the geometric pattern assumption does not hold in agriculture, consider fruit segmentation in horticulture.

In [32], the authors proposed RNNs as a method of reliably and flexibly incorporating spatio-temporal information. A key benefit witnessed by most spatio-temporal techniques was the layered structure of FCNs, which provide the opportunity to embed "feedback" layers in the network.

By embedding these "feedback" layers researchers can make use of the extra content and context provided by the multiple views (observations) of the same scene. This embedding acts on the system by biasing future outputs at $t+1$ based on the current output at t. Benefits of this method were found to hold when spatio-temporal dependencies exist between consecutive frames [28]. By comparing feed-forward and feed-backward networks [12] showed the later was able to increase the receptive fields of the layers. Ultimately, "feedback" layers provide richer feature maps to enable RNN-based systems to improve predictions.

Long Short-Term Memory (LSTM) [11] based architectures have also been shown to improve classification tasks by integrating temporal information [2,21]. These LSTM cells can be inserted into the network to augment performance, however, they add significant complexity to the network.

Another method used to improve the generalisability of networks is data augmentation. Kamilaris et al. [13] provided a comprehensive survey of early forms of data augmentation used by deep learning techniques. Generally, these methods include rotations, partitioning or cropping, scaling, transposing, channel jittering, and flipping [31]. Recent advances in data augmentation have lead to generative adversarial networks (GANs) which have the ability to generate synthetic data [4], enhancing the generalisability of the trained models. However, this adds further complexity to the pipeline as generating synthetic data can often be a time consuming exercise.

Generally, recent advances in machine learning have been made by using large labelled datasets such as ImageNet [3] which was used for object classification. These datasets only exist as labour is directed towards the particular task. This provides a major hurdle for deploying temporal approaches to novel domains such as agriculture.

To overcome the data requirements of spatio-temporal techniques, we make use of the partially labelled data of [27] and a newly captured dataset of sugar beets. Our proposed temporal data augmentation approach (see Sect. 3) generates virtual samples that only represent short-term temporal information, as such we do not explore LSTM-based approaches. This consists of dense (small spatial shifts between frames) temporal sequences that we augment and use to train a lightweight RNN architecture.

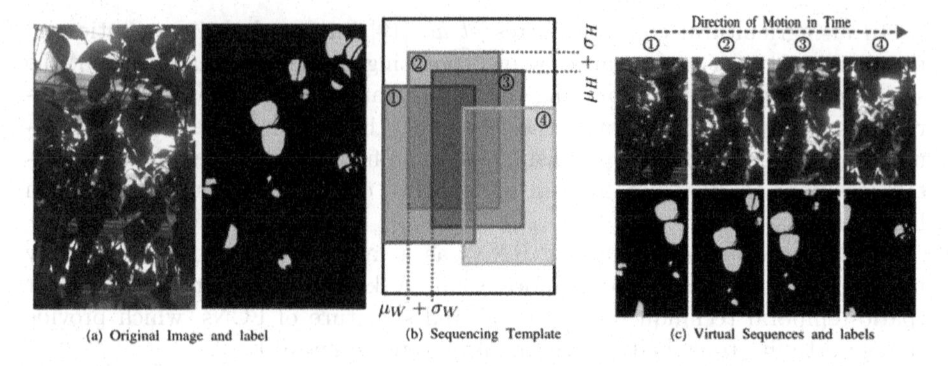

Fig. 2. An example virtual sequence generation. (a) Original image and label, (b) Sequencing template used for generating virtual frames with $N = 4$. Frame one is cropped at the position that will ensure N frames can be created without going beyond the border of the image, considering the means μ_W, μ_H and standard deviations σ_W, σ_H, (c) Generated virtual sequence with labels. (Color figure online)

3 Generating Virtual Spatio-Temporal Sequences

We propose that *virtual* sequences can be generated from an existing image and its annotation. The virtual sequences can simulate camera motion by performing consecutive crops and shifts as shown in Fig. 2. This has the advantage that the virtual sequence is fully annotated (per pixel) without having to perform laborious annotation or risk the propagation of noisy labels. While this approach is a simulation of camera motion and removes the natural occlusion witnessed in actual motion, it should reduce the requirement for mass annotation of large-scale datasets and provides fully labelled spatio-temporal data. Structuring motion in this manner (both the x and y direction) is important for simulating traditional temporal data on a robot where limited rotation occurs, for this reason we avoid other augmentation techniques such as up-down flipping and affine rotations. The aim of this data augmentation technique is to provide extra contextual information which enables a network to learn the relationship between consecutive frames and improve prediction accuracy. A limitation of this approach is that it assumes that the robot (camera) moves and that the scene is relatively static.

Virtual sequences are generated by employing a crop and shift technique based on manually obtained parameters. There are two sets of parameters that are needed. First, we need to estimate the movement parameters in the actual data (in image coordinate frames). Second, we need to define the number of frames N that will be generated for each *virtual* sequence. The number of frames N, sets the number of *virtual* samples to be simulated per image in the dataset, see Fig. 2. Once N is known we replicate the natural motion of the camera in the scene by cropping N fixed sized cropped images from the original image. Using the two sets of parameters, each cropped frame is then computed by,

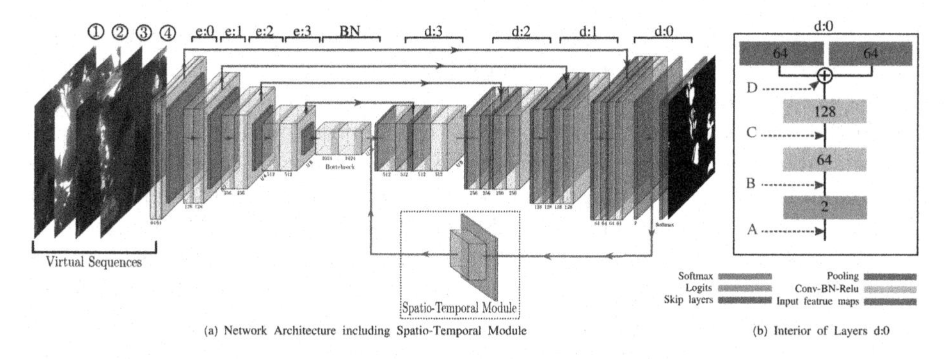

(a) Network Architecture including Spatio-Temporal Module (b) Interior of Layers d:0

Fig. 3. Network architecture. At the input, four sequences of virtually generated frames of a single image are passed to the network, separately. In each pass only one of the samples will be fed to the network (batch size = 1). The spatio-temporal module is feeding back 2 last layers (before the soft-max) to the layer *d:3-D*

$$\begin{aligned} W_n &= W_s + \mathcal{N}\left(\mu_w, \sigma_w^2\right) \lambda n, \\ H_n &= H_s + \mathcal{N}\left(\mu_h, \sigma_h^2\right) \gamma n. \end{aligned} \tag{1}$$

H_n, W_n denotes the new position of top-left corner of the cropped image in the original image, W_s, H_s represent the start position, and n is the frame index ranging from 0 to $N-1$. The "directions of travel" λ and γ are chosen between left-to-right ($\lambda = 1$), up-to-down ($\gamma = 1$) and right-to-left ($\lambda = -1$), down-to-up ($\gamma = -1$) for each annotated sample and are fixed for all frames within the sequence. This ensures no bias is introduced while training, considering the nature of motion in real data (platforms move in one direction). The directional bias is essential as the camera motion occurs in both directions for the BUP20 dataset (main motion in x direction), while SB20 only captured up-to-down motions (along the y axis).

4 Proposed Temporal Network

The primary research goal of this paper is to explore spatio-temporal relations of a CNN without the need for laborious manual labelling of temporal sequences. As such, we use a lightweight neural network structure based on the UNet [23] architecture; we also use UNet to produce baseline results for comparison.

From a spatio-temporal standpoint we implement feedback modules within the UNet architecture. We augment the UNet baseline architecture to allow different types of feedback within its current composition of layers. Figure 3-a outlines both the baseline system and the inclusion of the spatio-temporal module. While a number of techniques exist to provide feedback [12], we use layer concatenation to join the feedback with the feed forward channels. Concatenation was selected as it directly adds information to the feature map of the host layer, creating greater potential to learn direct relationships between the current and previous layers (frames).

In Fig. 3-a, all layers are assigned a specific name and number. Specifically, the n-th layer of the decoder (d) and encoder (e) are denoted to as *d:n* and *e:n* respectively and the bottleneck is denoted as BN. We access the sub-layers within the main layers of the network via the intervention points assigned with names ranging from A to D. For instance, layer *d:0* with its sub-divisions is depicted in Fig. 3-b, containing four intervention points. Each intervention point can be used as an extraction or insertion point for the spatio-temporal module.

A key complexity when using the different layers as a feedback to the network is the discrepancy between the two resolutions. The disparity between layer height and width when feeding back creates additional complexity and needs to be allowed for. For instance, in Fig. 3-a the feature maps at the sampling layer *d:0-A* with size of $2 \times H \times W$ are fed back to the first deconvolution layer after the bottleneck in *d:3-D* of the baseline architecture with size of $1024 \times H/16 \times W/16$. To alleviate this issue we propose two methods of modifying the feedback layer: 1) bi-linear interpolation (*Bi-linear*); and 2) $2D$ convolution re-sampling (*Conv*). The *Bi-linear* approach maintains the integrity of the feature map such that the re-sampled output is similar to the input; just down or up-sampled. By contrast, the *Conv* re-sampling adds the benefit of learning to re-sample as well as learning to transform the feature map. A limitation of the *Conv* approach is the added depth (complexity) required to produce the feedback feature map. The (*Conv*) re-sampling block consists of a 2D convolutional *block* with a 3×3 kernel (of stride 2 and padding 1) followed by batch normalization and a $ReLU$ activation function. In training and evaluation phases, at the start of a new temporal (*real* or *virtual*) sequence, the activation values of the spatio-temporal module are set back to $\mathbf{1}$ to avoid accumulating irrelevant information between non-overlapping frames.

5 Experimental Setup

We evaluate our proposed approach on two challenging agricultural datasets. The two datasets represent two contrasting scenarios, the first is a glasshouse environment with sweet pepper and the second is an arable farm with sugar beet. In both cases, the data was captured using robotic systems which allowed us to extract estimates of the motion information, we describe each dataset in more detail below.

When presenting results we compare algorithms with and without temporal information. First, we evaluate the impact of using the generated virtual sequences for still image segmentation. We use N virtual samples (per image) to train a non-recurrent network which we compare to a system trained on the full size images (baseline). This allows us to understand if increasing the number of samples at training, data augmentation using the N virtual samples, is leading to improved performance. Once again, we only consider crop and shift augmentation so we can directly compare the non-recurrent and recurrent models directly. Second, we evaluate the proposed *spatio-temporal* system using different feedback points, with a variable number of virtual samples N as well as against a

Fig. 4. Example image of dataset SB20 (left), same image with multi-class annotations representing different types of crops and weeds using different colors (middle), and (right) shows the crop weed annotation abstraction used in this paper.

classic RNN system where the last frame in the sequence is the only frame with annotations; only the output from the last frame has a label which can be used to produce a loss.

5.1 Datasets

The two datasets that we use contain video sequences captured from sweet pepper (BUP20) and sugar beet (SB20) fields respectively. Below we briefly describe each dataset.

BUP20 was captured under similar conditions to [8] and first presented in [27], however, it was gathered with a robot phenotyping platform. It contains two sweet pepper cultivar, *Mavera* (yellow) and *Allrounder* (red), and was recorded in a glasshouse environment. The dataset is captured from 6 rows, using 3 Intel RealSense D435i cameras recording RGB-D images as well as IMU and wheel odometry. Sweet pepper is an interesting yet challenging domain due to a number of facets. Two challenges when segmenting fruit are occlusions caused by leaves and other fruit, and the similarity between juvenile pepper and leaves. A sample image is shown in Fig. 2.

SB20 is a sugar beet dataset that was captured at a field in campus Klein-Altendorf of the University of Bonn. The data was captured by mounting Intel RealSense D435i sensor with a nadir view of the ground on a Saga robot [7]. It contains RGB-D images of crops (sugar beet) and eight different categories of weeds covering a range of growth stages, natural world illumination conditions, and challenging occlusions. The captured field is separated into three herbicide

Table 1. Dataset characteristics such as image size, frames per second (fps), number of images for the training, validation and evaluation sets as well as the estimated motion parameters (μ_w, σ_w, μ_h, σ_h).

		Image Size	fps	Train	Valid.	Eval.	μ_w	σ_w	μ_h	σ_h
1	BUP20	1280×720	15	124	62	93	5	10	1	3
2	SB20	640×480	15	71	37	35	2	3	5	7

treatment regimes (30%, 70%, 100%) which provides a large range of distribution variations of different classes. We only use super-categories: crop, weed, and background for semantic segmentation purposes while, the dataset provides multi-class annotations, the Fig. 4 shows an example annotated image of this dataset.

Both datasets consist of temporally sparse annotations, that is, the annotation of one image does not overlap with another image. As such, we can use this data to generate real temporal sequences, of arbitrary size N, where **only** the final frame in the sequence is labelled. This serves as the real-world temporal data used in our evaluations. The advantage of creating temporal sequences in this manner is that it includes natural occlusions, varying illumination, and motion blur. The disadvantage in this method over traditional sequences is that as N increases the total loss decreases by a factor of $1/N$. This is due to the impact of having only a single annotated spatio-temporal frame to calculate the loss compared to N annotated frames (i.e. the final frame compared to N frames).

Table 1 summarizes the information about the images sets provided by BUP20 and SB20 datasets. To derive the movement parameters of the *real* sequences, we used odometry data provided from the datasets. The estimated parameters express motion of the platforms in both the x and y directions in image coordinate frames, which are summarised in Table 1.

To facilitate the interpretation of our experimental results, we set certain hyper-parameters as constant. The *real* images of BUP20 dataset are of resized to 704×416 and its *virtual* images are generated with a of size 544×320, also we use SB20 images without resizing and make *virtual* sequences of size 544×416. The *still* image experiments are denoted by ***Still-**** and use the UNet model. The spatio-temporal (sequence) experiments are denoted by ***Temporal-**** and use the lightweight RNN model described in Sect. 4, which is based on the UNet model.

5.2 Implementation and Metrics

Our semantic segmentation network is based on the UNet model implemented in PyTorch. To train our network we use Adam [14] with a momentum of 0.8 and StepLR leaning rate scheduler with a decay rate of $\gamma = 0.8$, decreasing the learning rate every one hundred epochs, starting initially with a learning rate of 0.001. We train all models for a total of 500 epochs with a batch size of $B = 4$ using cross-entropy as our loss. Also, all models' weights are initialized before training with Xavier method [6]. For parameter selection we employ the validation set of the datasets and use the weighted mean intersection over union (mIoU) to evaluate the performance. For evaluation we also employ the mIoU metric as it provides the best metric for system performance [22].

6 Results and Evaluations

Here we present three studies. First, we evaluate the performance of still image systems and explore the impact that generating N cropped images, from a single

Table 2. The result of the UNet model when trained and evaluated using *still* images.

		BUP20	SB20
1	*Still-Real*	77.3	71.3
2	*Still-Virtual*	78.7	73.5

image, has on system performance (data augmentation). Second, we evaluate the performance of spatio-temporal models when trained using either *real* or *virtually* generated sequences. Third, we perform an ablation study to explore the impact of varying the number of frames N in the spatio-temporal sequences. In all cases, we utilize the validation set to select the best performing model. This can be achieved earlier than the maximum epoch value. All of the results that we present are on the *evaluation* set of *real* data (either still images or sequences), unless otherwise stated.

6.1 Still Image Systems

These experiments provide the baseline performance when using only *still* images to train the segmentation system. This allows us to investigate if the extra samples generated in the *virtual* sequences lead to improved performance through data augmentation. For the *Still-Real* experiments only the annotated images were used for training. The *Still-Virtual* experiment uses *virtual* samples with $N = 5$ (data augmentation) to train the same base-line network. The results in Table 2 highlight the benefits of the data augmentation in *Still-Virtual* over the *Still-Real*. There is an absolute performance improvement from *Still-Real* to *Still-Virtual* of 1.4 and 2.2 respectively for datasets BUP20 and SB20, which represents a relatively high improvement on the same network gained only through data augmentation.

6.2 Spatio-Temporal Systems

In these experiments we explore the performance implications of varying how the feedback is provided to our temporal model, as described in Sect. 4, and the utility of using *virtual* sequences to train the spatio-temporal model.

First, we investigate the impact of different insertion and extraction points for the spatio-temporal (RNN) model as well as the impact of using either the *Bi-linear* or *Conv* re-sampling methods. Table 3 summarises the results of how the feedback is provided to our spatio-temporal model[1].

From Table 3 two details are clear, first, *Conv* consistently outperforms its *Bi-linear* counterpart, and second BUP20 results are higher than SB20 across

[1] Initial empirical results found that insertion and extraction points in the encoder part of the network provided poor performance and so was not considered in further experiments.

Table 3. The outlines of different RNN models using feedback layers extracted from *d:0-B* and looped back to various layers; using Bi-linear and Covolutional down-sampling feedbacks. Outlines illustrate the performance of different networks trained with *virtual* sequences and evaluated on *real* samples.

		Feedback	Conv	Bi-linear
BUP20	1	d:0-B - d:0-D	81.9	78.9
	2	d:0-B - d:1-D	80.3	79.1
	3	d:0-B - d:2-D	81.1	78.9
	4	d:0-B - d:3-D	80.8	80.4
	5	d:0-B - BN	80.4	78.9
SB20	6	d:0-B - d:0-D	75.4	73.9
	7	d:0-B - d:1-D	76.2	75.1
	8	d:0-B - d:2-D	74.5	74.3
	9	d:0-B - d:3-D	76.0	75.5
	10	d:0-B - BN	75.5	75.1

the board. From the *Conv* versus *Bi-linear* perspective this shows that a learned representation as feedback is of greater benefit than a direct representation. Then the performance mismatch between the two feedback types also results in different layers performing best, in BUP20 it is *d:0-B-d:0-D* and *d:0-B-d:1-D* for SB20 (lines 1 and 7 respectively of Table 3). The variation in performance between the BUP20 and SB20 datasets can be attributed to two key factors. First, the resolution of SB20 is approximately half of the BUP20 dataset, meaning there is more fine-grained information available in BUP20. Second, SB20 has more diversity in the sub-classes: the multitude of weed types look significantly different particularly with respect to the crop, while BUP20 only varies the color of a sweet pepper (similarity between sub-classes). From hereon, all spatio-temporal models use these best performing architectures (see Table 3 - line 1 BUP20, line 7 SB20).

In Table 4 we present the results for several systems trained on *virtual* and *real* sequences, see Sect. 4 for more details on each system. It can be seen that all of the spatio-temporal systems outperform the still image systems. The worst

Table 4. Outlines the results of spatio-temporal models trained with *real* datasets, *virtual* and along with the models train with *virtual* sequences and fine-tuned with *real* temporal frames (line 3).

	Model	BUP20	SB20
1	*Temporal-Real*	78.2	73.9
2	*Temporal-Virtual*	81.9	76.2
3	*Fine-Tuned*	82.8	76.4

performing system uses only *real* sequences for training and achieves a performance of 78.2% and 73.9%, which yield an absolute improvement of 0.9 and 2.6 over the *Still-Real* equivalent, respectively for both BUP20 and SB20. This is similar to the performance improvement we achieved through data augmentation *Still-Virtual.*

Considerable performance improvements are obtained when using the *virtual* sequences for training the spatio-temporal models. The benefits of *Temporal-Virtual* is clearly visible when comparing to the *Still-Real* variants, with an absolute improvement of 4.6 (BUP20) and 4.9(SB20). Furthermore, the *Temporal-Virtual* system has a performance improvement of 3.7 (BUP20) and 2.3 (SB20) when compared to the *Temporal-Real* system. We attribute this increase in performance to the impact of using all images when calculating the loss compared to only the final image. These results outline the benefit of the contextual information supplied from the temporal sequences for improved semantic segmentation allowing for temporal error reduction in pixel-wise classification.

To further evaluate the performance of our temporal network we explore the potential to augment the *real* sequences by training on data from both directions, which led to a minor boost in performance of 79.5% (BUP20) and 74.8% (SB20) over the *Temporal-Real.* Overall, these evaluations outline the importance of having fully annotated sequences as *virtual* sequences outperforms the other experiments.

We also explore training a temporal model on *virtual* sequences and then fine-tuning with *real* sequences, referred to as *Fine-Tuned* in line 3 of Table 4. This fine-tuning trick results in the highest mIoU score, out-performing both *Temporal-Real* and *Temporal-Virtual* models . These results empirically support the use of *virtual* sequences with convolutional re-sampling for training spatio-temporal models.

6.3 Ablation Study; Varying the Number of Frames N in a Temporal Sequence

As an ablation study we explore the impact of varying the number of images in the temporal sequence. We only consider a maximum case of five images, that is $N = [2, 3, 4, 5]$, in alignment with previous experiments. The results in Table 5 provide two important insights. First, even for $N = 2$ (the smallest

Table 5. For a sequence containing 5 frames (0–4), same base model trained with *virtual* outlining different performances for various number of frames.

	Model	BUP20	SB20
1	*Temporal-Virtual* (N = 5)	81.9	76.2
2	*Temporal-Virtual* (N = 4)	81.3	75.7
3	*Temporal-Virtual* (N = 3)	81.2	74.9
4	*Temporal-Virtual* (N = 2)	80.9	74.0

sequence size) we achieve improved performance over using *still* images. This is an absolute performance improvement of 3.6 (BUP20) and 2.7 (SB20) compared to the same *Still-Real* systems in Table 2. It indicates that even a small temporal sequence is able to provide more robust information to the classifier, resulting in more accurate segmentation. Second, increasing the temporal field improves performance by incorporating previous predictions when performing pixel-wise classification. In our experiments $N = 5$ provided a suitable trade off between informative structure and temporal distance. But future work can utilise larger GPU RAM to allow exploration into larger temporal sequences.

7 Conclusion

In this paper we presented a novel approach to augment temporal data for training agriculture based segmentation models. By exploiting the relatively static scene and assuming that the robot (camera) moves we are able, with no extra annotation effort, to generate virtually labelled temporal sequences. This is applied to sparsely annotated images within a video sequence to generate *virtual* temporal sequences via structured crops and shifts. We also explore the potential of a lightweight RNN (spatio-temporal model) by varying the "feedback" layers within the UNet architecture. It was found that introducing a convolutional layer as part of the "feedback" led to improved performance, which we attribute to being able to learn a representation over the feature map (when compared to a Bi-linear transform). The validity of both our data augmentation technique and spatio-temporal (lightweight RNN) architecture was outlined by an increase in absolute performance of 4.6 and 4.9 when comparing two models, *Still-Virtual* and *Temporal-Virtual*, respectively for two datasets BUP20 and SB20. These results outline the ability for our technique to work regardless of the properties of the scene being surveyed by the robotic platform, making it dataset agnostic. We also showed empirically that increasing N, the number of frames, for our lightweight RNN increases performance, showing that temporal information does play a role in improving segmentation results. Finally, this paper showed that an uncomplicated approach to generating temporally annotated data from sparsely annotated images improves segmentation performance.

Acknowledgements. This work was funded by the Deutsche Forschungsgemeinschaft (DFG, German Research Foundation) under Germany's Excellence Strategy - EXC 2070 - 390732324.

References

1. Campos, Y., Sossa, H., Pajares, G.: Spatio-temporal analysis for obstacle detection in agricultural videos. Appl. Soft Comput. **45**, 86–97 (2016)
2. Cordts, M., et al.: The cityscapes dataset for semantic urban scene understanding. In: Proceedings of the IEEE Conference on Computer Vision and Pattern Recognition, pp. 3213–3223 (2016)

3. Deng, J., Dong, W., Socher, R., Li, L.J., Li, K., Fei-Fei, L.: Imagenet: a large-scale hierarchical image database. In: 2009 IEEE Conference on Computer Vision and Pattern Recognition, pp. 248–255. IEEE (2009)
4. Douarre, C., Crispim-Junior, C.F., Gelibert, A., Tougne, L., Rousseau, D.: Novel data augmentation strategies to boost supervised segmentation of plant disease. Comput. Electron. Agric. **165**, 104967 (2019)
5. Fawakherji, M., Youssef, A., Bloisi, D., Pretto, A., Nardi, D.: Crop and weeds classification for precision agriculture using context-independent pixel-wise segmentation. In: 2019 Third IEEE International Conference on Robotic Computing (IRC), pp. 146–152. IEEE (2019)
6. Glorot, X., Bengio, Y.: Understanding the difficulty of training deep feedforward neural networks. In: Proceedings of the Thirteenth International Conference on Artificial intelligence and Statistics, pp. 249–256. JMLR Workshop and Conference Proceedings (2010)
7. Grimstad, L., From, P.J.: Thorvald ii-a modular and re-configurable agricultural robot. IFAC-PapersOnLine **50**(1), 4588–4593 (2017)
8. Halstead, M., Denman, S., Fookes, C., McCool, C.: Fruit detection in the wild: the impact of varying conditions and cultivar. In: Proceedings of Digital Image Computing: Techniques and Applications (DICTA) (2020)
9. Halstead, M., McCool, C., Denman, S., Perez, T., Fookes, C.: Fruit quantity and ripeness estimation using a robotic vision system. IEEE Robot. Autom. Lett. **3**(4), 2995–3002 (2018)
10. He, Y., Chiu, W.C., Keuper, M., Fritz, M.: Std2p: rgbd semantic segmentation using spatio-temporal data-driven pooling. In: Proceedings of the IEEE Conference on Computer Vision and Pattern Recognition (CVPR) (July 2017)
11. Hochreiter, S., Schmidhuber, J.: Long short-term memory. Neural Comput. **9**(8), 1735–1780 (1997)
12. Jarvers, C., Neumann, H.: Incorporating feedback in convolutional neural networks. In: Proceedings of the Cognitive Computational Neuroscience Conference, pp. 395–398 (2019)
13. Kamilaris, A., Prenafeta-Boldú, F.X.: Deep learning in agriculture: a survey. Comput. Electron. Agric. **147**, 70–90 (2018)
14. Kingma, D.P., Ba, J.: Adam: A method for stochastic optimization. arXiv preprint arXiv:1412.6980 (2014)
15. Krizhevsky, A., Sutskever, I., Hinton, G.E.: Imagenet classification with deep convolutional neural networks. In: Advances in Neural Information Processing Systems, pp. 1097–1105 (2012)
16. LeCun, Y., Bottou, L., Bengio, Y., Haffner, P.: Gradient-based learning applied to document recognition. Proc. IEEE **86**(11), 2278–2324 (1998)
17. Lehnert, C., Sa, I., McCool, C., Upcroft, B., Perez, T.: Sweet pepper pose detection and grasping for automated crop harvesting. In: 2016 IEEE International Conference on Robotics and Automation (ICRA), pp. 2428–2434. IEEE (2016)
18. Long, J., Shelhamer, E., Darrell, T.: Fully convolutional networks for semantic segmentation. In: Proceedings of the IEEE Conference on Computer Vision and Pattern Recognition, pp. 3431–3440 (2015)
19. Lottes, P., Behley, J., Milioto, A., Stachniss, C.: Fully convolutional networks with sequential information for robust crop and weed detection in precision farming. IEEE Robot. Autom. Lett. **3**(4), 2870–2877 (2018)
20. Noh, H., Hong, S., Han, B.: Learning deconvolution network for semantic segmentation. In: Proceedings of the IEEE International Conference on Computer Vision (ICCV) (December 2015)

21. Pfeuffer, A., Schulz, K., Dietmayer, K.: Semantic segmentation of video sequences with convolutional lstms. In: 2019 IEEE Intelligent Vehicles Symposium (IV), pp. 1441–1447. IEEE (2019)
22. Rahman, M.A., Wang, Y.: Optimizing intersection-over-union in deep neural networks for image segmentation. In: Bebis, G., et al. (eds.) ISVC 2016. LNCS, vol. 10072, pp. 234–244. Springer, Cham (2016). https://doi.org/10.1007/978-3-319-50835-1_22
23. Ronneberger, O., Fischer, P., Brox, T.: U-Net: convolutional networks for biomedical image segmentation. In: Navab, N., Hornegger, J., Wells, W.M., Frangi, A.F. (eds.) MICCAI 2015. LNCS, vol. 9351, pp. 234–241. Springer, Cham (2015). https://doi.org/10.1007/978-3-319-24574-4_28
24. Shi, Z., Yang, Y., Hospedales, T.M., Xiang, T.: Weakly supervised image annotation and segmentation with objects and attributes (2017)
25. Simonyan, K., Zisserman, A.: Very deep convolutional networks for large-scale image recognition. arXiv preprint arXiv:1409.1556 (2014)
26. Slaughter, D., Giles, D., Downey, D.: Autonomous robotic weed control systems: a review. Comput. Electron. Agric. **61**(1), 63–78 (2008)
27. Smitt, C., Halstead, M., Zaenker, T., Bennewitz, M., McCool, C.: Pathobot: A robot for glasshouse crop phenotyping and intervention. arXiv preprint arXiv:2010.16272 (2020)
28. Tsuda, H., Shibuya, E., Hotta, K.: Feedback attention for cell image segmentation. arXiv preprint arXiv:2008.06474 (2020)
29. Wang, C., MacGillivray, T., Macnaught, G., Yang, G., Newby, D.: A two-stage 3d unet framework for multi-class segmentation on full resolution image. arXiv preprint arXiv:1804.04341 (2018)
30. Wang, J., Adelson, E.H.: Spatio-temporal segmentation of video data. In: Image and Video Processing II, vol. 2182, pp. 120–131. International Society for Optics and Photonics (1994)
31. Yu, X., Wu, X., Luo, C., Ren, P.: Deep learning in remote sensing scene classification: a data augmentation enhanced convolutional neural network framework. GIScience Remote Sens. **54**(5), 741–758 (2017)
32. Zhang, T., Zheng, W., Cui, Z., Zong, Y., Li, Y.: Spatial-temporal recurrent neural network for emotion recognition. IEEE Trans. Cybern. **49**(3), 839–847 (2018)

Weakly Supervised Segmentation Pretraining for Plant Cover Prediction

Matthias Körschens[1(✉)], Paul Bodesheim[1], Christine Römermann[1], Solveig Franziska Bucher[1], Mirco Migliavacca[2], Josephine Ulrich[1], and Joachim Denzler[1]

[1] Friedrich Schiller University Jena, Jena, Germany
{matthias.koerschens,paul.bodesheim,christine.roemermann, solveig.franziska.bucher,josephine.ulrich,joachim.denzler}@uni-jena.de
[2] Department Biogeochemical Integration, Max Planck Institute for Biogeochemistry, Jena, Germany
mmiglia@bgc-jena.mpg.de

Abstract. Automated plant cover prediction can be a valuable tool for botanists, as plant cover estimations are a laborious and recurring task in environmental research. Upon examination of the images usually encompassed in this task, it becomes apparent that the task is ill-posed and successful training on such images alone without external data is nearly impossible. While a previous approach includes pretraining on a domain-related dataset containing plants in natural settings, we argue that regular classification training on such data is insufficient. To solve this problem, we propose a novel pretraining pipeline utilizing weakly supervised object localization on images with only class annotations to generate segmentation maps that can be exploited for a second pretraining step. We utilize different pooling methods during classification pretraining, and evaluate and compare their effects on the plant cover prediction. For this evaluation, we focus primarily on the visible parts of the plants. To this end, contrary to previous works, we created a small dataset containing segmentations of plant cover images to be able to evaluate the benefit of our method numerically. We find that our segmentation pretraining approach outperforms classification pretraining and especially aids in the recognition of less prevalent plants in the plant cover dataset.

Keywords: Plant cover prediction · Biodiversity monitoring · Plant segmentation · Computer vision · Deep learning · Transfer learning · Neural network pretraining · Weakly supervised learning

1 Introduction

Analyzing the impact of environmental changes on plant communities is an essential part of botanical research. This way, we can find the causes and effects of such

Supplementary Information The online version contains supplementary material available at https://doi.org/10.1007/978-3-030-92659-5_38.

C. Bauckhage et al. (Eds.): DAGM GCPR 2021, LNCS 13024, pp. 589–603, 2021.
https://doi.org/10.1007/978-3-030-92659-5_38

changes and ways to counteract them. A prominent example of an environmental change investigated this way is climate change [22,23,30]. Other environmental aspects can be monitored like this as well, such as land-use [1,9] and insect abundance [33]. One possibility to monitor the effects of such changes on plants is to monitor the species diversity, specifically the species composition of plant communities. This is commonly done by the biologists directly in the field by estimating the so-called plant cover, which is defined as the percentage of soil covered by each plant species, disregarding any occlusion. Performing this task in an automated way based on automatically collected imagery would reduce the massive workload introduced by this recurring and laborious task, and enable an objective analysis of the data in high temporal resolution.

However, developing a correctly working system to perform plant cover prediction (PCP) is a difficult task due to multiple reasons. Firstly, the plant cover estimates are usually noisy due to human error and subjective estimations. Secondly, the plant cover is usually heavily imbalanced by nature, as plants always grow in strongly differing ratios in a natural environment. Thirdly, in addition to this, PCP is an ill-posed problem. This is primarily due to the fact that in the plant cover estimations used as annotation, occlusion is ignored. This makes it near impossible to, for example, train a convolutional neural network (CNN) well on this data alone, as, contrary to human vision, CNNs usually cannot inherently deal with occlusion. Therefore, the network often learns arbitrary features in the images to reduce the error during training in any way possible, which are mostly not the true features the network should use for a correct prediction. To counter this problem, Körschens *et al.* [17], who proposed a first solution for the task of plant cover prediction, suggested utilizing segmentation maps generated in a weakly supervised way by the network for visual inspection of the network's prediction. The segmentation maps make it possible to monitor what the network learned and which areas it uses to generate a prediction, increasing transparency of the system and its gathered knowledge. However, Körschens *et al.* could not solve the problem entirely in the paper, and their results also showed that, when the system is only trained on the plant cover data, the predictions are strongly biased towards the more prevalent plants in the dataset. We argue that these problems can be further alleviated by strong usage of pretraining on domain-similar datasets containing isolated plants, as with such datasets, we can directly control the data balance and have a better influence on the features the network learns. Using such a domain-similar dataset was also recently investigated by Körschens *et al.* [18], who demonstrated the correctness of this assumption on several standard network architectures.

Pretraining, for example the one in [18], is usually done by utilizing a backbone CNN serving as feature extractor, followed by global average pooling and a fully-connected classification layer. We argue that this training method results in the network focusing on the most prevalent features in the pretraining dataset, which might not be optimal since these features are not necessarily contained in the images of the target dataset. An example of this would be the blossom of the plants, which are usually the most discriminative features in isolated plant images but are comparably rare in simple vegetation images for PCP. To solve this problem, we suggest that encouraging the network to focus more on the entire plant instead of only the most discriminative parts by training on segmentation data would be beneficial. However, to the best of our knowledge, there are no comprehensive plant segmentation datasets publicly available. Therefore, we propose a system, which generates weakly supervised segmentations using a classification-trained network. The generated segmentations can then be used for segmentation-pretraining. To this end, we investigate the class activation mapping (CAM) method [48], which is often used in tasks like weakly supervised object localization [5,32,41,45,46,48] and weakly supervised semantic segmentation [12,15,34,37,40,43]. While most of these methods use global average pooling (GAP) as basis for their classification network, we found that pooling methods like global max pooling (GMP) or global log-sum-exp pooling (GLSEP), which is an approximation of the former, in parts generate better localizations which result in better segmentations. While similar functions have been investigated before for classification [44], and in another setting in weakly supervised object localization [28], to the best of our knowledge we are the first to apply in as feature aggregation method in conjunction with CAMs.

Furthermore, PCP can be viewed as a task effectively consisting of two parts: analysis of the visual and the occluded parts. Solving occlusion is heavily dependent on a good automated analysis of the visual part, as we can only complete the partially visible plants correctly if our analysis of the species in the visible parts is correct as well. Therefore, we will primarily focus on the correctness of the analysis of the visible parts. Analyzing the visible parts can be done, for example, by investigating segmentation maps as proposed by Körschens *et al.* [17]. However, they merely relied on visual inspection of the segmentations, as no ground-truth segmentations were available for quantitative evaluation. To enable the latter, we manually annotated several images from the plant cover dataset and can therefore also evaluate our approach quantitatively.

Hence, our contributions are the following: We introduce a novel pretraining pipeline, which converts existing classification data of a plant-domain dataset into segmentation data usable for pretraining on the plant cover task. Moreover, we investigate and compare multiple pooling approaches and their differences for generating the abovementioned segmentations. Lastly, we evaluate the segmentations of the final part of our system on the plant cover dataset quantitatively by utilizing a small set of manually annotated plant segmentations.

2 Related Work

2.1 Weakly Supervised Object Localization (WSOL)

Weakly supervised object localization is an established field in Computer Vision research. While there are different kinds of approaches, the most recent ones are based on the method of *Class Activation Mapping* of Zhou *et al.* [48]. In their paper, the authors propose to utilize the classification weights learned by a classification layer at the end of the network to generate a map containing class activations at each position of the last feature map, also called class activation map (CAM). To this end, the pooling layer is removed, and the fully connected layer is converted into a 1 by 1 convolutional layer. The generated CAM can then be thresholded and weakly supervised bounding boxes or segmentations generated. Multiple methods based on this approach tackle the problem by utilizing occluding data augmentation, e.g., by dropping parts of the images [5,32], or cutting and pasting parts of other images [41]. The aim of such augmentations is to prevent the network from relying too much on the most discriminative features and hence distribute the activations in the CAM more equally over the complete objects. In other approaches, this is done in more sophisticated ways, for example, by using adversarial erasing [45]. Choe *et al.* [4], however, showed recently that methods in this direction primarily gained performance improvements by indirectly tuning hyperparameters on the test set, resulting in almost no effective gain in performance on WSOL benchmarks in the last years. Nevertheless, recently, there have also been other approaches, which tackle the problem differently, for example, by modifying the way the CAMs are generated [26] or generating alternative maps for localizing objects [47].

We also base our method on CAMs. However, in contrast to the methods mentioned above, we also investigate changes to the base method by exchanging the pooling layer used during training. Specifically, utilizing global max pooling, or its approximation global log-sum-exp pooling, can potentially yield more benefits during the weakly supervised segmentation. This is because these methods do not depend on an averaging operation, potentially inhibiting good localization caused by dilution of activations during training.

2.2 Plant Analysis

The continuous developments of convolutional neural networks (CNNs) have also encouraged the development of automated plant analysis methods. These reach from simple plant species identification [2,10,19,39] over the detection of ripe fruit [7] and counting of agricultural plants [24,38] to the prediction of plant diseases [3]. However, the number of works concerned with plant cover determination is still small. The first work in this area was proposed by Kattenborn *et al.* [14], who analyzed the plant cover of several woody species, herbs and shrubs via UAV imagery. In their work, they utilized delineations in the images as training data for their custom CNN, which is not comparable with the data analyzed in this work. Nevertheless, previous works on plant cover analysis of herbaceous plants, specifically on the InsectArmageddon dataset by [36], were published by Körschens *et al.* [17,18]. They did several analyses with a custom network [17], as well as multiple established network architectures with different pretraining methods [18]. As the base of their approach, they model the problem as a weakly supervised segmentation approach, where pixel-wise probabilities for each plant are calculated, which are then aggregated into the final cover percentages. While we also utilize this basic approach, we go more into depth regarding the findings in [18]. Körschens *et al.* found that pretraining on a related dataset, albeit comparably small, is advantageous when tackling the

Fig. 1. Example images from the pretraining dataset. The plant species shown are from left to right: *Achillea millefolium* (Yarrow), *Centaurea jacea* (brown knapweed), *Plantago lanceolata* (ribwort plantain), *Trifolium pratense* (red clover), *Scorzoneroides autumnalis* (autumn hawkbit) and *Grasses*, which are not differentiated into different species. (Color figure online)

PCP problem. We, however, argue that the efficiency of pretraining could be massively improved when training on segmentation data. For regular classification training with global average pooling, the network primarily focuses on the most discriminative regions [48]. With segmentation data, however, the network is encouraged to focus on the full extent of the object instead of the few most discriminative parts. For this reason, we investigate a similar approach to [18] with a strongly different pretraining process, in which we initially generate weakly supervised segmentations instead of using classification data directly. Moreover, we also include a numerical evaluation of the prediction quality of the visible plants in the images.

3 Datasets

In our experiment we utilize two separate datasets: one for pretraining and one for the actual plant cover training.

3.1 Pretraining Dataset

The pretraining dataset we use in our experiments contains species-specific randomly selected images from the Global Biodiversity Information Facility[1] (GBIF) [8]. The dataset encompasses images of 8 different plant species in natural settings, which match the ones from the plant cover dataset explained in Sect. 3.2.

The plant species in the datasets and their respective abbreviations used in parts of this paper are *Achillea millefolium* (Ach_mil), *Centaurea jacea* (Cen_jac), *Lotus corniculatus* (Lot_cor), *Medicago lupulina* (Med_lup), *Plantago lanceolata* (Pla_lan), *Scorzoneroides autumnalis* (Sco_aut) and *Trifolium pratense* (Tri_pra).

The pretraining dataset comprises 6000 training images and 1200 validation images, which are evenly distributed across the classes, making the dataset balanced. Example images from the dataset are shown in Fig. 1.

[1] http://gbif.org.

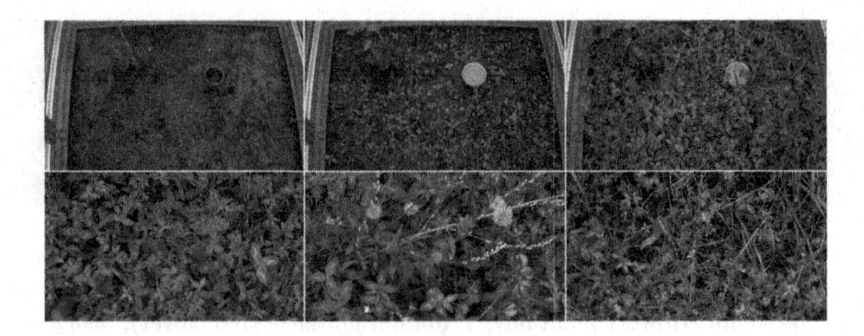

Fig. 2. Example images from the InsectArmageddon [17,36] dataset. The growth process of the plants, as well as other changes over the time, like the flowering process, are captured in the images.

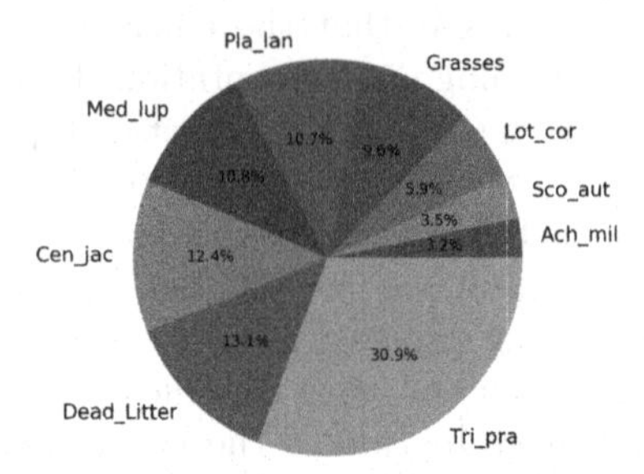

Fig. 3. The distribution of plant cover percentages over the different species from the InsectArmageddon dataset. *Trifolium pratense* takes almost a third of the dataset while *Achillea millefolium* is the least abundant plant with only 3.2% of the total cover. Figure taken from [18].

3.2 Plant Cover Dataset

We utilize the same dataset introduced by Ulrich *et al.* in [36] and Körschens *et al.* in [17]. As the latter refer to it as InsectArmageddon dataset due to its origin in the eponymous project[2], we will also refer to it this way. The dataset contains images from nine different plant species collected in enclosed boxes: so-called EcoUnits. Two cameras collected images in each of the 24 utilized EcoUnits over multiple months from above with a frequency of one image per day, hence also capturing the growth process of the plants. However, due to the laboriousness of the annotation of the images, only weekly annotations are available, leading to 682 images with annotations. Examples of the plants and their state of growth at different points in time are shown in Fig. 2.

The nine different plant species in the dataset are the same as introduced in Sect. 3.1 with the addition of *Dead Litter*. The latter was introduced to designate dead plant matter, which is usually indistinguishable regarding the original plant

[2] https://www.idiv.de/en/research/platforms_and_networks/idiv_ecotron/experiments/insect_armageddon.html.

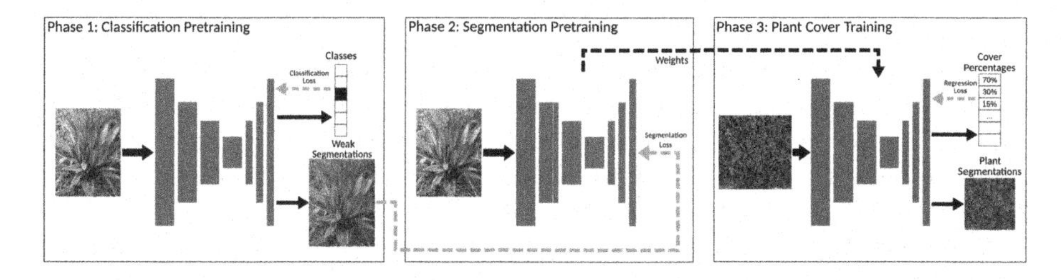

Fig. 4. Our proposed WSOL pretraining pipeline. The first network is trained on classification data and generates segmentation maps for the class in question based on the CAM. The second network is trained on this segmentation data and the trained weights are then utilized in the network used in plant cover training.

species. As seen in Fig. 3, the plants are in a heavily imbalanced long-tailed distribution, with *Trifolium pratense* spearheading the distribution and *Achillea millefolium* trailing it.

The plant cover annotations themselves are quantized into the so-called Schmidt-scale [27], so that the values can be estimated better. The scale contains the percentages 0, 0.5, 1, 3, 5, 8, 10, 15, 20, 25, 30, 40, 50, 60, 70, 75, 80, 90 and 100. For details on the data- and image-collection process, we refer to [6,17,35,36].

4 Method

In this section we will explain the general workflow of our method, followed by the pooling methods we use and the network head utilized during plant cover prediction. For better explainability, in the following we will view the network as consisting of three different parts: backbone, pooling layer and head. The backbone represents the feature extractor part of the network, and the head the task specific layer(s), which, in case of classification, is usually a fully connected layer that directly follows a pooling layer.

4.1 Pipeline

Our proposed pipeline can be divided into three distinct phases and an overview is shown in Fig. 4. The first phase consists of a simple classification training on the pretraining dataset. The network used during this training has the shape of a typical classification network: a backbone, followed by a pooling operation and a fully connected layer with softmax activation. Hence, a standard classification loss like categorical cross-entropy can be applied. After the training, we remove the pooling layer of the network and convert the fully-connected layer into a convolutional layer as shown in [48] to generate class activation maps (CAMs). We isolate the single CAM belonging to the class annotation of the image and apply a threshold to generate a discrete segmentation map. The threshold is, as described in [48], a value relative to the maximum activation, e.g., 0.2.

In the second phase of our method, we utilize the segmentations generated in the first phase to train another network consisting of only a backbone and a segmentation head, i.e., a simple pixel-wise classifier with sigmoid activation. For training this network, a segmentation loss, e.g., a dice loss [25], can be applied. Upon this network being trained, we can use the weights to initialize the network from phase three for transfer learning.

In the third phase of our approach, we use the network initialized with the segmentation weights to fine-tune on the PCP task. This can be done using a simple regression loss, e.g., the mean absolute error. Finally, the network can be used to generate plant cover predictions in addition to segmentation maps for the plants in the image.

4.2 Pooling Methods

As mentioned above, in our experiments, we will investigate three different pooling methods: global average pooling (GAP), global max pooling (GMP), and global log-sum-exp pooling (GLSEP). GAP is the pooling method usually used for classification training and, therefore, also in the CAM method. However, as shown in [48], networks trained in such a way focus primarily on the most discriminative features in the images. Moreover, as the averaging operation encourages the distribution of higher activations over greater areas in the images, the CAMs dilute and are not optimal for good localization, resulting in worse segmentations. This is not the case for GMP, however, Zhou *et al.* [48] argue that GMP is, in contrast, more prone to focusing on single points with high activations in the images. Therefore, we investigate GLSEP, as log-sum-exp is an approximation of the maximum function and, due to its sum-part, does not only focus on a single point in the image. Hence, GLSEP can also be viewed as a parameter-free compromise between GAP and GMP. While variations of such a pooling method have been investigated before [28,44], to the best of our knowledge, none have been investigated for WSOL in conjunction with CAMs.

4.3 Plant Cover Prediction Head

To take into account the relatively complex calculations done during cover prediction, we utilize the already established plant cover prediction head from [17], with the modification introduced in [18]. To summarize, with this network head the plant cover prediction is viewed as a pixel-wise classification problem, where the classes consist of the plant classes in the dataset in addition to a background class and an irrelevance class. The background class serves as indicator of regions relevant for the plant cover calculation, while not containing any relevant plants (e.g., the soil in the images or weeds not monitored in the experiment). The irrelevance class indicates regions, which are not relevant for the calculation (e.g., the walls of the EcoUnits). Due to possible occlusion between the plants, the plant classes are modeled as not mutually exclusive, while they are mutually exclusive with background and irrelevance. For the details on this approach, we refer to [17,18].

5 Experiments

5.1 Evaluation and Metrics

We evaluate the our method after the last step: the plant cover training. For the evaluation, we utilize the metrics introduced by Körschens *et al.* [17], i.e., the mean absolute error (MAE) and the mean scaled absolute error (MSAE). The latter was considered to enable a fairer comparison of two imbalanced plant species and is defined as the mean of the species-wise absolute errors divided by their respective species-wise mean cover percentage. The exact percentages used during calculation can be found in [17]. To evaluate the quality of the segmentations, we utilize the mean Intersection-over-Union (mIoU), which is commonly used to evaluate segmentation and object detection tasks:

$$mIoU = \frac{1}{N} \sum_{i=1}^{N} \frac{P_{pred}^{c_i} \cap P_{true}^{c_i}}{P_{pred}^{c_i} \cup P_{true}^{c_i}}, \tag{1}$$

with N being the number of classes, $P_{pred}^{c_i}$ the set of pixels predicted as class i, and $P_{true}^{c_i}$ the respective ground-truth counterpart. Since the original plant cover image dataset [17,36] did not contain any segmentations, we annotated 14 images containing large numbers of plants ourselves and used them for evaluation. It should be noted that these 14 images had no plant cover annotations, and hence are not seen during training. Several example images with our annotations can be found in the supplementary material.

5.2 Plant Cover Prediction

As mentioned above, our training is conducted in three separate phases, and we explain the setups in the following. In all three phases, we utilize the same networks architecture, which is a ResNet50 [11] with a Feature Pyramid Network (FPN) using a depth of 512 and the extraction layer P2, as shown in [20]. We choose this architecture because Körschens *et al.* [18] also investigated this one, among others. While in [18], a DenseNet121 [13] performed best, in our experiments we were not able to successfully train a DenseNet in the same setting and hence choose the ResNet50, as it achieved comparable performance in [18] when using an FPN.

Classification Pretraining. For classification pretraining, we utilize the standard cross-entropy loss and the Adam optimizer [16] with a learning rate of 1e-4, an L2 regularization of 1e-6, and a batch size of 12. The ResNet50 (without FPN) has been initialized with ImageNet weights [31]. We do not train for a fixed number of epochs, but monitor the accuracy on the validation set and reduce the learning rate by a factor of 0.1, if there were no improvements for four epochs. We repeat this process until there have been no improvements for six epochs, after which we end the training. Data augmentation is done by random rotations, random cropping, and random horizontal flipping. During the augmentation process, the images are resized such that the smaller image dimension has

Table 1. The results of our experiments with segmentation pretraining (SPT) in comparison to regular classification pretraining, considering different pooling methods. We evaluate the MAE, MSAE, total mIoU and the mIoU for plants only (without background and irrelevance classes). Best results are marked in bold font.

Pooling	SPT	MAE	MSAE	mIoU	mIoU (plants)
GAP	✗	5.23%	0.501	0.148	0.161
	✓	5.23%	0.501	**0.171**	**0.179**
GMP	✗	5.23%	0.501	0.144	0.157
	✓	**5.17%**	**0.494**	0.165	0.176
GLSEP	✗	5.24%	0.504	0.156	0.162
	✓	5.23%	0.503	0.161	0.174

a size of 512 pixels, then the images are cropped to a size of 448 × 448 pixels used for further processing.

Segmentation Pretraining. The setup for our segmentation pretraining is mostly the same as in the first phase. However, we utilize a dice loss during this training process and monitor the mIoU instead of the accuracy. It should be noted that we initialize the network again with ImageNet weights and do not use the weights from the first pretraining to have a fairer comparison between training using segmentations and training using class information only.

Plant Cover Training. During the plant cover training, we use the same setup as in [18], i.e., image sizes of 1536 × 768 px, a batch size of 1, Adam optimizer with a learning rate of 1e−5, a training duration of 40 epochs, and a simple horizontal flipping augmentation. For the loss function, we utilize the MAE and we run our experiments in a 12-fold cross-validation ensuring that images from the same EcoUnit are in the same subset, as done in [17,18].

Results. In our experiments, we compare the effect of training using the weakly supervised segmentations with training using only class labels. The latter corresponds to using the weights from the first phase of our method directly in the third phase, and in case of GAP, this is equivalent to the method shown in [18]. The results of these experiments are summarized in Table 1.

We notice that regarding MAE and MSAE, the results using segmentation pretraining always perform at least as good as with standard classification pretraining or even better, albeit often only by small amounts. Moreover, we can see that the top results for MAE and MSAE do not coincide with the best segmentations, as the top error values have been achieved using GMP segmentation pretraining (MAE of 5.17% and MSAE of 0.494), while the best segmentations, measured by mIoU, are achieved when using segmentation pretraining with GAP.

Table 2. Detailed results for the composition of the mIoU values from Table 1. Abbreviations used: SPT = Segmentation pretraining, DL = Dead Litter, BG = Background (relevant for calculation), IRR = Background (irrelevant for calculation), PO = Value for plants only (excludes BG and IRR); Top results per species are marked in bold font.

Pooling	SPT	Ach_mil	Cen_jac	Grasses	Lot_cor	Med_lup	Pla_lan	Sco_aut
GAP	✗	0.043 ±0.020	0.101 ±0.028	0.442 ±0.042	0.015 ±0.007	0.082 ±0.014	0.151 ±0.009	**0.001** ±0.001
GAP	✓	**0.099** ±0.033	**0.147** ±0.007	0.438 ±0.021	0.034 ±0.014	0.100 ±0.010	0.152 ±0.017	0.000 ±0.001
GMP	✗	0.024 ±0.020	0.097 ±0.021	0.435 ±0.024	0.010 ±0.003	0.062 ±0.011	0.151 ±0.016	**0.001** ±0.001
GMP	✓	0.095 ±0.028	0.135 ±0.010	0.429 ±0.034	0.035 ±0.016	0.110 ±0.017	0.159 ±0.017	**0.001** ±0.001
GLSEP	✗	0.024 ±0.020	0.102 ±0.017	**0.447** ±0.019	0.019 ±0.007	0.079 ±0.011	0.155 ±0.011	0.001 ±0.000
GLSEP	✓	0.064 ±0.032	0.141 ±0.017	0.411 ±0.034	**0.043** ±0.011	**0.112** ±0.016	0.167 ±0.015	0.000 ±0.000

Pooling	SPT	Tri_pra	DL	BG	IRR	Total	PO
GAP	✗	0.528 ±0.028	0.083 ±0.008	0.122 ±0.009	0.062 ±0.049	0.148 ±0.011	0.161 ±0.011
GAP	✓	**0.556** ±0.015	**0.088** ±0.009	**0.144** ±0.011	0.124 ±0.035	**0.171** ±0.007	**0.179** ±0.006
GMP	✗	0.550 ±0.013	0.086 ±0.006	0.125 ±0.010	0.043 ±0.031	0.144 ±0.004	0.157 ±0.005
GMP	✓	0.543 ±0.016	0.077 ±0.008	0.125 ±0.006	0.109 ±0.087	0.165 ±0.009	0.176 ±0.008
GLSEP	✗	0.551 ±0.016	0.082 ±0.010	0.122 ±0.006	**0.138** ±0.097	0.156 ±0.013	0.162 ±0.006
GLSEP	✓	0.539 ±0.013	0.086 ±0.006	0.133 ±0.011	0.076 ±0.055	0.161 ±0.009	0.174 ±0.008

5.3 Detailed Analysis of Segmentations

To have a more thorough insight into the effect of the segmentation pretraining (SPT), we also investigate the different IoU values for each plant species, shown in Table 2. It should be noted that due to the small size of many plants, precisely pinpointing their locations with CNNs is a difficult task, leading to relatively small IoU values overall. We see that the IoU values for the less abundant plants in the dataset, especially *Achillea millefolium*, *Centaurea jacea* and *Lotus corniculatus*, increase massively when applying SPT compared to only classification pretraining. Depending on the pooling strategy, the IoU increases by up to 250% for *Achillea millefolium* and *Lotus corniculatus*, and up to 150% for *Centaurea jacea*. This indicates that more relevant features for these species are included during SPT, confirming our intuition. In contrast to this, we also note that the IoU for *Grasses* and in parts for *Trifolium pratense* decreases. For the former, the reason might be that grasses are hard to segment automatically due to their thinness, leading to worse pretraining for these classes during SPT. To counter this problem, it might be possible to utilize a network with a higher output resolution, e.g., a deeper FPN layer. The decreasing IoU for *Trifolium pratense* can be attributed to the balancing effect in the predictions introduced by the SPT, as a more balanced dataset usually results in worse performance for the more dominant classes. Regarding the varying pooling strategies used during the classification pretraining, we notice that they perform differently, depending on the plant species. We attribute this to the structure of the different plant species. This means that thinner and smaller objects are more easily recognized after training with GLSEP, as this presumably generates features that are more focused on smaller areas. However, plants with large leaf areas are more easily recognizable by networks using the likely more unfocused features generated by networks with GAP due to the averaging operation. Finally, it should be noted that the low IoU for *Scorzoneroides autumnalis* is caused by the small abundance of this species in the segmentation dataset. Hence, no conclusion can be drawn for this plant at this point. Multiple example segmentations using our method can be found in the supplementary material.

To summarize, the segmentation pretraining proved to be superior to using only the standard classification pretraining. It consistently improves the MAE and MSAE over classification pretraining by small amounts and especially improves the quality of the segmentations in general, with more significant improvements for the less abundant plants in the dataset.

6 Conclusions and Future Work

In this work, we proposed a novel pretraining pipeline for plant cover prediction by generating segmentation maps via a weakly supervised localization approach and using these maps for an additional segmentation pretraining. In general, we demonstrated superior performance of our approach compared to only standard classification pretraining for identifying the plants in the visible parts of the image, with more significant improvements for the less abundant plants in the dataset. More specifically, we noticed that the recognition of the less abundant plants improved the most, while the detection of the most prevalent plants, i.e., grasses and *Trifolium pratense*, decreases slightly. This effect is likely caused by the increased training set balance that the segmentation pretraining introduces. We also observed that the investigated pooling strategies perform differently depending on the plant species, which is likely caused by the varying structures of each plant that are handled differently by the individual aggregation methods during training of the network.

Our approach offers multiple directions for further improvements. First, the quality of the segmentation maps generated after the classification pretraining in the first phase could be improved, for example, by applying further data augmentation approaches commonly used in weakly supervised object localization [5,32,41]. Second, the segmentation pretraining could be altered by using different kinds of augmentations, losses, or also different kinds of segmentation networks, e.g., a U-Net [29]. Third, as segmentation maps are generated in the training process, these maps could also be utilized for applying an amodal segmentation approach [21,42] to better deal with occlusions in the plant cover images. Lastly, the proposed approach might also potentially be applicable for fine-grained classification, as well as segmentation tasks, in general.

Acknowledgement. Matthias Körschens thanks the Carl Zeiss Foundation for the financial support. We would also like to thank Alban Gebler and the iDiv for providing the data for our investigations.

References

1. Aggemyr, E., Cousins, S.A.: Landscape structure and land use history influence changes in island plant composition after 100 years. J. Biogeogr. **39**(9), 1645–1656 (2012)
2. Barré, P., Stöver, B.C., Müller, K.F., Steinhage, V.: LeafNet: a computer vision system for automatic plant species identification. Eco. Inform. **40**, 50–56 (2017)
3. Chen, J., Chen, J., Zhang, D., Sun, Y., Nanehkaran, Y.: Using deep transfer learning for image-based plant disease identification. Comput. Electron. Agric. **173**, 105393 (2020)
4. Choe, J., Oh, S.J., Lee, S., Chun, S., Akata, Z., Shim, H.: Evaluating weakly supervised object localization methods right. In: Proceedings of the IEEE/CVF Conference on Computer Vision and Pattern Recognition, pp. 3133–3142 (2020)
5. Choe, J., Shim, H.: Attention-based dropout layer for weakly supervised object localization. In: Proceedings of the IEEE/CVF Conference on Computer Vision and Pattern Recognition, pp. 2219–2228 (2019)
6. Eisenhauer, N., Türke, M.: From climate chambers to biodiversity chambers. Front. Ecol. Environ. **16**(3), 136–137 (2018)
7. Ganesh, P., Volle, K., Burks, T., Mehta, S.: Deep orange: mask R-CNN based orange detection and segmentation. IFAC-PapersOnLine **52**(30), 70–75 (2019)
8. GBIF.org: Gbif occurrence downloads, 13 May 2020. https://doi.org/10.15468/dl.xg9y85, https://doi.org/10.15468/dl.zgbmn2, https://doi.org/10.15468/dl.cm6hqj, https://doi.org/10.15468/dl.fez33g, https://doi.org/10.15468/dl.f8pqjw, https://doi.org/10.15468/dl.qbmyb2, https://doi.org/10.15468/dl.fc2hqk, https://doi.org/10.15468/dl.sq5d6f
9. Gerstner, K., Dormann, C.F., Stein, A., Manceur, A.M., Seppelt, R.: Editor's choice: review: effects of land use on plant diversity-a global meta-analysis. J. Appl. Ecol. **51**(6), 1690–1700 (2014)
10. Ghazi, M.M., Yanikoglu, B., Aptoula, E.: Plant identification using deep neural networks via optimization of transfer learning parameters. Neurocomputing **235**, 228–235 (2017)
11. He, K., Zhang, X., Ren, S., Sun, J.: Deep residual learning for image recognition. In: Proceedings of the IEEE Conference on Computer Vision and Pattern Recognition, pp. 770–778 (2016)
12. Huang, Z., Wang, X., Wang, J., Liu, W., Wang, J.: Weakly-supervised semantic segmentation network with deep seeded region growing. In: Proceedings of the IEEE Conference on Computer Vision and Pattern Recognition, pp. 7014–7023 (2018)
13. Iandola, F., Moskewicz, M., Karayev, S., Girshick, R., Darrell, T., Keutzer, K.: DenseNet: implementing efficient convnet descriptor pyramids. arXiv preprint arXiv:1404.1869 (2014)
14. Kattenborn, T., Eichel, J., Wiser, S., Burrows, L., Fassnacht, F.E., Schmidtlein, S.: Convolutional neural networks accurately predict cover fractions of plant species and communities in unmanned aerial vehicle imagery. Remote Sens. Ecol. Conserv. **6**, 472–486 (2020)

15. Kim, B., Han, S., Kim, J.: Discriminative region suppression for weakly-supervised semantic segmentation. arXiv preprint arXiv:2103.07246 (2021)
16. Kingma, D.P., Ba, J.: Adam: a method for stochastic optimization. arXiv preprint arXiv:1412.6980 (2014)
17. Körschens, M., Bodesheim, P., Römermann, C., Bucher, S.F., Ulrich, J., Denzler, J.: Towards confirmable automated plant cover determination. In: Bartoli, A., Fusiello, A. (eds.) ECCV 2020. LNCS, vol. 12540, pp. 312–329. Springer, Cham (2020). https://doi.org/10.1007/978-3-030-65414-6_22
18. Körschens, M., Bodesheim, P., Römermann, C., Bucher, S.F., Ulrich, J., Denzler, J.: Automatic plant cover estimation with convolutional neural networks. arXiv preprint arXiv:2106.11154 (2021)
19. Lee, S.H., Chan, C.S., Wilkin, P., Remagnino, P.: Deep-plant: plant identification with convolutional neural networks. In: 2015 IEEE International Conference on Image Processing (ICIP), pp. 452–456. IEEE (2015)
20. Lin, T.Y., Dollár, P., Girshick, R., He, K., Hariharan, B., Belongie, S.: Feature pyramid networks for object detection. In: Proceedings of the IEEE Conference on Computer Vision and Pattern Recognition, pp. 2117–2125 (2017)
21. Ling, H., Acuna, D., Kreis, K., Kim, S.W., Fidler, S.: Variational amodal object completion. In: Advances in Neural Information Processing Systems 33 (2020)
22. Liu, H., et al.: Shifting plant species composition in response to climate change stabilizes grassland primary production. Proc. Natl. Acad. Sci. **115**(16), 4051–4056 (2018)
23. Lloret, F., Peñuelas, J., Prieto, P., Llorens, L., Estiarte, M.: Plant community changes induced by experimental climate change: seedling and adult species composition. Perspect. Plant Ecol. Evol. Syst. **11**(1), 53–63 (2009)
24. Lu, H., Cao, Z., Xiao, Y., Zhuang, B., Shen, C.: TasselNet: counting maize tassels in the wild via local counts regression network. Plant Methods **13**(1), 79 (2017)
25. Milletari, F., Navab, N., Ahmadi, S., V-Net: fully convolutional neural networks for volumetric medical image segmentation. In: Proceedings of the 2016 Fourth International Conference on 3D Vision (3DV), pp. 565–571
26. Muhammad, M.B., Yeasin, M.: Eigen-CAM: class activation map using principal components. In: 2020 International Joint Conference on Neural Networks (IJCNN), pp. 1–7. IEEE (2020)
27. Pfadenhauer, J.: Vegetationsökologie - ein Skriptum. IHW-Verlag, Eching, 2. verbesserte und erweiterte auflage edn. (1997)
28. Pinheiro, P.O., Collobert, R.: From image-level to pixel-level labeling with convolutional networks. In: Proceedings of the IEEE Conference on Computer Vision and Pattern Recognition, pp. 1713–1721 (2015)
29. Ronneberger, O., Fischer, P., Brox, T.: U-Net: convolutional networks for biomedical image segmentation. In: Navab, N., Hornegger, J., Wells, W.M., Frangi, A.F. (eds.) MICCAI 2015. LNCS, vol. 9351, pp. 234–241. Springer, Cham (2015). https://doi.org/10.1007/978-3-319-24574-4_28
30. Rosenzweig, C., et al.: Assessment of observed changes and responses in natural and managed systems. In: Climate Change 2007: Impacts, Adaptation and Vulnerability. Contribution of Working Group II to the Fourth Assessment Report of the Intergovernmental Panel on Climate Change, pp. 79–131 (2007)
31. Russakovsky, O., et al.: Imagenet large scale visual recognition challenge. Int. J. Comput. Vision **115**(3), 211–252 (2015). https://doi.org/10.1007/s11263-015-0816-y

32. Singh, K.K., Lee, Y.J.: Hide-and-seek: forcing a network to be meticulous for weakly-supervised object and action localization. In: 2017 IEEE International Conference on Computer Vision (ICCV), pp. 3544–3553. IEEE (2017)
33. Souza, L., Zelikova, T.J., Sanders, N.J.: Bottom-up and top-down effects on plant communities: nutrients limit productivity, but insects determine diversity and composition. Oikos **125**(4), 566–575 (2016)
34. Stammes, E., Runia, T.F., Hofmann, M., Ghafoorian, M.: Find it if you can: end-to-end adversarial erasing for weakly-supervised semantic segmentation. arXiv preprint arXiv:2011.04626 (2020)
35. Türke, M., et al.: Multitrophische biodiversitätsmanipulation unter kontrollierten umweltbedingungen im iDiv ecotron. In: Lysimetertagung, pp. 107–114 (2017)
36. Ulrich, J., et al.: Invertebrate decline leads to shifts in plant species abundance and phenology. Front. Plant Sci. **11**, 1410 (2020)
37. Wei, Y., Feng, J., Liang, X., Cheng, M.M., Zhao, Y., Yan, S.: Object region mining with adversarial erasing: a simple classification to semantic segmentation approach. In: Proceedings of the IEEE Conference on Computer Vision and Pattern Recognition, pp. 1568–1576 (2017)
38. Xiong, H., Cao, Z., Lu, H., Madec, S., Liu, L., Shen, C.: TasselNetv2: in-field counting of wheat spikes with context-augmented local regression networks. Plant Methods **15**(1), 150 (2019)
39. Yalcin, H., Razavi, S.: Plant classification using convolutional neural networks. In: 2016 Fifth International Conference on Agro-Geoinformatics (Agro-Geoinformatics), pp. 1–5. IEEE (2016)
40. Yao, Q., Gong, X.: Saliency guided self-attention network for weakly and semi-supervised semantic segmentation. IEEE Access **8**, 14413–14423 (2020)
41. Yun, S., Han, D., Oh, S.J., Chun, S., Choe, J., Yoo, Y.: CutMix: regularization strategy to train strong classifiers with localizable features. In: Proceedings of the IEEE/CVF International Conference on Computer Vision, pp. 6023–6032 (2019)
42. Zhan, X., Pan, X., Dai, B., Liu, Z., Lin, D., Loy, C.C.: Self-supervised scene de-occlusion. In: Proceedings of the IEEE/CVF Conference on Computer Vision and Pattern Recognition, pp. 3784–3792 (2020)
43. Zhang, B., Xiao, J., Wei, Y., Sun, M., Huang, K.: Reliability does matter: an end-to-end weakly supervised semantic segmentation approach. In: Proceedings of the AAAI Conference on Artificial Intelligence, vol. 34, pp. 12765–12772 (2020)
44. Zhang, B., Zhao, Q., Feng, W., Lyu, S.: AlphaMEX: a smarter global pooling method for convolutional neural networks. Neurocomputing **321**, 36–48 (2018)
45. Zhang, X., Wei, Y., Feng, J., Yang, Y., Huang, T.S.: Adversarial complementary learning for weakly supervised object localization. In: Proceedings of the IEEE Conference on Computer Vision and Pattern Recognition, pp. 1325–1334 (2018)
46. Zhang, X., Wei, Y., Kang, G., Yang, Y., Huang, T.: Self-produced guidance for weakly-supervised object localization. In: Ferrari, V., Hebert, M., Sminchisescu, C., Weiss, Y. (eds.) ECCV 2018. LNCS, vol. 11216, pp. 610–625. Springer, Cham (2018). https://doi.org/10.1007/978-3-030-01258-8_37
47. Zhang, X., Wei, Y., Yang, Y., Wu, F.: Rethinking localization map: towards accurate object perception with self-enhancement maps. arXiv preprint arXiv:2006.05220 (2020)
48. Zhou, B., Khosla, A., Lapedriza, A., Oliva, A., Torralba, A.: Learning deep features for discriminative localization. In: Proceedings of the IEEE Conference on Computer Vision and Pattern Recognition, pp. 2921–2929 (2016)

How Reliable Are Out-of-Distribution Generalization Methods for Medical Image Segmentation?

Antoine Sanner(✉), Camila González, and Anirban Mukhopadhyay

Technical University of Darmstadt, Karolinenpl. 5, 64289 Darmstadt, Germany
{antoine.sanner,camila.gonzalez,
anirban.mukhopadhyay}@GRIS.TU-DARMSTADT.DE

Abstract. The recent achievements of Deep Learning rely on the test data being similar in distribution to the training data. In an ideal case, Deep Learning models would achieve **Out-of-Distribution (OoD) Generalization**, i.e. reliably make predictions on out-of-distribution data. Yet in practice, models usually fail to generalize well when facing a shift in distribution. Several methods were thereby designed to improve the robustness of the features learned by a model through **Regularization-** or **Domain-Prediction-based** schemes. Segmenting medical images such as MRIs of the hippocampus is essential for the diagnosis and treatment of neuropsychiatric disorders. But these brain images often suffer from distribution shift due to the patient's age and various pathologies affecting the shape of the organ. In this work, we evaluate OoD Generalization solutions for the problem of hippocampus segmentation in MR data using both fully- and semi-supervised training. We find that no method performs reliably in all experiments. Only the **V-REx** loss stands out as it remains easy to tune, while it outperforms a standard U-Net in most cases.

Keywords: Semantic segmentation · Medical images · Out-of-Distribution Generalization

1 Introduction

Semantic segmentation of medical images is an important step in many clinical procedures. In particular, the segmentation of the hippocampus from MRI scans is essential for the diagnosis and treatment of neuropsychiatric disorders. Automated segmentation methods have improved vastly in the past years and now yield promising results in many medical imaging applications [11]. These methods can technically exploit the information contained in large datasets.

Supported by the Bundesministerium für Gesundheit (BMG) with grant [ZMVI1-2520DAT03A]. The final authenticated version of this manuscript will be published in Lecture Notes in Pattern recognition in the life and natural sciences.

C. Bauckhage et al. (Eds.): DAGM GCPR 2021, LNCS 13024, pp. 604–617, 2021.
https://doi.org/10.1007/978-3-030-92659-5_39

However, no matter how large a training dataset is or how good the results on the in-distribution data are, methods may fail on **Out-of-Distribution (OoD)** data. **OoD Generalization** remains crucial for the reliability of deep neural networks, as insufficient generalization may vastly limit their implementation in practical applications.

Distribution shifts occur when the data at test time is different in distribution from the training data. In the context of hippocampus segmentation, the age of the patient [13] and various pathologies [4] can affect the shape of the organ. Using a different scanner or simply using different acquisition parameters can also cause a distribution shift [5].

OoD Generalization alleviates this issue by training a model such that it generalizes well to new distributions at test time without requiring any further training. Several strategies exist to approach this. **Regularization-based** approaches enforce the learning of robust features across training datasets [2,9], which can reliably be used to produce an accurate prediction regardless of context. On the other hand, **Domain-Prediction-based** methods focus on harmonizing between domains [6,7] by including a domain predictor in the architecture.

In this work, we perform a thorough evaluation of several state-of-the-art OoD Generalization methods for segmenting the hippocampus. We find that no method performs reliably across all experiments. Only the Regularization-based **V-REx** loss stands out as it remains easy to tune, while its worst results remains relatively good.

2 Methods

The setting for Out-of-Distribution Generalization has been defined [2] as follows. Data is collected from multiple environments and the source of each data point is known. An *environment* describes a set of conditions under which the data has been measured. One can for instance obtain a different environment by using a different scanner or studying a different group of patients. These environments contain spurious correlations due for instance to dataset biases.

Only fully labeled datasets from a limited set of environments is available during training, and the goal is to learn a predictor, which performs well on all environments. As such, a model trained using this method will theoretically perform well on unseen but semantically related data.

2.1 Regularization-based Methods

Arjovsky et al. [2] formally define the problem of OoD Generalization as follows. Consider the datasets $D_e := \{(x_i^e, y_i^e)\}_{i=1}^{n_e}$ collected under multiple training environments $e \in \mathcal{E}_{tr}$. These environments describe the same pair of random variables measured under different conditions. The dataset D_e, from environment e, contains examples identically and independently distributed according to some probability distribution $P(X_e, Y_e)$. The goal is then to learn a predictor

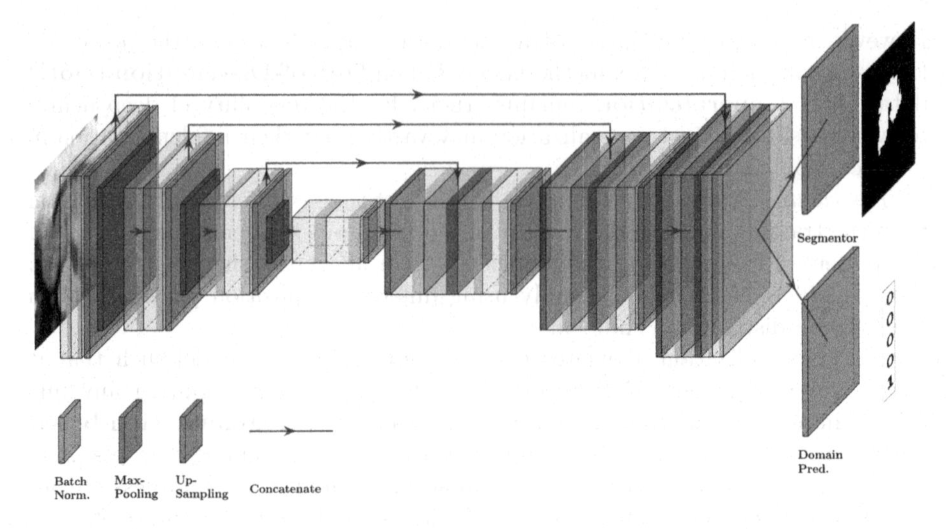

Fig. 1. Network architecture for the segmentation task with Domain-prediction mechanism. The U-Net generates features, which are then used by the segmentor for the main task and by the auxiliary head for domain prediction. This schema contains some simplifications: the segmentor and domain predictor are implemented respectively using a convolutional layer and a CNN. The architecture displayed is 2D, while the implementation is 3D.

$Y \approx f(X)$, which performs well across all unseen related environments $\mathcal{E}_{all} \supset \mathcal{E}_{tr}$. The objective is to minimize:

$$R^{OoD}(f) = \max_{e \in \mathcal{E}_{all}} \mathcal{R}^e(f)$$

where $\mathcal{R}^e(f) := \mathbb{E}_{X^e, Y^e}[\ell(f(X^e), Y^e)]$ is the risk under environment e.

IRMv1 [2] and Risk Extrapolation (**V-REx** and **MM-Rex**) [9] add regularization to the training loss to enforce strict equality between training risks by finding a data representation common to all environments. **IRM Games** [1] poses the problem as finding the Nash equilibrium of an ensemble game.

More precisely, Krueger et al. [9] propose to solve this problem by minimizing the following loss (V-REx):

$$\min_{\mathcal{X} \rightarrow \mathcal{Y}} \lambda \cdot \mathrm{Var}\{R^1(\Phi), ..., R^{n_e}(\Phi)\} + \frac{1}{|\mathcal{E}_{tr}|} \sum_{e \in \mathcal{E}_{tr}} R^e(\Phi)$$

where Φ is the invariant predictor, the scalar $\lambda \geq 0$ controls the balance between reducing average risk and enforcing the equality of risks, and Var stands for the variance between the risks across training environments. To the best of our knowledge, these approaches have not been evaluated yet on image segmentation tasks.

2.2 Domain-Prediction-based Methods

Given data from multiple sources, **Domain-Prediction-based** methods find a harmonized data representation such that all information relating to the source domain of the image is removed [6,7,14]. This goal can be achieved by appending another head to the network, which acts as a domain classifier. During training, the ability of the domain classifier to predict the domain is minimized to random chance, thus reducing the domain-specific information in the data representation. Domain-Prediction-based methods differ from Regularization-based methods in that they rather remove the need to annotate all images from a new target dataset to be able to train a model on new unlabeled data, and so allow leveraging non-annotated data.

As described by Dinsdale et al. [6], the network shown in Fig. 1 is a modified U-Net with a second head which acts as a domain predictor. The final goal is to find a representation that maximizes the performance on a segmentation task with input images $\mathbf{X_d} \in \mathbb{R}^{B_d \times W \times H \times D}$ and task labels $\mathbf{Y_d} \in \mathbb{R}^{B_d \times W \times H \times D \times C}$ while minimizing the performance of the domain predictor. This network is composed of an encoder, a segmentor, and a domain predictor with respective weights Θ_{repr}, Θ_{seg}, and Θ_{dp}.

The network is trained iteratively by minimizing the loss on the segmentation task L_s, the domain loss L_d, and the confusion loss L_{conf}. The confusion loss penalizes a divergence of the domain predictor's prediction from a uniform distribution and is used to remove source-related information from Θ_{repr}. It is also important that the segmentation loss L_s be evaluated separately for the data from each dataset to prevent the performance being driven by only one dataset, if their sizes vary significantly. In our work, the segmentation loss L_s takes the form of the sum of a Sorensen-Dice loss and a binary cross-entropy loss. The domain loss L_d is used to assess how much information remains in about the domains.

The Domain-Prediction method thus minimizes the total loss function:

$$\begin{aligned} L(\mathbf{X}, \mathbf{Y}, \mathbf{D}, \Theta_{repr}, \Theta_{seg}, \Theta_{dp}) = & \sum_{d \in \mathcal{E}_{tr}} L_s(\mathbf{X_d}, \mathbf{Y_d}; \Theta_{repr}, \Theta_{seg}) \\ & + \alpha \cdot L_d(\mathbf{X}, \mathbf{D}, \Theta_{repr}; \Theta_{dp}) \\ & + \beta \cdot L_{conf}(\mathbf{X}, \mathbf{D}, \Theta_{dp}; \Theta_{repr}) \end{aligned}$$

where α and β represent weights of the relative contributions for the different loss functions. $(\mathbf{X_d}, \mathbf{Y_d})$ corresponds to the labeled data available for the domain d, while $(\mathbf{X}, \mathbf{D})$ corresponds to all the images (labeled and unlabeled) available and their corresponding domain. This is interesting as training the domain predictor does not require labeled data and semi-supervised training can be used to find a harmonized data representation.

2.3 A Combined Method for OoD Generalization

The losses introduced in the context of Regularization-based OoD Generalization can be combined to the Domain-Prediction mechanism. We choose to apply the V-REx loss to the segmentor only. The intuition is that if the segmentor learns robust features through regularization, then there will be less source-related information in the extracted features and can make the Domain-Prediction part of the method easier. Besides, Domain Prediction schemes already require to compute the segmentation loss for each domain separately, which is also a requirement for Regularization-based methods. This method aims to minimize the following loss function:

$$\begin{aligned}
\min_{\mathcal{X}\to\mathcal{Y}} \lambda \cdot \mathrm{Var}\{&L_s(\mathbf{X_1}, \mathbf{Y_1}; \Theta_{repr}, \Theta_{seg}), ..., L_s(\mathbf{X_{n_d}}, \mathbf{Y_{n_d}}; \Theta_{repr}, \Theta_{seg})\} \\
&+ \sum_{d=1}^{n_d} L_s(\mathbf{X_d}, \mathbf{Y_d}; \Theta_{repr}, \Theta_{seg}) \\
&+ \alpha \cdot L_d(\mathbf{X}, \mathbf{D}, \Theta_{repr}; \Theta_{dp}) \\
&+ \beta \cdot L_{conf}(\mathbf{X}, \mathbf{D}, \Theta_{dp}; \Theta_{repr})
\end{aligned}$$

3 Experimental Setup

In the following, we present the datasets used in this work and detail the evaluation strategy used.

3.1 Datasets

We use a corpus of three datasets for the task of hippocampus segmentation from various studies. The *Decath* dataset [12] contains MR images of both healthy adults and schizophrenia patients. The *HarP* dataset [3] is the product of an effort from the European Alzheimer's Disease Consortium and Alzheimer's Disease Neuroimaging Initiative to harmonize the available protocols and contains MR images of senior healthy subjects and Alzheimer's disease patients. Lastly, the *Dryad* dataset [10] contains MR images of healthy young adults.

Since both *HarP* and *Dryad* datasets provide whole-head MR images at the same resolution, we crop each scan at two fixed positions, respectively for the right and left hippocampus. The shape of the resulting crops (64, 64, 48) fits every hippocampus in both datasets. Since segmentation instances from the *Decath* dataset are smaller, they are centered and zero-padded to the same shape. Similarly, as all the datasets do not provide the same number of classes in their annotation, we restrict the classes to "background" and "hippocampus".

3.2 Evaluation

We first train three instances of a 3D U-Net, using 3 layers for encoding, and decoding with no dropout, respectively on each of the datasets to assess how well the features learned on a dataset generalize to the other datasets. We then evaluate the proposed method with five-fold cross-validation. In the context of fully supervised training, a model is trained on two of the datasets and evaluated on the third dataset. 10% of the data in each fold is used for validation and guiding the training schedule. The dataset that is not used to train the model is considered entirely as test data. Due to the nature of the task, we need to find a hyper-parameter/data augmentation configuration that allows the model to generalize well on the test dataset no matter which pair of training dataset is used.

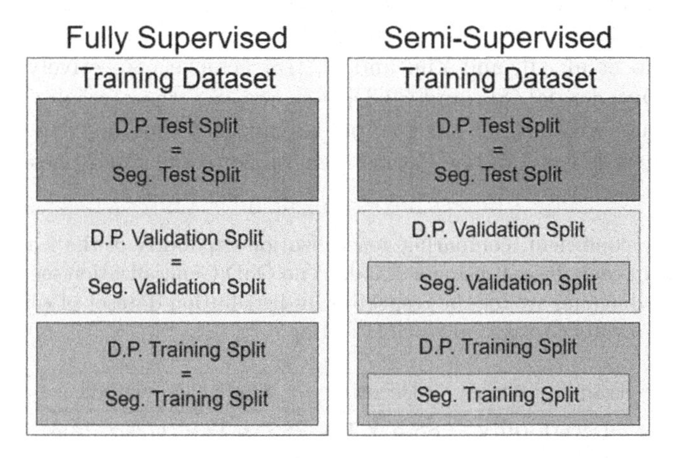

Fig. 2. Schema of testing/validation/training splits in the context of fully supervised and semi-supervised training. The test splits remain equal while the segmentor's training and validation splits are subsets of the splits for the domain predictor. "D.P." refers to the domain predictor and "Seg." to the segmentor.

For semi-supervised training, we train our models on all 3 datasets. The splits for the domain predictor are computed the same way as during fully supervised training. As for the segmentor, its training and validation splits are subsets of the domain predictor's respective splits, as shown in Fig. 2. The test splits for both heads remain equal. The splits are computed only once and reused for each method to reduce the variance during testing.

The network is implemented using Python 3.8 and PyTorch 1.6.0. Some light data augmentation is also used: *RandomAffine*, *RandomFlip*, *RandomMotion*. The batch size for each dataset is selected so that the training duration is the same with all datasets.

4 Results

In this section, we first evaluate the generality of the features learned by a standard U-Net on each dataset. We then compare the different methods in a fully supervised settings, and we outline their limitations. Finally, we assess whether Domain-Prediction-based methods can leverage their ability of training on unlabeled data.

4.1 U-Net Results on Each Dataset

We train three instances of a standard 3D U-Net respectively on each of the datasets, in order to assess how well features learned by a model on a given dataset generalize well to the other two datasets. The results can be seen in Table 1. Each of three trained models achieves a mean Dice score in the range of 84 to 90 with a low standard deviation below 2. As a comparison, Isensee et al. [8], Carmo et al. [4], and Zhu and al. [15] achieve respectively on Decath, HarP, and Dryad a 90, 86, and 89 Dice scores. So, these results are close to state-of-the-art, which is not our goal per se, but they should give a good insight on how features learned in one dataset generalize to the other ones.

Table 1. Dice coefficient, comparing generalization capability of the features learned by the U-Net in each dataset using a standard no OoD Generalization mechanism. The results on the diagonal are for the respective in-distribution dataset of each model and otherwise the results correspond to OoD testing.

Training dataset	Decath	HarP	Dryad
Decath **only**	$\mathbf{88.5 \pm 1.4}$	28.7 ± 12.0	17.1 ± 14.3
HarP **only**	76.5 ± 2.4	$\mathbf{84.6 \pm 1.4}$	80.0 ± 1.7
Dryad **only**	59.0 ± 4.8	57.9 ± 1.9	$\mathbf{85.2 \pm 1.9}$

The performance of the three models on OoD Generalization differ greatly. While the model trained on HarP already achieves great results, the other two models seem to struggle much more to generalize and attain a much lower average and higher standard deviation. So, training each of the three pairs of datasets provides a different setting and allows testing multiple scenarios.

4.2 Fully Supervised Training

For all dataset configuration, we use two datasets for training and the third one for OoD testing. The training datasets are also referred to as in-distribution datasets.

Training on Decath and HarP: In this first setup, we train all models on Decath and HarP. The results can be seen in Table 2. Training a U-Net only on HarP already learns features that generalize well, so we are interested in seeing how adding "lesser quality" data (Sect. 4.1) influences the results on Dryad. The results are shown in Table 2. If we first focus on the results on the in-distribution datasets, we observe that all methods achieve state-of-the-art results. All methods achieve satisfying results on the out-of-distribution dataset with the combined method having a lead. The domain prediction accuracy reaches near random results for all Domain-Prediction-based methods, which indicates that the domain-identifying information is removed.

Table 2. Dice coefficients and domain prediction accuracy (when applicable). A **bold** dataset name denotes that the dataset is used for training.

Method	**Decath**	**HarP**	Dryad	Dom. Pred. Acc
U-Net	89.5 ± 0.4	85.7 ± 0.7	81.2 ± 1.5	–
U-Net + V-REx	89.4 ± 0.6	85.3 ± 1.2	81.7 ± 0.5	–
Domain-Prediction	$\mathbf{90.0 \pm 0.4}$	$\mathbf{86.8 \pm 1.3}$	82.8 ± 2.5	48.6 ± 6.2
Combined	$\mathbf{88.9 \pm 0.7}$	85.2 ± 0.8	$\mathbf{84.6 \pm 0.4}$	56.1 ± 10.9

Training on Decath and Dryad: This training setup is perhaps the most interesting one of the three, as we have seen that training a model on either of Decath and Dryad does not yield good generalization results. As shown in Table 3, all methods yield state-of-the-art results on both in-distribution datasets. Surprisingly, all methods perform well on HarP, although the regular U-Net to a lesser degree. Regarding domain prediction accuracy, the Domain-Prediction method reaches once more near random chance results. The domain prediction accuracy is particularly high for the combined method, since these results stem from an intermediary training stage where the segmentor and domain predictor are trained jointly.

Table 3. Dice coefficients and domain prediction accuracy (when applicable). A **bold** dataset name denotes that the dataset is used for training.

Method	**Decath**	HarP	**Dryad**	Dom. Pred. Acc
U-Net	88.7 ± 0.7	69.8 ± 2.4	89.2 ± 0.5	–
U-Net + V-REx	89.3 ± 0.2	$\mathbf{73.7 \pm 1.9}$	89.9 ± 0.3	–
Domain-Prediction	$\mathbf{89.6 \pm 0.4}$	72.9 ± 2.4	$\mathbf{90.5 \pm 0.3}$	56.5 ± 11.5
Combined	$\mathbf{89.6 \pm 0.2}$	73.2 ± 1.5	$\mathbf{90.5 \pm 0.3}$	90.0 ± 6.3

Training on HarP and Dryad: In this last setup, we train the models on HarP and Dryad and test them on Decath. The results can be seen in Table 4. All methods perform almost equally well on in-distribution datasets. However, they differ vastly on the first column, with the U-Net + V-REx method being the clear winner. The regular U-Net and the Domain-Prediction method perform similarly, with the latter having a smaller variance. Removing domain-related information from the U-Net features does not seem to be enough as both Domain-Prediction methods achieve a near random domain prediction accuracy, which indicates that domain-identifying information has been removed, but the combined method vastly underperforms.

Table 4. Dice coefficients and domain prediction accuracy (when applicable). The usual data augmentation scheme is used. A **bold** dataset name denotes that the dataset is used for training.

Method	Decath	**HarP**	**Dryad**	Dom. Pred. Acc
U-Net	65.6 ± 18.1	84.9 ± 0.8	89.8 ± 0.7	–
U-Net + V-REx	**78.1 ± 2.6**	85.3 ± 0.8	90.1 ± 0.3	–
Domain-Prediction	69.4 ± 6.4	**86.3 ± 0.8**	**90.6 ± 0.6**	54.4 ± 7.0
Combined	46.2 ± 10.6	84.2 ± 1.1	89.7 ± 0.9	52.6 ± 9.3

If we observe the segmentation masks predicted by models from different folds of the same cross-validation, we can see in Fig. 3 that while the first prediction matches the ground truth nicely, the second prediction fails to fully recognize the hippocampus and is very noisy. What is troublesome is that both models achieve Dice scores of 86% and 90% respectively for the test splits of in-distribution datasets. The domain prediction accuracy is reduced to random chance without causing any significant drop in performance on the segmentation task. As such, the Domain-Prediction method reduces domain information in the model features as expected. Reducing domain prediction accuracy does not seem to be enough however, as the combined method also achieves low domain detection results, but fares worse than a regular U-Net on the test dataset. Overall, we have no way of predicting whether a model is usable on new data by looking at the metrics on the validation data.

It is worth noting that by adding some *RandomBiasField* and *RandomNoise* to the data augmentation scheme, both Domain-Prediction-based methods perform significantly better, as can be seen in Table 5. The results on the training datasets remain stable with a decrease in standard deviation. The Dice scores on Decath, however, significantly improve. However, this change cannot be justified by just looking at the metrics on validation data. Furthermore, this scheme yields significantly worse results when training on any other dataset pair, which is why it is not used anywhere else.

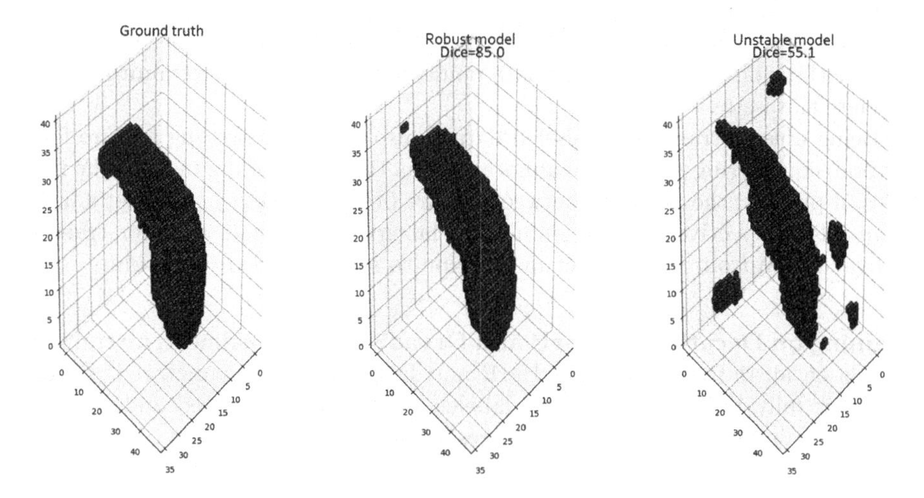

Fig. 3. Segmentation mask prediction for an instance from Decath. Both models are trained using the Domain-Prediction-based method but with different training data folds and achieve respectively 78.9% and 59.9% Dice score on the out-of-distribution dataset Decath. Dice scores on the test splits of the in-distribution datasets are within 1% of each other.

Table 5. Dice coefficients and domain prediction accuracy (when applicable) for a different data augmentation scheme. A **bold** dataset name denotes that the dataset is used for training.

Method	Decath	**HarP**	**Dryad**	Dom. Pred. Acc
Domain-Prediction	**75.1 $\pm$ 1.7**	**84.2 $\pm$ 0.9**	**90.0 $\pm$ 0.3**	57.8 $\pm$ 7.6
Combined	74.1 $\pm$ 4.3	82.3 $\pm$ 1.2	89.0 $\pm$ 0.6	61.2 $\pm$ 3.8

Discussion. OoD Generalization methods seem to be quite limiting in the sense that they require the user to find a hyper-parameter configuration that works well in all setups. The results on the validation data are too good to allow us to fine tune the hyper-parameters for each method in each of the settings. As such, no method tested gives reliable results, although the U-Net + V-REx method always performs at least on par, and often better than, the reference U-Net.

We also observe that **we have no way of predicting whether a model is usable on new data by looking at the metrics on the validation data**. The Dice scores on the validation data are not representative of how well a model generalizes on out-of-distribution data. A near random guessing domain prediction accuracy is also not sufficient, as seen in Sect. 4.2.

4.3 Semi-supervised Training

During the fully supervised training, we noted that Domain-Prediction-based methods trained using HarP and Dryad can fail to fully recognize the hippocampus and produce very noisy segmentations. The main goal of these experiments is to test whether adding a few labeled data points from Decath to the training set can alleviate this issue.

Being able to train on unlabeled data is also a very interesting feature for a method, considering that image annotation is a time-consuming task. In the next titles, if present, the numbers between square brackets next to a dataset's name mean refer respectively to the number of labeled data points used for training and validation from this dataset in addition to the unlabeled data points.

Training on Decath[5,5], HarP and Dryad: In this experiment, we train the models on all datasets but only HarP and Dryad are fully annotated. Decath is almost fully unlabeled. Only 10 labeled points are considered and are split between training and validation. The results are shown in Table 6.

Table 6. Dice coefficients and domain prediction accuracy (when applicable). A **bold** dataset name denotes that the dataset is used for training. If present, the numbers between brackets next to an in-distribution dataset's name refer respectively to the number of data points used for training and validation from this dataset.

Method	**Decath[5, 5]**	**HarP**	**Dryad**	Dom. Pred. Acc
U-Net	$\mathbf{88.1 \pm 0.6}$	$\mathbf{85.9 \pm 1.0}$	$\mathbf{90.5 \pm 0.5}$	–
U-Net + V-REx	87.8 ± 0.8	85.7 ± 1.0	$\mathbf{90.5 \pm 0.4}$	–
Domain-Prediction	87.4 ± 0.7	85.4 ± 1.4	90.0 ± 0.5	56.6 ± 5.2
Combined	86.9 ± 1.7	85.8 ± 1.1	90.3 ± 0.6	57.2 ± 1.5

All methods perform similarly on the segmentation task and reach state-of-the-art results on all datasets. The Domain-Prediction-based methods decrease the domain prediction accuracy, but it remains far better than random chance (55% vs 33%). During training, sudden variations in metrics for the segmentation tasks were observed even at learning rates as low as $5 \cdot 10^{-5}$. However, the results achieved here are only slightly worse than the one reached during fully-supervised training. As such, we expect that refining this tuning only leads to marginal gains on the segmentation task. Overall, no method here is able to outperform the others and yield significant improvement over using a standard U-Net.

Training on Decath[6,4], HarP[6,4] and Dryad[6,4]: From what we have seen in Sect. 4.3, training models on two fully labeled datasets and one partially labeled one does not allow discriminating between methods. Instead, we can consider that we have multiple dataset available, but only a few labeled instances for each of them. With only 18 instances being used for training, we want to see whether one method outperforms the others.

As shown in Table 7, all methods perform almost equally well. The results for the U-Net and U-Net + V-REx methods are impressive, considering that

Table 7. Dice coefficients label and domain prediction accuracy (when applicable). A **bold** dataset name denotes that the dataset is used for training. If present, the numbers between parentheses next to an in-distribution dataset's name refer respectively to the number of data points used for training and validation from this dataset.

Method	**Decath[6, 4]**	**HarP[6, 4]**	**Dryad[6, 4]**	Dom. Pred. Acc
U-Net	86.7 ± 0.7	80.2 ± 1.4	87.3 ± 1.2	–
U-Net + V-REx	**86.9 ± 0.5**	**80.7 ± 1.0**	**87.5 ± 1.0**	–
Domain-Prediction	82.9 ± 3.3	75.7 ± 4.8	85.6 ± 1.5	56.1 ± 4.4
Combined	85.6 ± 0.5	78.4 ± 2.0	86.4 ± 0.7	54.7 ± 3.7

there is only a slight drop in performance now that we are training these models using far fewer labeled data points. The results for the Domain-Prediction-based methods are slightly worse, and the domain prediction accuracy remains relatively high for both methods. These results are coherent with the ones obtained during the previous experiment as the same hyper-parameter combination is used, although the combined method achieves better results than the standard Domain-Prediction method on the second column.

Discussion. Overall, no method reliably outperforms the standard U-Net. Domain-Prediction-based methods actually performs worse than the reference in the last experiment. Either removing domain related information does not scale well with the number of training domains, or a much finer hyper-parameter tuning is required to allow these methods to leverage their ability to train on unlabeled data. The feasibility of this tuning can also be put in question due to the increase in instability observed during training.

5 Conclusion

In this work, we evaluate a variety of methods for **Out-of-Distribution Generalization** on the task of hippocampus segmentation. In particular, we evaluate **Regularization-based** methods which regularize the performance of a model across training environments. We also consider a **Domain-Prediction-based** method, which adds a domain classifier to the architecture, its goal being to reduce the prediction accuracy of this classifier. Lastly, we explore how well a method uniting both approaches performs.

We compare these methods in a fully supervised setting, and we observe the limitations of OoD Generalization methods. No method performs reliably in all experiments. Only the **V-REx** loss stands out as it remains easy to tune, while its worst results remains close to the reference U-Net.

To gauge the ability of Domain-Prediction-based methods to train on unlabeled data, we subsequently evaluate all methods in a semi-supervised settings, using the minimum number of labeled images required to retain a stable training trajectory. The model trained with a V-REx loss maintains impressive results despite the lack of training data, while Domain-Prediction-based methods show their limits when training on an increased number of domains.

The combined method achieves good results in some settings, but suffers from instability in others. Domain-Prediction-based methods have a lot of potential, but require a lot of fine-tuning and can cause the training to slow down.

In the future, we wish to evaluate these methods on another corpus of datasets. This different setting will allow us to evaluate whether the results observed here still hold, especially whether Domain-Prediction-based methods can make use of training on unlabeled data during semi-supervised training. We also wish to confirm whether using the V-REx loss remains a good option in another setting, as it achieves similar or better results than a standard loss on this corpus of datasets.

References

1. Ahuja, K., Shanmugam, K., Varshney, K.R., Dhurandhar, A.: Invariant risk minimization games (2020). http://arxiv.org/abs/2002.04692
2. Arjovsky, M., Bottou, L., Gulrajani, I., Lopez-Paz, D.: Invariant risk minimization. http://arxiv.org/abs/1907.02893
3. Boccardi, M., et al.: Training labels for hippocampal segmentation based on the EADC-ADNI harmonized hippocampal protocol **11**(2), 175–183 (2015). https://doi.org/10.1016/j.jalz.2014.12.002, https://linkinghub.elsevier.com/retrieve/pii/S155252601402891X
4. Carmo, D., Silva, B., Yasuda, C., Rittner, L., Lotufo, R.: Hippocampus segmentation on epilepsy and Alzheimer's disease studies with multiple convolutional neural networks (2020). http://arxiv.org/abs/2001.05058
5. Castro, D.C., Walker, I., Glocker, B.: Causality matters in medical imaging. Nat. Commun. **11**(1) (2020). https://doi.org/10.1038/s41467-020-17478-w
6. Dinsdale, N.K., Jenkinson, M., Namburete, A.I.L.: Deep learning-based unlearning of dataset bias for MRI harmonisation and confound removal. bioRxiv (2020). https://doi.org/10.1101/2020.10.09.332973, https://www.biorxiv.org/content/early/2020/12/14/2020.10.09.332973
7. Ganin, Y., Lempitsky, V.: Unsupervised domain adaptation by backpropagation (2015). http://arxiv.org/abs/1409.7495
8. Isensee, F., et al.: nnU-net: self-adapting framework for U-net-based medical image segmentation (2018). http://arxiv.org/abs/1809.10486
9. Krueger, D., et al.: Out-of-distribution generalization via risk extrapolation (REx) (2020). http://arxiv.org/abs/2003.00688
10. Kulaga-Yoskovitz, J., et al.: Multi-contrast submillimetric 3 tesla hippocampal subfield segmentation protocol and dataset **2**(1), 150059 (2015). https://doi.org/10.1038/sdata.2015.59, http://www.nature.com/articles/sdata201559
11. Litjens, G., et al.: A survey on deep learning in medical image analysis. Med. Image Anal. **42** (2017). https://doi.org/10.1016/j.media.2017.07.005
12. Simpson, A.L., et al.: A large annotated medical image dataset for the development and evaluation of segmentation algorithms (2019). http://arxiv.org/abs/1902.09063

13. Xu, Y., et al.: Age effects on hippocampal structural changes in old men: the HAAS. NeuroImage **40**(3), 1003–1015 (2008) https://doi.org/10.1016/j.neuroimage.2007.12.034, https://www.sciencedirect.com/science/article/pii/S105381190701141X
14. Xue, Y., Feng, S., Zhang, Y., Zhang, X., Wang, Y.: Dual-task self-supervision for cross-modality domain adaptation. In: Martel, A.L., et al. (eds.) MICCAI 2020. LNCS, vol. 12261, pp. 408–417. Springer, Cham (2020). https://doi.org/10.1007/978-3-030-59710-8_40
15. Zhu, H., et al.: Dilated dense u-net for infant hippocampus subfield segmentation **13**, 30 (2019) https://doi.org/10.3389/fninf.2019.00030, https://www.frontiersin.org/article/10.3389/fninf.2019.00030/full

[illegible]

3D Modeling and Reconstruction

Clustering Persistent Scatterer Points Based on a Hybrid Distance Metric

Philipp J. Schneider(✉) and Uwe Soergel

Institute for Photogrammetry, University of Stuttgart, Geschwister-Scholl-Str. 24D, 70174 Stuttgart, Germany
{philipp.schneider,uwe.soergel}@ifp.uni-stuttgart.de

Abstract. Persistent Scatterer Interferometry (PSI) is a powerful radar-based remote sensing technique, able to monitor small displacements by analyzing a temporal stack of coherent synthetic aperture radar images. In an urban environment it is desirable to link the resulting PS points to single buildings and their substructures to allow an integration into building information and monitoring systems. We propose a distance metric that, combined with a dimension reduction, allows a clustering of PS points into local structures which follow a similar deformation behavior over time. Our experiments show that we can extract plausible substructures and their deformation histories on medium sized and large buildings. We present the results of this workflow on a relatively small residential house. Additionally we demonstrate a much larger building with several hundred PS points and dozens of resulting clusters in a web-base platform that allows the investigation of the results in three dimensions.

Keywords: Persistent scatterer interferometry · Clustering · Distance metric

1 Introduction

The remote sensing technique persistent scatterer interferometry (PSI) [8,9] allows monitoring of deformations in large areas such as entire cities. This differential interferometric synthetic aperture radar method is established and well understood. High resulting SAR images lead to millions of persistent scatterer (PS) points in the covered area. The most interesting result of PSI is a deformation time series that describes the points movement, along the satellites line-of-sight, over time with millimeter accuracy. Today the subsequent analysis of deformation processes is often limited to visual inspection of PS point clouds superimposed onto an orthophoto. This is not suitable in order to gain full insight of the 3d distribution and motion patterns of the PS points, especially in an urban environment. In addition, often the deformation time series for each point are condensed to a linear trend. This motion model neglects dynamic characteristics of the PS points and makes it difficult to identify groups of points

C. Bauckhage et al. (Eds.): DAGM GCPR 2021, LNCS 13024, pp. 621–632, 2021.
https://doi.org/10.1007/978-3-030-92659-5_40

that, for example, show similar seasonal deformation behavior. Furthermore, the assignment of PS points to single buildings and later to building substructures is desirable, since this can open the PS-technique for automated per building risk assignment and its integration into long term building information management systems.

In this paper, we propose a clustering approach for PS points based on an initial non-linear dimension reduction and a new distance metric. The resulting clusters show a similar deformation behavior and segment the building in its substructures. The underlying assumption hereby is that PS points on rigid structures show a correlated behavior in their deformation histories. We exploit this fact to cluster them and thereby find groups of points that represent redundant measurements of the underlying deformation process. The spatial distribution of these groups gives an insight into the internal static structure of the building.

Depending on the size of the investigated building, the resulting clusters, their spatial distribution and related time series are complex to visualize. Therefore, we present an exemplary web-based platform which can be used by decision makers, such as civil engineers, to inspect the PS points and the resulting clusters superimposed with a three-dimensional representation of the building.

Previous clustering approaches by [18,20] and [4] have in common that they do not use a correlation based distance metric to describe the similarity between the PS points. [14] show that such PS-clusters can be confirmed by in-situ ground based techniques and introduce a correlation-based distance metric. [15] suggest an initial dimension reduction to counteract problems in high-dimensional clustering.

In the following, we briefly introduce our data set, explain PSI technique and the resulting PS point cloud. We explain the reverse geocoding approach we use to transfer the building labels into the radar geometry in detail. The main focus of this work is the newly proposed distance metric, which combines a correlation based distance with an Euclidean one. Finally, we show results of our approach on an exemplary, relative small building. In the appendix, we provide a link to a web-based three-dimensional presentation of the before shown results, along with another, much larger building.

2 Methods and Data Set

2.1 Synthetic Aperture Radar and Airborne Laser Scanning Data

The here used synthetic aperture radar (SAR) data have been acquired by the German X-Band SAR satellite TerraSAR-X (TSX). The corresponding slant range - azimuth resolution for the "High Resolution Spotlight 300 Mhz acquisition mode" is $0.6\,\mathrm{m} \times 1.1\,\mathrm{m}$ [1]. The 132 images were captured during a 4 years time span (September 2016 to October 2020) with an 11 days repeat cycle. For the interferogram generation a master image in November 2018 was chosen.

We are using sovereign airborne laser scanning (ALS) data (40 Points/m^2, flight altitude: 1000 m) to represent the 3D structure of buildings and to derive a digital surface model (DSM) from this point cloud. The DSM serves as a reference elevation model for the PSI processing and the label transformation, as well as for the visualization of the PS-clusters on the buildings.

2.2 Study Site

We choose two exemplary buildings to demonstrate our clustering approach. *Building 1* is a relatively small apartment building, which suffers from damage [17] due to an underground tunnel construction. *Building 2* one is a large commercial complex, that was build in 2014. No underground construction is carried out here, but harmless post-construction settlements can be expected here.

2.3 Persistent Scatterer Interferometry

Persistent scatterer interferometry (PSI) is an advanced Differential Interferometric SAR (DInSAR) technique. The main idea of this algorithm is the detection of temporally coherent pixels in a stack of co-registered SAR images. By analyzing the phase history of such pixels in each image of the stack, relative to a master image, the line-of-sight (LOS) movement evolution and a 3d position of this scatterer can be estimated [8,9]. [6] give a good overview over the basics and the capabilities of PSI algorithms. For more detailed insights we highly recommend reading this article.

The results presented here were obtained using ENVI's SARscape software [13] with a temporal coherence threshold of 0.7, which is a good trade of between point quality and density [19].

PSI works well for dense urban areas, since man-made structures especially metal parts located at house façades and roofs act as retro reflectors [16].

Since PSI is analyzing time series of multiple SAR images, the displacement history of each scatterer is one of the results. For every PS point we obtain its relative displacement $d(t)$ as a time series with a measurement for each of the $M = 132$ SAR acquisitions (see Eq. (1)). The accuracy for each measurement can be better than 2 mm [11].

The deformation time series represent the most advanced PSI product and is the base for the later clustering approach. As [10] and [5] have shown, PSI time series, derived from high-resolution SAR data, are able to reveal the annual movements of buildings. They confirm thermal expansion of buildings up to several millimeters in amplitude over the year. We exploit this fact for our clustering, under the assumption that each segment of a building exhibits a characteristic movement behavior.

2.4 Building Footprints and Label Transformation

We use *OpenStreetMap* (OSM) building footprints to assign the extracted PS points to single building entities. We apply a *Reverse Geocoding* approach to transfer the footprints from WGS84 coordinate system to the radar image's range azimuth geometry. The main idea hereby is to utilize an existing pipeline [13] to geocode two lookup tables (LuTs). The LuTs contain a unique identifier for each range and azimuth cell. After geocoding, these LuTs are used to transfer labels, respectively building footprints, into the master images range-azimuth geometry. This allows precise assignment for each PS point, without the need of the estimated UTM-coordinates. An overview over the entire workflow is given in Fig. 1. Results are shown in Fig. 2. This approach is also suitable to transfer the labels of training data into range-azimuth geometry, e.g. for land classification applications.

The semantic representation of a building as a polygon in OSM is arguably not suitable for a structural analysis. Often, building complexes, as presented in the results in Fig. 5, are divided into several units, even though they are connected and need to be regarded as a single structure. This can be simply overcome by merging the individual components before the analysis. On the other hand, some buildings consist obviously of several independent units but are listed as a single OSM building entity (see Building 2 in Appendix 5). In practice, this does not affect the results too much, since the here proposed distance metric does consider the actual distances of building parts by it-self. The downside is that a larger amount of points has to be considered in the analysis, which has negative effects on the run time.

2.5 Deformation Space

We treat the deformation histories of each PS point as points in a M-dimensional space, with a dimension for each acquisition date. The example in Fig. 3 shows the embedding of such a (exemplary) deformation history for $M = 3$. Each point $d_n \in \mathbb{R}^M$ is defined by the M measurements d:

$$d_n = \left[d_1^n \; d_2^n \; \dots \; d_{M-1}^n \; d_M^n\right] . \tag{1}$$

In order to have a metric for the Euclidean distance we also use the Cartesian coordinates of the PS points. For the following steps we define the coordinate tuple for each point as:

$$X_n = \left[x_n \; y_n \; z_n\right] . \tag{2}$$

2.6 Distance Metric

We use a combined distance D to describe the similarity of two PS points d_a and d_b in the M-dimensional deformation space. D is composed of the correlation distance D_C which is "1- minus the sample correlation" (see Eq. (4)) and the normalized Euclidean distance D_E between the PS points coordinates X_a and X_b. The normalization is achieved by dividing all distances by the maximum (See Eq. (6)). We combine D_E and D_C as follows:

$$D = \sqrt{D_C^2 + D_E^2 \cdot \lambda} . \tag{3}$$

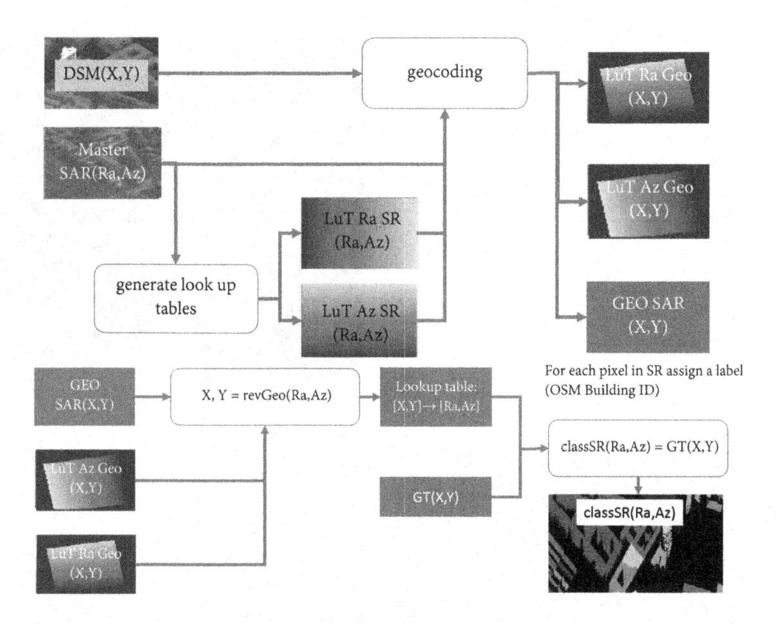

Fig. 1. Workflow for Reverse Geocoding: We generate Lookup tables (LuTs) for Range and Azimuth based on the dimensions and acquisition parameters of the master image. The LuTs are geocoded, using the master image's SAR properties and a precise digital surface model (DSM) derived from the ALS point cloud. The geocoded LuTs allow a mapping of the ground truth (GT) from UTM coordinates to slant-range geometry: $X, Y \rightarrow Ra, Az$, including SAR characteristic distortion properties like foreshortening and layover (Fig. 2).

The weight factor λ allows control over the cluster size. In all our experiments we fix it to $\lambda = 1$.

The correlation distance D_C considers two points (d_a and d_b) as close if their deformation behavior is correlated:

$$D_C = 1 - corr(d_a, d_b) \tag{4}$$

$$= 1 - \frac{(d_a - \overline{d_a})(d_b - \overline{d_b})^T}{\sqrt{(d_a - \overline{d_a})(d_a - \overline{d_a})^T} \cdot \sqrt{(d_b - \overline{d_b})(d_b - \overline{d_b})^T}} \tag{5}$$

where $\overline{d_a} = \frac{1}{M} \sum_m^M d_m^a$
$\overline{d_b} = \frac{1}{M} \sum_m^M d_m^b$
and $d_{a/b}$ as in Eq. (1).

The normalized Euclidean distance D_E is the length of the line segment between the two points X_a and X_b divided by the maximal overall distance between all points:

$$D_E = \frac{\sqrt{(x_a - x_b)^2 + (y_a - y_b)^2 + (y_a - y_b)^2}}{D_{max}} \tag{6}$$

Fig. 2. Exemplary Scene and results from the Reverse Geocoding. **Top left:** OpenStreetMap labels. **Bottom left:** Digital surface model. **Top right:** SAR image in slant-range geometry. **Bottom right:** Labels in slant-range geometry.

where D_{max} is the maximal Euclidean distance between all other points, with X_a defined as in Eq. (2).

This hybrid definition of the distance groups points that show a similar deformation behavior and are not far in Euclidean space.

2.7 Uniform Manifold Approximation and Projection

As [2] have shown, dimension reduction can drastically improve the performance of a following clustering and minimizes the need of hyper parameter tuning. We are aware that such dimensionality reduction might lead to an artificial split of bigger clusters. This is acceptable trade off since we are interested in reliably finding clusters on a city wide scale and can therefore not do hyper parameter tuning for each individual building. One could tackle this issue by regrouping clusters as shown in [15], however the authors of these previous studies don't suggest a large benefit.

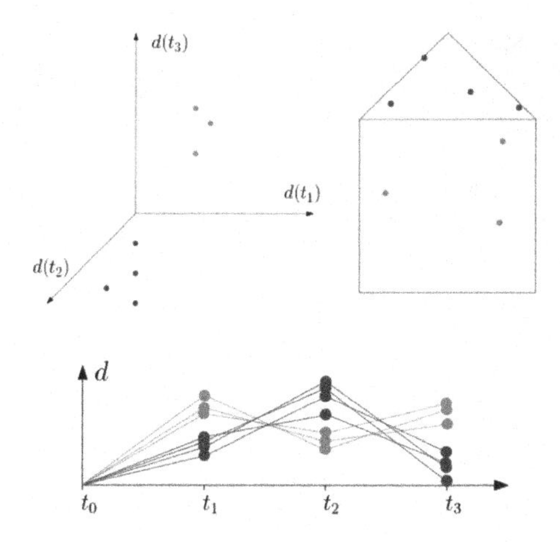

Fig. 3. Representations of PS points. **Top right**: on building. **Bottom**: deformation time series. **Top left**: in deformation space. The assumption is that points which lay on a rigid structure e.g. the roof show similar deformation behavior and therefore form clusters in deformation space.

The non-linear dimension reduction technique Uniform Manifold Approximation and Projection (UMAP) [12] is used to reduce the $(M+3)$-dimensions drastically to a two-dimensional space. We are using the distance metric D as defined in Eq. (3) to model the relationship of the PS points in $(M+3)$-dimensions.

For the UMAP hyper parameter, we use random initialization, with minimum neighbor number of 5 and a minimum distance of 0.3. We found that an iteration for 5000 epochs reliable converges towards a stable embedding.

2.8 Clustering Workflow

Each point in the embedded result is characterized by the core distance (CD) density estimator [3]. We exclude all points that have a bigger CD than twice the median overall CD. The remaining points are clustered with the density-based spatial clustering of applications with noise (DBSCAN) approach [7] with $minPts{=}5{+}\lceil \frac{N}{1000} \rceil$ (N is the total number of PS points on the building). The hyper parameter ϵ is set to the overall median of CD. Finally, we remove points from each cluster if their correlation distance D_C to the mean of the cluster is bigger than 0.3. A schematic workflow with exemplary interim results is presented in Fig. 4.

3 Experiments

To evaluate our proposed workflow, we choose a building complex that suffers known damage from underground construction activities. We can assign 185 PS points to the building. The UMAP plot with the resulting clusters are presented

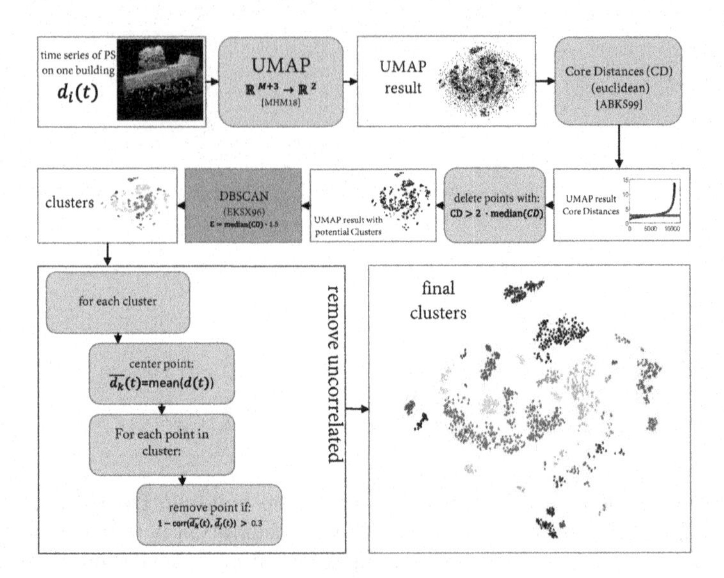

Fig. 4. Schematic workflow of the clustering process. The $(M + 3)$ dimensional PS points on a single building are embedded into a 2D space via UMAP. For all points in this embedding the core distance CD is estimated. Points that have a CD greater than twice the overall median are then excluded. DBSCAN is then performed on the remaining points (minPts $= 5$, $\varepsilon = median(CD) \cdot 1.5$). For each point in each cluster the distance to the cluster's center of gravity in the original M-dimensional deformation space is then calculated using the correlation based distance D_C. Based on this metric points are excluded from a cluster if their correlation distance is greater than a empirical threshold $(D_C = 0.3)$ [15].

in Fig. 5, along with the PS points on the building and the corresponding time series. The clustering workflow shows that the deformation behavior can be grouped in several areas that follow a unique deformation behavior.

In the time series in Fig. 5, we can observe a relief rupture (Fig. 6) in May 2019, that coincides with press reports about compression injections due to an ongoing tunnel project under this area [17]. The time series also indicate a stabilization of these structures over the period of the following year.

We also investigated a much larger building in the same data set. The resulting 69 clusters along with their time series are presented in the supplementary material (see Appendix in Sect. 5). Here we can observe the annual temperature oscillation along with post-construction subsidence in the order of several millimeters per year.

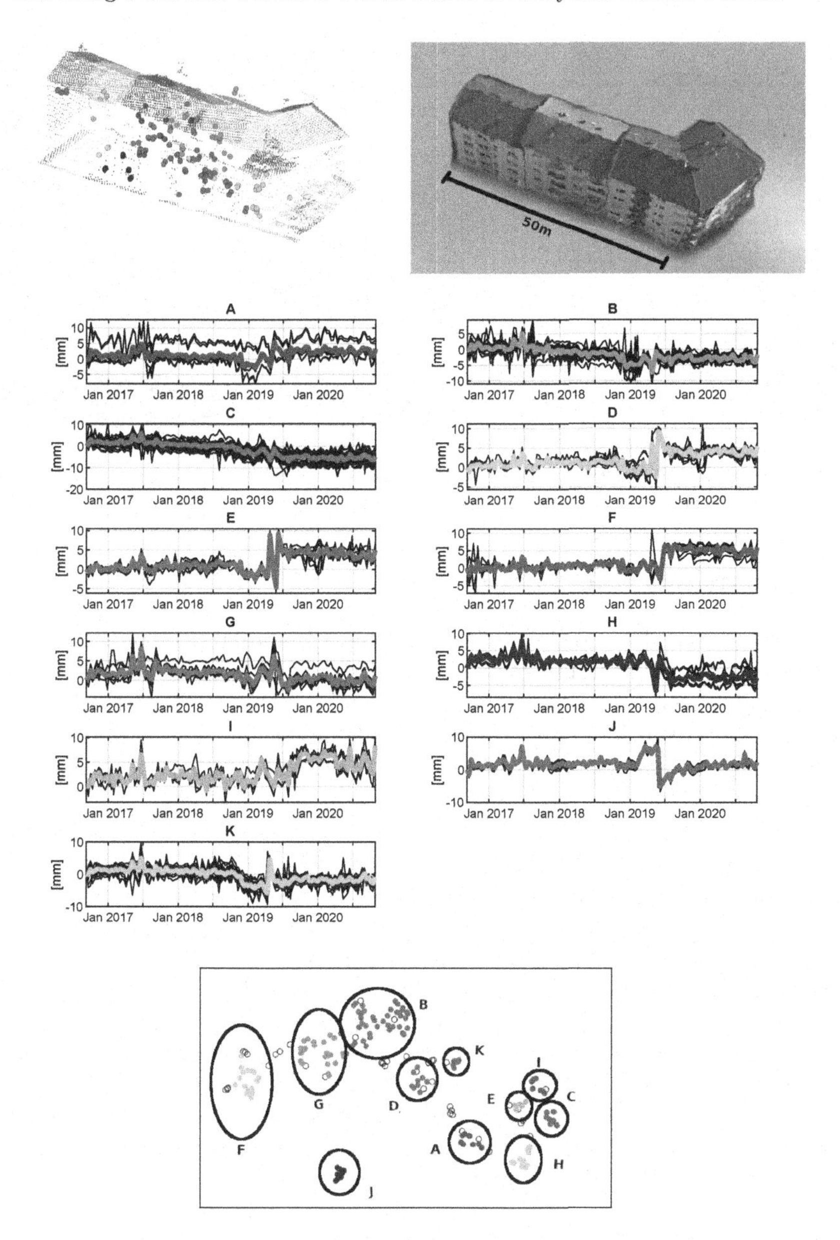

Fig. 5. Left: First from the top: Spatial distribution of 144 PS points on a small urban building complex, super imposed with an ALS-point cloud. The colors indicate the membership to the extracted clusters. **Second from the top**: textured mesh representation of the same building complex. **The time series A to K** show the deformation behavior for each of the extracted cluster. The bold, colored graph represents the centroid of the cluster in deformation space as defined in Fig. 3. **Bottom**: UMAP's low-dimensional embedding of all PS points. The color and letter correspond with the time series and spatial distribution above. Points that were excluded during the clustering process (see Sect. 2.8) are drawn in white.

Fig. 6. Cracks in the building started to appear at the same time (May 2019) as the observed release rupture in the time series (Fig. 5).

4 Conclusions and Future Work

We propose a new distance metric to describe the similarity of persistent scatterers on single buildings. The metric considers the similarity of the deformation histories and the spatial distance of the points. This allows to group the PS points into clusters that show an equal deformation behavior but does suppress groups of points that are distributed too far on the building. Our experiments show that this metric combined with a dimension reduction and clustering approach can reliably extract plausible groups of PS points that are locally connected and show the same deformation behavior over time. The visualization of those clusters, their spatial distribution and the related time series is exemplary shown in a web-based tool, that allows decision makers an insight into deformation events, beyond the traditional two dimensional superposition with an orthophoto.

Further experiments need to include more semantic information of the buildings to link the extracted clusters to specific structures. This allows an integration of the persistent scatterer interferometry technique into building information modeling systems. Another long term aim of this work is an automated city-wide monitoring approach. By extracting clusters and analyzing the deformation patterns relative to each other, structural stress could be observed. This could lead to a per-building risk and damage assignment.

Acknowledgements. The SAR data were provided by the *German Aerospace Center* (DLR) through the proposal LAN0634. We would like to thank the *State Office for Spatial Information and Land Development Baden-Württemberg* (LGL) for providing citywide ALS/Mesh data and orthophotos.

5 Appendix

As supplementary material, we provide an online visualization of the above presented results. We also provide a secondary, much larger building from the same data set. Those visualizations allow an three-dimensional investigation of the achieved results. A colorized ALS point cloud is shown together with the PS points. The clusters are color coded and correspond with the extracted time series on the right hand side.

Building 1: https://ifpwww.ifp.uni-stuttgart.de/philipp/gcpr2021/building1/
Building 2: https://ifpwww.ifp.uni-stuttgart.de/philipp/gcpr2021/building2/

References

1. Airbus: TerraSAR-X Image Product Guide - Basic and Enhanced Radar Satellite Imagery, February 2020. https://www.intelligence-airbusds.com/files/pmedia/public/r459_9_20171004_tsxx-airbusds-ma-0009_tsx-productguide_i2.01.pdf, October 2017, https://www.intelligence-airbusds.com/files/pmedia/public/r459_9_20171004_tsxx-airbusds-ma-0009_tsx-productguide_i2.01.pdf
2. Allaoui, M., Kherfi, M.L., Cheriet, A.: Considerably improving clustering algorithms using UMAP dimensionality reduction technique: a comparative study. In: El Moataz, A., Mammass, D., Mansouri, A., Nouboud, F. (eds.) ICISP 2020. LNCS, vol. 12119, pp. 317–325. Springer, Cham (2020). https://doi.org/10.1007/978-3-030-51935-3_34
3. Ankerst, M., Breunig, M.M., Kriegel, H.P., Sander, J.: Optics. ACM SIGMOD Rec. **28**(2), 49–60 (1999). https://doi.org/10.1145/304181.304187
4. Costantini, M., et al.: Automatic detection of building and infrastructure instabilities by spatial and temporal analysis of insar measurements. In: IGARSS 2018–2018 IEEE International Geoscience and Remote Sensing Symposium, pp. 2224–2227 (2018)
5. Crosetto, M., Monserrat, O., Cuevas-González, M., Devanthéry, N., Luzi, G., Crippa, B.: Measuring thermal expansion using X-band Persistent Scatterer Interferometry. ISPRS J. Photogramm. Remote. Sens. **100**, 84–91 (2015)
6. Crosetto, M., Monserrat, O., Cuevas-González, M., Devanthéry, N., Crippa, B.: Persistent scatterer interferometry: a review. ISPRS J. Photogramm. Remote Sens. **115**, 78–89 (2016). https://doi.org/10.1016/j.isprsjprs.2015.10.011
7. Ester, M., Kriegel, H.P., Sander, J., Xu, X.: A density-based algorithm for discovering clusters in large spatial databases with noise. In: Proceedings of the Second International Conference on Knowledge Discovery and Data Mining. KDD 1996, pp. 226–231. AAAI Press (1996). http://dl.acm.org/citation.cfm?id=3001460.3001507
8. Ferretti, A., Prati, C., Rocca, F.: Permanent scatterers in SAR interferometry. IEEE Trans. Geosci. Remote Sens. **39**(1), 8–20 (2001). https://doi.org/10.1109/36.898661
9. Ferretti, A., Prati, C., Rocca, F.: Nonlinear subsidence rate estimation using permanent scatterers in differential SAR interferometry. IEEE Trans. Geosci. Remote Sens. **38**, 2202–2212 (2000). https://doi.org/10.1109/36.868878
10. Gernhardt, S., Adam, N., Eineder, M., Bamler, R.: Potential of very high resolution SAR for persistent scatterer interferometry in urban areas. Ann. GIS **16**, 103–111 (2010)
11. Maccabiani, J., Nicodemo, G., Peduto, D., Ferlisi, S.: Investigating building settlements via very high resolution SAR sensors. In: Proceedings of the Fifth International Symposium on Life-Cycle Civil Engineering. IALCCE 2016, p. 8, January 2017. https://doi.org/10.1201/9781315375175-332

12. McInnes, L., Healy, J., Saul, N., Großberger, L.: UMAP: uniform manifold approximation and projection. J. Open Source Softw. **3**(29), 861 (2018). https://doi.org/10.21105/joss.00861
13. SARMAP: Sarscape: Technical description. SARMAP: Purasca, May 2020. http://www.sarmap.ch/tutorials/PS_v553.pdf
14. Schneider, P.J., Khamis, R., Soergel, U.: Extracting and evaluating clusters in dinsar deformation data on single buildings. ISPRS Ann. Photogramm. Remote Sens. Spat. Inf. Sci. **V-3-2020**, 157–163 (2020). https://doi.org/10.5194/isprs-annals-v-3-2020-157-2020
15. Schneider, P.J., Soergel, U.: Segmentation of buildings based on high resolution persistent scatterer point clouds. ISPRS Ann. Photogramm. Remote Sens. Spat. Inf. Sci. **V-3-2021**, 65–71 (2021). https://doi.org/10.5194/isprs-annals-V-3-2021-65-2021, https://www.isprs-ann-photogramm-remote-sens-spatial-inf-sci.net/V-3-2021/65/2021/
16. Schunert, A., Schack, L., Soergel, U.: Matching persistent scatterers to buildings. ISPRS - Int. Arch. Photogramm. Remote Sens. Spat. Inf. Sci. **XXXIX-B7**, 79–84 (2012). https://doi.org/10.5194/isprsarchives-xxxix-b7-79-2012
17. SWR: Risse in Hauswänden - verursacht durch Stuttgart 21. [YouTube video], May 2019. https://www.youtube.com/watch?v=CawL7bCnNpg. Accessed 25 May 2021
18. Tanaka, T., Hoshuyama, O.: Persistent scatterer clustering for structure displacement analysis based on phase correlation network. In: 2017 IEEE International Geoscience and Remote Sensing Symposium (IGARSS). IEEE, July 2017. https://doi.org/10.1109/igarss.2017.8128030
19. Yang, C.H.: Spatiotemporal change detection based on persistent scatterer interferometry: a case study of monitoring urban area (2019). https://doi.org/10.18419/OPUS-10423
20. Zhu, M., et al.: Detection of building and infrastructure instabilities by automatic spatiotemporal analysis of satellite SAR interferometry measurements. Remote Sens. **10**(11), 1816 (2018). https://doi.org/10.3390/rs10111816

CATEGORISE: An Automated Framework for Utilizing the Workforce of the Crowd for Semantic Segmentation of 3D Point Clouds

Michael Kölle(✉), Volker Walter, Ivan Shiller, and Uwe Soergel

Institute for Photogrammetry, University of Stuttgart, Geschwister-Scholl-Str. 24D, 70174 Stuttgart, Germany
{michael.koelle,volker.walter,ivan.shiller, uwe.soergel}@ifp.uni-stuttgart.de

Abstract. This paper discusses the CATEGORISE framework meant for establishing a supervised machine learning model without i) the requirement of training labels generated by experts, but by the crowd instead and ii) the labor-intensive manual management of crowdsourcing campaigns. When crowdworking is involved, quality control of results is essential. This control is an additional overhead for an expert diminishing the attractiveness of crowdsourcing. Hence, the requirement for an automated pipeline is that both quality control of labels received and the overall employment process of the crowd can run without the involvement of an expert. To further reduce the number of necessary labels and by this human labor (of the crowd), we make use of Active Learning. This also minimizes time and costs for annotation. Our framework is applied for semantic segmentation of 3D point clouds. We firstly focus on possibilities to overcome the aforementioned challenges by testing different measures for quality control in context of real crowd campaigns and develop the CATEGORISE framework for full automation capabilities, which leverages the *microWorkers* platform. We apply our approach to two different data sets of different characteristics to prove the feasibility of our method both in terms of accuracy and automation. We show that such a process results in an accuracy comparable to that of Passive Learning. Instead of labeling or administrative responsibilities, the operator solely monitors the progress of the iteration, which runs and terminates (using a proper stopping criterion) in an automated manner.

Keywords: Crowdsourcing · Labeling · Automation · Active Learning · Semantic segmentation · Random Forest · 3D point clouds

1 Introduction

At latest since the emergence of Convolutional Neural Networks (CNNs), it has become clear that supervised machine learning (ML) systems are severely

C. Bauckhage et al. (Eds.): DAGM GCPR 2021, LNCS 13024, pp. 633–648, 2021.
https://doi.org/10.1007/978-3-030-92659-5_41

hindered by the lack of labeled training data. To boost the development of such systems, many labeled benchmark data sets were crafted both in the domain of imagery [7,17] and 3D point clouds [8,28]. However, these data sets might be insufficient for new tasks, for instance in remote sensing (e.g., airborne vs. terrestrial systems). Although labeled data sets are also present in the remote sensing domain [18,24,27], due to the rapid development of new sensor types and system design (e.g., for airborne laser scanning (ALS): conventional ALS [24], UAV laser scanning [18] often enriched by imaging sensors, single photon LiDAR [23]) labeled data might be quickly out of date requiring new labeling campaigns. Although transfer learning might help to reduce the amount of task-specific ground truth data (GT) by building upon GT from another domain, it is often necessary to generate one's own training data [25].

Such a labeling process is typically carried out by experts [18,24], which is both time-consuming and cost-intensive. Hence, the idea is to outsource this tedious task to others in order to free experts from such duties in the sense of crowdsourcing. In this context many platforms for carrying out crowd campaigns (such as *Amazon Mechanical Turk* [5] or *microWorkers* [12]) have emerged. In addition to such crowdsourcing platforms, also services which offer to take over the complete labeling process come into focus (such as Google's *Data Labeling Service* [9]). Although the expert loses control of the labeling campaign (outsourcing vs. crowdsourcing), the justification of the latter is to avoid time-consuming campaign management (hiring, instructing, checking and paying crowdworkers). Hence, for crowdsourcing to remain competitive, the aforementioned tasks ought to be automated.

Another major challenge of employing crowdworkers is quality control of results received from the crowd. Walter & Soergel [33] have shown that data quality varies significantly. Therefore an employer either needs to check results manually (which might become even more labor-intensive than the actual labeling task) or rely on proper means for quality control. This problem is most pronounced in context of paid crowdsourcing, where often the sole aim of crowdworkers is to make money as fast as possible and there might be even malicious crowdworkers. In this regard, motivation and consequently quality of work in paid crowdsourcing differs significantly from volunteered crowdsourcing (or volunteered geographic information to be precise). In case of the latter, workers are intrinsically motivated and aim to contribute to a greater cause, which is for example freely available map data in case of *OpenStreetMap* [4]. Nevertheless, paid crowdsourcing could already be successfully used for annotation of airborne imagery [33], detecting and describing trees in 3D point clouds [32] or labeling of individual points according to a specified class catalog [15]. To minimize labeling effort (for the crowd), a common approach is Active Learning (AL).

2 Related Work on Active Learning

AL aims on selecting only most informative instances justifying manual annotation effort [29]. In Mackowiak et al. [22] querying such instances (2D image subsets) is accomplished by combining both predictive uncertainty and expected human annotation effort for deriving a semantic segmentation of 2D imagery. Luo et al. [21] transferred the AL idea to the semantic segmentation of mobile mapping point clouds relying on a sophisticated higher order Markov Random Field. However, only few works focus on ALS point clouds, such as Hui et al. [14], who apply an AL framework for iteratively refining a digital elevation model. For semantic segmentation of ALS data, Li & Pfeifer [19] introduce an artificial oracle by propagating few available class labels to queried points based on their geometric similarity.

For exceeding the limits of automatically answering the query of the AL loop, Lin et al. [20] define an AL regime for the semantic segmentation of ALS point clouds relying on the *PointNet++* [26] architecture, where labels are given by an omniscient oracle. The inherent problem of employing CNN approaches in AL is that usually the majority of points does not carry a label and cannot contribute to the loss function. Often, this problem is circumvented by firstly performing an unsupervised segmentation for building subsets of points, which are to be completely annotated by the oracle [13,20]. Although such a procedure drastically reduces the amount of necessary labels, the oracle is still asked to deliver full annotations (of subsets) requiring a lot of human interaction. In Kölle et al. [16] this issue is directly addressed by excluding unlabeled points from the loss computation while still implicitly learning from them as geometric neighbors of labeled points. Additionally, the authors found that AL loops utilizing the state-of-the-art SCN [10] architecture can result in more computational effort due to relearning (or at least refining) features in every iteration step and might converge slower compared to conventional feature driven classifiers. Hence, a CNN design might not be optimal for AL.

In most of the aforementioned works it is assumed that labels of selected primitives are received by an omniscient oracle, which is a naive assumption, regardless of whether an expert or the crowd is labeling [33]. Consequently, for fully relieving experts from labeling efforts and to form a feasible hybrid intelligence [31] or human-in-the-loop system [2], integration of crowdsourced labeling into the AL procedure in an automated manner is required.

Our contribution can be summarized as follows: We develop a framework referred to as CATEGORISE (**C**rowd-based **A**c**t**iv**e** Learnin**g** f**or** Po**i**nt **Se**mantics), which is tailored for 3D point annotation, but can be easily transferred to other tasks as well. This includes a detailed discussion of i) possibilities for automated quality control tested in various crowd campaigns (Sect. 3.1), ii) measures for automation of the crowd management (Sect. 3.2) and iii) a suitable intrinsic quality measure for the AL loop to enable an operator to monitor the training progress of the machine (Sect. 3.3). Please note that in contrast to related work in this domain [15,16], which mainly focuses on how to employ AL in a crowd-based scenario for semantic segmentation of point clouds, the focus of this paper lies in the automation and enables running such AL loops incorporating real crowdworkers as if the annotation is a subroutine of a program.

3 The CATEGORISE Framework

As aforementioned the backbone of our framework is to combine crowdsourcing with AL to iteratively point out only the subset of points worth labeling. Starting from an initial training data set (see Sect. 5.2), a first classifier C is trained and used to predict on the remaining training points. In our case, we apply a Random Forest (RF) classifier [3] (features are adopted from Haala et al. [11]). Predicted a posteriori probabilities $p(c|x)$ (that point x belongs to class c) are then used to determine samples the classifier is most uncertain about (i.e., the classifier would benefit from knowing the actual label). This sampling score can be derived via entropy E:

$$x_E = \underset{x}{\operatorname{argmax}} \left(-\sum_c p(c|x) \cdot \log p(c|x) \right) \tag{1}$$

In order to also consider imbalanced class occurrences, we further rely on a weighting function, which is derived based on the total number of points n_T currently present in the training data set and the number of representatives of each class n_c at iteration step i: $w_c(i) = n_T(i)/n_c(i)$. For avoiding sampling of points which are similar in terms of their representation in feature space (in context of pool-based AL) and to boost the convergence of the iteration, we adapt the recommendation of Zhdanov [35]. Precisely, we apply a k-means clustering in feature space and sample one point from each cluster (number of clusters equals number of points n_{AL} to be sampled) in each iteration step.

To especially account for the employment of real crowdworkers, we rely on a sampling add-on proposed by Kölle et al. [16], which aims on reducing the interpretation uncertainty of points situated on class borders, where the true class is hard to tell even by experts, referred to as *RIU* (*Reducing Interpretation Uncertainty*). Precisely, in each case, we use a point with highest sampling score as seed point but select an alternative point within a distance of d_{RIU} (in object space) instead. Combining these sampling strategies yields to an optimal selection of points both in context of informativeness and crowd interpretability crucial for our framework.

3.1 Automation of Quality Control

Such a fully automated framework is only applicable if the operator can trust labels received from the crowd to be used for training a supervised ML model. Since results from crowdworkers might be of heterogeneous nature [33], quality control is of high importance. Although interpretation of 3D data on 2D screens requires a distinct spatial imagination, in Kölle et al. [15] it was already shown that crowdworkers are generally capable of annotating 3D points. Within the present work, we aim to analyze in which way the performance of crowdworkers for 3D point annotation can be further improved. Quality control measures can be categorized as i) *quality control on task designing* and ii) *quality improvement after data collection* [34]. In case of labeling specific selected points, which can be thought of as categorization task, one realization of the latter can be derived from the phenomenon of the *wisdom of the crowd* [30]. This means that aggregating answers of many yields to a result of similar quality compared to one given by a single dedicated expert. In our case *wisdom of the crowd* can be translated to simple majority vote (MV) of class labels given by a group of crowdworkers (i.e., a crowd oracle). Consequently, this raises the question of how many crowdworkers are necessary to get results sufficient to train a ML model of desired quality. Detailed discussion of experimental set up can be found in Sect. 5.1.

We would like to stress that to clarify this question it is insufficient to run a labeling campaign multiple times and vary the number n of crowdworkers employed since results would be highly prone to individual acquisitions, which might be extraordinary good or bad (especially for small n). To derive a more general result, we ran the campaign only once and each point was covered by a total of k crowdworkers ($k \geq n$). From those k acquisitions, for each n (i.e., the number of crowdworkers required; range is $[1, k]$) we derive all possible combinations:

$$n_{comb} = \binom{n}{k} = \frac{n!}{(n-k)! \cdot k!} \tag{2}$$

For each n, acquisitions for each combination were aggregated via MV and evaluated according to Overall Accuracy (OA) and classwise mean F1-score. Afterwards, quality measures of each combination (for a specific n) were averaged. Consequently, our quality metrics can be considered as typical result when asking for labels of n crowdworkers.

However, a drawback of accomplishing quality control by *wisdom of the crowd* is increased costs due to multiple acquisitions. Therefore, it is beneficial to also employ *quality control on task designing* [34]. In our case, this is realized by including check points in our tasks. Precisely, in addition to labeling queried AL points, each crowdworker is asked to label a specific number of check points with known class label. Those additional data points can then be used to filter and reject results of low quality. Hence, labels from: i) crowdworkers who did not understand the task, ii) crowdworkers who are not capable of dealing with this kind of data or even from iii) malicious crowdworkers (i.e., who try to maximize their income by randomly selecting labels in order to quickly finish the task) can be filtered. This poses the question of the right number of check points to be included and the consequent impact to labeling accuracy (analyzed in Sect. 5.1).

3.2 Automation of Crowd Management

To realize a truly automated process, we need to avoid any engagement between operator (i.e., employer) and crowdworkers. Within our framework (visualized in Fig. 1(a)), we draw on the crowd of *microWorkers*, which also handles the payment of crowdworkers by crediting salaries to the *microWorkers* account of the crowdworker (avoiding to transfer money to individual bank accounts, which

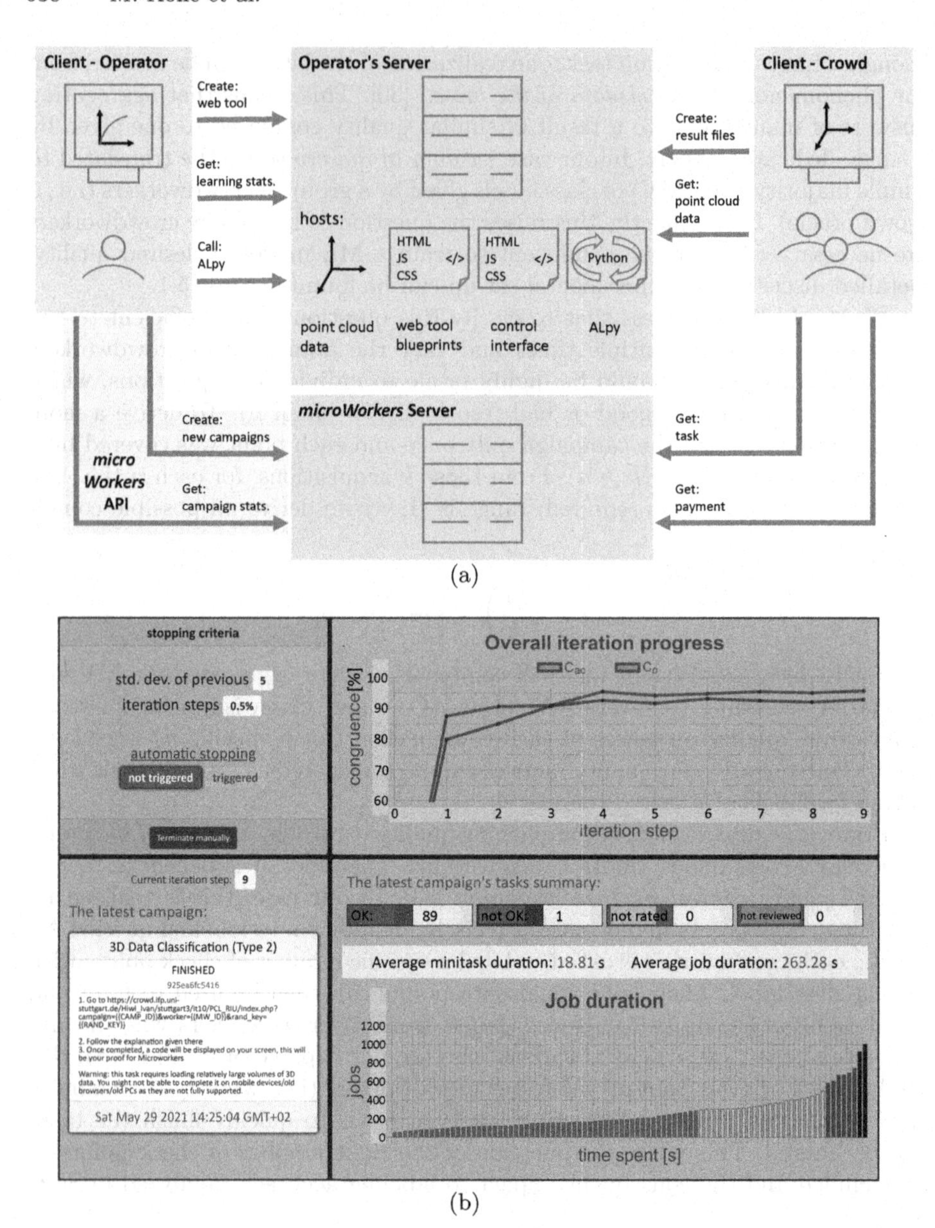

Fig. 1. Architecture of the CATEGORISE framework (a) and the interface used by the operator (b). The latter is designed so that the operator may monitor and control the complete AL run.

would be laborious and would cause fees). Crowd campaigns can be prompted in an automated manner by leveraging the *microWorkers* API. Simultaneously, a respective web tool is set up (by feeding parameters to custom web tool blueprints) and input point clouds to be presented to crowdworkers are prepared (all point cloud data is hosted on the operator's server). An exemplary tool is visualized in Fig. 2(a). When a crowdworker on *microWorkers* accepts a task, he uses the prepared web tool and necessary point cloud data is transferred to him. After completion of the task, results are transmitted to the operator's server (via php) and the crowdworker receives the payment through *microWorkers*.

Meanwhile, by usage of our control interface (see Fig. 1(b)) hosted on the operator's server, the operator can both request the state of an ongoing crowd campaign from the *microWorkers*' server (e.g., number of jobs completed, respective ratings and avg. task duration (see Sect. 3.1), etc.) and the current training progress from the operator's server in order to monitor the overall progress. The control interface and web tools are implemented in Javascript (requests are handled with AJAX). As soon as all points of one iteration step are labeled, the evaluation routine is called (which is implemented in python) and the AL iteration continues (provided the stopping criterion is not met).

3.3 Automated Stopping of the AL Loop

When we recall our aim of an automated framework, it is crucial that it not only runs in an automated manner but also stops automatically when there is no significant quality gain anymore (i.e., the iteration converges). Therefore, our aim is to find an effective measure of quality upon which we can build our stopping criterion for the AL loop and which does not need to resort to GT data. Inspired by the approach of Bloodgood & Vijay-Shanker [1], we accomplish this by determining congruence of predicted labels (for the distinct test set) from the current iteration step to the previous one (i.e., we compute the relative amount of points for which the predicted class label has not changed). In addition to this overall congruence C_o, to sufficiently account for small classes, we further derive a classwise congruence value by first filtering points currently predicted as c and check whether this class was assigned to those points in the previous iteration step as well. These individual class scores can be averaged to get an overall measure C_{ac} equally sensitive for each class. For actually stopping the iteration, we assume that the standard deviation of congruence values of the previous n_{stop} iteration steps (counted from the current one) converges towards 0, which means that change in predictions stays almost constant (i.e., only classes of few most demanding points change).

4 Data Sets

All our tests were conducted on both ISPRS' current benchmark data sets, one being the well-known Vaihingen 3D (V3D) data set [24] captured in August 2008 and the other one being the recently introduced Hessigheim 3D (H3D) data set [18] acquired in March 2018. V3D depicts a suburban environment covering an area of 0.13 km^2 described by about 1.2 M points. We colorized the points by orthogonal projection of colors from an orthophoto received from Cramer

[6]. Color information is used both for deriving color based features and for presenting point cloud data to crowdworkers. H3D is an UAV laser scanning data set and consists of about 126 M points covering a village of an area of about 0.09 km^2. The class catalog of both data sets can be seen in Table 1. Please note that in order to avoid labeling mistakes of crowdworkers solely due to ambiguous class understanding, in case of V3D class *Powerline* was merged with class *Roof* and classes *Tree*, *Shrub* and *Fence* were summarized to class *Vegetation*. In case of H3D, class *Shrub* was merged with *Tree* (to *Vegetation*), *Soil/Gravel* with *Low Vegetation*, *Vertical Surface* with *Façade* and *Chimney* with *Roof*.

5 Results

Within this section, we first discuss our experiments for determining proper measures for quality control (Sect. 5.1), which constitute the basis for conducting our crowd-based AL loops presented in Sect. 5.2.

5.1 Impact of Quality Control Measures to Label Accuracy

To determine measures for quality control (see Sect. 3.1) a total of 3 crowd campaigns were conducted for H3D. These campaigns are dedicated to i) analyze labeling accuracy w.r.t the number of multiple acquisitions when quality control is done by MV only, ii) explore the impact of including check points and iii) derive the optimal number of multiple acquisitions when combining both check points and MV (i.e., realizing *quality control on task designing* and *quality improvement after data collection*). For each campaign, we randomly selected 20 points per class and organized those in jobs of 6 points each (one point per class), which results in a total of 20 jobs. Points were randomly shuffled within each job to avoid that crowdworkers realize a pattern of point organization. Each job was processed by $k = 20$ different crowdworkers, who used the web tool visualized in Fig. 2(a). In addition to the point to be labeled, we extract a 2.5 D neighborhood having a radius of 20 m (trade-off between a large enough subset for feasible interpretation and required loading time, limited by the available bandwidth) to preserve spatial context.

Figure 2(b) depicts the labeling accuracy of the crowd w.r.t. the number of acquisitions used for MV (please note that results are averaged over all n_{comb} combinations; see Eq. 2). In addition to the OA and classwise F1-scores, we further derive entropy from relative class votes to gain a measure of uncertainty for crowd labels. This first campaign shows that MV leads to almost perfect results in the long run proving the concept of the *wisdom of the crowd*. However, this requires a lot of multiple acquisitions and thus causes increased costs. F1-scores of most classes converge from about $n = 10$ acquisitions on. However, most classifiers are capable to cope with erroneous labels to some extent. In Kölle et al. [16] it was shown that about 10 % of label errors only marginally harm the performance of a classifier. Considering those findings, a significantly smaller number for n of about 5 is actually required.

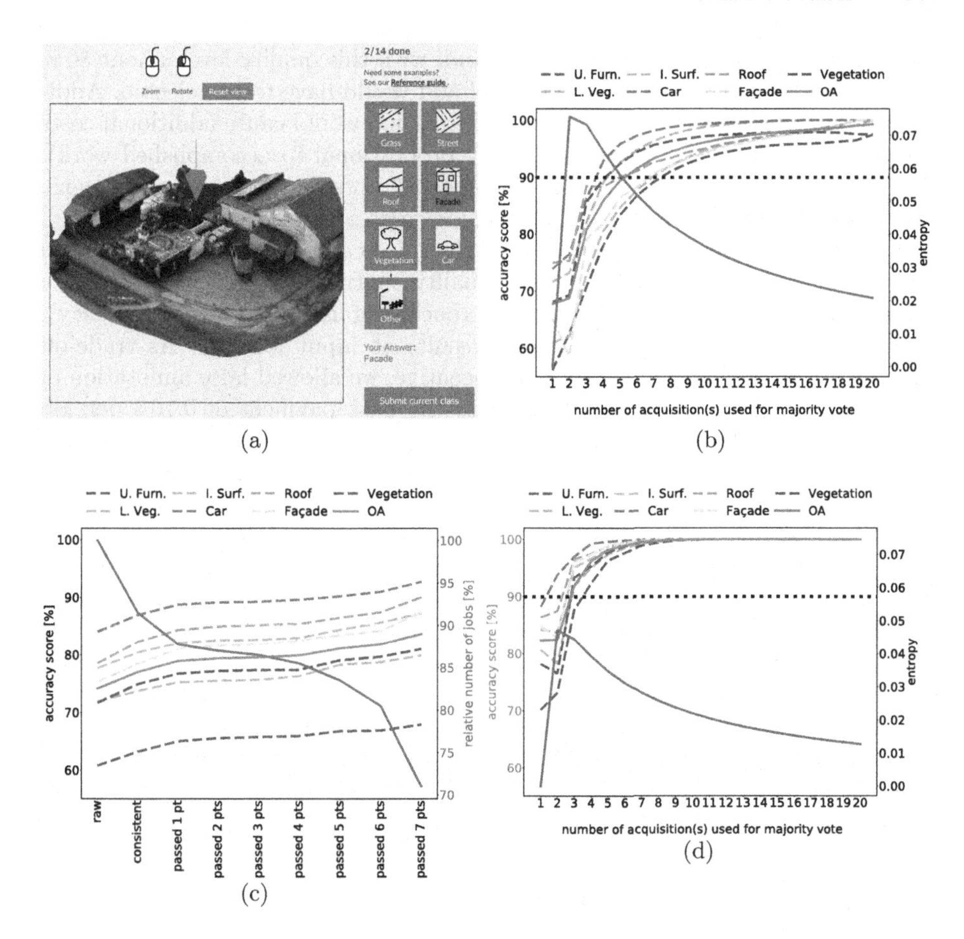

Fig. 2. Developed web tool used by crowdworkers for labeling 3D points (a) and derived results. We compare the result of pure MV (b) to the result of the same task when adding check points (d). The quality improvement by check points is displayed in (c). (web tool can be tried out at https://crowd.ifp.uni-stuttgart.de/DEMO/index.php).

Nevertheless, we aim to further save costs by introducing check points (see Sect. 3.1). We dedicated the second campaign to determining the optimal number of check points. Precisely, we added the same 7 check points to each task and showed the first and last check point twice in order to check consistency of given labels. Crowdworkers were informed about presence of check points but without giving further details or resulting consequences. In post-processing, we used these check points to gradually filter results which do not meet a certain quality level (see Fig. 2(c)). In this context, quality level *consistent* means that the check point presented twice was labeled identically but not necessarily correct. Correctness however, is assumed for level *passed 1 pt* (following quality levels additionally incorporate correctness of more than one check point). We can see that OA can be improved by about 10 percentage points (pps) when enforcing the highest quality level. On the other hand, with this quality level, about 30 % of jobs would not pass our quality control and would have to be rejected. Additionally, an extra labeling effort of 8 points per job would cause additional costs (since crowdworkers should be paid fairly proportional to accomplished work). Therefore, we decided to use quality level *passed 3 pts* for our future campaigns, which offers a good trade-off between accuracy gain and number of jobs rejected.

Considering these findings, we posted the third campaign, which differs from the previous ones as it combines both quality control strategies (MV & check points). By using a total of 3 check points (one being used twice for consistency), we aim on receiving only high-quality results as input for MV. As trade-off between error tolerance and additional incentive, we allowed false annotation of one point but offered a bonus of 0.05 $ to the base payment of 0.10 $ per job (which is also the base payment for campaign 1 and 2) when all check points are labeled correctly. Results obtained are displayed in Fig. 2(d). We observed that adding check points drastically boosts convergence of accuracy. Using check points and relying on results from 10 crowdworkers leads to an even better result than considering 20 acquisitions without check points (see Fig. 2(b) vs. (d)). This holds true for all classes with class *Urban Furniture* having the worst accuracy and class *Car* having top accuracy (overall trend is identical to the first campaign). Class *Urban Furniture* is of course difficult for interpretation since it actually serves as class *Other* [18], which makes unique class affiliation hard to determine. If we again accept a labeling OA of about 90% for training our classifier, 3 acquisitions are sufficient (2 less than for the first campaign). This offers to significantly minimize costs in case of larger crowd campaigns where many hundred points are to be annotated.

5.2 Performance of the AL Loop

Finally, we employ the CATEGORISE framework for conducting a complete AL loop for both the V3D and H3D data set. For setting up the initial training set, often random sampling is pursued. Since this might lead to severe undersampling of underrepresented classes (such as *Car*), we launch a first crowd campaign where a total of 100 crowdworkers are asked to select one point for each class. Since this kind of job cannot be checked at first (due to lack of labeled reference data), we present selected points of each crowdworker to another one for verification. Points which are tagged false are discarded. For both data sets we conduct a total of 10 iteration steps, sample $n_{AL} = 300$ points in each step and parametrize all RF models by 100 binary decision trees with maximum depth of 18.

Performance of the Crowd within the AL Loop. Figure 3 (*top row*) visualizes the accuracy of crowd labeling (obtained from MV from 3 acquisitions using quality level *passed 3 pts*) throughout the iteration process for both presenting queried points to the crowd oracle $\mathcal{O}_C$ and points which were selected by adapting the query function by *RIU*. Please note that OA is used as quality

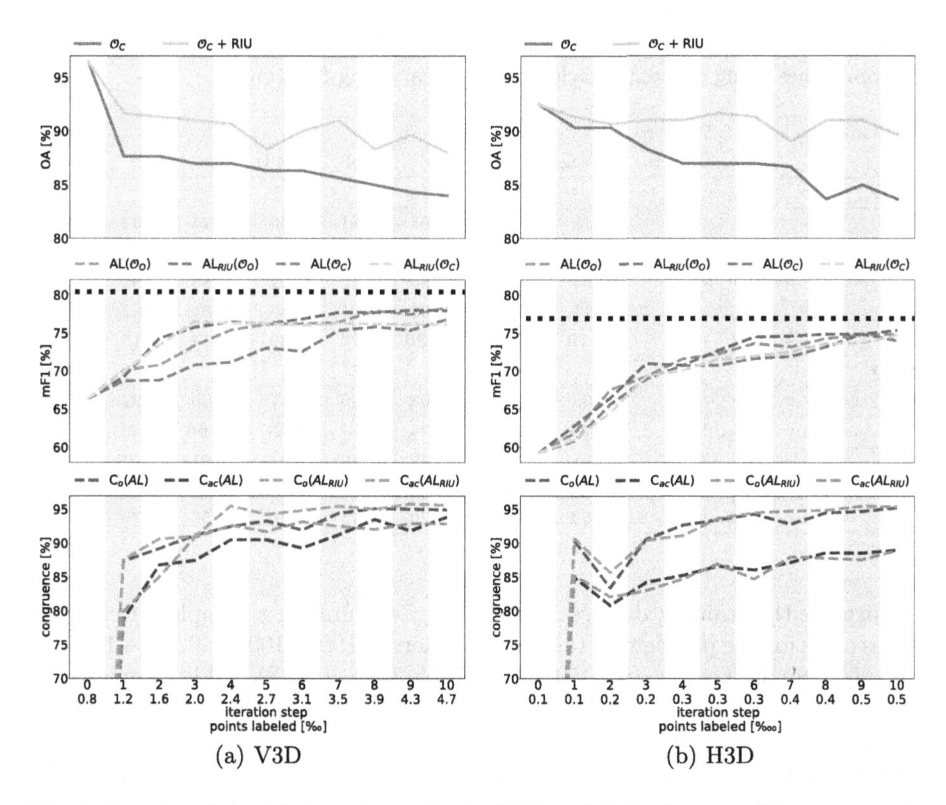

Fig. 3. Results of the AL loop for both the V3D and H3D data set. We compare the OA achieved by the crowd for labeling (*top*) to the mean F1-score of the classifier's performance (*middle*). *Dotted black line* represents the baseline result of PL. As intrinsic measure to describe training progress, we rely on predictive congruence (*bottom*) evaluated for our crowd-based runs (using $\mathcal{O}_C$).

metric since AL tends to sample points in an imbalanced manner w.r.t. class affiliation, so that the mean F1-score would only poorly represent actual quality. In both cases, we can observe a negative trend for labeling accuracy (V3D & H3D). This is due to our AL setting where we start with points easy to interpret (i.e., points freely chosen by the crowd), for which the crowd yields top accuracies consequently. From the first iteration step on, selection of points is handled by the classifier. In the beginning it might select points which are also easy to interpret (since such points might just be missing in the training set so far) and then it continues with points which are more and more special (and typically harder for interpretation). However, *RIU* is capable to at least alleviate this problem. Using $d_{RIU} = 1.5\,\text{m}$ leads to an OA which is about 4 pps higher in case of V3D and up to 6 pps for H3D. In case of the latter, this effect especially improves quality for later iteration steps when sampling points more complex for interpretation.

Table 1. Comparison of accuracies reached both for V3D and H3D for PL and various AL approaches using different oracle types and sampling functions.

Method	Oracle	F1-score [%]							[%]	
		U. Furn.	L. Veg.	I. Surf.	Car	Roof	Façade	Veg.	mF1	OA
V3D										
PL	–	–	82.25	91.28	65.64	94.81	62.30	86.24	80.42	88.11
AL	$\mathcal{O}_O$	–	79.35	90.70	66.00	**93.14**	**57.86**	**83.06**	**78.35**	**85.89**
	$\mathcal{O}_O + RIU$	–	80.67	91.13	67.80	91.12	55.22	81.83	77.96	85.29
	$\mathcal{O}_C$	–	80.11	90.71	66.05	91.25	49.87	82.91	76.82	85.03
	$\mathcal{O}_C + RIU$	–	**81.10**	**91.24**	**68.26**	88.76	45.39	82.14	76.15	84.19
H3D										
PL	–	41.14	90.68	85.26	51.63	92.96	83.77	93.05	76.92	88.16
AL	$\mathcal{O}_O$	33.93	**90.31**	82.70	56.34	88.33	79.73	**92.66**	74.86	**86.65**
	$\mathcal{O}_O + RIU$	**36.97**	89.91	**83.84**	55.42	**90.05**	79.61	91.91	**75.39**	86.60
	$\mathcal{O}_C$	33.37	88.34	78.14	**57.40**	88.89	**79.83**	92.07	74.01	85.15
	$\mathcal{O}_C + RIU$	34.56	89.59	81.81	56.20	89.18	79.35	92.56	74.75	86.26

Forming the training data set within AL (i.e., running a complete iteration) causes costs for the payment of the crowdworkers of 190 \$ (100 pts·0.10 \$+100 pts·3 rep.·0.15 \$+10 it. steps·(n_{AL}/10 pts per job)·3 rep.·0.15 \$) and is completed in about 5 days (approx. 11 h · 10 it. steps + approx. 16 h for initialization = 126 h). Compared to this, estimating total expenses of Passive Learning (PL) is hard to conduct since it is on one hand determined by external factors such as the skills and salary of the annotator, the required software, the required hardware and on the other hand by the complexity of the scene and the targeted class catalog etc.

Performance of the Machine within the AL Loop. Our RF classifier relies on these crowd-provided labels within training. We compare the results of simulated AL runs using an omniscient oracle $\mathcal{O}_O$ to the respective runs using a real crowd oracle $\mathcal{O}_C$ each with and without *RIU* and to the baseline result of PL. Figure 3 (*middle row*) & Table 1 present the accuracies achieved when predicting on the unknown and disjoint test set of each data set. For V3D, we achieve a result ($AL_{RIU}(\mathcal{O}_C)$) which only differs by about 4 pps from the result of PL both in OA and mean F1-score, while this gap is only about 2 pps for H3D. In case of V3D, relying on $AL_{RIU}(\mathcal{O}_C)$ results in a significantly higher increase of mean F1-score in early iteration steps compared to $AL(\mathcal{O}_C)$ (which is beneficial when labeling budget is limited) and performs close to the optimal corresponding AL run relying on $AL_{RIU}(\mathcal{O}_O)$. For the run without *RIU*, the performance of the RF is diminished, which is due to the lower labeling accuracy of the crowd. We would also like to underline the effectiveness of *RIU* even when labels are given by $\mathcal{O}_O$, which demonstrates that this strategy also helps to receive a more generalized training set by avoiding to sample only most uncertain points. In case of H3D, all AL runs perform similarly well with $AL_{RIU}(\mathcal{O}_C)$ marginally outperforming $AL(\mathcal{O}_C)$. For both data sets, Table 1 demonstrates that while classes with a high in-class variance such as *Urban Furniture* and *Façade* (note that façade furniture also belongs to this class) suffer in accuracy whereas underrepresented classes such as *Car* yield better results than the PL baseline.

Terminating the AL Loop. As mentioned in Sect. 3.3, we need to provide a criterion for deciding about stopping the iteration. Congruence values derived for this purpose are displayed in Fig. 3 (*bottom row*). For the initialization, congruence is 0 due to the lack of a previous prediction. Generally, congruence curves correspond well to the test results for our AL iterations. For instance, consider accuracy values for $AL_{RIU}(\mathcal{O}_C)$ (V3D), which reach close-to-final accuracy level in the third iteration step. Therefore, from the fourth iteration step on (one step offset) intrinsic congruence values $C_o(AL_{RIU})$ & $C_{ac}(AL_{RIU})$ have also reached a stable level. We would like to stress that AL runs having a close to linear increase in accuracy ($AL(\mathcal{O}_C)$ for V3D and $AL(\mathcal{O}_C)/AL_{RIU}(\mathcal{O}_C)$ for H3D) show a similar behavior in congruence. All congruence measures flatten with increasing number of iteration steps. To avoid early stopping by a too strict stopping criterion, we set $n_{stop} = 5$ (Sect. 3.3), which indeed leads to std. dev. values close to 0 for $AL_{RIU}(\mathcal{O}_C)$ (V3D), which obviously can be stopped earlier compared to the other runs (0.4 % vs. ca. 1.4 % for the other runs at it. step 10).

In case of H3D, the decrease of congruence in the second iteration step is noteworthy indicating that the predictions have changed significantly. In other words, we assume that the newly added training points have a positive impact to the training process, which is rather unstable at this point, i.e., the iteration should not be stopped here at all. The explanation for this effect is that labels of the initialization (iteration step 0) lead to a level of prediction which is slightly improved in iteration step 1 by adding unknown but comparably similar points (w.r.t. their representation in feature space), which were just missing so far (in fact mostly façade points were selected). The second iteration step then leads to the addition of a greater variety of classes so that accuracy increased significantly, causing changed class labels of many points and by this a drop in congruence. However, one inherent limitation of our intrinsic congruence measure is that it can only detect whether a stable state of training was achieved, which does not necessarily correspond to accuracy.

6 Conclusion

We have shown that the combination of crowdsourcing and AL allows to automatically train ML models by generating training data on the fly. The peculiarity of our CATEGORISE framework is that although there are humans (i.e., crowdworkers) involved in the process, the pipeline can be controlled just like any program, so that carrying out a labeling campaign can be considered as subroutine of this program. This requires automated quality control, automation of crowd management and an effective stopping criterion all addressed within this work. Although the framework was designed for annotation and semantic segmentation of 3D point clouds, it can be easily adapted to other categorization tasks (e.g., for imagery) by i) adapting the web tool used by crowdworkers (mainly concerning the data viewer) and ii) minor changes for the AL loop (as long as individual instances such as points or pixels are to be classified).

References

1. Bloodgood, M., Vijay-Shanker, K.: A method for stopping active learning based on stabilizing predictions and the need for user-adjustable stopping. In: Proceedings of the Thirteenth Conference on Computational Natural Language Learning (CoNLL-2009), pp. 39–47. Association for Computational Linguistics, Boulder, June 2009. https://www.aclweb.org/anthology/W09-1107
2. Branson, S., et al.: Visual recognition with humans in the loop. In: Daniilidis, K., Maragos, P., Paragios, N. (eds.) ECCV 2010. LNCS, vol. 6314, pp. 438–451. Springer, Heidelberg (2010). https://doi.org/10.1007/978-3-642-15561-1_32
3. Breiman, L.: Random forests. Mach. Learn. **45**(1), 5–32 (2001). https://doi.org/10.1023/A:1010933404324
4. Budhathoki, N.R., Haythornthwaite, C.: Motivation for open collaboration: crowd and community models and the case of OpenStreetMap. Am. Behav. Sci. **57**(5), 548–575 (2012). https://doi.org/10.1177/0002764212469364
5. Buhrmester, M., Kwang, T., Gosling, S.D.: Amazon's mechanical turk: a new source of inexpensive, yet high-quality, data? Perspect. Psychol. Sci. **6**(1), 3–5 (2011). https://doi.org/10.1177/1745691610393980
6. Cramer, M.: The DGPF-test on digital airborne camera evaluation - overview and test design. Photogrammetr. - Fernerkundung - Geoinf. **2010**(2), 73–82 (2010). https://doi.org/10.1127/1432-8364/2010/0041
7. Deng, J., Dong, W., Socher, R., Li, L.J., Li, K., Li, F.F.: ImageNet: a large-scale hierarchical image database. In: CVPR 2009, pp. 248–255 (2009). https://doi.org/10.1109/CVPR.2009.5206848
8. Geiger, A., Lenz, P., Urtasun, R.: Are we ready for autonomous driving? The KITTI vision benchmark suite. In: 2012 IEEE Conference on Computer Vision and Pattern Recognition, pp. 3354–3361 (2012). https://doi.org/10.1109/CVPR.2012.6248074
9. Google: AI platform data labeling service [WWW Document] (2021). https://cloud.google.com/ai-platform/data-labeling/docs. Accessed 2 June 2021
10. Graham, B., Engelcke, M., v. d. Maaten, L.: 3D semantic segmentation with submanifold sparse convolutional networks. In: CVPR 2018, pp. 9224–9232 (2018)
11. Haala, N., Kölle, M., Cramer, M., Laupheimer, D., Mandlburger, G., Glira, P.: Hybrid georeferencing, enhancement and classification of ultra-high resolution uav lidar and image point clouds for monitoring applications. ISPRS Ann. Photogr. Remote Sens. Spat. Inf. Sci. **V-2-2020**, 727–734 (2020). https://doi.org/10.5194/isprs-annals-V-2-2020-727-2020
12. Hirth, M., Hoßfeld, T., Tran-Gia, P.: Anatomy of a crowdsourcing platform - using the example of microworkers.com. In: IMIS 2011, pp. 322–329. IEEE Computer Society, Washington (2011). http://dx.doi.org/10.1109/IMIS.2011.89
13. Hou, J., Graham, B., Nießner, M., Xie, S.: Exploring data-efficient 3D scene understanding with contrastive scene contexts. ArXiv abs/2012.09165 (2020). http://arxiv.org/abs/2012.09165
14. Hui, Z., et al.: An active learning method for DEM extraction from airborne LiDAR point clouds. IEEE Access **7**, 89366–89378 (2019)

15. Kölle, M., Walter, V., Schmohl, S., Soergel, U.: Hybrid acquisition of high quality training data for semantic segmentation of 3D point clouds using crowd-based active learning. ISPRS Ann. Photogr. Remote Sens. Spat. Inf. Sci. **V-2-2020**, 501–508 (2020). https://www.isprs-ann-photogramm-remote-sens-spatial-inf-sci.net/V-2-2020/501/2020/
16. Kölle, M., Walter, V., Schmohl, S., Soergel, U.: Remembering both the machine and the crowd when sampling points: active learning for semantic segmentation of ALS point clouds. In: Del Bimbo, A., et al. (eds.) ICPR 2021. LNCS, vol. 12667, pp. 505–520. Springer, Cham (2021). https://doi.org/10.1007/978-3-030-68787-8_37
17. Krizhevsky, A.: Learning multiple layers of features from tiny images. Technical report TR-2009, University of Toronto, Toronto (2009)
18. Kölle, M., et al.: The Hessigheim 3D (H3D) benchmark on semantic segmentation of high-resolution 3D point clouds and textured meshes from UAV lidar and multi-view-stereo. ISPRS Open J. Photogr. Remote Sens. **1**, 100001 (2021). https://doi.org/10.1016/j.ophoto.2021.100001
19. Li, N., Pfeifer, N.: Active learning to extend training data for large area airborne LiDAR classification. ISPRS - Int. Arch. Photogr. Remote Sens. Spat. Inf. Sci. **XLII-2/W13**, 1033–1037 (2019). https://doi.org/10.5194/isprs-archives-XLII-2-W13-1033-2019
20. Lin, Y., Vosselman, G., Cao, Y., Yang, M.Y.: Active and incremental learning for semantic ALS point cloud segmentation. ISPRS J. Photogramm. Remote. Sens. **169**, 73–92 (2020). https://doi.org/10.1016/j.isprsjprs.2020.09.003
21. Luo, H., et al.: Semantic labeling of mobile LiDAR point clouds via active learning and higher order MRF. TGRS **56**(7), 3631–3644 (2018)
22. Mackowiak, R., Lenz, P., Ghori, O., Diego, F., Lange, O., Rother, C.: CEREALS - Cost-Effective REgion-based Active Learning for Semantic Segmentation. In: BMVC 2018 (2018). http://arxiv.org/abs/1810.09726
23. Mandlburger, G., Lehner, H., Pfeifer, N.: A comparison of single photon and full waveform lidar. ISPRS Ann. Photogr. Remote Sens. Spat. Inf. Sci. **IV-2/W5**, 397–404 (2019). https://doi.org/10.5194/isprs-annals-IV-2-W5-397-2019
24. Niemeyer, J., Rottensteiner, F., Soergel, U.: Contextual classification of lidar data and building object detection in urban areas. ISPRS J. Photogramm. Remote. Sens. **87**, 152–165 (2014). https://doi.org/10.1016/j.isprsjprs.2013.11.001
25. Penatti, O.A.B., Nogueira, K., dos Santos, J.A.: Do deep features generalize from everyday objects to remote sensing and aerial scenes domains? In: 2015 IEEE Conference on Computer Vision and Pattern Recognition Workshops (CVPRW), pp. 44–51 (2015). https://doi.org/10.1109/CVPRW.2015.7301382
26. Qi, C.R., Yi, L., Su, H., Guibas, L.J.: PointNet++: deep hierarchical feature learning on point sets in a metric space. In: NIPS 2017, pp. 5105–5114. Curran Associates Inc., USA (2017). http://dl.acm.org/citation.cfm?id=3295222.3295263
27. Roscher, R., Volpi, M., Mallet, C., Drees, L., Wegner, J.D.: Semcity toulouse: a benchmark for building instance segmentation in satellite images. ISPRS Ann. Photogr. Remote Sens. Spat. Inf. Sci. **V-5-2020**, 109–116 (2020). https://doi.org/10.5194/isprs-annals-V-5-2020-109-2020
28. Roynard, X., Deschaud, J.E., Goulette, F.: Paris-Lille-3D: a large and high-quality ground-truth urban point cloud dataset for automatic segmentation and classification. Int. J. Robot. Res. **37**(6), 545–557 (2018). https://doi.org/10.1177/0278364918767506
29. Settles, B.: Active learning literature survey. Computer Sciences Technical report 1648, University of Wisconsin-Madison (2009)

30. Surowiecki, J.: The Wisdom of Crowds. Anchor (2005)
31. Vaughan, J.W.: Making better use of the crowd: how crowdsourcing can advance machine learning research. Journ. Mach. Learn. Res. **18**(193), 1–46 (2018). http://jmlr.org/papers/v18/17-234.html
32. Walter, V., Kölle, M., Yin, Y.: Evaluation and optimisation of crowd-based collection of trees from 3D point clouds. ISPRS Ann. Photogr. Remote Sens. Spat. Inf. Sci. **V-4-2020**, 49–56 (2020). https://doi.org/10.5194/isprs-annals-V-4-2020-49-2020
33. Walter, V., Soergel, U.: Implementation, results, and problems of paid crowd-based geospatial data collection. PFG **86**, 187–197 (2018)
34. Zhang, J., Wu, X., Sheng, V.S.: Learning from crowdsourced labeled data: a survey. Artif. Intell. Rev. **46**(4), 543–576 (2016). https://doi.org/10.1007/s10462-016-9491-9
35. Zhdanov, F.: Diverse mini-batch active learning. CoRR abs/1901.05954 (2019). http://arxiv.org/abs/1901.05954

Zero-Shot Remote Sensing Image Super-Resolution Based on Image Continuity and Self Tessellations

Rupak Bose(✉), Vikrant Rangnekar, Biplab Banerjee, and Subhasis Chaudhuri

Indian Institute of Technology, Bombay, Mumbai, India

Abstract. The goal of zero-shot image super-resolution (SR) is to generate high-resolution (HR) images from never-before-seen image distributions. This is challenging, especially, because it is difficult to model the statistics of an image that the network has never seen before. Despite deep convolutional neural networks (CNN) being superior to traditional super-resolution (SR) methods, little attention has been given to generating remote sensing scene-based HR images which do not have any prior ground truths available for training. In this paper, we propose a framework that harnesses the inherent tessellated nature of remotely images using continuity to generate HR images that tackle atmospheric and radiometric condition variations. Our proposed solution utilizes self tessellations to fully harness the image heuristics to generate an SR image from a low resolution (LR) input. The salience of our approach lies in a two-fold data generation in a self-preservation case and a cascaded attention sharing mechanism on the latent space for content preservation while generating SR images. By learning a mapping from LR space to SR space while keeping the content statistics preserved helps in better quality image generation. The attention sharing between content and tessellations aids in learning the overall big picture for super-resolution without losing an eye on the main image to be super-resolved. We showcase our results with the generated images given the low resolution (LR) input images in zero-shot cases comparable to state-of-the-art results on EuroSAT and PatternNet datasets with metrics of SSIM and PSNR. We further show how this architecture can be leveraged for non-remote sensing (RS) applications.

Keywords: Super-resolution · Remote sensing · Attention sharing

1 Introduction

Generating an SR image from a given LR image in a smooth end-to-end fashion is what is expected from an automated super-resolving framework. Such a framework becomes highly desirable as on-demand detailed image generation would make efficient storage use by allowing HR images to be downsampled

C. Bauckhage et al. (Eds.): DAGM GCPR 2021, LNCS 13024, pp. 649–662, 2021.
https://doi.org/10.1007/978-3-030-92659-5_42

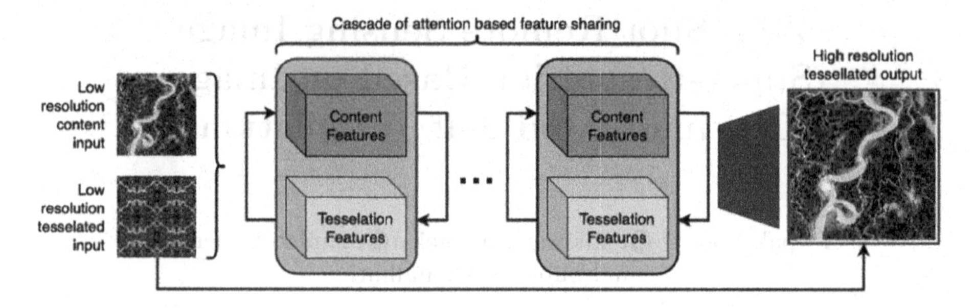

Fig. 1. A schematic of our proposed model. A cascade of attention based feature sharing modules for generating super-resolved images. The attention-sharing-based interaction helps the content statistics trickle down into the tessellation features.

and stored. Also, it would mean lower bandwidth consumption while transmission if images at the end-user have access to an SR framework. Apart from having upsides in traditional computer vision, bringing super-resolution to the remote sensing domain can attempt to address various problems like generating high-resolution spectro-spatial bands from a low spatial-high spectral band configuration (Fig. 1).

Given that multi-spectral images (MSI) are defined by multi-resolution bands, it is favorable to bring all the bands to a common higher resolution to better understand the features. Super-resolution in the remote sensing domain differs from traditional approaches as the objects in the satellite image are very small compared to the huge scale of a satellite image. Here, we leverage the fact that most remote sensing scenes are a kind of tessellation by nature in some sense. Tessellations are those structures whose individual components are repeated to form a pattern as a representative of a whole sample and can be found repeating throughout the sample at various scales. If we can learn the properties of the smaller clusters, the generation of features for larger clusters would be a simple task. Some of the challenges with satellite images include different atmospheric conditions for different images, diverse shape generations from low-resolution images, and various spectral signatures present in the image.

The traditional method of increasing satellite image resolution is by pan-sharpening where the 1st principal component is replaced by a panchromatic (PAN) image and then inverse principle component analysis is done which brings all the bands to the PAN image resolution. Methods like average interpolation, bi-linear, and bi-cubic interpolation are fast methods for upscaling but they lack behind in generating sharp features while upscaling. Advance deep learning methods like convolutional neural networks (CNN) based architectures [3,8,16] and generative adversarial networks [7,13] perform well at hallucinating the details while upscaling and generating sharp high-resolution images in the traditional setting.

Super-resolution being an inherently difficult problem, some of the above-mentioned problems can be tackled if the framework is shifted to zero-shot and the learning is based on self patterns [11,12]. The scene-based radiometric corrections can be reduced if the scene stays unchanged in the training phase. Better robust features can be generated for true context while generating features in the testing phase. Thus, even though zero-shot has many potential advantages, it has been hugely neglected in the satellite images domain.

Given that the traditional methods perform well, they miss out on addressing a few problems. Pan-sharpening does increase the resolution, but it can't be used to generate resolutions higher than the PAN image. It also fails if the bands have co-related noise as an inherent property. The interpolating methods tend to generate smooth images as it performs a weighted average of the values depending on the nature of interpolation and generates the intermediate values. Deep learning-based models require a huge amount of data and time with the scenes being from a relatively similar environment to generate comparable results. Also building a unified model for super-resolution on satellite images is a challenge as different sensors output images having different configurations and different image properties.

To this end, we propose a cascaded attention sharing-based model which aims to address the aforementioned problems faced by the existing architectures. We use the zero-shot framework on satellite images to increase their resolution. This eliminates the atmospheric condition problem as it is trained on the same atmospheric conditions as the original image is. And the requirement of huge data for training models is eliminated as we just require the original image in the training phase for the super-resolution task. Attention sharing helps robust feature learning in a bottom-up approach. Also with the reduction in training times, we can attempt super-resolution as an on-the-fly method. Our contributions in this paper would be:

- We propose two unique methods based on image continuity, i.e., internal and external tessellations, using self-image continuity without loss of generality for data generation in a zero-shot setting.
- We also propose a novel cascade of attention sharing networks on the latent space that helps trickle-down content statistics into the super-resolution domain while upsampling to intermediate scales.
- We showcase our model's performance on popular datasets like UC Merced land-use, EuroSAT rgb, and PatternNet datasets in terms of PSNR, SSIM and cosine similarity.
- We propose an extension to a true zero-shot case in terms of internal tessellation for efficient image size invariant scaling of up to $16\times$ and for external tessellation of up to $8\times$ scaling for generalized uses.

2 Related Works

Apart from the traditional methods, the recent trends for super-resolution are centred around training on low-high resolution training pairs and testing the model on a test low-resolution pair. The task being generative, CNN based architectures, especially generative adversarial networks (GANs) and auto-encoders show promising results. Skip connections play a key role in generating a colour accurate high-resolution image by transferring data between encoder and decoder is well demonstrated by RedNet [9] and Deep Memory Connected Network [14]. It shows that a better flow of information across the encoder-decoder network is as important as having a deep convolutional model. Residual connections enhance the generated feature maps and this is utilized well in the super-resolution of Sentinel-2 images [6]. The performance of GANs is proved by D-SRGAN [2]. It shows promising results by generating high-resolution DEMs from low-resolution DEMs without the need for extra data. However, D-SRGAN does not perform uniformly on all terrains. Flatter terrains produce better results than rugged terrains. Super-Resolution increases the performance of detection tasks by enhancing objects can be seen in vehicle detection algorithm [5] and object detection algorithm [10].

Given the merits of zero-shot super-resolution, the vision community has been harnessing its usefulness. [12] demonstrates meta-transfer learning for exploiting internal image properties for a faster SR method. [1] propose a depth guided methodology and learning-based downsampling to leverage internal statistics for producing SR images. [11] uses internal recurrence of information to train the model. This shows us that internal learning can give us a more robust instance-based learning for SR tasks. However, these methods don't effectively utilize the internal pattern-based tessellations to the fullest extent. This in turn makes the model miss out on the bigger picture.

3 Methodology

3.1 Tessellation Based Data Generation

Tessellations are patterns generated due to structural repetitive cycles. As tessellations are uniform in all directions, the statistics such as mean and variance of the tessellated image is largely similar to that of the content image. This comes to our advantage when used to generate random augmented batches for training the network as it encourages the model to learn different patterns emerging with the same statistical distributions. Thus a robust learning approach is established when mapping an LR image to an HR image. To utilize these properties, we propose two methods to produce tessellation based HR/LR pairs based on internal and external cycles.

Internal Cycle: It is an algorithm to build HR/LR pairs using internal patterns. Here we sample smaller scale images from the original images itself to be tessellated for generation of a larger image. This larger image is then down-scaled to form a low resolution image. As shown in (Fig. 2. a.), we take a random patch from the content image and tessellate it to generate a high-resolution image and use bicubic down-sampling to downscale the generated tessellated image. Looping through the random patches of the content images generated a set of HR/LR pairs.

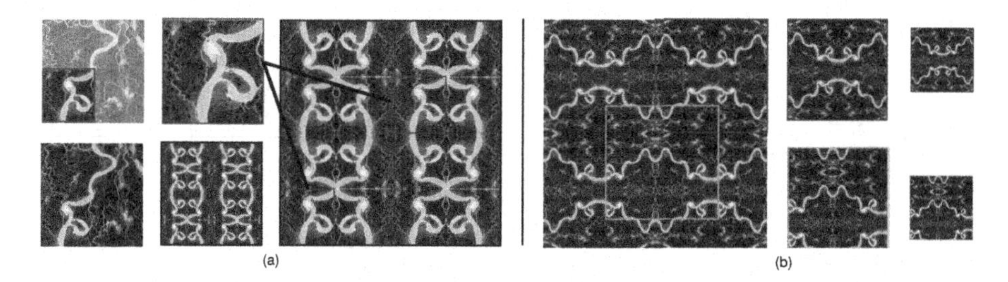

Fig. 2. The data generation techniques. (a) The internal tessellation based HR and LR image pair generation. An internal patch is selected and stitched with continual pattern generating augmentations for required scaling. (b) The external tessellation based HR-LR pair generation based on sliding window operations on the continued image.

External Cycle: It is an algorithm to build HR/LR pairs using external patterns. Here we tessellate the original image directly to generate a larger image. The high resolution images are then sampled from the larger tessellated image and down-scaled to generate low resolution images. As shown in (Fig. 2. b.), we take tessellate the content image to generate an $8\times/16\times$ image. Then random HR patches are obtained from this $8\times/16\times$ image and bicubic down-sampling is applied to obtain LR images.

3.2 Proposed Architecture

Super-resolution is a data-driven process and to make the process efficient, it is highly desirable to understand what features play an important role and how much its contribution is towards generating the final output. To achieve this, an advanced mechanism called self-attention can be used with soft weights. This soft self-attention pushes the weights between 0 and 1 in a continual fashion.

To this end, we propose a cascaded latent space attention sharing network (Fig. 3) for both the content stream and tessellation stream to model mapping from LR to HR while learning to super-resolve tessellations as the content features are preserved. The continual interaction of the latent space not only helps joint feature learning but also helps reach a common ground that highlights feature importance for both content as well as tessellation streams. The highlighting is done through the lower level feature to the higher level features. This helps in efficient feature migration while super resolving in the testing phase.

HR and LR images share similar statistical distributions. Thus it is important to selectively highlight features not only at the deeper levels but also in the initial levels too. The inter-connectivity of attention weights ensures cross-content feature preservation. To this end, we propose a cross attention weights based module at upscaled latent embeddings. In this mechanism (Fig. 4) the attention weights are self-attention soft weights. For a given feature, the attention is calculated using Global Max Pooling and Global Average Pooling followed by a dense layer with 'sigmoid' activation. The 'sigmoid' activation assigns soft weights $\in$

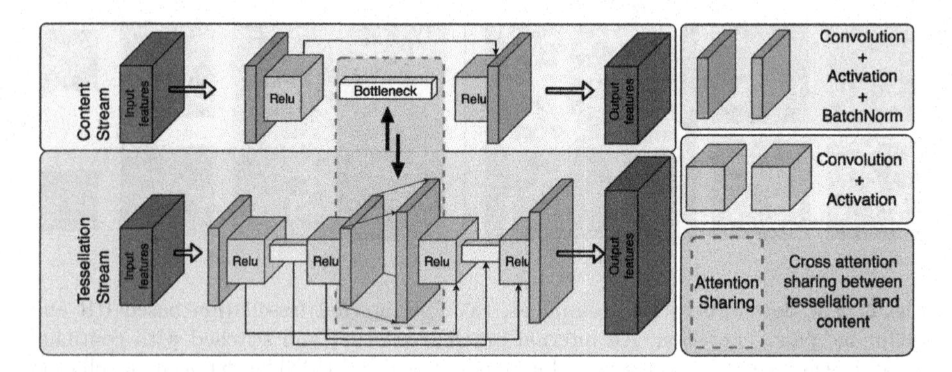

Fig. 3. The overall architecture of our model. The auto-encoder based building blocks with attention sharing at the upscaling latent space for increased interactions between content and tessellation streams.

[0,1]. These self-attention scores are then point-wise multiplied with cross features followed by residual self skip connections to enhance cross attention based self feature highlighting. This sharing mechanism is run in a cascaded manner. A stack of attention sharing mechanisms ensures attention scores from the lower level based filters to deeper latent features. In this way, the super-resolution happens for the tessellation stream, whereas the features of the content stream are constantly preserved.

The building blocks of the overall architecture are auto-encoder(AE) style modules with skip connections around a bottleneck. The AE has the initial layers of 2D Convolutions with activation of 'relu' followed by batch normalization. The internal layers are of 2D convolutions with activation of 'relu'. The output of AE module is upscaled using transposed 2D convolutions and AE module is applied on this upscaled feature. The activation of the bottleneck layer is kept 'tanh' for an even distribution of values in the latent space.

It is in this upscaling space, the attention score sharing happens between content and tessellation streams. The cascaded connections interaction points share attention scores from these features to exchange cross-feature information. The bottleneck ensures relevant information being filtered in due to non-linearity based dimensionality reduction. The skip connections allow for a smoother gradient flow.

4 Experimental Setup

4.1 Training Protocol

The model is trained for 500 epochs with a learning rate of 5×10^{-4}. The loss for both the content as well as HR image is mean squared error. Mean squared error [Eq. 1] takes into account the squared pixel-wise difference between the generated image and ground truth across all channels. Higher is the mean squared error, the more dissimilar the generated images and ground truth is. Given $\hat{y}$ is the

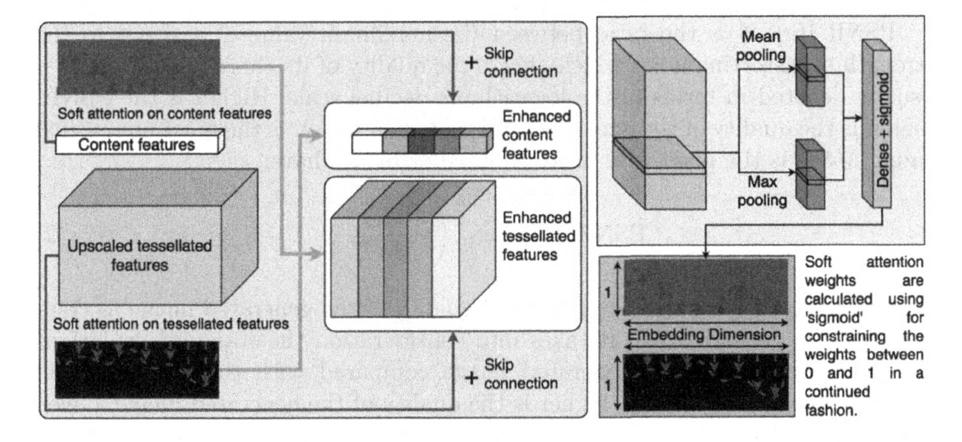

Fig. 4. The attention sharing mechanism. The soft self-attention weights are generated using 'sigmoid' activation on max and mean pooling. These attention scores are exchanged and cross-latent features are highlighted.

generated pixel value and y is the ground truth, (m, n) are image height and image width respectively, we have:

$$\mathcal{L}_{\mathcal{MSE}} = \frac{1}{mn}\sum_{i=1}^{m}\sum_{j=1}^{n}(\hat{y}_{ijk} - y_{ijk})^2 \tag{1}$$

We use internal tessellation based image generation for training purposes. For a given image, we generate tessellation data with a continual sliding window-based patch which is further extended 3 times with using rotation augmentation of 90°, 180° and 270°. The content image is kept constant with data augmentation with rotation of 90°, 180° and 270°. After every epoch, the generated data is shuffled and then fed into the network to randomize learning and lower the memorization of parameters. The model is tested on randomly sampled 100 images per class and images are generated per instance basis for quality monitoring.

4.2 Accuracy Metrics

We use three standardized accuracy metrics popularly used for super-resolution tasks: Peak Signal to Noise Ratio (PSNR), Structural Similarity Index Measurement (SSIM) and cosine similarity. These metrics are used to quantify how close the produced images are to the ground truth in terms of signal quality and visual perception. Each of the metrics have characteristics that convey image quality.

PSNR [Eq. 2] is the ratio between the maximum value of a signal to the strength of distorting noise which affects the quality of its characterization. It is usually denoted in terms of the logarithmic decibel scale. Higher is the PSNR, higher is the quality of the generated image. Given MAX is the maximum signal value, MSE is the pixel-wise mean squared error, we have:

$$\text{PSNR} = 10\log_{10}\left(\frac{MAX^2}{MSE}\right) \tag{2}$$

SSIM [Eq. 3] measures the perceived quality of the generated image as compared to the ground truth. It takes into consideration the standard deviations along with means of the generated image compared with the ground truth. Higher is the value of SSIM, higher is the quality of the generated image. Given $\hat{y}$ is the generated pixel value and y is the ground truth with μ and σ as mean and standard deviation, C_1 and C_2 are constants, we have:

$$\text{SSIM}(\hat{\text{y}}, \text{y}) = \frac{(2\mu_{\hat{y}}\mu_y + C_1) + (2\sigma_{\hat{y}y} + C_2)}{(\mu_{\hat{y}}^2 + \mu_y^2 + C_1)(\sigma_{\hat{y}}^2 + \sigma_y^2 + C_2)} \tag{3}$$

Cosine similarity [Eq. 4] computes the angular nearness between the generated images and the ground truth. The nearer the predicted image is to the ground truth, the angle (θ) between them tends to zero. Thereby, the angular similarity, $\cos(\theta)_{\theta \to 0} \to 1$. Thus a value closer to 1 is desirable. More closer is it's value to 1, higher is the image quality. Given $\hat{y}$ is the generated pixel value and y is the ground truth, we have:

$$\cos(\hat{y}, y) = \frac{\hat{y} \cdot y}{\|\hat{y}\|\|y\|} = \frac{\sum_{i=1}^{n} \hat{y}_i y_i}{\sqrt{\sum_{i=1}^{n} (\hat{y}_i)^2}\sqrt{\sum_{i=1}^{n} (y_i)^2}} \tag{4}$$

5 Results

5.1 Datasets

We experimentally validate our framework on the of the most popular remote sensing datasets having high spatial resolution: EuroSAT (MS) [4], PatternNet [17] and UC Merced land-use [15] dataset. The EuroSAT (MS) dataset consists of 10 classes having 27000 images of 13 spectral channels of the Sentinel-2 satellite. Each image consists of 3 channels with 64 × 64 pixels per channel. We use the RGB subset of EuroSAT (MS). The PatternNet dataset is a collection of 30400 images from 38 classes, 800 images per class. The images are collected from Google earth imagery. UC Merced is a land-use dataset spanning 21 classes having 100 images per class. It was extracted from large images from the USGS National Map Urban Area Imagery collection.

As seen in Table 1, our model outperforms other models in terms of PSNR and cosine similarity and MZSR marginally outperforms us only in the SSIM domain. One of the causes for this can be diverse changes in color gamut along

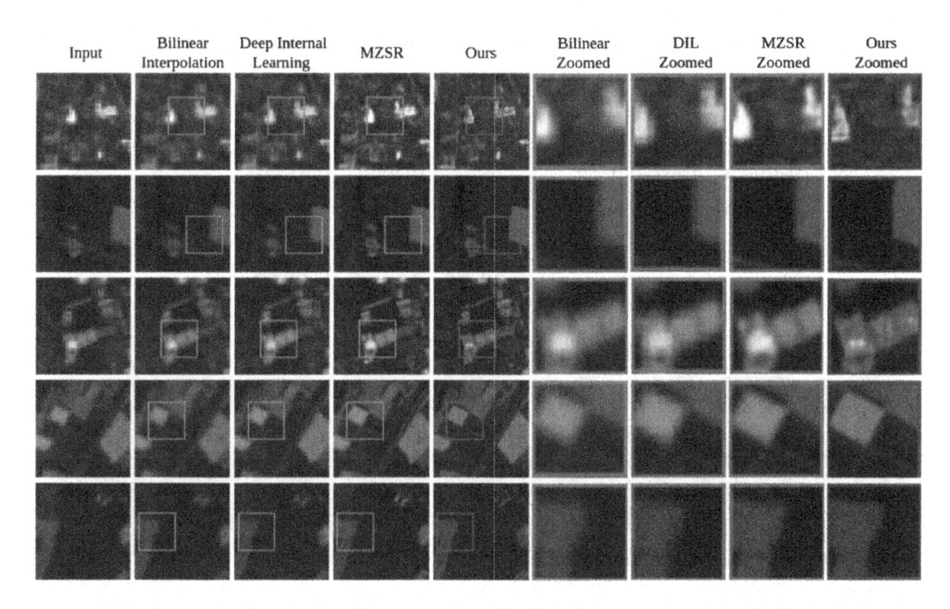

Fig. 5. A visual depiction of comparative model outputs on UC Merced dataset. It can be clearly seen in the zoomed boxes that our proposed model generates a crisper SR image as compared to others.

Table 1. Quantitative performance of compared models in terms of PSNR, SSIM and cosine similarity on the UC Merced dataset for 2× scaling.

Model	PSNR	SSIM	Cosine Similarity
Bi-linear Interpolation	28.10	0.9754	0.9623
Deep internal Learning [11]	30.14	0.9867	0.9806
Meta Transfer ZSR [12]	29.92	**0.9916**	0.9824
Ours	**31.27**	0.9911	**0.9891**

Table 2. Quantitative performance of compared models in terms of PSNR, SSIM and cosine similarity on the EuroSAT dataset for 2× scaling.

Model	PSNR	SSIM	Cosine Similarity
Bi-linear Interpolation	29.87	0.9777	0.9865
Deep internal Learning [11]	31.34	0.9832	0.9913
Meta Transfer ZSR [12]	31.88	0.9802	0.9932
Ours	**32.54**	**0.9834**	**0.9941**

the edges of the super-resolved images. By basic visual inspection, it can be seen that our model generates images and acts as a de-blurring model also (Fig. 5).

Table 3. Quantitative performance of compared models in terms of PSNR, SSIM and cosine similarity on the PatternNet dataset for 2× scaling.

Model	PSNR	SSIM	Cosine Similarity
Bi-linear Interpolation	28.04	0.9689	0.9733
Deep internal Learning [11]	28.86	0.9721	0.9876
Meta Transfer ZSR [12]	29.78	0.9763	0.9885
Ours	**30.27**	**0.9789**	**0.9913**

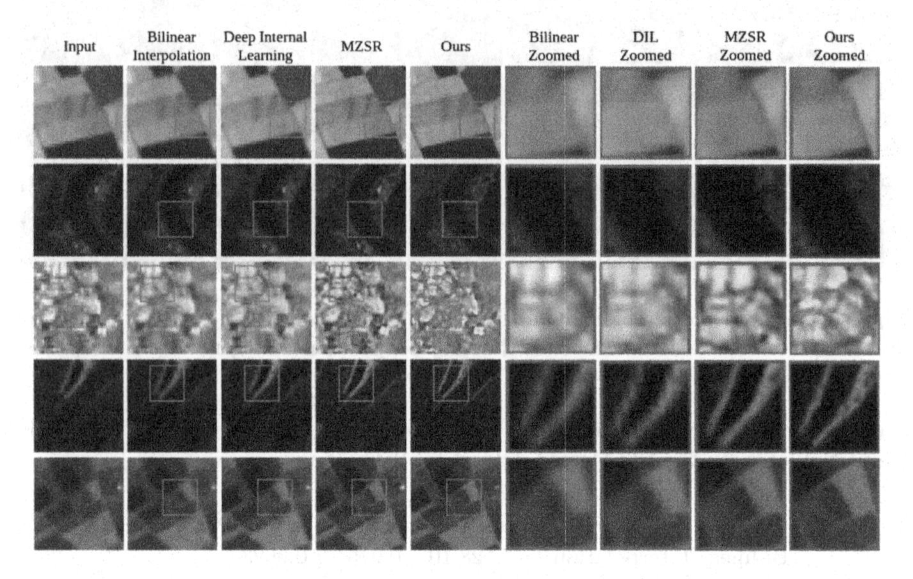

Fig. 6. A visual depiction of comparative model outputs on EuroSAT dataset. It can be clearly seen in the zoomed boxes that our proposed model generates a crisper SR image as compared to others.

As seen in Table 2, our frame outperforms other models in terms of PSNR and cosine similarity. Deep internal learning comes closer to our model by a fraction only in the SSIM domain. In this dataset, our model handles upscaling and de-blurring in a balanced mode to generate high fidelity images (Fig. 6).

As seen in Table 3, even though our model outperforms other models in terms of PSNR, cosine similarity and MZSR [12], it proves to be a difficult dataset to handle. This is due to a large number of high-frequency components present in considerable classes of urban zones. By basic visual inspection (Fig. 7), it can be seen that our model generates images and handles high-frequency components well.

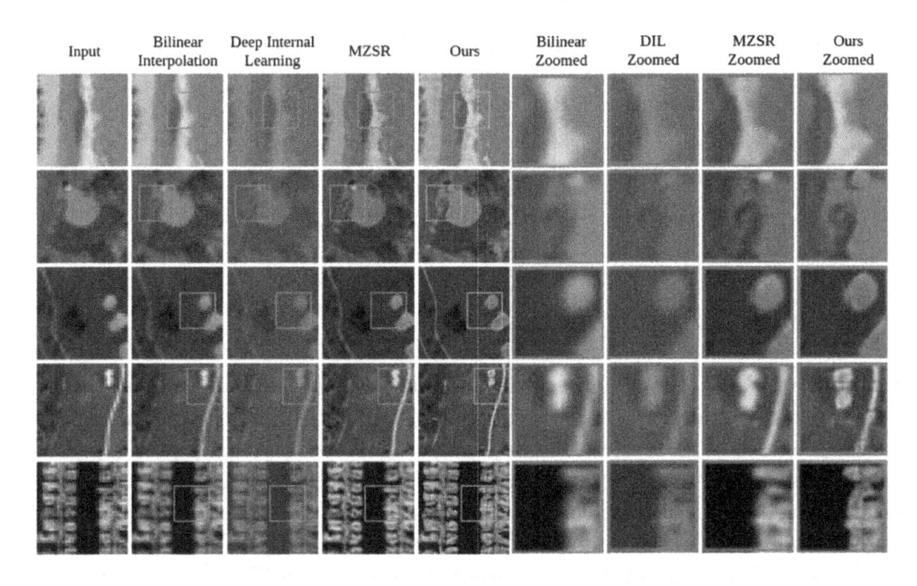

Fig. 7. A visual depiction of comparative model outputs on PatternNet dataset. It can be clearly seen in the zoomed boxes that our proposed model generates a crisper SR image as compared to others.

5.2 Ablation Studies

For ablation studies, we run the network without the attention score sharing to see its impact. Ablation is also performed on the number of modules used in the cascade. This is to validate the learning effectiveness of the attention sharing in the latent space as well as how deep the attention sharing needs to go.

Table 4. Ablation without attention score sharing for 2× scaling.

Dataset	PSNR	SSIM	Cosine Similarity
UC Merced	30.12	0.9807	0.9762
EuroSAT	31.04	0.9719	0.9852
PatternNet	29.20	0.9701	0.9877

From Table 4 and Table 5, we can infer that the attention score sharing plays a crucial role in generating images that are highly coherent with the ground truth. The absence of score sharing reduces the performance of the model. Also, increasing the number of modules in the stack does boost performance. But after a certain number of modules, the performance reaches a plateau and further increase of modules.

Table 5. Ablation on module number in stack for EuroSAT dataset for 2× scaling.

Module No	PSNR	SSIM	Cosine Similarity
4	30.89	0.9763	0.9889
6	32.54	0.9834	0.9941
8	32.50	0.9825	0.9924

6 Extended Application Results

We showcase **8K** (Fig. 8) some of the results with no prior SR images available as reference. To maintain the zero-shot scenario, we train the model on remotely sensed images and test on a different image, i.e. telescopic image and a person's image. The photos are highly scaled to get an 8k image. Given that current devices have patch-based selective upsampling when being zoomed on, a sliding window approach is used to generate the SR images from LR images. The artifacts arise due to the patch being stitched back to give the whole zoomed-out picture given the limited size of the window.

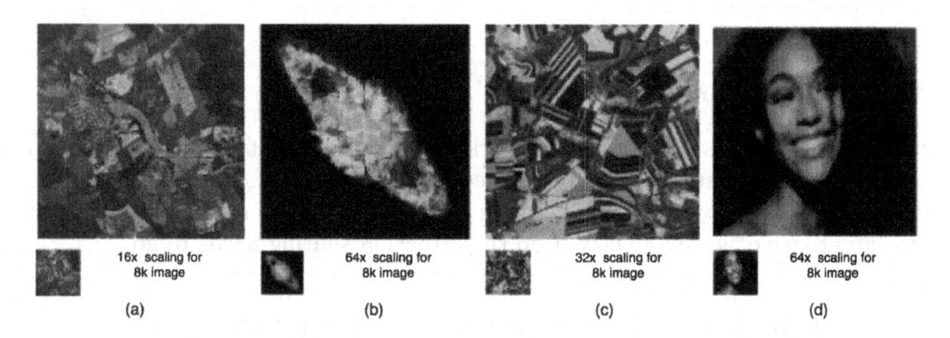

Fig. 8. The model that trained on the smaller figure of (a) enhanced (a) as well as (b) in the testing phase. Similarly, the model that trained on the smaller figure of (c) enhanced (c) as well as (d) in the testing phase. The artifacts are due to stitching of grid sampled images in the low resolution input.

7 Conclusion and Future Work

We introduce a tessellation based zero-shot super-resolution framework that utilizes instance-based statistics and image continuity. This helps in the efficient generation of high fidelity super-resolved images in cases with no prior references. The cascaded attention sharing network aids in selective highlighting of features throughout the feature space interactions which helps in building robust content-based HR images with superior quality. The auto-encoder style format aids in capturing high-frequency signals which in turn outputs crisp and sharp HR images. We also show extended efficient upscaling results that can be applied to non RS domains. These tessellations being based on image continuity opens up new avenues in traditional computer vision areas as well as remotely sensed domains. It can help in efficient information storage and transmission by utilizing highly compressed images that can be restored at the consumer end.

Acknowledgements. The authors would like to thank the members of PerceptX labs[1] for their valuable feedback that contributed to this research.

References

1. Cheng, X., Fu, Z., Yang, J.: Zero-shot image super-resolution with depth guided internal degradation learning. In: Vedaldi, A., Bischof, H., Brox, T., Frahm, J.-M. (eds.) ECCV 2020. LNCS, vol. 12362, pp. 265–280. Springer, Cham (2020). https://doi.org/10.1007/978-3-030-58520-4_16
2. Demiray, B.Z., Sit, M., Demir, I.: D-SRGAN: dem super-resolution with generative adversarial networks. SN Comput. Sci. **2**(1), 1–11 (2021). https://doi.org/10.1007/s42979-020-00442-2
3. Dong, C., Loy, C.C., He, K., Tang, X.: Image super-resolution using deep convolutional networks (2015)
4. Helber, P., Bischke, B., Dengel, A., Borth, D.: EuroSAT: a novel dataset and deep learning benchmark for land use and land cover classification (2019)
5. Ji, H., Gao, Z., Mei, T., Ramesh, B.: Vehicle detection in remote sensing images leveraging on simultaneous super-resolution. IEEE Geosci. Remote Sens. Lett. **17**(4), 676–680 (2020). https://doi.org/10.1109/LGRS.2019.2930308
6. Lanaras, C., Bioucas-Dias, J., Galliani, S., Baltsavias, E., Schindler, K.: Super-resolution of sentinel-2 images: learning a globally applicable deep neural network. ISPRS J. Photogramm. Remote Sens. **146**, 305–319 (2018). https://doi.org/10.1016/j.isprsjprs.2018.09.018
7. Ledig, C., et al.: Photo-realistic single image super-resolution using a generative adversarial network (2017)
8. Lim, B., Son, S., Kim, H., Nah, S., Lee, K.M.: Enhanced deep residual networks for single image super-resolution (2017)
9. Müller, M.U., Ekhtiari, N., Almeida, R.M., Rieke, C.: Super-resolution of multi-spectral satellite images using convolutional neural networks. ISPRS Ann. Photogramm. Remote Sens. Spat. Inf. Sci. **V-1-2020**, 33–40 (2020). https://doi.org/10.5194/isprs-annals-v-1-2020-33-2020
10. Rabbi, J., Ray, N., Schubert, M., Chowdhury, S., Chao, D.: Small-object detection in remote sensing images with end-to-end edge-enhanced GAN and object detector network (2020)
11. Shocher, A., Cohen, N., Irani, M.: "zero-shot" super-resolution using deep internal learning (2017)
12. Soh, J.W., Cho, S., Cho, N.I.: Meta-transfer learning for zero-shot super-resolution (2020)
13. Wang, X., et al.: ESRGAN: enhanced super-resolution generative adversarial networks. In: Leal-Taixé, L., Roth, S. (eds.) ECCV 2018. LNCS, vol. 11133, pp. 63–79. Springer, Cham (2019). https://doi.org/10.1007/978-3-030-11021-5_5

[1] PerceptX labs: https://sites.google.com/view/perceptx/home.

14. Xu, W., Xu, G., Wang, Y., Sun, X., Lin, D., Wu, Y.: High quality remote sensing image super-resolution using deep memory connected network. In: IGARSS 2018–2018 IEEE International Geoscience and Remote Sensing Symposium, July 2018. https://doi.org/10.1109/igarss.2018.8518855
15. Yang, Y., Newsam, S.: Bag-of-visual-words and spatial extensions for land-use classification. In: ACM SIGSPATIAL International Conference on Advances in Geographic Information Systems (ACM GIS) (2010)
16. Zhang, J., Wang, Z., Zheng, Y., Zhang, G.: Cascaded convolutional neural network for image super-resolution. In: Sun, X., Zhang, X., Xia, Z., Bertino, E. (eds.) ICAIS 2021. CCIS, vol. 1422, pp. 361–373. Springer, Cham (2021). https://doi.org/10.1007/978-3-030-78615-1_32
17. Zhou, W., Newsam, S., Li, C., Shao, Z.: PatternNet: a benchmark dataset for performance evaluation of remote sensing image retrieval. ISPRS J. Photogramm. Remote Sens. **145**, 197–209 (2018). https://doi.org/10.1016/j.isprsjprs.2018.01.004

A Comparative Survey of Geometric Light Source Calibration Methods

Mariya Kaisheva(✉) and Volker Rodehorst

Bauhaus-Universität Weimar, Weimar, Germany
{mariya.kaisheva,volker.rodehorst}@uni-weimar.de

Abstract. With this survey paper, we provide a comprehensive overview of geometric light source calibration methods developed in the last two decades and a comparison of those methods with respect to key properties such as dominant lighting cues, time performance and accuracy. In addition, we discuss different light source models and propose a corresponding categorization of the calibration methods. Finally, we discuss the main application areas of light source calibration and seek to inspire a more unified approach with respect to evaluation metrics and data sets used in the research community.

Keywords: Geometric light source calibration · Image-based illumination estimation · Light source models · Survey

1 Introduction

Image formation is influenced by three main independent factors: properties of the observed scene, illumination conditions, and viewing perspective. Image-based 3D reconstruction, appearance acquisition, physically-based rendering, camera and light source calibration are some of the major computer vision and computer graphics research areas related to these factors. In this paper, we focus on the aspect of illumination. In particular, the paper reviews and categorizes methods for light source calibration.

Geometric light source calibration aims at the reconstruction of spatial properties of a single or multiple light sources, such as location, orientation and area size of emissive surfaces. Depending on the used light source model, different subsets of those geometric properties can be estimated (see Subsect. 1.2). In contrast, additional illumination properties such as color and intensity are referred to as photometric, and can be estimated via photometric calibration [57]. In the scope of this paper, we focus on the review of geometric properties of lighting.

While some of the earliest work on light direction detection dates back to the early 80s [60], this survey paper includes publications within the time span of the past two decades providing a comprehensive reference on recent light calibration techniques.

C. Bauckhage et al. (Eds.): DAGM GCPR 2021, LNCS 13024, pp. 663–680, 2021.
https://doi.org/10.1007/978-3-030-92659-5_43

In the following subsection, we feature common applications for light source calibration and point out its importance. After a summary of key concepts and common assumptions in Subsect. 1.3, the core of the paper is dedicated to the specific subcategories of light source calibration (Sects. 2 to 4).

1.1 Applications

The knowledge about the lighting conditions in a given scene is of crucial importance for different applications in the field of visual computing. In the field of computer vision, light source calibration is necessary for the precise estimation of 3D surface geometry from photometric methods such as shape-from-shading [25], photometric stereo [75] or shape-from-shadows [24]. While it is true that a significant research effort has been put into the development of the so-called uncalibrated photometric stereo methods, the problem still remains challenging and at least some approximate lighting information is still required [50]. Calibrated photometric reconstruction methods are still considered more accurate [64] and can benefit from practical and versatile light calibration techniques.

The reconstruction of material properties, for example in the form of Bidirectional Reflectance Distribution Functions (BRDFs), which is another major computer vision task, also requires information about the light sources. In the past, classical light calibration methods have been used in the context of BRDF reconstruction [21]. In more recent times, the integration of the light estimation into the overall material reconstruction process [26] is more common.

Another use case for estimating lighting conditions is the creation of photorealistic visual effects. One the one hand, film and entertainment industry often employs the blending of real and virtual objects. This integration process inevitably requires knowledge about the real lighting in order to correctly shade the virtual objects [41]. On the other hand, light source calibration can also facilitate scene relighting [48], allowing for the convincing modification of the visual appearance of purely non-virtual objects.

1.2 Lighting Models

There are two main approaches to modeling illumination: parametric and non-parametric [4]. The former approach represents individual light sources of a specific type, while the latter gives an overall approximation of a complex lighting environment. Figure 1 shows a schematic overview of the different parametric lighting models described in this section.

Among the most commonly used parametric light source models are those representing a point light source and a directional or parallel one. While truly parallel lighting can be achieved under lab conditions, for example with the help of an optical collimator [44], directional light is often approximated by a distant point light source. A typical example for this approximation is the sunlight coming to Earth from such a long distance that it can be modeled as a point light source at infinity, or simply a parallel light source. At a smaller scale, strong light sources placed at large distances relative to the size of the

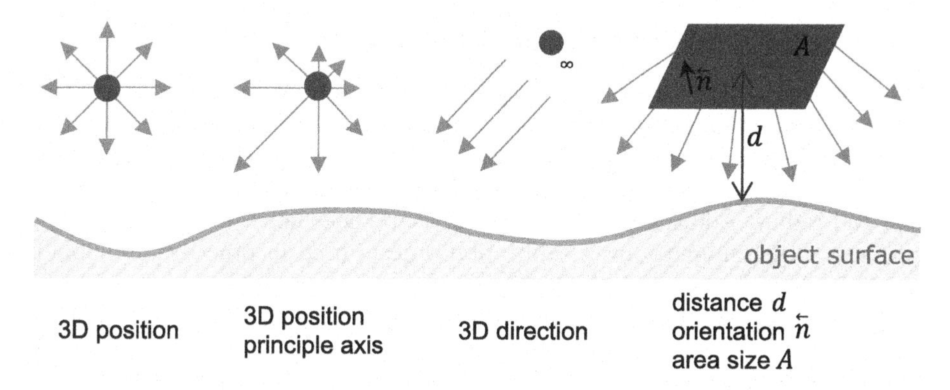

Fig. 1. From left to right this figure visualizes the following light source models with their respective geometric parameters listed below: isotropic point, non-isotropic point, parallel and area light source.

illuminated objects are also adequately approximated by parallel light sources [82], which are geometrically fully characterized by a directional 3D vector.

However, in close-range parallel light approximation cannot be applied. Near light sources are instead modeled as point lights, for which the light direction is different for each illuminated surface point. The calibration of point light sources means finding their position in space relative to a chosen coordinate system. In recent years, the increasing use of near light sources has led to additional distinction between isotropic and non-isotropic point light sources. The latter are characterized by an angular light intensity fall-off. The direction of the strongest light intensity is referred to as their principle optical axis and should be determined in addition to their 3D position. The area light source [80] is another parametric model that can be used to represent near lights of non-negligible size. It is defined by its location, orientation and area size.

In addition, several works [36,64,73,82] have conceptualized more general light source models, which can be used for the simultaneous calibration of different light source models. The idea behind such general models is that basic light source models can be understood as a special case of the general model, in which certain parameters are degenerated. These general models and the corresponding calibration methods are discussed in more detail in Sect. 4.

In relation to the non-parametric lighting modeling, spherical harmonics are frequently used as basis functions [22,62] for modeling complex illumination. They are a particularly convenient representation due to their low memory requirements. A popular alternative is the usage of so-called environment maps, also referred to as incident light maps. Although Sect. 2.2 gives an overview of light estimation methods for the main non-parametric lighting models, the focus of this paper falls on the calibration of parametrically modeled light sources.

1.3 General Overview of Light Calibration Methods

Many light calibration techniques require geometric information about the imaged scene. Frequently, the necessary geometric information comes from an explicit calibration object. Spheres [1,38,61,76,80] are often chosen because their geometry can be easily represented implicitly with known or derived [74] position and radius. Additionally, their convex shape does not cause self-shadowing artifacts. Other simple convex calibration objects used in practice are cuboids [6,52,65,73] and planes [40,45,59,67].

However, the use of an explicit calibration object is not always desirable or even possible, for example when working with pre-existing archive video footage. Arbitrary objects present in the scene can serve as a practical alternative to calibration targets. Yet, in highly constraint cases, only approximate geometry can be extracted. In [56], Nishino and Nayar demonstrate illumination estimation from reflections in the eyes of movie actors by adopting a geometric eye model based on average human anatomy. For a geometric approximation of an arbitrary object, Lopez-Moreno et al. [43] derive surface normals from object silhouettes, assuming that surface normals along silhouette are perpendicular to the viewing direction. In other cases, scene geometry can be reconstructed via 3D scanners [78], multi-view stereo [19] or RGB-D sensors [7,27,31,46]. Table 3 provides an overview of technical requirements such as the type of a calibration object, or the number of required input images, imposed by different calibration methods.

Other common assumptions include working with a calibrated camera with linear or known radiometric response function and the presence of achromatic (white) light sources [32,39,43]. A few methods [10,59,74], however, use the same calibration object to jointly calibrate the camera and the light source. Often, for simplification purposes, an orthographic camera model is assumed [12]. In addition, light source calibration usually takes place without the presence of participating (scattering) media such as fog or murky water. In specialized applications like underwater photometric stereo [18,53], the participating media causes complex light attenuation and scattering effects, which have to be modeled in addition to the standard light source calibration.

Depending on the underlying lighting model, the light calibration methods can be divided into two main categories: methods for distant light (Sect. 2) sources, which estimate the direction of the incident lighting, and methods for near light sources (Sect. 3) that compute the actual 3D position of a light source. Tables 1 and 2 include an overview of calibration methods for distant and near light sources.

Finally, it is worth noting that most light calibration techniques rely on one of three lighting cues: *diffuse reflectance*, *specular reflectance*, or *cast shadows*. The behavior of light rays reflected off a purely diffuse or specular surface are well described by the Lambertian cosine law and the law of reflection, respectively. Therefore, in combination with known geometry, these basic reflectance models prove very useful to light calibration. Similarly, cast shadows caused by a known occluder also provide valuable calibration information.

2 Methods for Distant Light Sources

2.1 Single and Multiple Light Sources

Among the earliest light source calibration methods are those assuming that there is a single distant light source influencing the imaged scene, which is the simplest calibration scenario. Some mixed reality applications [6,30] focus on modeling only one dominant distant illumination source, which often results in sufficiently plausible visual output at low computational cost. However, for other applications the calibration of all distinct light sources could be necessary. Since methods designed to recover the direction of multiple distant light sources can trivially be applied also in case of a single light source, in this section we focus on multi-source calibration and reference single-source methods [10,30,48,52] in Table 1. The notation (s) used in the table indicates results for synthetic data. The entries in the table are sorted in chronological order.

Table 1. Methods for calibration of directional light sources

reference paper	lighting cue	multiple light sources	additional photometric calibration	real time	accuracy (avg. geom. err.)
Zhang & Yang [79]	diffuse	yes	intensity	no	0.5° to 1.5° (s)
Zhou & Kambhamettu [81]	specular	yes	intensity	no	3°
Wang & Samaras [72]	diffuse, shadows	yes	no	no	4° (s)
Masselus et al. [48]	diffuse	no	intensity	n.a.	3° to 4°
Li et al. [39]	diffuse, specular, shadows	yes	intensity	no	0.49°
Mori et al. [52]	diffuse	no	color	no	n.a.
Lagger & Fua [35]	specular	yes	color, intensity	no	0.72° to 2.50°
Cao & Foroosh [10]	shadows	no	no	no	0.1° to 2.3° (s)
Wong et al. [74]	specular	yes	no	no	0.07° to 1.52° (s)
Lee & Jung [37]	specular	yes	intensity	yes	n.a.
Lopez-Moreno et al. [43]	diffuse	yes	intensity	no	10° to 20°
Chen et al. [12]	learned	yes	intensity	no	3.7° to 6.3°
Kán and Kaufmann [30]	learned	no	no	yes	38.3° to 41°

Zhang and Yang [79] introduced the concept of critical points observed on the surface of a Lambertian sphere with constant albedo. The critical points in the image plane are conjugate to object points for which the surface normal is perpendicular to a given light direction. By grouping multiple critical points that belong to the same cut-off curve, and therefore to the same light source, an initial illumination direction is estimated. Finally, a least-squares optimization is used to determine the exact light source direction and its intensity. Wang and Samaras [72] propose an extension of the method in [79] for Lambertian objects of arbitrary known shape. To detect the critical points, the chosen scene objects are first mapped onto the surface of a virtual sphere. Since this mapping is generally not dense, in order to compensate for inaccuracies due to missing data points, the authors propose to also consider information from cast shadows.

An alternative method that uses the specular highlights on the surface of a sphere is proposed by Zhou and Kambhamettu [81]. In this work, the sphere is imaged by a calibrated stereo camera set-up, which provides enough information to automatically measure its location and radius from the input images. To reduce the influence of noise, the final lighting direction is computed as the average of the directions computed from the two stereo images. Similar to [61], corresponding highlight spots between the images are automatically found based on the image space coordinates of the centroid for each highlight. Also using a mirror sphere, Lee and Jung [37] present Augmented Reality (AR) application capable of tracking multiple dynamic light sources in real-time.

Instead of using only a single lighting cue, Li et al. [39] show that a combination of diffuse, specular and shadow cues can increase calibration robustness in the presence of strong texture variance. Their method simulates the effects of hypothetical lighting directions sampled over a tessellated hemisphere and looks for highest agreement with cues detected in the input image. It is assumed that all scene light sources are achromatic and that the geometry is known. For the calibration of multiple small area light sources without the assumption of achromatic light, Lagger and Fua [35] use multiple images of a moving object with known geometry. The gliding behavior of specular highlights on the surface of the object, allows for a robust light source detection even in textured regions.

Lopez-Moreno et al. [43] present an interesting approach based on image space silhouettes that can be applied in the absence of scene geometry information. However, their method requires user interaction for outlining an appropriate convex silhouette of a chosen scene object.

In recent years, deep-learning based approaches to light source calibration have emerged. Kán and Kaufmann [30] estimate a single dominant lighting direction from RGB-D sensor data. The model of Chen et al. [12] uses multiple images of a static object of general shape and reflectance and poses the calibration task as a classification of predefined number of discretized lighting directions.

2.2 Complex Environment Lighting

This section provides a brief overview of methods that generate more general representation of lighting conditions. We refer to these as methods for estimation of environment lighting. A typical assumption here is that there exists a large number of distant, i.e. parallel light sources. This assumption is practical in cases of uncontrolled lighting environments in which the precise calibration of each individual light source is infeasible due to high numbers and complexity.

With this argumentation, Sato et al. [65] introduce an approach that approximates the distribution of a complex natural illumination. A shadow caster of known shape is used as calibration object. With help of the observed shadow image, a system of linear equations is formed in order to determine the weights of n uniformly spaced imaginary distant point lights, where n is a free parameter and determines the number of sampling points over the surface of a geodetic hemisphere. Takai et al. [69] also use discretization over a geodetic hemisphere and shadow analysis. However, instead of cast shadows on planar surface, they analyze self-shadows on the surface of a special calibration object shaped as an open cube. The approach allows for shadow sampling on multiple surfaces of different orientation, which leads to more stable results.

After the introduction of image-based lighting by [13], the acquisition of environment lighting with the means of High Dynamic Range (HDR) photography established itself as an industry practice [70]. HDR images in the form of spherical panoramas, referred to as environment maps, can be acquired via an omnidirectional camera or with help of a mirror sphere. When using environment maps, each pixel in the map can be interpreted as describing the incident radiance of an individual distant light source. However, depending on the particular use case and the resolution of the map, this strategy can prove computationally infeasible. Therefore, a reduced number of light sources can be extracted from the initial environment map, for example through clustering [14] or image segmentation [17]. Instead of a single map, Unger et al. [70] capture HDR-video and show how real-time AR applications benefit from spatio-temporal variance in the estimated illumination.

Lombardi and Nishino [42] simultaneously recover object surface reflectance and natural illumination using an iterative Bayesian optimization. The method only requires a single input image to generate the so-called illumination map, which is similar to an environment map. Many other authors, among which [15,71], have also adopted the assumption of distant natural illumination and show joint reflectance and illumination estimation.

An alternative to the explicit environment maps is modeling illumination as a linear combination of a small number of basis functions. A benefit of this representation is its memory efficiency. Alldrin and Kriegman [4] conceptualize a custom planar calibration object that allows the capturing of a uniform distant lighting as a set basis functions in the frequency domain. The evaluation of their method suggests that it is best suited for recovering the low frequencies of the scene illumination. Spherical harmonics are also frequently used as basis functions to store and represent environment lighting [22,33,46,63].

Since the methods summarized in this section produce rather an approximate estimation of complex lighting environments than precise calibration, we do not include a formal accuracy evaluation. For more detailed overview of methods dedicated to capturing environment lighting, we would like to refer the interested readers to the works of Kronander et al. [34] and Alhakamy and Tuceryan [2].

3 Methods for Near Light Sources

In this section we discuss the calibration of both isotropic and non-isotropic near point light sources. The modeling and calibration of such light sources has gained a lot of attention in the recent years because they often represent real lighting conditions more accurately than the distant light model.

3.1 Isotropic Point Light Sources

There is an abundant body of literature on the calibration of isotropic point light sources. While earlier works tend to be mostly probe-based, more recent works estimate illumination conditions as a part of the general scene reconstruction process. All methods discussed in this section are also summarized in Table 2. Note that the table is subdivided based on the used evaluation metrics as follows: absolute distance error, relative distance error, absolute angular error, no explicit accuracy measurement. The entries in each subgroup are ordered chronologically.

Direct Observations and Known Calibration Objects. In case a given light source is directly visible in at least two different views of a calibrated camera, classical triangulation can be applied for its localization [1]. For example, in the context of an AR application, Frahm et al. [17] use a moving multi-camera system consisting of a fisheye and a video camera. The video camera captures the scene, while the fisheye camera oriented upwards images all present light sources. After saturation-based image segmentation, the image sequence from the fisheye camera is used for direct triangulation.

In the absence of direct light source observations, triangulation from observed specular highlights can be used instead. This method applies the standard rule of specular reflection to compute the reflected rays for triangulation. Therefore, at least two corresponding highlight observations per light source in the scene are necessary. Since the intersection of rays in 3D is not always possible, some authors take instead the middle point of the shortest line connecting the two rays [41]. Equivalently, a least-squares method can also be used to find the point in space closest to all reflected rays [1,61].

One option for the acquisition of multiple corresponding specular responses is to use a stereo camera setup in combination with a single reflective calibration object. Instead of relying on multiple camera views, Powell et al. [61] apply triangulation by using a stereo pair of specular spheres with known relative position, while Shen and Cheng [67] use two different known orientations of a mirror plane. For increased computational robustness, Lensch et al. [38] use six steel balls in combination with precise positioning of the spheres through meticulous hardware calibration. Ackermann et al. [1] investigate the influence of increasing the number of calibration spheres, and propose a reverse formulation of the same problem that minimizes the reprojection error. However, little difference in accuracy between the standard and the so-called backward calibration method has been observed.

In contrast, motivated by practicality, Aoto et al. [5] introduce a single-view single-object method, which relies on the double reflections on the surface of a hollow transparent shiny sphere. Bunteong and Chotikakamthorn [8] also propose a single-view single-object method which combines specular highlights and cast shadows from a custom shaped calibration target.

Similarly to [67], Liu et al. [40] prefer to use a shiny planar object, in this case a flat monitor display, to triangulate the specular reflections. The main argumentation for their choice is to improve calibration accuracy using more stable geometric calibration of the used light probe, i.e. the shiny display surface. They argue that localizing the center of a calibration sphere is more difficult than estimating the orientation of plane.

Probeless and Optimization-Based Approaches. Provided with known arbitrary geometry and assuming a predefined parametric shading model, several authors [23,26,27,78] have proposed optimization approaches for the simultaneous estimation of reflectance parameters and light source positions. Estimating the 3D position of a single point light source at a time, Hara et al. [23] adopt the dichromatic reflectance model [66] and assume that the specular component can be expressed by the Torrance-Sparrow [49] model. Xu and Wallace [78] increase the number of calibrated light sources up to three and additionally estimate their intensities. Their approach uses a stereo-view setup and applies the Blinn-Phong shading model. Innmann et al. [26] recover parameters of the Disney BRDF [9] model and achieve a more accurate light position estimation compared to optimization under the assumption of a simple diffuse shading model. In their framework, a mirror sphere is used to generate an initial guess.

While the works discussed above rely on detected specularities, Boom et al. [7] treat scene geometry as Lambertian and reconstruct the position of a single light source via iterative inverse rendering technique. They apply superpixel segmentation to find homogeneous albedo regions. Like in [7], Kasper and Heckman [32] also rely on geometry from RGB-D sensors. They observe that strong specular highlights are usually lost during reconstruction and therefore also assume diffuse surfaces. They employ a non-linear optimization approach to determine the most likely light source candidates among thousands of potential point light sources distributed in a volume around the reconstructed scene and avoid making any assumptions about object albedo.

Calibration techniques based on cast shadows are presented in [28,29,55]. In an earlier work Jiddi et al. [28], first segment scene geometry into occluding objects and ground plane. In an iterative process, synthesized shadow maps are compared against shadowing information from the input images in order to identify optimal light source positions among multiple evenly distributed candidates. Later, in [29], a combination of specular and shadow cues helps to recover both light source position and color even in the presence of strong texture variations. Similarly, Nieto et al. [54] also utilize specular and shadow information. Unlike [29], they apply a continuous optimization avoiding discretized light source locations. However, they also assume known material properties.

Papadhimitri and Favaro [58] were first to present a solution for uncalibrated close-range photometric stereo, which allows for the simultaneous reconstruction of surface geometry, albedo, and light source position and intensity without any specific assumptions about object reflectance or lighting distribution.

Recently, Marques et al. [47] presented a learning-based approach for light position estimation. The method is fast and practical. However, the accuracy of the results as well as the duration of training process strongly depends on the complexity of the available training data. Like in [32], a set of possible light source positions is determined in advance, but in this case the potential lights are distributed over the surface of a virtual hemisphere instead of a cubic volume.

3.2 Non-isotropic Light Sources

The calibration of non-isotropic light sources is of particular importance for the accurate application of close-range photometric stereo. Due to multiple advantages ranging from energy efficiency to adjustable color temperature [77], the light-emitting diodes (LEDs), which behave as non-isotropic point light sources, are frequently employed in close-range photometric stereo setups. A detailed optical model of a single LED source can be found in [51].

Xie et al. [76] present an optimization-based techniques for the calibration of LED light sources. A reference sphere, which exhibits both diffuse and specular reflectance properties, is used as a calibration object. The observed specular highlight is assumed to indicate the optical axis of the light source. For the computation of the light source position, only the diffuse reflectance components are used after their extraction from the input image.

Park et al. [59] introduced a method for simultaneous geometric and photometric calibration of a non-isotropic point light source. Their method requires multiple-view images of a plane with known orientation. The light source is rigidly paired with the moving camera. It is assumed that the radiant intensity distribution of the non-isotropic light source is radially symmetric with respect to the optical axis of the light source. This leads to bilaterally symmetric shading on a Lambertian plane with respect a 2D line on this plane. A RANSAC algorithm [16] is used to detect the symmetry line. By detecting the symmetry line in multiple images taken under different viewing angles, the optical axis of the light source is recovered.

Nie et al. [54] present a two-stage calibration method, which recovers first the position of the light source and then its optical axis. During the first stage, position of the light source is estimated using multiple specular spheres for triangulation of reflected light rays. In the second stage, the optical axis is determined using a Lambertian plane with known orientation. Similar to [59], the work discusses the existence of a symmetry line. However, instead of searching for this line, Nie et al. focus on directly detecting the brightest point along it, which they prove to be a unique point in the image.

A practical single-image calibration method was recently proposed by Ma et al. [45]. Both position and orientation of the light source are simultaneously estimated with help of a custom-made planar calibration object. The orientation

Table 2. Methods for calibration of isotropic point light sources

reference paper	lighting cue	multiple light sources	additional photom. calibration	real time	accuracy (avg. geom. err.)
Lensch et al. [38]	specular	no	no	no	2 cm
Hara et al. [23]	specular, diffuse	no	no	no	3 to 4 cm
Aoto et al. [5]	specular	yes	no	no	2.3 cm
Ackermann et al. [1] forward calib.	specular	yes	no	no	1.3 to 13.0 cm
Ackermann et al. [1] backward calib.	specular	yes	no	no	0.9 to 8.2 cm
Papadhimitri & Favaro [58]	diffuse	yes	intensity	no	3.85 to 4.79 cm
Jiddi et al. [27]	specular	no	no	no	16 cm
Jiddi et al. [28]	shadow	yes	intensity	near	17 cm
Kasper & Heckman [32]	diffuse	yes	intensity	no	8 to 20 cm
Jiddi et al. [29]	specular, shadow	yes	color	near	13 cm
Powell et al. [61]	specular	yes	no	no	6% in magnitude 9% in direction
Zhou & Kambhamettu [80]	specular	yes	no	no	4%
Xu & Wallace [78]	specular	yes	intensity	no	0.32% to 38.44%
Liu et al. [41]	specular	yes	no	yes	14.9%
Bunteong & Chotikakamthorn [8]	specular, shadow	yes	intensity	no	4.17% to 8.16%
Innmann et al. [26]	specular	yes	no	no	5%
Shen & Cheng [67]	specular	no	intensity	no	0.04° to 0.47°
Boom et al. [7]	diffuse	no	intensity	no	10°
Nieto et al. [55]	specular, shadow	yes	intensity	no	0.09° to 0.21°
Frahm et al. [17]	direct LS observation	yes	no	No	n.a.
Marques et al. [47]	learned	no	no	no	n.a.

of the plane is automatically estimated based on circular markers positioned in one of its corners. Afterwards, the calibration process is formulated as an iterative optimization problem.

4 Unified Frameworks

In this section we review works that focus on methods for unified geometric light source calibration covering at least two different light source types. Typically, unified approaches define more complex mathematical models, which generalize the established light source models.

Weber and Cipolla [73] introduce a general model for point light sources and a practical single-image calibration method which requires a homogeneous diffuse calibration object with known convex geometry. The described mathematical model is applicable both for isotropic and non-isotropic point light sources as well as for distant ones. In practice, however, the authors present an evaluation of their calibration methods only on isotropic point light sources. The proposed procedure allows the calibration of a single light source.

Extending upon the work of Langer et al. [36], Zhou and Kambhamettu [82] develop a General Light source Model (GLM) and present a framework for both geometric and photometric light source calibration. The geometric calibration relies on specular highlights on the surface of a calibration sphere imaged via well calibrated stereo camera setup. The proposed GLM is geometrically defined by two parameters: the source region A defined on a plane, and the set of directions Q of all rays emitted by a given light source.

Table 3. Overview of technical requirements

requirement category		reference paper
number of input images	single	[1, 5–8, 23, 30, 39, 41, 43, 45, 47, 48, 52, 72–74, 76, 79]
	multiple	[10, 12, 17, 26–29, 32, 35, 37, 38, 40, 54, 55, 58, 59, 61, 64, 67, 78, 80–82]
type of calibration object	sphere	[1, 5, 26, 37, 38, 41, 48, 54, 61, 74, 76, 79–82]
	cuboid	[6, 52, 73]
	plane	[40, 54, 59, 67]
	custom or learned	[8] (spherical object with a surface discontinuity), [45] (custom plane), [47] (human hands), [64] (small shadow casters)
	arbitrary object	[7, 10, 12, 23, 26–30, 32, 35, 39, 55, 72, 78]
	none	[17, 58]
requires geometric information	yes	[1, 6–8, 23, 26–30, 32, 35, 37, 39, 55, 72, 73, 76, 78]
	partially	[5, 10, 38, 41, 43, 47, 48, 54, 61, 79]
	no	[12, 17, 40, 45, 58, 59, 64, 67, 74, 80–82]

Recently, Santo et al. [64] have proposed a light calibration technique based on moving cast shadows, which is suitable for both distant and near point light sources. The choice of a light model in a given scene is done automatically, exploiting the different number of degrees-of-freedom between distant and near point light sources. The authors use small pins distributed over a movable diffuse plane as shadow casters. Since knowing their position is not required, the custom calibration target can be easily reproduced. The motion of the plane is tracked with the help of ArUco markers [20]. Multiple light sources can be calibrated simultaneously.

5 Conclusion and Outlook

With this work we propose a general classification of methods for geometric light source calibration and summarize their main application areas. The diverse areas of application of calibrated illumination show the importance of both accurate and at the same time practical calibration techniques. Although, researchers have achieved good results in either of these directions, a combination of both accuracy, ease of use and computational speed is still an open challenge.

In order to better support future light calibration research, there is a necessity for publicly available data sets with reliable ground truth information covering a diverse range of lighting sources. We observe that the majority of scientific publication on the topic of light calibration evaluate the accuracy of the proposed methods completely autonomously without quantitative comparison with other related works. Often the evaluation is mainly qualitative and focused around the performance of a specific added task. For example, the visual fidelity of an AR scene or the photometric reconstruction error are used to indirectly judge the light calibration quality. Some of the more prominent public data sets of images and light calibration data include Light Stage Data Gallery [11], Gourd&Apple [3] and DiLiGenT [68] and were originally composed for applications in object relighting and photometric stereo. Other data sets originated in the context of AR research [7,43]. However, to the best of our knowledge, currently there is no specific data set designed for the sake of light source calibration. Such data set should provide an accurate ground truth and large diversity of lighting conditions, also including scenes with multiple light sources of different type.

Acknowledgments. This work was partially supported by the Thüringer Aufbaubank (TAB), the Free State of Thuringia and the European Regional Development Fund (EFRE) under project number 2019 FGI 0026.

References

1. Ackermann, J., Fuhrmann, S., Goesele, M.: Geometric point light source calibration. In: Bronstein, M., Favre, J., Hormann, K. (eds.) Vision, Modeling & Visualization. The Eurographics Association (2013)
2. Alhakamy, A., Tuceryan, M.: Real-time illumination and visual coherence for photorealistic augmented/mixed reality. ACM Comput. Surv. **53**(3), 1–34 (2020)
3. Alldrin, N., Zickler, T., Kriegman, D.: Photometric stereo with non-parametric and spatially-varying reflectance. In: IEEE Conference on Computer Vision and Pattern Recognition, pp. 1–8. IEEE (2008)
4. Alldrin, N., Kriegman, D.: A planar light probe. In: IEEE Computer Society Conference on Computer Vision and Pattern Recognition (CVPR 2006), vol. 2, pp. 2324–2330 (2006)

5. Aoto, T., Taketomi, T., Sato, T., Mukaigawa, Y., Yokoya, N.: Position estimation of near point light sources using a clear hollow sphere. In: Proceedings of the 21st International Conference on Pattern Recognition (ICPR 2012), pp. 3721–3724 (2012)
6. Arief, I., McCallum, S., Hardeberg, J.Y.: Realtime estimation of illumination direction for augmented reality on mobile devices. In: Color and Imaging Conference, vol. 2012, pp. 111–116. Society for Imaging Science and Technology (2012)
7. Boom, B., Orts-Escolano, S., Ning, X., McDonagh, S., Sandilands, P., Fisher, R.: Point light source estimation based on scenes recorded by a RGB-D camera. In: British Machine Vision Conference, Bristol (2013)
8. Bunteong, A., Chotikakamthorn, N.: Light source estimation using feature points from specular highlights and cast shadows. Int. J. Phys. Sci. **11**, 168–177 (2016)
9. Burley, B., Studios, W.D.A.: Physically-based shading at disney. In: ACM SIGGRAPH, vol. 2012, pp. 1–7 (2012)
10. Cao, X., Foroosh, H.: Camera calibration and light source orientation from solar shadows. Comput. Vis. Image Underst. **105**(1), 60–72 (2007)
11. Chabert, C.F., et al.: Relighting human locomotion with flowed reflectance fields. In: ACM SIGGRAPH 2006 Sketches, pp. 76–es (2006)
12. Chen, G., Han, K., Shi, B., Matsushita, Y., Wong, K.Y.K.K.: Self-calibrating deep photometric stereo networks. In: IEEE/CVF Conference on Computer Vision and Pattern Recognition (CVPR), pp. 8731–8739 (2019)
13. Debevec, P.: Rendering synthetic objects into real scenes: bridging traditional and image-based graphics with global illumination and high dynamic range photography. In: Proceedings of the 25th Annual Conference on Computer Graphics and Interactive Techniques, SIGGRAPH 1998, pp. 189–198. Association for Computing Machinery, New York (1998)
14. Debevec, P.: A median cut algorithm for light probe sampling. In: ACM SIGGRAPH 2005 Posters, SIGGRAPH 2005, pp. 66–es. Association for Computing Machinery, New York (2005)
15. Dong, Y., Chen, G., Peers, P., Zhang, J., Tong, X.: Appearance-from-motion: recovering spatially varying surface reflectance under unknown lighting. ACM Trans. Graph. (TOG) **33**(6), 1–12 (2014)
16. Fischler, M.A., Bolles, R.C.: Random sample consensus: a paradigm for model fitting with applications to image analysis and automated cartography. Commun. ACM **24**(6), 381–395 (1981)
17. Frahm, J.M., Koeser, K., Grest, D., Koch, R.: Markerless augmented reality with light source estimation for direct illumination. In: The 2nd IEE European Conference on Visual Media Production CVMP 2005, pp. 211–220 (2005)
18. Fujimura, Y., Iiyama, M., Hashimoto, A., Minoh, M.: Photometric stereo in participating media considering shape-dependent forward scatter. In: IEEE/CVF Conference on Computer Vision and Pattern Recognition, pp. 7445–7453 (2018)
19. Furukawa, Y., Hernández, C.: Multi-view stereo: a tutorial. Found. Trends. Comput. Graph. Vis. **9**(1–2), 1–148 (2015)
20. Garrido-Jurado, S., Muñoz-Salinas, R., Madrid-Cuevas, F., Marín-Jiménez, M.: Automatic generation and detection of highly reliable fiducial markers under occlusion. Pattern Recogn. **47**(6), 2280–2292 (2014)
21. Goldman, D.B., Curless, B., Hertzmann, A., Seitz, S.M.: Shape and spatially-varying BRDFs from photometric stereo. IEEE Trans. Pattern Anal. Mach. Intell. **32**(6), 1060–1071 (2010)

22. Gruber, L., Richter-Trummer, T., Schmalstieg, D.: Real-time photometric registration from arbitrary geometry. In: IEEE International Symposium on Mixed and Augmented Reality (ISMAR), pp. 119–128 (2012)
23. Hara, K., Nishino, K., Ikeuchi, K.: Light source position and reflectance estimation from a single view without the distant illumination assumption. IEEE Trans. Pattern Anal. Mach. Intell. **27**, 493–505 (2005)
24. Hatzitheodorou, M.: Shape from shadows: a Hilbert space setting. J. Complex. **14**(1), 63–84 (1998)
25. Horn, B.K.: Shape from shading: a method for obtaining the shape of a smooth opaque object from one view. Technical report, Massachusetts Institute of Technology (1970)
26. Innmann, M., Süßmuth, J., Stamminger, M.: BRDF-reconstruction in photogrammetry studio setups. In: IEEE Winter Conference on Applications of Computer Vision (WACV), pp. 3346–3354 (2020)
27. Jiddi, S., Robert, P., Marchand, E.: Reflectance and illumination estimation for realistic augmentations of real scenes. In: IEEE International Symposium on Mixed and Augmented Reality (ISMAR-Adjunct), pp. 244–249 (2016)
28. Jiddi, S., Robert, P., Marchand, E.: Estimation of position and intensity of dynamic light sources using cast shadows on textured real surfaces. In: 25th IEEE International Conference on Image Processing (ICIP), pp. 1063–1067 (2018)
29. Jiddi, S., Robert, P., Marchand, E.: Detecting specular reflections and cast shadows to estimate reflectance and illumination of dynamic indoor scenes. IEEE Trans. Vis. Comput. Graph. 1 (2020, online). https://doi.org/10.1109/tvcg.2020.2976986
30. Kán, P., Kafumann, H.: DeepLight: light source estimation for augmented reality using deep learning. Vis. Comput. **35**(6), 873–883 (2019)
31. Karaoglu, S., Liu, Y., Gevers, T., Smeulders, A.W.M.: Point light source position estimation from RGB-D images by learning surface attributes. IEEE Trans. Image Process. **26**(11), 5149–5159 (2017)
32. Kasper, M., Heckman, C.: Multiple point light estimation from low-quality 3D reconstructions. In: 2019 International Conference on 3D Vision (3DV), pp. 738–746 (2019)
33. Knorr, S.B., Kurz, D.: Real-time illumination estimation from faces for coherent rendering. In: IEEE International Symposium on Mixed and Augmented Reality (ISMAR), pp. 113–122. IEEE (2014)
34. Kronander, J., Banterle, F., Gardner, A., Miandji, E., Unger, J.: Photorealistic rendering of mixed reality scenes. Comput. Graph. Forum **34**(2), 643–665 (2015)
35. Lagger, P., Fua, P.: Using specularities to recover multiple light sources in the presence of texture. In: 18th International Conference on Pattern Recognition (ICPR 2006), vol. 1, pp. 587–590 (2006)
36. Langer, M., Zucker, S.: What is a light source? In: Proceedings of IEEE Computer Society Conference on Computer Vision and Pattern Recognition, pp. 172–178 (1997)
37. Lee, S., Jung, S.K.: Estimation of illuminants for plausible lighting in augmented reality. In: International Symposium on Ubiquitous Virtual Reality, pp. 17–20 (2011)
38. Lensch, H.P.A., Kautz, J., Goesele, M., Heidrich, W., Seidel, H.P.: Image-based reconstruction of spatial appearance and geometric detail. ACM Trans. Graph. **22**(2), 234–257 (2003)
39. Li, Y., Lin, Lu, H., Shum, H.Y.: Multiple-cue illumination estimation in textured scenes. In: Proceedings Ninth IEEE International Conference on Computer Vision, vol. 2, pp. 1366–1373 (2003)

40. Liu, C., Narasimhan, S., Dubrawski, A.: Near-light photometric stereo using circularly placed point light sources. In: IEEE International Conference on Computational Photography (ICCP), pp. 1–10 (2018)
41. Liu, Y., Kwak, Y.S., Jung, S.K.: Position estimation of multiple light sources for augmented reality. In: Park, J., Stojmenovic, I., Jeong, H., Yi, G. (eds.) Computer Science and its Applications. LNEE, vol. 330, pp. 891–897. Springer, Heidelberg (2015). https://doi.org/10.1007/978-3-662-45402-2_126
42. Lombardi, S., Nishino, K.: Reflectance and natural illumination from a single image. In: Fitzgibbon, A., Lazebnik, S., Perona, P., Sato, Y., Schmid, C. (eds.) ECCV 2012. LNCS, vol. 7577, pp. 582–595. Springer, Heidelberg (2012). https://doi.org/10.1007/978-3-642-33783-3_42
43. Lopez-Moreno, J., Garces, E., Hadap, S., Reinhard, E., Gutiérrez, D.: Multiple light source estimation in a single image. In: Computer Graphics Forum, vol. 32 (2013)
44. Luo, T., Wang, G.: Compact collimators designed with point approximation for light-emitting diodes. Light. Res. Technol. **50**(2), 303–315 (2018)
45. Ma, L., Liu, J., Pei, X., Hu, Y., Sun, F.: Calibration of position and orientation for point light source synchronously with single image in photometric stereo. Opt. Express **27**(4), 4024–4033 (2019)
46. Mandl, D., et al.: Learning lightprobes for mixed reality illumination. In: IEEE International Symposium on Mixed and Augmented Reality (ISMAR), pp. 82–89 (2017)
47. Marques., B.A.D., Drumond., R.R., Vasconcelos., C.N., Clua., E.: Deep light source estimation for mixed reality. In: Proceedings of the 13th International Joint Conference on Computer Vision, Imaging and Computer Graphics Theory and Applications - GRAPP, pp. 303–311. SciTePress (2018)
48. Masselus, V., Dutré, P., Anrys, F.: The free-form light stage. In: Debevec, P., Gibson, S. (eds.) Eurographics Workshop on Rendering. The Eurographics Association (2002)
49. Meister, G., Wiemker, R., Monno, R., Spitzer, H., Strahler, A.: Investigation on the torrance-sparrow specular BRDF model. In: IGARSS 1998. Sensing and Managing the Environment. IEEE International Geoscience and Remote Sensing. Symposium Proceedings (Cat. No.98CH36174), vol. 4, pp. 2095–2097 (1998)
50. Mo, Z., Shi, B., Lu, F., Yeung, S.K., Matsushita, Y.: Uncalibrated photometric stereo under natural illumination. In: IEEE/CVF Conference on Computer Vision and Pattern Recognition, pp. 2936–2945 (2018)
51. Moreno, I., Avendaño-Alejo, M., Tsonchev, R.: Designing light-emitting diode arrays for uniform near-field irradiance. Appl. Opt. **45**, 2265–2272 (2006)
52. Mori, K., Watanabe, E., Watanabe, K., Katagiri, S.: Estimation of object color, light source color, and direction by using a cuboid. Syst. Comput. Jpn. **36**, 1–10 (2005)
53. Murez, Z., Treibitz, T., Ramamoorthi, R., Kriegman, D.: Photometric stereo in a scattering medium. In: IEEE International Conference on Computer Vision (ICCV), pp. 3415–3423 (2015)
54. Nie, Y., Song, Z., Ji, M., Zhu, L.: A novel calibration method for the photometric stereo system with non-isotropic led lamps. In: IEEE International Conference on Real-time Computing and Robotics (RCAR), pp. 289–294 (2016)
55. Nieto, G., Jiddi, S., Robert, P.: Robust point light source estimation using differentiable rendering. CoRR abs/1812.04857 (2018). http://arxiv.org/abs/1812.04857
56. Nishino, K., Nayar, S.K.: Eyes for relighting. ACM Trans. Graph. (TOG) **23**(3), 704–711 (2004)

57. Ohno, Y.: NIST measurement services: photometric calibrations, vol. 250–37. Special Publication (NIST SP), National Institute of Standards and Technology, Gaithersburg (1997)
58. Papadhimitri, T., Favaro, P.: Uncalibrated near-light photometric stereo. In: Proceedings of the British Machine Vision Conference. BMVA Press (2014)
59. Park, J., Sinha, S.N., Matsushita, Y., Tai, Y.W., Kweon, I.S.: Calibrating a non-isotropic near point light source using a plane. In: IEEE Conference on Computer Vision and Pattern Recognition, pp. 2267–2274 (2014)
60. Pentland, A.P.: Finding the illuminant direction. J. Opt. Soc. Am. **72**(4), 448–455 (1982)
61. Powell, M.W., Sarkar, S., Goldgof, D.: A simple strategy for calibrating the geometry of light sources. IEEE Trans. Pattern Anal. Mach. Intell. **23**(9), 1022–1027 (2001)
62. Ramamoorthi, R., Hanrahan, P.: A signal-processing framework for inverse rendering. In: Proceedings of the 28th Annual Conference on Computer Graphics and Interactive Techniques, SIGGRAPH 2001, pp. 117–128. Association for Computing Machinery, New York (2001)
63. Richter-Trummer, T., Kalkofen, D., Park, J., Schmalstieg, D.: Instant mixed reality lighting from casual scanning. In: IEEE International Symposium on Mixed and Augmented Reality (ISMAR), pp. 27–36 (2016)
64. Santo, H., Waechter, M., Lin, w.y., Sugano, Y., Matsushita, Y.: Light structure from pin motion: Geometric point light source calibration. Int. J. Comput. Vis. **128**, 1889–1912 (2020)
65. Sato, I., Sato, Y., Ikeuchi, K.: Illumination from shadows. IEEE Trans. Pattern Anal. Mach. Intell. **25**(3), 290–300 (2003)
66. Shafer, S.A.: Using color to separate reflection components. Color Res. Appl. **10**(4), 210–218 (1985)
67. Shen, H.L., Cheng, Y.: Calibrating light sources by using a planar mirror. J. Electron. Imaging **20**, 013002 (2011)
68. Shi, B., Wu, Z., Mo, Z., Duan, D., Yeung, S.K., Tan, P.: A benchmark dataset and evaluation for non-lambertian and uncalibrated photometric stereo. In: Proceedings of the IEEE Conference on Computer Vision and Pattern Recognition, pp. 3707–3716 (2016)
69. Takai, T., Maki, A., Matsuyama, T.: Self shadows and cast shadows in estimating illumination distribution. In: 4th European Conference on Visual Media Production, pp. 1–10 (2007)
70. Unger, J., Kronander, J., Larsson, P., Gustavson, S., Ynnerman, A.: Temporally and spatially varying image based lighting using HDR-video. In: 21st European Signal Processing Conference (EUSIPCO 2013), pp. 1–5 (2013)
71. Wang, T.Y., Ritschel, T., Mitra, N.: Joint material and illumination estimation from photo sets in the wild. In: International Conference on 3D Vision (3DV), pp. 22–31 (2018)
72. Wang, Y., Samaras, D.: Estimation of multiple directional light sources for synthesis of mixed reality images. In: 10th Pacific Conference on Computer Graphics and Applications, pp. 38–47. IEEE Computer Society (2002)
73. Weber, M., Cipolla, R.: A practical method for estimation of point light-sources. In: Proceedings of BMVC 2001, vol. 2, pp. 471–480 (2001)
74. Wong, K.-Y.K., Schnieders, D., Li, S.: Recovering light directions and camera poses from a single sphere. In: Forsyth, D., Torr, P., Zisserman, A. (eds.) ECCV 2008. LNCS, vol. 5302, pp. 631–642. Springer, Heidelberg (2008). https://doi.org/10.1007/978-3-540-88682-2_48

75. Woodham, R.J.: Photometric method for determining surface orientation from multiple images, pp. 513–531. MIT Press (1989)
76. Xie, L., Song, Z., Huang, X.: A novel method for the calibration of an led-based photometric stereo system. In: IEEE International Conference on Information and Automation (ICIA), pp. 780–783 (2013)
77. Xie, L., Song, Z., Jiao, G., Huang, X., Jia, K.: A practical means for calibrating an led-based photometric stereo system. Opt. Lasers Eng. **64**, 42–50 (2015)
78. Xu, S., Wallace, A.M.: Recovering surface reflectance and multiple light locations and intensities from image data. Pattern Recogn. Lett. **29**(11), 1639–1647 (2008)
79. Zhang, Y., Yang, Y.H.: Multiple illuminant direction detection with application to image synthesis. IEEE Trans. Pattern Anal. Mach. Intell. **23**(8), 915–920 (2001)
80. Zhou, W., Kambhamettu, C.: Estimation of the size and location of multiple area light sources. In: International Conference on Pattern Recognition, vol. 4, pp. 214–217. IEEE Computer Society (2004)
81. Zhou, W., Kambhamettu, C.: Estimation of illuminant direction and intensity of multiple light sources. In: Heyden, A., Sparr, G., Nielsen, M., Johansen, P. (eds.) ECCV 2002, Part IV. LNCS, vol. 2353, pp. 206–220. Springer, Heidelberg (2002). https://doi.org/10.1007/3-540-47979-1_14
82. Zhou, W., Kambhamettu, C.: A unified framework for scene illuminant estimation. Image Vis. Comput. **26**(3), 415–429 (2008)

Quantifying Point Cloud Realism Through Adversarially Learned Latent Representations

Larissa T. Triess[1,2(✉)], David Peter[1], Stefan A. Baur[1], and J. Marius Zöllner[2,3]

[1] Mercedes-Benz AG, Research and Development, Stuttgart, Germany
larissa.triess@daimler.com
[2] Karlsruhe Institute of Technology, Karlsruhe, Germany
[3] Research Center for Information Technology, Karlsruhe, Germany

Abstract. Judging the quality of samples synthesized by generative models can be tedious and time consuming, especially for complex data structures, such as point clouds. This paper presents a novel approach to quantify the realism of local regions in LiDAR point clouds. Relevant features are learned from real-world and synthetic point clouds by training on a proxy classification task. Inspired by fair networks, we use an adversarial technique to discourage the encoding of dataset-specific information. The resulting metric can assign a quality score to samples without requiring any task specific annotations.

In a series of experiments, we confirm the soundness of our metric by applying it in controllable task setups and on unseen data. Additional experiments show reliable interpolation capabilities of the metric between data with varying degree of realism. Further, we demonstrate how the local realism score can be used for anomaly detection in point clouds.

Keywords: Point cloud · Metric · Adversarial training · Local features

1 Introduction

Generative models, such as generative adversarial networks (GANs), are often used to synthesize realistic training data samples to improve the performance of perception networks. Assessing the quality of such synthesized samples is a crucial part of the process which is usually done by experts, a cumbersome and time consuming approach. Though a lot of work has been conducted to determine the quality of generated images, very little work is published about how to quantify the realism of LiDAR point clouds. Visual inspection of such

Supplementary Information The online version contains supplementary material available at https://doi.org/10.1007/978-3-030-92659-5_44.

C. Bauckhage et al. (Eds.): DAGM GCPR 2021, LNCS 13024, pp. 681–696, 2021.
https://doi.org/10.1007/978-3-030-92659-5_44

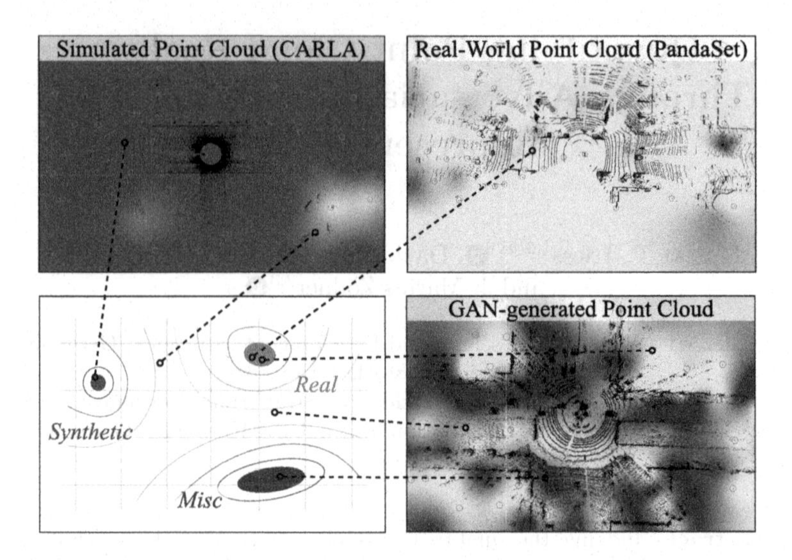

Fig. 1. Proposed Approach: The realism measure has a tripartite understanding of the 3D-world (bottom left). The other images show the color-coded metric scores at discrete query point locations (gray circles). Local regions in the simulated sample (top left) are largely predicted as being of synthetic origin (blue), while regions in the real-world sample (top right) are predicted to be realistic (green). The bottom right shows a GAN-generated sample which has large areas with high distortion levels that neither appear realistic nor synthetic. The metric therefore assigns high misc scores (red). (Color figure online)

data is expensive and not reliable given that the interpretation of 3D point data is rather unnatural for humans. Because of their subjective nature, it is difficult to compare generative approaches with a qualitative measure. This paper closes the gap and introduces a quantitative evaluation for LiDAR point clouds.

In recent years, a large amount of evaluation measures for GANs emerged [4]. Many of them are image specific and cannot be applied to point clouds. Existing work on generating realistic LiDAR point clouds mostly relies on qualitative measures to evaluate the generation quality. Alternatively, some works apply annotation transfer [32] or use the *Earth Mover's* distance as an evaluation criterion [5]. However, these methods require either annotations associated with the data or a matching target, i.e. Ground Truth, for the generated sample. Both are often not feasible when working with large-scale data generation.

The generation capabilities of the models are often directly graded by how much the performance of perception networks changes after being trained with the generated data. This procedure serves as an indication for realism, but cannot solely verify realism and fidelity [32,39]. The improved perception accuracy is primarily caused by the larger quantity of training data. Only measuring the performance difference in downstream perception tasks is therefore not an adequate measure for the realism of the data itself.

This paper proposes a novel learning-based approach for robust quantification of LiDAR point cloud quality. The metric is trained to learn relevant features via a proxy classification task. To avoid learning global scene context, we use hierarchical feature set learning to confine features locally in space. This locality aspect additionally enables the detection and localization of anomalies and other sensor effects within the scenery. To discourage the network from encoding dataset-specific information, we use an adversarial learning technique which enables robust quantification of unseen data distributions. The resulting metric does not require any additional annotations.

2 Related Work

2.1 Point Cloud Quality Measures

A decisive element to process unordered point sets is the ability to operate in a permutation-invariant fashion, such that the ordering of points does not matter. *Earth Mover's* distance (EMD) and *Chamfer* distance (CD) are often used to compare such sets. EMD measures the distance between the two sets by attempting to transform one set into the other, while CD measures the squared distance between each point in one set and its nearest neighbor in the other set. Both metrics are often used to measure the reconstruction of single object shapes, like those from ShapeNet [1]. Caccia et al. [5] use the metrics as a measure of reconstruction quality on entire scenes captured with a LiDAR scanner. This procedure is only applicable to paired translational GANs or supervised approaches, because it requires a known target, i.e. Ground Truth to measure the reconstruction error. It is therefore not useable for unpaired or unsupervised translation on which many domain adaptation tasks are based.

In previous work, we use the mean opinion score (MOS) testing to verify the realism of generated LiDAR point clouds [39]. It was previously introduced in [22] to provide a qualitative measure for realism in RGB images. In contrast to [22], where untrained people were used to determine the realism, [39] requires LiDAR experts for the testing process to assure a high enough sensor domain familiarity of the test persons.

Some domain adaptation GANs are used to improve the performance of downstream perception tasks by training the perception network with the generated data [32,36]. The generation capabilities of the GAN are derived by how much the perception performance on the target domain changes when trained with the domain adapted data in addition or instead of the source data. These capabilities include the ability to generate samples that are advantageous for perception network training but might not be suitable to quantify the actual generation quality. Inspecting example images in state-of-the-art literature [5,32] and closely analyzing our own generated data, we observe that the GAN-generated point clouds are usually very noisy, especially at object boundaries (see bottom right in Fig. 1). Therefore, we find that solely analyzing GAN generation performance on downstream perception tasks is not enough to claim realistic generation properties.

2.2 GAN Evaluation Measures

The GAN objective function is not suited as a general quality measure or as a measure of sample diversity, as it can only measure how well the generator and the discriminator perform relative to the opponent. Therefore, a considerable amount of literature deals with how to evaluate generative models and propose various evaluation measures. The most important ones are summarized in extensive survey papers [4,25,40]. They can be divided into two major categories: qualitative and quantitative measures.

Qualitative evaluation [9,15,20,24,26,37,42] uses visual inspection of a small collection of examples by humans and is therefore of subjective nature. It is a simple way to get an initial impression of the performance of a generative model but cannot be performed in an automated fashion. The subjective nature makes it difficult to compare performances across different works, even when a large inspection group, such as via Mechanical Turk, is used. Furthermore, it is expensive and time-consuming.

Quantitative evaluation, on the other hand, is performed over a large collection of examples, often in an automated fashion. The two most popular quantitative metrics are the Inception Score (IS) [31] and the Fréchet Inception Distance (FID) [18]. In their original form, these and other ImageNet-based metrics [8,17,44] are exclusively applicable to camera image data, as they are based on features learned from the ImageNet dataset [11] and can therefore not be directly applied to LiDAR scans.

Therefore, [35] proposes Fréchet point cloud distance (FPD), which can measure the quality of GAN-generated 3D point clouds. Based on the same principle as FID, FPD calculates the 2-Wasserstein distance between real and fake Gaussian measures in the feature spaces extracted by PointNet [7]. In contrast to our method, FPD requires labels on the target domain to train the feature extractor. Further, it is only possible to compare a sample to one particular distribution and therefore makes it difficult to obtain a reliable measure on unseen data.

There also exists a number of modality independent GAN metrics, such as distribution-based [2,16,30,38], classification-based [23,33,41], and model comparison [21,27,43] approaches. However, we do not further consider them in this work, since their focus is not on judging the quality of individual samples, but rather on capturing sample diversity and mode collapse in entire distributions.

2.3 Metric Learning

The goal of deep metric learning is to learn a feature embedding, such that similar data samples are projected close to each other while dissimilar data samples are projected far away from each other in the high-dimensional feature space. Common methods use siamese networks trained with contrastive losses to distinguish between similar and dissimilar pairs of samples [10]. Thereupon, triplet loss architectures train multiple parallel networks with shared weights to achieve the feature embedding [12,19]. This work uses an adversarial training technique to push features in a similar or dissimilar embedding.

2.4 Contributions

The aim of this paper is to provide a reliable metric that gives a quantitative estimate about the realism of generated LiDAR data. The contributions of this work are threefold. First and foremost, we present a novel way to learn a measure of realism in point clouds. This is achieved by learning hierarchical point set features on a proxy classification task. Second, we utilize an adversarial technique from the fairness-in-machine-learning domain in order to eliminate dataset-specific information. This allows the measure to be used on unseen datasets. Finally, we demonstrate how the fine-grained local realism score can be used for anomaly detection in LiDAR scans.

3 Method

3.1 Objective and Properties

The aim of this work is to provide a method to estimate the level of realism for arbitrary LiDAR point clouds. We design the metric to learn relevant realism features directly from distributions of real-world data. The output of the metric can then be interpreted as a distance measure between the input and the learned distribution in a high dimensional space.

Regarding the discussed aspects of existing point cloud quality measures and GAN measures, we expect a useful LiDAR point cloud metric to be:

Quantitative: The realism score is a quantitative measure that determines the distance of the input sample to the internal representation of the learned realistic distribution. The score has well defined lower and upper bounds that reach from 0 (unrealistic) to 1 (realistic).

Universal: The metric has to be applicable to any LiDAR input and therefore must be independent from any application or task. This means no explicit ground truth information is required.

Transferable: The metric must give a reliable and robust prediction for all inputs, independent on whether the data distribution of the input sample is known by the metric or not. This makes the metric transferable to new and unseen data.

Local: The metric should be able to compute spatially local realism scores for smaller regions within a point cloud. It is also expected to focus on identifying the quality of the point cloud properties while ignoring global scene properties as much as possible.

Flexible: Point clouds are usually sets of unordered points with varying size. Therefore, it is crucial to have a processing that is permutation-invariant and independent of the number of points to process.

Fast: Speed is not the most important property, but a fast computation time allows the metric to run in parallel to the training of a neural network for LiDAR

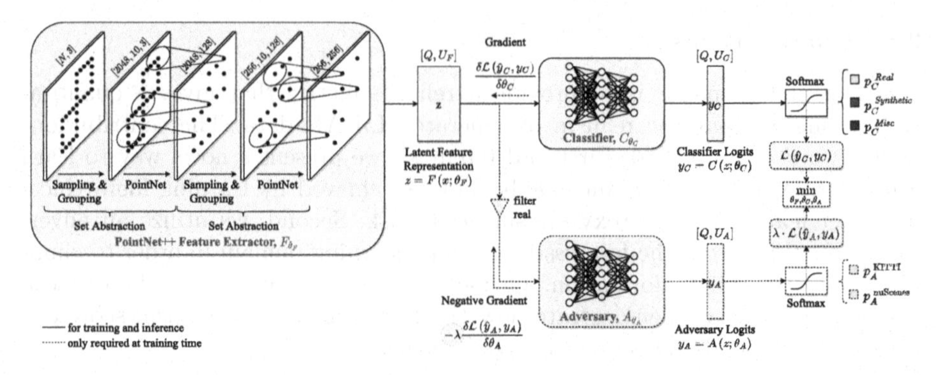

Fig. 2. Architecture: The feature extractor F_{θ_F} uses hierarchical feature set learning from PointNet++ [28] to encode information about each of the Q query points and their nearest neighbors. The neighborhood features z are then passed to two identical networks, the classifier C_{θ_C} and the adversary A_{θ_A}. The classifier outputs probability scores $\mathbf{p}_C$ for each category – *Real*, *Synthetic*, *Misc* – while the adversary outputs probability scores $\mathbf{p}_A$ for each dataset. For both network outputs a multi-class cross-entropy loss is minimized. To let the classifier perform as good as possible while the adversary should perform as bad as possible, the gradient is inverted between the adversary input and the feature extractor [3,29]. The input to the adversary is limited to samples from the *Real* category. λ is a factor that regulates the influence of the two losses, weighting the ratio of accuracy versus fairness. In our experiments we use a factor of $\lambda = 0.3$. For more information we refer to the supplementary material. (Color figure online)

data generation. This enables monitoring of generated sample quality over the training time of the network.

We implement our metric in such a way that the described properties are fulfilled. To differentiate the metric from a GAN discriminator, we want to stress that a discriminator is not *transferable* to unseen data.

3.2 Architecture

Figure 2 shows the architecture of our approach. The following describes the components and presents how each part is designed to contribute towards achieving the desired metric properties. The underlying idea of the metric design is to compute a distance measure between different data distributions of *realistic* and *unrealistic* LiDAR point cloud compositions. The network learns features indicating realism from data distributions by using a proxy classification task. Specifically, the network is trained to classify point clouds from different datasets into three categories: *Real*, *Synthetic*, *Misc*. The premise is the possibility to divide the probability space of LiDAR point clouds into those that derive from real-world data (*Real*), those that derive from simulations (*Synthetic*), and all the others (*Misc*), e.g. distorted or randomized data. Refer to Fig. 1 for an impression. By acquiring the prior information about the tripartite data distribution, the metric does not require any target information or labels for inference.

The features are obtained with hierarchical feature set learning, explained in Sect. 3.2. Section 3.2 outlines our adversarial learning technique.

Feature Extractor. The left block in Fig. 2 visualizes the PointNet++ [28] concept of the feature extractor F_{θ_F}. We use two abstraction levels, sampling 2048 and 256 query points, respectively, with 10 nearest neighbors each. Keeping the number of neighbors and abstraction levels low limits the network to only encode information about *local* LiDAR-specific statistics instead of global scenery information. On the other hand, the high amount of query points helps to cover many different regions within the point cloud and guarantees the *local* aspect of our method. We use a 3-layer MLP in each abstraction level with filter sizes of $[64, 64, 128]$ and $[128, 128, 256]$, respectively. This results in the neighborhood features, a latent space representation $z = F(x, \theta_F)$ of size $[Q, U_F]$ with $U_F = 256$ features for each of the $Q = 256$ query points. The features z are then fed to a densely connected classifier C_{θ_C} (yellow block). It consists of a hidden layer with 128 units and 50% dropout, and the output layer with U_C units.

As the classifier output, we obtain the probability vector $\mathbf{p}_{C,q} = \text{softmax}(y_C) \in [0, 1]^{U_C}$ per query point q. The vector has $U_C = 3$ entries for each of the categories *Real*, *Synthetic* and *Misc*. The component $p_{C,q}^{Real}$ quantifies the degree of realism in each local region q. The scores $\mathbf{S} = \frac{1}{Q} \sum_q \mathbf{p}_{C,q}$ for the entire scene are given by the mean over all query positions. Here, S^{Real} is a measure for the degree of realism of the entire point cloud. A score of 0 indicates low realism while 1 indicates high realism.

Fair Categorization. To obtain a *transferable* metric network, we decided to leverage a concept often used to design fair network architectures [3,29]. The idea is to force the feature extractor to encode only information into the latent representation z that is relevant for realism estimation. This means, we actively discourage the feature extractor from encoding information that is specific to the distribution of a single dataset. In other words, using fair networks terminology [3], we treat the concrete dataset name as a sensitive attribute. With this procedure we can improve the generalization ability towards unknown data.

To achieve this behavior, we add a second output head, the adversary A_{θ_A}, which is only required at training time (see orange block in Fig. 2). Its layers are identical to the one of the classifier, except for the number of units in the output layer. Following the designs proposed in [3,29], we train all network components by minimizing the losses for both heads, $\mathcal{L}_C = \mathcal{L}(y_C, \hat{y}_C)$ and $\mathcal{L}_A = \mathcal{L}(y_A, \hat{y}_A)$, but reversing the gradient in the path between the adversary input and the feature extractor. The goal is for C to predict the category y_C and for A to predict the dataset y_A as good as possible, but for F to make it hard for A to predict y_A. Training with the reversed gradient results in F encoding as little information as possible for predicting y_A. The training objective is formulated as

$$\min_{\theta_F, \theta_C, \theta_A} \mathcal{L}\Big(C\big(F(x; \theta_F); \theta_C\big), \hat{y}_C\Big) + \mathcal{L}\Big(A\big(J_\lambda[F(x; \theta_F)]; \theta_A\big), \hat{y}_A\Big) \quad (1)$$

with θ being the trainable variables and J_λ a special function such that

$$J_\lambda[F] = F \quad \text{but} \quad \nabla J_\lambda[F] = -\lambda \cdot \nabla F \quad . \tag{2}$$

The factor λ determines the ratio of accuracy and fairness.

In contrast to the application examples from related literature [3,29], our requested attribute, the category, can always be retrieved from the sensitive attribute, the specific dataset. Therefore, we consider only the data samples coming from the *Real* category in the adversary during training, indicated by the triangle in Fig. 2. Considering samples from all categories leads to unwanted decline in classifier performance (see ablation study in Sect. 4.5). Our filter extension forces the feature extractor to encode only common features within the *Real* category, but keep all other features in the other categories. In principal, it is possible to use a separate adversary for each category, but this enlarges the training setup unnecessarily since the main focus is to determine the realism of the input.

4 Experiments

Section 4.1 and Sect. 4.2 present the datasets and baselines used in our experiments. We provide extensive evaluations to demonstrate the capability and applicability of our method. First, we present the classification performance on completely unseen data distributions (Sect. 4.3). Then, we show that the metric provides a credible measure of realism when applied to generated point clouds (Sect. 4.4). In Sect. 4.5 we show the feature embedding capabilities of the fair training strategy in an ablation study. Section 4.6 shows that our metric is capable to interpolate the space between the training categories in a meaningful way. As an important application, Sect. 4.7 demonstrates how our method can be used for anomaly detection in point clouds. In the end, we elaborate on the limitations of our method (Sect. 4.8). Additional results, visualizations, and implementation details are provided in the supplementary material.

4.1 Datasets

We use two different groups of datasets. While the datasets of Table 1 are used to train the metric network, the ones in Table 2 are exclusively used to evaluate the metric. With the strict separation of training and evaluation datasets, additionally to the training and test splits, we can demonstrate that our method is a useful measure on unknown data distributions. In both cases alike, the datasets are divided into three categories: *Real*, *Synthetic*, *Misc*.

Publicly available real-world datasets are used in *Real* for training (KITTI, nuScenes), and evaluation (PandaSet).

For *Synthetic*, we use the CARLA simulator where we implement the sensor specifications of a Velodyne HDL-64 sensor to create ray-traced range measurements. GeomSet is the second support set in this category. Here, simple geometric

Table 1. Training Datasets: The table lists the datasets that are used as support sets and shows the number of samples used to train the metric network.

Category	Support set	Train split
Real	KITTI [14]	18,329
	nuScenes [6]	28,130
Synthetic	CARLA [13]	106,503
	GeomSet	18,200
Misc	Misc 1, 2, 3	∞

Table 2. Evaluation Datasets: The table lists the datasets that are used to evaluate the performance of the metric. The train split states the number of samples used to train the up-sampling networks.

Category	Dataset	Train split
Real	PandaSet [34]	7,440
Misc	Misc 4	–

objects, such as spheres and cubes are randomly scattered on a ground plane in three dimensional space and ray-traced in a scan pattern.

Finally, we add a third category, *Misc*, to allow the network to represent absurd data distributions, as they often occur during GAN trainings or with data augmentation. Therefore, *Misc* contains randomized data that is generated at training/evaluation time. Misc 1 and Misc 2 are generated by linearly increasing the depth over the rows or columns of a virtual LiDAR scanner, respectively. Misc 3 is a simple Gaussian noise with varying standard deviations. Misc 4 is created by setting patches of varying height and width of the LiDAR depth projection to the same distance. Misc 4 is only used for evaluation. Varying degrees of Gaussian noise are added to the Euclidean distances of Misc $\{1,2,4\}$.

It is certainly possible to add more support sets to each category, for example by including more datasets or by augmenting existing ones. However, we want to stress that our method is a reliable metric even when using just a small number of support sets, as demonstrated by the experiments below. In addition to the training data listed in the tables, we use 600 samples from a different split of each dataset to obtain our evaluation results. No annotations or additional information are required to train or apply the metric, all operations are based on the xyz coordinates of the point clouds.

4.2 Baselines and Evaluation

We evaluate our metric by applying it to the generated outputs of different versions of LiDAR up-sampling networks. The task of up-sampling offers the possibility to compare against other distance metrics by exploiting the high-resolution Ground Truth. Furthermore, it renders the difficulty for the networks to synthesize realistic high-resolution LiDAR outputs and is therefore a suitable testing candidate for our realism metric.

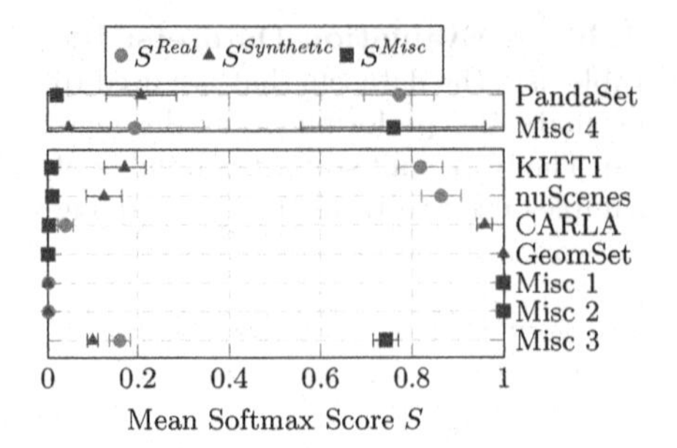

Fig. 3. Metric results: Shown is the metric output S for all three categories on different datasets. The upper part shows the unseen datasets, PandaSet and Misc 4. They are both correctly classified as *Real* and *Misc*, respectively. The results of the support sets are depicted in the lower part. The text coloring indicates the correct class.

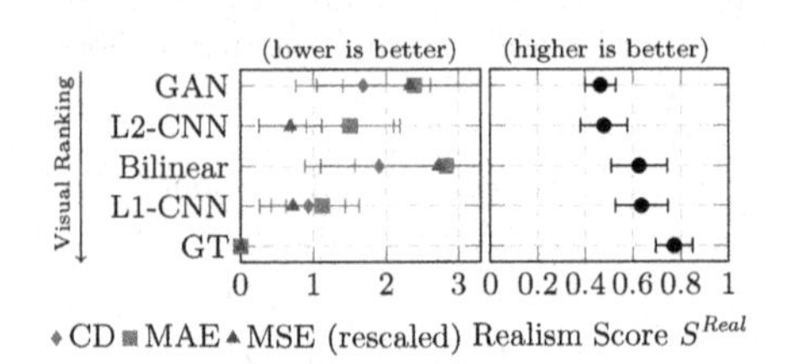

Fig. 4. Metric scores for up-sampling methods: The diagram lists four methods to perform 4× LiDAR scan upsampling and the high-resolution *Ground Truth* (GT) data. The left side shows different baseline measures, i.e. reconstruction errors, whereas the right side shows our realism score. Methods are ordered by their respective human visual judgment ratings.

As a baseline, we compute the *Chamfer* distance (CD) between the high-resolution prediction and the Ground Truth and compare against our metric results and human visual judgment. Since we follow the procedure proposed in [39], where the data is represented by cylindrical depth projections, we can compute additional distance measures. Specifically, we compute the mean squared error (MSE) and mean absolute error (MAE) from the dense point cloud representation between the reconstruction and the target.

4.3 Metric Results

We run our metric network on the test split of the evaluation datasets, as well as on the support sets. Figure 3 shows the mean of the metric scores S for each of the three categories. The observation is that our method correctly identifies the unknown PandaSet data as being realistic by outputting a high *Real* score. Misc 4 is also correctly classified, as indicated by the high *Misc* scores.

4.4 Metric Verification

In this section we demonstrate how our quality measure reacts to data that was generated by a neural network and compare it to traditional evaluation measures, as introduced in Sect. 4.2.

We compare bilinear interpolation to two convolutional neural network (CNN) versions and one GAN. The CNNs are trained with $\mathcal{L}_1$ and $\mathcal{L}_2$ loss, respectively. For the GAN experiments, we adapt SRGAN [22] from which the

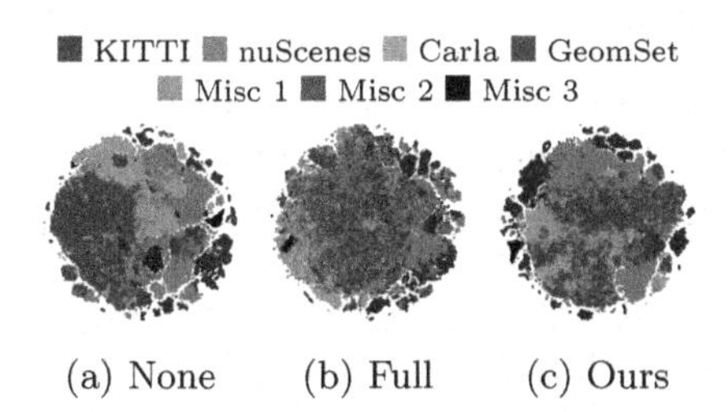

Fig. 5. Learned feature embedding: Shown are the t-Distributed Stochastic Neighbor Embedding (t-SNE) plots for the feature embeddings z of three versions of the metric network. (a) represents the learned features when the metric is trained without an adversary. (b) visualizes when an adversary was used for training as in related literature. Our approach is depicted by (c), where the adversary only focuses on data samples from the *Real* category.

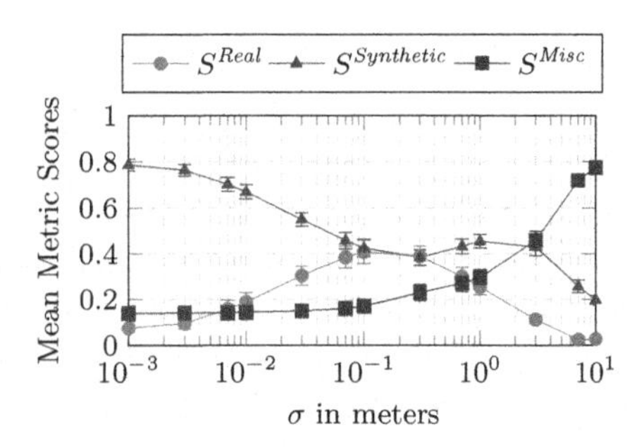

Fig. 6. Point cloud distortion: Increasing levels of Gaussian noise are applied to the distance measurements of CARLA point clouds. For small σ of a few centimeters, the synthetic data appears more realistic. For higher σ, the noise is dominant, as indicated by decreasing *Real* and *Synthetic* scores.

generator is also used in the CNN trainings. We conduct the experiments for 4× up-sampling and compare the predictions of our metric to the qualitative MOS ranking in [39].

Figure 4 shows the realism score on the right side and the baseline metrics on the left. The realism score for the original data (GT), is displayed for reference and achieves the best results with the highest realism score and lowest reconstruction error. The methods are ranked from top to bottom by increasing realism as perceived by human experts which matches the ordering by mean opinion score [39]. In general, the baseline metrics do not show clear correlation to the degree of realism and struggle to produce a reasonable ordering of the methods. Our realism score, on the other hand, sorts the up-sampling methods according to human visual judgment. These results align with the ones in [39], where the MOS study showed that a low reconstruction error does not necessarily imply a good generation quality.

The GAN obtains the lowest realism score which aligns with visual inspections of the generated data. An example scene is visualized in the bottom right of Fig. 1 and further images are provided in the supplementary material. We attribute the high noise levels and the low generation quality to two facts: First, due to the limited size of PandaSet, the GAN could only be trained with $\sim 7k$ samples, which is very small for these data complexities. Second, the SRGAN architecture is not specifically optimized for LiDAR point clouds.

4.5 Ablation Study

The proposed approach uses the fairness setup with a filter extension to embed features related to realism and omit dataset-specific information. To demonstrate the advantage of the adversarial feature learning and the necessity of our extension, we train two additional metric networks, one without the adversary and one with the adversary but without the extension.

Figure 5 shows plots of the t-SNE of the neighborhood features z, extracted from the three different metric versions. Each support set is represented by a different color. If two points or clusters are close, it means their encoding is more similar to those with higher spacing.

The t-SNE plots show that KITTI and nuScenes form two separable clusters when trained without the adversary (Fig. 5a). In our approach (Fig. 5c), they are mixed and form a single cluster. This shows that our proposed approach matches features from KITTI and nuScenes to the same distribution, whereas the network without the fair setup does not. Consequently, the fair setup enforces the feature extractor to only encode the overlap of the real dataset features.

Figure 5b visualizes the encoding of the metric trained with the adversary but without the filter extension. Clusters are not clearly separable, even those of support sets from different categories. This indicates impaired classification capabilities and justifies the use of the filter extension.

4.6 Feature Continuity

An important property of a metric is to represent a continuous feature space from which we can obtain reasonable realism estimates at any point within the learned tripartite distribution. To demonstrate this continuity, we take the CARLA test split and add normally distributed noise with zero mean and varying standard deviation σ.

Figure 6 shows the mean metric scores S over a wide range of additive noise levels applied to CARLA data. Notably, at low noise levels, the *Real* score increases. This reflects the fact that ideal synthetic data needs a certain level of range noise in order to appear more realistic. On the other hand, at high noise levels, the data barely possesses any structure, as indicated by high *Misc* and low *Real* and *Synthetic* scores.

We interpret the smooth transitions between the states as an indication for a disentangled latent representation within our metric network. Further, it shows the necessity for all three categories when using real-world and synthetic data.

4.7 Anomaly Detection

Given the locality aspect of our method, we are able to find anomalies or regions with differing appearance within a single scan. Figure 7 shows three examples where our metric outputs the lowest *Real* score within all test scans and one example where a part of the horizontal field of view is distorted with noise. The method can successfully identify anomalies within the support sets, i.e. KITTI

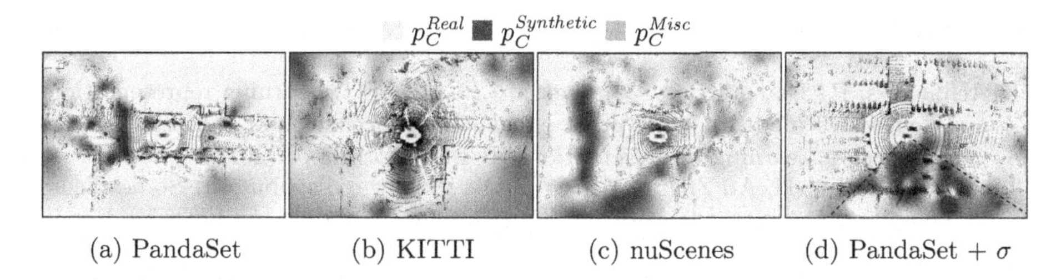

(a) PandaSet (b) KITTI (c) nuScenes (d) PandaSet + σ

Fig. 7. Localization of anomalies: Example scenes with low *Real* scores. The colors represent interpolated p_C scores which are discrete values located at the query points (gray circles). In (a), the purple area marks a road section with extreme elevation changes. In the lower half of (b), the metric highlights seemingly floating branches of a huge tree, that enter the LiDAR field-of-view from above. (c) shows an unusual scene in a dead end road with steep hills surrounding the car. (d) illustrates a PandaSet sample where the region indicated by dashed lines has been manually distorted with additive Gaussian noise. (Color figure online)

and nuScenes. To a lesser extent, the method is also capable of identifying similar unusual constellations in completely unseen data (PandaSet). Weird sensor effects, such as the region altered with additive Gaussian noise in the PandaSet scan, are also detected by the metric (purple areas).

4.8 Limitations

LiDAR data is sparse and our approach is dependent on measurements. Wherever there is no measurement data, the metric cannot give information about the data quality. This limits the ability for reliable prediction of very sparse data, for example at high distances (>100 m) or when information is lost in advance to metric score computation. On the other hand, this enables the processing of point clouds that are not a full 360° scan of the world.

5 Conclusion

This paper presented a novel metric to quantify the degree of realism of local regions in LiDAR point clouds. In extensive experiments, we demonstrated the reliability and applicability of our metric on unseen data. Through adversarial learning, we obtain a feature encoding that is able to adequately capture data realism instead of focusing on dataset specifics. Our approach provides reliable interpolation capabilities between various levels of realism without requiring annotations. The predictions of our method correlate well with visual judgment, unlike reconstruction errors serving as a proxy for realism. In addition, we demonstrated that the local realism score can be used to detect anomalies.

Acknowledgment. The research leading to these results is funded by the German Federal Ministry for Economic Affairs and Energy within the project "KI Delta Learning" (Förderkennzeichen 19A19013A).

References

1. Achlioptas, P., Diamanti, O., Mitliagkas, I., Guibas, L.: Learning representations and generative models for 3D point clouds. In: Proceedings of the International Conference on Learning Representations (ICLR) Workshops (2018)
2. Arora, S., Risteski, A., Zhang, Y.: Do GANs learn the distribution? Some theory and empirics. In: Proceedings of the International Conference on Learning Representations (ICLR) (2018)
3. Beutel, A., Chen, J., Zhao, Z., Chi, E.H.: Data decisions and theoretical implications when adversarially learning fair representations. In: Workshop on Fairness, Accountability, and Transparency in Machine Learning (2017)
4. Borji, A.: Pros and cons of GAN evaluation measures. In: Computer Vision and Image Understanding (CVIU), pp. 41–65 (2019)
5. Caccia, L., van Hoof, H., Courville, A., Pineau, J.: Deep generative modeling of LiDAR data. In: Proceedings of IEEE International Conference on Intelligent Robots and Systems (IROS), pp. 5034–5040 (2019)
6. Caesar, H., et al.: nuScenes: a multimodal dataset for autonomous driving. In: Proceedings of IEEE Conference on Computer Vision and Pattern Recognition (CVPR), pp. 11618–11628 (2020)
7. Charles, R.Q., Su, H., Kaichun, M., Guibas, L.J.: PointNet: deep learning on point sets for 3D classification and segmentation. In: Proceedings of IEEE Conference on Computer Vision and Pattern Recognition (CVPR), pp. 77–85 (2017)
8. Che, T., Li, Y., Jacob, A.P., Bengio, Y., Li, W.: Mode regularized generative adversarial networks. In: Proceedings of the International Conference on Learning Representations (ICLR) (2017)
9. Chen, X., Duan, Y., Houthooft, R., Schulman, J., Sutskever, I., Abbeel, P.: InfoGAN: interpretable representation learning by information maximizing generative adversarial nets. In: Advances in Neural Information Processing Systems (NIPS) (2016)
10. Chicco, D.: Siamese neural networks: an overview. In: Cartwright, H. (ed.) Artificial Neural Networks. Methods in Molecular Biology, vol. 2190, pp. 73–94. Springer, New York (2021). https://doi.org/10.1007/978-1-0716-0826-5_3
11. Deng, J., Dong, W., Socher, R., Li, L.J., Li, K., Fei-Fei, L.: ImageNet: a large-scale hierarchical image database. In: Proceedings of IEEE Conference on Computer Vision and Pattern Recognition (CVPR), pp. 248–255 (2009)
12. Dong, X., Shen, J.: Triplet loss in siamese network for object tracking. In: Ferrari, V., Hebert, M., Sminchisescu, C., Weiss, Y. (eds.) ECCV 2018. LNCS, vol. 11217, pp. 472–488. Springer, Cham (2018). https://doi.org/10.1007/978-3-030-01261-8_28
13. Dosovitskiy, A., Ros, G., Codevilla, F., Lopez, A., Koltun, V.: CARLA: an open urban driving simulator. In: Proceedings of the 1st Annual Conference on Robot Learning, pp. 1–16 (2017)
14. Geiger, A., Lenz, P., Stiller, C., Urtasun, R.: Vision meets robotics: the KITTI dataset. Int. J. Robot. Res. (IJRR) **32**(11), 1231–1237 (2013)
15. Goodfellow, I., et al..: Generative adversarial nets. In: Advances in Neural Information Processing Systems (NIPS) (2014)
16. Gretton, A., Borgwardt, K.M., Rasch, M.J., Schölkopf, B., Smola, A.: A kernel two-sample test. J. Mach. Learn. Res. (JMLR) **13**, 723–773 (2012)
17. Gurumurthy, S., Sarvadevabhatla, R.K., Babu, R.V.: DeLiGAN: generative adversarial networks for diverse and limited data. In: Proceedings of IEEE Conference on Computer Vision and Pattern Recognition (CVPR), pp. 4941–4949 (2017)

18. Heusel, M., Ramsauer, H., Unterthiner, T., Nessler, B., Hochreiter, S.: GANs trained by a two time-scale update rule converge to a local nash equilibrium. In: Advances in Neural Information Processing Systems (NIPS), pp. 6629–6640 (2017)
19. Hoffer, E., Ailon, N.: Deep metric learning using triplet network. In: Feragen, A., Pelillo, M., Loog, M. (eds.) SIMBAD 2015. LNCS, vol. 9370, pp. 84–92. Springer, Cham (2015). https://doi.org/10.1007/978-3-319-24261-3_7
20. Huang, X., Li, Y., Poursaeed, O., Hopcroft, J., Belongie, S.: Stacked generative adversarial networks. In: Proceedings of IEEE Conference on Computer Vision and Pattern Recognition (CVPR), pp. 1866–1875 (2017)
21. Im, D.J., Kim, C.D., Jiang, H., Memisevic, R.: Generating images with recurrent adversarial networks. arXiv.org (2016)
22. Ledig, C., et al.: Photo-realistic single image super-resolution using a generative adversarial network. In: Proceedings of IEEE Conference on Computer Vision and Pattern Recognition (CVPR), pp. 105–114 (2017)
23. Lehmann, E.L., Romano, J.P.: Testing Statistical Hypotheses. Springer, Heidelberg (2006)
24. Lin, Z., Khetan, A., Fanti, G., Oh, S.: PacGAN: the power of two samples in generative adversarial networks. In: Advances in Neural Information Processing Systems (NIPS), pp. 324–335 (2018)
25. Lucic, M., Kurach, K., Michalski, M., Gelly, S., Bousquet, O.: Are GANs created equal? A large-scale study. In: Advances in Neural Information Processing Systems (NIPS), pp. 698–707 (2018)
26. Mathieu, M.F., Zhao, J.J., Zhao, J., Ramesh, A., Sprechmann, P., LeCun, Y.: Disentangling factors of variation in deep representation using adversarial training. In: Advances in Neural Information Processing Systems (NIPS), pp. 5047–5055 (2016)
27. Olsson, C., Bhupatiraju, S., Brown, T., Odena, A., Goodfellow, I.: Skill rating for generative models. arXiv.org (2018)
28. Qi, C.R., Yi, L., Su, H., Guibas, L.J.: PointNet++: deep hierarchical feature learning on point sets in a metric space. In: Advances in Neural Information Processing Systems (NIPS) (2017)
29. Raff, E., Sylvester, J.: Gradient reversal against discrimination: a fair neural network learning approach. In: Proceedings of IEEE International Conference on Data Science and Advanced Analytics (DSAA), pp. 189–198 (2018)
30. Richardson, E., Weiss, Y.: On GANs and GMMs. In: Advances in Neural Information Processing Systems (NIPS), pp. 5852–5863 (2018)
31. Salimans, T., et al.: Improved techniques for training GANs. In: Advances in Neural Information Processing Systems (NIPS), pp. 2234–2242 (2016)
32. Sallab, A.E., Sobh, I., Zahran, M., Essam, N.: LiDAR sensor modeling and data augmentation with GANs for autonomous driving. In: Proceedings of the International Conference on Machine learning (ICML) Workshops (2019)
33. Santurkar, S., Schmidt, L., Madry, A.: A classification-based study of covariate shift in GAN distributions. In: Proceedings of the International Conference on Machine learning (ICML), pp. 4480–4489 (2018)
34. Scale AI: PandaSet (2020). https://pandaset.org
35. Shu, D., Park, S.W., Kwon, J.: 3D point cloud generative adversarial network based on tree structured graph convolutions. In: Proceedings of the IEEE International Conference on Computer Vision (ICCV), pp. 3858–3867 (2019)
36. Sixt, L., Wild, B., Landgraf, T.: RenderGAN: generating realistic labeled data. Front. Robot. AI **5**, 6 (2018)

37. Srivastava, A., Valkov, L., Russell, C., Gutmann, M.U., Sutton, C.: VEEGAN: reducing mode collapse in GANs using implicit variational learning. In: Advances in Neural Information Processing Systems (NIPS) (2017)
38. Tolstikhin, I.O., Gelly, S., Bousquet, O., Simon-Gabriel, C.J., Schölkopf, B.: AdaGAN: boosting generative models. In: Advances in Neural Information Processing Systems (NIPS), pp. 5424–5433 (2017)
39. Triess, L.T., Peter, D., Rist, C.B., Enzweiler, M., Zöllner, J.M.: CNN-based synthesis of realistic high-resolution LiDAR data. In: Proceedings of IEEE Intelligent Vehicles Symposium (IV), pp. 1512–1519 (2019)
40. Xu, Q., et al.: An empirical study on evaluation metrics of generative adversarial networks. arXiv.org (2018)
41. Yang, J., Kannan, A., Batra, D., Parikh, D.: LR-GAN: layered recursive generative adversarial networks for image generation. In: Proceedings of the International Conf. on Learning Representations (ICLR) (2017)
42. Zhang, H., et al.: StackGAN: text to photo-realistic image synthesis with stacked generative adversarial networks. In: Proceedings of the IEEE International Conference on Computer Vision (ICCV), pp. 5908–5916 (2017)
43. Zhang, Z., Song, Y., Qi, H.: Decoupled learning for conditional adversarial networks. In: Proceedings of the IEEE Winter Conference on Applications of Computer Vision (WACV), pp. 700–708 (2018)
44. Zhou, Z., et al.: Activation maximization generative adversarial nets. In: Proceedings of the International Conference on Learning Representations (ICLR) (2018)

Full-Glow: Fully Conditional Glow for More Realistic Image Generation

Moein Sorkhei[1(✉)], Gustav Eje Henter[1], and Hedvig Kjellström[1,2]

[1] KTH Royal Institute of Technology, Stockholm, Sweden
{sorkhei,ghe,hedvig}@kth.se
[2] Silo AI, Stockholm, Sweden

Abstract. Autonomous agents, such as driverless cars, require large amounts of labeled visual data for their training. A viable approach for acquiring such data is training a generative model with collected real data, and then augmenting the collected real dataset with synthetic images from the model, generated with control of the scene layout and ground truth labeling. In this paper we propose *Full-Glow*, a fully conditional Glow-based architecture for generating plausible and realistic images of novel street scenes given a semantic segmentation map indicating the scene layout. Benchmark comparisons show our model to outperform recent works in terms of the semantic segmentation performance of a pretrained PSPNet. This indicates that images from our model are, to a higher degree than from other models, similar to real images of the same kinds of scenes and objects, making them suitable as training data for a visual semantic segmentation or object recognition system.

Keywords: Conditional image generation · Generative models · Normalizing flows

1 Introduction

Autonomous mobile agents, such as driverless cars, will be a cornerstone of the smart society of the future. Currently available datasets of labeled street scene images, such as Cityscapes [6], are an important step in this direction, and could, e.g., be used for training models for semantic image segmentation. However, collecting such data poses challenges including privacy intrusions, the need for accurate crowd-sourced labels, and the requirement to cover a huge state-space of different situations and environments. Another approach – especially useful to gather data representing dangerous situations such as collisions with pedestrians – is to generate training images with known ground-truth labeling using game engines or other virtual worlds, but this approach requires object and state-space variability to be manually engineered into the system.

Supplementary Information The online version contains supplementary material available at https://doi.org/10.1007/978-3-030-92659-5_45.

C. Bauckhage et al. (Eds.): DAGM GCPR 2021, LNCS 13024, pp. 697–711, 2021.
https://doi.org/10.1007/978-3-030-92659-5_45

A viable alternative to both these approaches is to augment existing datasets with synthetically-generated novel datapoints, produced by generative image models trained on the existing data. This builds on recent applications of generative models for a variety of tasks such as image style transfer [15] and modality transfer in medical imaging [36].

Among currently-available deep generative approaches, GANs [10] are probably the most widely used in image generation, owing to their achievements in synthesizing realistic high-resolution output with novel and rich detail [3,16]. Auto-regressive architectures [26,27] are usually computationally demanding (not parallelizable) and not feasible for generating higher-resolution images. Image samples generated by early variants of VAEs [20,33] tended to suffer from blurriness [44], although the realism of VAE output has improved in recent years [32,40].

This article considers normalizing flows [7,8], a different model class of growing interest. With recent improvements such as *Glow* [19], flows can generate images with a quality that approaches that produced by GANs. Flows have also achieved competitive results in other tasks such as audio and video generation [17,21,30]. Flow-based models exhibit several benefits compared to GANs: 1) stable, monotonic training, 2) learning an explicit representation useful for down-stream tasks such as style transfer, 3) efficient synthesis, and 4) exact likelihood evaluation that could be used for density estimation.

In this paper, we propose a new, fully conditional Glow-based architecture called *Full-Glow* for generating plausible street scene images conditioned on the structure of the image content (i.e., the segmentation mask). We show that, by using this model, we are able to synthesize moderately high-resolution images that are consistent with the given structure but differ substantially from the existing ground-truth images. A quantitative comparison against previously proposed Glow-based models [23,36] and the popular GAN-based conditional image-generation model pix2pix [15], finds that our improved conditioning allows us to synthesize images that achieve better semantic classification scores under a pre-trained semantic classifier. We also provide visual comparisons of samples generated by different models.

The remainder of this article is laid out as follows: Sect. 2 presents prior work in street-scene generation and image-to-image translation, while Sect. 3 provides technical background on normalizing flows. Our proposed fully-conditional architecture is then introduced in Sect. 4 and validated experimentally in Sect. 5.

2 Related Work

Synthetic Data Generation. Street-scene image datasets such as Cityscapes [6], CamVid [4], and the KITTI dataset [9] are useful for training vision systems for street-scene understanding. However, collecting and labeling such data is costly, resource demanding, and associated with privacy issues. An effective alternative that allows for ground-truth labels and scene layout control is synthetic data generation using game engines [34,35,39]. Despite these advantages, images generated by game engines tend to differ significantly from real-world images and may not always act as a replacement for real data. Moreover, game engines generally only synthesize objects from pre-generated assets or recipes, meaning that variation has to be hand-engineered in. It is therefore difficult and costly to obtain diverse data in this manner. Data generated from approaches such as ours address these shortcomings while maintaining the benefits of ground-truth labeling and scene layout control.

Image-to-Image Translation. In order to generate images for data-augmentation of supervised learning tasks, it is necessary to condition the image generation on an input, such that the ground-truth labeling of the generated image is known. For street-scene understanding, this conditioning takes the form of per-pixel class labels (a *segmentation mask*), meaning that the augmentation task can be formulated as an image-to-image translation problem. GANs [10] have been employed for both paired and unpaired image-to-image translation problems [15,45]. While GANs can generate convincing-looking images, they are known to suffer from mode collapse and low output diversity [11]. Consequentially, their value in augmenting dataset diversity may be limited.

Likelihood-based models, on the other hand, explicitly aim to learn the probability distribution of the data. These models generally prefer sample diversity, sometimes at the expense of sample quality [8], which has been linked to the mass-covering property of the likelihood objective [25,37]. Like for GANs [3], perceived image quality can often be improved by reducing the entropy of the distribution at synthesis time, relative to the distribution learned during training, cf. [19,40]. Flow-based models are a particular class of likelihood-based model that have gained recent attention after an architecture called Glow [19] demonstrated impressive performance in unconditional image generation. Previous works have applied flow-based models for image colorization [1,2], image segmentation [23], modality transfer in medical imaging [36], and generating point clouds and image-to-image translation [31].

So far, Glow-based models proposed for image-to-image translation [23,31,36] have only considered low-resolution tasks. Although the results are promising, they do not assess the full capacity of Glow for generating realistic image detail, for example in street scenes. High-resolution street-scene synthesis has been performed by the GAN-based model pix2pixHD [41] on a GPU with very high memory capacity (24 GB). In the present work, we synthesize moderately high resolution street scene images using a GPU with lower memory capacity (11–12 GB). We extend previous works on Glow-based models by introducing a fully conditional architecture, and also by modeling high-resolution street-scene images, which is a more challenging task than the low-resolution output considered in prior work.

3 Flow-Based Generative Models

Normalizing flows [28] are a class of probabilistic generative models, able to represent complex probability densities in a manner that allows both easy sampling and efficient training based on explicit likelihood maximization. The key idea is to use a sequence of *invertible* and *differentiable* functions/transformations which (nonlinearly) transform a random variable $\mathbf{z}$ with a simple density function to another random variable $\mathbf{x}$ with a more complex density function (and vice versa, thanks to invertibility):

$$\mathbf{x} \overset{\mathbf{f}_1}{\longleftrightarrow} \mathbf{h}_1 \overset{\mathbf{f}_2}{\longleftrightarrow} \mathbf{h}_2 \overset{\mathbf{f}_3}{\longleftrightarrow} \dots \overset{\mathbf{f}_K}{\longleftrightarrow} \mathbf{z}. \tag{1}$$

Each component transformation $\mathbf{f}_i$ is called a *flow step*. The distribution of $\mathbf{z}$ (termed the *latent*, *source*, or *base distribution*) is assumed to have a simple parametric form, such as an isotropic unit Gaussian. Similar to in GANs, the generative process can be formulated as:

$$\mathbf{z} \sim p_{\mathbf{z}}(\mathbf{z}), \tag{2}$$
$$\mathbf{x} = \mathbf{g}_{\theta}(\mathbf{z}) = \mathbf{f}_{\theta}^{-1}(\mathbf{z}) \tag{3}$$

where $\mathbf{z}$ is sampled from the base distribution and $\mathbf{g}_{\theta}$ represents the cumulative effect of the parametric invertible transformations in Eq. (1). The log-density function of $\mathbf{x}$ under this transformation can be written as:

$$\log p_{\mathbf{x}}(\mathbf{x}) = \log p_{\mathbf{z}}(\mathbf{z}) + \sum_{i=1}^{K} \log \left| \det \frac{d\mathbf{h}_i}{d\mathbf{h}_{i-1}} \right| \tag{4}$$

using the *change-of-variables theorem*, where $\mathbf{h}_0 \triangleq \mathbf{x}$ and $\mathbf{h}_K \triangleq \mathbf{z}$. Equation (4) can be used to compute the exact dataset log-likelihood (not possible in GANs) and is the sole objective function for training flow-based models.

The central design challenge of normalizing flows is to create expressive invertible transformations (typically parameterized by deep neural networks) where the so-called *Jacobian log-determinant* in Eq. (4) remains computationally feasible to evaluate. Often, this is achieved by designing transformations whose Jacobian matrix is triangular, making the determinant trivial to compute. An important example is NICE [7]. NICE introduced the *coupling layer*, which is a particular kind of flow nonlinearity that uses a neural network to invertibly transform half of the elements in $\mathbf{h}_k$ with respect to the other half. RealNVP [8] improved on this architecture using more general invertible transformations in the coupling layer and by imposing a hierarchical structure where the flow is partitioned into *blocks* that operate at different resolutions. This hierarchy allows using smaller $\mathbf{z}$-vectors at the initial, smaller resolutions, speeding up computation, and has lately been used by other prominent image-generation systems [32,40]. Glow [19] added *actnorm* as a replacement for batchnorm [14] and introduced invertible 1×1 convolutions to more efficiently mix variables in between the couplings.

A number of Glow-based architectures have been proposed for conditional image generation. In these models, the goal is to learn a distribution over the target image $\mathbf{x}_b$ conditioned on the source image $\mathbf{x}_a$. C-Glow [23] is based on the standard Glow architecture from [19], but makes all sub-steps inside the Glow conditional on the raw conditioning image $\mathbf{x}_a$. The Dual-Glow [36] architecture instead builds a generative model of both source and target image together. It consists of two Glows where the base variables $\mathbf{z}_a$ of the source-image Glow determine the Gaussian distribution of the corresponding base variables $\mathbf{z}_b$ of the target-image Glow through a neural network. Because of the hierarchical structure of Glow, several different conditioning networks are used, one for each block of flow steps. C-Flow [31] described a similar structure of side-by-side Glows, but kept the Gaussian base distributions in the two flows independent. Instead, they used the latent variables $\mathbf{h}_{a,i}$ at every flow step i of the target-domain Glow to condition the transformation in the coupling layer at the corresponding level in the source-domain Glow. Compared to the raw image-data conditioning in C-Glow, Dual-Glow and C-Flow simplify the conditional mapping task at the different levels since the source and target information sit at comparable levels of abstraction.

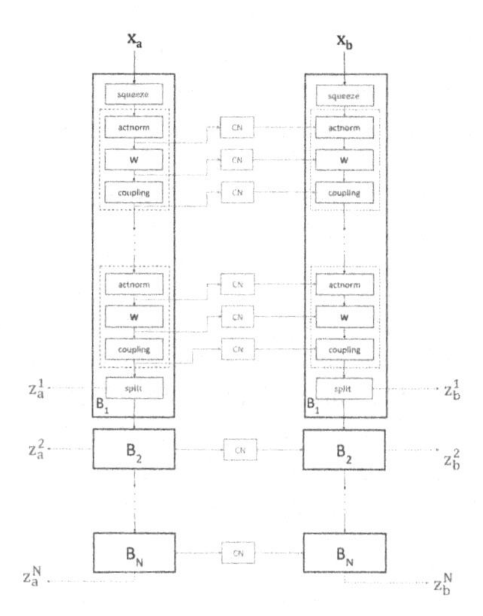

Fig. 1. The proposed architecture, where all substeps have been made conditional by inserting conditioning networks. $\mathbf{x}_a$ and $\mathbf{x}_b$ are paired images in the source and target domains, respectively.

4 Fully Conditional Glow for Scene Generation

This section introduces our new, fully conditional Glow architecture for image-to-image translation, which combines key innovations from all three previous architectures, C-Glow, Dual-Glow, and C-Flow: Like Dual-Glow and C-Flow (but unlike C-Glow) we use two parallel stacks of Glow, so that we can leverage conditioning information at the relevant level of hierarchy and not be restricted to always use the raw source image as input. In contrast to Dual-Glow and C-Flow (but reminiscent of C-Glow), we introduce conditioning networks that allow *all* operations in the target-domain Glow conditional on the source-domain information. The resulting architecture is illustrated in Fig. 1. Because of its fully conditional nature, we dub this architecture *Full-Glow*.

In our proposed architecture, not only is the coupling layer conditioned on the output of the corresponding operation in the source Glow, but the actnorm and the 1×1 convolutions in the target Glow are also connected to the source Glow. In particular, the parameters of these two operations in each target-side step are generated by conditioning networks CN built from convolutional layers followed by fully connected layers. These networks also enable us to exploit other side information for conditioning, for instance by concatenating the side information with the other input features of each conditioning network. We experimentally show that making the model fully conditional indeed allows for learning a better conditional distribution (measured with lower conditional bits-per-dimension) and more semantically meaningful images (measured using a pre-trained semantic classifier).

We will now describe the architecture of the fully-conditional target-domain Glow in more detail. We describe the computational statistical inference ($\mathbf{f}_\theta$), but every transformation is invertible for synthesis ($\mathbf{g}_\theta$) given the conditioning image $\mathbf{x}_a$.

Conditional Actnorm. The shift $\mathbf{t}$ and scale $\mathbf{s}$ parameters of the conditional actnorm are computed as follows:

$$\mathbf{s}, \mathbf{t} = \mathrm{CN}\left(\mathbf{x}_{\text{act}}^{\text{source}}\right) \tag{5}$$

where $\mathbf{x}_{\text{act}}^{\text{source}}$ is the output of the corresponding actnorm in the source Glow. For initializing the actnorm conditioning network (CN), we set all parameters of the network except those of the output layer to small, random values. Similar to [19,42], the weights of the output layer are initialized to 0 and the biases are initialized such that the output activations of the target side after applying actnorm have mean 0 and std of 1 per channel for the first batch of data, similar to the scheme used for initialising actnorm in regular Glow.

Conditional 1×1 *Convolution.* Like Glow, we represent the convolution kernel W using an LU decomposition for easy log-determinant computation, but we have conditioning networks generate the $\mathbf{L}$, $\mathbf{U}$ matrices and the $\mathbf{s}$ vector:

$$\mathbf{L}, \mathbf{U}, \mathbf{s} = \mathrm{CN}\left(\mathbf{x}_{\mathbf{W}}^{\text{source}}\right) \tag{6}$$

where $\mathbf{x}_{\mathbf{W}}^{\text{source}}$ is the output of the corresponding 1×1 convolution in the source Glow, $\mathbf{L}$ is a lower triangular matrix with ones on the diagonal, $\mathbf{U}$ is an upper triangular matrix with zeros on the diagonal, and $\mathbf{s}$ is a vector. Initialization again follows [19]: we first sample a random rotation matrix $\mathbf{W}_0$ per layer, which we factorise using the LU decomposition as $\mathbf{W}_0 = \mathbf{P}\mathbf{L}_0\left(\mathbf{U}_0 + \mathrm{diag}(\mathbf{s}_0)\right)$. The conditioning network is then set up similar as for the actnorm, with weights and biases set so that its outputs on the first batch are constant and equivalent to the sampled rotation matrix $\mathbf{W}_0$. The permutation matrix $\mathbf{P}$ remains fixed throughout the optimization.

Conditional Coupling Layer. The conditional coupling layer resembles that of C-Flow (except that we use the whole source coupling output rather than half of it), where the network in the coupling layer takes input from both source and target sides:

$$\mathbf{x}_1^{\text{target}}, \mathbf{x}_2^{\text{target}} = \texttt{split}(\mathbf{x}^{\text{target}}) \tag{7}$$

$$\mathbf{o}_1, \mathbf{o}_2 = \mathrm{CN}\left(\mathbf{x}_2^{\text{target}}, \mathbf{x}^{\text{source}}\right) \tag{8}$$

$$\mathbf{s} = \mathrm{Sigmoid}(\mathbf{o}_1 + \mathbf{2}) \tag{9}$$

$$\mathbf{t} = \mathbf{o}_2 \tag{10}$$

where the `split` operation splits the input tensor along the channel dimension, $\mathbf{x}^{\text{source}}$ is the output of the corresponding coupling layer in the source Glow, and $\mathbf{x}^{\text{target}}$ is the output of the preceding 1×1 convolution in the target Glow. $\mathbf{s}$ and $\mathbf{t}$ are the affine coupling parameters. The conditioning network inputs are concatenated channel-wise.

The objective function for the model has the same form as that of Dual-Glow:

$$\frac{1}{N}\left[-\sum_{n=1}^{N} \lambda \log p_\theta\left(\mathbf{x}_a^{(n)}\right) - \sum_{n=1}^{N} \log p_\phi\left(\mathbf{x}_b^{(n)} \mid \mathbf{x}_a^{(n)}\right)\right] \tag{11}$$

where $\boldsymbol{\theta}$ are the parameters of the source Glow and $\boldsymbol{\phi}$ are the parameters of the target Glow. We note that there is one model (and term) for unconditional image generation in the source domain, coupled with a second model (and term) for conditional image generation in the target domain. With the tuning parameter λ set to unity, Eq. 11 is the joint likelihood of the source-target image pair $(\mathbf{x}_a, \mathbf{x}_b)$, and puts equal emphasis on learning to generate (and to normalize/analyze) both images. In the limit $\lambda \rightarrow \infty$, we will learn an unconditional model of source images only. Using a λ below 1, however, helps the optimization process instead put more importance on the conditional distribution, which is our main priority in image-to-image translation. This "exchange rate" between bits of information in different domains is reminiscent of the tuning parameter in the information bottleneck principle [38].

5 Experiments

This section reports our findings from applying the proposed model to the Cityscapes dataset from [6]. Each data instance is a photo of a street scene that has has been segmented into objects of 30 different classes, such as road, sky, buildings, cars, and pedestrians. 5000 of these images come with fine per-pixel class annotations of the image, a so called *segmentation mask*. We used the data splits provided by the dataset (2975 training and 500 validation images), and trained a number of different models to generate street-scene images conditioned on their segmentation masks.

A common way to evaluate the quality of images generated based on the Cityscapes dataset is to apply well-known pre-trained classifiers such as FCN [22] and (here) PSPNet [43] to synthesized images (as done by [15,41]). The idea is that if a synthesized image is of high quality, a classifier trained on real data should be able to successfully classify different objects in the synthetic image, and thus produce an estimated segmentation mask that closely agrees with the ground-truth segmentation mask. For likelihood-based models we also consider the conditional bits per dimension (BPD), $-\log_2 p(\mathbf{x}_b|\mathbf{x}_a)$, as a measure of how well the conditional distribution learned by the model matches the real conditional distribution, when tested on held-out examples.

Implementation Details. Our main experiments were performed on images from the Cityscapes data down-sampled to 256×256 pixels (higher than C-Flow [31] that uses 64×64 resolution). The Full-Glow model was implemented in PyTorch [29] and trained using the Adam optimizer [18] with a learning rate of 10^{-4} and a batch size of 1. The conditioning networks (CN) for the actnorm and 1×1 convolution in our model consisted of three convolutional layers followed by four fully connected layers. The CN for the coupling layer had two convolutional layers. Network weights were initialized as described in Sect. 4. We used $\lambda = 10^{-4}$ in the objective function Eq. 11. Training was consistently stable and monotonic; see the loss curve in the supplement. Our implementation could be found at: https://github.com/MoeinSorkhei/glow2.

5.1 Quantitative Comparison with Other Models

We compare the performance of our model against C-Glow [23] and Dual-Glow [36] (two previously proposed Glow-based models) and pix2pix [15] a widely used GAN-based model for image-to-image translation.

Since C-Glow was proposed to deal with low-resolution images, the authors exploited deep conditioning network in their model. We could not use equally deep conditioning networks in this task because the images we would like to generate are of higher resolution (256×256). To enable valid comparisons, we trained two versions of the their model. In the first version, we allowed the conditioning networks to be deeper while the Glow itself is shallower (3 Blocks each with 8 Flows). In the second version, the Glow model is deeper (4 Blocks and each with 16 Flows) but the conditioning networks are shallower. More details about the models and their hyper-parameters can be found in the supplementary material. Note that the Glow models in C-Glow version 2, Dual-Glow, and our model are all equally deep (4 Blocks and each having 16 Flows). All models, including Full-Glow, were trained for $\sim$45 epochs using the same training procedure described earlier.[1]

We sampled from each trained model 3 times on the validation set, evaluated the synthesized images using PSPNet [43], and calculated the mean and standard deviation of the performance (denoted by $\pm$). The metrics used for evaluation are mean pixel accuracy, mean class accuracy, and mean intersection over union (IoU), as formulated in [22]. The mean pixel accuracy essentially computes mean accuracy over all the pixels of an image (which could easily be dominated by the sky, trees, and large objects that are mostly classified correctly.) Mean class accuracy, however, calculates the accuracy over the pixels of each class, and then takes average over different classes (where all classes are treated equally). Finally, mean class IoU calculates for each class the intersection over union for the objects of that class segmented in the synthesized image compared against the objects in the ground-truth segmentation. Optimally, this number should be 1, signifying complete overlap between segmented and ground-truth objects.

Quantitative results of applying each model to the Cityscapes dataset in the label $\rightarrow$ photo direction could be seen in Table 1. The results show that street scene images generated by Full-Glow are of higher quality from the viewpoint of semantic segmentation. The noticeable difference in classification performance confirms that the objects in the images generated by our model are more easily *distinguishable* by the off-the-shelf semantic classifier. We attribute this to the fact that making the model fully conditional

Table 1. Comparison of different models on the Cityscapes dataset for label $\rightarrow$ photo image synthesis.

Model	Cond. BPD	Mean pixel acc.	Mean class acc.	Mean class IoU
C-Glow v.1 [23]	2.568	35.02 ± 0.56	12.15 ± 0.05	7.33 ± 0.09
C-Glow v.2 [23]	2.363	52.33 ± 0.46	17.37 ± 0.21	12.31 ± 0.24
Dual-Glow [36]	2.585	71.44 ± 0.03	23.91 ± 0.19	18.96 ± 0.17
pix2pix [15]	—	60.56 ± 0.11	22.64 ± 0.21	16.42 ± 0.06
Our model	**2.345**	$\mathbf{73.50 \pm 0.13}$	$\mathbf{29.13 \pm 0.39}$	$\mathbf{23.86 \pm 0.30}$
Ground-truth	—	*95.97*	*84.31*	*77.30*

[1] We used these repositories 1, 2, 3 to obtain the official implementations of C-Glow, Dual-Glow, and pix2pix. We did not find any official implementation for C-Flow.

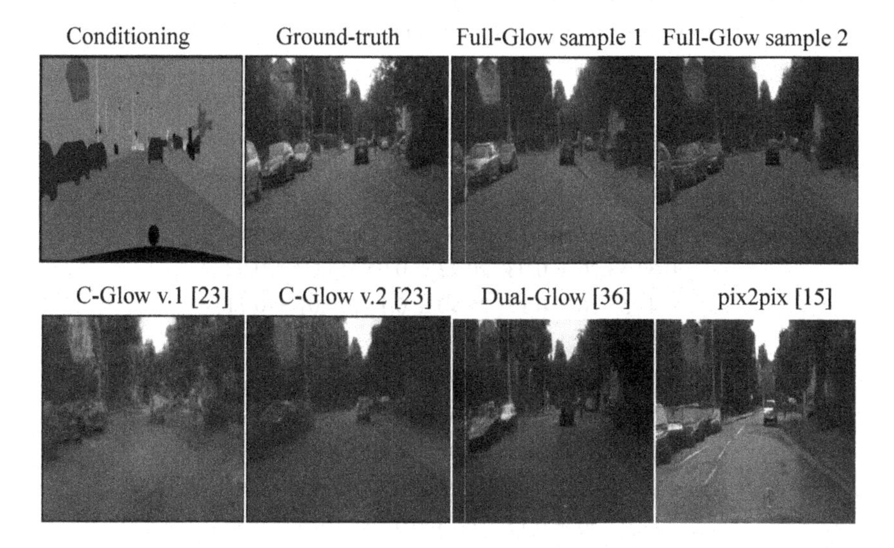

Fig. 2. Visual samples from different models. Samples from likelihood-based models are taken with temperature 0.7. Please zoom in to see more details.

enables the target Glow to exploit the information available in the source image and to synthesize an image that follows the structure most.

5.2 Visual Comparison with Other Models

It is interesting to see how samples generated by different models are different visually. Figure 2 illustrates samples from different models given the same condition. An immediate observation is that C-Glow v.1 [23] (which has deeper conditioning networks but shallower Glow) is essentially unable to generate any meaningful image. Dual-Glow [36], however, is able to generate plausible images. Samples generated by pix2pix [15] exhibit vibrant colors (especially for the buildings) but the important objects (such as cars) that constitute the general structure of the image are sometimes distorted. We believe this is the reason behind scoring low with the semantic classifier. Respecting the structure seems to be more important than having vibrant colors in order to get higher classification accuracy. Different samples taken from our model show the benefit of flow-based models in synthesizing different images every time we sample. Most of the difference can be seen in the colors of the objects such as cars.

Generally, the samples generated by likelihood-based models appear dimmed to some extent. This is in contrast with GAN-based samples, which often have realistic colors. This is probably related to the fundamental difference in the optimization process of the two categories. GAN-based models tend to collapse to regions of data where only plausible samples could have come from, and they might not have support over other data regions [11] – also seen as lack of diversity in their samples. In contrast, likelihood-based models try to learn a distribution that has support over wider data regions while maximizing the probability of the available datapoints. The latter app-

Table 2. Effect of temperature T evaluated using a pre-trained PSPNet [43]. Each column lists the mean over repeated image samples.

T	Pixel acc.	Class acc.	Class IoU
0.0	27.83 ± 0	12.48 ± 0	7.69 ± 0
0.1	29.47 ± 0.04	13.15 ± 0.05	8.33 ± 0.01
0.2	34.97 ± 0.13	15.83 ± 0.13	10.65 ± 0.01
0.3	43.36 ± 0.08	20.22 ± 0.05	14.69 ± 0.02
0.4	55.26 ± 0.12	25.90 ± 0.15	20.01 ± 0.11
0.5	69.83 ± 0.01	31.66 ± 0.16	25.82 ± 0.12
0.6	79.38 ± 0.05	34.90 ± 0.28	29.37 ± 0.23
0.7	82.14 ± 0.09	35.51 ± 0.23	29.98 ± 0.21
0.8	$\mathbf{83.52 \pm 0.04}$	$\mathbf{35.94 \pm 0.19}$	$\mathbf{30.67 \pm 0.17}$
0.9	82.27 ± 0.05	35.24 ± 0.37	29.85 ± 0.36
1.0	73.50 ± 0.13	29.13 ± 0.39	23.86 ± 0.30

roach seems to result in generating samples that are diverse but have somewhat muted colors (especially with lower temperatures).

5.3 Effect of Temperature

As noted above, likelihood-based models such as Glow [19] generally tend to *overestimate* the variability of the data distribution [25,37], hence occasionally generating implausible output samples. A common way to circumvent this issue is to reduce the diversity of the output at generation time. For flows, this can be done by reducing the standard deviation of the base distribution by a factor T (known as the temperature). While $T = 1$ corresponds to sampling from the estimated maximum-likelihood distribution, reducing T generally results in the output distribution becoming concentrated on a core region of especially-probable output samples. Similar ideas are widely used not only in flow-based models (cf. [19]) but also in other generative models such as GANs, VAEs, and Transformer-based language models [3,5,13,40].

We investigated the effect of temperature by evaluating the performance of the model on samples generated at different temperatures (instead of $T = 1$ as in previous experiments). We sampled on the validation set 3 times with the trained model and evaluated using the PSPNet semantic classifier as before. The results are reported in Table 2 and suggest that the optimal temperature is around 0.8 for this task. That setting strikes a compromise where colors are vibrant while object structure is well-maintained, enabling the classifier to well understand the objects in the synthesized image. Also note the small standard deviation at lower temperatures, which agrees with our expectation that inter-sample variability would be small at low temperature. Example images generated at different temperatures are provided in the supplementary material.

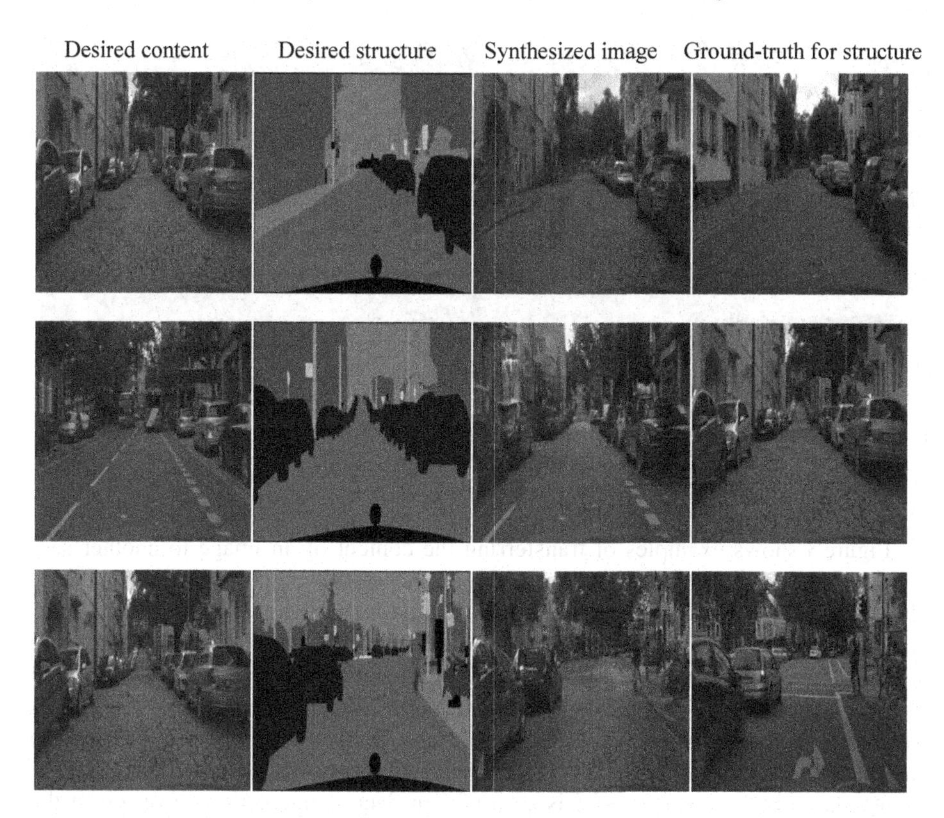

Fig. 3. Examples of applying a desired content to a desired structure. Please zoom in to see more details.

5.4 Content Transfer

Style transfer in general is an interesting application for synthesizing new images, which incorporates transferring the style of an image into a new structure (condition). In this experiment, we try to transfer the *content* of a real photo to a new structure (segmentation). Previous work [1,31] has performed similar experiments with flow-based models but either on a different dataset or in very low resolution (64×64). We demonstrate how the learned representation enables us to synthesize an image given a desired content and a desired structure in relatively high resolution while maintaining the details of the content.

Suppose $\mathbf{x}_b^1$ is the image with the desired content (with $\mathbf{x}_a^1$ being its segmentation) and $\mathbf{x}_a^2$ is the new segmentation to which we are interested to transfer the content. We can take the following steps ($\mathbf{g}_\theta(.)$ and $\mathbf{f}_\phi(.)$ are the forward functions in the source and target Glows respectively):

1. Extract representation of the desired content: $\mathbf{z}_{\mathrm{b}}^1 = \mathbf{f}_\phi\left(\mathbf{x}_{\mathrm{b}}^1 \mid \mathbf{g}_\theta\left(\mathbf{x}_{\mathrm{a}}^1\right)\right)$.
2. Apply the content to the new segmentations: $\mathbf{x}_{\mathrm{b}}^{\mathrm{new}} = \mathbf{f}_\phi^{-1}\left(\mathbf{z}_{\mathrm{b}}^1 \mid \mathbf{g}_\theta\left(\mathbf{x}_{\mathrm{a}}^2\right)\right)$.

Fig. 4. Higher resolution samples of our model taken with temperature 0.9. Left: conditioning, middle-right: samples 1, 2, 3. Please zoom in to see more details.

Fig. 5. Higher resolution samples of our model taken with temperature 0.9. Left: conditioning, middle-right: samples 1, 2, 3. Please zoom in to see more details.

Figure 3 shows examples of transferring the content of an image to another segmentation. We can see that the model is able to successfully apply the content of large objects such as buildings, trees, cars to the desired image structure. So often a given content and a given structure may not agree with each other. For instance, when there are cars in the content which are missing in the segmentation or vice versa. This kind of mismatch is quite common for content transfer on Cityscapes images as these images have a lot of objects placed in different positions. In such a case, the model tries to respect the structure as much as possible while filling it with the given content. The results show that content transfer is so useful in data augmentation since, given the desired content, the model can fill the structure with coherent information which makes the output image much realistic. This technique could practically be applied to any content and any structure (provided that they do not mismatch completely), hence enabling one to synthesize many more images.

5.5 Higher-Resolution Samples

In order to see how expressive the model is at even higher resolutions, we trained the proposed model on 512×1024 images. Example output images from the trained model are provided in Figs. 4 and 5. It is known that higher temperatures show more diversity but the structures become somewhat distorted. We chose to sample with temperature 0.9 for these higher-resolution images. The diversity between multiple samples is especially obvious looking at the cars. Additional high-resolution samples are available in the supplementary material.

6 Conclusions

In this paper, we proposed a fully conditional Glow-based architecture for more realistic conditional street-scene image generation. We quantitatively compared our method against previous work and observed that our improved conditioning allows for generating images that are more interpretable by the semantic classifier. We also used the architecture to synthesize higher-resolution images in order to better show diversity of samples and the capabilities of Glow at higher resolutions. In addition, we demonstrated how new meaningful images could be synthesized based on a desired content and a desired structure, which is a compelling option for high-quality data augmentation.

Future Work. While our results are promising, further work remains to be done in order to close the gap between synthetic images and real-world photographs, for instance by adding self-attention [12] and by leveraging approaches that combine the strong points of Full-Glow with the advantages of GANs, e.g., following [11,24]. That said, we believe the quality of the synthetic images is already at a level where it also is worth exploring their utility in training systems for autonomous cars and other mobile agents, which remains to be observed in future works.

References

1. Ardizzone, L., Kruse, J., Lüth, C., Bracher, N., Rother, C., Köthe, U.: Conditional invertible neural networks for diverse image-to-image translation. In: Akata, Z., Geiger, A., Sattler, T. (eds.) DAGM GCPR 2020. LNCS, vol. 12544, pp. 373–387. Springer, Cham (2021). https://doi.org/10.1007/978-3-030-71278-5_27
2. Ardizzone, L., Lüth, C., Kruse, J., Rother, C., Köthe, U.: Guided image generation with conditional invertible neural networks. arXiv preprint arXiv:1907.02392 (2019)
3. Brock, A., Donahue, J., Simonyan, K.: Large scale GAN training for high fidelity natural image synthesis. In: International Conference on Learning Representations (2019)
4. Brostow, G.J., Fauqueur, J., Cipolla, R.: Semantic object classes in video: a high-definition ground truth database. Pattern Recogn. Lett. **30**(2), 88–97 (2009)
5. Brown, T.B., et al.: Language models are few-shot learners. In: Neural Information Processing Systems, pp. 1877–1901 (2020)
6. Cordts, M., et al.: The Cityscapes dataset for semantic urban scene understanding. In: IEEE Conference on Computer Vision and Pattern Recognition (2016)
7. Dinh, L., Krueger, D., Bengio, Y.: NICE: non-linear independent components estimation. arXiv preprint arXiv:1410.8516 (2014)
8. Dinh, L., Sohl-Dickstein, J., Bengio, S.: Density estimation using Real NVP. In: International Conference on Learning Representations (2017)
9. Geiger, A., Lenz, P., Stiller, C., Urtasun, R.: Vision meets robotics: the KITTI dataset. Int. J. Robot. Res. **32**(11), 1231–1237 (2013)
10. Goodfellow, I., et al.: Generative adversarial nets. In: Neural Information Processing Systems (2014)
11. Grover, A., Dhar, M., Ermon, S.: Flow-GAN: combining maximum likelihood and adversarial learning in generative models. In: AAAI Conference on Artificial Intelligence (2018)
12. Ho, J., Chen, X., Srinivas, A., Duan, Y., Abbeel, P.: Flow++: improving flow-based generative models with variational dequantization and architecture design. arXiv preprint arXiv:1902.00275 (2019)
13. Holtzman, A., Buys, J., Du, L., Forbes, M., Choi, Y.: The curious case of neural text degeneration. In: International Conference on Learning Representations (2020)
14. Ioffe, S., Szegedy, C.: Batch normalization: accelerating deep network training by reducing internal covariate shift. arXiv preprint arXiv:1502.03167 (2015)
15. Isola, P., Zhu, J.Y., Zhou, T., Efros, A.A.: Image-to-image translation with conditional adversarial networks. In: IEEE Conference on Computer Vision and Pattern Recognition (2017)
16. Karras, T., Laine, S., Aila, T.: A style-based generator architecture for generative adversarial networks. In: IEEE Conference on Computer Vision and Pattern Recognition (2019)

17. Kim, S., Lee, S., Song, J., Kim, J., Yoon, S.: FloWaveNet: a generative flow for raw audio. In: International Conference on Machine Learning (2019)
18. Kingma, D.P., Ba, J.: Adam: a method for stochastic optimization. arXiv preprint arXiv:1412.6980 (2014)
19. Kingma, D.P., Dhariwal, P.: Glow: generative flow with invertible 1×1 convolutions. In: Neural Information Processing Systems (2018)
20. Kingma, D.P., Welling, M.: Auto-encoding variational Bayes. In: International Conference on Learning Representations (2014)
21. Kumar, M., et al.: VideoFlow: a flow-based generative model for video. arXiv preprint arXiv:1903.01434 (2019)
22. Long, J., Shelhamer, E., Darrell, T.: Fully convolutional networks for semantic segmentation. In: IEEE Conference on Computer Vision and Pattern Recognition (2015)
23. Lu, Y., Huang, B.: Structured output learning with conditional generative flows. In: AAAI Conference on Artificial Intelligence (2020)
24. Lucas, T., Shmelkov, K., Alahari, K., Schmid, C., Verbeek, J.: Adaptive density estimation for generative models. In: Neural Information Processing Systems, pp. 11993–12003 (2019)
25. Minka, T.: Divergence measures and message passing. Technical report, MSR-TR-2005-173, Microsoft Research, Cambridge, UK (2005)
26. van den Oord, A., Kalchbrenner, N., Espeholt, L., Vinyals, O., Graves, A., Kavukcuoglu, K.: Conditional image generation with pixelCNN decoders. In: Neural Information Processing Systems (2016)
27. van den Oord, A., Kalchbrenner, N., Kavukcuoglu, K.: Pixel recurrent neural networks. arXiv preprint arXiv:1601.06759 (2016)
28. Papamakarios, G., Nalisnick, E., Rezende, D.J., Mohamed, S., Lakshminarayanan, B.: Normalizing flows for probabilistic modeling and inference. arXiv preprint arXiv:1912.02762 (2019)
29. Paszke, A., et al.: Automatic differentiation in PyTorch. In: NIPS 2017 Workshop Autodiff (2017)
30. Prenger, R., Valle, R., Catanzaro, B.: WaveGlow: a flow-based generative network for speech synthesis. In: IEEE International Conference on Acoustics, Speech and Signal Processing (2019)
31. Pumarola, A., Popov, S., Moreno-Noguer, F., Ferrari, V.: C-Flow: conditional generative flow models for images and 3D point clouds. In: IEEE Conference on Computer Vision and Pattern Recognition (2020)
32. Razavi, A., van den Oord, A., Vinyals, O.: Generating diverse high-fidelity images with VQ-VAE-2. In: Neural Information Processing Systems (2019)
33. Rezende, D.J., Mohamed, S., Wierstra, D.: Stochastic backpropagation and approximate inference in deep generative models. In: International Conference on Machine Learning (2014)
34. Richter, S.R., Vineet, V., Roth, S., Koltun, V.: Playing for data: ground truth from computer games. In: Leibe, B., Matas, J., Sebe, N., Welling, M. (eds.) ECCV 2016. LNCS, vol. 9906, pp. 102–118. Springer, Cham (2016). https://doi.org/10.1007/978-3-319-46475-6_7
35. Ros, G., Sellart, L., Materzynska, J., Vazquez, D., Lopez, A.M.: The SYNTHIA dataset: a large collection of synthetic images for semantic segmentation of urban scenes. In: IEEE Conference on Computer Vision and Pattern Recognition (2016)
36. Sun, H., et al.: DUAL-GLOW: conditional flow-based generative model for modality transfer. In: IEEE International Conference on Computer Vision (2019)
37. Theis, L., van den Oord, A., Bethge, M.: A note on the evaluation of generative models. In: International Conference on Learning Representations (2016)

38. Tishby, N., Pereira, F.C., Bialek, W.: The information bottleneck method. In: Proceedings of the Allerton Conference on Communication, Control and Computing, vol. 37, pp. 368–377 (2000)
39. Tsirikoglou, A., Kronander, J., Wrenninge, M., Unger, J.: Procedural modeling and physically based rendering for synthetic data generation in automotive applications. arXiv preprint arXiv:1710.06270 (2017)
40. Vahdat, A., Kautz, J.: NVAE: a deep hierarchical variational autoencoder. arXiv preprint arXiv:2007.03898 (2020)
41. Wang, T.C., Liu, M.Y., Zhu, J.Y., Tao, A., Kautz, J., Catanzaro, B.: High-resolution image synthesis and semantic manipulation with conditional GANs. In: IEEE Conference on Computer Vision and Pattern Recognition (2018)
42. Zhang, H., Dauphin, Y.N., Ma, T.: Fixup initialization: residual learning without normalization. In: International Conference on Learning Representations (2019)
43. Zhao, H., Shi, J., Qi, X., Wang, X., Jia, J.: Pyramid scene parsing network. In: IEEE Conference on Computer Vision and Pattern Recognition (2017)
44. Zhao, S., Song, J., Ermon, S.: Towards deeper understanding of variational autoencoding models. arXiv preprint arXiv:1702.08658 (2017)
45. Zhu, J.Y., Park, T., Isola, P., Efros, A.A.: Unpaired image-to-image translation using cycle-consistent adversarial networks. In: IEEE International Conference on Computer Vision (2017)

Multidirectional Conjugate Gradients for Scalable Bundle Adjustment

Simon Weber(✉), Nikolaus Demmel, and Daniel Cremers

Technical University of Munich, Munich, Germany
{sim.weber,nikolaus.demmel,cremers}@tum.de

Abstract. We revisit the problem of large-scale bundle adjustment and propose a technique called Multidirectional Conjugate Gradients that accelerates the solution of the normal equation by up to 61%. The key idea is that we enlarge the search space of classical preconditioned conjugate gradients to include multiple search directions. As a consequence, the resulting algorithm requires fewer iterations, leading to a significant speedup of large-scale reconstruction, in particular for denser problems where traditional approaches notoriously struggle. We provide a number of experimental ablation studies revealing the robustness to variations in the hyper-parameters and the speedup as a function of problem density.

Keywords: Large-scale reconstruction · Bundle adjustment · Preconditioned conjugate gradients

1 Introduction

The classical challenge of image-based large scale reconstruction is witnessing renewed interest with the emergence of large-scale internet photo collections [2]. The computational bottleneck of 3D reconstruction and structure from motion methods is the problem of large-scale bundle adjustment (BA): Given a set of measured image feature locations and correspondences, BA aims to jointly estimate the 3D landmark positions and camera parameters by minimizing a non-linear least squares reprojection error. More specifically, the most time-consuming step is the solution of the normal equation in the popular Levenberg-Marquardt (LM) algorithm that is typically solved by Preconditioned Conjugate Gradients (PCG).

In this paper, we propose a new iterative solver for the normal equation that relies on the decomposable structure of the competitive block Jacobi preconditioner. Inspired by respective approaches in the domain-decomposition literature, we exploit the specificities of the Schur complement matrix to enlarge the search-space of the traditional PCG approach leading to what we call *Multidirectional Conjugate Gradients* (MCG). In particular our contributions are as follows:

- We design an extension of the popular PCG by using local contributions of the poses to augment the space in which a solution is sought for.

C. Bauckhage et al. (Eds.): DAGM GCPR 2021, LNCS 13024, pp. 712–724, 2021.
https://doi.org/10.1007/978-3-030-92659-5_46

(a) *Final-1936* from BAL dataset

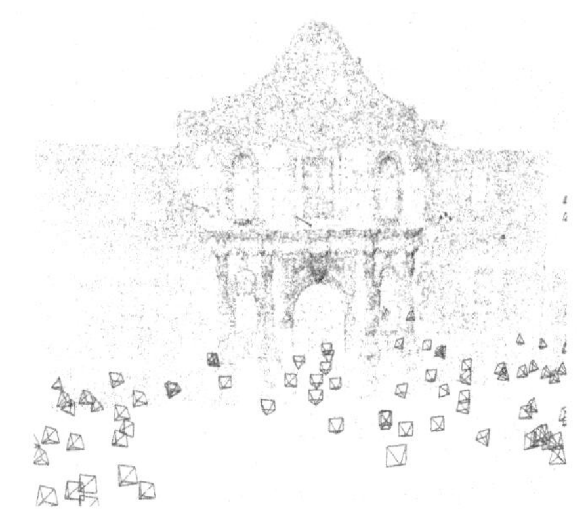

(b) *Alamo* from 1dSfM dataset

Fig. 1. (a) Optimized 3D reconstruction of a *final* BAL dataset with 1936 poses and more than five million observations. For this problem MCG is 39% faster than PCG and the overall BA resolution is 16% faster. (b) Optimized 3D reconstruction of Alamo dataset from 1dSfM with 571 poses and 900000 observations. For this problem MCG is 56% faster than PCG and the overall BA resolution is 22% faster.

- We experimentally demonstrate the robustness of MCG with respect to the relevant hyper-parameters.
- We evaluate MCG on a multitude of BA problems from BAL [1] and 1dSfM [20] datasets with different sizes and show that it is a promising alternative to PCG.
- We experimentally confirm that the performance gain of our method increases with the density of the Schur complement matrix leading to a speedup for solving the normal equation of up to 61% (Fig. 1).

2 Related Work

Since we propose a way to solve medium to large-scale BA using a new iterative solver that enlarges the search-space of the traditional PCG, in the following we will review both scalable BA and recent CG literature.

Scalable Bundle Adjustment

A detailed survey of the theory and methods in BA literature can be found in [17]. Sparsity of the BA problem is commonly exploited with the Schur complement matrix [5]. As the performance of BA methods is closely linked to the resolution of the normal equations, speed up the solve step is a challenging task. Traditional direct solvers such as sparse or dense Cholesky factorization [11] have been outperformed by inexact solvers as the problem size increases and are therefore frequently replaced by Conjugate Gradients (CG) based methods

[1,6,18]. As its convergence rate depends on the condition number of the linear system a preconditioner is used to correct ill-conditioned BA problems [14]. Several works tackle the design of performant preconditioners for BA: [9] proposed the band block diagonals of the Schur complement matrix, [10] exploited the strength of the coupling between two poses to construct cluster-Jacobi and block-tridiagonal preconditioners, [7] built on the combinatorial structure of BA. However, despite these advances in the design of preconditioners, the iterative solver itself has rarely been challenged.

(Multi-preconditioned) Conjugate Gradients

Although CG has been a popular iterative solver for decades [8] there exist some interesting recent innovations, e.g. flexible methods with a preconditioner that changes throughout the iteration [13]. The case of a preconditioner that can be decomposed into a sum of preconditioners has been exploited by using Multi-Preconditioned Conjugate Gradients (MPCG) [4]. Unfortunately, with increasing system size MPCG rapidly becomes inefficient. As a remedy, Adaptive Multi-Preconditioned Conjugate Gradients have recently been proposed [3,15]. This approach is particularly well adapted for domain-decomposable problems [12]. While decomposition of the reduced camera system in BA has already been tackled e.g. with stochastic clusters in [19], to our knowledge the decomposition *inside* the iterative solver has never been explored. As we will show in the following, this modification gives rise to a significant boost in performance.

3 Bundle Adjustment and Multidirectional Conjugate Gradients

We consider the general form of bundle adjustment with n_p poses and n_l landmarks. Let x be the state vector containing all the optimization variables. It is divided into a pose part x_p of length $d_p n_p$ containing extrinsic and eventually intrinsic camera parameters for all poses (generally $d_p = 6$ if only extrinsic parameters are unknown and $d_p = 9$ if intrinsic parameters also need to be estimated) and a landmark part x_l of length $3n_l$ containing the 3D coordinates of all landmarks. Let $r(x) = [r_1(x), ..., r_k(x)]$ be the vector of residuals for a 3D reconstruction. The objective is to minimize the sum of squared residuals

$$F(x) = \|r(x)\|^2 = \sum_i \|r_i(x)\|^2 \tag{1}$$

3.1 Least Squares Problem and Schur Complement

This minimization problem is usually solved with the Levenberg Marquardt algorithm, which is based on the first-order Taylor approximation of $r(x)$ around the current state estimate $x^0 = (x_p^0, x_l^0)$:

$$r(x) \approx r^0 + J\Delta x \tag{2}$$

where

$$r^0 = r\left(x^0\right), \tag{3}$$

$$\Delta x = x - x^0, \tag{4}$$

$$J = \frac{\partial r}{\partial x} \mid_{x=x^0} \tag{5}$$

and J is the Jacobian of r that is decomposed into a pose part J_p and a landmark part J_l. An added regularization term that improves convergence gives the damped linear least squares problem

$$\min_{\Delta x_p, \Delta x_l} \left(\|r^0 + \left(J_p \; J_l \right) \begin{pmatrix} \Delta x_p \\ \Delta x_l \end{pmatrix} \|^2 + \lambda \| \left(D_p \; D_l \right) \begin{pmatrix} \Delta x_p \\ \Delta x_l \end{pmatrix} \|^2 \right) \tag{6}$$

with λ a damping coefficient and D_p and D_c diagonal damping matrices for pose and landmark variables. This damped problem leads to the corresponding normal equation

$$H \begin{pmatrix} \Delta x_p \\ \Delta x_l \end{pmatrix} = - \begin{pmatrix} b_p \\ b_l \end{pmatrix} \tag{7}$$

where

$$H = \begin{pmatrix} U_\lambda & W \\ W^\top & V_\lambda \end{pmatrix}, \tag{8}$$

$$U_\lambda = J_p^\top J_p + \lambda D_p^\top D_p, \tag{9}$$

$$V_\lambda = J_l^\top J_l + \lambda D_l^\top D_l, \tag{10}$$

$$W = J_p^\top J_l, \; b_p = J_p^\top r^0, \tag{11}$$

$$b_l = J_l^\top r^0 \tag{12}$$

As the system matrix H is of size $(d_p n_p + 3n_l)^2$ and tends to be excessively costly for large-scale problems [1], it is common to reduce it by using the Schur complement trick and forming the reduced camera system

$$S\Delta x_p = -\widetilde{b} \tag{13}$$

with

$$S = U_\lambda - W V_\lambda^{-1} W^\top, \tag{14}$$
$$\widetilde{b} = b_p - W V_\lambda^{-1} b_l \tag{15}$$

and then solving (13) for Δx_p and backsubstituting Δx_p in

$$\Delta x_l = -V_\lambda^{-1}\left(-b_l + W^\top \Delta x_p\right) \tag{16}$$

3.2 Multidirectional Conjugate Gradients

Direct methods such as Cholesky decomposition [17] have been studied for solving (13) for small-size problems, but this approach implies a high computational cost whenever problems become too large.

A very popular iterative solver for large symmetric positive-definite system is the CG algorithm [16]. Since its convergence rate depends on the distribution of eigenvalues of S it is common to replace (13) by a preconditioned system. Given a preconditioner M the preconditioned linear system associated to

$$S \Delta x_p = -\widetilde{b} \tag{17}$$

is

$$M^{-1} S \Delta x_p = -M^{-1}\widetilde{b} \tag{18}$$

and the resulting algorithm is called Preconditioned Conjugate Gradients (PCG) (see Algorithm 1). For block structured matrices as S a competitive preconditioner is the block diagonal matrix $D(S)$, also called block Jacobi preconditioner [1]. It is composed of the block diagonal elements of S. Since the block S_{mj} of S is nonzero if and only if cameras m and j share at least one common point, each diagonal block depends on a unique pose and is applied to the part of conjugate gradients residual r_i^j that is associated to this pose. The motivation of this section is to enlarge the conjugate gradients search space by using several local contributions instead of a unique global contribution.

Adaptive Multidirections

Local Preconditioners. We propose to decompose the set of poses into N subsets of sizes $l_1, \ldots, l_N$ and to take into consideration the block-diagonal matrix $D_p(S)$ of the block-jacobi preconditioner and the associated residual r^p that correspond to the l_p poses of subset p (see Fig. 2(a)). All direct solves are performed inside these subsets and not in the global set. Each local solve is treated as a separate preconditioned equation and provides a unique search-direction. Consequently the conjugate vectors $Z_{i+1} \in \mathbb{R}^{d_p n_p}$ in the preconditioned conjugate gradients (line 10 in Algorithm 1) are now replaced by conjugate matrices

Algorithm 1. Preconditioned Conjugate Gradients

1: x_0, $r_0 = -\widetilde{b} - Sx_0$, $Z_0 = D(S)^{-1}r_0$, $P_0 = Z_0$, ϵ;
2: **while** $i <$ imax **do**
3: $\quad Q_i = SP_i$;
4: $\quad \Delta_i = Q_i^\top P_i$; $\gamma_i = P_i^\top r_i$; $\alpha_i = \frac{\gamma_i}{\Delta_i}$;
5: $\quad x_{i+1} = x_i + \alpha_i P_i$;
6: $\quad r_{i+1} = r_i - \alpha_i Q_i$;
7: $\quad$ **if** $r_{i+1} < \epsilon * r_0$ **then**
8: $\quad\quad$ break
9: $\quad$ **end if**
10: $\quad Z_{i+1} = D(S)^{-1} r_{i+1}$;
11: $\quad \Phi_i = Q_i^\top Z_{i+1}$; $\beta_i = \frac{\Phi_i}{\Delta_i}$;
12: $\quad P_{i+1} = Z_{i+1} - \beta_i P_i$;
13: **end while**
14: return x_{i+1};

(a) Decomposed preconditioned CG residuals

(b) Enlarged search-space

Fig. 2. (a) Block-Jacobi preconditioner $D(S)$ is divided into N submatrices $D_p(S)$ and each of them is directly applied to the associated block-row r^p in the CG residual. (b) Up to a τ-test the search-space is enlarged. Each iteration provides N times more search-directions than PCG.

$Z_{i+1} \in \mathbb{R}^{d_p n_p \times N}$ whose each column corresponds to a local preconditioned solve. The search-space is then significantly enlarged: N search directions are generated at each inner iteration instead of only one. An important drawback is that matrix-vector products are replaced by matrix-matrix products which can lead to a significant additional cost. A trade-off between convergence improvement and computational cost needs to be designed.

Adaptive τ-Test. Following a similar approach as in [15] we propose to use an adaptive multidirectional conjugate gradients algorithm (MCG, see Algorithm 2) that adapts automatically if the convergence is too slow. Given a threshold $\tau \in \mathbb{R}^+$ chosen by the user, a τ-test determines whether the algorithm sufficiently reduces the error (case $t_i > \tau$) or not (case $t_i < \tau$). In the first case a global block Jacobi preconditioner is used and the algorithm performs a step of PCG; in the second case local block Jacobi preconditioners are used and the search-space is enlarged (see Fig. 2(b)).

Algorithm 2. Multidirectional Conjugate Gradients

1: x_0, $r_0 = -\widetilde{b} - Sx_0$, $Z_0 = D(S)^{-1}r_0$, $P_0 = Z_0$, ϵ;
2: **while** $i <$ imax **do**
3: $\quad Q_i = SP_i$;
4: $\quad \Delta_i = Q_i^\top P_i$; $\gamma_i = P_i^\top r_i$; $\alpha_i = \Delta_i^\dagger \gamma_i$;
5: $\quad x_{i+1} = x_i + P_i\alpha_i$;
6: $\quad r_{i+1} = r_i - Q_i\alpha_i$;
7: $\quad$ **if** $r_{i+1} < \epsilon * r_0$ **then**
8: $\quad\quad$ break
9: $\quad$ **end if**
10: $\quad t_i = \frac{\gamma_i^\top \alpha_i}{r_{i+1}^\top D(S)^{-1} r_{i+1}}$;
11: $\quad$ **if** $t_i < \tau$ **then**
12: $$Z_{i+1} = \begin{pmatrix} D_1(S)^{-1}r_{i+1}^1 & & 0 \\ 0 & \cdots & \cdots \\ \cdots & \cdots & 0 \\ 0 & & D_N(S)^{-1}r_{i+1}^N \end{pmatrix};$$
13: $\quad$ **else**
14: $\quad\quad Z_{i+1} = D(S)^{-1}r_{i+1}$;
15: $\quad$ **end if**
16: $\quad \Phi_{i,j} = Q_j^\top Z_{i+1}$; $\beta_{i,j} = \Delta_j^\dagger \Phi_{i,j}$ for $j = 0, ..., i$;
17: $\quad P_{i+1} = Z_{i+1} - \sum_{j=0}^{i} P_j\beta_{i,j}$;
18: **end while**
19: return x_{i+1};

Optimized Implementation. Besides matrix-matrix products two other changes appear. Firstly an $N \times N$ matrix Δ_i must be inverted (or pseudo-inverted if Δ_i is not full-rank) each time $t_i < \tau$ (line 4 in Algorithm 2). Secondly a full reorthogonalization is now necessary (line 16 in Algorithm 2) because of numerical errors while $\beta_{i,j} = 0$ as soon as $i \neq j$ in PCG.

To improve the efficiency of MCG we do not directly apply S to P_i (line 3 in Algorithm 2) when the search-space is enlarged. By construction the block S_{kj} is nonzero if and only if cameras k and j observe at least one common point. The trick is to use the construction of Z_i and to directly apply the non-zero blocks S_{jk}, i.e. consider only poses j observing a common point with k, to the column in Z_i associated to the subset containing pose k and then to compute

$$Q_i = SZ_i - \sum_{j=0}^{i-1} Q_j\beta_{i,j} \tag{19}$$

To get t_i we need to use a global solve (line 10 in Algorithm 2). As the local block Jacobi preconditioners $\{D_p(S)\}_{p=1,\dots,N}$ and the global block Jacobi preconditioner $D\,(S)$ share the same blocks it is not necessary to derive all local solves to construct the conjugates matrix (line 12 in Algorithm 2); instead it is more efficient to fill this matrix with block-row elements of the preconditioned residual $D\,(S)^{-1}\,r_{i+1}$.

As the behaviour of CG residuals is *a priori* unknown the best decomposition is not obvious. We decompose the set of poses into $N-1$ subsets of same size and the last subset is filled by the few remaining poses. This structure presents the practical advantage to be very easily fashionable and the parallelizable block operations are balanced.

4 Experimental Evaluations

4.1 Algorithm and Datasets

Levenberg-Marquardt (LM) Loop. Starting with damped parameter 10^{-4} we update λ according to the success or failure of the LM loop. Our implementation runs for at most 25 iterations, terminating early if a relative function tolerance of 10^{-6} is reached. Our evaluation is built on the LM loop implemented in [19] and we also estimate intrinsics parameters for each pose.

Iterative Solver Step. For a direct performance comparison we implement our own MCG and PCG solvers in C++ by using Eigen 3.3 library. All row-major-sparse matrix-vector and matrix-matrix products are multi-threaded by using 4 cores. The tolerance ϵ and the maximum number of iterations are set to 10^{-6} and 1000 respectively. Pseudo-inversion is derived with the pseudo-inverse function from Eigen.

Datasets. For our evaluation we use 9 datasets with different sizes and heterogeneous Schur complement matrix densities d from BAL [1] and 1dSfM [20] datasets (see Table 1). The values of N and τ are arbitrarily chosen and the robustness of our algorithm to these parameters is discussed in the next subsection.

We run experiments on MacOS 11.2 with Intel Core i5 and 4 cores at 2 GHz.

4.2 Sensitivity with τ and N

In this subsection we are interested in the solver runtime ratio that is defined as $\frac{t_{MCG}}{t_{PCG}}$ where t_{MCG} (resp. t_{PCG}) is the total runtime to solve all the linear systems (12) with MCG (resp. PCG) until a given BA problem converges. We investigate the influence of τ and N on this ratio.

Sensitivity with τ. We solve BA problem for different values of τ and for a fixed number of subsets N given in Table 1. For each problem a wide range of values supplies a good trade-off between the augmented search-space and the additional computational cost (see Fig. 3). Although the choice of τ is crucial it does not require a high accuracy. That confirms the tractability of our solver with τ.

Sensitivity with N. Similarly we solve BA problem for different values of N and for a fixed τ given in Table 1. For each problem a wide range of values supplies a good trade-off between the augmented search-space and the additional computational cost (see Fig. 4). That confirms the tractability of our solver with N.

Table 1. Details of the problems from BAL (prefixed as: F for *final*, L for *Ladybug*) and 1dSfM used in our experiments. d is the density of the associated Schur complement matrix, N is the number of subsets, τ is the adaptive threshold that enlarges the search-direction space.

Names	Poses	Points	Projections	d	N	τ
Piazza del Popolo	335	37,609	195,016	0.57	33	10
Metropolis	346	55,679	255,987	0.50	34	6
F-394	394	100,368	534,408	0.94	131	6
Montreal	459	158,005	860,116	0.60	22	3
Notre-Dame	547	273,590	1,534,747	0.77	45	10
Alamo	571	151,085	891,301	0.77	47	10
L-646	646	73,484	327,297	0.25	64	2
F-871	871	527,480	2,785,977	0.40	43	2.5
F-1936	1,936	649,673	5,213,733	0.91	121	3.3

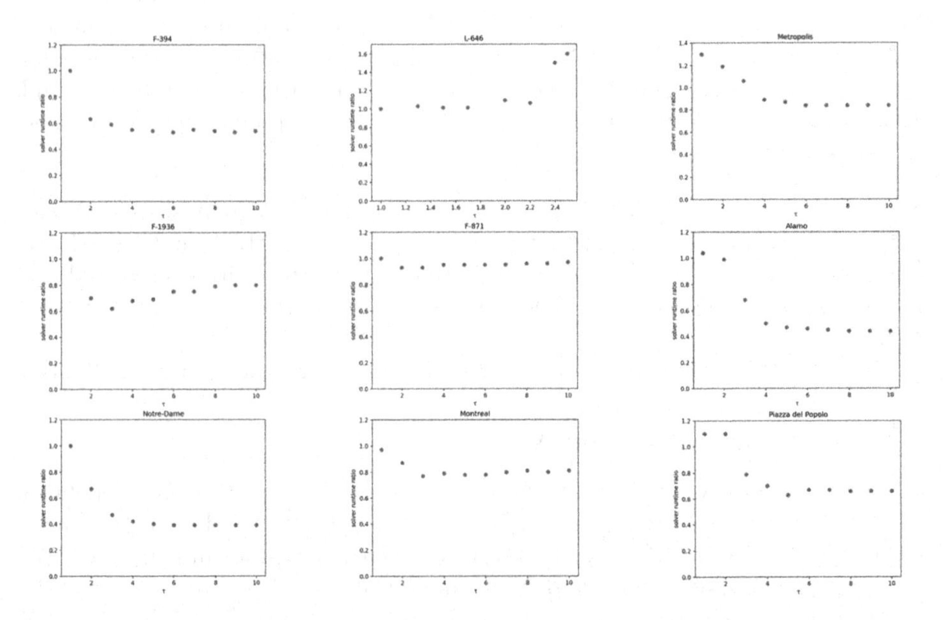

Fig. 3. Robustness to τ. The plots show the performance ratio as a function of τ for a number of subsets given in Table 1. The wide range of values that give similar performance confirms the tractability of MCG with τ.

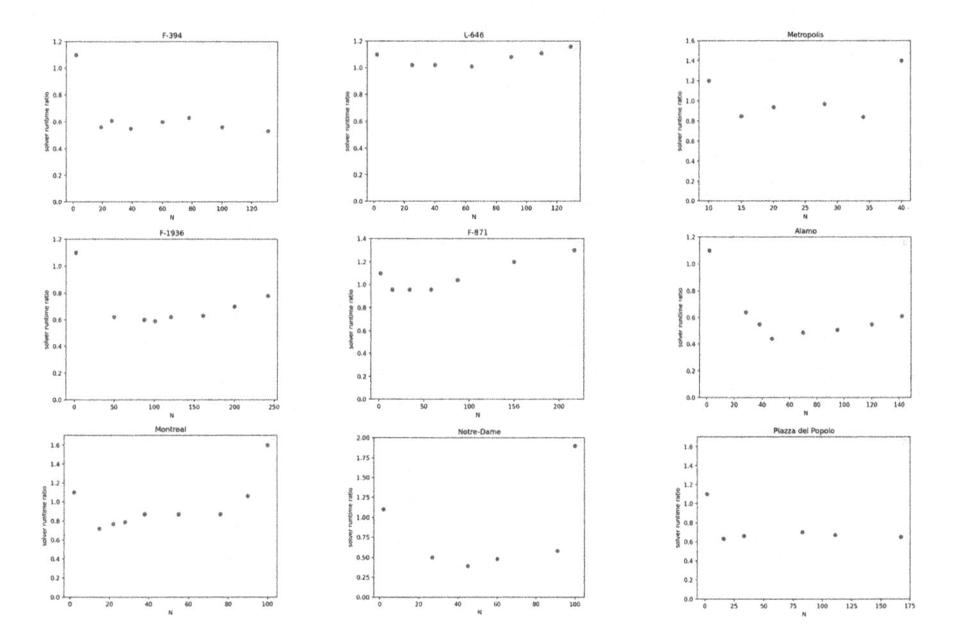

Fig. 4. Robustness to the number of subsets N. The plots show the performance ratio as a function of N for τ given in Table 1. The wide range of values that give similar performance confirms the tractability of MCG with N.

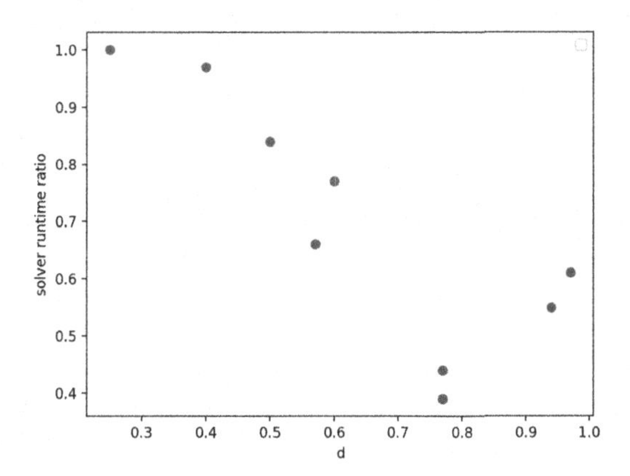

Fig. 5. Density effect on the relative performance. Each point represents a BA problem from Table 1 and d is the density of the Schur complement matrix. Our solver competes PCG for sparse Schur matrix and leads to a significant speed-up for dense Schur matrix.

4.3 Density Effect

As the performance of PCG and MCG depends on matrix-vector product and matrix-matrix product respectively we expect a correlation with the density of the Schur matrix. Figure 5 investigates this intuition: MCG greatly outperforms PCG for dense Schur matrix and is competitive for sparse Schur matrix.

4.4 Global Performance

Figures 6 and 7 present the total runtime with respect to the number of BA iterations for each problem and the convergence plots of total BA cost for *F-1936* and *Alamo* datasets, respectively. MCG and PCG give the same error at each BA iteration but the first one is more efficient in terms of runtime. Table 2 summarizes our results and highlights the great performance of MCG for dense Schur matrices. In the best case BA resolution is more than 20% faster than using PCG. Even for sparser matrices MCG competes PCG: in the worst case MCG presents similar results as PCG. If we restrict our comparison to the linear system solve steps our relative results are even better: MCG is up to 60% faster than PCG and presents similar results as PCG in the worst case.

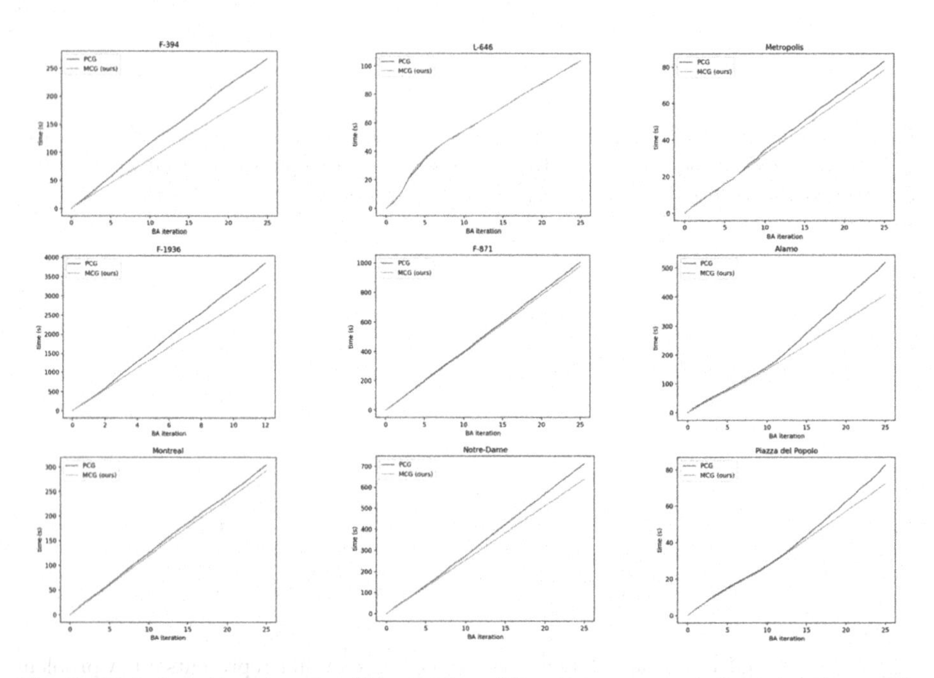

Fig. 6. Global runtime to solve BA problem. The plots represent the total time with respect to the number of BA iterations. For almost all problems the BA resolution with MCG (orange) is significantly faster than PCG (blue). (Color figure online)

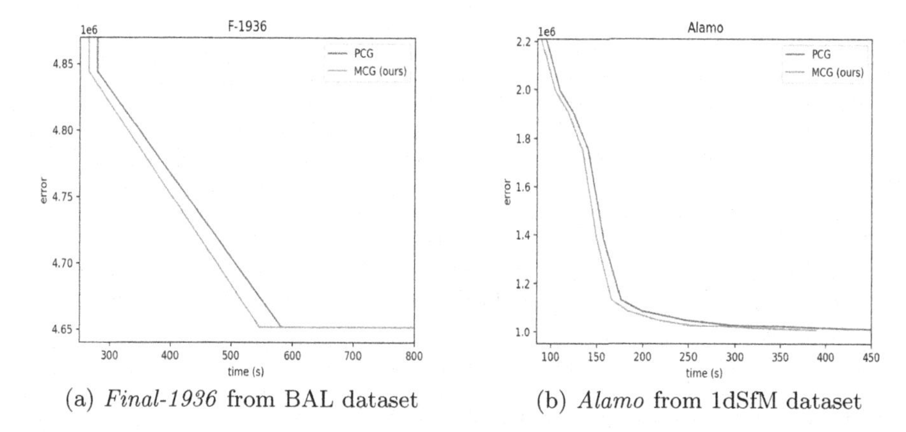

(a) *Final-1936* from BAL dataset (b) *Alamo* from 1dSfM dataset

Fig. 7. Convergence plots of (a) *Final-1936* from BAL dataset and (b) *Alamo* from 1dSfm dataset. The y-axes show the total BA cost.

Table 2. Relative performances of MCG w.r.t. PCG. d is the density of the associated Schur complement matrix. MCG greatly outperforms PCG (up to 61% faster) for dense Schur matrix and competes PCG for sparse Schur matrix. The global BA resolution is up to 22% faster.

Name	Solver runtime ratio	Global runtime ratio	d
Notre-Dame	**0.39**	**0.86**	0.77
Alamo	**0.44**	**0.78**	0.77
F-394	**0.55**	**0.80**	0.94
F-1936	**0.61**	**0.84**	0.97
Piazza del Popolo	**0.66**	**0.87**	0.57
Montreal	**0.77**	0.96	0.60
Metropolis	**0.84**	0.92	0.50
F-871	0.93	0.97	0.40
L-646	1.01	1.00	0.25

5 Conclusion

We propose a novel iterative solver that accelerates the solution of the normal equation for large-scale BA problems. The proposed approach generalizes the traditional preconditioned conjugate gradients algorithm by enlarging its search-space leading to a convergence in much fewer iterations. Experimental validation on a multitude of large scale BA problems confirms a significant speedup in solving the normal equation of up to 61%, especially for dense Schur matrices where baseline techniques notoriously struggle. Moreover, detailed ablation studies demonstrate the robustness to variations in the hyper-parameters and increasing speedup as a function of the problem density.

References

1. Agarwal, S., Snavely, N., Seitz, S.M., Szeliski, R.: Bundle adjustment in the large. In: Daniilidis, K., Maragos, P., Paragios, N. (eds.) ECCV 2010. LNCS, vol. 6312, pp. 29–42. Springer, Heidelberg (2010). https://doi.org/10.1007/978-3-642-15552-9_3
2. Agarwal, S., Snavely, N., Simon, I., Seitz, S.M., Szeliski, R.: Building Rome in a day. In: International Conference on Computer Vision (ICCV) (2009)
3. Bovet, C., Parret-Fréaud, A., Spillane, N., Gosselet, P.: Adaptive multipreconditioned FETI: scalability results and robustness assessment. Comput. Struct. **193**, 1–20 (2017)
4. Bridson, R., Greif, C.: A multipreconditioned conjugate gradient algorithm. SIAM J. Matrix Anal. Appl. **27**(4), 1056–1068 (2006). (electronic)
5. Brown, D.C.: A solution to the general problem of multiple station analytical stereo triangulation. RCA-MTP data reduction technical report no. 43 (or AFMTC TR 58–8), Patrick Airforce Base, Florida (1958)
6. Byrod, M., Aström, K., Lund, S.: Bundle adjustment using conjugate gradients with multiscale preconditioning. In: BMVC (2009)
7. Dellaert, F., Carlson, J., Ila, V., Ni, K., Thorpe, C.E.: Subgraph-preconditioned conjugate gradients for large scale slam. In: IROS, pp. 2566–2571 (2010)
8. Hestenes, M.R., Stiefel, E.: Methods of conjugate gradients for solving linear systems. J. Res. Nat. Bur. Standards **49**(409–436), 1952 (1953)
9. Jeong, Y., Nister, D., Steedly, D., Szeliski, R., Kweon, I.-S.: Pushing the envelope of modern methods for bundle adjustment. In: CVPR, pp. 1474–1481 (2010)
10. Kushal, A., Agarwal, S.: Visibility based preconditioning for bundle adjustment. In: CVPR (2012)
11. Lourakis, M., Argyros, A.: Is Levenberg-Marquardt the most efficient optimization algorithm for implementing bundle adjustment. In: International Conference on Computer Vision (ICCV), pp. 1526–1531 (2005)
12. Mandel, J.: Balancing domain decomposition. Comm. Numer. Methods Eng. **9**(3), 233–241 (1993)
13. Notay, Y.: Flexible conjugate gradients. SIAM J. Sci. Comput. **22**(4), 1444–1460 (2000)
14. Saad, Y.: Iterative Methods for Sparse Linear Systems. SIAM, PHiladelphia (2003)
15. Spillane, N.: An adaptive multipreconditioned conjugate gradient algorithm. SIAM J. Sci. Comput. **38**(3), A1896–A1918 (2016)
16. Trefethen, L., Bau, D.: Numerical Linear Algebra. SIAM, Philadelphia (1997)
17. Triggs, B., McLauchlan, P.F., Hartley, R.I., Fitzgibbon, A.W.: Bundle adjustment — a modern synthesis. In: Triggs, B., Zisserman, A., Szeliski, R. (eds.) IWVA 1999. LNCS, vol. 1883, pp. 298–372. Springer, Heidelberg (2000). https://doi.org/10.1007/3-540-44480-7_21
18. Wu, C., Agarwal, S., Curless, B., Seitz, S.: Multicore bundle adjustment. In: CVPR, pp. 3057–3064 (2011)
19. Zhou, L., et al.: Stochastic bundle adjustment for efficient and scalable 3D reconstruction. In: Vedaldi, A., Bischof, H., Brox, T., Frahm, J.-M. (eds.) ECCV 2020. LNCS, vol. 12360, pp. 364–379. Springer, Cham (2020). https://doi.org/10.1007/978-3-030-58555-6_22
20. Wilson, K., Snavely, N.: Robust global translations with 1DSfM. In: Fleet, D., Pajdla, T., Schiele, B., Tuytelaars, T. (eds.) ECCV 2014. LNCS, vol. 8691, pp. 61–75. Springer, Cham (2014). https://doi.org/10.1007/978-3-319-10578-9_5

Author Index

Ahmadi, Alireza 574
Akata, Zeynep 235
Alamri, Faisal 467
Amini, Arash 530
Anzén, William 559
Ardizzone, Lynton 79
Argus, Max 250

Banerjee, Biplab 649
Baur, Stefan A. 681
Behnke, Sven 530
Bodesheim, Paul 48, 142, 589
Boll, Bastian 498
Bose, Rupak 649
Brandt, Felix 545
Brissman, Emil 206
Brox, Thomas 250
Brust, Clemens-Alexander 159
Bucher, Solveig Franziska 589

Chaudhari, Prashant 328
Chaudhuri, Subhasis 649
Cheema, Noshaba 313
Chicca, Elisabetta 297
Chli, Margarita 515
Cremers, Daniel 3, 712

Danelljan, Martin 206
Das, Anurag 235
Demmel, Nikolaus 712
Dengel, Andreas 361
Denzler, Joachim 48, 142, 159, 589
Despinoy, Fabien 282
Dörr, Laura 545
Duhme, Michael 265
Dutta, Anjan 467

Eldesokey, Abdelrahman 222

Fan, Yue 63
Farha, Yazan Abu 282
Felsberg, Michael 206, 222
Fischer, Max 559
Francesca, Gianpiero 282
Frolov, Stanislav 361

Gall, Juergen 282
González, Camila 604
Gonzalez-Alvarado, Daniel 453
Gowda, Shreyank N. 191
Gurevych, Iryna 405

Haefner, Bjoern 3
Halstead, Michael 574
Harb, Robert 18
He, Yang 235
Hees, Jörn 361
Henter, Gustav Eje 697
Herzog, Robert 343
Huang, Huili 174

Ihrke, Ivo 313
Illgner-Fehns, Klaus 313

Jaiswal, Sunil Prasad 313
Johnander, Joakim 206

Kaisheva, Mariya 663
Karayil, Tushar 361
Keller, Frank 191
Kim, Kiyoon 191
Kirschner, Matthias 439, 484
Kjellström, Hedvig 697
Klein, Franz 421
Knöbelreiter, Patrick 18
Köhler, Thomas 328
Kölle, Michael 633
Korsch, Dimitri 142
Körschens, Matthias 589
Köthe, Ullrich 79
Kugele, Alexander 297
Kukleva, Anna 63

Lee, Haebom 343

Maffra, Fabiola 515
Mahajan, Shweta 421
Maier, Andreas 328
Mascaro, Ruben 515
McCool, Chris 574
Memmesheimer, Raphael 265
Migliavacca, Mirco 589
Möllenhoff, Thomas 3
Mukhopadhyay, Anirban 604
Müller, Jens 79

Naumann, Alexander 545
Nesterov, Vitali 376

Parbhoo, Sonali 376
Paulus, Dietrich 265
Penzel, Niklas 159
Periyasamy, Arul Selvam 530
Peter, David 681
Pfeiffer, Jonas 405
Pfeiffer, Michael 297
Pfeil, Thomas 297
Pototzky, Daniel 439, 484
Pouls, Martin 545

Quéau, Yvain 3

Rangnekar, Vikrant 649
Raue, Federico 361
Reimers, Christian 48, 159
Rexilius, Jan 343
Riess, Christian 328
Robinson, Andreas 222
Rodehorst, Volker 663
Rohrbach, Marcus 191
Römermann, Christine 589
Roozbahani, M. Mahdi 174
Roth, Stefan 33, 405, 421
Roth, Volker 376
Rother, Carsten 79, 343
Runge, Jakob 48

Sainudiin, Raazesh 559
Samarin, Maxim 376
Sanner, Antoine 604
Schiele, Bernt 63, 235
Schirrmacher, Franziska 328
Schmidt-Thieme, Lars 439, 484
Schmier, Robert 79
Schmitt, Jannik 33
Schneider, Philipp J. 621
Schnörr, Christoph 453, 498
Schulze, Henning 392
Sevilla-Lara, Laura 191
Sharma, Avneesh 361
Sharma, Shivam 313
Shekhovtsov, Alexander 111, 127
Shiller, Ivan 633
Sitenko, Dmitrij 498
Slusallek, Philipp 313
Soergel, Uwe 621, 633
Sorkhei, Moein 697
Souri, Yaser 282
Steitz, Jan-Martin O. 405
Sultan, Azhar 484

Taray, Samim Zahoor 313
Teixeira, Lucas 515
Triess, Larissa T. 681

Ulrich, Josephine 589

von Koch, Christian 559

Waibel, Alexander 392
Walter, Volker 633
Wang, Shiming 515
Weber, Simon 712
Wieczorek, Aleksander 376
Wieser, Mario 376

Xian, Yongqin 235

Yaman, Dogucan 392
Yanush, Viktor 111
Ye, Zhenzhang 3

Zeilmann, Alexander 453
Zimmermann, Christian 250
Zöllner, J. Marius 681

Printed in the United States
by Baker & Taylor Publisher Services